Handbook of hygiene control in the food industry

Related titles:

Poultry meat processing and quality
(ISBN 978-1-85573-727-3)
To ensure the continued growth and competitiveness of the poultry meat industry, it is essential that poultry meat quality is maintained during all stages of production and processing. This authoritative collection reviews how quality can be maintained at key points in the supply chain, from breeding and husbandry to packaging and refrigeration.

Understanding pathogen behaviour: Virulence, stress response and resistance
(ISBN 978-1-85573-953-6)
Pathogens respond dynamically to their environment. Understanding their behaviour is critical to ensuring food safety. This authoritative collection summarises the key research on pathogen virulence, stress response and resistance. It reviews the behaviour of individual pathogens and evidence of resistance to particular preservation techniques.

Improving the safety of fresh fruit and vegetables
(ISBN 978-1-85573-956-7)
Fresh fruit and vegetables have been identified as a significant source of pathogens and chemical contaminants. As a result, there has been a wealth of research on identifying and controlling hazards at all stages in the supply chain. *Improving the safety of fresh fruit and vegetables* reviews this research and its implications for food processors.

Details of these books and a complete list of Woodhead's titles can be obtained by:

- visiting our web site at www.woodheadpublishing.com
- contacting Customer Services (e-mail: sales@woodheadpublishing.com; fax: +44 (0) 1223 893694; tel.: +44 (0) 1223 891358 ext. 130; address: Woodhead Publishing Limited, Abington Hall, Granta Park, Great Abington, Cambridge CB21 6AH, UK)

If you would like to receive information on forthcoming titles in this area, please send your address details to: Francis Dodds (address, tel. and fax as above; email: francis.dodds@woodheadpublishing.com). Please confirm which subject areas you are interested in.

Woodhead Publishing Series in Food Science, Technology and Nutrition:
Number 116

Handbook of hygiene control in the food industry

Edited by
H. L. M. Lelieveld, M. A. Mostert and J. Holah

CRC Press
Boca Raton Boston New York Washington, DC

WOODHEAD PUBLISHING LIMITED
Oxford Cambridge New Delhi

Published by Woodhead Publishing Limited, Abington Hall, Granta Park
Great Abington, Cambridge CB21 6AH, UK
www.woodheadpublishing.com

Woodhead Publishing India Private Limited, G-2, Vardaan House, 7/28 Ansari Road, Daryaganj, New Delhi – 110002, India
www.woodheadpublishingindia.com

Published in North America by CRC Press LLC, 6000 Broken Sound Parkway, NW Suite 300, Boca Raton FL 33487, USA

First published 2005, Woodhead Publishing Limited and CRC Press LLC
Reprinted 2008, 2010

British Library Cataloguing in Publication Data
A catalogue record for this book is available from the British Library.

Library of Congress Cataloging in Publication Data
A catalog record for this book is available from the Library of Congress.

Woodhead Publishing ISBN: 978-1-85573-957-4 (print)
Woodhead Publishing ISBN: 978-1-84569-053-3 (online)
CRC Press ISBN: 978-0-8493-3439-9
CRC Press order number: WP3439
ISSN 2042-8049 (print)
ISSN 2042-8057 (online)

Printed in the United Kingdom by Lightning Source UK Limited

Contents

Contributor contact details

(* = main contact)

Chapter 1

Professor S. Notermans
Obrechtlaan 17
3723 KA Bilthoven
The Netherlands

E-mail: S.Notermans@wxs.nl

Chapter 2

Professor M. H. Zwietering* and
Dr E. D. van Asselt
Laboratory of Food Microbiology
Wageningen University
PO Box 8129
6700 EV Wageningen
The Netherlands

E-mail: Marcel.Zwietering@wur.nl

Chapter 3

Dr G. Wirtanen* and Ms S. Salo
VTT Biotechnology
PO Box 1500
Espoo
FIN-02044 VTT
Finland

E-mail: gun.wirtanen@vtt.fi
satu.salo@vtt.fi

Chapter 4

Ir A. J. van Asselt and
Dr M. C. te Giffel*
NIZO Food Research
Kernhemseweg 2
PO Box 20
6710 BA Ede
The Netherlands

E-mail: meike.te.giffel@nizo.nl
arjan.van.asselt@nizo.nl

Chapter 5

Dr D. Burfoot
Silsoe Research Institute
Wrest Park
Silsoe
Bedford MK45 4HS
UK

E-mail: dean.burfoot@bbsrc.ac.uk

Chapter 6

Professor Lynn J. Frewer and Dr Arnout R. H. Fischer*
Marketing and Consumer Behaviour Group
Wageningen University
Hollandseweg 1
6706 KN Wageningen
The Netherlands

E-mail: Lynn.frewer@wur.nl
arnout.fischer@wur.nl

Chapters 7 and 10

Mr D. J. Graham
Graham Sanitary Design Consulting Limited
14318 Aitken Hill Court
Chesterfield MO 63017
USA

E-mail: grahamdj@prodigy.net

Chapter 8

Dr John Holah
Campden & Chorleywood Food Research Association
Chipping Campden GL55 6LD
UK

E-mail: j.holah@campden.co.uk

Chapter 9

Dr Brigitte Carpentier
AFSSA (French Food Safety Agency)
Laboratoire d'étude et de recherches sur la qualité des aliments et sur les procédés agro-alimentaires
23 avenue du Général de Gaulle
F-94706 Maisons-Alfort cedex
France

Tel: 33(0)149 77 26 46
Fax: 33(0)149 77 26 40
E-mail: b.carpentier@afssa.fr

Chapter 11

Professor A. Friis* and Dr B. B. B. Jensen
Biocentrum
Technical University of Denmark
Building 221
Soltofts Plads
2800 Lyngby
Denmark

E-mail: af@biocentrum.dtu.dk
bbb@biocentrum.dtu.dk

Chapters 12 and 35

Dr A. P. M. Hasting
Tony Hasting Consulting
37 Church Lane
Sharnbrook
Bedford
UK

E-mail: tony.hasting@virgin.net

Chapter 13

Ing. Karel Mager
Quest International
PO Box 2
NL-1400 CA Bussum
The Netherlands

E-mail: karel.mager@questintl.com

Chapter 14

Ir C. J. de Koning
CFS b.v. (Convenience Food Systems)
PO Box 1
5760 AA Bakel
The Netherlands

E-mail: cees.de.koning@cfs.com

Chapter 15

Dr L. Uiterlinden*
GTI Process Solutions BV
PO Box 845
NL-5201 AV 's- Hertogenbosch
The Netherlands

E-mail: luiterlinden@gti-group.com

Dr H. M. J. van Eijk
The Food Processing Group
Unilever R&D Vlaardingen
PO Box 114
NL-3130 AC Vlaardingen
The Netherlands

Chapter 16

Mr F. T. Schonrock
11302 Alms House Court
Fairfax Station
Virginia
USA

E-mail: ftracy1@cox.net

Chapter 17

Dr H. Hoogland
Unilever R&D Vlaardingen
PO Box 114
NL-3130 AC Vlaardingen
The Netherlands

E-mail: hans.hoogland@unilever.com

Chapter 18

Dr R. Stahlkopf
Tuchenhagen GmbH
Am Industrie Park 2-10
D-21514 Buchen
Germany

E-mail:
stahlkopf.ralf@tuchenhagen.de

Chapter 19

Dr M. Bücking*
Environmental and Food Analysis
Fraunhofer IME
Auf dem Aberg 1
57392 Schmallenberg
Germany

E-mail:
mark.buecking@ime.fraunhofer.de

Dr J. E. Haugen
Matforsk AS
Osloveien 1
N-1430 As
Norway

E-mail: john-erik.haugen@matforsk.no

Chapter 20

Dr I. H. Huisman*
Nutricia
PO Box 1
2700 MA Zoetermeer
The Netherlands

E-mail: i_huisman@hotmail.com

Dr E. Espada Aventín
Unilever R&D Vlaardingen
PO Box 114
NL-3130 AC Vlaardingen
The Netherlands

Chapter 21

Professor J. Ralph Blanchfield MBE
17 Arabia Close
Chingford
London E4 7DU
UK

E-mail: jralphb@easynet.co.uk

Chapter 22

Professor R. H. Schmidt
Department of Food Science
University of Florida
Gainsville
Florida 32611-0370
USA

E-mail: rschmidt@ifas.ufl.edu

Chapter 23

Dr René Crevel
Safety & Environmental Assurance Centre
Unilever R&D Colworth
Bedford MK44 1LQ
UK

E-mail: rene.crevel@unilever.com

Chapter 24

Dr Laura Raaska
VTT Biotechnology
PO Box 1500
Espoo
FIN-02044 VTT
Finland

E-mail: Laura.Raaska@vtt.fi

Chapter 25

Dr E.U. Thoden van Velzen* and Dr Ir. L.J.S. Lukasse
Agrotechnology and Food Innovations BV
Wageningen University and Research Centre
Bornsesteeg 59
6708 PD Wageningen
The Netherlands

E-mail: ulphard.thodenvanvelzen@wur.nl

Chapter 26

Professor E. Shaaya
Department of Food Science
ARO
The Volcani Center
Bet Dagan 50-250
Israel

E-mail: vtshaaya@agri.gov.il

Chapter 27

Knuth Lorenzen
GEA Tuchenhagen Dairy Systems GmbH
Am Industriepark 2-10
D-21514 Buchen
Germany

Tel: 49(0) 4155 49 2427
Fax: 49(0) 4155 49 2764
E-mail: lorenzen.knuth@tuchenhagen.de

Chapter 28

Mr L. Keener
International Product Safety Consultants, Inc.
4021 W. Bertona Street
Seattle
WA 98199-1934
USA

E-mail: lkeener@aol.com

Chapter 29

Professor P. J. Fryer* and Dr G. K. Christian
Centre for Formulation Engineering
Department of Chemical Engineering
University of Birmingham
Birmingham B15 2TT
UK

E-mail: p.j.fryer@bham.ac.uk

Chapter 30

Dr S. Salo
VTT Biotechnology
PO Box 1500
Espoo
FIN-02044 VTT
Finland

E-mail: satu.salo@vtt.fi

Chapter 31

Dr L. Fielding* and Mr R. Bailey
School of Applied Sciences
University of Wales Institute Cardiff
Llandaff Campus
Western Avenue
Cardiff CF5 2YB
UK

E-mail: lfielding@uwic.ac.uk

Chapter 32

Dr Ing A. Grasshoff
Federal Dairy Research Centre
PO Box 60 69
D-24121 Kiel
Germany

E-mail: grasshoff@bafm.de

Chapter 33

Dr J. Lundén*, Professor J. Björkroth and Professor H. Korkeala
Department of Food and Environmental Hygiene
Faculty of Veterinary Medicine
University of Helsinki
Finland

E-mail: janne.lunden@helsinki.fi

Chapter 34

Professor Joanna Verran
Department of Biological Sciences
Manchester Metropolitan University
Chester Street
Manchester M1 5GD
UK

E-mail: j.verran@mmu.ac.uk

Chapter 36

Professor C. Griffith
School of Applied Sciences
University of Wales Institute Cardiff
Llandaff Campus
Western Avenue
Cardiff CF5 2YB
UK

E-mail: cgriffith@uwic.ac.uk

Chapter 37

Dr H. Miettinen
VTT Biotechnology
Espoo
PO Box 1500
FIN-02044 VTT
Finland

E-mail: hanna.miettinen@vtt.fi

Chapter 38

Dr Jean-Yves Maillard
Welsh School of Pharmacy
Cardiff University
Redwood Building
King Edward VII Avenue
Cardiff CF10 3XF
UK

E-mail: MaillardJ@Cardiff.ac.uk

Chapter 39

Dietmar Rosner
Training & Service Manager
Food & Beverage Division
Ecolab GmbH & Co. OHG
Reisholzer Werftstrasse 38–42
40589 Duesseldorf
Germany

E-mail: dietmar.rosner@ecolab.com

Chapter 40

Dr P. Overbosch
Kraft Foods R&D
Bayerwaldstrasse 8
D-81737 Munich
Germany

E-mail: poverbosch@krafteurope.com

Preface

Following the publication of *Hygiene in Food Processing*, the editors have focussed in this book on how current best practice in hygiene may be further improved. The food related illnesses reported daily in surveys of the European and American food safety authorities, for example, show that, in many instances, such improvements are highly desirable. We hope therefore that this book will not only reach those who are *now* responsible for product quality and safety in food companies, and for the design, building and installation of food plants, but particularly also to those who will assume such responsibility in the future. Students in food science, food technology, food engineering, microbiology and food chemistry may benefit from using this handbook, since much of the information needed in practice in the food industry – in its widest sense – is, in most cases, not part of the courses they follow.

The book starts with an introduction discussing the history of hygiene. This chapter discusses the first origins of hygiene as a concept thousands of years ago. It demonstrates very clearly why hygiene is so important and why, even today, people die because of not complying with basic hygiene requirements. To be able to decide on measures to control product safety, it is essential to understand the risks associated with product safety. Part I therefore is devoted to the range of microbiological risks in food processing. Risk perception is one of the most important determinants of consumer behaviour in the hygienic handling and consumption of food. It is also important because factors influencing consumer behaviour may be very similar to those affecting the behaviour of employees in the food chain. Part I therefore includes a chapter discussing consumer risk perception since an understanding of such behaviour may help to devise effective measures to reduce risks or eliminate undue hazards.

Part II is devoted to improving the design of production facilities: buildings, equipment and equipment components. It covers areas not covered in previous books on the subject such as requirements for electrical installations and sensors. Understanding risks in food production and hygienic design requirements, however, will not guarantee hygienic production. Inadequate management is often a factor in food safety incidents. Part III therefore discusses risk management and control, covering such areas as good manufacturing practice (GMP) and standard operating procedures (SOPs) in relation to processing, cleaning and sanitation. It also covers ways of monitoring the effectiveness of hygiene in food processing.

If you bought this book to address issues that you want to solve, we hope that you will find the answers or, at least, where to go next to find the answers. In case you come across issues that you feel are important but have not been addressed, we invite you to contact us so that we may take this into account in future editions of the book.

In Part I, chapter 2 provides a general introduction to what the following chapters cover. Parts II and III include brief introductions to the main themes that follow.

Huub Lelieveld
Tineke Mostert
John Holah

1

Introduction

S. Notermans and S. C. Powell, Lancashire Postgraduate School of Medicine and Health, UK and E. Hoornstra, TNO Nutrition and Food Research, The Netherlands

1.1 Introduction: the evolution of food hygiene

The art of healing is almost as old as people themselves. Instincts, needs and experiences taught humans the art of healing. Throughout history, medicine and hygiene have been counterparts in healing and preventing diseases. Both disciplines have mostly gone hand in hand with improving human health. This introductory chapter starts with the early aspects of hygiene and, where necessary, interfaces between healing and preventing diseases will be discussed. After the recognition of germs as causative agent of diseases, the significance of hygiene developed rapidly and is now considered as the cornerstone of safe food production.

1.1.1 The origin of the hygiene concept

Hygeia, the goddess of health

In Greek mythology, Asclepius, son of Apollo and referred to as the god of medicine or healing, was a healer who became a Greek demigod, and was a famous physician. He was the most important among the Greek gods and heroes who were associated with health and curing disease. Shrines and temples of healing, known as Asclepieia, were erected throughout Greece where the sick came to worship and sought cures for their ills. Among the children of Asclepius the best known are his daughters Hygeia and Panacea. Hygeia became the goddess of healing and she focused on the healing power of cleanliness. She introduced and promoted the idea of washing patients with soap and water. She had lots of hospital shrines and played an important role in the cult of Asclepius as a giver of health. At the beginning she was the goddess of corporal well-being. Later she was also connected to mental health; the aphorism '*mens sana in corpore sano*' applies to this, 'a healthy mind in a healthy body'. Her sister was faced, like her father, with healing by medicines.

Hygeia was celebrated in many places in the Greek and Roman world. She was sung about and represented by many artists from the 4th century BC until the end of the Roman period. Statues of Hygeia were made by well-known masters such as Skopias, Tomotheos and Bryaxis. The name of Hygeia has survived in the word hygiene and its components. Her sacred snake together with the rod of Asclepius is the sign for medicine.

Hippocrates (460–377 BC)

Hippocrates, the most famous doctor in ancient Greece, was called the Father of Medicine. Hippocrates based medicine on objective observation and deductive reasoning. His medical school and sanatorium on the island of Kos developed principles and methods in curing that have been used ever since. Hippocrates and his followers elaborated an entirely rational system that was based on the classification of the symptoms of different diseases. He taught that medicine should build the patient's strength through diet and hygiene, resorting to more drastic treatment only when necessary. All historians agree that he taught validly concerning epidemics, fever, epilepsy, fractures, the difference between malignant and benign tumours, health in general and, most of all, the importance of hygiene, the healing power of food and the need for high ethical values in the practice of medicine. He laid utmost stress on hygiene and diet, but used herbal remedies and surgery when necessary.

An overview of the work of Hippocrates is presented in the book *Magni Hippocratis Coi Opera Omnia* (Hollier, 1623). It contains everything that had been ascribed to Hippocrates up to the 17th century.

Other hygiene measures

Over many millennia, humankind has learned how to select edible plant and animal species, and how to produce, harvest and prepare them for food purposes. This was mostly done on the basis of trial and error and from long experience. Many of the lessons learned, especially those relating to adverse effects on human health are reflected in various religious taboos, which include a ban on eating specific items, such as pork, in the Jewish and Muslim religions (Tannahill, 1973). Other taboos showed a more general appreciation of food hygiene. In India, for example, religious laws prohibited the consumption of certain 'unclean' foods, such as meat cut with a sword, or sniffed by a dog or cat, and meat obtained from carnivorous animals (Tannahill, 1973). Most of these food safety requirements were established thousands of years ago when religious laws were likely to have been the only ones in existence. The introduction of control measures in civil law was of a much later date.

The re-emergence of hygiene

In the Middle Ages folk-medicine developed rapidly. Medicinal plants, animal parts and minerals were used to get rid of disease symptoms. Later, surgery was used as a cure. At the beginning of the 1800s the excesses of doctors and the cottage industry of drugs led to general loathing and ridicule of the medical profession by the public in the USA and Europe. For at least a century strychnine was the best remedy the profession had for palsy and paralysis. It was used to kill rats, cats and dogs. But when given as medicine, it was tonic, a nerving, a remedy for palsied people. It was standard medical practice to withhold water from the ill, and thousands of patients literally died of dehydration. Alcohol was a foundation of the many bitters that were sold to the people as tonics, as it was the chief ingredient in many of the patent nostrums sold. Remedies were sold against alcoholism that were chiefly alcohol. In addition to drugging their patients to death, physicians have frequently bled them to death. Bleeding was employed in wounds and head injuries that resulted in unconsciousness. Not only were pregnant mothers bled, but physicians also drew blood from blue babies. In these days patients were bled, blistered, purged, vomited, narcotised, mercurialised or alcoholised into chronic invalidism or into the grave. The death rate was high and the sick person who recovered without sequelae was so rare as to be negligible. In that time hygiene was very poor as well. Physicians not only frowned upon but opposed bathing. Surgeons performed operations without washing their hands, and operating rooms of hospitals were veritable pig-sties. Physicians would go from the post-mortem room directly to the delivery room and assist in the birth of a child without washing their hands. Child-bed fever was a very common disease and the death rate from it was very high.

This is the time when the revolt, 'hygiene', re-emerged. Out of the contradictions, confusions, chaos and delusions called the science of medicine grew a need for new thoughts, and a crusade for health reform developed.

The 'Natural Hygiene' concept

One of the first pioneers was Isaac Jennings (see the book *Awakening our self-healing body* by Arthur Michael Baker (1994)). In 1822, after having practised medicine for 20 years and being thoroughly discouraged with the results, Jennings begins to administer placebos of bread pills, starch powders and coloured water tonics to patients, while instructing them in healthy living. Jennings and physiologist/minister Sylvester Graham started to educate citizens with failures and contradictions of current medical practice and theory. Graham developed a significant following of Grahamites in response to his eloquent lectures and writings. To the temperance movement he offered a vegetarian diet as a cure for alcoholism. He also advocated sexual restraint and hygiene measures such as bathing.

The truths proclaimed by Jennings and Graham found immediate and widespread acceptance. After becoming fully convinced of the correctness of his 'Do-Nothing Cure' and the 'No-Medicine Plan', Jennings announced his discovery to the world, but he was misunderstood. The work of Jennings in the USA was continued by many others. People were taught to bathe, to eat more fruit and vegetables, to ventilate their homes, to get exercise and sunshine. Hygiene became so popular, that traditional medicine finally had to adopt parts of the 'Natural Hygiene' concept. Later, when it became clear that 'germs' were the cause of many diseases, the new 'hygiene' was incorporated with the drug usage of medicine and the word hygiene got the meaning it has today.

Hygienic developments in Europe

In the middle of the 19th century two people laid the foundation of modern hygiene: the Hungarian physician Semmelweis and the British surgeon Lister. Both introduced hygienic methods that are still essential in modern society.

Ignác Fülöp Semmelweis (1818–1865) was a Hungarian physician who demonstrated that puerperal fever[1] (also known as 'childbed fever') was contagious and that its incidence could be drastically reduced by enforcing appropriate hand-washing behaviour by medical care-givers. Semmelweis made this discovery in 1847 while working in the Maternity Department of the Vienna Lying-in Hospital. He realised that the number of cases of puerperal fever was much larger in clinic 1 where students did both post-mortem examinations with human cadavers in the autopsy rooms and midwifery in the maternity rooms. In clinic 1 the average death rate amounted to 9.92%. In clinic 2 where students were involved in midwifery and not allowed to do autopsies, the average death rate was much less at around 3.38%. After testing a few hypotheses, Semmelweis found that the number of cases was drastically reduced if the doctors washed their hands carefully before dealing with a pregnant woman (see Table 1.1). Risk was especially high if they had been in contact with corpses before they treated the women. The germ theory of disease had not been developed at the time. Thus, Semmelweis concluded that some unknown 'cadaveric material' caused childbed fever. Since the cadaverous matter could not been removed from the doctors' hands merely by washing them with soap and water Semmelweis started experiments with different chemicals. Finally he prescribed the additional use of chlorinated lime. Due to this the death rate caused by puerperal fever deceased to zero (see Table 1.1).

Semmelweis lectured publicly about his results in 1850. However, the reception by the medical community was cold, if not hostile. His observations went against the current scientific opinion of that time, which blamed diseases on an imbalance of the basic 'humours' in the body. It was also argued that even if his findings were correct, washing one's hands each time before treating a pregnant woman, as Semmelweis advised, would be too much work. Nor were doctors eager to admit that they had caused so many deaths. Semmelweis spent 14 years developing his ideas and lobbying for their acceptance, culminating in a book he wrote in 1861 (Semmelweis, 1861). The book received poor reviews,

1. A serious form of septicaemia contracted by a woman during childbirth or abortion (usually attributable to unsanitary conditions); formerly widespread but now uncommon.

Table 1.1 The effect of hand-washing and hand-washing with chlorinated lime on maternal death caused by puerperal fever

Year/period	Maternal death rate in (%)	
	Medical students (clinic 1)	Midwife students (clinic 2)
1841–1846	9.92	3.38
May 1847	12.42	
Introduction of hand-washing		
June 1847	2.38	
July 1847	1.20	
August 1847	1.89	
Introduction of chlorinated lime hand-washing		
October 1847	1.27	1.33
March 1848	0	0
August 1848	0	0

and he responded with polemic. His failure to convince his fellow doctors was not helped by his poor ability to communicate. In 1865, he suffered a nervous breakdown and was committed to an insane asylum where he soon died from blood poisoning. Only after Dr Semmelweis's death was the germ theory of disease developed. He is now recognised as a pioneer of antiseptic policy and prevention of nosocomial disease.

Joseph Lister (Lord Lister, 1827–1912) introduced antiseptic surgery. By the middle of the 19th century, post-operative sepsis infection accounted for the death of almost half of patients undergoing major surgery. A common report by surgeons was: operation successful but patient died. For many years he had explored the inflammation of wounds, at the Glasgow Infirmary. These observations led him to consider that infection was not due to bad air alone, and that 'wound sepsis' was a form of decomposition. When, in 1865, Louis Pasteur suggested that decay in wounds was caused by living organisms in the air, which on entering matter caused it to ferment, Lister made the connection with wound sepsis. As a meticulous researcher and surgeon, Lister recognised the relationship between Pasteur's research and his own. He considered that microbes in the air were likely to be causing the putrefaction and must be destroyed before they entered the wound. In 1864 Lister had heard that 'carbolic acid' was being used to treat sewage in Carlisle (UK), and that fields treated with the effluent became free of parasites causing disease in cattle. Even before the work of Pasteur on fermentation and putrefaction, Lister had been convinced of the importance of scrupulous cleanliness and the usefulness of deodorants in the operating room. In 1865 he began spraying a carbolic acid solution during surgery to kill germs. In the end, Lister gives Semmelweis his due by saying 'Without Semmelweis, my achievements would be nothing'. Through Pasteur's researches, he realised that the formation of pus was due to bacteria, so he proceeded to develop his antiseptic surgical methods. The immediate success of the new treatment led to its general adoption, with results of such beneficence as to make it rank as one of the great discoveries of the age. Lister also began to clean wounds and dress them using a solution of carbolic acid. He was able to announce at a British Medical Association meeting, in 1867, that his wards at the Glasgow Royal Infirmary had remained clear of sepsis for nine months. German surgeons were also beginning to practise antiseptic surgery, which involved keeping wounds free from microorganisms by the use of sterilised instruments and materials.

The 1870s were some of the happiest years of Lister's life, largely because of the German experiments with antisepsis during the Franco-German War. His clinics were crowded with visitors and eager students. Lister made a triumphal tour of the leading surgical centres in Germany in 1875. Here he met Robert Koch who demonstrated in 1878 the usefulness of steam for sterilising surgical instruments and dressings.

1.1.2 Foodborne diseases and hygiene since 1850

Foodborne diseases

Public health concern with foodborne diseases emerged around the 1880s. This was after microorganisms had been found to be infectious agents. Koch and his assistants devised the techniques for culturing bacteria outside the body, and formulated the rules for showing whether or not a bacterium is the cause of a disease (Koch, 1883). Before that time two types of illness with foodstuffs were recognised: one associated with ageing, and the other with foods normally not causing illness and apparently incapable of adulteration such as meat and fish. The last type had long been associated with decomposition; in the early 19th century it was thought to be due to chemical poisons, later to ptomaines,[2] or putrefactive alkaloids (Dewberry, 1959). Uncooked fruit and vegetables were also associated with upset stomachs, but here illness was generally attributed to unripeness or acidity (Hardy, 1999).

It was not until the late 1880s that the generic term 'food poisoning' emerged: before this, and still occasionally for decades thereafter, episodes were usually described by the precise item of food involved: 'cheese poisoning', 'meat poisoning', 'pork-pie poisoning', etc. Despite Robert Koch's identification of specific organisms causing foodborne diseases, such as anthrax, in 1876 (Koch, 1876) the above terms for food poisoning remain in use and examples have been described by, among others, Durham (1898) and Peckham (1923–24) who reported on respectively outbreaks of meat poisoning and pork-pie poisoning.

2. Food poisoning, erroneously believed to be the result of ptomaine ingestion. The word ptomaine was invented by the Italian chemist Selmi for the basic substances produced in putrefaction. They belong to several classes of chemical compounds and are any of various amines (such as putrescine or cadaverine) formed by the action of putrefactive bacteria.

The 1880s were also the decade when bacterial food poisoning displaced ptomaine poisoning. In the late 1870s, German researchers had begun to draw attention to connections between septic and pyaemic[3] diseases in animals used for food and meat poisoning outbreaks. In the early 1880s they started to investigate meat poisoning outbreaks bacteriologically (described by Ostertag, 1902).

Hard historical information about the incidence rate of foodborne infections in the 19th century is lacking. One of the reasons is that food poisoning was not notifiable. (In the UK food poisoning became notifiable in 1939.) There are, however, two indicators for the behaviour of food infections in cities: typhoid and epidemic diarrhoea. Typhoid emerged in the UK as a major urban hazard in the 1830s and was largely water-borne. The role of human carriers and of contaminated foodstuffs is likely to have been significant. Death rates from typhoid fell rapidly between 1870 and 1885, as urban water supplies were improved, but then stabilised until early in the 20th century (Greenwood, 1935). With the discovery of the human carrier and of foodstuffs as a vehicle of infection, and with greater (hygienic) care of patients, death rates fell rapidly and disappeared around 1920. Epidemic diarrhoea contributed even more to food poisoning. The term encompasses infant diarrhoea, the condition responsible for some 30% of infant mortality before 1901. Huck's local studies (Huck, 1994) showed that rising infant mortality was closely associated with the growth of industrial towns in the early 19th century. Other contemporary studies found that infant death was only the visible tip of the iceberg of extensive familial episodes of diarrhoea (Woods *et al.*, 1988) which emphasised a very high degree of multiple infections in households. It was Ballard (1887, cited by Hardy, 1999) who linked the infections with contaminated foodstuffs. When the first bacteriological analyses of epidemic diarrhoea came to performed, the leading contenders for causation came from bacteria belonging to the family of Salmonellae (Niven, 1909–1910). In 1888 the German Gärtner (1888) discovered and described the *Salmonella* bacterium, which he named *Bacillus enteritidis*. He demonstrated the presence of the organism in a slaughtered cow that had caused gastroenteritis in the people who had eaten its meat. The discovery of other such organisms quickly followed in the pioneering bacteriological laboratories of the 1890s and actually the identification of specific agents of disease became a competitive game. Despite new isolation and identification techniques, the bacteriology of food poisoning and infection appeared to be an immensely complicated subject, partly because of the number of different organisms apparently involved in the process, and partly because of the vexed questions of their nature and natural habit. For example, questions about whether *Salmonella* was a natural inhabitant of the intestinal tract of human and animals, was *Salmonella* present in the flesh of the animal or was it present only in diseased animals needed to be solved.

3. The invasion of bloodstream by pyogenic (pus-forming) organisms.

Identification of agents involved in foodborne diseases and the aetiological research of foodborne diseases began at the end of the 19th century when the work of Van Ermengem served to clarify the aetiology of botulism in humans (Van Ermengem, 1897). Later milestones in this category included the recognition of *Clostridium perfringens* as a foodborne pathogen in 1943 (McClane, 1979) and *Bacillus cereus* in the 1950s (Kramer and Gilbert, 1989). Human infections with *Listeria monocytogenes* were well known by the 1940s and foodborne transmission was suspected (Rocourt and Cossart, 1997), but it was not until the occurrence of an outbreak in Canada in 1981 that proper evidence was obtained. In this case, illness followed the consumption of contaminated coleslaw (Farber and Peterkin, 2000). Since then, numerous foodborne outbreaks have been reported in different countries, and prevention of listeriosis has become a major challenge for the food industry. Around 1980–1985 *Salmonella* Enteritidis re-emerged via the internal contamination of chicken eggs. At the same time a new emerging pathogenic started to emerge: *Escherichia coli* O157: H7 (Willshaw *et al.*, 2001). This organism causes haemorrhagic colitis. Some victims, particularly the very young, may develop haemolytic uremic syndrome (HUS) which is characterised by renal failure and haemolytic anaemia. From 0 to 15% of haemorrhagic colitis victims may develop HUS. The disease can lead to permanent loss of kidney function.

Although there were enormous developments in foodborne disease research at the beginning of the 20th century the reporting of incidents remained low. Unless one or more deaths were involved, or an outbreak was on a considerable local scale, incidents of gastroenteritis rarely came to the knowledge of the authorities. Savage and Bruce White (1925) complained after a case in which an elderly woman died as a result of eating canned salmon. Investigations were not carried out. Difficulties in reporting and adequate investigations were acknowledged obstacles to a fuller understanding of the nature of and factors in bacterial food poisoning. Before the Second World War, most food poisoning incidents undoubtedly remained hidden. To improve the situation in 1939 the UK established the 'Emergency Public Health Laboratory Services' – a network of 19 provincial and 10 metropolitan laboratories, whose services were to be available free of charge to medical officers of health if required for the investigation and control of infectious diseases.

The Second World War is generally seen as a seminal event in history of food poisoning. During and after the war, there was a rapid extension of mass catering, both in terms of feeding large numbers of people in canteens and restaurants, and in mass production of prepared foodstuffs. This resulted in many new problems. Egg-borne *Salmonella* infections received widespread publicity when incidents were traced to the use of bulk imports of American powdered eggs (Hardy, 1999). Trade in both human and animal foodstuffs became internationalised and opened many European countries to a large number of exotic *Salmonella* types from all over the world.

After the introduction of compulsory notification, foodborne diseases acquired a statistical profile. After an uncertain start notifications began to

Table 1.2 Microorganisms detected in patients with symptoms of acute enteritis and controls (de Wit *et al.*, 2001)

	Patients (N = 857)		Controls (N = 574)	
	No.	%	No.	%
Salmonella spp.	33	3.9	1	0.2
Campylobacter spp.	89	10.4	3	0.5
Yersinia spp.	6	0.7	6	1.1
Shigella spp.	1	0.1	0	0.0
VTEC	4	0.5	3	0.6
Rotavirus	45	5.3	8	1.4
Adenovirus	19	2.2	2	0.4
Astrovirus	13	1.5	2	0.4
Norwalk-like viruses	43	5.0	6	1.1
Sapporo-like viruses	5	2.1	1	0.2
Parasites	64	7.4	26	4.5
Total	**322**	**37.6**	**58**	**10.1**

rise steadily. For example in the UK the notifications increased in a 10 year period (1941–1951) from an initial couple of hundred to over 3000 a year.

As indicated by Hardy (1999) in her historical overview of food poisoning in Britain, the history of foodborne disease is one of social and scientific change, but is not simply of an increasing preference for foodstuffs prepared outside the home rather than within it. Rather, it is the story of how social and scientific change has gradually exposed unchanging economies of time and hygiene that most people have always made in their everyday lives. However, information about foodborne diseases is still not complete. A current problem is that although most countries have mandatory systems for notifying foodborne diseases, the information provided is generally poor and there is a dramatic underreporting. This came to light after modern analyses were used, including sentinel and population studies. It became clear that in developed countries on average 10 000–20 000 persons per 1 000 000 population suffer yearly from a foodborne disease (de Wit *et al.*, 2001; Fitzgerald *et al.*, 2004). In addition in about 37.5% of the cases investigated in sentinel studies a causative organism was identified (see Table 1.2).

Hygiene

Following the discovery, around 1880, that food can be an important source of disease-causing organisms, investigations started to concentrate on the reservoirs and routes of transmission of pathogens. The research of Buchanan (cited by Oddy and Millar, 1985) revealed an association between infant diarrhoea, refuse tips and flies. Further elucidation of reservoirs and routes of transmission stimulated the British public health authorities to include this emerging field in preventive medicine. As an example, the health authorities began extensive anti-fly campaigns, both through public education and by tackling the breeding grounds of the flies themselves. At much the same time, attention began to focus on the presence of pathogenic bacteria in the intestines of animals, as a source of food contamination, and foods of animal origin, as routes of transmission to humans.

Savage (1909) observed that faecal contamination of food must be very common. Milk, in particular, was suspected to be a vehicle of infection. Theodor Escherich, a German paediatrician, who devoted his efforts to improving childcare, particularly in relation to infant hygiene and nutrition, was the first to make a plea for heat-processing of milk to prevent infant diarrhoea (Escherich, 1890). After that time, the heating processes used for food began to improve. Real progress was made when Esty and Meyer (1922) developed the concept of process-performance criteria for heat treatment of low-acid, canned food-products to reduce the risk of botulism. Later, many other foods subjected to heat treatment were controlled in the same manner. An outstanding example is the work of Enright *et al.* (1956, 1957), who established performance criteria for the pasteurisation of raw milk that provided an appropriate level of protection against *Coxiella burnetii*, the causative agent of Q fever. Studies on the agent responsible for tuberculosis had been carried out earlier. These are early examples of the use of risk-assessment principles in deriving process criteria for control purposes.

The recognition of animal reservoirs of *Salmonella* served to reinforce the perceived complexity of the food poisoning problem. It was known that the key to preventing typhoid lay in blocking the routes by which the causative bacteria might pass from animals to humans and then among the human population. From the public health viewpoint, it became clear that there were several elements in the food poisoning situation: firstly, there was the animal-health aspect, with veterinary, slaughterhouse and culinary factors to consider; then, there was the matter of personal hygiene, which involved toilet and hand-washing habits. The question of developing suitable legislation was also apparent and, finally, there was the bacteriological aspect, with the need for more extensive laboratory provision to help in unravelling evidence from the field. Based on the need to improve hygiene in slaughterhouses, the USA was one of the first countries to introduce a Meat Inspection Act in 1906. This brought the following reforms to the processing of cattle, sheep, horses, swine and goats destined for human consumption:

- all animals were required to pass an inspection by the US Drug Administration prior to slaughter;
- all carcasses were subject to a post-mortem inspection;
- standards of cleanliness were established for slaughterhouses and processing plants.

In the UK, it was recognised that legislation alone was not sufficient to protect consumers against foodborne diseases, and the health authorities became aware of the need for public education to achieve cleaner food supplies. Food handling practices were very poor. Some examples from the 1920s were described by Porter (1924–1925) and included the following:

- Glass washing:
  - it was common for glasses to be dipped only in dirty water before being re-used.
- Personal cleanliness among food handlers:
  - food handlers regularly licked their fingers when dealing with wrapping paper;
  - they blew into paper bags to open them;
  - butchers often failed to wash their hands after eviscerating animals;
  - the habit of fingering the nose and/or mouth, while serving food, was common.

When wrapped bread was introduced into the UK, the innovation proved unpopular among housewives. One reason was that the wrappers became dirty and people failed to realise that, without wrappers, the dirt would be on the bread (Hardy, 1999). Hand-washing facilities were mostly unavailable and, where present, were rarely used initially. Toilet paper, too, was accepted reluctantly and, when it became available, the quality was very poor (Whitebread, 1926). Only when a new Food and Drug Act was introduced in the UK in 1938, was it necessary to use hygienic conditions and practices in handling, wrapping and delivering food, and adequate hand-washing facilities were required for food handlers.

A clear breakthrough in public health was the processing and disposal of domestic and sewage wastes, in conjunction with the purification of water supplies to ensure that any pathogens present were not passed to consumers via drinking water. Also, sanitary microbiologists were appointed to inspect food processing and eating establishments to ensure that proper food-handling procedures were followed. These made a significant contribution to the development of appropriate hygiene standards.

1.2 Definitions of hygiene

In ancient times, it was clear that diseases could be overcome, either by actively curing (Asclepius) or through the power of cleanliness (Hygeia). Curing diseases with the use of medicines was traditionally the role of the physician. Preventing diseases, on the other hand, became the domain of the hygienist. The first definitions of 'hygiene' are derived from the work of the goddess Hygeia:

- 'healing through cleanliness';
- 'the science dealing with the preservation and promotion of health'.

In the course of time, medicines became the principal means of curing diseases. However, because of the many failures during the 18th and 19th centuries, hygiene re-emerged as the key discipline. In the USA, the 'Natural Hygiene' movement came into being. The main objective of this science-based movement was not to treat the effect, but to remove the cause of a disease (treating the

Table 1.3 Definitions of food hygiene in current use.

- Conditions and practices that preserve the quality of food to prevent contamination and foodborne illnesses.

http://www.nlm.nih.gov/medlineplus/ency/article/002434.htm

- All measures necessary to ensure the safety and wholesomeness of foodstuffs.

EU's General Food Hygiene Directive (Anon., 1993).

- All conditions and measures necessary to ensure the safety and suitability of food at all stages of the food chain.

Codex Alimentarius Commission (CAC/RCP, 2003)

- The measures and conditions necessary to control hazards and ensure fitness for human consumption of a foodstuff, taking into account its intended use

Environmental Health Journal, 2000, 108/9; http://www.ehj-online.com/archive/2000/september/sept10.html; Council Directive 93/43, 1993.

effect without addressing the root cause was then the usual practice of medicine). Natural Hygiene addresses all aspects of living: the environment, food, work, home, economics, spirituality, psychology, politics, etc. and those other factors that positively influence health and well-being.

Following the recognition of germs as the principal causes of disease at the end of the 19th century, hygiene measures rapidly became established. By the beginning of the 20th century, it had become clear that preventive measures were the only way to produce safe food, and the discipline of food hygiene was born. Current definitions of 'food hygiene' are presented in Table 1.3.

Based on these definitions, it can be concluded that the concept involves all necessary measures to produce safe and healthy food. Any means to prevent contamination, decontaminate food (such as pasteurisation) and measures to improve wholesomeness and fitness for consumption are considered to be part of the hygiene concept. Various factors are contributory, such as personal hygiene and hygienic design of facilities, equipment, etc., as well as activities relating to cleaning and disinfection of food premises and hygienic disposal of waste, which are referred to as 'sanitation'.

1.2.1 Personal hygiene

Personal hygiene is of great importance for the maintenance of health in general. Human beings are natural carriers of many microorganisms and sources include the hair, skin, mucous membranes, digestive tract, wounds, infections and clothing. Good personal hygiene is primarily directed towards preventing both disease and discomfort. Hand-washing, dental care, avoidance of spitting, daily showering, etc., as well as clean living, play an important part. Disposal of waste is also important. All these measures are preventive in character and are readily carried out.

1.2.2 Hygienic design of facilities and equipment

Hygienic design of food production facilities, processing equipment, etc., is a most important factor in ensuring that food is safe and wholesome. Poorly designed farms, factories and equipment can easily result in contamination of food products and lead to food poisoning incidents. Furthermore, design deficiencies may result in losses of product due to spoilage, increased cleaning costs and reduced production time. These aspects are also of possible environmental concern. Therefore, it is essential that both manufacturers and users of food processing equipment are aware of hygienic design principles and requirements such as those described in EU Directives 98/37/EC and 93/43/EEC, and Hygienic Design DIN EN 1672/2 (1997). Hygienic production of food thus depends upon a combination of food processing procedures and hygienic design of buildings and equipment, in full compliance with legislation.

1.2.3 Sanitation

Sanitation is a term for the hygienic disposal or recycling of waste materials, particularly human excrement. In consequence, sanitation is an important public health measure that is essential for the prevention of disease. In the USA, there is a particular focus on the concept of food sanitation, which may be defined as 'the hygienic practices designed to maintain a clean and wholesome environment for food production, preparation and storage' (Marriot, 1999). This second definition links hygiene more specifically with maintaining a clean working environment for food processing. Even here, hygiene requirements extend beyond the practice of cleaning itself to incorporate those elements that make effective cleaning possible and allow control of insects and other pests. In the microbiological sense, sanitation is defined as 'a cleaning and disinfection process that results in a 99–99.9% reduction in the number of vegetative bacteria present'.

1.3 Sources of food contamination[4]

There are three main types of food contaminant:

- microbiological;
- chemical;
- physical.

Foods can become contaminated during growth and harvesting of raw materials, storage and transport to the factory, and processing into finished products. The final product may then become (re-)contaminated during subsequent storage and transport to shops, and during storage and preparation by the consumer. The main sources of contamination are the environment, animals and people. The main transmission routes (vectors) of contamination are contaminated surfaces,

4. Partly based on Lelieveld (2003).

air, water, people and pests. Processing, packaging material and equipment, and transport vehicles may also act as vectors. Contact between food material and an inert surface leaves residual food debris that favours the growth of microorganisms. Over time, these can multiply to significant numbers and become endemic in a processing plant. Chemical contamination may also result from contact with surfaces, if they are not adequately rinsed after cleaning and disinfection procedures. Lubricants, often unavoidable in equipment with moving parts, may also contribute to chemical contamination (Steenaard *et al.*, 2002). Non-contact surfaces, such as floors, walls, ceilings, overhead beams and equipment supports, are potential reservoirs of microbial contamination and can also be a source of physical and chemical contaminants (e.g. from flaking plaster and its associated chemicals). They need to be designed so that they are durable and can be cleaned effectively.

Animals are important reservoirs of microorganisms, and slaughter animals introduce large numbers of microorganisms into the processing plant. Among them are many so-called zoonotic pathogens that are present on the skin and in the gastrointestinal and respiratory tracts. Pathogens carried on hands are also a major source of contamination (Taylor and Holah, 2000).

Air can be a significant medium for the transfer (vector) of contaminants to food products (Brown, 1996). Unless the air is filtered, microorganisms will be present, and air may also carry 'light' foreign bodies, such as dust, straw-type debris and insects. Chemical taints can enter the production area through airborne transmission. Water is used in the food industry as an ingredient, a processing aid and for cleaning. Its use as an ingredient or processing aid can give rise to both microbial and chemical contamination, so it is important to use water of a high microbiological and chemical quality (i.e. potable quality). Water used in hand-washing facilities poses a potential problem, as does that from condensation of steam or water vapour, leaking pipes and drains, and rainwater. Stagnant water is particularly hazardous, since microbial levels can increase rapidly under favourable conditions. The water used in cleaning programmes also needs to be of adequate quality (Holah, 1997; Dawson, 1998, 2000). Personnel can transfer enteric and respiratory pathogens to food, e.g. via aerosol droplets from coughing near the processing line (Guzewich and Ross, 1999). People can equally be vectors of physical contaminants, such as hair or fingernail fragments, earrings, plasters and small personal belongings.

Pests, such as birds, insects and rodents, are potentially a major contamination problem, and particular care needs to be taken to prevent their entry into food production areas. Buildings must be designed to keep them out. Floors, ceilings and walls should not allow insects and other invertebrates the chance to live and breed.

1.3.1 Microbial contaminants

Pathogenic microorganisms are the major safety concern for the food industry. The vast majority of outbreaks of food-related illness are due to microbial pathogens, rather than to chemical or physical contaminants. As they are generally undetectable by the unaided human senses (i.e. they do not usually cause colour changes or produce 'off'-flavours or taints in the food) and they are capable of rapid growth under favourable storage conditions, much time and effort are spent in controlling and/or eliminating them. Even if the microbes in a food are ultimately destroyed by cooking, they may have already produced toxins, so it is vital to prevent contamination through the use of hygienic practices. Like microbial pathogens, spoilage organisms can either be present naturally or gain access to food. Although not a food safety concern, increased levels of spoilage organisms will usually mean a reduction in the length of time that the food remains fit to eat. This can affect product quality and thus influence the consumer's perception of the product.

Growth of microorganisms depends on a number of factors, such as temperature, humidity/water activity (a_W), pH, availability of nutrients, presence or absence of oxygen and inhibitory compounds such as preservatives. Different organisms require different conditions for optimal growth (e.g. some grow only in the absence of oxygen, others prefer either warm or cool conditions). Bacterial growth is by the simple division of one cell into two (binary fission), and their number will increase exponentially under favourable conditions. The influence of factors such as temperature, oxygen, pH and a_W on microbial activity may be interdependent. Microbes generally become more sensitive to oxygen, pH and a_W at temperatures near growth minima or maxima. Often, bacteria grow at higher pH and a_W, and at a lower temperature under anaerobic conditions than they do aerobically. Organisms that grow at lower temperatures are usually aerobic and generally have a high a_W requirement. Lowering a_W by adding salt or excluding oxygen from foods (such as meat) that are held at a chill temperature dramatically reduces the growth rate of spoilage microbes.

Normally, some microbial growth occurs when any one of the factors that controls the growth rate is at a limiting level. If more than one factor becomes limiting, microbial growth is drastically curtailed or even completely prevented. Effective control of pathogenic and spoilage bacteria thus depends on a thorough understanding of the growth conditions favouring particular organisms. This understanding can be used to minimise contamination of incoming raw materials, to inactivate bacteria during processing and prevent decontaminated food from becoming recontaminated. It is also important to know where and how microorganisms can become established, if growth conditions are favourable. They are particularly attracted to surfaces that provide a stable environment for survival and growth. Surfaces exposed to the air are always vulnerable unless frequently and effectively cleaned and disinfected. However, surfaces within closed equipment may also be vulnerable. There are usually places in processing lines, even when correctly designed, where some product residues remain longer than is desirable. Even if 'dead' areas have been 'designed out', some product will attach to equipment surfaces, despite the possibility of fast-moving liquids. Microbes may reside on such surfaces long enough to multiply, and contaminate the product. The problem is exacerbated when a process includes dead spaces where product can stagnate.

As an example, if a single cell of *Escherichia coli* is trapped in a dead space filled with 5 ml of a slightly viscous low-acid food product at a temperature of approx. 25 °C, it could take less than 24 hours for the number of microbial cells to increase to 0.2×10^9 per ml, assuming they double every 40 minutes (Lelieveld, 2000). If 1 ml per hour is washed out from the dead space by the passing product, then the product would be contaminated with 200 million *E. coli* cells per hour, by the end of the first day's production. If the production capacity of the line is 5×10^6 ml per hour, the average level of *E. coli* contamination would be 200/5 = 40 per ml. Many traditional process lines have much larger (often very contaminated) dead spaces and growth-rates can be higher if conditions permit.

Microbes may also penetrate through very small leaks. There is considerable evidence that they can pass through microscopic openings very rapidly and that pressure differences may retard, but not prevent, passage, even if the pressure difference is as high as 0.5 bar. The bacterium *Serratia marcescens* may move at a speed of 160 mm per hour (Schneider and Dietsch, 1974). Motile bacteria may propel themselves against the flow of liquid through a leak. Whether motile or not, they may also penetrate by forming a biofilm on the surface. Studies on the migration of microorganisms through microscopic channels show that passage can occur through holes a few micrometres in diameter in a metal plate of 0.1 mm thickness (Brénot *et al.*, 1995).

When attracted to a surface, microbes are deposited, attach and initiate growth. As they grow and multiply, the newly formed cells attach to each other, as well as to the surface, forming a growing colony. When this mass of cells becomes large enough to entrap debris, nutrients and other microorganisms, a microbial biofilm is established (IFT, 1994). Biofilms form in two stages. First, an electrostatic attraction occurs between the surface and the microbe. The process is reversible at this stage. The next phase occurs when the organism forms an extracellular polysaccharide, which firmly attaches the cell to the surface. The cell then multiplies, forming micro-colonies and, ultimately, the biofilm (Notermans *et al.*, 1991). These films are very difficult to remove during cleaning operations (Firstenberg *et al.*, 1979). Microorganisms that appear to be more difficult to remove because of biofilm formation include the pathogens *Staphylococcus aureus* and *Listeria monocytogenes* (Notermans, 1979). Current information suggests that heat treatment is more effective than the application of chemical sanitisers, and Teflon appears to be easier to clear of biofilm than stainless steel (Marriott, 1999).

Biofilm development may take place on any type of surface and is difficult to prevent, if conditions sustain microbial growth. Many organisms, including a number of pathogens (*Listeria monocytogenes, Salmonella* Typhimurium, *Yersinia enterocolitica, Klebsiella pneumoniae, Legionella pneumophila* and *Staphylococcus aureus*) form biofilms, even under hostile conditions, such as the presence of disinfectants. Adverse conditions even stimulate microorganisms to grow in biofilms (van der Wende *et al.*, 1989; van der Wende and Characklis, 1990). Thermophilic bacteria (such as *Streptococcus thermophilus*) can form a biofilm in the cooling section of a milk pasteuriser, sometimes within 5 hours, resulting in massive contamination of the pasteurized product (up to 10^6 cells per ml) (Driessen and Bauman, 1979; Langeveld *et al.*, 1995). On metal (including stainless steel) surfaces, biofilms may also enhance corrosion, leading to the development of microscopic holes. Such pinholes allow the passage of microbes and thus may cause contamination of the product. Like other causes of fouling, biofilms will also affect heat transfer in heat exchangers. On temperature probes, biofilms may seriously affect heat transfer and thereby the accuracy of the measurement. Reducing the effectiveness of heat treatment may itself help to stimulate further bacterial growth. On conveyor belts and on the surfaces of blanching equipment, for example, biofilms may contaminate cooked or washed products, which are assumed to have been made pathogen-free by the temperature treatment received.

Biofilms may be much more difficult to remove than ordinary soil. If the cleaning procedure used is not capable of removing the biofilm completely, decontamination of the surface by either heat or chemicals may fail, since a biofilm dramatically increases the resistance of the embedded organisms (IFT, 1994). It is therefore imperative that product contact-surfaces are well cleaned before disinfection. Krysinski *et al.* (1992) studied the effects of a variety of cleaning and sanitising compounds on *L. monocytogenes*, which was allowed to attach to stainless steel and plastic material used in conveyor belts over a period of 24 hours. They found that sanitisers alone had little effect on the attached organisms, even when the exposure time was increased to 10 mins. Unattached cells, on the other hand, showed a 5-log reduction in numbers within 30 seconds. In general, acidic quaternary ammonium compounds, chlorine dioxide and peracetic acid were the most effective sanitisers for eliminating attached cells. Least effective were chlorine, iodophors and neutral quaternary ammonium compounds. When the attached organisms were exposed to cleaning compounds prior to treatment with sanitisers, the bacteria were readily inactivated.

1.4 Hygiene control measures in food processing

Hygiene in food processing started with the introduction of general measures, including cleaning and disinfection, prevention of recontamination and treatment of food products to kill any microbial pathogens present. Heat treatment was introduced into food processing even before the underlying causes of foodborne illness were known. It was Nicholas Appert in France and Peter Durand in England who introduced canning of food and the use of thermal processing around 1800. However, neither Appert nor Durand understood why thermally processed foods did not spoil and remained safe to eat (Hartman, 1997). Then, Louis Pasteur showed that certain bacteria were either associated with food spoilage or caused specific diseases. Based on Pasteur's findings, commercial heat treatment of wine was first used in 1867 to destroy any undesirable microorganisms, and the process was described as 'pasteurisation'. This process was also recommended by Escherich (1890) to decontaminate milk.

In the course of time, it became clear that the effects of certain antimicrobial treatments were predictable. Two historical examples were the setting of performance criteria for destroying spores of *Clostridium botulinum* in low-acid, canned foods by Esty and Meyer (1922) and the process criteria for *Coxiëlla burnetii* in milk pasteurisation, as determined by Enright *et al.* (1957). Further research resulted in predictions relating to many other processes, such as acidification, drying and the use of curing agents in meat products, on both pathogenic and spoilage organisms. Such knowledge ushered in a new era in safe food production. This era is characterised by the division of hygiene measures into specific practices that are controllable and other general measures, the effects of which are largely unpredictable at present.

1.4.1 General hygiene practices

One of the first safety systems developed by the food industry was that involving the application of good manufacturing practice (GMP), as a supplement to end-product testing. GMP covers all aspects of production, from starting materials, premises and equipment to the training of staff, and the WHO has established detailed guidelines. GMP also provide a framework for hygienic food production, which is often referred to as good hygienic practice (GHP). The establishment of GHP is the outcome of long practical experience, and has the following major components:

- **Design of premises and equipment**. This includes the location and layout of the premises to avoid hygiene hazards and facilitate safe food production. Food processing and handling equipment should always be designed with hygiene in mind, including ease of cleaning.
- **Control of the production process**. Control measures are applied throughout the supply chain and cover factors such as raw materials, packaging and process water, as well as the product itself. Key aspects include management and supervision of the process as a whole, as well as appropriate recording systems.
- **Plant maintenance and cleaning**. Both processing equipment and the fabric of the building should be maintained in good order. Suitable programmes need to be developed for plant cleaning and disinfection, and their effectiveness monitored routinely. Systems are also needed for pest control and management of waste.
- **Personal hygiene**. Staff are required to maintain high standards of personal hygiene in relation to wearing of protective clothing, hand-washing and general behaviour. Visitors must also be strictly controlled in these respects. The health status of personnel should be monitored regularly and any illness or injuries recorded.

- **Transportation.** Requirements should be established for the use and maintenance of transport vehicles, including their cleaning and disinfection. Vehicle usage should be managed and supervised.
- **Product information and consumer awareness.** It is important that the final product is suitably labelled and that the consumer is provided with all relevant information on product handling and storage, including a 'use-by' date. Labelling should also indicate the batch and origin of the product, so that full traceability is possible.
- **Staff training.** In relation to food hygiene and safety, all personnel should receive appropriate training and be made fully aware of their individual responsibilities. Such training should be repeated and updated as required.

The GHP concept is largely subjective and its benefits tend to be qualitative rather than quantitative. It has no direct relationship to the safety status of the product, but its application is considered to be a necessary preventive measure in producing safe food. Those hygiene measures that have a predictable outcome and are subject to control can be incorporated in the Hazard Analysis Critical Control Point (HACCP) concept. This concept seeks, among other things, to avoid reliance on microbiological testing of the end-product as a means of controlling food safety. Such testing may fail to distinguish between safe and unsafe batches of food and is both time-consuming and relatively costly. However, effective application of the HACCP concept depends upon GHP being used.

1.4.2 HACCP

The HACCP concept is a systematic approach to the identification, assessment and control of hazards in a particular food operation. It aims to identify problems before they occur and establish measures for their control at stages in production that are critical to ensuring the safety of food. Control is based on scientific knowledge and is proactive, since remedial action is taken in advance of problems occurring. The key aspects fall into four main categories:

- quality of the raw materials used;
- the type of process used, which may include heat treatment, irradiation, high-pressure technology, etc.;
- product composition, including addition of, e.g., salt, acids or other preservatives;
- storage conditions, involving storage temperature and time, gas packaging etc.

The effects of the last three categories on the hygienic condition of the end-product are predictable and relatively easy to determine. Effective management of these categories allows all food safety requirements to be met. In doing so, it is necessary to define criteria for process performance, product composition and storage conditions. The setting of such criteria is the task of the risk manager, and use of the HACCP concept is the managerial tool that ensures that the criteria will be met in practice.

In a review of the historical background, Barendsz (1995) and Untermann *et al.* (1996) described the development of the HACCP approach, which began in the 1960s. The concept arose from a collaboration between the Pillsbury Company, the US Army Natick Research and Development Laboratories and the US National Aeronautics and Space Administration. The original purpose was to establish a system of safe food production for use in human space travel. At that time, the limitations of end-product testing were already appreciated and therefore more attention was given to controlling the processes involved in food production and handling. When first introduced at a congress on food protection (Department of Health, Education and Welfare, 1972), the concept involved three principles: (i) hazard identification and characterisation; (ii) identification of critical control points (CCPs); and (iii) monitoring of the CCPs.

Many large food companies started to apply HACCP principles on a voluntary basis and, in 1985, the US National Academy of Science recommended that the system should be used. Further support came from the ICMSF (1988), which extended the concept to six principles. They added specification of criteria, corrective action and verification. In 1989, the US National Advisory Committee on Microbiological Criteria for Foods added a further principle: the establishment of documentation concerning all procedures and records appropriate to the principles and their application. Use of the HACCP system was given an international dimension by the Codex Alimentarius Commission (CAC), which published details of the principles involved and their practical application (CAC, Committee on Food Hygiene (1991). In 1997, the CAC laid down the 'final' set of principles and clarified the precise meaning of the different terms (CAC, Committee on Food Hygiene, 1997):

- General principles of food hygiene (Alinorm 97/13. Appendix II).
- HACCP system and guidelines for its application (Alinorm 97/13A, Appendix II).
- Principles for the establishment and application of microbiological criteria for foods (Alinorm 97/13A, Appendix III).

The full HACCP system, as described in Alinorm 97/13, is shown in Table 1.4. The document also gives guidelines for practical application of the HACCP system. By 1973, the Food and Drug Administration (FDA) had made the use of HACCP principles mandatory for the production of low-acid canned foods (FDA, 1973) and, in 1993, the system became a legal requirement for all food products in the European Union (Directive 93/43).

It was Notermans *et al.* (1995) who first made a plea for the principles of quantitative risk assessment to be used in setting critical limits at the critical control points (CCPs) (process performance, product and storage criteria). It was their opinion that only when the critical limits are defined in quantitative terms can the level of control at CCPs be expressed realistically. At the International Association of Food Protection (IAFP) meeting in 2001, Buchanan *et al.* (2001)

Table 1.4 The seven principles of the HACCP system (CAC, Committee on Food Hygiene, 1997)

Principle	Activity
1. Conduct a hazard analysis	List all potential hazards associated with each step, conduct a hazard analysis, and consider any measures to control identified hazards
2. Determine the critical control points (CCPs)	Determine critical control points (CCPs)
3. Establish critical limit(s)	Establish critical limits for each CCP
4. Monitoring	Establish a system of monitoring for each CCP
5. Establish corrective actions	Establish the corrective action to be taken when monitoring indicates that a particular CCP is not under control
6. Establish verification procedures	Establish procedures for verification to confirm that the HACCP system is working effectively
7. Establish documentation and record keeping	Establish documentation concerning all procedures and records appropriate to these principles and their application

also favoured the use of these principles and suggested that food safety objectives should encompass end-product criteria, which are related to the criteria used in processing. New developments in the HACCP system concern the verification process. These involve verifying the criteria and/or food safety objectives set and use of a probabilistic approach to assessing risk reduction, thus providing information on the degree of control obtained.

1.5 Future trends

1.5.1 Improving information on foodborne diseases

As indicated earlier, present information is far from complete and, in 50–60% of cases of acute enteritis, a causative agent is not detected (de Wit *et al.*, 2001) In order to define better the burden of such diseases, novel techniques should be developed to test for unsuspected pathogens. For this purpose, a multifactorial approach is advocated and should include a study of the aetiology of unsuspected foodborne agents and their epidemiology, the risk factors involved, identification of virulence genes, demographic factors, clinical characteristics, etc. Knowledge of the relevant risk factors and their contribution to the problem is particularly important for the development of appropriate intervention strategies, and this aspect also needs to have an international dimension.

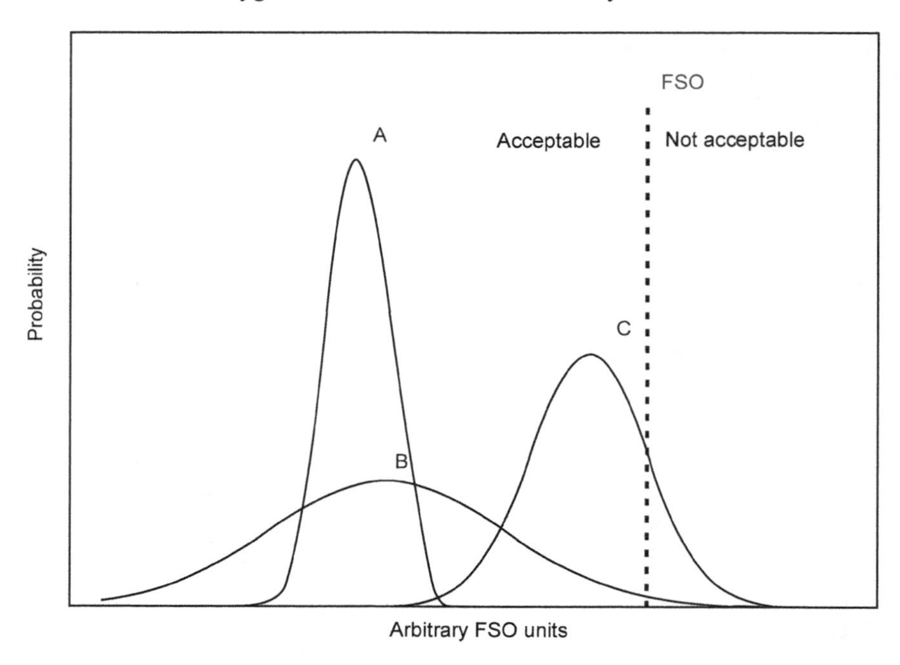

Fig. 1.1 HACCP-verification based on a probabilistic approach. The Food Safety Objective (FSO) is set as a criterion that separates 'acceptable' and 'unacceptable' products.

1.5.2 Assessment of process performance

Verification of HACCP involves the establishment of procedures to confirm that the HACCP system is working effectively. However, this stage is still in its infancy. Currently, verification is limited to demonstrating that controls are operating as intended and no proper data are collected. Instead, it is possible to determine the effects of control measures by carrying out a risk assessment. The principles of such an approach are in given in Fig. 1.1. Values to the left of the food safety objective (FSO[5]) are considered to be acceptable and values to the right are unacceptable.

Instead of 'single-point estimates' that result from the performance of a particular process verification data are presented in a probabilistic way. A single point estimate does not provide any information on the probability of exceeding the FSO. The curves A, B and C are so-called 'probability distribution curves' that are based on three levels of process performance. It can now be seen that, in some cases, the FSO is exceeded. The process performance values expressed in curves B and C are unacceptable because a substantial proportion of the product

5. An FSO may be a criterion or a target. When the criterion or target is met, an appropriate level of protection will be obtained at the time of consumption.

is beyond the FSO. Scenario B shows that the average is well within the target, but because of the large variation in part of the process, the FSO will be exceeded. Curve A is an example of an acceptable curve: the product meets the required FSO and the relatively small standard deviation of the curve indicates that the process is under control while this is not the case for curve B.

Another drawback of the present verification process is that food production is subject to unobserved changes. However, HACCP is based only on existing knowledge and, therefore, it is recommended that consumer complaints are also considered in the process of verification.

1.5.3 Further development of hygiene control

From long experience, it has become clear that certain hygiene controls are very effective in reducing foodborne disease, and the effects of certain measures, such as heating the product, have a predictable outcome. Thus, they have been incorporated eventually in the HACCP system. However, there are still a large number of important measures that contribute to food safety but their effects are neither quantifiable nor properly understood. Examples include the effects of cleaning and disinfection, steps to prevent cross-contamination in food processing and hand-washing and other aspects of personal hygiene. On the other hand, microorganisms may sometimes become established unexpectedly in processing equipment and food production facilities, thus increasing contamination of the product. In this case, the usual process parameters are controlled, but other, unknown factors are having an effect. Clearly, more information is needed on the factors that affect product safety and those that have little or no effect.

1.5.4 Changing pattern of microbial hazards

Society is increasingly confronted with microbial problems that are not susceptible to control by traditional measures. This may involve new hazards, including viral contamination of food and the occurrence of bacteria resistant to antibiotics and disinfectants. Many of these problems arise from the introduction of new technologies, new methods of producing raw food-materials and socio-economic changes in society, including overcrowding, increased travelling and global food production and trade. Foodborne disease continues at a high level, despite increasing attention to food hygiene, and with no alternative strategy available. This situation is an important challenge to modern society and requires a degree of foresight that goes well beyond present concepts of hygiene control. There is a similar problem with the availability of potable water. In developing countries, more than one billion people have no access to a basic water supply and 2.4 billion have no proper sanitation. The developed world has problems too in this respect, with climate change leading to water shortages in many areas. Can all these problems be overcome by technology?

1.5.5 Building hygiene into the system

A new research area that aims to improve general hygiene involves nanotechnology. This technology is a promising means of developing processes that are inherently hygienic. For example, coatings based on nanotechnology can make the environment more hygienic by preventing bacterial attachment to surfaces (ceilings, floors and walls of processing facilities, conveyor belts, etc.) and/or bacterial proliferation on these surfaces. Coatings have already been developed and successfully applied to prevent fouling of, for example, windows, water closets and tiles.

Another example concerns photocatalytic oxidation technology (www.cuhk.edu.hk/ipro/pressrelease/021007e.htm). The first application was developed by Professor Jimmy Yu Chai-mei of the Department of Chemistry, Hong Kong University, and involves the deposition of a uniform, nanometre-thick titanium dioxide coating on a solid substrate. The coating exhibits strong photocatalytic activity when exposed to visible light that results in the emission of local ultraviolet irradiation. As a result, it can oxidise most organic and inorganic pollutants, and kill bacteria such as *Escherichia coli* and *Vibrio cholerae* within seconds. This leads to a very attractive and safe technology for water treatment. The new treatment system has proved to be more effective than conventional UV irradiation, and it is said to be suitable for producing drinking water and treating industrial or agricultural wastewater and seawater. A similar air-purification system can be installed in hospitals, offices, schools, restaurants and homes. Thus, modern technology can do much to protect society from pathogenic agents, but this takes no account of one important factor: natural disease resistance. Without such resistance, human beings will continue to be highly vulnerable and require ever more protection from pathogens.

1.6 References

ANON. (1993), 'Council directive 93/94.EEC of 14 June 1993 on the hygiene of foodstuffs', *Off. J. Eur. Comm*, **L 157**, 1–11.

BAKER A M (1994), *Awaking our Self-healing Body – A solution to the health care crisis*, Self Health Care Systems, Los Angeles, p. 5.

BARENDSZ A W (1995), 'Kwaliteitsmanagement: HACCP de ontbrekende schakel' in *HACCP, A Practical Manual*, Keesing Noordervliet, Houten.

BRÉNOT O, DALEBOUT A and ODEN C (1995), personal communication.

BROWN K L (1996), *Guidelines on Air Quality Standards for the Food Industry*, Guideline No. 12, Campden & Chorleywood Food Research Association, Chipping Campden.

BUCHANAN R L *et al.* (2001), 'Moving beyond HACCP – Risk management and food safety objectives', in Symposium Abstracts IAFP 88th Annual Meeting, Minneapolis.

CAC, COMMITTEE ON FOOD HYGIENE (1991), *Draft Principles and Applications of the Hazard Analysis Critical Control Point (HACCP) System*. Alinorm 93/13, Appendix VI. Food and Agriculture Organization, World Health Organization, Rome.

CAC, COMMITTEE ON FOOD HYGIENE (1997), *Hazard Analysis Critical Control Point (HACCP) and Guidelines for its Application.* Alinorm 97/13. Food and Agriculture Organization, World Health Organization, Rome.

CAC, COMMITTEE ON FOOD HYGIENE/RCP (2003), *Recommended International Code of Practice General Principles of Food Hygiene.* CAC/RCP 1-1969, Rev. 4-2003.

COUNCIL DIRECTIVE 93/43/EEC (1993) of 14 June 1993 on the hygiene of foodstuffs. Official Journal L175, 19/07/1993, pp. 0001–0011.

DAWSON D (1998), *Water Quality for the Food Industry: an introductory manual,* Campden & Chorleywood Food Research Association, Chipping Campden.

DAWSON D (2000), *Water Quality for the Food Industry: management and microbiological issues*, Campden & Chorleywood Food Research Association, Chipping Campden.

DEPARTMENT OF HEALTH, EDUCATION AND WELFARE (1972), Proceedings '*National Conference on Food Protection*', US Governmental Printing Office, Washington, DC.

DEWBERRY E B (1959), 'Food poisoning', in *Food-borne Infection and Intoxication*, Leonard Hill, London, pp 6–7.

DE WIT M A S, KOOPMANS M P G, KORTBEEK L M, VAN LEEUWEN N J, BARTELDS A I M and VAN DUYNHOVEN Y T H P (2001), 'Gastroenteritis in sentinel general practices, The Netherlands', *Emerging Infectious Dis*, **7**, No. 1, January–February.

DRIESSEN F M and BOUMAN S (1979), 'Growth of thermoresistant streptococci in cheese milk pasteurizers – experiment with a model pasteurizer', *Voedingsmiddelen-technologie*, **12**, 34–37.

DURHAM H E (1898), 'The present knowledge of outbreaks of meat poisoning', *Br Med J*, **I**, 1797–1801.

ENRIGHT J B, SADLER W W and THOMAS R C (1956), 'Observations on the thermal inactivation of the organism of Q fever in milk', *J Milk Food Technol*, **10**, 313–318.

ENRIGHT J B, SADLER W W and THOMAS R C (1957), 'Thermal inactivation of *Coxiella burnetii* and its relation to pasteurisation of milk', Public Health Service Publication No. 517. United States Government Printing Office, Washington, DC.

ESCHERICH T (1890). 'Ueber Milchsterilisirung zum Zwecke der Säuglingsernährung mit Demonstration eines neuen Apparates', *Berliner klinische Wochenschrift*, **27**, 1029–1033.

ESTY J R and MEYER K F (1922), 'The heat resistance of spores of *Bacillus botulinus* and allied anaerobes', *XI J Inf Dis*, **31**, 650–663.

FARBER J M and PETERKIN P I (2000), '*Listeria monocytogenes*', in Lund B M, Baird-Parker T C and Gould G W, *The Microbiological Safety and Quality of Food* – Vol. I, Aspen Publishers Inc, Gaithersburg, MD.

FDA (1973), '*Acidified foods and low acid foods in hermetically sealed containers*' in Code of US Federal Regulations, Title 21, 1 Parts 113 and 114 (renumbered since 1973), FDA, Washington, DC.

FIRSTENBERG-EDEN R, NOTERMANS S, THIEL F, HENSTRA S and KAMPELMACHER E H (1979), 'Scanning electron microscopic investigations into attachments of bacteria to teats of cows', *J Food Prot*, **42**, 305–309.

FITZGERALD M, SCALLAN E, COLLINS C, CROWLEY D, DALY L, DEVINE M, IGOE D, QUIGLEY T and SMYTH B (2004), 'Results of the first population based telephone survey of acute gastroenteritis in Northern Ireland and the Republic of Ireland', *Eurosurveillance Weekly*, 22 April, Volume 8, Issue 17.

GÄRTNER E (1888), 'Über die Fleischuergiftung in Frankenhausen a. Kyffh. und der Erreger derselben', *Korrespondenzblatt des Allgemeinen arztlichen Vereins von Thuringen*, **17**, 573–600.

GREENWOOD M (1935), *Epidemics and Crowd Diseases*, Williams and Norgate, London, p. 157.

GUZEWICH J and ROSS P (1999), 'Evaluation of risks related to microbiological contamination of ready-to-eat food by food preparation workers and the effectiveness of interventions to minimise those risks', Food and Drug Administration White Paper, FDA, CFSAN, in http://cfsan.fda.gov/~ear/.

HARDY A (1999), 'Food, hygiene, and the laboratory. A short history of food poisoning in Britain, circa 1850–1950', *Social History Med*, **12**, 293–311.

HARTMAN P A (1997), 'The evolution of food microbiology', in Doyle M P, Beuchat L R and Montville T J, *Food Microbiology: Fundamentals and frontiers*, ASM Press, Washington, 3–13.

HOLAH J T (1997), 'Microbiological control of food industry process waters: Guidelines on the use of chlorine dioxide and bromine as alternatives to chlorine'. Guideline No.15, Campden & Chorleywood Food Research Association, Chipping Campden.

HOLLIER J (1623), *Opera omnia practica*, Genevae.

HUCK P (1994), 'Infant mortality in nine industrial parishes in Northern England, 1813–1836', *Population Studies*, **48**, 513–526.

ICMSF (The International Commission on Microbiological Specifications of Foods) (1988), *Micro-organisms in Foods. Application of the hazard analysis critical control point (HACCP) system to ensure microbiological safety and quality*. Blackwell Scientific Publications, Oxford.

IFT (1994), 'Microbial attachment and biofilm formation: a new problem for the food industry?', *Food Technol*, **48**, 107–114.

KOCH R (1876), 'Die Aetiologie der MilzbrandKrankheit, begrundet auf die Entwicklungsgeschichte des Bacillus Anthracis' (The etiology of anthrax, based on the life history of *Bacillus anthracis*), *Beitrage zur Biologie der Phlanzen*, **2**, 277–310.

KOCH R (1883), 'New methods for the detection of microorganisms in soil, air and water', report at XI German Congress of Physicians in Berlin.

KRAMER J M and GILBERT R J (1989), *Bacillus cereus* and other Bacillus species. In Doyle M P, *Foodborne Bacterial Pathogens*, Marcel Dekker, Inc., New York.

KRYSINSKI E P, BROWN L J and MARCHISELLO T J (1992), 'Effect of cleaners and sanitizers on *Listeria monocytogenes* attached to product contact surfaces', *J Food Protect*, **55**, 246–251.

LANGEVELD L P M, MONTFORT-QUASIG R M G E VAN, WEERKAMP A H *et al.* (1995), 'Adherence, growth and release of bacteria in a tube heat exchanger for milk', *Netherl Milk & Dairy J*, **49**, 207–220.

LELIEVELD H L M (2000), 'Hygienic design of factories and equipment', in Lund B M *et al.* (eds), *The Microbiology of Food*, Aspen Publishers Inc., Gaithersburg, MD.

LELIEVELD H L M (2003), 'Sources of contamination', in Lelieveld H L M, *Hygiene in Food Processing*, Woodhead Publishing Ltd, Cambridge, pp. 61–75.

MARRIOT N (1999) *Principles of Food Sanitation*, fourth edition, Aspen Publishers, Inc., Gaithersburg, MD.

McCLANE B C (1979), '*Clostridium perfringens*', in Doyle M P, Beuchat L R and Montville T J, *Food Microbiology: fundamentals and frontiers*, ASM Press, Washington, DC, pp. 305–326.

NIVEN J (1909–1910), 'Summer diarrhoea and enteric fever', *Proc R Soc Med*, **3.2**, 133.
NOTERMANS S (1979), 'Attachment of bacteria to meat surfaces'. *Antonie van Leeuwenhoek*, **45**, 324–325
NOTERMANS S, DORMANS J A M A and MEAD G C (1991), 'Contribution of surface attachment to the establishment of microorganisms in food processing plants – a review', *Biofouling*, **5**, 21–36.
NOTERMANS S, GALLHOFF G, ZWIETERING M H and MEAD G C (1995), 'Identification of critical control points in the HACCP system with a quantitative effect on the safety of food products', *Food Microbiol*, **12**, 93–98.
ODDY D J and MILLAR S (1985), *Diet and Health in Modern Brittain*, Croom Helm, London, pp. 147–148.
OSTERTAG R (1902), *Handbook of Meat Inspection*, Washington, DC.
PECKHAM K F (1923–24), 'An outbreak of pork pie poisoning at Derby', *J Hygiene*, **22**, 69–76
PORTER C (1924–25), 'Cleanliness in food handling: impression of American methods', *J Sanitary Inst*, **XLV**, 289.
ROCOURT J and COSSART P (1997), '*Listeria monocytogenes*', in Doyle M P, Beuchat L R and Montville T J, *Food Microbiology, Fundamentals and frontiers*. ASM Press, Washington, DC.
SAVAGE W G (1909), 'Further report on the presence of Gaertner group of organisms in animal intestines', Medical Officer's Annual Report, Local Government Board, XXVIII, p. 479.
SAVAGE W H and BRUCE WHITE P (1925), '*Food poisoning*', Medical Research Council, Special Report Series no. 92.
SCHNEIDER W R and DIETSCH R N (1974), 'Velocity measurements of motile bacteria by use of a videotape recording technique', *Appl Microbiol*, **27**, 283–284.
SEMMELWEIS I P (1861), 'Die Ätiologie, der Begriff und die Prophylaxis des Kindbettfiebers', Pest-Wien-Leipzig, 1861. Translated into English by Murphy F R (1941), 'The etiology, the concept and prophylaxis of childbed fever', *Medical Classics* **5**, 350–773.
STEENAARD P, MAAS H, VAN DEN BOGAARD J, PINCHIN R and DE BOER M (2002), 'Production and use of food-grade lubricants', EHEDG Doc. 23, CCFRA Technology, Campden.
TANNAHILL R (1973), *Food in History*, Stein and Day Publishers, New York.
TAYLOR J H and HOLAH J T (2000), 'Hand hygiene in the food industry: a review', Review 18, Campden & Chorleywood Food Research Association, Chipping Campden.
UNTERMANN F, JAKOB P and STEPHAN R (1996), '35 Jahre HACCP-System. Von NASA-Komzept bis zu den Definitionen des Codex Alimentarius', *Fleischwirtschaft* **76**, 589–594.
VAN DER WENDE E and CHARACKLIS W G (1990), 'Biofilms in potable water distribution systems', in McFeters G A, *Drinking Water Microbiology: Progress and recent developments*, Brock/Springer Series in Contemporary Bioscience, Springer-Verlag, New York, pp. 249–268.
VAN DER WENDE E, CHARACKLIS W G AND SMITH D B (1989), 'Biofilms and bacterial drinking water quality', *Water Res*, **23**, 1313–1322.
VAN ERMENGEM E (1897), 'Ueber einem neuen anaeroben *Bacillus* und seine Beziehungen zum Botulismus', *Z Hyg Infectionskrankh*, **26**, 1–56. English translation (1979), *Rev Infect Dis*, **1**, 701–719.
WHITEBREAD F G (1926), 'Faecal organisms carriers', *Safe Med*, **34**, 734.

WILLSHAN G A, CHEASTY T, SMITH H R, O'BRIEN S J and ADAK G K (2001), 'Verocytotoxin-producing *Escherichia coli* (VTEC) O157 and other VTEC from human infections in England and Wales: 1995–1998', *J Med Microbiol*, **50**, 135–142.

WOODS R I, WOODWARD J and WATTERSON P (1988), 'The cause of rapid infant mortality decline in England and Wales, 1861–1921, part 1', *Population Studies*, **42**, 343–366.

Part I

Risks

2

The range of microbial risks in food processing

M. H. Zwietering and E. D. van Asselt, Wageningen University, The Netherlands

2.1 Introduction: the risk of microbial foodborne disease

Accurate estimates of foodborne diseases are difficult to make because of underreporting. If estimates are made, however, one can see that orders of magnitude of the various organisms differ so much that even with these uncertainties, global conclusions remain appropriate. It is estimated that in the USA there is a chance of about 1 in 3 per year of getting a foodborne illness, a chance of 1 in 800 of getting hospitalised, and a chance of 1 in 55 000 of dying (Mead *et al.*, 1999). It is useful to compare this estimate with other causes of death, to put it in perspective (Table 2.1). From this comparison one can conclude that although foodborne diseases are not among the most relevant causes of death, they are relevant and more important than minor causes such as lightning.

2.1.1 Microorganisms responsible for foodborne diseases

Comparing the various organisms can also show relevant trends (Table 2.2), we can see that in the USA noro-viruses, *Campylobacter* and *Salmonella* are responsible for more than 90% of the diseases, and for deaths *Toxoplasma* and *Listeria* are also of relevance. So only five organisms give rise to more than 90% of the problem (Mead *et al.*, 1999). That does not mean that one should not look at other organisms, since these are only estimates, which may change over time.

The comparison of different countries will result in different outcomes, because of real differences in risks since other products, procedures, habits are used, but also because of different surveillance systems. The contributions of

Table 2.1 Estimated chance of dying (per year) by different causes*

P dying	Number per million	Cause
1:115	8800	Total
1:10 000	100	Infectious disease
1:10 000	100	Suicide
1:15 000	75	Traffic accident
1:55 000	18	Foodborne disease
1:200 000	6	Drowning
1:10 000 000	0.1	Natural disasters
1:20 000 000	0.05	Lightning

* All data (except foodborne diseases, which are from the USA; Mead *et al.* (1999)) are based on Dutch statistical data.

Table 2.2 Relative importance of various causes for disease, hospitalisation and death (Mead *et al.*, 1999)

	Illness (%)	Hospital (%)	Death (%)
Bacillus cereus	0.198	0.014	0
Botulism	0.00042	0.076	0.246
Brucella	0.0056	0.100	0.306
Campylobacter	**14.2**	**17.3**	**5.7**
Clostridium perfringens	1.8	0.064	0.360
Escherichia coli	1.3	4.6	4.3
Listeria monocytogenes	0.018	3.8	**27.5**
Salmonella non-typhoidal	**9.7**	**25.7**	**30.4**
Shigella	0.649	2.0	0.790
Staphylococcus	1.3	2.9	0.107
Streptococcus	0.369	0.586	0
Vibrio	0.038	0.203	1.7
Yersinia enterolitica	0.628	1.8	0.126
Cryptosporidium parvum	0.217	0.327	0.365
Cyclospora cayetanensis	0.106	0.025	0.021
Giardia lamblia	1.4	0.822	0.055
Toxoplasma gondii	0.814	4.1	**20.7**
Trichinella spiralis	0.00038	0.0069	0.0086
Noro-viruses	**66.9**	**32.9**	**6.8**
Rotavirus	0.282	0.822	0
Astrovirus	0.282	0.205	0
Hepatitis A	0.030	0.891	0.460
Total	**100**	**100**	**100**

Bold figures represent the most important causative agents (values larger than 5%).

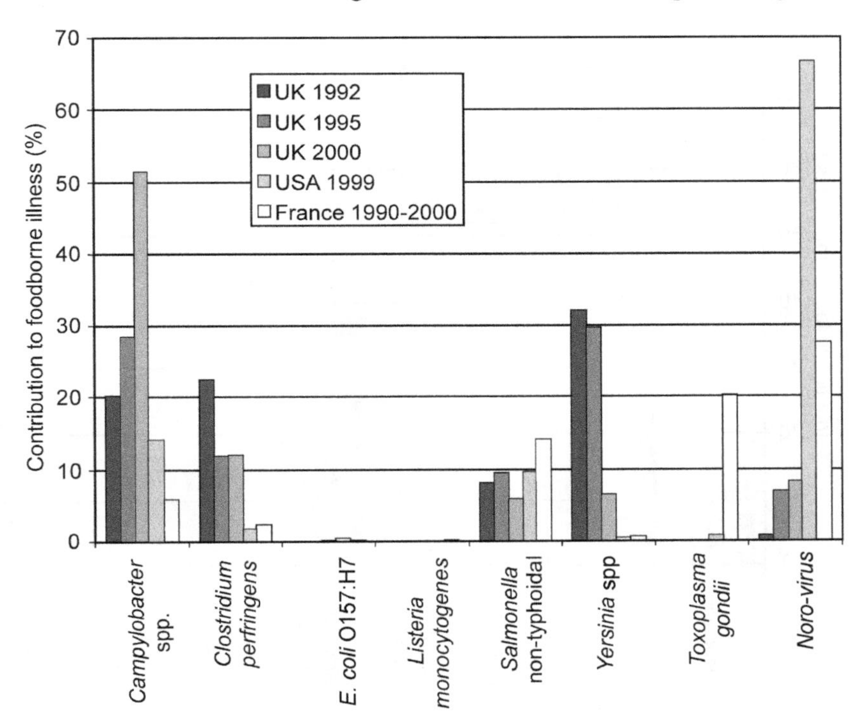

Fig. 2.1 The contribution of various pathogens to the total number of foodborne illnesses in percentages. No data available for *Toxoplasma gondii* in the UK.

various pathogens to the total number of foodborne illnesses and deaths are given in Figs 2.1 and 2.2 respectively, for surveys from the USA (Mead *et al.*, 1999), the UK (Adak *et al.*, 2002) and France (Vaillant *et al.*, 2004).

As described before, there is underreporting in the number of documented foodborne illnesses. To account for this underreporting, various surveys use an underreporting factor to obtain a realistic estimate of the real number of illnesses based on documented cases. The USA survey is based on surveillance systems and uses underreporting factors that are either based on literature studies or are estimated based on, for example, expert opinions (Mead *et al.*, 1999). The UK survey (Adak *et al.*, 2002) uses underreporting factors that are based on data from laboratory reports and the incidence rate in the population in 1995 found in a study to infectious intestinal disease. This factor is then applied to data from laboratory reports in other years. Which pathogens are most important in a country, therefore, also depends on the underreporting factor used (Table 2.3).

It can be seen that in all three countries noro-virus, *Campylobacter* and *Salmonella* are very important pathogens for causing foodborne illnesses. In the UK, *Clostridium perfringens* and *Yersinia* spp. are also in the top five of the most important pathogens. However, in these cases much higher underreporting factors are used than in the USA. Foodborne deaths in the three countries are

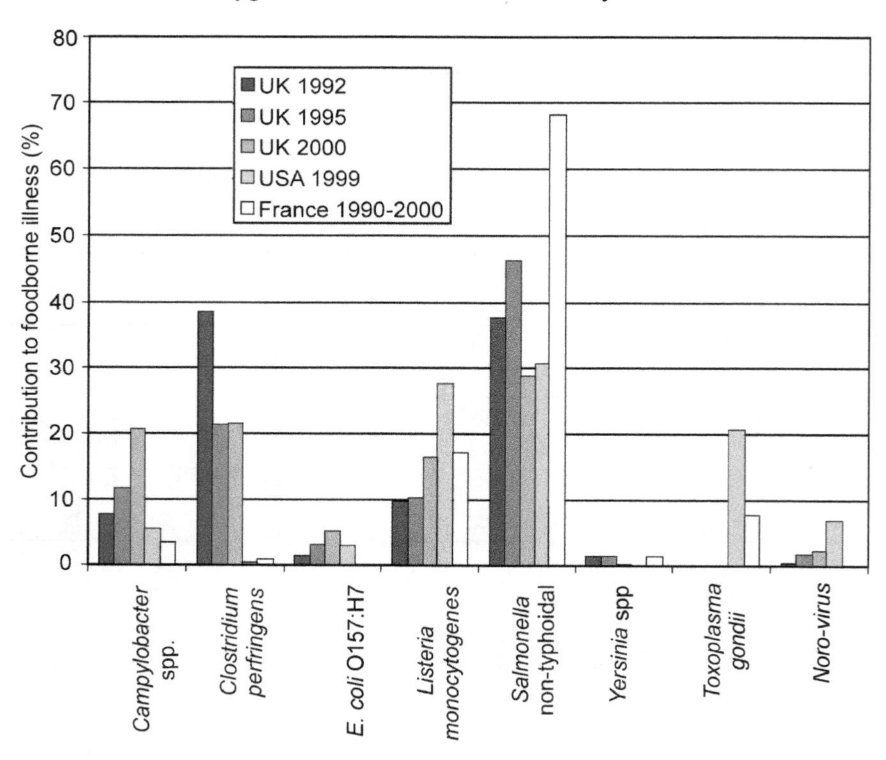

Fig. 2.2 The contribution of various pathogens to the total number of foodborne deaths in percentages. No data available for *Toxoplasma gondii* in the UK and for noro-virus in France.

Table 2.3 Ranking of pathogens according to their contribution to the overall number of foodborne illnesses or deaths together with the underreporting factor used in the UK (Adak *et al.*, 2002), USA (Mead *et al.*, 1999) and France (Vaillant *et al.*, 2004)

Microorganism	Ranking illness			Ranking death			Underreporting factor		
	UK[a]	USA	France	UK[a]	USA	France	UK[a]	USA	France
Campylobacter spp.	1	2	4	3	5	4	10	38	–
C. perfringens	2	4	5	2	7	6	364	38	–
L. monocytogenes	7	8	8	4	2	2	2	2	–
E. coli O157:H7	6	7	7	5	6	7	2	20	–
Salmonella spp.	5	3	3	1	1	1	4	38	–
Yersinia spp.	4	6	6	7	8	5	1254	38	–
Toxoplasma gondii	?	5	2	?	3	3	?	15	–
Noro-virus	3	1	1	6	4	?	276	?	–

[a] Data from 2000.
? Unknown.
– No under-reporting factor used.

mainly caused by *Salmonella* and *L. monocytogenes*. In the UK, *Cl. perfringens* is the second most important pathogen causing foodborne deaths, but again the underreporting factor is much higher than in the USA.

2.1.2 Related products

Once an important organism has been identified, it is important to identify its transmission route. One can follow three approaches:

- identify foods in outbreaks (by case-control, typing, ...);
- comparing types in cases/foods (sero-, phage-typing);
- quantitative risk assessment.

The first method is straightforward in that if a certain food/organism combination is suspected in a case or an outbreak, one can clearly establish a link between the food product and the case/outbreak by typing. Foods suspected in incidents are usually investigated by food inspection services and the WHO has collected these data for various countries in the world (Fig. 2.3) (Rocourt *et al.*, 2003). It can be seen that the products that contribute most to the number of foodborne illnesses are meat and meat products and eggs and egg products. This is probably caused by the presence of *Salmonella* and *Campylobacter* spp., which are known to be an important cause of foodborne outbreaks (see previous

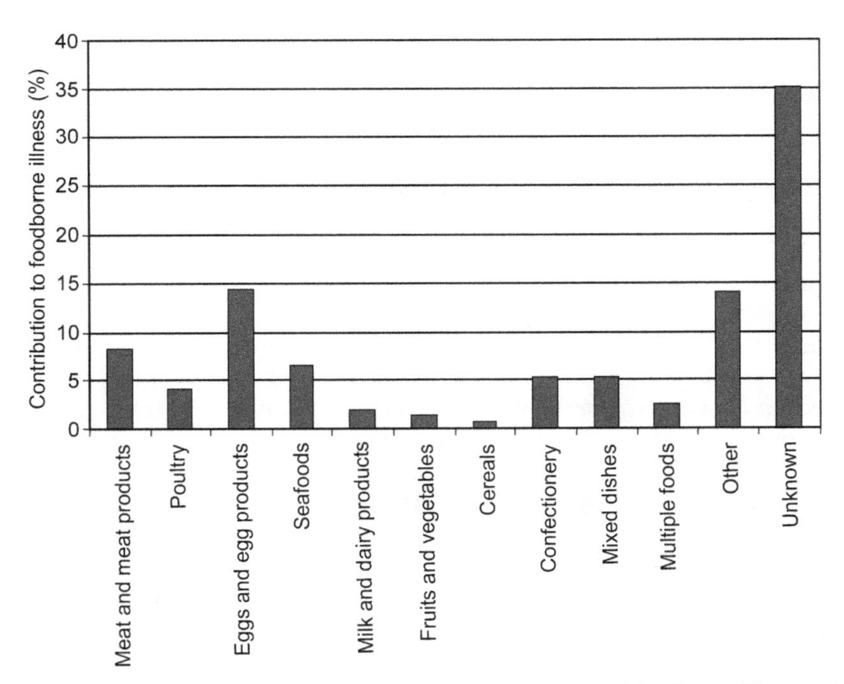

Fig. 2.3 Contribution of various food products to the number of foodborne illnesses for various countries in 1998–2001 (Rocourt *et al.*, 2003).

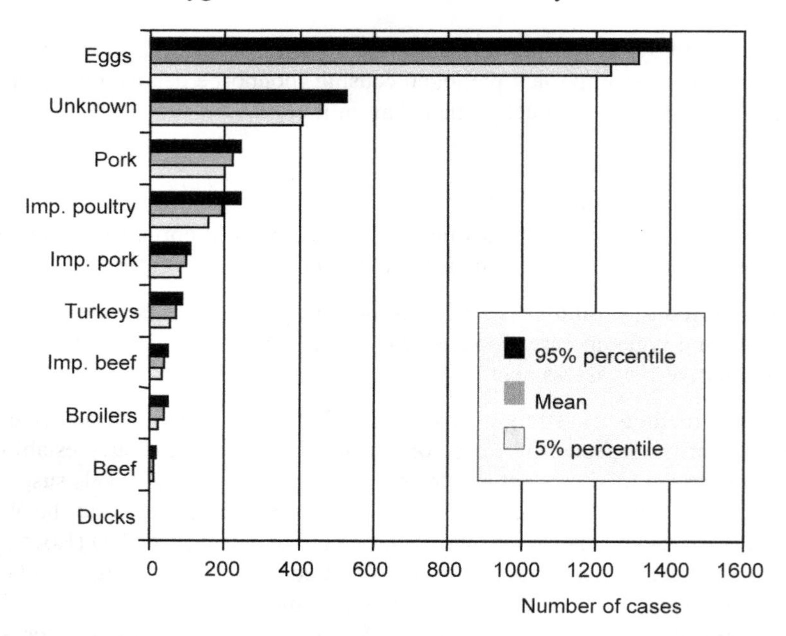

Fig. 2.4 Estimated number of sporadic and domestic cases of human salmonellosis in Denmark in 1999 attributed to different animal-food sources (Hald *et al.*, 2002).

section). Important considerations are that sporadic cases will often go undetected, since the focus with this method is on larger incidents. Moreover, it is often difficult to gather all relevant details of an outbreak, because of the time between the moment of consumption and the investigation.

An example of the second approach can be found in a study by Hald *et al.* (2002) who compared various types of salmonellae in patients with types in various food products, resulting in estimates of the contribution of these food products to the disease burden (Fig. 2.4). An example of the third approach concerns the use of quantification to determine the importance of various routes of contamination. Evers *et al.* (2004) determined the contribution in the exposure of *Campylobacter* of various routes. This type of quantification can effectively help to target interventions. The very extended FDA/FSIS (Food and Drug Administration/Food Safety Inspection Service) risk assessment (HHS/ USDA, 2003) concerning the relative risks of *Listeria monocytogenes* in ready-to-eat foods resulted in the identification of the most important product groups. Deli meats were identified as largely the most important source, followed unexpectedly by pasteurised milk. In this product, the risk per serving is not very high, but owing to the large number of consumed units for the per annum risk, and thus the number of cases in a year, it is a relevant source. The low prevalence of *Listeria* in this product is probably caused by recontamination (HHS/USDA, 2003).

2.2 The control of food safety

When a relevant hazard has been determined in a certain food product or a relevant contamination pathway has been determined, one should investigate interventions to control this hazard for this route of contamination. For this control, the HACCP system (Hazard Analysis Critical Control Points) can be used. In this system, the potential hazards are first determined, before deciding where these hazards can be controlled (CCPs, critical control points), what limits one should set, how one can monitor to ensure these limits are obeyed, what should be done if one is off-limit, how one can verify if everything is under control, and this all should be documented.

HACCP is not all that is needed (Fig. 2.5). Basic hygiene must be under control (GHP, good hygienic practice; GMP, good manufacturing practice). When this basic hygiene is not under control, HACCP will not be effective. Quantitative risk analysis (QRA) can help in setting limits, for example for specific organisms or for specific steps in the process, such as pasteurisation. Another important aspect is that apart from setting up the system, procedures should be followed strictly. There are always reasons to change certain things, but if one does not follow procedures strictly, the system will not be effective. Therefore, certification and ISO are important. Additionally, it is important that personnel are well educated, in order to prevent stupid errors. Continuous training and education is, therefore, also a relevant aspect.

With these structured systems, safety can be controlled, but zero risk is unattainable. One can, however, reduce risks by intelligent interventions. International organisations are moving more and more towards quantitative risk analysis. Also many recently created food safety authorities more and more frequently set quantitative objectives. If one wants, for example, to reduce the number of food infections by 30% in five years, it is clear, given the first tables, that one should focus on the most important causative organisms. Even if the

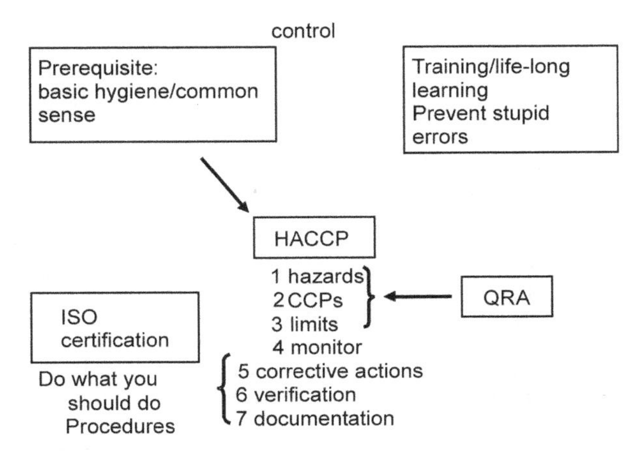

Fig. 2.5 Overview of food safety control systems.

whole *E. coli* problem could be reduced to zero, an overall reduction of 30% of total infections would not be achieved. These quantitative approaches give more transparency and more flexibility to control food safety problems in a chain at the location where it is the most effective. This can be well illustrated with the food safety objective (FSO) approach proposed by the International Commission on Microbiological Specifications for Foods (ICMSF, 2002).

2.3 Using food safety objectives to manage microbial risks

Definitions of a food safety objective (FSO) and related terms are provided in Table 2.4. The principle of FSO is very simple, the initial level (H_o) minus the sum of all reductions (R) plus the sum of all growth (G) and recontamination (C) must be smaller than the FSO, a limit set by governments:

$$H_o - \Sigma R + \Sigma G + \Sigma C < FSO \tag{2.1}$$

This FSO is the maximum frequency and/or concentration of a hazard in a food at the time of consumption that provides or contributes to the appropriate level of protection (ALOP) (CAC, 2004). This clearly shows the philosophy: a risk of zero does not exist. It is much better to set an appropriate level and perform actions to achieve this objective than to have one's head in the sand. Of course, this appropriate level is not something that remains the same forever; it can be changed for societal, political or technical reasons.

The FSO is a maximal concentration that will result in a certain, appropriate level of cases. In this respect, definitions and the correct reporting of units are crucial. If one talks about concentration (organisms per gram) or dose (organisms per consumption, for example 100 g), there is a difference of a factor of 100. The use of disease cases per consumption or per year can also easily differ by a factor of 100 if for a certain food product 100 units are consumed per year. It is also important to report whether a case is defined as infection, disease or death.

Table 2.4 Definitions of a food safety objective (FSO) and related terms

Food safety objective (FSO): The maximum frequency and/or concentration of a hazard in a food at the time of consumption that provides or contributes to the appropriate level of protection (ALOP).

Performance objective (PO): The maximum frquency and/or concentration of a hazard in a food at a specified step in the food chain before the time of consumption that provides or contributes to an FSO or ALOP, as applicable.

Performance criteria (PC): The effect in frequency and/or concentration of a hazard in a food that must be achieved by the application of one or more control measures to provide or contribute to a PO or an FSO.

Source: Codex Alimentarius Commission (2004): ALINORM 04/27/13; Appendix III (p. 83).

Apart from the fact that it is difficult to estimate the number of cases based on an FSO (or the other way around, to derive an FSO from an ALOP), it is also very difficult to set a specific ALOP. It is difficult both to determine what is appropriate and also to 'distribute' disease cases over various transmission routes. For example, one can set a level for campylobacteriosis (as the public health goal that is the result of food transmission and other sources), and needs to determine what the FSO for food products should be. In that case one should select what the specific ALOP will be for food transmission of this organism, since the FSO influences only the food transmitted part of all the cases but not any other sources that can cause campylobacteriosis. Secondly, an additional problem is that it is often not the product itself that gives the risk, but the fact that the product cross-contaminates other products, via utensils, surfaces or hands. The fact that it is difficult to set an ALOP does not mean that it should not be done. It is much better to do it directly based on the current state of knowledge and data than to wait until all information is available, since this will never be the case. However, if new information does become available, one should evaluate whether the level should be changed.

2.3.1 Distribution over the chain

A positive aspect about this concept is that once an FSO has been set, the objectives can be distributed over the whole chain from primary production to consumption. A performance objective (PO) can be set for every link in the chain, so that in total the FSO is achieved. This has the great advantage that the most efficient distribution of the objectives over the chain can be found: one has the flexibility to do more in the first stage, or in the last stage, or both. If the PO has been set for one stage, this can again be distributed over various process steps. This defines the performance criterion (PC), for example for a reduction step (pasteurisation) a 6 log reduction is necessary. With this criterion, one can then define process criteria that will attain this reduction (e.g. 72 °C, 15 s). This is indicated in Fig. 2.6, which shows the relation with HACCP and critical limits. The advantage of this concept is that one has the flexibility to change limits in one stage, as long as one equalises this in another. For example, a process criterion can be changed so that only 5 log reductions are achieved if this factor of 10 is balanced in another process step, or even in another stage in the chain.

One of the problems in setting FSOs, and in relating FSOs to ALOPS, is the fact that it is not the setting of a limit that determines the health burden, but that in many cases extreme levels are determining. This can be illustrated by a very large survey published by Gombas *et al.* (2003) in which 31 700 ready-to-eat foods were sampled for *Listeria*. Of these samples, 1.8% were contaminated with *Listeria* (577 samples). Only 2 out of the 577 positive samples (0.006% of the 31 700 products) contained more than 10^5 organisms per gram. If we determine the total exposure of all *Listeria* in these products, these two samples alone represented 97.5% of the total exposure in the 31 700 products, because of

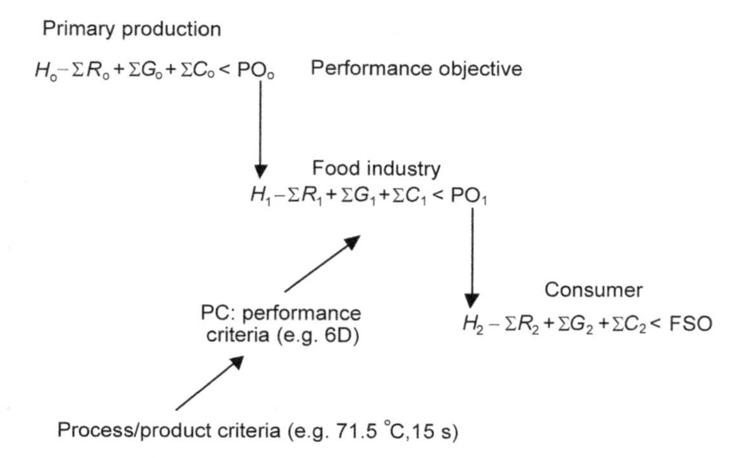

Fig. 2.6 FSO: link limits with end result.

their high contamination level. These samples are largely above every FSO that should be set. Therefore one could argue that it is then not so important where to set the limit, but how one controls the compliance, and especially how one can detect and, more importantly, prevent these low-frequency, extreme levels.

2.3.2 Quantitative methods

To estimate the values in the FSO equation one can use microbiological methods or use quantitative microbiology. Characteristic numbers (Zwietering, 2002) showing the change in log numbers can supply the necessary numbers for the equation in a direct way for every stage in the chain, with the first characteristic number, the step characteristic (*SC*):

$$SC = \frac{kt}{\ln(10)} \text{ for growth } (G) \text{ or inactivation } (R) \tag{2.2}$$

in which k is the specific growth rate or inactivation rate (depending on the temperature and other factors) and t is the time.

It should be noted that *SC* is only 'condition' dependent, i.e. the effect of a heat treatment remains the same whether the initial level of microorganisms is 10^3 organisms/g or 1 organism/g, e.g. a 6D reduction. Therefore, growth and inactivation are 'additive' on a logarithmic scale. If growth and inactivation processes are considered to follow first order kinetics, it is possible to express a process without recontamination as:

$$N = N_0 \exp(k_1 t) \exp(k_2 t) \exp(k_3 t) \exp(k_4 t) \ldots \tag{2.3}$$

with k the specific growth or inactivation rate, depending on the actual conditions in the stage.

On a log scale these kinetics become additive:

$$\log(N) = \log(N_0) + \frac{k_1 t}{\ln(10)} + \frac{k_2 t}{\ln(10)} + \frac{k_3 t}{\ln(10)} + \frac{k_4 t}{\ln(10)}$$
$$= H_0 + SC_1 + SC_2 + SC_3 + SC_4 \quad (2.4)$$

If, for example, SC_2 is an inactivation, and the other three growth, $\Sigma G = SC_1 + SC_3 + SC_4$ and $\Sigma R = SC_2$. In principle, the outcome will be equal if process steps are interchanged. It does not matter if first a 4 log growth and then a 6 log reduction takes place, or first a 6 log reduction and then 4 log growth: in both cases the result will be an overall 2 log reduction. This can also be seen from the fact that in eqn 2.2 the effect is dependent only on k and not on the actual level.

There are three exceptions:

1. If within growth the stationary phase is reached, but this is generally not the case for pathogens (and should not be).
2. If the number of organisms in a product unit becomes smaller than 1. Even in that case for large numbers of product units and proportional dose–response relations without threshold, this does not have an overall effect on the outcome of the risk estimate.
3. History effects may make stages interdependent.

In order to incorporate contamination in the calculations, one can use the second characteristic number, the contamination characteristic (*CC*):

$$CC = \log\left(\frac{N_{in} + R_c}{N_{in}}\right) \text{ for (re)contamination } (C) \quad (2.5)$$

in which N_{in} is the numbers entering the stage and R_c is the (re)contamination rate (in colony-forming units/g).

CC is not only condition dependent but also state dependent, depending on the number of entering microorganisms. Contamination is 'additive' on a linear scale and not on a logarithmic scale.

For a case where in all stages of the process both growth or inactivation and contamination can take place, one gets:

$$N = \{[[(N_0 + R_{c1})\exp(k_1 t) + R_{c2}]\exp(k_2 t) + R_{c3}]\exp(k_3 t) + R_{c4}\}\exp(k_4 t)\ldots \quad (2.6)$$

In this case the final effect can be totally different if contamination occurs at stage 1, 2, 3 or 4 (for example before or after pasteurisation). This can also be seen from eqn 2.5 where the characteristic number depends on the recontamination level (R_c) and on the actual state (N_{in}). A recontamination with 10 cells per gram is much more important if the actual concentration is 1 cfu/g than if it is already 100 cfu/g. This is illustrated in the following example (Fig. 2.7). An imaginary production process is chosen and *Staphylococcus aureus* is selected as the pathogen that can be present in the product. During the first production step (mixing) growth is an important factor causing an increase of more than 1 log cfu/g ($GC = 1.36$). When contamination takes place in the next

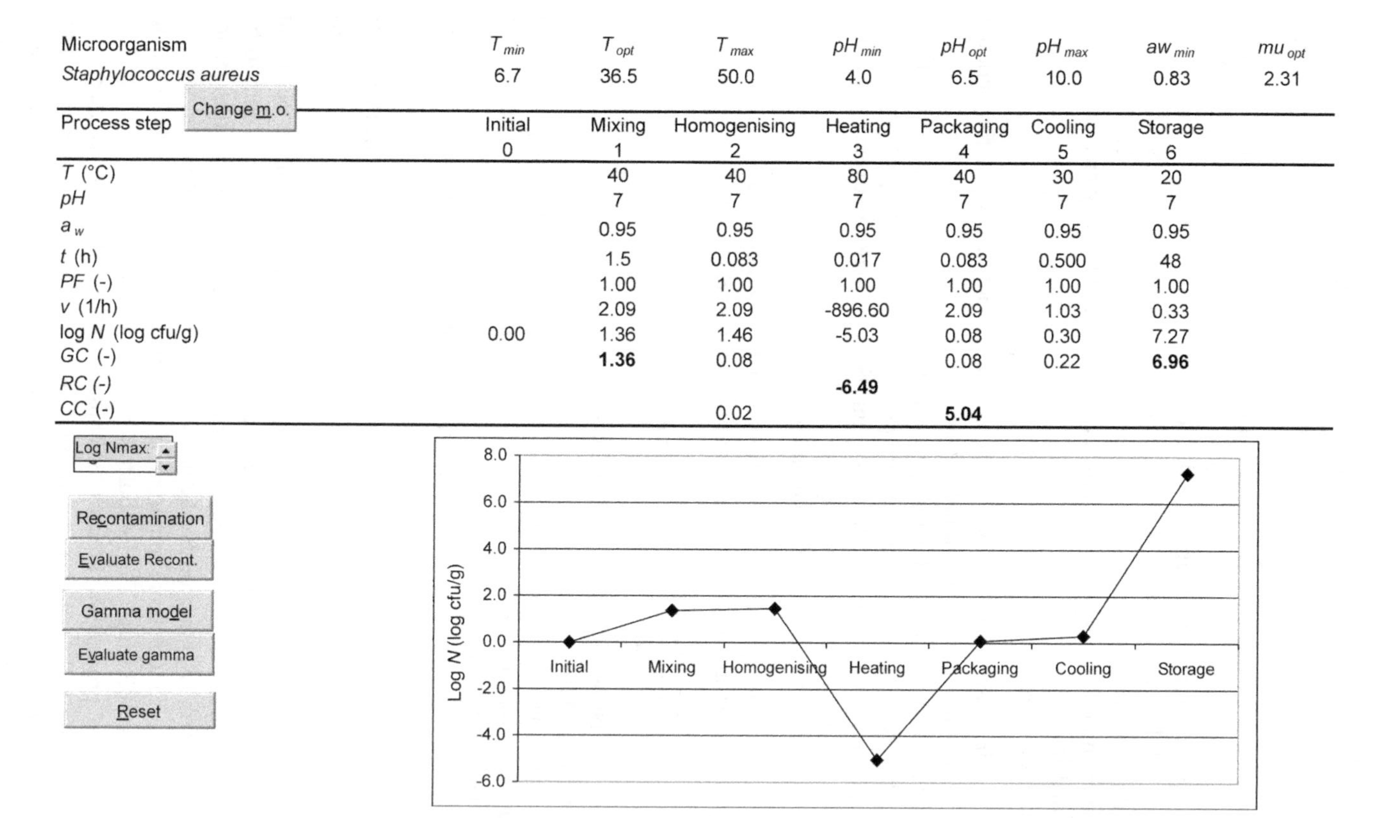

Microorganism	T_{min}	T_{opt}	T_{max}	pH_{min}	pH_{opt}	pH_{max}	aw_{min}	mu_{opt}
Staphylococcus aureus	6.7	36.5	50.0	4.0	6.5	10.0	0.83	2.31

Process step	Initial 0	Mixing 1	Homogenising 2	Heating 3	Packaging 4	Cooling 5	Storage 6
T (°C)		40	40	80	40	30	20
pH		7	7	7	7	7	7
a_w		0.95	0.95	0.95	0.95	0.95	0.95
t (h)		1.5	0.083	0.017	0.083	0.500	48
PF (-)		1.00	1.00	1.00	1.00	1.00	1.00
v (1/h)		2.09	2.09	-896.60	2.09	1.03	0.33
log N (log cfu/g)	0.00	1.36	1.46	-5.03	0.08	0.30	7.27
GC (-)		**1.36**	0.08		0.08	0.22	**6.96**
RC (-)				**-6.49**			
CC (-)			0.02		**5.04**		

Fig. 2.7 Evaluation of a production process contaminated with *Staph. aureus* with characteristic numbers *GC* (growth characteristic), *RC* (reduction characteristic) and *CC* (contamination characteristic). Bold numbers indicate changes in number of organisms larger than 1 log.

step (homogenising) with 1 cfu/g, this does not give a high increase in total number of *Staph. aureus* cells, since the concentration after the mixing step was already 1.36 log cfu/g (or 23 cfu/g). At the heating step, there is a large reduction of cells ($RC = -6.49$ log units). When contamination takes place after this heating step (packaging) with 1 cfu/g, this becomes a very important step ($CC = 5$ log units). Because of the heating step, almost all *Staph. aureus* cells are inactivated and additional contamination, although at a low level, thus causes a high increase in concentration. When growth is possible during storage, the product can end up with a high number of bacteria ($GC = 7$ log units). This example shows that the relevance of recontamination strongly depends on the number of microorganisms already present on the product and thus on the process stage.

In the whole process $H_0 = 0$ (or 1 cfu/g), $\Sigma G = 1.36 + 0.08 + 0.08 + 0.22 + 6.96 = 8.7$, $\Sigma R = -6.49$, $\Sigma C = 0.02 + 5.04 = 5.06$, resulting in an exposure of $0 + 8.7 - 6.49 + 5.06 = 7.27$ log cfu/g (see log N in the storage step).

2.3.3 Quantification of recontamination

Growth and inactivation can be modelled using various predictive models, such as the first order models as presented previously. Contamination, however, is more difficult to quantify. Nevertheless, attempts should be made to incorporate this factor in the FSO equation so that the relevance of contamination can be compared with growth and inactivation. Recontamination can take place at several stages in a production process. Examples are through biofilm formation in process lines, contaminated equipment via air, or at consumer level where cross-contamination can occur. A way to obtain cross-contamination in the kitchen is the use of the same cutting board to cut chicken followed by preparation of a salad. Cross-contamination then depends on the transfer rates of microorganisms from one surface to the next. Transfer rates from chicken to stainless steel vary between 0 and 10% with a mean of 1.6% for *Salmonella* and 2.4% for *Campylobacter*. Transfer from stainless steel to cucumber has a larger variation (between 0 and 100%) with a mean of 34.8% for *Salmonella* and 42.5% for *Campylobacter* (Kusumaningrum *et al.*, 2004). This means that when a chicken is contaminated with *Campylobacter* at a concentration of 4 log cfu/cm^2 (or 10^4 cfu/cm^2), the mean number of microorganisms on the stainless steel surface will be 2.4 log cfu/cm^2 (or 240 cfu/cm^2 = 2.4% of 10 000). The cucumber salad will then be contaminated with 2 log cfu/cm^2 (or 102 cfu/cm^2 since 42.5% of 240 = 102). This means that when around 50% is transferred, on a log scale this means that both surfaces end up with around the same concentration (initially 2.4 log (240 cfu/cm^2), after transfer 2.13 log (140 cfu/cm^2) left on the surface (57.5 % of 240 cfu/cm^2) and 2 log on the cucumber (42.5% of 240 cfu/cm^2)).

There are several models available to quantify the various recontamination routes in the production process (den Aantrekker *et al.*, 2002). A relatively simple model to quantify recontamination via the air was developed by Whyte (1986):

$$R_c = C_{air} v_s At / W \qquad (2.7)$$

where R_c is the contamination rate (cfu/g), C_{air} is the concentration of micro-organisms in the air (cfu/m^3), v_s is the settling velocity (m/s), A is the exposed product area (m^2), t is the exposure time (s), and W is the weight of the product (g).

For example, during the production of sliced meat, the product with A = 140 cm^2 and W = 17 g is exposed to the air for 45 s, resulting in a contamination level of 4 cfu/product or 0.2 cfu/g (C_{air} is 3.39 log cfu/m^3 and v_s is −2.59 log m/s) (den Aantrekker *et al.*, 2003). This means that when the product is sterile, contact with contaminated air causes an increase in concentration with 4 cfu/product.

Although these simple models do not incorporate all factors that may be of relevance, they can be used to provide an indication of the importance of air contamination compared with the initial contamination of the product and possible growth and inactivation during the production process.

2.4 Conclusions

Quantitative methods can help to determine which microorganisms give the highest contribution to the number of foodborne illnesses. This helps to decide which pathogen(s) one should focus on in order to reduce the number of illnesses. The same accounts for the products that are related to foodborne illnesses. Quantitative microbiology can be used in controlling food safety in a farm-to-fork approach. Using such models, one can try various approaches to obtain the same FSO.

More work needs to be done to incorporate recontamination in predictive models. The models available at the moment, although simple, can already be applied to determine the importance of (re)contamination compared with growth and inactivation of pathogens.

2.5 References

ADAK, G. K., LONG, S. M. and O'BRIEN, S. J. (2002), 'Trends in indigenous foodborne disease and deaths, England and Wales: 1992 to 2000', *Gut*, **51** (6), 832–841.

CAC (CODEX ALIMENTARIUS COMMISSION) (2004). *ALINORM 04/27/13* Appendix III, p. 83.

DEN AANTREKKER, E. D., BOOM, R. M., ZWIETERING, M. H. and VAN SCHOTHORST, M. (2002), 'Quantifying recontamination through factory environments – a review', *Int. J. Food Microbiol.*, **80** (2), 117–130.

DEN AANTREKKER, E. D., BEUMER, R. R., VAN GERWEN, S. J. C., ZWIETERING, M. H., VAN SCHOTHORST, M. and BOOM, R. M. (2003), 'Estimating the probability of recontamination via the air using Monte Carlo simulations', *Int. J. Food Microbiol.*, **87** (1-2), 1–15.

EVERS, E. G., FELS, H. J., NAUTA, M. H., SCHIJVEN, J. F. and HAVELAAR, A. H. (2004), *The relative importance of* Campylobacter *transmission routes based on human exposure estimates*, Bilthoven, the Netherlands, RIVM.

GOMBAS, D. E., CHEN, Y. H., CLAVERO, R. S. and SCOTT, V. N. (2003), 'Survey of *Listeria monocytogenes* in ready-to-eat foods', *J. Food Protect.*, **66** (4), 559–569.

HALD, T., VOSE, D. and WEGENER, H. C. (2002), 'Quantifying the contribution of animal-food sources to human salmonellosis in Denmark in 1999', in *Foodborne zoonoses: a co-ordinated food chain approach*, Bilthoven, the Netherlands, 83–86.

HHS/USDA (2003), *Quantitative assessment of relative risk to public health from foodborne* Listeria monocytogenes *among selected categories of ready-to-eat foods*, http://www.foodsafety.gov/~dms/lmr2-toc.html

ICMSF (2002) *Microorganisms in foods – microbiological testing in food safety management*, New York, USA, Kluwer Academic/Plenum Publishers.

KUSUMANINGRUM, H. D., VAN ASSELT, E. D., BEUMER, R. R. and ZWIETERING, M. H. (2004), 'A quantitative analysis of cross-contamination of *Salmonella* and *Campylobacter* via domestic kitchen surfaces', *J. Food Protect.*, **67**, 1892–1903.

MEAD, P. S., SLUTSKER, L., DIETZ, V., MCGAIG, L. F., BRESEE, J. S., SHAPIRO, C., GRIFFIN, P. M. and TAUXE, R. V. (1999), 'Food-related illness and death in the United States', *Emerg. Infect. Dis.*, **5** (5), 607–625.

ROCOURT, J., MOY, G., VIERK, K. and SCHLUNDT, J. (2003), *Present state of foodborne disease in OECD countries*, Geneva, Switzerland, World Health Organization.

VAILLANT, V., DE VALK, H. and BARON, E. (2004), *Morbidité et mortalité – dues aux maladies infectieuses d'origine alimentaire en France*, Saint-Maurice, France, Institut de Veille Sanitaire. http://www.invs.sante.fr

WHYTE, W. (1986), 'Sterility assurance and models for assessing airborne bacterial contamination', *J. Parent. Sci. Techn.*, **40** (5), 188–197.

ZWIETERING, M. H. (2002), 'Quantification of microbial quality and safety in minimally processed foods', *Int. Dairy J.*, **12** (1–3), 263–271.

3

Biofilm risks

G. Wirtanen and S. Salo, VTT Biotechnology, Finland

3.1 Introduction: biofilm formation and detection

This chapter deals with biofilm formation, sampling and detection methods, pathogens in biofilms, persistent and non-persistent microbes, prevention of biofilm formation and biofilm removal as well as future trends in biofilm control in the food industry. Microbes that inhabit contact and environmental sites in food processing are mostly harmful because microbial communities in the wrong places lead to contamination of surfaces and of the product produced in the process (Wirtanen, 1995). Documented biofilms have been almost entirely composed of bacteria, and the types of bacterial biofilms particularly related to pathogens are detailed in Section 3.2. There are, however, very few published studies concerning yeast biofilms in food processing. Storgårds *et al.* (1997) studied the tendency of spoilage yeasts isolated from brewery samples to form biofilms. This study showed that the slow-growing strains covered tested surfaces with 2–4% biofilm in 10 days; fast biofilm producers had already covered the whole surface in 2 days. In addition to the problems in food industry, biofilm formation also causes problems in food-related systems, e.g. industrial water systems as well as the paper and packaging industry (Bryers, 2000; Alakomi *et al.*, 2002). On the positive side, however, biofilms have also been applied successively in food-related processes, e.g. in brewing and in water treatment (Kronlöf, 1994; Zottola and Sasahara, 1994; Wong and Cerf, 1995; Bryers, 2000).

3.1.1 Factors affecting biofilm formation

In order to be able to survive hostile environmental factors such as heat and chemicals, microbes in microcolonies have a tendency to form protective extracellular matrices, which mainly consist of polysaccharides and glycoproteins, and are called biofilms (Wirtanen, 1995). The microcolony formation is the first stage in biofilm formation, which occurs under suitable conditions on any surface – both inert and living. Microbes can start up this formation when there is water or moisture available (Bryers, 2000). Physical parameters such as fluid flow rate, charge, hydrophobicity and micro-topography of the surface material affect the attachment of cells to the surface. Cells must overcome the energy-intensive repulsion barrier, which affects the particle surfaces (van Loosdrecht *et al.*, 1989). Bacteria with pili could conceivably overcome this barrier to achieve micro-colonisation and biofilm formation (Zottola & Sasahara, 1994). It has been found that temperatures below 50 °C promote biofilm formation (Miller & Bott, 1982).

In the food industry, equipment design plays the most important role in combating biofilm formations. The choice of materials and their surface treatments as well as roughness, e.g. grinding and polishing, are important factors for inhibiting the formation of biofilm and making surfaces easier to clean. Treating surface materials so that they reject biofilms can be performed actively to remove or passively to retard biofilm reoccurrence. The cleanliness of surfaces, training of personnel and good manufacturing and design practices are the most important tools in combating biofilm problems in the food industry (Holah & Timperley, 1999; Wirtanen, 2002).

3.1.2 Biofilm formation on food processing surfaces

It is also important to remember that about 85–96% of a biofilm consists of water, which means that only 2–5% of the total biofilm volume is detectable on dry surfaces (Costerton *et al.*, 1981). Biofilm can generally be produced by any microbes under suitable conditions, although some microbes naturally have a higher tendency to produce biofilm than others. A biofilm consists of microbial cell clusters with a network of internal channels or voids in the extracellular polysaccharide and glycoprotein matrix (Carpentier & Cerf, 1993). This allows nutrients and oxygen to be transported from the bulk liquid to the cells (Stoodley *et al.*, 1994; Kostyál, 1998).

It has been suggested that the mechanisms of microbial attachment and biofilm build-up occur in two-, three-, five- and eight-step processes (Wirtanen, 1995; Bryers, 2000). The two-step process is divided into reversible and irreversible biofilm formation. The reversible phase involves the association of cells near to but not in contact with the surface. Cells associated with the surface synthesise exopolymers, which irreversibly bind the cells to the surface. Characklis (1981) described biofilm build-up using the following five steps: transportation of cells to a wetted surface, absorption of the cells into a conditioning film, adhesion of microbial cells to the wetted surface, reaction of the cells in the biofilm and detachment of biofilm from the surface. Bryers and Weightman (1995) divided the biofilm build-up into the following eight steps: preconditioning of the surface by macromolecules, transport of cells to the surface, reversible and irreversible adsorption to the surface, cell replication, transport of nutrients and metabolism, production of extracellular polymers and, finally, detachment.

3.1.3 Sampling and detection of biofilm formation in food processing sites

Methods for studying biofilm formation include microbiological, chemical, microscopical and molecular biological methods (Wirtanen, 1995; Holah & Timperley, 1999; Wirtanen *et al.*, 2000a,b; Salo *et al.*, 2000, 2002; Maukonen *et al.*, 2003). Practical methods for assessing microbes and organic soil on processing surfaces are needed to establish the optimal cleaning frequency of the equipment. Hygiene monitoring is currently based on conventional cultivation using swabbing, rinsing or contact plates. Surface sampling can be improved by wetting the surface in advance. In methods that use swabs, sponges or something similar, the detachment of surface-bound microbes is a limiting factor. In the cultivation of biofilm microbes, it is important for the sample to be detached and mixed properly. Agitation used too forcefully in the detachment of the biofilm from the surface may harm the cells, making them unable to grow on the agar plates, whereas insufficient mixing may result in clumps and inaccurate results. Ultrasonics detaches about ten times the number of cells from the surface compared with swabbing (Wirtanen *et al.*, 2000b). In biofilm detection the planktonic cell counts of processing fluids should be interpreted with caution because they are not always representative of the sessile organisms found on surfaces, especially in badly designed equipment and process lines. Organisms from extreme environments are difficult to culture and therefore standard plate counts do not give accurate estimates. The choice of agar and incubation conditions during the cultivation is governed by the characteristics of the microbes that are considered to be the most important.

Conventional culturing techniques are used to measure the number of viable cells able to grow on the chosen agar at given circumstances. The plates and slides are usually incubated at 25–30 °C for 2–3 days. The agars are either nutrient agars, which may contain tryptose, yeast, glucose and agar-agar, or selective agars based on growth inhibitors, e.g. nutritional, antibiotic or acidic compounds. The international standard methods for the detection and enumeration of spoilage and pathogenic microbes are based on culturing techniques (van Netten & Kramer, 1992; Salo *et al.*, 2000).

Impedance techniques can be used to enumerate microorganisms directly on surfaces as the increase in conductance and capacitance due to the metabolic activity of the microbes in the sample leads to a decrease in the impedance. The measurement of the change in impedance value at suitable time intervals provides an impedance curve and thus the detection time of microbial growth in the sample (Firstenberg-Eden, 1986). The detection time depends on the number of microbes in the sample. Results are achieved more rapidly with impedance measurements than with cultivation. Impedance measurement is used in the food industry to control product quality and to assess the effect of cleaning agents and disinfectants (Holah *et al.*, 1990; Flint *et al.*, 1997; Wirtanen *et al.*, 1997).

The chemical methods used in the assessment of biofilm formation are indirect methods based on the utilisation or production of specific compounds, e.g. organic carbon, oxygen, polysaccharides and proteins, or on the biofilm microbial activity, e.g. living cells and ATP (adenosine 5′-triphosphate) content (Characklis *et al.*, 1982). ATP measurement is a luminescence method based on the luciferine–luciferase reaction. The ATP content of the biofilm is proportional to the number of living cells in the biofilm and provides information about their metabolic activity. Kinetic data obtained for freely suspended cells should not be used to assess immobilised biomass growth, e.g. biofilm. The ATP method is insensitive and therefore not suitable for hygiene measurements in equipment where absolute sterility is needed, because with most of the reagents used today a count of at least 10^3 bacterial cells is needed to obtain a reliable ATP value (Wirtanen, 1995; Lappalainen *et al.*, 2000).

Important tools in modern biotechnology-related research are based on microscopical techniques. One advantage of microscopical analysis is that it can measure surface-adhered cells, rather than cells that have been detached from the surface. Various microscopical techniques for studying cell adhesion and biofilm formation on surface materials are available including: epifluorescence, scanning and transmission electron microscopy, Fourier transformation infrared spectrometry, quartz-crystal microbalance and infrared spectroscopy as well as confocal laser scanning and atomic force microscopying techniques. Fluorescence is a type of luminescence in which light is emitted from molecules for a short period of time following the absorption of light. Fluorescence occurs when an excited electron returns to a lower-energy orbit and emits a photon of light. Many different fluorochromes have been used for the staining of microbes in food samples, biofilms and environmental samples (Wirtanen, 1995; Kostyál, 1998).

Flow cytometry using fluorescent probes is a direct optical technique for the measurement of functional and structural properties of individual cells in a cell population. The cells are forced to flow in single file along a rapidly moving fluid stream through a powerful light source. This technique has been used to determine the viability of protozoa, fungi and bacteria. It measures the viability of a statistically significant number of organisms (5000–25 000 cells per sample). The advantages of flow cytometry are accuracy, speed, sensitivity and reproducibility (Wirtanen *et al.*, 2000b).

In the food industry, the first step is to identify the biofilm problems in a particular process or site. Subsequently, it is important to use the best possible methods for isolation and detection of the biofilm for further characterisation in the laboratory using molecular biology and biochemical methods. These methods can be utilised in the detection and identification of microbes in two ways by performing identification either directly from sample material or indirectly from pure cultures obtained from the samples. The two major techniques applied in the molecular detection and identification of bacteria are the polymerase chain reaction and the hybridisation technique (Maukonen *et al.*, 2003).

3.2 Pathogens in biofilms

It is somewhat alarming to know that pathogens such as *Escherichia coli* O157:H7, *Listeria monocytogenes, Salmonella* Typhimurium, *Campylobacter jejuni* and *Yersinia enterocolitica* can easily produce biofilms or be part of biofilm communities that cause severe disinfection and cleaning problems on surfaces in the food industry (Somers *et al.*, 1994; Griffiths, 2003; Stopforth *et al.*, 2003b). According to a study by Peters *et al.* (1999) pathogens were isolated from biofilm communities. In this study *Listeria* spp. were found in 35% of food contact sites and 42% of environmental sites, with *Staphylococcus aureus* being present in a total of 7% and 8%, respectively. Joseph *et al.* (2001) have reported pathogenic bacteria such as *Klebsiella* spp., *Campylobacter* spp. and enterohaemorrhagic *E. coli* in biofilms.

In laboratory studies, specific properties of pathogens in biofilms have been studied, and it has been found that biofilm cells of *Listeria* were more resistant than planktonic cells to disinfectants containing, e.g., chlorine, iodine, quaternary ammonium and anionic acid compounds (Wirtanen, 1995; Chae & Schraft, 2000; Lundén, 2004; Wirtanen & Salo, 2004). Chae and Schraft (2000) used 13 *L. monocytogenes* strains in biofilm studies on glass surfaces at static conditions of 37 °C for up to 4 days. After 3 h incubation bacterial cells from all 13 strains had attached themselves to the glass slides and they formed biofilms within 24 hours. Two poultry isolates of *Salmonella* were used to study biofilm formation on three commonly used food contact surfaces viz. plastic, concrete and stainless steel. Both isolates, i.e. *Salmonella* Weltevreden and *Salmonella* FCM 40, showed similar patterns in the biofilm formation with the greatest growth on plastic followed by concrete and stainless steel (Joseph *et al.*, 2001). In the following chapters there are more examples of the biofilm formation capability of some Gram-negative and Gram-positive pathogenic bacteria.

3.2.1 *Salmonella* biofilms

Salmonella is a genus within the family *Enterobacteriaceae* in which approximately 2200 serotypes are recognised. Some of these strains are specifically adapted to hosts and largely restricted to them, e.g. *S.* Typhi in man and *S.* Dublin in cattle. *Salmonella* is a non-spore-forming rod-shaped, motile Gram-negative bacterium with non-motile exceptions such as *S.* Gallinarum and *S.* Pullorum (Price & Tom, 2003b). *Salmonella* serotypes are traditionally named as if they were separate species but, because of their genetic similarity, a single species, *S. enterica*, has been proposed, with food-poisoning serotypes mostly classified subspecies, also named *enterica* (Mead, 1993). The growth range for salmonellae is 5–47 °C at pH 4.0–9.0, with optimum growth at 35–37 °C and pH 6.5–7.5. Salmonellae are not particularly salt-tolerant, although growth can occur in the presence of 4% sodium chloride. The lower limit of water activity (a_w) permitting growth is 0.93 (Mead, 1993).

Foods commonly associated with the disease include raw meats, poultry, eggs, milk and dairy products (Price & Tom, 2003b). Milk-borne salmonellosis is common in parts of the world where milk is neither boiled nor pasteurised. It occurs, but much less frequently, in developed countries where the main products implicated are pasteurised milk, powdered milk and certain cheeses (Mead, 1993). Formation of a biofilm by *Salmonella* on various types of surfaces used in the food processing industry has been reported by several groups (Mafu *et al.*, 1990; Helke *et al.*, 1993; Ronner & Wong, 1993; Joseph *et al.*, 2001). These studies have shown that *Salmonella* spp. can form biofilms on food contact surfaces and that the cells in biofilms are much more resistant to sanitisers compared to planktonic cells (Ronner & Wong, 1993; Joseph *et al.*, 2001; Stepanovic *et al.*, 2003). Mokgatla and co-workers (1998) studied the resistance of *Salmonella* sp. isolated from a poultry abattoir and found out that it will grow in the presence of in-use concentrations of hypochlorous acid. The presence of *Pseudomonas fluorescens* in the biofilm resulted in the increased resistance of *S.* Typhimurium to chlorine (Leriche & Carpentier, 1995).

3.2.2 *Escherichia coli* biofilms

Escherichia coli is a Gram-negative, rod-shaped bacterium. Because many microbes from faeces are pathogenic in animals and humans, the presence of the intestinal bacterium *E. coli* in water and foods indicates a potential hygiene hazard. Most strains of *E. coli* are harmless. However, a few strains with well-characterised traits are known to be associated with pathogenicity (Venkitanarayanan & Doyle, 2003). Those of greatest concern in water and foods are the intestinal pathogens, which are classified into five major groups: the enterohaemorrhagic *E. coli* (EHEC), the enterotoxigenic *E. coli* (ETEC), the enteroinvasive *E. coli* (EIEC), the enteropathogenic *E. coli* (EPEC) and the enteroaggregative *E. coli* (EAEC). Growth can occur at 7–46 °C with the maximal growth rate at 35–37 °C. The minimum a_w for growth ranges from 0.94 and 0.97. The optimum pH for growth is approximately 7.0, with a minimum and maximum pH for growth of 4.5 and 9.0. EHEC has been shown to grow poorly at temperatures of 44 °C (Venkitanarayanan & Doyle, 2003).

Escherichia coli has been isolated from a large number of foods and drinks, e.g. fermented meat sausage, dairy products, vegetables, meat, poultry and fish products, water and apple cider. These agents can cause diarrhoeal outbreaks (Junkins & Doyle, 1993; Venkitanarayanan & Doyle, 2003). Unpasteurised milk is a common vehicle of *E. coli* O157:H7 transmission to humans (Dontorou *et al.*, 2004). *E. coli* can also survive for extended periods of time in several acidic foods, e.g. cheese and yogurt. Acid-adapted *E. coli* O157:H7 has shown enhanced survival and prevalence in biofilms on stainless steel surfaces (Stopforth *et al.*, 2003a,b; Venkitanarayanan & Doyle, 2003). In a hygiene survey performed in the food industry by Holah *et al.* (2002), microbial strains, e.g. *E. coli* and *L. monocytogenes*, were found either on surfaces or in products or in both, and some of these strains were persistent. Faille *et al.* (2002, 2003) found out that *E. coli* cells were poorly adhered to surfaces. The cells were embedded in the organic matrix of the biofilm, which shows that the structure of the biofilm formed affects the way in which the surfaces should be cleaned. Oulahal-Lagsir *et al.* (2003) showed in their studies that proteolytic and glycolytic enzyme treatment together with ultrasonics enhance the removal of *E. coli* biofilm from stainless steel soiled with milk. These findings correspond with results obtained in the food industry.

3.2.3 *Campylobacter* biofilms

Campylobacter spp. are microaerophilic, very small, curved and thin Gram-negative rods (Price & Tom, 2003a). Growth can occur in a microaerophilic atmosphere containing 3–15% oxygen and 3–5% carbon dioxide at 30–48 °C with a maximal growth rate at 42–43 °C. The minimum a_w for growth is above 0.987. The optimum pH for growth is approximately 6.5–7.5, with a minimum at 4.9 and a maximum at 9.0 (Stern & Kazami, 1989; Roberts *et al.*, 1996; van Vliet & Ketley, 2001). *C. coli* and *C. laridis* can grow at 30.5 °C while *C. jejuni* cannot. *C. laridis* tolerates slightly more salt than *C. jejuni* or *C. coli* and ceases growing in the presence of 2.5% sodium chloride (Roberts *et al.*, 1996). Illness can be caused by ingestion of as few as 500–800 cells in milk. Since the infective dose is rather low and the food in many cases may contain only a few cells, liquid enrichment methods are normally required before plating on selective media in order to detect contamination with *C. jejuni* or *C. coli.* Successful detection of these organisms requires incubation at 42 °C under microaerophilic conditions (Roberts *et al.*, 1996).

In laboratory tests *Campylobacter* has been shown to form a biofilm in optimum circumstances on stainless steel and glass beads in 2 days (Somers *et al.*, 1994; Dykes *et al.*, 2003). In studies performed by de Beer *et al.* (1994) biofilms are shown to form zones with low oxygen content in aerobic surroundings and *Campylobacter* spp. can therefore more easily survive in biofilms. Trachoo *et al.* (2002) showed that viable *C. jejuni* cells grown on polyvinyl chloride surfaces decreased with time and the greatest reduction occurred on surfaces without a pre-existing biofilm. The number of viable *C. jejuni* determined by using a direct viable count was greater than by using culturing techniques, which indicates that *C. jejuni* cells can form a viable but non-culturable state within the biofilm. Both determination methods showed that biofilms enhance the survival of *C. jejuni* during a 7-day period at 12 °C and 23 °C (Trachoo *et al.*, 2002). Taking the resistance of the viable but non-culturable *C. jejuni* cells into account is important in the optimisation of cleaning and decontamination procedures, especially in those food industrial processes in which raw meat products are processed (Rowe *et al.*, 1998; Trachoo & Frank, 2002). Organic soil, e.g. food residues, or moisture improve the survival of campylobacter on surfaces (Humphrey *et al.*, 1995; Kusamaningrum *et al.*, 2003). Boucher and co-workers (1998) showed that *C. jejuni* survived very well on wooden surfaces because the pores in the wood protect the cells from oxygen.

3.2.4 *Listeria monocytogenes* biofilms

Listeria monocytogenes is a facultatively anaerobic Gram-positive, non-spore-forming short rod that is widely distributed in nature (El-Kest & Marth, 1988; Griffiths, 2003). It is a non-host specific pathogen (El-Kest & Marth, 1988; Lundén, 2004). Listeriosis may occur sporadically or epidemically. The organism has been isolated from raw milk, mastitic milk and pasteurised milk. Foodstuffs associated with listeriosis outbreaks also include cold-smoked and gravad rainbow trout products, sliced cold meat, soft cheese, butter, ice-cream and coleslaw. Examples of epidemic sources are: coleslaw in Canada 1981, unpasteurised milk in the USA 1983, Mexican-style soft cheese in USA 1985, pork product in France 1992, chocolate milk in the USA 1994, soft cheese in Swizerland 1995, rainbow trout in Sweden 1997, corn in Italy 1997, hot dogs in the USA 1998–99 and butter in Finland 1999 (Lyytikäinen *et al.*, 2000; Weinstein & Ortiz, 2001). Treated wastewater can also be a source of *L. monocytogenes*. Of the 13 different *L. monocytogenes* serotypes only three (1/2a, 1/2b and 4b) have been predominantly implicated in human diseases (Chae & Schraft, 2000). It has been reported that healthy people can be carriers of *L. monocytogenes* (El-Kest & Marth, 1988). *L. monocytogenes* is able to grow in many environments, at a low oxygen tensions, in high salt concentrations and over a wide range of pH (5–9.5) and temperatures (3–45 °C) with an optimum at 30 °C (Griffiths, 2003; Lundén, 2004). The bacterium can survive for a limited time in up to 25% salt at 4 °C (El-Kest & Marth, 1988).

Hygiene monitoring in the food processing industry is important because *L. monocytogenes*, in particular, can colonise and form biofilms in food processing environments and on surfaces (Husu *et al.*, 1990; Eklund *et al.* 1995; Autio *et al.*, 1999; Miettinen *et al.*, 1999, 2001; Lyytikäinen *et al.*, 2000; Aarnisalo *et al.*, 2003; Lundén, 2004; Miettinen & Wirtanen, 2005; Wirtanen & Salo, 2004). *Listeria* sources in processing plants are conveyor belts, cutters, slicers, brining and packaging machines, coolers and freezers as well as floors and drains (Wirtanen, 2002; Lundén, 2004; Wirtanen & Salo, 2004). *L. monocytogenes* has been found to form biofilms on common food contact surfaces such as plastic, polypropylene, rubber as well as stainless steel and also on glass (Mafu *et al.*, 1990; Helke *et al.*, 1993; Ronner & Wong, 1993; Wirtanen, 1995; Chae & Shraft, 2000; Borucki *et al.*, 2003; Lundén, 2004).

3.2.5 *Staphylococcus aureus* biofilms

Staphylococcus aureus is a Gram-positive, aerobic, non-spore-forming catalase postitive rod. It is ubiquitous in the mucous membrane and skin of most warm-blooded animals. Nasal and skin carriage are frequent vehicles in the transportation of *S. aureus*. It is an opportunistic pathogen causing infections via open wounds, for example (Roberts *et al.*, 1996). The growth temperature for this bacterium is 7–48 °C, with an optimum around 37 °C. Growth has been demonstrated over the pH range 4–10, with an optimum at 6–7. The lower limit of a_w

permitting growth is 0.83. It readily produces enterotoxins, which are not destroyed in heat treatment (Roberts *et al.*, 1996).

Staphylococcus aureus is a pathogen that can also affect dairy products. Its occurrence in sour milk products such as yoghurt is worthwhile investigating as it is present in relatively high numbers in raw milk (Benkerroum *et al.*, 2002). According to studies by Benkerroum *et al.* (2002), staphylococci grew rapidly during the initial fermentation. Similar behaviour by *S. aureus* has previously been reported both in yoghurt and cheese (Ahmed *et al.*, 1983; Attaie *et al.*, 1987). Elvers *et al.* (1999) isolated *S. aureus* in a total of 7% of food contact sites and 8% of environmental sites from 10 small and medium sized enterprises producing high risk foods in their study, which was performed for the Ministry of Agriculture, Fisheries and Food (now DEFRA) in the UK. The source of *S. aureus* almost always originated from food handlers or from utensils previously contaminated by humans (Elvers *et al.*, 1999; Peters *et al.*, 1999). A survey revealed that *S. aureus* was involved in 15% of the recorded foodborne illnesses caused by dairy products in eight developed countries whereas *L. monocytogenes* was involved in 22% (Benkerroum *et al.*, 2002). It is resistant to drying and may also colonise complex food-processing equipment, which is left in wet conditions (Bolton *et al.*, 1988). It can also be found in the dust in ventilation systems (Roberts *et al.*, 1996). Resistance to oxidative disinfectants has mainly been associated with biofilm formation (Bolton *et al.*, 1988). Luppens *et al.* (2002) showed that *S. aureus* biofilm formed on stainless steel, polystyrene and glass in a nutrient flow needed concentrations of benzalkonium chloride that were 50 times higher and concentrations of hypochlorite that were 600 times higher to achieve 4 log killing of *S. aureus* compared with cells in suspensions. Supporting results were obtained by Møretrø *et al.* (2003a).

3.2.6 *Bacillus cereus* biofilms

Bacillus cereus is a Gram-positive, aerobic, spore-forming rod, normally present in soil, dust, and water (Jay, 1996). It can also grow well anaerobically. Cells of *B. cereus* are large and motile. The growth temperature for this bacterium is 4–50 °C, with an optimum around 28–35 °C. Growth has been demonstrated over the pH range 4.9–9.3 (Jay, 1996; Granum, 2003; Shelef, 2003; Svensson *et al.*, 2004). The organism elaborates a number of toxins with distinct diarrhoeal and emetic syndromes (Shelef, 2003). *B. cereus* occurs extensively in the environment but despite the fact that it is a common contaminant in raw milk, food poisoning outbreaks caused by dairy products contaminated with *B. cereus* have been rare (Wirtanen *et al.*, 2002; Svensson *et al.*, 2004).

In a dairy product survey, Wong (1998) showed that *B. cereus* was found in 52% of ice-creams, 35% of soft ice-creams, 29% of milk powders, 17% of fermented milks and 2% of pasteurised milks and fruit-flavoured milks. Svensson *et al.* (1999) found indications of a prolonged contamination problem caused by mesophilic *B. cereus* strains early in the production chain of one dairy plant. Additional contamination of milk by the *B. cereus* biofilm was shown to occur in the filling machine. Different *Bacillus* spp., and among them *B. cereus*, have been found on liquid packaging boards and blanks and these could thus be an additional source of biofilms containing *Bacillus* spp. (Svensson *et al.*, 2004). Furthermore, spores of *B. cereus* are reported to possess a pronounced ability to adhere to the surface of stainless steel, which is commonly used in food processing. Both *B. cereus* and *B. subtilis* biofilms were detected on stainless steel and Teflon surfaces, and removal from stainless steel was more difficult than from Teflon because of surface roughness (Wirtanen *et al.*, 1996). Te Giffel and co-workers (1997) showed that spores of *B. cereus* adhered, germinated and multiplied on the stainless steel surfaces of a tube heat exchanger. The cells of *B. cereus* were isolated from the individual tubes after cleaning. The attachment of *B. cereus* in process lines may act as a continual source of post-pasteurisation contamination (Elvers *et al.*, 1999). Lindsay (2001) found that the biofilms of food spoilage *Bacillus* and *Pseudomonas* species attach themselves to liquid food processing equipment surfaces and cells in biofilms, even if treated with an in-use concentration of sanitisers, manage to survive and grow. This phenomenon is even stronger when mixed biofilms are involved. The attached *B. cereus* cells may subsequently form a biofilm on a stainless steel surface and present a major problem for the food industry (Peng *et al.*, 2002).

3.2.7 *Clostridium perfringens* biofilms

Clostridium perfringens is a spore-forming, Gram-positive, anaerobic, non-motile rod which forms large, regular, round and slightly opaque and shiny colonies on the surface of agar (Brynestad & Granum, 2002). There are five types of *C. perfringens*: A, B, C, D and E, which produce different types of toxins (Labbé, 2003). *C. perfringens* can grow between 10 °C and 52 °C, with a maximum of 45 °C for most strains (Brynestad & Granum, 2002). It is often a cause of human food poisoning due to its ability to grow over a wide temperature range. Its spores can also survive several food processing procedures. Spores of some strains are resistant to temperatures of 100 °C for more than 1 h (Labbé, 2003).

Clostridium perfringens food poisoning from new food sources, because the bacterium is so adaptable and prolific, has helped to show how our perceptions and understanding of safe food change with new knowledge (Foster, 1997). *C. perfringens* can be found as part of the normal flora of the intestinal tracts of both animals and humans, as well as in soil, clothing and skin. It has been found in virtually all environments tested, including water, milk and dust (Brynestad & Granum, 2002). In view of its widespread presence in moist soil, its presence in air and dust in kitchens, catering and food processing environments is not surprising (Labbé, 2003). *C. perfringens* serotypes commonly associated with human illness have been found on recently slaughtered carcasses. Other foods contaminated with *C. perfringens* are poultry, fish, vegetables and dairy products (Roberts *et al.*, 1996). As with other agents of human food poisoning, the number of outbreaks of food poisoning attributable to *C. perfringens* is under-reported (Labbé, 1993).

3.2.8 *Mycobacterium* biofilms

The genus *Mycobacterium* contains approximately 50 species, which are divided into rapid growers, slow growers and the human leprosy bacillus (Collins & Grange, 1993). Mycobacteria are weakly Gram-positive, non-motile, slender, non-spore-forming, rod-shaped, aerobic and free-living in soil and water (Payeur, 2000). They do not produce appreciable amounts of toxin substances and do not cause food poisoning (Collins & Grange, 2003). Mycobacteria are widely distributed in nature and have been isolated from natural and piped waters, wet soil, mud, compost, grasses, vegetables, unpasteurised milk and butter. They have also been isolated from domestic water pipes from which they readily enter drinking water (Collins & Grange, 2003). *M. tuberculosis, M. africanum, M. bovis, M. bovis* BCG and *M. microti* are collectively referred to as the *M. tuberculosis* complex because these organisms cause tuberculosis (Payeur, 2000). Infections in humans and animals may be caused by most of the slowly growing mycobacteria such as *M. avium, M. intracellulare*, *M. scofulaceum*, *M. kansasii*, *M. marinum*, *M. simiae*, *M. ulcerans* and *M. xenopi*. The only rapidly growing pathogenic species are *M. chelonae* and *M. fortuitum*. The principal source of these infections seems to be water (Payeur, 2000).

A pilot plant pasteuriser was used to examine the heat resistance of *M. avium* subsp. *paratuberculosis* (*M. paratuberculosis*) during high temperature short time (HTST) pasteurisation using raw milk samples under various time and temperature conditions. Results indicated that low numbers of *M. paratuberculosis* may also survive extreme HTST treatments (Hammer *et al.*, 2002). Torvinen *et al.* (2004) studied 16 drinking water distribution systems in Finland for growth of mycobacteria by sampling water from waterworks and in different parts of the systems. In the experimental part, mycobacterial colonisation as biofilms on polyvinyl chloride tubes was studied. The isolation frequency of mycobacteria increased from 35% at the waterworks to 80% in the systems, and the number of mycobacteria in the positive samples increased from 15 to 140 cfu/l, respectively. The densities of mycobacteria in the developing biofilms were highest at the distal sites of the system. Over 90% of the mycobacteria isolated from water deposits belonged to *M. lentiflavum*, *M. tusciae*, *M. gordonae* and a previously unclassified group of mycobacteria. Dailloux *et al.* (2003) investigated the ability of *M. xenopi* to colonise an experimental drinking water distribution system. *M. xenopi* was found to be present in the biofilm within an hour of introduction. After 9 weeks, it was constantly present in all outlet water samples (1–10 cfu/100 ml) and in biofilm samples (10^2–10^3 cfu/cm^2). Biofilms may be considered to be the reservoirs for the survival of *M. xenopi*. Gao *et al.* (2002) studied the survival of *M. paratuberculosis* in 7 regular batch and 11 HTST pasteurisation experiments using raw milk or ultra-heat treatment (UHT) milk samples spiked with *M. paratuberculosis*. *E. coli* and *M. bovis* BCG strains were used as controls. No survivors were detected from any of the slants or broths corresponding to the 7 regular batch, but survivors were detected in 2 of the 11 HTST experiments. No survivors were detected after heat treatment for 15 min at 63 °C. These results indicate that *M. paratuberculosis* may survive HTST pasteurisation (Gao *et al.*, 2002).

3.3 Biofilms and microbial contamination in food processing

Prolonged or persistent contamination of some *Listeria monocytogenes* strains, which means that they have caused food plant contamination for long periods of up to several years, has been reported in several food industry areas, e.g. meat, poultry, fish, dairy and fresh sauces (Miettinen *et al.*, 1999; Borucki *et al.*, 2003; Rørvik *et al.*, 2003; Lundén, 2004). *Escherichia coli* and *Salmonella* isolates are also known to be persistent in food and fish feed factories (Holah *et al.*, 2004; Møretrø *et al.*, 2003b). Persistent *L. monocytogenes* plant contamination appears to be the result of the interaction of several different factors. Properties that influence survival, including enhanced adherence to food contact surfaces and adaptation to disinfectants, in addition to such predisposing factors in the processing line as complex processing machines and poor zoning may lead to persistent plant contamination (Lundén, 2004). The eradication of persistent contamination of *L. monocytogenes* has been shown to be difficult but not impossible. Targeted and improved sanitation has led to successful eradication (Miettinen *et al.*, 1999).

In studies performed by Lundén (2004), persistent *L. monocytogenes* strains were observed to adhere to stainless steel surfaces in higher cell numbers than non-persistent strains after short contact times. Such enhanced adherence increases the likelihood of the survival of the persistent strains due to increased resistance against prevention methods and may have an effect on the initiation of persistent plant contamination. If the adherence period of strains was prolonged then the adherence level of non-persistent strains was close to the adherence level of persistent strains (Lundén, 2004). The initial resistance of persistent and non-persistent *L. monocytogenes* strains to disinfectants varied, and the increase in resistance was similar for persistent and non-persistent strains. The concentrations of disinfectants used at food processing plants were not reached in the studies performed by Lundén (2004). Also Holah *et al.* (2002) reported in their studies that the resistance of persistent strains of *L. monocytogenes* and *E. coli* found in the food industry to the most commonly used disinfectants were not significantly different from the laboratory control strain. A study carried out by Earnshaw and Lawrence (1998) concluded that it is unlikely that the strains that persisted in the poultry processing environment did so by means of plasmid-mediated resistance to the commercial disinfectants used.

3.4 Prevention of biofilm formation and biofilm removal

Harmful microbes may enter the manufacturing process and reach the end-product in several ways, e.g. through raw materials, air in the manufacturing area, chemicals employed, process surfaces or factory personnel (Lelieveld *et al.*, 2003; Maukonen *et al.*, 2003). Once a biofilm is formed, either on food contact or environmental surfaces, it can be a source of contamination for foods passing through the same processing line. For example, *Listeria monocytogenes* is difficult to remove from the factory environment once it has become a part of the house microbiota (Lundén, 2004). Therefore, it is especially important for the persistent growth of pathogenic and harmful microbes to be prevented in the food processing line using all available means (Wirtanen, 1995; Joseph *et al.*, 2001; Lundén, 2004). In the food industry, equipment design and the choice of surface materials are important in fighting microbial biofilm formation. Attention should also be paid to the quality of additives and raw materials as well as the processing water, steam and other additives, because using poor quality materials leads to the easy spoiling of the process (Wirtanen & Salo, 2004). The aim of microbial control in a process line is two-fold: to reduce or limit the number of microbes in liquids and products and to reduce or limit their activity and to prevent and control the formation of biofilms on surfaces. At present the most efficient means for limiting the growth of microbes are good production hygiene, the rational running of the process line, and the well-designed use of cleaning and decontamination processes (Alakomi *et al.*, 2002; Wirtanen & Salo, 2004). The cleanliness of surfaces, the training of personnel and good manufacturing and design practices are important in combating biofilm problems in the food industry.

3.4.1 Hygienic equipment design

Several conferences and literature reviews have shown that the design of the equipment and process line in the food processing and packaging industry are important in preventing biofilm formation to improve the process and production hygiene (Wimpenny *et al.*, 1999; Wirtanen *et al.*, 1999; Bryers, 2000; Gilbert *et al.*, 2001; Alakomi *et al.*, 2002; Wirtanen, 2002; McBain *et al.*, 2003; Maukonen *et al.*, 2003; Lundén, 2004; Wirtanen & Salo, 2004). The most significant laws regarding the food industry are the EU directive 98/37/EU and machine standard EN 1672-2:1997. EN 1672 draws particular attention to dead spaces, corners, crevices, cracks, gaskets, seals, valves, fasteners and joints owing to their ability to harbour microorganisms that can subsequently endure adverse/harmful process conditions (Lelieveld *et al.*, 2003; Wirtanen & Salo, 2004). Equipment that causes problems in food processing and packaging includes slicing and cutting equipment, filling and packing machines, conveyors, plate heat exchangers and tanks with piping. These types of equipment can cause contamination through spoilage microbes and pathogens as they are difficult to clean, e.g. the pathogen *Listeria monocytogenes* is often associated with harbourage in poorly designed equipment.

3.4.2 Biofilm removal

The elimination of biofilms is a very difficult and demanding task, because many factors affect the detachment, such as temperature, time, mechanical forces and chemical forces. Sanitation, i.e. cleaning and disinfection, is carried out in food processing plants in order to produce safe products with an acceptable shelf-life and quality. The key to the effective cleaning of a food plant is the understanding of the type and nature of the soil and of the microbial growth on the surfaces to be removed. The intelligent integration of decontamination programmes in the manufacturing are essential to achieve both successful cleaning and business profit (Lelieveld *et al.*, 2003). Lelieveld as early as 1985 wrote that there is a trend towards longer production runs with short intervals for sanitation, because the sanitation should be performed as cost-effectively and safely as possible.

The mechanical and chemical power, temperature and contact time in the cleaning regime should be carefully chosen to achieve an adequate cleaning effect (Wirtanen, 1995). An efficient cleaning procedure consists of a sequence of rinses and detergent and disinfectant applications in various combinations of temperature and concentration, finally letting the equipment and process lines dry in well-ventilated areas. The basic task of detergents is to reduce the interfacial tensions of soils so that the soil attached to surfaces, for example biofilm, becomes miscible in water. The effect of the surfactants is increased by the mechanical effect of turbulent flow or water pressure, or by abrasives, for example salt crystals. Prolonged exposure of the surfaces to the detergent makes removal more efficient. Detergents to be used in the cleaning of open systems are formulated to be effective at temperatures in the range 35–50 °C. In closed systems the detergents are formulated to be used at temperatures in the range of 55–80 °C (Troller, 1993; Wirtanen *et al.*, 2002).

Elimination of biofilms in open systems is performed as follows: gross soil should be removed by dry methods, e.g. brushing, scraping or vacuuming, and, if the process is wet, the visible soil can be rinsed off with low-pressure water. The effective elimination of biofilms from open systems is achieved by dismantling the equipment in the process line and cleaning is then carried out using either foam or gel. Foams are most effective in situations where contact with the soil for an extended contact time is necessary. The surfactants, which suspend the adhered particles and microbes from the surfaces in the water, are added to increase the cleaning effect, which is also increased by using water of sufficient volume at the correct temperature and pressure. The dismantled equipment and utensils should thereafter be stored on racks and tables, not on the floor. The cleaning is mostly carried out in combination with a final disinfection, because viable microbes on the surfaces are likely to harm production (Troller, 1993; Wirtanen *et al.*, 2000a).

In the cleaning regime for closed processes, pre-rinsing with cold water is carried out to remove loose soil. Cleaning-in-place (CIP) treatment is normally performed using hot cleaning solutions, but cold solutions can also be used in the processing of fat-free products. The warm alkaline cleaning solution, normally 1% sodium hydroxide, is heated to 75–80 °C and the cleaning time is 15–20 min. The equipment is rinsed with cold water before the acid treatment is performed at approximately 60–70 °C for 5 min. The effect of chlorine-based agents can be divided into three phases: loosening of the biofilm from the surface, breakage of the biofilm and the disinfective effect of the active chlorine. The cleaning solutions should not be re-used in processes in order to achieve total sterility because the reused cleaning solution can contaminate the equipment. Single-phase CIP is more commonly used nowadays because the processing industry wants to save time. In single-phase cleaning procedures the time it takes to carry out one cleaning process, normally the acid treatment, and a rinsing step can be saved (Costerton *et al.*, 1985; Wirtanen *et al.*, 2000a). The photobacterial test can be used to test that rinsing has been performed properly (Lappalainen *et al.*, 2003).

3.5 Future trends

The food and drink industry is the leading manufacturing sector in Europe with production representing 13% of the total of all industrial manufacturers in the EU and with an annual turnover of about €700 billion. Three main employers in the EU employ more than 4 million people. This position illustrates the major economic role of the food and drink industry, a very diverse sector that is characterised by the variety of its activities, its elaborated products and structures.

Microbiological and chemical issues will be especially important for the safe production of feed, food and packaging material in the future. A number of outbreaks in recent years have seriously damaged the European consumer's trust in food safety and therefore knowledge of product safety, including equipment hygiene, is of the utmost importance both for the product manufacturer and for the consumer. Development of optimal pathogen management strategies requires knowledge of pathogen contamination routes, the consumer, how the food becomes a vehicle for disease transmission and the differentiation of risks and hazards. Hazards in the food industry can be of microbial, e.g. biofilm formation, biological, chemical, physical or informational origin. The function of risk assessment is to give objective and relevant information about specified risks. An important problem in risk assessment at the manufacturing level is that a more quantitative systematic approach should be used: the risk assessment procedure should be based on scientific knowledge and performed in a team that has the knowledge and experience needed to perform the reliable evaluation of risks (Wirtanen & Raaska, 2004). Therefore, reliable monitoring systems, which can provide information about microbial growth on-line, directly and in real time, are required within the process. The methods should be based on optical and electro-chemical measurements, ion mobility and infrared techniques as well as bioluminescence. The successful transfer of these techniques for on-line monitoring of food quality and process cleanliness should be based on microbial reference methods. This means that the threshold values for detected amounts of contaminants must be very low (Maukonen *et al.*, 2003). The following topics are of interest: (1) exploring pathogen physiology/ecology with emphasis on the understanding of survival of and resistance towards processing and in pathogen–host interactions; (2) exploring virulence traits with the emphasis on understanding pathogenicity and infectivity; (3) identifying specific microbial characteristics to assist in the identification of pathogenic microbes in the food environment under investigation; and (4) assistance in risk assessment carried out by governments and food safety management in industry (Vaughan, 2004).

3.6 Sources of further information and advice

The food and drink industry should offer a wide range of safe, wholesome and nutritious food and drink products to 450 million consumers in an enlarged Europe. At a time when quality is being subjected to evaluation by the market and is not addressed through regulatory prescriptions, the production of safe food products is being subjected to great stress. Any food safety obligation must be respected by all the links in the food chain including farmers and animal feed producers. Regulation 178/2002 confirms the new approach to food safety – from the farm to the fork – which implies close cooperation between all those involved in the food chain. The International Food Standard, the British Retail Consortium Standard, the Danish Standard, the Dutch Standard and the soon to be adopted ISO 22000 are all tools for assessing manufacturers in producing safe food in a secure environment with a documented and effective quality management (Wirtanen & Raaska, 2005). The choice of various standards is influenced by many factors, such as availability of advisers and retailers (Zagorc, 2004).

Furthermore, the European Hygienic Engineering and Design Group (EHEDG) is currently producing a guideline on hygienic systems integration. This coming EHEDG guideline has the task of linking and supporting current guidelines on hygienic design regarding specific equipment and hygienic tests. It can be viewed as both vertical and horizonal guidelines. The most fundamental EHEDG guidelines in hygienic integration are: Document 8 'Hygienic equipment design criteria', Document 10 'Hygienic design of closed equipment for the processing of liquid food', Document 13 'Hygienic design of equipment for open processing', Document 22 'General hygienic design criteria for the safe processing of dry particulate materials' and Document 26 'Hygienic engineering of plants for the processing of dry particulate materials'. Neither the EN1672-2 nor the HACCP standards are replaced by this guideline (Steenstrup *et al.*, 2004).

3.7 References

AARNISALO K, AUTIO T, SJÖBERG A-M, LUNDÉN J, KORKEALA H and SUIHKO M-L (2003) Typing of *Listeria monocytogenes* isolates originating from the food processing industry with automated ribotyping and pulsed-field gel electrophoresis, *J Food Prot*, **66**, 249–255.

AHMED A A-H, MUSTAFA M K and MARTH E (1983) Growth and enterotoxin production by *Staphylococcus aureus* in whey from the manufacture of Domiati cheese, *J Food Prot*, **46**, 235–237.

ALAKOMI H-L, KUJANPÄÄ K, PARTANEN L, SUIHKO M-L, SALO S, SIIKA-AHO M, SAARELA M, MATTILA-SANDHOLM T and RAASKA L (2002) *Microbiological problems in paper machine environments, VTT Research Notes 2152*, Espoo, Otamedia Oy.

ATTAIE R, WHALEN R J, SHAHANI K M and AMER M A (1987) Inhibition of *Staphylococcus aureus* during production of Acidophilus yoghurt, *J Food Prot*, **50**, 224–228.

AUTIO T, HIELM S, MIETTINEN M, SJÖBERG A-M, AARNISALO K, BJÖRKROTH J, MATTILA-SANDHOLM T and KORKEALA H (1999) Sources of *Listeria monocytogenes* contamination in a cold-smoked rainbow trout processing plant detected by pulsed-field gel electrophoresis typing, *Appl Environ Microbiol*, **65**, 150–155.

BEER DE D, STOODLEY P, ROE F and LEWANDOWSKI Z (1994) Effects of biofilm structure on oxygen distribution and mass transport, *Biotechnol Bioeng*, **43**, 1131–1138.

BENKERROUM N, OUBEL H and BEN MIMOUN L (2002) Behavior of *Listeria monocytogenes* and *Staphylococcus aureus* in yoghurt fermented with a bacteriocin-producing thermophilic starter, *J Food Prot*, **65**, 799–805.

BOLTON K J, DODD C E R, MEAD G C and WAITES W M (1988) Chlorine resistance of strains of *Staphylococcus aureus* isolated from poultry processing plants, *Lett Appl Microbiol*, **6**, 31–34.

BORUCKI M K, PEPPIN J D, WHITE D, LOGE F and CALL D R (2003) Variation in biofilm formation among strains of *Listeria monocytogenes*, *Appl Environ Microbiol*, **69**, 7336–7342.

BOUCHER S N, CHAMBERLAIN A H L and ADAMS M R (1998) Enhanced survival of *Campylobacter jejuni* in association with wood, *J Food Prot*, **61**, 26–30.

BRYERS J ed. (2000) *Process Analysis and Applications*, New York: John Wiley-Liss Inc.

BRYERS J and WEIGHTMAN A (1995) The Centre for Biofilm Engineering: an international resource in managing complex biological systems, *SIM News*, **45**, 103–111.

BRYNESTAD S and GRANUM P E (2002) *Clostridium perfringens* and food-borne infections, *Int J Microbiol*, **74**, 195–202.

CARPENTIER B and CERF O (1993) Biofilms and their consequences, with particular reference to hygiene in the food industry, *J Appl Bacteriol*, **75**, 499–511.

CHAE M S and SCHRAFT H (2000) Comparative evaluation of adhesion and biofilm formation of different *Listeria monocytogenes* strains, *Int J Food Microbiol*, **62**, 103–111.

CHARACKLIS W G (1981) Fouling biofilm development: a process analysis, *Biotechnol Bioeng*, **23**, 1923–1960.

CHARACKLIS W G, TRULEAR M G, BRYERS J D and ZELVER N (1982) Dynamics of biofilm processes: methods, *Water Res*, **16**, 1207–1216.

COLLINS C H and GRANGE J M (1993) Mycobacteria, in Macrae M, Robinson R K and Sadler M J, *Encyclopaedia of Food Science, Food Technology and Nutrition*, London, Academic Press, vol. 5, 3187–3191.

COLLINS C H and GRANGE J M (2003) Mycobacteria, in Caballero B, Trugo L C and Finglas

P M, *Encyclopedia of Food Sciences and Nutrition*, 2nd edition, London, Academic Press, vol. 6, 4067–4072.

COSTERTON J W, IRVIN R T and CHENG K-J (1981) The bacterial glycocalyx in nature and disease, *Ann Rev Microbiol*, **35**, 299–324.

COSTERTON J W, MARRIE T J and CHENG K-J (1985) Phenomena of bacterial adhesion, in Savage D C and Fletcher M *Bacterial Adhesion*, New York, Plenum Press, 3–43.

DAILLOUX M, ALBERT M, LAURAIN C, ANDOLFATTO S, LOZNIEWSKI A, HARTEMANN P and MATHIEU L (2003) *Mycobacterium xenopi* and drinking water biofilms, *Appl Environ Microbiol*, **69**, 6946–6948.

DONTOROU A, PAPADOPOULOU C, FILIOUSSIS G, APOSTOLOU I, ECONOMOU V, KANSOUZIDOU A and LEVIDIOTOU S (2004) Isolation of a rare *Escherichia coli* O157:H7 strain from farm animals in Greece, *Comp Immunol Microbiol Infect Dis*, **27**, 201–207.

DYKES G A, SAMPATHKUMAR B and KORBER D R (2003) Planktonic or biofilm growth affects survival, hydrophobicity and protein expression of a pathogenic *Campylobacter jejuni* strain, *Int J Food Microbiol*, **89**, 1–10.

EARNSHAW A M and LAWRENCE L M (1998) Sensitivity to commercial disinfectants, and the occurrence of plasmids within various *Listeria monocytogenes* genotypes isolated from poultry products and the poultry processing environment, *J Appl Microbiol*, **84**, 642–648.

EKLUND M W, POUSKY F T, PARANJPYE R N, LASHBROOK L C, PETERSON M E and PELROY G A (1995) Incidence and sources of *Listeria monocytogenes* in cold-smoked fishery products and processing plants, *J Food Prot*, **58**, 502–508.

EL-KEST S and MARTH E H (1988) *Listeria monocytogenes* and its inactivation by chlorine: a review, *Lebensm Wiss Technol*, **21**, 346–351.

ELVERS K T, PETERS A C and GRIFFITH C J (1999) Development and control of biofilms in the food industry in Wimpenny J, Gilbert P, Walker J, Brading M and Bayston R *Biofilms – the good, the bad and the ugly*, Cardiff, BioLine, 139–145.

FAILLE C, FONTAINE F, LELIEVRE C and BENEZECH T (2003) Adhesion of *Escherichia coli, Citrobacter freundii* and *Klebsiella pneumoniae* isolated from milk: Consequence on the efficiency of sanitation procedures, *Water Sci Technol*, **44**, 225–231.

FAILLE C, JULLIEN C, FONTAINE F, BELLON- FONTAINE M-N, SLOMIANNY C and BENEZECH T (2002) Adhesion of *Bacillus* spores and *Escherichia coli* cells to inert surfaces: Role of surface hydrophobicity, *Can J Microbiol*, **48**, 728–738.

FIRSTENBERG-EDEN R (1986) Electrical impedance for determining microbial quality of foods in Pierson M D and Stern N J *Foodborne microorganisms and their toxins: developing methodology*, New York, Marcel Dekker Inc., 129–144.

FLINT S H, BROOKS J D and BREMER P J (1997) Use of the Malthus conductance growth analyser to determine numbers of thermophilic streptococci on stainless steel, *J Appl Microbiol*, **83**, 335–339.

FOSTER E M (1997) Historical overview of key issues in food safety, *Emerg Infect Dis*, **3**, 481–482.

GAO A, MUTHARIA L, CHEN S, RAHN K and ODUMERU J (2002) Effect of pasteurization on survival of *Mycobacterium paratuberculosis* in milk, *J Dairy Sci*, **85**, 3198–3205.

GIFFEL TE M C, BEUMER R R, LANGEVELD L P M, ROMBOUTS F M and TE GIFFEL M C (1997) The role of heat exchangers in the contamination of milk with *Bacillus cereus* in dairy processing plants, *Int J Dairy Technol*, **50** (2), 43–47.

GILBERT P, ALLISON D, BRADING M, VERRAN J and WALKER J (2001) *Biofilm community interactions: chance or necessity?*, Cardiff, BioLine.

GRANUM P E (2003) *Bacillus*, in Caballero B, Trugo L C and Finglas P M *Encyclopedia of*

Food Sciences and Nutrition, London, Academic Press, vol. 1, 359–365.

GRIFFITHS M W (2003) *Listeria*, in Caballero B, Trugo L C and Finglas P M *Encyclopedia of Food Sciences and Nutrition*, London, Academic Press, vol. 6, 3562–3573.

HAMMER P, KIESNER C, WALTE H-G, KNAPPSTEIN K and TEUFEL P (2002) Heat resistance of *Mycobacterium avium* spp. *paratuberculosis* in raw milk tested in a pilot plant pasteurizer, *Kieler-Milchwirtschaftliche-Forschungsberichte*, **54**, 275–303.

HELKE D M, SOMERS E B and WONG A C L (1993) Attachment of *Listeria monocytogenes* and *Salmonella typhimurium* to stainless steel and Buna-N in the presence of milk and individual milk components, *J Food Prot*, **56**, 479–484.

HOLAH J and TIMPERLEY A (1999) Hygienic design of food processing facilities and equipment in Wirtanen G, Salo S and Mikkola A *30th R^3-Nordic contamination contol symposium, VTT Symposium 193*, Espoo, Libella Painopalvelu Oy, 11–39.

HOLAH J T, HIGGS C, ROBINSON S, WORTHINGTON D and SPENCELEY H (1990) A conductance-based surface disinfection test for food hygiene, *Lett Appl Microbiol*, **11**, 255–259.

HOLAH J T, TAYLOR J H, DAWSON D J and HALL K E (2002) Biocide use in the food industry and the disinfectant resistance of persistent stains of *Listeria monocytogenes* and *Escherichia coli*, *Soc Appl Microbiol Symp Ser*, **31**, 111S–120S.

HOLAH J T, BIRD J and HALL K E (2004) The microbial ecology of high risk, chilled food factories; evidence for persistent *Listeria* spp. and *Escherichia coli* strains. *J. Appl Microbiol*, **97**, 68–77.

HUMPHREY T, MASON M and MARTIN K (1995) The isolation of *Campylobacter jejuni* from contaminated surfaces and its survival in diluents, *Int J Food Microbiol*, **26**, 295–303.

HUSU J R, SEPPÄNEN J T, SIVELÄ S K and RAURAMAA A L (1990) Contamination of raw milk by *Listeria monocytogenes* on dairy farms, *J Vet Med*, **37**, 268–275.

JAY J M (1996) *Modern Food Microbiology*, 5th edition, New York, Chapman & Hall, 13–66, 469–477.

JOSEPH B, OTTA S K, KARUNASAGAR I and KARUNASAGAR I (2001) Biofilm formation by *Salmonella* spp. on food contact surfaces and their sensitivity to sanitizers, *Int. J. Food Microbiol*, **64**, 367–372.

JUNKINS A and DOYLE M P (1993) *Enterobacteriaceae*, in Macrae M, Robinson R K and Sadler M J *Encyclopaedia of Food Science, Food Technology and Nutrition*, 2nd edition, London, Academic Press, vol. 3, 1613–1618.

KOSTYÁL E (1998) *Removal of Chlorinated Organic Matter from Wastewaters, Chlorinated Ground, and Lake Water by Nitrifying Fluidized-bed Biomass*, Helsinki, Hakapaino Oy.

KRONLÖF J (1994) *Immobilized Yeast in Continuous Fermentation of Beer*. VTT Publications 167. Espoo, The Technical Research Centre of Finland.

KUSAMANINGRUM H D, RIBOLDI G, HAZELEGER W C and BEUMER R R (2003) Survival of foodborne pathogens on stainless steel surfaces and cross-contamination to foods, *Int J Food Microbiol*, **85**, 227–236.

LABBÉ R G (1993) *Clostridium*, in Macrae M, Robinson R K and Sadler M J *Encyclopaedia of Food Science, Food Technology and Nutrition*, London, Academic Press, vol. 5, 1043–1051.

LABBÉ R G (2003) *Clostridium*, in Caballero B, Trugo L C and Finglas P M *Encyclopedia of Food Sciences and Nutrition*, 2nd edition, London, Academic Press, vol. 3, 1398–1401.

LAPPALAINEN J, LOIKKANEN S, HAVANA M, KARP M, SJÖBERG A-M and WIRTANEN G (2000) Microbial testing methods for detection of residual cleaning agents and disinfectants – Prevention of ATP bioluminescence measurement errors in the food industry, *J Food Prot*, **63**, 210–215.

LAPPALAINEN J, SALO S and WIRTANEN G (2003) Detergent and disinfectant residue testing with photobacteria in Wirtanen G and Salo S *34th R^3 – Nordic contamination control symposium*, Espoo, Otamedia Oy, 151–159.

LELIEVELD H L M (1985) Hygienic design and test methods, *J Soc Dairy Technol*, **38**, 14–16.

LELIEVELD H L M, MOSTERT M A, HOLAH J and WHITE B (2003) *Hygiene in Food Processing*, Cambridge, Woodhead Publishing Limited.

LERICHE V and CARPENTIER B (1995) Viable but nonculturable *Salmonella typhimurium* in single- and binary-species biofilms in response to chlorine treatment, *J Food Prot*, **58**, 1186–1191.

LINDSAY D (2001) *Ecophysiology, Biofilm Formation and Sanitizer Susceptibility of Food Spoilage Bacteria with Emphasis on Selected Bacillus Species*, Johannesburg, University of the Witwatersrand.

LOOSDRECHT VAN M C M, LYKLEMA J, NORDE W and ZEHNDER A J B (1989) Bacterial adhesion: a physicochemical approach, *Microb Ecol*, **17**, 1–15.

LUNDÉN J (2004) *Persistent* Listeria monocytogenes *Contamination in Food Processing Plants*, Helsinki, Yliopistopaino.

LUPPENS S B I, REIJ M W, VAN DER HEIJDEN R W L, ROMBOUTS F M and ABEE T (2002) Development of a standard test to assess the resistance of *Staphylococcus aureus* biofilm cells to disinfectants, *Appl Environ Microbiol*, **68**, 4194–4200.

LYYTILÄINEN O, AUTIO T, MAIJALA R, RUUTU P, HONKANEN-BUZALSKI T, MIETTINEN M, HATAKKA M, MIKKOLA J, ANTTILA V-J, JOHANSSON T, RANTALA L, AALTO T, KORKEALA H and SIITONEN A (2000) An outbreak of *Listeria monocytogenes* serotype 3a from butter in Finland, *J Infect Dis*, **181**, 1838–1841.

MAFU A A, ROY D, GOULET J and MAGNY P (1990) Attachment of *Listeria monocytogenes* to stainless steel, glass, polypropylene, and rubber surfaces after short contact times, *J Food Prot*, **53**, 742–746.

MAUKONEN J, MÄTTÖ J, WIRTANEN G, RAASKA L, MATTILA-SANDHOLM T and SAARELA M (2003) Methodologies for the characterization of microbes in industrial environments: a review, *J Ind Microbiol Biotechnol*, **30**, 327–356.

MCBAIN A, ALLISON D, BRADING M, RICKARD A, VERRAN J and WALKER J (2003) *Biofilm Communities: Order from Chaos?*, Cardiff, BioLine.

MEAD G C (1993) *Salmonella*, in Macrae M, Robinson R K and Sadler M J *Encyclopaedia of Food Science, Food Technology and Nutrition* 2nd edition, London, Academic Press, vol. 6, 3981–3985.

MIETTINEN H and WIRTANEN G (2005) Prevalence and location of *Listeria monocytogenes* in farmed rainbow trout, *Int J Food Microbiol*, submitted.

MIETTINEN M K, BJÖRKROTH K J and KORKEALA H J (1999) Characterization of *Listeria monocytogenes* from an ice-cream plant by serotyping and pulsed-field gel electrophoresis, *Int J Food Microbiol*, **46**, 187–192.

MIETTINEN H, AARNISALO K, SALO S and SJÖBERG A-M (2001) Evaluation of surface contamination and the presence of *Listeria monocytogenes* in fish processing factories, *J Food Prot*, **64**, 635–639.

MILLER P C and BOTT T R (1982) Effects of biocide and nutrient availability on microbial contamination of surfaces in cooling-water systems, *J Chem Technol Biotechnol*, **32**, 538–546.

MOKGATLA R M, PROZEL V S and GOUWS P A (1998) Isolation of *Salmonella* resistant to hypochlorous acid from a poultry abattoir, *Lett Appl Microbiol*, **27**, 379–382.

MØRETRØ T, HERMANSEN L, HOLCK A, SIDHU M S, RUDI K and LANGSRUD S (2003a) Biofilm formation and the presence of the intercellular adhesion locus ica among staphylococci from food and food processing environments, *Appl Environ Microbiol*, **69**, 5648–5655.

MØRETRØ T, MIDTGAARD E S, NESSE L L and LANGSRUD S (2003b) Susceptibility of *Salmonella* isolated from fish feed factories to disinfectants and air-drying at surfaces, *Vet Microbiol*, **94**, 207–217.

NETTEN VAN P and KRAMER J M (1992) Media for the detection and enumeration of *Bacillus cereus* in foods: A review, *Int J Food Microbiol*, **17**, 85–99.

OULAHAL-LAGSIR N, MARTIAL-GROS A, BONNEAU M and BLOM L J (2003) '*Escherichia coli* – milk' biofilm removal from stainless steel surfaces: Synergism between ultrasonic waves and enzymes, *Biofouling*, **19**, 159–168.

PAYEUR J B (2000) *Mycobacterium*, in Robinson R K, Batt C A and Patel P D *Encyclopedia of Food Microbiology*, London, Academic Press, 1500–1511.

PENG J S, TSAI W C and CHOU C C (2002) Inactivation and removal of *Bacillus cereus* by sanitizer and detergent, *Int J Food Microbiol*, **77**, 11–18.

PETERS A C, ELVERS K T and GRIFFITH C J (1999) Biofilms in the food industry: Assessing hazards and risks to health in Wimpenny J, Gilbert P, Walker J, Brading M and Bayston R *Biofilms – The Good, the Bad and the Ugly*, Cardiff, BioLine.

PRICE R P and TOM P D (2003a) Compendium of fish and fishery product processing methods, hazards and controls, Chapter 11: *Campylobacter* spp., www.seafood.ucdavis.edu/ HACCP/ Compendium/Chapt11.htm, 21 October 2004.

PRICE R P and TOM P D (2003b) Compendium of fish and fishery product processing methods, hazards and controls, Chapter 17: *Salmonella*, www.seafood.ucdavis.edu/ HACCP/ Compendium/Chapt17.htm, 21 October 2004.

ROBERTS T A, BAIRD-PARKER A C and TOMPKIN R B (1996) *Microorganisms in Foods 5 – Characteristics of microbial pathogens*, London, Blackie Academic & Professional, 112–125.

RONNER A B and WONG A C L (1993) Biofilm development and sanitizer inactivation of *Listeria monocytogenes* and *Salmonella typhimurium* on stainless steel and Buna-N rubber, *J Food Prot*, **56**, 750–758.

RØRVIK L M, AASE B, ALVESTAD T and CAUGANT D A (2003) Molecular epidemiological survey of *Listeria monocytogenes* in broilers and poultry products, *J Appl Microbiol*, **94**, 633–640.

ROWE M T, DUNSTALL G, KIRK R, LOUGHNEY C F, COOKE J L and BROWN S R (1998) Development of an image system for the study of viable but non-cultural forms of *Campylobacter jejuni* and its use to determine their resistance to disinfectants, *Food Microbiol*, **15**, 491–498.

SALO S, LAINE A, ALANKO T, SJÖBERG A-M and WIRTANEN G (2000) Validation of the microbiological methods Hygicult dipslide, contact plate and swabbing in surface hygiene control: A Nordic collaborative study, *J AOAC Int*, **83**, 1357–1365.

SALO S, ALANKO T, SJÖBERG A-M and WIRTANEN G (2002) Validation of Hygicult E dipslides in surface hygiene control: A Nordic collaborative study, *J AOAC Int*, **85**, 388–394.

SHELEF L A (2003) *Bacillus*, in Caballero B, Trugo L C and Finglas P M *Encyclopedia of Food Sciences and Nutrition*, 2nd edition, London, Academic Press, vol. 1, 358–365.

SOMERS E B, SCHOENI J L and WONG A C L (1994) Effect of trisodium phosphate on biofilm and planktonic cells of *Campylobacter jejuni*, *Escherichia coli* O157:H7, *Listeria monocytogenes* and *Salmonella typhimurium, Int J Food Microbiol*, **22**, 269–276.

STEENSTRUP L D, COCKER R and FRIIS A (2004) The integrated approach to hygienic engineering in Wirtanen G and Salo S *DairyNET – Hygiene control in Nordic dairies, VTT Publication 545*, Espoo, Otamedia Oy, 199–204.

STEPANOVIC S, CIRKOVIC I, MIJAC V and SVABIC-VLAHOVIC M (2003) Influence of the incubation temperature, atmosphere and dynamic conditions on biofilm formation by *Salmonella* spp, *Food Microbiol*, **20**, 339–343.

STERN N L and KAZMI S U (1989) *Campylobacter jejuni*, in Doyle M P *Foodborne Bacterial Pathogens*, New York, Marcel Dekker Inc., 71–110.

STOODLEY P, DE BEER D and LEWANDOWSKI Z (1994) Liquid flow in biofilm systems, *Appl Environ Microbiol*, **60**, 2711–2716.

STOPFORTH J D, SAMELIS J, SOFOS J N, KENDALL P A and SMITH G C (2003a) Influence of extended acid stressing in fresh beef decontamination runoff fluids on sanitizer resistance of acid-adapted *Escherichia coli* O157:H7 in biofilms, *J Food Prot*, **66**, 2258–2266.

STOPFORTH J D, SAMELIS J, SOFOS J N, KENDALL P A and SMITH G C (2003b) Influence of organic acid concentration on survival of *Listeria monocytogenes* and *Escherichia coli* O157:H7 in beef carcass wash water and on model equipment surfaces, *Food Microbiol*, **20**, 651–660.

STORGÅRDS E, PIHLAJAMÄKI O and HAIKARA A (1997) Biofilms in the brewing process – a new approach to hygiene management, in *Proceedings of the 26th Congress of European Brewery Convention*, Maastricht, 24–29 May 1997, 717–724.

SVENSSON B, EKELUND K, OGURA H and CHRISTIANSSON A (2004) Characterisation of *Bacillus cereus* isolated from milk silo tanks at eight different dairy plants, *Int Dairy J*, **14**, 17–27.

SVENSSON B, ENEROTH Å, BRENDEHAUG J and CHRISTIANSSON A (1999) Investigation of *Bacillus cereus* contamination sites in a dairy plant with RAPD-PCR, *Int Dairy J*, **9**, 903–912.

TORVINEN E, SUOMALAINEN S, LEHTOLA M J, MIETTINEN I T, ZACHEUS O, PAULIN L, KATILA M-L and MARTIKAINEN P J (2004) Mycobacteria in water and loose deposits of drinking water distribution system in Finland, *Appl Environ Microbiol*, **70**, 1973–1981.

TRACHOO N and FRANK J F (2002) Effectiveness of chemical sanitizers against *Campylobacter jejuni* containing biofilms, *J Food Prot*, **65**, 1117–1121.

TRACHOO N, FRANK J F and STERN N J (2002) Survival of *Campylobacter jejuni* in biofilms isolated from chicken houses, *J Food Prot*, **65**, 1110–1116.

TROLLER J A (1993) *Sanitation in Food Processing*, San Diego, Academic Press Inc.

VAUGHAN E E (2004) Future of omics technologies in food safety in Raspor P, Smole Možina S and Cencič A *New Tools for Improving Microbial Food Safety and Quality – Biotechnology and molecular biology approaches*, Ljubljana, Slovenian Microbiological Society, 410.

VENKITANARAYANAN K S and DOYLE M P (2003) *Escherichia coli*, in Caballero B, Trugo L C and Finglas P M *Encyclopedia of Food Sciences and Nutrition*, 2nd edition, London, Academic Press, vol. 4, 2149–2152.

VLIET VAN A M H and KETLEY J M (2001) Pathogenesis of enteric *Campylobacter* infection, *J Appl Microbiol*, **90**, 45S–56S.

WEINSTEIN K B and ORTIZ J (2001) *Listeria monocytogenes*, http://www.emedicine.com / med/topic1312.htm, 26 October 2004.

WIMPENNY J, GILBERT P, WALKER J, BRADING M and BAYSTON R (1999) *Biofilms – The Good, the Bad and the Ugly*, Cardiff, BioLine.

WIRTANEN G (1995) *Biofilm Formation and its Elimination from Food Processing Equipment*, VTT Publications 251, Espoo, VTT Offsetpaino.

WIRTANEN G (2002) *Equipment Hygiene in the Food Processing Industry – hygiene problems and methods of controlling* Listeria monocytogenes, VTT Publications 480, Espoo, Otamedia Oy. In Finnish.

WIRTANEN G and RAASKA L (2004) Needs for qualitative and quantitative risk assessment in the future, in Raspor P, Smole Možina S and Cenciè A *New Tools for Improving Microbial Food Safety and Quality – Biotechnology and molecular biology approaches*, Ljubljana, Slovenian Microbiological Society, 411.

WIRTANEN G and RAASKA L (2005) Food safety regulations, standards and guidelines in Europe, in *36th R^3-Nordic Symposium & the 5th European Patenteral Conference*, Linköping, May 23–25, Genarp, R^3-Nordic Association, 151–160.

WIRTANEN G and SALO S (2004) *DairyNET – hygiene control in Nordic dairies*, VTT Publication 545, Espoo, Otamedia Oy.

WIRTANEN G, HUSMARK U and MATTILA-SANDHOLM T (1996) Microbial evaluation of the biotransfer potential from surfaces with *Bacillus* biofilms after rinsing and cleaning procedures in closed food-processing systems, *J Food Prot*, **59**, 727–733.

WIRTANEN G, SALO S, MAUKONEN J, BREDHOLT S and MATTILA-SANDHOLM T (1997) *NordFood – sanitation in dairies*, VTT Publications 309, Espoo, VTT Offsetpaino.

WIRTANEN G, SALO S and MIKKOLA A (1999) *30th R^3-Nordic Contamination Control Symposium*, VTT Symposium 193. Espoo, Libella Painopalvelu Oy.

WIRTANEN G, SAARELA M and MATTILA-SANDHOLM T (2000a) Biofilms – impact on hygiene in food industries in Bryers J *Biofilms II: Process analysis and applications*, New York, John Wiley-Liss Inc., 327–372.

WIRTANEN G, STORGÅRDS E, SAARELA M, SALO S and MATTILA-SANDHOLM T (2000b) Detection of biofilms in the food and beverage industry, in Walker J, Surman S and Jass J *Industrial Biofouling*, Chichester, John Wiley & Sons, Ltd., 175–203.

WIRTANEN G, LANGSRUD S, SALO S, OLOFSON U, ALNÅS H, NEUMAN M, HOMEID J P and MATTILA-SANDHOLM T (2002) *Evaluation of Sanitation Procedures for Use in Dairies*, VTT Publication 481, Espoo, Otamedia Oy.

WONG A C L (1998) Biofilms in food processing environments, *J Dairy Sci*, **81**, 2765–2770.

WONG A C L and CERF O (1995) Biofilms: Implications for hygiene monitoring of dairy plant surfaces, *Bull Int Dairy Fed*, **302**, 40–50.

ZAGORC T (2004) How to achieve harmonized standards for safer food production?, in Raspor P, Smole Možina S and Cenciè A *New Tools for Improving Microbial Food Safety and Quality – Biotechnology and molecular biology approaches*, Ljubljana, Slovenian Microbiological Society, 412.

ZOTTOLA E A and SASAHARA K C (1994) Microbial biofilms in the food processing industry – should they be a concern?, *Int J Food Microbiol*, **23**, 125–148.

4

Pathogen resistance to sanitisers

A. J. van Asselt and M. C. te Giffel, NIZO Food Research, The Netherlands

4.1 Introduction: disinfection methods

In the food industry worldwide millions of tonnes of safe and healthy food are produced every year, by many people using a large amount of equipment. In producing food, the equipment used gets soiled by both product and microorganisms. In order to avoid recontamination of the fresh product due to fouled surfaces, each piece of equipment or processing line needs to be cleaned and disinfected at regular intervals. Therefore, cleaning and disinfection are important unit-operations that are carried out in each food factory on a regular basis. Within the dairy industry, for example, cleaning and disinfection is carried out on a daily basis, sometimes several times a day. For condiments the frequency differs per batch of product; however, the equipment is cleaned and disinfected usually after 8–16 hours operation. In the beverage industry, because of the acid character of fruit juices and soft drinks, cleaning and disinfection is applied after 60–100 hours of production.

Disinfection is defined as the treatment of surfaces/equipment using physical or chemical means such that the amount of microorganisms present is reduced to an acceptable level (Krop, 1990; Donhauser *et al.*, 1991). Prior to disinfecting, cleaning of the surface is necessary to remove organic compounds adhered to the surface. Without proper cleaning, disinfection is useless, as remaining product will inactivate the disinfecting agent and microorganisms present will survive the disinfecting treatment. In practice 90–95% of the microorganisms present are removed by an efficient cleaning protocol (Krop, 1990). Disinfection reduces the amount of remaining microorganisms. This means that, in general, a disinfected surface/piece of equipment is not sterile and means that disinfection is not equal to sterilisation where viable microorganisms can no longer be detected.

Disinfection can be performed by using physical (steam, ultraviolet, irradiation) or chemical methods. In general, physical methods are preferred as they are very reliable and leave no residues behind. However, physical methods cannot always be applied owing to restrictions such as temperature, safety of personnel and design of the equipment. In those cases chemical disinfectants are used (Krop, 1990).

In this chapter the mode of action of the main disinfectants, the behaviour/response of pathogenic bacteria towards chemical disinfectants and some future developments are discussed. The effect of physical methods is not discussed.

4.2 Factors influencing the effectiveness of cleaning and disinfection

A wide range of disinfectants is available that can be divided in the following groups (see also Table 4.1):

- halogen-releasing agents (HRA);
- quaternary ammonium compounds (QAC);
- peroxygens;
- alcohols;
- aldehydes;
- (bis)phenols;
- biguanides.

Each of the different groups has its own applications within the food industry and its own restrictions in use. It is important to realise what the proposed effect of a disinfectant is on a target-organism and what possible protection mechanisms are present within the organism. In the following sections, the different compounds, their mode of action and their applications are discussed.

Table 4.1 Disinfectants and their mode of action

Biocide	Mode of action	Target
Halogen-releasing agents	Halogenation/oxidation	Nucleic acids, proteins
Quaternary ammonium compounds (QACs)	Electrostatic (ionic) interaction	Cell surface, enzymes, proteins
Peroxygens	Oxidation	Lipids, proteins, DNA
Alcohols (ethanol)	Protein denaturation	Plasma membrane
Aldehydes	Alkylation reaction	Cell wall
(bis)Phenols	Penetration/partition phospholipids bilayer	Phospholipid bilayer
Biguanides	Electrostatic (ionic) interaction	Cytoplasmic membrane (bacteria)/plasma membrane (yeasts)

4.2.1 Halogen-releasing agents (HRA)

Chlorine-based compounds are the most frequently applied HRAs. They include sodium hypochlorite, chlorine dioxide, and the *N*-chloro compounds such as sodium dichloroisocyanurate (NaDCC). A very cheap and frequently applied formulation is an aqueous solution of sodium hypochlorite producing hypochlorous acid (HClO) (Krop, 1990; McDonnell and Russell, 1999) (Table 4.3 on page 77). HClO is the active component and results in the inactivation of all types of microorganisms such as bacteria, viruses and spores (Sofos and Busta, 1999). Another applied form of chlorine is chlorine dioxide (ClO_2). It is synthesised by the reaction of chlorine and sodium hypochlorite. However, chlorine dioxide is much more unstable than a standard hypochlorous solution and decomposes chlorine into gas at temperatures higher than 30 °C when exposed to light (Beuchat, 1998). This can lead to dangerous situations as high concentrations of chlorine gas are explosive (Speek, 2002; Codex, 2003). However, when the solution is kept cool and protected from light the disinfectant can be kept stable at concentrations up to $10\,g\,l^{-1}$ (Erco Worldwide, 2004).

Mode of action of hypochlorous acid

Although the exact mode of action is not known, the main disinfecting effect of chlorine is caused by oxidative activity. In particular, nucleic acids and proteins are destroyed, resulting in irreversible changes and disruption of DNA-protein synthesis (Krop, 1990). The mechanism of killing of spores differs owing to their thick proteinaceous coat. Therefore higher concentrations are needed than for inactivation of vegetative cells. Young and Setlow (2003) concluded that hypochlorite affects spore germination possibly because of the severe damage to the spore's inner membrane. For spore suspensions, Young and Setlow (2003) showed that a concentration of $50\,mg\,l^{-1}$ during 10 min at room temperature is sufficient to achieve 4 decimal reductions of *Bacillus subtilis* spores. A concentration of $50\,mg\,l^{-1}$ resulted in 1 decimal reduction of *B. cereus* spores after 1.5 min (Wang *et al.*, 1973). These results show that the minimal inhibitory concentration can vary per species.

Mode of action of chlorine dioxide

Chlorine dioxide (ClO_2), if applied properly, appears to be 2.5 times more oxidative than sodium hypochlorite (Speek, 2002; Rodgers *et al.*, 2004), and is effective against bacteria, viruses and spores (Hoxey and Thomas, 1999). The action of chlorine dioxide involves disruption of the cell's protein synthesis and membrane permeability control mechanism. It produces no harmful by-products as trihalomethans, nor does it react with ammonia. After treatment with chlorine dioxide, spores of *Bacillus subtilis* can undergo the initial steps in spore germination but the process stops because of membrane damage (Young and Setlow, 2003). An aqueous chlorine dioxide treatment of alfalfa seeds inoculated with *E. coli* for 10 min at a concentration of $25\,mg\,l^{-1}$ resulted in approximately 1 log reduction of the microorganism (Singh *et al.*, 2003). Compared with standard chlorine solutions (sodium hypochlorite) a concentration of $3\,mg\,l^{-1}$ chlorine dioxide has the same inactivating effect on *E. coli* O157:H7 and *L. monocytogenes* as $200\,mg\,l^{-1}$ of chlorine when applied for decontamination of fruit surfaces (Rodgers *et al.*, 2004).

Iodine

Iodine is widely used for sanitising food processing equipment and surfaces. Iodine is less reactive than chlorine and less affected by the presence of organic matter but also has disadvantages such as staining human skin, plastic parts of equipment, and also has a relatively high price as compared with chlorine (Krop, 1990; Hugo and Russell, 1999). Solutions of 15% active chlorine are commercially available for €0.20–0.30 per kg whereas a 6% solution of iodine in 70% ethanol costs approximately €400 per kg (Boom Chemicals). Iodine is applied in three possible formulations: ethanol-iodine, aqueous iodine solutions and iodophores. The iodophores are most frequently applied and have high solubility in water, produce no vapour (below 50 °C), are less corrosive to stainless steel than chlorine-containing solutions, and are generally effective against Gram-negative and Gram-positive vegetative cells, yeasts, moulds and viruses (Bernstein, 1990; Beuchat, 1998). Bacterial spores (*B. cereus*, *B. subtilis* and *C. botulinum* type) are more resistant to iodophors (*D*-values are 10–100 times higher) and higher concentrations are necessary to achieve inactivation.

Mode of action of iodine

Similar to chlorine, the exact mode of action of iodine is not known. Iodine penetrates into microorganisms and attacks specific groups of proteins, nucleotides and fatty acids in a way comparable to chlorine (McDonnell and Russell, 1999). The effective concentration of iodine is approximately $100\,mg\,l^{-1}$ which is as effective as $300\,mg\,l^{-1}$ of chlorine (Krop, 1990).

4.2.2 Quaternary ammonium compounds (QACs)

QACs can be divided in two main subgroups (Mohr and Duggal, 1997; Reuter, 1998):

- tri-alkylbenzyl-ammonium compounds (e.g. benzalkonium chloride);
- tetra-alkyl-ammonium compounds (e.g. didecyldimethyl-ammonium chloride).

QACs combine antimicrobial properties with surface-active properties and are therefore useful for hard surface cleaning and deodorisation (McDonnell and Russell, 1999). Compared with chlorine they are more expensive but have the advantage of having residual action. QACs remain active on surfaces for approximately 1 day (e.g. fish industry) and therefore discourage further bacterial growth (Tatterson and Windsor, 2001). This adherence to the surface also has disadvantages. Removing the disinfectant from the surface by flushing with water becomes difficult, resulting in possible residues in the product (Kraemer, 1998).

Table 4.2 Efficacy of quaternary ammonium compounds on different infectious agents

Infectious agent	Efficacy	Comments	Source
Bacteria			Russell (1995)
Gram-positive	+		
Gram-negative	+	MIC higher than Gram +	
Spores	−	Sporostatic	Russell (1990)
Viruses			Quinn and Markey (1999)
Lipid	+		
Small non-lipid	−		
Non-lipid	+/−		
Mycobacteria	−		Russell (1996)
Yeast/moulds	+	Moulds more resistant	Russell (1999c)

+, effective, − ineffective, +/−, limited efficacy

In general QACs are effective against vegetative bacteria but have greatest effectiveness against Gram-positive bacteria. Yeast and moulds can be inactivated to some extent but higher concentrations are necessary (Krop, 1990; Bernstein, 1990) (see Table 4.2). QACs are most effective in the range of pH 6 and 10 (Beuchat, 1998), which limits their applicability in acid environments.

Mode of action

The principal actions of QACs are lowering of surface tension, inactivation of enzymes and denaturation of cell proteins. As a result of adsorption of QACs onto the microorganism's surface, the cell's permeability is changed dramatically. This results in leakage of intracellular low-molecular compounds, degradation of proteins and nucleic acids, and cell wall lysis by autolytic enzymes (McDonnell and Russell, 1999). The concentration applied depends on the type of microorganisms present in the product, the processing system and the environment. Concentrations typically used are in the range between 150 and 250 $mg\,l^{-1}$ of active Quaternary Ammonium (QA) (Bernstein, 1990; Beuchat, 1998). Allerberger and Dierich (1988) showed a bactericidal effect on *E. coli* at a concentration of 100 $mg\,l^{-1}$. Low concentrations (0.0005% w/v = 5 $mg\,l^{-1}$) of benzalkonium chloride are sporostatic, inhibiting outgrowth but not germination. QACs are not sporicidal (Russell, 1990).

4.2.3 Peroxygens

Hydrogen peroxide and peracetic acid are the main representatives of the group of peroxygens. Hydrogen peroxide is widely applied within the food industry and is commercially available in concentrations varying between 3% and 90% w/v, with 35% routinely used in the food industry (McDonnell *et al.*, 2002). It is applied for sterilising packaging material prior to filling (Mohr and Duggal, 1997), sterilising contact lenses and sterilising the surface of fruit and vegetables. Hydrogen peroxide is both bactericidal and sporicidal (Hugo and Russell, 1999), in general a concentration of 6% is bactericidal. Peroxygens are generally more active against Gram-positive bacteria than Gram-negative bacteria (Russell, 1990; McDonnell and Russell, 1999). To achieve a sporicidal effect, concentrations between 10 and 30% are necessary. Peracetic acid is commercially available in 15% solutions as a mixture of water, hydrogen peroxide and acetic acid and acts faster than hydrogen peroxide. It has a broad spectrum of efficacy against viruses, bacteria, yeast and spores (Bernstein, 1990). Compared with hydrogen peroxide, the activity of peracetic acid is hardly influenced by organic matter (Russell, 1990; McDonnell and Russell, 1999). Disadvantages are that peroxygens corrode on tools and equipment and are aggressive to, e.g., human tissues (Reuter, 1998). However the development and use of anticorrosives has reduced this concern (Marquis *et al.*, 1995).

Mode of action

The mode of action of peroxygens is based on free-radical oxidation (e.g. hydroxyl radicals) of essential cell components such as lipids, proteins and DNA (McDonnell and Russell, 1999). Peracetic acid not only attacks the proteins in the cell wall but also migrates into the cell and disrupts inner cell components as well (Donhauser *et al.*, 1991).

4.2.4 Alcohols

The most widely used alcohols for disinfection are: ethyl-alcohol (ethanol, alcohol), isopropyl alcohol (isopropanol, propane-2-ol) and *n*-propanol, the latter especially in Europe (Mohr and Duggal, 1997; McDonnell and Russell, 1999). In food production areas, alcohols are particularly used for the decontamination of hard surfaces of equipment (e.g. filling machines). The most effective concentration is between 60 and 70% v/v (Mohr and Duggal, 1997). The concentrations to achieve reduction of growth or complete inactivation are higher than for chorine solutions or organic acids. Alcohols are quick reacting, have a broad spectrum of antimicrobial activity and inhibit growth of vegetative bacteria, viruses and fungi. Spores are rather resistant against the effects of alcohol; however, a combination of 70% v/v concentration with temperatures up to 65 °C results in inactivation of spores, for example *Bacillus subtilis* spores (Setlow *et al.*, 2002). Compared with other disinfectants the concentrations applied are much higher (50–100 times) and in fact alcohols are only effective if used as the substance itself, instead of a low-concentration solution. This property makes alcohol more expensive in use compared with chlorine and QACs, and therefore is not frequently applied on a large, industrial scale but is used mostly for applications such as small, difficult to reach spots in equipment, temperature probes and quick wipe-downs of working surfaces and scales.

Mode of action

The general mode of action for inactivation of microorganisms by alcohols is by denaturation of proteins (Schlegel, 1993), with the primary site of action being the cell (plasma) membrane. As a result of deterioration of the plasma membrane, the cell wall starts to leak essential cell components such as ions (Ca^{2+}) and low molecular weight solutes such as peptides and amino acids. Therefore, the mode of action and its effect on the metabolism of the microorganism depends very much on the concentration. Moulds and actinomycetes are most susceptible to alcohols and are inhibited at 4% (v/v) whereas most bacteria can still grow at these concentrations (Kalathenos and Russell, 2003). Application of 5.5% (v/v) shows a bacteriostatic effect on *E. coli*, but in order to kill this microorganism concentrations of 22.2% or higher are necessary (Allerberger and Dierich, 1988). Yeasts are able to grow at higher alcohol concentrations (8–12% v/v), which is not surprising since they are responsible for the production of beer and wine (*Saccharomyces cerevisiae*). Spores are affected by ethanol. Setlow and co-workers (2002) showed that the spore coat can be permeabilized. Consequently, ethanol in combination with other components or with high temperature (> 65 °C) is more effective than ethanol itself in activating spores.

4.2.5 Aldehydes

Two aldehyde compounds are mainly used for disinfecting, glutaraldehyde and formaldehyde. Aldehydes are active against a wide range of bacteria, viruses, moulds and spores, are easily removed from surfaces and are (bio) degradable (Mohr and Duggal, 1997). However, the activity of aldehydes is very easily influenced by remaining (protein) fouling, which necessitates sufficient cleaning prior to disinfecting. From a toxicological point of view, aldehydes do not cause problems for humans when used within the prescribed concentrations (Mohr and Duggal, 1997). On the other hand, it is possible that formaldehyde can have mutagenic effects (McDonnell and Russell, 1999).

Mode of action

The mode of action of glutaraldehyde involves a strong association with the outer layers of bacterial cells (Denyer and Stewart, 1998; McDonnell and Russell, 1999). The cell's chemical reaction with glutaraldehyde results in metabolic and replicative inhibition (Denyer and Stewart, 1998). The way formaldehyde reacts is most probably the same. Concerning processing conditions, an alkali environment is more favourable than an acid environment as more reactive sites will be formed on the cell surface. Applied concentrations vary between 0.08 and 1.6% (w/w) for inactivating *E. coli*. For a sporicidal effect, a solution of 2% is normally sufficient.

4.2.6 Bisphenols

Bisphenols are hydroxy halogenated derivatives of diphenyl methane, diphenyl ether and diphenyl sulphide, and are active against bacteria, fungi and algae.

Triclosan and hexachlorophene are the most widely used (McDonnell and Russell, 1999). Triclosan, a derivative of diphenyl ether, is known as an ingredient in some medicated soaps and hand-cleansing gels and toothpastes, and is effective against staphylococci (Hugo and Russell, 1999). It is currently applied as antimicrobial layer in packaging material (Vermeiren *et al.*, 2002; Chung *et al.*, 2003) and conveyer belts (Quantex Laboratories, 2001; Stekelenburg and Hartog, 2002). Unfortunately, depending on the impurity of the starting material, Triclosan can contain concentrations of dioxin and dibenzofurans, both substances highly toxic to humans (Quantex Laboratories, 2001). Therefore, it is of great importance that the origin and way of production are known prior to application in food production areas. Hexachlorophene has been used in soaps as well; in 1972 it was restricted in use by the US Food and Drug Administration (FDA) to levels less than 0.1%. Nowadays, application as a surgical scrubber in case of certain infections is permitted (Spectrum Laboratories).

Mode of action

The exact mode of action is unknown so far but it is suggested that Triclosan affects the cytoplasmic membrane. However, current research shows that Triclosan inhibits one specific enzyme of the fatty acid synthesis of *E. coli*. This increases the risk of resistance against Triclosan as one mutation of a gene can result in a decreased efficacy of the disinfectant (Sixma, 2001). Hexachlorophene affects bacteria by inducing leakage, causing protoplast lysis and inhibiting respiration.

4.2.7 Biguanides

The group of biguanides is represented by chlorhexidine, alexidine and polymeric biguanides (McDonnell and Russell, 1999; Hugo and Russell, 1999). Chlorhexidine is probably the most widely applied biocide in hand-washing and oral products such as mouthwash, mouth spray and throat-lozenges (Sixma, 2001) and is bacteriostatic at concentrations of 0.0001 $mg\,l^{-1}$ as well as bactericidal at concentrations of 0.002 $mg\,l^{-1}$ (Russell, 1991). Chlorhexidine has a broad spectrum of activity and is pH-dependent (higher efficacy at alkaline rather than acid pH); its efficacy is greatly reduced by the presence of organic matter. High concentrations of chlorhexidine cause coagulation of intracellular constituents (Russell, 1990; McDonnell and Russell, 1999). Chlorhexidine is only sporicidal at elevated temperatures (>0.005 $mg\,l^{-1}$ at 70 °C) and is in general more sporostatic; it has little effect on the germination of the spore but does not prevent the outgrowth of the spore (Russell, 1991; Gorman *et al.*, 1987). Alexidine and the polymeric biguanides are used only on a small scale. The polymeric biguanides are used in particular by the food industry and also for the disinfection of swimming pools. An example is poly(hexamethylene biguanide) hydrochloride (PHMB) which is the main active ingredient of Vantocil, which is widely used in the food industry, hospitals, nursing homes and consumer households (Avecia, 2004).

Table 4.3 Summary of disinfecting agents

Biocide	Application	Bactericidal	Sporicidal	Comments
Halogen-releasing agents	50–250 mg l^{-1}	>10 mg l^{-1}	>50 mg l^{-1}	Chlorine cheap Iodine expensive Influenced by organic substances
Quaternary ammonium compounds	150–250 mg l^{-1}	>100 mg l^{-1}	No	Residual action (approx 1 day), neutral, non-aggressive
Peroxygens	3–90%	>6%	10–30%	More effective as mixture with acetic acid
Alcohols (ethanol)	20–70% (w/v)	>22% (w/v)	60–70% (w/v)	Not for large industrial application
Aldehydes	0.8–16 mg l^{-1}	<10 mg l^{-1}	20 mg l^{-1}	
Bisphenols	2–20 mg kg^{-1}	>10 mg l^{-1}	No	
Biguanides (chlorhexidine)	>150 mg l^{-1}	1–60 mg l^{-1}	—	Applied in hand-washing and oral products

Mode of action

In principle chlorhexidine attacks the outer cell layer but not sufficiently to induce lysis or cell death. However, after crossing the cell wall it damages the cytoplasmic membrane (bacteria) or plasma membrane (yeast) (McDonnell and Russell, 1999). Polymeric biguanide appears to have a non-specific mode of attack against cell membranes resulting in quick cell death.

Summary

The different effective concentrations for the biocides are summarised in Table 4.3. It is obvious that, depending on the type of application or type and metabolic state of the microorganism, the proper disinfectant must be chosen.

4.3 Strategies for optimisation of cleaning and disinfection

Resistance development as a result of cleaning and disinfection is not (yet) a matter of major concern for the food industry. However the food industry (and also the pharmaceutical industry) has to realise that the current processes of cleaning and disinfecting need to be carried out properly in order to avoid development of resistance. Even a short-term exposure to sub-lethal concentrations of QACs causes cellular changes of *Listeria monocytogenes* (Lundén *et al.*, 2003). In addition, recirculation of product in the process chain (re-work) implies a possible risk as (remaining) microorganisms are exposed a second time to a cleaning and disinfecting step. This might induce the development of resistant mutants of the spoilage microorganisms.

For the application of cleaning and disinfecting agents the following issues are important:

- use of appropriate product;
- application of correct processing conditions;
- influence of neutralising components;
- monitoring.

4.3.1 Use of the appropriate product

Application of the right type of agent is important to achieve the desired chemical effect. With respect to disinfectants it is necessary that a product with the proper spectrum of activity is chosen. For example, to inactivate spores the application of alcohols or QACs is useless as those agents are not sporicidal (Russell, 1990). Another point is that some solutions (e.g. chlorine solutions) act very aggressively towards metal surfaces and polymer seals. This results in corrosion of the materials providing bacteria with places where they are able to survive cleaning and disinfecting procedures (Kraemer, 1998).

4.3.2 Application of the right cleaning/disinfecting conditions

The combination of concentration, mechanical action, time and temperature is of major importance for efficient cleaning and disinfection. The applied concentration should not be higher or lower than the advised concentration. Too high concentrations can lead to insolubility and increased corrosiveness. Concerning protein fouling, it is known that too high concentrations of alkali (>0.5%) result in polymerisation of the protein and form a rubbery layer (Bird, 1994; Jeurnink and Brinkman, 1994). These kinds of rubbery layers obstruct and prevent the penetration of cleaning and disinfecting solution into the fouling, resulting in a decreased fouling removal rate (Jeurnink *et al.*, 1996). Concerning starch fouling, the concentrations of alkali needed vary between 9 and 20% (w/v) (Bird, 1994), which is quite different from those for dairy processes. Therefore, the applied concentration depends on the type of fouling.

An increase in temperature results in increased efficiency. However, for dairy processes at temperatures above 80 °C the opposite effect can be achieved, as proteins coagulate, resulting in an increase of fouling instead of a decrease. In addition, for all processes, cleaning at temperatures above 80 °C results in higher energy consumption use without extra cleaning benefit and can lead to damage to the equipment (corrosion). An optimal working temperature therefore is around 70 °C. In combination with the 0.5% alkaline solution (for dairy environments) this is sufficient to inactivate any vegetative pathogenic microorganism (Jeurnink *et al.*, 1996). In the case of membrane systems even lower temperatures (40–60 °C) are advised, owing to the rather vulnerable composition of the membranes and its modules (Shorrock *et al.*, 1998).

Contact time is the third important parameter of disinfection processes. The longer the contact time, the greater the number of microorganisms inactivated. In most cases there is a direct link between contact time and concentration. There are various models predicting the inactivation of a disinfectant but not all of them are easy to use (e.g. too many unknown parameters). In general, the simple Chick-Watson (1908) log-linear model is used (Lambert and Johnston, 2000; Kamase *et al.*, 2003; Cho *et al.*, 2003):

$$\log\left(\frac{N_1}{N_0}\right) = -kC^n t \tag{4.1}$$

where N_1 = number of surviving microorganisms, N_0 = initial number of microorganisms, k = disinfection rate constant, C = disinfectant concentration, n = dilution coefficient and t = contact time.

The dilution coefficient (n) differs per type of disinfectant. For example, for QACs $n = 1$, which implies that by halving the concentration the contact time (t) is doubled. For ethanol $n = 10$ which implies an efficiency reduction by a factor of 2^{10} (= 1024) when halving the concentration (Krop, 1990).

The effect of mechanical action is obvious; the more mechanical energy is put into the removal of the fouling the more efficiently the fouling will be removed, thus a more efficient cleaning process is obtained (Gibson *et al.*, 1999). However, there is a limit, as too much mechanical action (e.g. by using metal scrubbing devices) may cause damage to the cleaned object/surface.

The final effect depends on the right combination of the conditions discussed. However, it remains possible to choose for different combinations of conditions as long as 'the sum' of the conditions will be the same, e.g. a reduction of the concentration can be compensated by an increase in time or mechanical action (Krop, 1990).

4.3.3 Influence of neutralising components

Prior to disinfecting, the equipment or surface to be treated should not contain any components that can inactivate the disinfectant. Organic matter (e.g. food residues, milk stone, blood) are well known for their neutralising effect. In general these organic materials interfere by reacting with the biocide, leaving a reduced concentration of antimicrobial agent for attack on microorganisms. In addition to organic materials, surface-active agents and metal ions can act as an interfering substrate (Russell, 1999a).

4.3.4 Monitoring

As shown, a lot of characteristics concerning the application of disinfectants and the inactivation of microorganisms are known. But knowledge does not guarantee appropriate control of the process. Thorough analysis of available data is necessary to make the right decision with regard of type of disinfectant, process conditions and required effect. Monitoring devices to analyse cleaning and disinfection processes, and databases containing inactivation kinetics of relevant microorganisms in combination with predictive knowledge can be a great help in optimising relevant processes.

With regard to monitoring, OPTICIP, a monitoring device to make and optimise cleaning-in-place (CIP) procedures can be applied (van Asselt and te Giffel, 2002). A typical cleaning procedure of an evaporator before and after optimisation is shown in Fig. 4.1. The system monitors the removal of organic and inorganic fouling off-line in combination with the in-line measurement of parameters such as temperature, flow, conductivity and valve settings. The turbidity of the cleaning solution is a measure for the removal of organic fouling. The calcium concentration is a measure for the removal of inorganic fouling. Conductivity measurement is used for separation of the various cleaning phases and gives an indication of the concentration of the cleaning solution used. Sharp slopes between subsequent phases indicate that rinsing and cleaning phases are properly separated (van Asselt *et al.*, 2002). More simplified systems are also available. Johnson-Diversey introduced 'Shurlogger', a real-time CIP monitoring system based on flow, temperature and conductance (Dodd, 2003). However, the fouling removal is not taken into account. Therefore, this system gives less detailed analyses compared to OPTICIP.

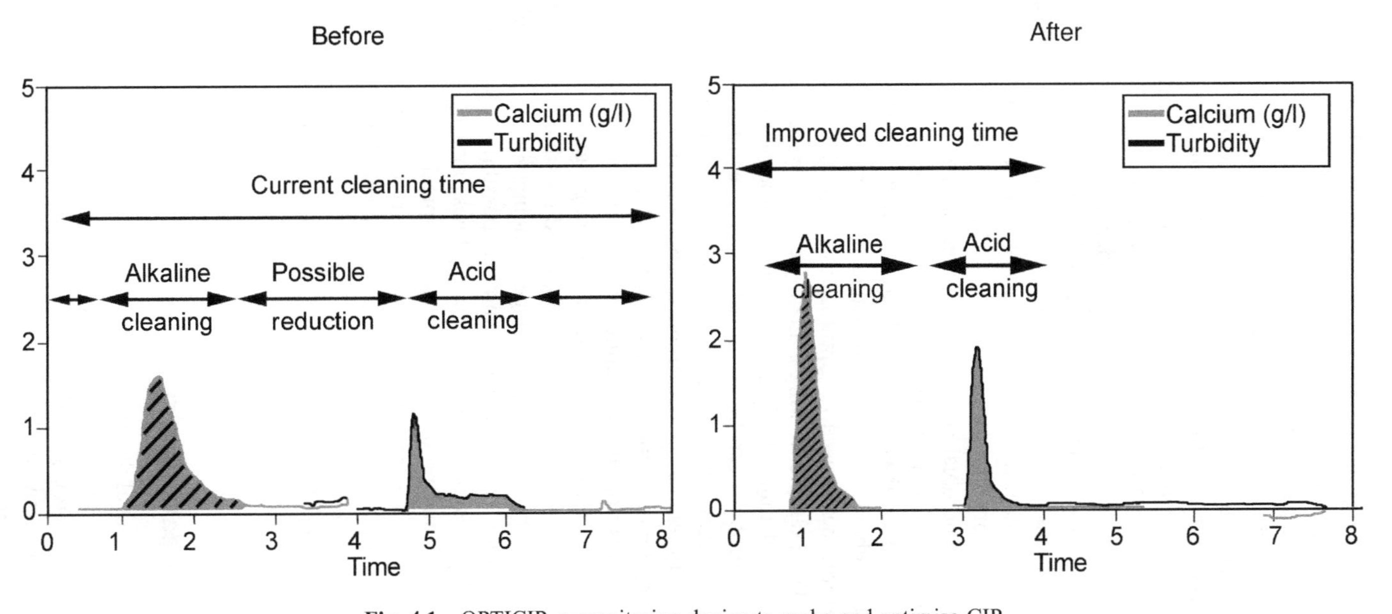

Fig. 4.1 OPTICIP, a monitoring device to make and optimise CIP.

A quick monitoring device is the application of ATP as measurement for remaining microorganisms and/or organic substances. The principle is based on the fact that every organic cell contains ATP as energy carrier. The reaction of the enzyme luciferase with ATP results in the emission of light that can be measured by a specific light-measuring device. The more light is emitted the more ATP was present and the more the surface or liquid was contaminated with microorganisms or organic matter. It is even possible to differentiate between microbial and organic ATP. A disadvantage of this method is that the detection limit is relatively high. The minimum concentration of microorganisms is approx. 10^3–10^4 cfu ml^{-1} (Moore *et al.*, 2001) before this method becomes reliable whereas this amount is already crossing the limit of levels of contamination. Thus, measuring ATP is suitable for a quick inventory of the cleanliness of equipment or rinsing water. The method is not applicable to determine the antimicrobial activity of disinfecting agents.

A different way of optimising cleaning and disinfecting processing concerns the combination of databases and predictive modelling. NIZO Premia™ is an example that combines research knowledge with predictive modelling. It is a software platform that is used for optimisation of product properties or process performance. For example, fouling is mainly caused by denaturation of proteins and precipitation of minerals. The denaturation process of β-lactoglobulin (an important whey protein) can be described as a consecutive set of reactions (de Jong, 1996). This knowledge can be used to predict the fouling behaviour in heat exchangers of different dairy-type products. By predicting the amount of fouling produced, the optimum running time for heat exchangers can be determined. In addition the composition of the fouling layer is known which makes it possible to choose the right cleaning procedures (cleaning agents, temperature, etc.). After optimisation with NIZO Premia it appeared possible to reduce the amount of fouling by 50–80%, resulting in longer running times and higher process efficiency (de Jong *et al.*, 2002). Another possibility is using predictive modelling for the design of new processing lines, making the effects visible concerning fouling and product properties. A typical example is the development of a new type of evaporator at a Dutch dairy company where the use of NIZO Premia resulted in 70% less energy use compared with standard designed evaporators (Vissers *et al.*, 2002).

Thus, predictive modelling is a powerful tool to analyse and optimise critical processes within the food industry.

4.4 Types of pathogen response

When applying chemical disinfectants in a process or on process equipment it is important to know how microorganisms/pathogens may respond. Like every other organism, microorganisms protect themselves against all kinds of influence from the environment. Some of the protection mechanisms are intrinsic (natural property) but others are acquired (mutation or acquisition of plasmids) during the evolution of the organism. A general overview of relevant disinfectants and their mode of action is given in Table 4.1.

The possible mode of actions of the applied disinfecting treatments and pathogen response as discussed in the previous section are discussed below.

4.4.1 Target area of disinfectant

- Cell membrane and its outer layers, breaking down results in quick cell death/inactivation of the microorganism (Todar, 2001).
- Damage to enzymes + important metabolic processes; some heavy metals (e.g. copper, silver, mercury) act as poisons to enzymes. Added as salts or organic combinations they bind to SH groups of enzymes and cause changes in the structure (tertiary and quaternary) of these proteins (Schlegel, 1993).
- Affecting the synthesis of proteins in the target organism results in growth prohibition (Todar, 2001; Schlegel, 1993).
- Inhibition of DNA synthesis or breakage of the DNA strands, resulting in the blockage of cell growth (McDonnell and Russell, 1999).

4.4.2 Pathogen response

Adding disinfectants will result in increased stress on the bacteria and their metabolism. In principle they have three ways of responding to disinfectants:

- alteration of the target;
- reduction of target access;
- inactivation of the disinfectant.

As disinfectants have a broad spectrum of activity, it is not likely that the alteration of the target will work. The two other mechanisms seem to be possible and a combination of resistance mechanisms is also one of the possibilities (Chapman, 1998). The fact that microorganisms show this kind of behaviour is caused by either intrinsic or acquired resistance (Russell, 1995; McDonnell and Russell, 1999).

4.4.3 Intrinsic resistance

This type of resistance is defined as a natural chromosomally controlled property of a bacterial cell to circumvent the action of a disinfectant. It is demonstrated especially by Gram-negative bacteria and bacterial spores (Russell, 1991; McDonnell and Russell, 1999). Bacterial spores, the genera *Bacillus* and *Clostridium* in particular, are the most resistant, e.g *Cl. perfringens* and *B. cereus* (Russell, 1995). The exact mechanism of sporicidal action is not fully understood; however, as the prime target area of biocides lies within the spore it is expected that, owing to the different layers of the spore, the penetration of biocides is limited.

4.4.4 Acquired resistance

Acquired, non-plasmid-encoded resistance occurs when bacteria are exposed to gradually increasing concentrations of a certain biocide. Acquired, plasmid-encoded resistance is, in most cases, some form of resistance against metal-based biocides (silver, copper or mercury) (Chapman, 1998; Russell, 1999b; McDonnell and Russell, 1999). However, a recent study investigated the resistance of *Salmonella* against hypochlorous acid (concentrations up to 28 mg l^{-1}) and indicated an emerging problem for the food industry (Mokgatla *et al.*, 2002). Normally, a chlorine concentration of 10 mg l^{-1} is sufficient to inactivate vegetative, non-spore-forming microorganisms (Krop, 1990). This type of resistance might be caused by remaining organic substances (partly) inactivating the chlorine solution. It does show that, when applying certain disinfectants, it is important to apply to correct concentrations of disinfectant in combination with a clean surface in order to achieve efficient inactivation of microorganisms. Therefore, this kind of resistance appears to be unstable and could also be considered as pseudo-resistance (Heinzel, 1998). Pseudo-resistance occurs when bacteria appear to be resistant to a certain kind of biocide, but when placed in a biocide-free environment the resistance disappears. A few reasons are known to cause this apparent resistance:

- use of an inefficient product (i.e. disinfectant with limited spectrum of activity);
- incorrect use of the disinfectant (not according to the conditions recommended by the supplier);
- insufficient contact (time) with the surface to be treated;
- insufficient availability of the reactive agent.

It is obvious that these reasons may lead to survival of bacteria. Although it is not considered to be microbial resistance, it is probably the most widespread form of perceived resistance (Heinzel, 1998). In addition, it is even thought possible that some microorganisms are able to use the intended disinfectant as a source of energy: instead of being inactivated, they start to grow.

4.5 Predicting microbial resistance

Predicting pathogen resistance against current disinfectants would be very useful for application in food factories and hospitals. Compared with antibiotics, the mode of action of preservatives/disinfectants is less well understood. Antibiotics normally have one specific group or subgroup of bacteria as target microorganisms whereas disinfectants attack bacteria in general (Russell, 1991). Therefore, it is rather difficult to determine the exact effect on microorganisms beforehand. However, the mechanisms of action of disinfectants become more and more clear allowing the effect they have on microorganisms to be predicted. Whether microorganisms will survive disinfection in practice depends on more than one factor. At least 15 factors appear to influence the possible resistance of a microbial strain (Baquero *et al.*, 1998). A model to predict the effect of a single disinfectant in an unspecified environment might therefore be difficult to realise. Thus, testing under practical conditions remains necessary to determine the effect of a certain disinfectant. Although not every detail is known, it is possible to determine whether a disinfectant will be effective based on the following information:

- type of bacteria – metabolic state;
- revival of injured cells/biodiversity of microorganisms;
- influence of remaining organic matter/biofilms;
- processing conditions: temperature, pH.

4.5.1 Type of microorganisms – metabolic state

The metabolic state of microorganisms is important in determining the possible effect of the disinfectant. With regard to vegetative cells, Gram-negative microorganisms appear more resistant than Gram-positive microorganisms owing to the composition of the cell wall. The cell wall of Gram-positives contains fewer lipids compared with Gram-negatives (Russell, 1999a). Bacterial spores are highly resistant to chemical and physical agents, which is mainly due to the spore coat and spore cortex (Bloomfield and Arthur, 1994; Setlow *et al.*, 2002). For chemical agents, sporicidal concentrations are in most cases 10 times (or more) higher than bactericidal concentrations (Russell, 1990). In the case of phenols, organic acids, QACs, biguanides, organomercurials (e.g. methyl-Hg, ethyl-Hg) and alcohols used at high concentrations the agents have no sporicidal effect (Russell, 1990).

4.5.2 Revival of injured cells

Another aspect is the difference in cell damage after treatment with disinfectants or physical agents. This implies that a certain amount of cells can revive. However, this revival does not strictly indicate a resistance mechanism but is due to the statistical variance of the protection systems of the microorganism. It is important to realise that sensitivity also varies within the defined species and resistance is defined as the tolerance of a disinfectant that exceeds the natural variance (Heinzel, 1998; Russell, 1991).

4.5.3 Processing conditions: pH, temperature, concentration

pH is an important factor as it can modify the practical application of the disinfectant used (Russell, 1991). For example for chlorine the pH needs to be in the range between 5 and 8 in order to be effective as hypochlorous acid. Below pH 5 chlorine gas is produced and above pH 8 ClO^- ions are produced which are, apart from the acute toxicity of chlorine gas, not active as a disinfectant (Krop, 1990). Similar effects are known for other disinfectants. Therefore, knowing the pH of the environment makes it possible to predict whether a disinfectant will be active. In addition to pH (as discussed earlier), processing parameters such as temperature, concentration and application time are important factors concerning (pseudo)resistance of microorganisms. For example, when the temperature of a solution of formaldehyde is increased by 10 °C the effectiveness is increased between three and five times (Krop, 1990; Russell, 1999a). Concerning concentration, when the concentration of the applied disinfectant is too low, the disinfectant only works bacteriostatically instead of bactericidally. This implies that as soon as the disinfectant is used up the bacteria start growing again. For example chlorhexidine is bacteriostatic at $0.0001\,mg\,l^{-1}$ and bactericidal at $0.002\,mg\,l^{-1}$ (Russell, 1990). When concentrations are too high the disinfectant will act faster but the question is whether that is strictly necessary. When this is not the case it will cost money and may be dangerous for the environment (personnel, equipment).

4.5.4 Residual organic matter/biofilms

The guideline for cleaning and disinfection is that disinfection can be effective only when the equipment or surface is properly cleaned prior to the disinfection (Krop, 1990). This can be explained by the fact that remaining organic matter will inactivate the disinfectant and microorganisms will not be affected (Kraemer, 1998). A second reason is that organic compounds act as a protective layer for the microorganisms. This is also the case when microorganisms have formed a biofilm where, as a result of nutrient limitation, a reduced growth rate makes the specific microorganisms less susceptible to disinfectants (Brown and Gilbert, 1993; Luppens, 2002). The fact that microorganisms can form biofilms, implying a change in their growth characteristics, can also result in resistance against disinfectants for the following reasons (Brown and Gilbert, 1993):

- exclusion/influencing of the disinfectant by the glycocalyx (a slimy layer surrounding the cell);
- chemical reaction of the glycocalyx with disinfecting agents;
- limited availability of key nutrients results in decreased growth rate;
- the attachment to surfaces, causing depressing of genes associated with sessile (directly to the substrate) existence, which coincidentally affects antimicrobial susceptibility.

These effects can be regarded as pseudo-resistance as the effect will end as soon as the biofilm no longer exists.

4.6 Future trends

As microorganisms evolve and adapt to disinfecting strategies, the development of more effective cleaning and disinfecting strategies and new tools to monitor the efficiency of these strategies will continue. The following trends can be distinguished.

4.6.1 Disinfecting agents – total service

The producers of disinfectants are continuously working on new formulations and new active components and new total service concepts to serve their customers. Ecolab, a producer of cleaning and sanitising solutions (www.ecolab.com), offers a complete farm to fork approach concerning the food safety of the products of their customers called ECO-SHIELD. Other suppliers such as Johnson Diversey and Alconox offer the same kind of total service concepts. By offering these kinds of product, a great responsibility lies with the manufacturers of disinfectants to prescribe the right concentrations and procedures for application in order to avoid an increase in (pseudo)resistant pathogenic microorganisms in factory environments.

4.6.2 Incorporation of disinfectants

Where possible, disinfectants become integrated with, e.g., processing equipment, packaging material or sanitary devices (Stekelenburg and Hartog, 2002; Chung *et al.*, 2003). The advantage is that growth of (pathogenic) microorganisms is continuously inhibited as long as the disinfectant remains active. The disadvantages are a decreased activity in time as a result of biological breakdown or uptake by the environment. Another issue is that there is a risk that personnel will become negligent with regard to factory hygiene, resulting in an unwanted change of attitude.

4.6.3 Objective monitoring tools

Process monitoring will become more and more common sense. Currently it is possible to monitor on-line physical and chemical parameters such as flow, conductivity, pH, temperature, turbidity, concentration and pressure. Developments are ongoing for new sensors such as Isfets (ion selective transistors) used for specific ion concentrations (van Asselt *et al.*, 2002) or biosensors based on oxygen yeast cells used for the determination of ethanol in beverages (Rotariu and Bala, 2003). For monitoring microorganisms a range of on-line monitoring devices such as flow cytometry (e.g. Bactoscan) and ATP could be applicable. However, the main issue for these methods is the detection limit, which is in most cases higher than 10^3 cfu ml^{-1} (http://www.foss.dk/). This implies that the method is currently useful only in emergencies to stop the process (i.e. that once the method generates a positive signal, the process cannot be changed or optimised, only stopped) and not as a monitoring device. It is expected that the accuracy of the methods will improve, but to what extent will depend on the demands from market and government.

4.6.4 Genomics

A relatively new development in the study of microorganism is genomics. Since the first microbial genome sequence was published in 1995, genomics caused a revolution in the way people think about microorganisms. One of the main application of genomics is industrial strain development, e.g. in order to provide a certain microorganism with a gene that produces a specific flavour or functional property. Information available on the genome sequences may be used to determine the cell response to different stress situations such as high temperatures, high pressure, osmotic shock or disinfectants (Wells and Bennik, 2003; Abee and Wouters, 1999). Screening techniques (e.g. DNA microarray) enable the screening of large amounts of microorganisms on the specific properties and select the microorganisms containing that property. Concerning pathogenic microorganisms, a possible application could be the screening on pathogenicity or response towards disinfectant agents. This approach will, based on comparison between disinfectant resistant versus disinfectant-sensitive strains, allow the determination of disinfectant efficacy or critical concentration.

4.7 Sources of further information and advice

EHEDG
Guidelines and test methods http://www.ehedg.org/f_guidelines.htm (6 August 2004)

European Union
Guidelines http://europa.eu.int/eur-lex/nl/search/search_lif.html (28 July 2004)
European biocide guideline 98/8/EG:
http://europa.eu.int/servlet/portail/RenderServlet?search=DocNumber&lg=nl&nb_docs=25&domain=Legislation&coll=&in_force=NO&an_doc=1998&nu_doc=8&type_doc=Directive (28 July 2004)

United States – Food Drug Administration
Environmental Protection Agency: http://www.epa.gov (28 July 2004)
Pesticides: http://www.epa.gov/pesticides/factsheets/alpha_fs.htm (28th July 2004)

FAO/WHO
Codex alimentarius; Codex Committee on Food Additives and Contaminants: 'Code of practice on the safe use of active chlorine' (currently in preparation, currently at step 3 of 6).

4.8 References

ABEE, T. and WOUTERS, J. A. (1999), Microbial stress response in minimal processing, *International Journal of Food Microbiology*, **50**, 65–91.

ALLERBERGER, F. and DIERICH, M. P. (1988), Effects of disinfectants on bacterial metabolism evaluated by microcalorimetric investigations, *Zentralblatt für Bakteriologie, Mikrobiologie und Hygiene*, **187**, 166–179.

AVECIA (2004), http://www.avecia.com/biocides/products/vantocil/vantocil_folder.pdf, 30 March 2004.

BAQUERO, F., NEGRI, M.-C., MOROSINI, M.-I. and BLÁZQUEZ, J. (1998), Antibiotic-Selective Environments, *Clinical Infectious Diseases*, **27**, S5–S11.

BERNSTEIN, M. (1990), The chemistry of disinfectants, in Romney, A. J. D. *CIP: Cleaning in Place*, Society of Dairy Technology, Huntingdon, pp. 30–40.

BEUCHAT, L. R. (1998), *Surface Decontamination of Fruits and Vegetables Eaten Raw: a review*, Georgia, USA, Food Safety Unit World Health Organization.

BIRD, M. R. (1994), Cleaning agent concentration and temperature optima in the removal of food based deposition, in Fryer, P. J., Hasting, A. P. M. and Jeurnink, T. J. M., eds, *Fouling and Cleaning in Food Processing*. European Commission, Jesus College, Cambridge, UK.

BLOOMFIELD, S. F. and ARTHUR, M. (1994), Mechanisms of inactivation and resistance of spores to chemical biocides, *Journal of Applied Bacteriology*, **76**, 91S–104S.

BROWN, M. R. W. and GILBERT, P. (1993), Sensitivity of biofilms to antimicrobial agents, *Journal of Applied Bacteriology*, **74**, 87S–97S.

CHAPMAN, J. S. (1998), Characterizing bacterial resistance to preservatives and disinfectants, *International Biodeterioration & Biodegradation*, **41**, 241–245.

CHO, M., CHUNG, H. and YOON, J. (2003), Disinfection of water containing natural organic matter by using ozone-initiated radical reactions, *Applied Environmental Microbiology*, **69**, 2284–2291.

CHUNG, D., PAPADAKIS, S. E. and YAM, K. L. (2003), Evaluation of a polymer coating containing triclosan as the antimicrobial layer for packaging materials, *International Journal of Food Science Technology*, **38**, 165–169.

CODEX (2003), *Proposed Draft Code of Practice on the Safe Use of Active Chlorine*, The Hague, Codex Alimentarius Commission.

DE JONG, P. (1996), *Modelling and Optimisation of Thermal Processes in the Dairy Industry*, Delft University of Technology, Delft.

DE JONG, P., TE GIFFEL, M. C., STRAATSMA, H. and VISSERS, M. M. M. (2002), Reduction of fouling and contamination by predictive kinetic models, *International Dairy Journal*, **12**, 285–292.

DENYER, S. P. and STEWART, G. S. A. B. (1998), Mechanisms of action of disinfectants, *International Biodeterioration & Biodegradation*, **41**, 261–268.

DODD, T. (2003), Cleaning records and CIP optimization, *International Journal of Dairy Technology*, **56**, 247–247.

DONHAUSER, S., WAGNER, D. and GEIGER, E. (1991), Zur Wirkung von Desinfektionsmitteln in der Brauerei, *Brauwelt*, **131**, 604, 606, 609, 612, 614, 616.

ERCO WORLDWIDE (2004), http://www.clo2.com/factsheet/factindex.html, 4 August 2004.

GIBSON, H., TAYLOR, J. H., HALL, K. E. and HOLAH, J. T. (1999), Effectiveness of cleaning techniques used in the food industry in terms of the removal of bacterial biofilms, *Journal of Applied Microbiology*, **87**, 41–48.

GORMAN, S. P., JONES, D. S. and LOFTUS, A. M. (1987), The sporicidal activity and inactivation of chlorhexidine gluconate in aqueous and alcoholic solution, *Journal of Applied Bacteriology*, **63**, 183–188.

HEINZEL, M. (1998), Phenomena of biocide resistance in microorganisms, *International Biodeterioration & Biodegradation*, **41**, 225–234.

HOXEY, E. V. and THOMAS, N. (1999), Gaseous sterilization, in Russell, A. D., Hugo, W. B. and Ayliffe, G. A. J. *Principles and Practice of Disinfection, Preservation and Sterilization*, Blackwell Science Ltd, Oxford.

HUGO, W. B. and RUSSELL, A. D. (1999), Types of antimicrobial agents, in Russell, A. D., Hugo, W. B. and Ayliffe, G. A. J. *Principles and Practice of Disinfection, Preservation and Sterilization*, Blackwell Science Ltd, Oxford.

JEURNINK, T. J. M. and BRINKMAN, D. W. (1994), The cleaning of heat exchangers and evaporators after processing milk or whey, *International Dairy Journal*, **4**, 347–368.

JEURNINK, T. J. M., WALSTRA, P. and DE KRUIF, C. G. (1996), Mechanisms of fouling in dairy processing, *Netherlands Milk and Dairy Journal*, **50**, 407–426.

KALATHENOS, P. and RUSSELL, N. J. (2003), Ethanol as a food preservative, in Russell, N. J. and Gould, G. W. *Food Preservatives*, Kluwer Academic/Plenum Publishers, New York, pp. 196–217.

KAMASE, Y., MURAKAMI, H., TAKAHASHI, R., TAKAOKA, R. and NAKAMURA, Y. (2003), Development of medical disinfector using ozone, *IHI Engineering Review*, **36**, 131–134.

KRAEMER, J. (1998), Cleaning and disinfection, *Mitteilungen aus dem Gebiete der Lebensmitteluntersuchung und Hygiene*, **89**, 14–20.

KROP, J. J. P. (1990), *Reiniging en Desinfectie*, Bolsward, Agrarische Hogeschool Friesland.

LAMBERT, R. J. W. and JOHNSTON, M. D. (2000), Disinfection kinetics: a new hypothesis and model for the tailing of log-survivor/time curves, *Journal of Applied Microbiology*, **88**, 907–913.

LUNDÉN, J., AUTIO, T., MARKKULA, A., HELLSTROM, S. and KORKEALA, H. (2003), Adaptive and cross-adaptive responses of persistent and non-persistent *Listeria monocytogenes* strains to disinfectants, *International Journal of Food Microbiology*, **82** (3), 265–272.

LUPPENS, S. B. I. (2002), *Suspensions or Biofilms and Other Factors that Affect Disinfectant Testing on Pathogens*, Wageningen University, Wageningen.

MARQUIS, R. E., RUTHERFORD, G. C., FARACI, M. M. and SHIN, S. Y. (1995), Sporicidal action of peracetic acid and protective effects of transition metal ions, *Journal of Industrial Microbiology*, **15**, 486–492.

MCDONNELL, G. and RUSSELL, A. D. (1999), Antiseptics and disinfectants: activity, action, and resistance, *Clinical Microbiology Reviews*, **12**, 147–179.

MCDONNELL, G., GRIGNOL, G. and ANTLOGA, K. (2002), Vapor phase hydrogen peroxide decontamination of food contact surfaces, *Dairy, Food and Environmental Sanitation*, **22**, 868–873.

MOHR, M. and DUGGAL, S. (1997), Zielgerichte Sauberkeit; Teil 2, *Lebensmitteltechnik*, **29**, 60–62.

MOKGATLA, R. M., GOUWS, P. A. and BROZEL, V. S. (2002), Mechanisms contributing to hypochlorous acid resistance of a *Salmonella* isolate from a poultry-processing plant, *Journal of Applied Microbiology*, **92**, 566–573.

MOORE, G., GRIFFITH, C. and FIELDING, L. (2001), A comparison of traditional and recently developed methods for monitoring surface hygiene within the food industry: a laboratory study, *Dairy, Food and Environmental Sanitation*, **21**, 478–488.

QUANTEX LABORATORIES (2001), http://www.quantexlabs.com/triclosan.htm, 17 August 2004.

QUINN, P. J. and MARKEY, B. K. (1999), Viricidal activity of biocides part B; Activity against veterinary viruses, in Russell, A. D., Hugo, W. B. and Ayliffe, G. A. J. *Principles and Practice of Disinfection, Preservation and Sterilization*, Blackwell Science Ltd, Oxford.

REUTER, G. (1998), Disinfection and hygiene in the field of food of animal origin,

International Biodeterioration & Biodegradation, **41**, 209–215.

RODGERS, S. L., CASH, J. N., SIDDIQ, M. and RYSER, E. T. (2004), A comparison of different chemical sanitizers for inactivating *Escherichia coli* O157:H7 and *Listeria monocytogenes* in solution and on apples, lettuce, strawberries, and cantaloupe, *Journal of Food Protection*, **67**, 721–731.

ROTARIU, L. and BALA, C. (2003), New type of ethanol microbial biosensor based on a highly sensitive amperometric oxygen electrode and yeast cells, *Analytical Letters*, **36**, 2459–2471.

RUSSELL, A. D. (1990), Bacterial spores and chemical sporicidal agents, *Clinical Microbiology Reviews*, **3**, 99–119.

RUSSELL, A. D. (1991), Mechanisms of bacterial resistance to non-antibiotics: food additives and food and pharmaceutical preservatives, *Journal of Applied Bacteriology*, **71**, 191–201.

RUSSELL, A. D. (1995), Mechanisms of bacterial resistance to biocides, *International Biodeterioration & Biodegradation*, **36**, 247–265.

RUSSELL, A. D. (1996), Activity of biocides against mycobacteria, *Journal of Applied Bacteriology*, **81**, 87S–101S.

RUSSELL, A. D. (1999a), Factors influencing the efficacy of antimicrobial agents, in Russell, A. D., Hugo, W. B. and Ayliffe, G. A. J. *Principles and Practice of Disinfection, Preservation and Sterilization*, Blackwell Science Ltd, Oxford.

RUSSELL, A. D. (1999b), Bacterial resistance to disinfectants: present knowledge and future problems, *Journal of Hospital Infection*, **43 Suppl. S**, S57–S68.

RUSSELL, A. D. (1999c), Antifungal activity of biocides, in Russell, A. D., Hugo, W. B. and Ayliffe, G. A. J. *Principles and Practice of Disinfection, Preservation and Sterilization*, Blackwell Science Ltd, Oxford.

SCHLEGEL, H. G. (1993), *General Microbiology*, Cambridge University Press, Cambridge.

SETLOW, B., LOSHON, C. A., GENEST, P. C., COWAN, A. E., SETLOW, C. and SETLOW, P. (2002), Mechanisms of killing spores of *Bacillus subtilis* by acid, alkali and ethanol, *Journal of Applied Microbiology*, **92**, 362–375.

SHORROCK, C. J., BIRD, M. R. and HOWELL, J. A. (1998), Yeast deposit removal from polymeric microfiltration membrane, in Wilson, D. I., Fryer, P. J. and Hasting, A. P. M., eds, *Fouling and Cleaning in Food Processing '98*. European Commission, Cambridge.

SINGH, N., SINGH, R. K. and BHUNIA, A. K. (2003), Sequential disinfection of *Escherichia coli* O157:H7 inoculated alfalfa seeds before and during sprouting using aqueous chlorine dioxide, ozonated water, and thyme essential oil, *Lebensmittel Wissenschaft und Technologie*, **36**, 235–243.

SIXMA, J. J. (2001), *Disinfectants in Consumer Products*, Health Council of the Netherlands, The Hague.

SOFOS, J. N. and BUSTA, F. F. (1999), Chemical food preservatives, in Russel, A. D., Hugo, W. B. and Ayliffe, G. A. J. *Principles and Practice of Disinfection, Preservation and Sterilization*, Blackwell Science Ltd, Oxford.

SPECTRUM LABORATORIES, http://www.speclab.com/compound/c70304.htm, 17 August 2004.

SPEEK, A. J. (2002), *Onderzoek naar het toepassen van decontaminatie – en desinfectiemiddelen in de groenten en fruit verwerkende industrie*, Keuringsdienst van Waren Noordwest, Amsterdam.

STEKELENBURG, F. K. and HARTOG, B. J. (2002), Efficacy testing of antimicrobial agents, *International Food Hygiene*, **12**, 5, 7.

TATTERSON, I. N. and WINDSOR, M. L. (2001), *Cleaning in the Fish Industry*, Torry Research Station, Aberdeen.

TODAR, K. (2001), *The Control of Microbial Growth*, University Wisconsin-Madison, Madison.

VAN ASSELT, A. J. and TE GIFFEL, M. C. (2002), Opti-Cip optimaliseert en valideert CIP-reiniging, *Voedingsmiddelentechnologie*, **35**, 84–85.

VAN ASSELT, A. J., VAN HOUWELINGEN, G. and TE GIFFEL, M. C. (2002), Monitoring system for improving cleaning efficiency of cleaning-in-place processes in dairy environments, *Food and Bioproducts Processing*, **80**, 276–280.

VERMEIREN, L., DEVLIEGHERE, F. and DEBEVERE, J. (2002), Effectiveness on some recent antimicrobial packaging concepts, *Food Additives and Contaminants*, **19**, 163–171.

VISSERS, M. M. M., DE JONG, P. and DE WOLFF, J. J. (2002), Nieuwe indampertechniek resulteert in 70 procent energiebesparing, *Voedingsmiddelentechnologie*, **35**, 16.

WANG, M. Y., COLLINS, E. B. and LOBBEN, J. C. (1973), Destruction of psychrotrophic strains of *Bacillus* by chlorine, *Journal of Dairy Science*, **56**, 1253–1257.

WELLS, J. M. and BENNIK, M. H. J. (2003), Genomics of food borne bacterial pathogens, *Nutrition Research Reviews*, **16**, 21–35.

YOUNG, S. B. and SETLOW, P. (2003), Mechanisms of killing of *Bacillus subtilis* spores by hypochlorite and chlorine dioxide, *Journal of Applied Microbiology*, **95**, 54–67.

5

Aerosols as a contamination risk

D. Burfoot, Silsoe Research Institute, UK

5.1 Introduction

Aerosols consist of particles dispersed in air. The particles may be liquid droplets or solid particles or include both types of matter. The aerosols of most concern in food premises are those that include microorganisms. Aerosols may enter production areas through many routes including doorways, hatches, drains and any other opening that connects low- and high-care areas (Burfoot *et al.*, 2001). Aerosols can come from many sources, including raw materials, people, packaging, and moving or rotating equipment. Holah *et al.* (1995) showed that cleaning operations are major sources of aerosols that may include microorganisms. Cleaning operations such as boot washing, tray washing, equipment cleaning and floor scrubbing are all potential sources of aerosol.

The best approach to reducing contamination via the airborne route is to restrict the generation of aerosols. Once particles are airborne it is difficult to control the movements of every particle because various mechanisms, such as advection (air movement), turbulent dispersion, gravity and thermal convection affect the particle motions. However, correct specification and implementation of air-handling equipment can ensure that the majority of the airborne particles do not contaminate exposed foods. Such systems rely on three approaches: (i) using sufficient air exchange rates and filtration to remove the particles from the air; (ii) providing sufficient air to maintain a positive pressure in the high-care area and restrict the flow of air from low-care areas; and (iii) ensuring air flows do not create lower-pressure regions near to doorways, hatches and other openings that can lead to contamination entering from nearby low-care areas. These approaches have been developed significantly in sectors that use clean rooms, such as those for the manufacture of electronics and medical devices.

Food production areas differ from such environments in that they are often wet, include many more sources of aerosol, including more people, and the food sector is more concerned with microbial contaminants rather than total particle contaminants. Nonetheless the general approaches identified above are adopted by the food industry but often at a lower level of control than in other sectors.

It is interesting to compare the particle concentrations found in clean rooms and high-care food production areas. Clean rooms are classified according to various standards (Möller, 1999) with the Federal Standard 209E being one of the most commonly used. In this standard, each cubic foot of air in a Class 100 environment would not contain more than 100 particles of 0.5 μm diameter or larger. Similarly, the air in a Class 100 000 environment, often known as a 'white room', would not contain more than 100 000 particles per cubic foot with diameters of $\geq$0.5 μm. These concentrations would usually be measured without operators in the room. In comparison, Burfoot and Brown (2004a) found particle concentrations up to 230 000 ft^{-3} (8 100 000 m^{-3}) near to an operating boot scrubber in a high-care sandwich assembly area and down to 9000 ft^{-3} (330 000 m^{-3}) during a period of no activity in the same area. Particle concentrations up to 600 000 ft^{-3} (21 000 000 m^{-3}) were measured in a chilled dessert filling area.

Filtration is one of the major factors in controlling the concentration of airborne particles in rooms. A large range of filters and classification schemes have been developed and many of those of relevance to the high-care/risk chilled food industry are described in a guidance document produced by the Campden & Chorleywood Food RA (1996). Generally, air-handling systems for some high-care areas would be fitted with F9 filters, some high-care areas and high-risk areas with H11 filters, whereas clean rooms would use H13 or even higher levels of filtration. F9 filters remove almost all particles of 1 μm diameter and above, H11 filters remove particles of 0.5 μm particles and above, and H13 filters remove almost all particles above 0.3 μm.

5.2 Factors affecting aerosol contamination

The risk of food contamination in high-care environments depends on many factors including the rate of generation of airborne particles, particle sizes and speeds, the number of particles that include organisms, the direction of the air flow in the room and the exposure time and surface area of the food. These factors are important because they control the distance travelled, flight time and spatial distribution of concentration of the organisms. Methods are available to measure each of these important factors. In the remainder of this chapter, the term 'particle' will be used to refer to both droplets of liquid and solid particles, as often an aerosol may contain both.

5.2.1 Droplet generation, size and speed

Most of the methods of measuring droplet generation, size and speed are based on laser technology. Phase-Doppler analysers have been used to measure the sizes of very small particles, up to 40 μm, at high concentration near to cleaning operations (Burfoot *et al.*, 2003a). These devices utilise two laser beams and light is scattered at the intersection of the beams. Analysis of the scattering can be used to assess the particle size distribution, flux and velocity. Image-based systems use a laser and high-speed camera and they are used to measure the size and velocity of particles beyond the size range studied with a phase-Doppler analyser. Air particle counters are commonly used in clean rooms and these are also based on laser technology. They extract a sample of air for analysis rather than analysing the aerosol *in situ*. Further information on particle size analysers is given by Mitchell (1995).

5.2.2 Number of organisms

Settle plates can be used to assess the deposition of airborne organisms. Other equipment is available from various manufacturers that allows measurements of the concentrations of airborne organisms. Most of those used in the food sector rely on the impaction of the organisms onto solid media in a Petri dish as factory air is drawn across the dish. Calibrations are then used to convert the number of organisms on the dish to an airborne concentration. The Andersen (1958) sampler is widely used for research as, by using multiple dishes arranged in stages, it can provide information on particle size distribution. Crook (1995) describes various samplers, though those constructed from glass would generally be considered unsuitable for use in chilled food production areas.

5.2.3 Air flow

The speed and direction of the air flow in the production area are important as they can move contamination around a factory. Air speed can be measured using hot wire anemometers while vane anemometers provide speed data and some indication of flow direction, although care is needed. Both of these devices can be used in conjunction with a 'windicator' such as a small length of freely hanging fabric filaments to indicate the direction of the air flow (Burfoot *et al.*, 2001). Smoke tests may be applicable in some areas. Ultrasonic anemometers are relatively expensive but allow the measurement of air speed, direction and fluctuation.

5.3 Aerosol generation

Burfoot *et al.* (2003a) examined the generation of particles by four cleaning operations: low-pressure hosing (100 psi (689×10^3 N/m^2), hose type), boot scrubbing (mechanical walk through), hand-washing and floor scrubbing (mechanical with brushes, squeegee and vacuum). Hosing was found to produce a very high particle flux of 144 000 particles per square centimetre per second below 40 μm diameter. This flux, which was measured 15 cm from the impact

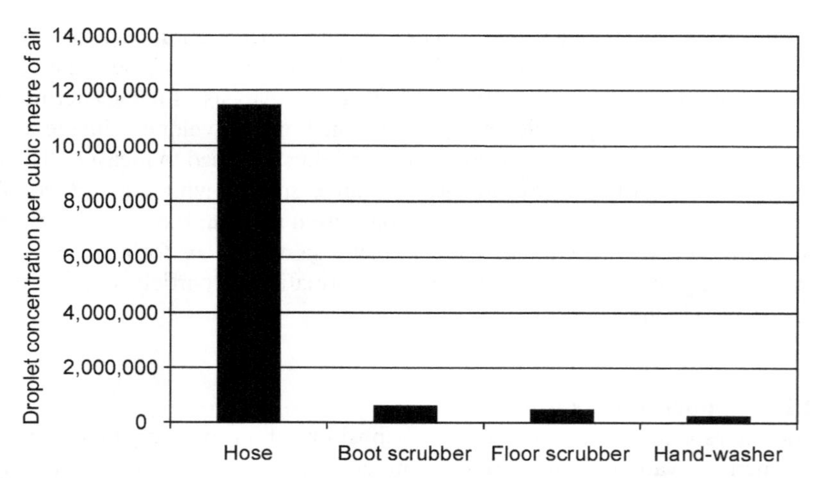

Fig. 5.1 Concentration of droplets above 1 μm diameter measured in air around 1 m from four cleaning operations (taken from Burfoot *et al.* (2003b)).

point of the water jet on a surface, led to particularly large increases in the number of airborne droplets nearby. Figure 5.1 shows the particle concentrations measured about 1 m from the various cleaning operations, including the hose, relative to a zero background count. The hose led to around 12 million particles per cubic metre whereas the concentrations resulting from the other cleaning operations were around 20 times lower. All the cleaning operations produced particles over a wide size range. with most of the particles being small and the number decreasing with size: Fig. 5.2 shows the trend for a hose.

Burfoot and Brown (2004a) report on the number of organisms in the air in four food factory environments: high-risk sandwich assembly area, chilled dessert filling area, chilled pie and quiche production and a changing area connected to a high-care meat production room. The concentration of airborne organisms measured in each area varied from 42 m^{-3} in the sandwich area to 2508 m^{-3} in the changing area. The number of particles containing an organism relative to the total number of particles varied from 1 in 200 near to staff during hand-washing to 1 in 30 000 in periods of inactivity in a well-designed production area. The highest concentrations of organisms and total particles were found near to cleaning operations. The greatest ratios of organisms to total particles were found next to cleaning operations and next to staff.

5.4 Aerosol dispersal

Large particles, above 100 μm, can settle near to the cleaning operation. Medium sized particles may settle near to the cleaning operation or evaporate to become smaller particles, below 20 μm, which can disperse easily around the production area. The airborne particles of most interest in chilled food factories are those

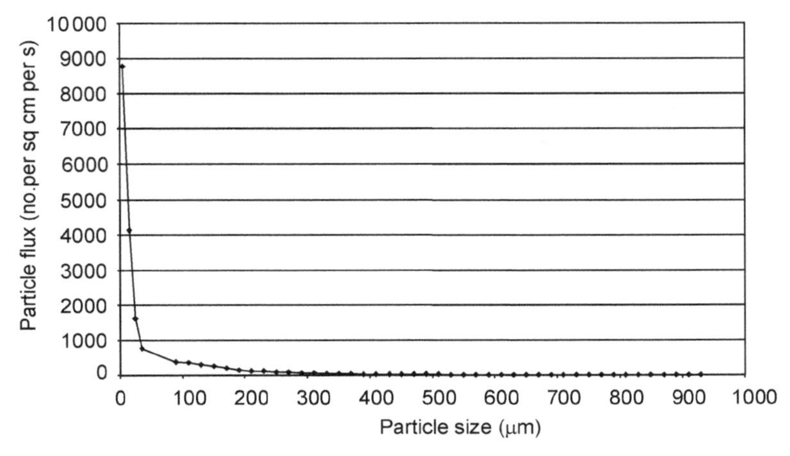

Fig. 5.2 Particle sizes produced by a low-pressure (100 psi/0.69 MPa) trigger-type hose (taken from Burfoot *et al.*, 2003b).

containing bacteria; such particles have a diameter of 1 μm or more. Tests have been carried out in which surfaces were smeared with a solution of *Bacillus subtilis* var. *globigii* and then cleaned and the airborne dispersal of the organisms around a room was detected using settle plates (Burfoot and Brown, 2004b). These tests showed that contamination is easily spread by hosing (Fig. 5.3a), with contamination travelling many metres. In this example, plates directly in front of the hose became so wet that they could not be used to assess microbial contamination. It is expected that the counts on such plates would have been very high. Much of the contamination from a boot scrubber (Fig. 5.3b) settled within 2 m and from hand-washing (Fig. 5.3c) within 1 m of the sink. Contamination from a floor scrubber with a vacuum was very low and detected only next to the scrubber (Fig. 5.3d). In all cases, much of the contamination fell close to the cleaning operation but there was always some contamination spread throughout the room, as evidenced by the low counts away from the main sources in Fig. 5.3. This spread results from the dispersal of small particles.

Measurements in factories and controlled environment rooms and the use of computer models have led to some important conclusions:

- The risk of product contamination is greatest when the direction of the air flow is from a source of contamination towards the food.
- The smaller the particle the greater the flight time and the distance it may travel.
- Generally, in high-care production areas, less than 1% of the particles generated will settle. Most will be removed by the filtration system or escape through doorways and hatches.
- The temporal change in concentration of very small particles, around 1 μm diameter, is affected by the air exchange rate, the efficiency of the filtration, the rate of generation of particles and the leakage of the room.

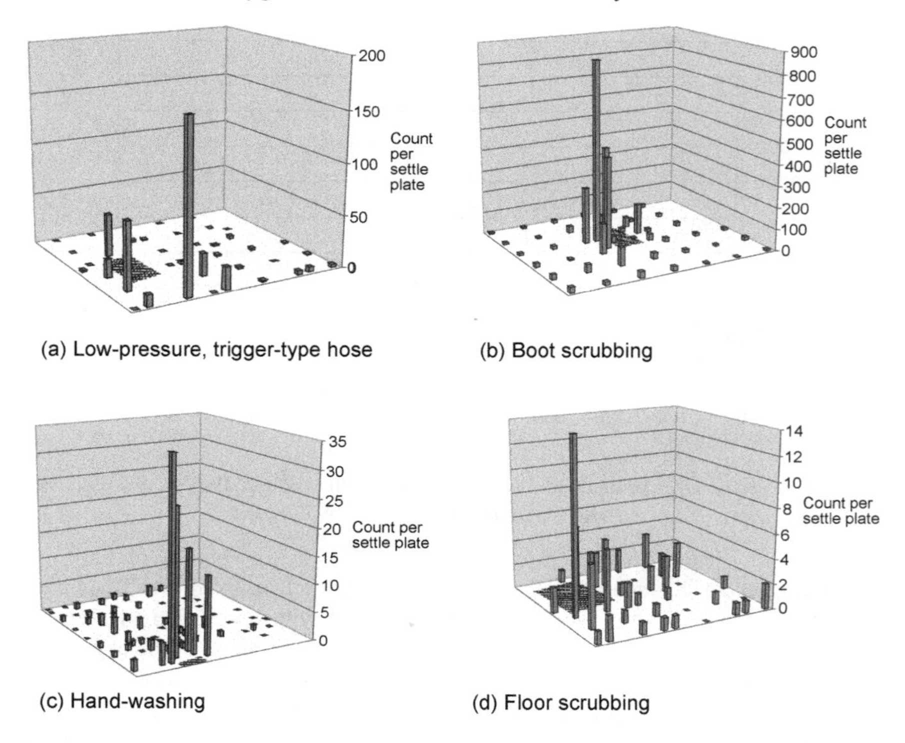

Fig. 5.3 Number of organisms on settle plates located around a room ($11.25 \times 10.25\,m^2$ floor) after using four types of cleaning operation. Data from Burfoot and Brown (2004b). The hatched areas show the position of the equipment on the floor of the room.

- Increasing the air change rate or the filtration efficiency reduces the clearance time of the particles. Clearance times of 10–30 min are typical for most high-care areas.

5.5 Ways to reduce the risk from airborne contamination

Methods to reduce the risk of food contamination from the airborne route fall into five categories: factory design and factory operation, equipment design and equipment operation, and monitoring. Much has been written about the hygienic design and operation of chilled food factories including guidelines from the UK Chilled Food Association (1997, 2001), and the various contributions elsewhere in this book. Here we concentrate on the management of the air and the design and operation of open cleaning operations.

5.5.1 Management of the air

There are many requirements for the correct management of the air in a high-care area.

- Ensure that adequate air is blown in to the room to maintain a positive pressure.
- Ensure that air extraction is not so great that it reduces the pressure in the high-care area below that in neighbouring rooms.
- Allow for air leakage from the high-care area at the design stage. Consider what will be the effect of opening doorways, etc., once the factory is in operation.
- Provide additional air if further air extracts or hoods are subsequently added to a production area following the introduction of new equipment, for example.
- Try not to position air extracts in high-care areas close to hatches or doorways connecting to low-care areas.
- Keep the return side of ceiling mounted chillers away from the hatches and doorways leading to low-care areas.
- Design for the air-handling and delivery system to be accessible and easily cleaned and maintained. The system should not be a source of contamination.
- Think about future changes to the factory, such as expansion.

5.5.2 Design and operation of open cleaning operations

Earlier data have shown that cleaning operations can be major sources of airborne contamination. Good design and operating procedures, such as the following examples, can help to minimise their impact:

- Poorly cleaned equipment can become a significant source of contamination. Cleaning equipment should be cleaned and sanitised according to a defined schedule. An area for cleaning should be provided outside the production area.
- Hosing creates high concentrations of aerosol and the use of this practice during production should be discouraged. If cleaning is essential during production, methods that produce the least generation of aerosols should be considered, for example using a 'scraper blade' or cloth may be adequate and produce far less aerosol than hosing.
- Compressed air lines are sometimes used to dislodge contamination and this can also generate aerosols (Holah *et al.*, 2004). The use of this practice during production should also be discouraged.
- Avoid areas in equipment design where water could collect, for example, in the reel casing of a retractable hose.
- Provide facilities for the disposal of water from cleaning operations such as the wash water from the tank of a mechanical floor scrubber.

These are just examples that illustrate the general principles of good design and operation of open cleaning operations. Obviously, factory layout and operation can also have a very significant impact on the dispersal of aerosols and again good practice is essential. Ensuring that cleaning is well away from production and that both product and packaging are exposed for only short periods are examples of good practice. Obviously, since 'deep cleaning' operations (thorough cleaning normally applied after production) produce high concentrations of aerosol, it is essential that aerosols are allowed to settle or be removed by the air-handling system after such cleaning. Ideally, the air-handling system would not be used during deep cleaning or otherwise bacteria, made airborne during the cleaning operations, could be collected on the cooling coils of the refrigeration system. Operating the air-handling system at full extract after deep cleaning is good practice if possible.

5.6 Future trends

The development of the high-care and high-risk chilled food sectors has seen an increase in the use of zoning in factories and more recently, for some factories, the use of clean room technologies. Both of these topics are covered elsewhere in this book. Having increasingly cleaner environments closest to the food is a good concept. The question arises as to the length scale on which the graduation or zoning is carried out. Currently, in most chilled food factories in the UK, manufacturing short shelf-life products, the main zones are high-care/risk areas and low-care areas. Some factories are, in addition, installing localised air delivery systems that provide air at an even higher quality, than is usual, directly towards the food. Burfoot *et al.* (2000) show a number of different designs that have been considered for this purpose. These include the direction of clean air vertically or horizontally towards the foods or the circulation of clean air around the foods. Localised air delivery has been found to reduce the airborne contamination of foods. A further advantage of these systems is that they could provide a potential energy saving if cold air is supplied locally allowing the factory to be run at a higher temperature (Burfoot *et al.*, 2004). Also, by maintaining the food temperature they reduce the need to cool the products after they leave the production area. However, for this approach to provide high energy savings, most ingredients need to have been cooled prior to entering the high-care/risk area. The use of localised air delivery is beginning to be applied but many see that such approaches are restricted to products where an extension of shelf-life is a major goal. For products such as prepared salads that have significant microbial load, or products that have a very short shelf-life due to quality degradation rather than microbial spoilage, providing ultra-clean air close to the product probably has less application than in the case of other products such as sliced ham.

5.7 Sources of further information and advice

There are many sources of information and advice relating to airborne contamination and air handling. Other chapters in this book clearly provide associated information. The engineering research and food research organisations are sources of information, such as Silsoe Research Institute (www.sri.bbsrc.ac.uk) and the Campden and Chorleywood Food Research Association (www.campden.co.uk) in the UK. Recognised trade and professional bodies are also important sources of information such as the Heating and Ventilating Contractors' Association (www.hvca.org.uk) in the UK and the American Society of Heating Refrigeration and Air Conditioning Engineers (www.ashrae.org) in the USA. The ASHRAE handbooks and standards are particularly useful. The Internet provides links to many manufacturers and suppliers of equipment for measuring the sizes and concentrations of particles and air speeds.

5.8 References

ANDERSEN, A.A. (1958) New sampler for the collection, sizing and enumeration of viable airborne particles. *Journal of Bacteriology*, **76**, 471–484.

BURFOOT, D. and BROWN, K. (2004a) A relationship between the airborne concentration of particles and organisms in chilled food factories. Paper 9 on CD ROM *Proceedings of the International Conference on Engineering and Food, ICEF9*, 7–11 March 2004, Montpellier, France.

BURFOOT, D. and BROWN, K. (2004b) Reducing food contamination via the airborne route. Paper 8 on CD ROM *Proceedings of the International Conference on Engineering and Food, ICEF9*, 7–11 March 2004, Montpellier, France.

BURFOOT, D., BROWN, K., XU, Y., REAVELL, S.V. and HALL, K. (2000) Localised air delivery systems in the food industry, *Trends in Food Science and Technology*, **11**, 410–418.

BURFOOT, D., BROWN, K., DUKE, N., NEWTON, K., HALLIGAN, A., MORGAN, W. and SAINTER, J. (2001) *Best practice guidelines on air flows in high-care and high-risk areas*. Report from Silsoe Research Institute, Silsoe, Bedford, UK.

BURFOOT, D., REAVELL, S.V., TUCK, C. and WILKINSON, D. (2003a) Generation and dispersion of droplets from cleaning equipment used in the chilled food industry. *Journal of Food Engineering*, **58**, 343–353.

BURFOOT, D. *et al.* (2003b) *Guidance to reduce food contamination from cleaning operations*. Report from Silsoe Research Institute, Silsoe, Bedford, UK.

BURFOOT, D., REAVELL, S., WILKINSON, D. and DUKE, N. (2004) Localised air delivery to reduce energy use in the food industry. *Journal of Food Engineering*, **62**, 23–28.

CAMPDEN & CHORLEYWOOD FOOD RESEARCH ASSOCIATION (1996) *Guidelines on air quality standards for the food industry*. CCFRA Guideline No. 12, CCFRA, Chipping Campden, UK.

CHILLED FOOD ASSOCIATION (1997) *Guidelines for good hygienic practice in the manufacture of chilled foods*, Third Edition. Chilled Food Association, Kettering, UK.

CHILLED FOOD ASSOCIATION (2001) *High risk area best practice guidelines*, Second Edition. Chilled Food Association, Kettering, UK.

CROOK, B. (1995) Inertial samplers: Biological perspectives. In Cox, C.S., Wathes, C.M. (Eds.) *Bioaerosols Handbook*. CRC Press Inc., Boca Raton, Florida, pp. 247–267.

HOLAH, J., HALL, K.E., HOLDER, J., ROGERS, S.J., TAYLOR, J. and BROWN, K.L. (1995) *Airborne microorganism levels in food processing environments*. Campden R&D Report No. 12 from Campden & Chorleywood Food Research Association, Chipping Campden, UK.

HOLAH, J., MIDDLETON, K.E. and SMITH, D.L. (2004) *Cleaning issues in dry production environments*. Confidential R&D Report No. 192 from Campden & Chorleywood Food Research Association, Chipping Campden, UK.

MITCHELL, J.P. (1995) Particle size analysers: Practical procedures and laboratory techniques. In Cox, C.S., Wathes, C.M. (Eds.) *Bioaerosols Handbook*. CRC Press Inc., Boca Raton, Florida, pp. 177–246.

MÖLLER, Å.L. (1999) International standards for the design of clean rooms. In Whyte, W. (Ed.) *Cleanroom Design*. John Wiley & Sons, Chichester, UK, pp. 21–50.

6

Consumer perceptions of risks from food

L. J. Frewer and A. R. H. Fischer, Wageningen University, The Netherlands

6.1 Introduction

Unlike some other public health problems (for example, poor nutrition), the health outcomes of food poisoning are acute and measurable, primarily as a consequence of the ready identification of causal agents (Hayward, 1997). As a consequence, microbial food contamination represents a public health problem that, in theory, is amenable to influence by effective risk management. In practice, the incidence of foodborne diseases continues to remain a significant public health problem. Interestingly, there is some evidence that the reduction of microbiological risks has remained a low public priority relative to other food-related risks for several decades (Hall, 1971; Lee, 1989; Sparks and Shepherd, 1994), although this pattern is not invariable (see for example, Buzby and Skees, 1994; Lynch and Lin, 1994). The observation that this reduction represents a low consumer priority for risk mitigation in itself does not explain why consumers continue to experience illness. This is because consumer *behaviour* related to food preparation must also be taken into account, since the proper hygienic food preparation practices by the consumer could eliminate many of the risks associated with food safety.

Food safety objectives have been introduced in order to promote public health objectives through a reduction of the number of cases of foodborne illnesses. Generally speaking, it is difficult, if not impossible, to legislate for consumer behaviour. Inappropriate storage, food preparation and cross-contamination may occur, resulting in illness, even though products met food safety objectives at the point of sale. The goal of improving public health can be obtained only through implementation of appropriate and effective information interventions.

There has been much recent discussion about at what point in the food chain these food safety objectives should be set, and whether these should be benchmarked at the point of sale of food products, or at the point of consumption. From the perspective of public health, it is far more useful to set food safety objectives at the point of consumption, as the least controllable part of the food chain is within the domestic environment. However, public health is ultimately contingent on the adopted safety level of food preparation practices by the consumer. The setting of food safety objectives at the point of consumption has currently been agreed by a recent meeting of the Codex Alimentarius (Codex Committee on General Principles, 2004). This implies that more effective information provision must be developed to optimise domestic hygiene practices relevant to food preparation.

Thus it is important to conduct research in order to understand any potential barriers to the adoption of healthy food hygiene practices by consumers, and to apply this understanding to the implementation of effective intervention strategies specifically focusing on influencing consumer behaviour. To this end, it is essential that an understanding of consumer risk perceptions associated with food safety be developed, and linked to actual consumer behaviours when preparing food. It is also important to understand individual differences in perceptions and behaviours, as some groups of the population may take greater risks than others. This may be particularly problematic when considering risk vulnerability, where some groups in the population may be more at risk than others.

This chapter aims to briefly summarise what is known about consumer risk perceptions, and apply this to understanding why consumers undertake potentially risky behaviours. For a more extensive review of the literature in this area, the reader is referred to Hansen *et al.* (2003). The issue of individual perceptions and behaviours will also be addressed. Existing research examining consumers and domestic food hygiene practices will be examined, and recommendations for future research identified. Finally, risk communication insights regarding the development and implementation of best practice regarding information interventions will be provided.

6.2 Risk perceptions of consumers are not the same as technical risk assessments

Individual responses to risks are driven by perceptions or beliefs about risks, and these may apparently bear little relationship to technical risk estimates. Indeed, consumer responses to different hazards cannot be understood in isolation of the wider context in which different hazards are embedded. A good starting point for understanding consumer risk perception is provided by the psychometric paradigm developed by Paul Slovic and his co-workers (see for example, Fischhoff *et al.*, 1978). Research within the psychometric paradigm has indicated that psychological factors determine individual responses to different risks. These include, for example, whether the risk is perceived by individuals to be *involuntary* (i.e. in terms of personal exposure), *catastrophic* (i.e. affecting large numbers of people at the same time) or *unnatural* (i.e. technological in nature). These psychological factors increase the threat value of some hazards, and reduce the same factor in others. The perceived benefits associated with a particular hazard may, under certain circumstances, offset perceived risk (Alhakami and Slovic, 1994). Flynn *et al.* (1994) have used the psychometric approach to explain the apparent differences between lay and expert perceptions of risk.

This points at a weighing of risk factors in a complex multidimensional and potentially holistic way. In general, lay perceptions are often richer and more complex than perception held by experts, involving more constructs (albeit psychological in origin) and multidimensionality (Flynn *et al.*, 1994). For example, consider the case of voluntary *versus* involuntary exposure to a particular hazard such as radiation. Most individuals are more tolerant of the potential risks of both medical and natural radiation compared with the risks they associate with the nuclear industry, because of the following:

- Artificial radiation adds risk to a situation where it was not present before. In comparison, natural radiation is tolerable as it represents part of the natural order (Frewer, 1999a). Thus public negativity to the nuclear industry may not be equal to their enthusiasm regarding attempts to mitigate the risks of natural radiation.
- Medical radiation is perceived to have a benefit to the population generally. This may not be the case for the nuclear industry, where financial reward is perceived by the public to accrue to company shareholders, but the risks accrue to the general public and the environment.

In very broad terms, it may be useful to distinguish between two categories of potential hazard, those related to *technology* and those related to *lifestyle* choices (Miles *et al.*, 2004). Perceptions of technology risks are shaped by perceptions that the risks are out of control, are unnatural and are somehow adding unnecessarily to the risk environment. Much activity in the area of technology acceptance has, in the past, focused on aligning public views with those of experts in order to align the two perspectives (Frewer, 1999b). More recently, there has been increased emphasis on getting the public involved in the debate about how to manage and commercialise technological innovations (Renn *et al.*, 1995).

6.2.1 Optimistic bias

In contrast to technological risks, where the public estimates the risks as higher than experts, lifestyle hazards are associated with high levels of *optimistic bias* or *unrealistic optimism* (Weinstein, 1980). People tend to rate their own personal risks from a particular lifestyle hazard as being less when compared with an 'average' member of society, or indeed compared with someone else with similar demographic characteristics (for a review in the food area see Miles and

Scaife, 2003). In the area of food risks, optimistic biases are much greater for lifestyle hazards (such as food poisoning contracted in the home, or illness experienced as a consequence of inappropriate dietary choices) compared with technologies applied to food production (such as food irradiation or genetic modification of food). At the same time, people perceive that they know more about the risks associated with lifestyle choices when compared with other people, and are in greater control over their personal exposure to specific hazards. This is not the case for perceptions of personal knowledge about, and control over, technology-related food risks. In consequence, this optimistic bias means that a barrier to effective risk communication about lifestyle risks can be identified. People perceive that information about risk reduction is directed towards other individual consumers who are at more risk from the hazard, and who also have less control about their personal exposure to the associated risks, and possess less knowledge regarding self-protective behaviours. It has been well established that people exhibiting optimistic bias may not take precautions to reduce their risk from a hazard (Perloff and Fetzer, 1986; Weinstein, 1987, 1989).

The importance of optimistic bias, and approaches to reducing the disparity between perceived risk to the self and to other people, have been reviewed in detail elsewhere (Miles and Scaife, 2003). A brief summary of issues relevant to optimistic bias and food poisoning will be provided here.

In general, research into optimistic bias within the food domain has focused on two broad areas (Miles and Scaife, 2003). The first addresses comparative risk judgements for negative health outcomes associated with food choices, and the second focuses on risk factors associated with specific behaviours. Both are likely to be relevant to food safety and consumers. This is because, in part, consumers are likely to compare their own risks of food poisoning with individuals they perceive to be more vulnerable than themselves. They may also over-estimate the efficacy of their own health-protective behaviours.

Optimistic bias is *reduced* for hazards perceived to occur more frequently (Weinstein, 1987), or which have been experienced by individuals (Weinstein, 1987; Lek and Bishop, 1995). Increased perceptions of personal control *increase* optimistic bias (Weinstein, 1987; Hoorens and Buunk, 1993; Lek and Bishop, 1995). Similarly, if an individual can identify a stereotypical 'at risk' individual, who is unlike themselves, optimistic bias is increased (Weinstein, 1980); where an individual perceives the stereotype to be rather similar to themselves, optimistic bias is decreased (Lek and Bishop, 1995). Welkenhuysen *et al.* (1996) report that optimistic bias is not related to the perceived severity of the hazard nor (contrary to some public health policy approaches) to an individual's knowledge about the hazard and associated risks.

Why do people exhibit optimistic bias for some types of hazard? Motivational explanations assume that people are motivated to make judgements about risks that promote psychological well-being through removing threat to self-esteem by inducing anxiety (Weinstein, 1989). In contrast, cognitive explanations have tended to place emphasis on systematic biases in human information processing of incoming risk information, for example through inability to adopt the perspective of others, or comparison with vulnerable stereotypes (Weinstein, 1980). Kunda (1990) has argued that motivational goals may influence how information about a risk is processed. One might argue, as a consequence, that high levels of optimistic bias might therefore act as a motivational cue, or heuristic, to prevent people processing information related to the risks associated with a particular hazard.

Some empirical research has attempted to determine how to reduce optimistic bias. This includes increasing perceived accountability associated with an individual's risk judgement. It can be achieved through providing information about actual risk-taking behaviours (McKenna and Myers, 1997), or through making people compare themselves with an individual similar to themselves (Harris *et al.*, 2000), or an individual similar to the receiver of the risk information (Alicke *et al.*, 1995). Data show that there has been varied success in reducing optimistic bias through cognitive approaches (Miles and Scaife, 2003) although the dual-processing approach described later in this chapter may offer a theoretical approach to combining cognitive and motivational approaches to reducing optimistic bias in the area of food safety.

6.3 Risk perception and barriers to effective risk communication

Clearly, perception of risk will influence attitudes towards microbiological risks and food-handling practices (Frewer, 2001). Optimistic bias is likely to act as a barrier to attempts to mitigate public health problems associated with food hygiene. An additional barrier is associated with attitudes to food technologies introduced to alleviate problems associated with microbiological risks. One consequence of public concern about food technology is that novel food processing technologies, such as food irradiation (Bruhn, 1995) or high-pressure processing, may not be acceptable to consumers (Frewer *et al.*, 2004).

Research conducted within the psychometric paradigm has demonstrated that microbiological food risks tend to be moderately dreaded by consumers, but also perceived to be highly familiar, which reduces their threat potential (Fife-Schaw and Rowe, 2000). A further factor to consider in the area of public perception of microbial risk is that some consumer concerns are very specific to particular hazard domains, and this is very much the case in relation to food poisoning (Miles and Frewer, 2001). Qualitative research has confirmed the optimistic bias effect. The results indicated that respondents were maintaining optimistic biases regarding their own risks from food hygiene through comparing themselves with individuals perceived to be more 'at risk' than they themselves. Respondents also invariably perceived that they know and apply optimal food hygiene practices. They also reported that microbial risks were the frequent subject of media 'hype' and exaggeration (and thus discountable as potentially having a negative effect on health), confined to certain product categories such as eggs.

The same concerns were not expressed for other hazard types. For example, in the case of BSE, concern about animal welfare dominated perceptions. Genetic modification of food was linked to concern about the environment and the potential for unintended effects. Neither BSE nor genetic modification was associated with perceptions that were apparently optimistically biased.

In addition, it is important to remember that individual differences in risk perceptions may be quite extensive (Barnett and Breakwell, 2001). Affective or emotional factors, such as 'worry', may influence perceived risk (Baron *et al.*, 2000). Personality correlates such as 'anxiety' may also be influential (Bouyer *et al.*, 2001). Differences in perceptions of risk and benefit associated with various hazards exist between different *countries* and *cultures*, between different *individuals*, and even within different individuals at different times and within different contexts (Burger *et al.*, 2001). For example, women are typically reporting higher risk perceptions than men across a range of different health and environment hazards (Dosman *et al.*, 2001). This may result in greater risk-taking behaviour being exhibited by men more generally.

6.4 Developing an effective risk communication strategy

In the case of communication about food-handling practices, the ultimate goal is to improve public health through persuading consumers to adopt more appropriate domestic hygiene practices. As a consequence, communicators need to understand how the public perceives risk and hazards to facilitate the structuring of risk-related messages in such a way that consumers change their attitudes about the risks. If we adopt the social psychological idea that attitudes are the proximal causes for behaviour (Ajzen, 1991), changing attitudes should also lead to changes in the risky behaviour. Therefore, the various models of attitude change may provide insights into how this may be accomplished. In *persuasion* research it was noted that to process the supplied information as fully as possible, a lot of cognitive effort is required (Cacioppo and Petty, 1982). It was found that when not much cognitive effort was applied to processing the information, attitudes changed in a different way from when more effort was made.

6.4.1 Dual-process models

This realisation lead to the construction of dual-processing models of persuasion, such as the elaboration likelihood model (Cacioppo *et al.*, 1986). The elaboration likelihood model posits that long-term attitude change will occur only if the person receiving the message carefully and thoughtfully assesses its arguments, following what is described as the central route to information processing. When there is no motivation or cognitive ability to process the information, the communication will not be processed in such an elaborate way and will follow a peripheral route to processing. The peripheral route to processing is based on an individual assessing a cognitive or affective (emotional) cue associated with the persuasive message. This permits them to decide whether and how to process the information, and whether and how the arguments contained in the information can be assessed as to their merits, without recourse to complex processing of the information (Petty and Cacioppo, 1986). If an attitude change is the consequence of such peripheral processing, it is likely to result in temporary attitude shifts that are also more susceptible to counter-persuasion, leading to less predictable behaviour.

A central theme in all dual-process theories is that elaborate processing of all arguments is relatively costly in cognitive resources, leading to, for example, fatigue. So the main aim of these models is to give an idea when these costly processes are applied and when and how the simpler peripheral solutions are conducted. The elaboration likelihood model assumes that someone wants to base decisions on a solution that is as good as possible (accuracy motivation), therefore it assumes that consumers embrace the central route to persuasion unless the motivational or ability demands are not met (Petty and Wegener, 1999). Another dual-process model, the heuristic systematic model (Chen and Chaiken, 1999), is very similar but has two major differences. Firstly, it assumes the central route (or systematic processing as Chen and Chaiken name it) will not be chosen when heuristics (peripheral processes) lead to satisfactory solutions. This sparing use of cognitive resources is often labelled as the *cognitive miser* assumption. Although the assumptions underlying the selection of the processes are different, both models similarly define the effect of motivation and availability of cognitive resources on the processing of information.

A second more structural difference lies in the assumption of the heuristic systematic model that heuristics are used throughout the process unless there is a need for cognition; which means that in reality often a mix of heuristic and systematic processes occurs. The elaboration likelihood model, on the other hand, assumes that processing is heuristic only when cognitive processing is not possible at all, so that either the central or the peripheral route to attitude change is taken. It should be noted that heuristic cues can lead to the central processing of information, thus accounting for a sequential mix of the processing modes (Petty and Wegener, 1999).

For risk communication about hygiene-related food safety issues to be successful in the long run, it would be best to design the communication strategy in such a way that it enforces the central or systematic process to run its course. So the question of risk communication with regard to the dual process approach is: what sort of information should be given, or in what way should the information be supplied to influence the selection of either the central or the peripheral mode of information processing by the message receiver?

6.4.2 Communicating information following the dual-process approach

If information is highly relevant to the person receiving the information, motivation to process this information elaborately is likely to be high (Fazio and

Towles-Schwein, 1999). However to conform to *availability of resources* demand, the arguments contained in the message need to be salient and of high quality, otherwise the consumer might want to take care of the arguments but cannot (Wood *et al.*, 1985; Areni, 2003). The quality of arguments is shown to be a necessary precondition to process information if the targeted consumer is motivated to process information following the central route. Therefore whenever risks are communicated, considerable care should be taken to design these high quality arguments.

Motivating consumers to follow the central route of processing

In this chapter we will focus on motivation of consumers rather than on the effect of message quality. So whenever there is a high relevance of the information and the information itself is well structured, the arguments will be weighed and used by the consumer. So to achieve the central processing of information and the accompanying lasting attitude change, it would be useful to be able to motivate consumers and, of course, to know how to supply the information. One way of motivating people to process information is by increasing their level of fear (Kruglanski and Freund, 1983). Although research with fear as motivator has been conducted, in general little research on the relation between emotions and persuasion attempts following dual-process models has been conducted. The use of fear might lead to some unforeseen side effects, as we will discuss at the end of the following section on peripheral processing of information.

Peripheral processing of information

When the information is of low relevance, there is no intrinsic motivation to process the arguments elaborately. McGuire (1985) has reported that the extent to which a source is perceived to possess expertise may act as a cue that increases the likelihood of persuasion occurring. In these cases the impact of the arguments are probably mediated by peripheral cues. Factors such as expertise may act as such a referential cue as to the quality of the arguments. So if the information is derived from an expert source, and the conclusions are taken into account without going into the actual arguments, a change in attitude might follow. Trust in the information source providing the information may also act as a peripheral cue as to the merits of the messages contained (Petty and Cacioppo, 1986). There is, however, some evidence that, in the case of communication about microbial food safety, information source characteristics are less influential than message relevance in influencing risk perceptions associated with food poisoning (Frewer *et al.*, 1997).

Another process that follows peripheral rather then central processes might be called the affect heuristic (Finucane *et al.*, 2000). This emotion-related heuristic implies that when one is feeling good, risk perception will be perceived as far lower and benefits as higher, so a good mood probably infers the use of positive rather then negative information. Alternatively a bad mood should enforce the processing of negative information, which might be one of the specific functions of fear in persuasion (Lerner and Keltner, 2001). The exact effects of emotion induction as heuristic are, however, not well understood as of today. Fear can be a cue to take account of negative information but could also lead to despair (no use of information at all) or realisation that fear might be a bad councillor (Meijnders *et al.*, 2001). As mentioned above, fear may also be used to stimulate the awareness of personal relevance, so triggering a central rather than a peripheral processing of the subsequent arguments. This may complicate matters even further when the aims of fear and the message are considered as one. The fear should be aimed at avoiding the risks, not at avoiding the risk communication. So although emotions seem a powerful cue for peripheral processing of information as well as a potential trigger for central processing, owing to the limited knowledge of their exact working, it is hard to predict their exact effect. Understanding the effects of emotions on attitudes and behaviour is currently one of the major research areas in social psychology.

It is not always clear whether information will be processed following the central of the peripheral route, or alternatively whether it will be processed systematically or heuristically, or even a mix thereof. Therefore, risk communication effort should ensure that the message conveyed in the logical arguments (for systematic processing) and in the cues (trust, expertise, layout and wording, etc.) is in concert. If this is not the case, perfectly valid arguments might be disregarded or perhaps even worse, carefully built images of trustworthiness and expertise might be lastingly damaged (Chaiken and Maheswaran, 1994).

6.4.3 Tailored information campaigns

Following the dual process approach, as the personal relevance increases, the likelihood that information will be systematically processed will increase. One approach to effective risk communication may focus on segmenting the population according to their information needs, and developing specific information with high levels of personal relevance to specific groups of respondents. Information is more likely to result in attitude change (and subsequent behaviour change) if perceived personal relevance is high (Petty and Cacioppo, 1986). An example is provided by another area of public health, that of HIV transmission, in the late 1980s and early 1990s. Information developed by the medical authorities focused on cause of the illness in terms of viral transmission, whereas the risk information would have been both more salient to the population, and more effective in preventing disease, if it had focused on people's behaviours (Fischhoff *et al.*, 1993).

The problem with such an approach is that it is resource intensive, as research first needs to be conducted in order to identify individual differences with respect to people's perceptions and behaviours, and then tailored information needs to be delivered using delivery mechanisms preferred by different respondents.

6.5 Application of combined consumer behaviour – food safety studies

Food safety consumer studies often focus on measures of self-reported behaviour or attitudes towards food safety rather than actual observations of consumer behaviour and what this might imply for the incidence of food poisoning. Clayton and colleagues have attempted to validate self-reported behaviours by comparing these data with observational data (Clayton *et al.*, 2002). The results indicate that some important actions such as hand-washing were more frequently reported than actually enacted by respondents. Comparison of observational data with safety protocols such as Hazard Analysis and Critical Control Points (HACCP) did, in fact, indicate that consumer behaviours were verifiable against microbial contamination (Griffith and Worsfold, 1994, Worsfold and Griffith, 1995; Griffith *et al.*, 1998). However, few research studies reported in two recent review papers on consumer behaviour and relation to food safety, studied consumer attitudes and risk perception as well as consumer behaviour observations at the same sample (Redmond and Griffith, 2003a,b). To our knowledge, the relevant cognitive representations of consumers, resulting consumer behaviours and microbial contamination have not been studied simultaneously. There is some convergence of results across different studies. However, a fundamental understanding of what consumer behaviours and activities result in what levels of microbial contamination, how these behaviours and activities vary among individuals, and the role of human information processing and affect (i.e. emotion such as anxiety or fear) in developing effective communication strategies, remains largely unexplored. It is suggested that the only effective way to understand the relationship between these different areas is therefore to integrate social and natural sciences, which may indicate the need for a new research agenda in this area.

6.6 The need for more intensive cooperation between natural and social scientists

To be able to tailor information campaigns to individual information needs, much more detailed information on risk-related attitudes and behaviours is needed, as well as what the consequences of these are for individual health outcomes. In the case of a national campaign targeted at population level audiences, average risk levels to consumers are generally applied, and it is unlikely that individual consumers (possibly those most at risk) will attend to the information contained in risk messages. In contrast, tailored or targeted campaigns must focus on the information needs of groups or segments in the population. To be able to design a successful campaign, realistic estimates of risks should be communicated, along with any uncertainties about these risk estimates (if they exist). Failure to do so may have a negative impact on trust in the information source (Frewer *et al.*, 1996). Thus when targeting distinct groups the relation between specific behaviour and specific risks should be known.

And, of course, if we wish to know more about the specific behaviour of different consumers, we also need to know more about the specific psychological attitudes, beliefs and values of these consumers. In other words, the outcomes of the risk predictions developed by microbiologists should be communicated to the target group of consumers in a way that fits the values and motivation of that target group.

These requirements can be met only when knowledge from different disciplines is combined. To assess and predict the specific food safety risks, food safety experts and, in the case of microbial hygiene food, microbiologists are needed. To be able to predict consumer behaviour based on attitudes, values and beliefs of specific groups, and to develop targeted information strategies, consumer psychologists play an important role. Subsequently, the impact of risk communication on consumer health must be assessed by food microbiologists. This implies close cooperation between consumer psychologists and microbiologists specifically, or social and natural scientists more generally.

Before illustrating these ideas by a current research initiative, we would like to mention that cooperation requires effort from all researchers involved and is therefore not a simple thing to accomplish. A precondition for cooperation is that researchers from both social science disciplines and natural sciences are willing to cooperate with each other. This implies a willingness to accept the research paradigms and methods used in the different disciplines, and requires effort to avoid jargon and communicate in a way that can be understood by the partners.

Ongoing research is currently developing these ideas further (Fischer *et al.*, 2005). The research combines contemporary insights from both risk perception and communication theories directed towards reducing risky behaviours. It is argued that three elements should be addressed from a psychological point of view if people are to adopt healthy domestic food hygiene practices following risk communication. Due account must be taken of the following psychological factors:

- the resistance against attitudes change invoked by optimistic bias;
- the limitations in motivation and mental capacity of consumers in processing information; and
- the observation that information processing by consumers follows an experiential and affect-driven solving strategy rather than one of formal logic.

Taking these psychological factors into account the next question is: how do consumer perceptions and attitudes relate to actual risks resulting from inappropriate consumer behaviours? At this stage, it is important to analyse the technical risks associated with specific domestic food hygiene practices across different consumer groups. Therefore behavioural observations and microbiological research into finished meals will be combined. The outcome of this study, analysed by adopting a microbiological approach developed from HACCP might provide the necessary inputs to design a quantitative mathematical risk assessment (QMRA) (see Nauta, 2002). This QMRA might then be able to generate the necessary information that can be used to develop a targeted communication strategy.

By taking due account of the attitude change theories with regard to dual-processing of information, it should be possible to understand what cues associated with different messages will motivate consumers to read and process the risk information. For example, it may be possible to enhance information processing by using emotions or affective factors.

Finally, in the case of consumer-based food safety objectives it is important to validate the impact of changed consumer behaviour on microbial contamination, by conducting additional microbiological measurements, implying further cooperation between the natural and social sciences.

So, for a comprehensive understanding of the effects of consumer behaviour with regard to hygiene-related food safety practices and to understand the effectiveness of information interventions aimed at those consumers and the subsequent changes in behaviour, a close cooperation between natural and social scientists is required.

6.6.1 Implications beyond consumers

Up to this point the discussion in this chapter has focused on the consumer. In part, this is because most research into human behaviour and food safety has had the same focus. This may be because consumer behaviour is the only part of the food chain that cannot be enforced to comply with food safety standards. Thus understanding consumer behaviour, and developing interventions to reduce risky practices, may be the only way to improve public health associated with food safety. However, it is likely that professional workers in the food industry (for example, in the catering sector) are bound by the same psychological factors as consumers. After all, workers, as highly skilled as they may be, are humans like all of us. Thus the provision to workers in the catering sector with a large and possibly complicated safety manual will not guarantee that the rules and guidelines contained in the manual are followed. Food industry and catering workers not only have to follow these rules, but they also have to comply to the production standards set by their employer and regulatory bodies.

If the company has a good safety policy, this might go a long way in generating an adequate level of worker motivation towards compliance. However, if the regulations are too complex, or inappropriately presented or described, their correct implementation might lie beyond the cognitive capabilities of the employees involved in food preparation, especially in a stressful or time-limited situation (Wickens and Hollands, 2000). This might be the case especially for the hotel and catering industry, which is often under considerable time pressure and in which the staff often lacks formal training.

In the manufacturing industry in general (for example, within the field of modern aviation) a lot of effort is spent on 'human factors': interfaces and procedures are specifically designed to accommodate the operator's cognitive potential even in situations of extreme stress, in order to prevent the potentially catastrophic results of human error in these industries. These efforts were undertaken after the occurrence of some serious safety incidents, and resulted from the need to protect both the public and the employee.

Similar insights relating to equipment operation and procedure design have, to our knowledge, not been extended to the food industry, whether food processing plants or to catering and hotel businesses. Arguably, the applicable safety standards in the food industry that are constructed without taking human factors into account are unlikely to result in optimal levels of safety for employees and consumers. Some of the approaches, procedures and information interventions adopted in the human factors literature generally, and consumer risk psychology literature specifically, may be usefully applied to improve safety in the food production and catering sectors.

6.7 Conclusions

Simply applying legislative reforms to small sectors of the food chain is unlikely to have a major impact on public health unless consumer behaviour is also addressed. It is difficult, if not impossible, to legislate for consumer behaviour in the home. The development of an effective and targeted communication strategy is likely to be the only way to produce improvement in public health in the food safety area.

Understanding the risk perception of consumers is an essential first step in predicting and, possibly, changing their behaviour with regard to hygiene-related food safety practices.

- Risk perceptions result in involuntary, potentially catastrophic and unnatural risks (among others) being perceived as more a focus of consumer concern and anxiety than similarly assessed risks that are seen as voluntary, non-catastrophic and natural. For this reason, food hygiene may not be a priority for many consumers. For the same reason, technological processes developed to mitigate food safety risks may not be acceptable to some consumers.
- People tend to regard their own risks from microbiological foodborne illnesses as lower than that of the general population. This leads to the rejection of risk information since the targeted individual does not perceive it as directed at him or her, but to the vulnerable other person.

Risk communication relies on understanding consumer risk attitudes (and how to change these attitudes) in order to influence behaviour. At present, this area merits further empirical investigation, but a theoretical perspective exists within social psychology that may provide a useful basis from which to develop an effective communication strategy. This is likely to entail targeted communication approaches focusing on the information needs of particular consumers, and build on current knowledge of motivation and cognitive capacity in human information processing theory, to ensure that people change their attitudes and adopt appropriate behaviours with respect to improving domestic food hygiene practices.

In order to target communication at specific groups, social science investigation of food safety behaviour should be integrated with natural sciences research investigating what consumer behaviours actually increase risks associated with different microbial hazards. Taken together, understanding what hazardous practices are conducted by consumers in the kitchen, and why they are doing it, should provide the basis for an effective information strategy that will deliver real benefits to public health.

A final note we would like to make is, that although this chapter, and much of the research reviewed in it, has focused on consumers and consumer behaviour, it is conceivable that similar theoretical approaches will play an important role influencing hygiene or more general food safety-related behaviour of workers in the food industry (for example, factory workers and employees in the catering sector), and might promote more effective working practices. After all industry workers are human beings, just as are consumers, rather than machines.

6.8 References

AJZEN, I. (1991) The theory of planned behavior. *Organizational Behavior and Human Decision Processes,* **50**, 179–211.

ALHAKAMI, A. S. & SLOVIC, P. (1994) A psychological study of the inverse relationship between perceived risk and perceived benefit. *Risk Analysis,* **14**, 1085–1096.

ALICKE, M. D., KLOTZ, M. L., BREITENBECHER, D. L., YURAK, T. J. & VREDENBURG, D. S. (1995) Personal contact, individuation, and the better-than-average effect. *Journal of Personality and Social Psychology,* **68**, 804–825.

ARENI, C. S. (2003) The effects of structural and grammatical variables on persuasion: An elaboration likelihood model perspective. *Psychology and Marketing,* **20**, 349–375.

BARNETT, J. & BREAKWELL, G. M. (2001) Risk perception and experience: Hazard personality profiles and individual differences. *Risk Analysis,* **21**, 171–178.

BARON, J., HERSHEY, J. C. & KUNREUTHER, H. (2000) Determinants of priority for risk reduction: The role of worry. *Risk Analysis,* **20**, 413–427.

BOUYER, M., BAGDASSARIAN, S., CHAABANNE, S. & MULLET, E. (2001) Personality correlates of risk perception. *Risk Analysis,* **21**, 457–466.

BRUHN, C. M. (1995) Consumer attitudes and market response to irradiated food. *Journal of Food Protection,* **58**, 175–181.

BURGER, J., GAINES, K. F. & GOCHFELD, M. (2001) Ethnic differences in risk from mercury among Savannah River fishermen. *Risk Analysis,* **21**, 533–544.

BUZBY, J. C. & SKEES, J. R. (1994) Consumers want reduced exposure to pesticides on food. *Food Review,* **17**, 19–22.

CACIOPPO, J. T. & PETTY, R. E. (1982) The need for cognition. *Journal of Personality and Social Psychology,* **42**, 116–131.

CACIOPPO, J. T., PETTY, R. E., KAO, C. F. & RODRIGUEZ, R. (1986) Central and peripheral routes to persuasion: An individual difference perspective. *Journal of Personality and Social Psychology,* **51**, 1032–1043.

CHAIKEN, S. & MAHESWARAN, D. (1994) Heuristic processing can bias systematic processing: Effects of source credibility, argument ambiguity, and task importance on attitude judgment. *Journal of Personality and Social Psychology,* **66**, 460–473.

CHEN, S. & CHAIKEN, S. (1999) The heuristic-systematic model in its broader context. In Chaiken, S. & Trope, Y. (Eds.) *Dual Process Theories in Social Psychology.* New York, Guilford Press.

CLAYTON, D. A., GRIFFITH, C. J., PRICE, P. & PETERS, A. C. (2002) Food handlers' beliefs and self-reported practices. *International Journal of Environmental Health Research,* **12**, 25–39.

CODEX COMMITTEE ON GENERAL PRINCIPLES (2004) ALINORM 04/27/13. Appendix III, http://www.codexalimentarius.net/web/codex/codex27_en.htm.

DOSMAN, D. M., ADAMOWICZ, W. L. & HRUDEY, S. E. (2001) Socioeconomic determinants of health- and food safety-related risk perceptions. *Risk Analysis,* **21**, 307–318.

FAZIO, R. H. & TOWLES-SCHWEIN, T. (1999) The MODE model of Attitude-Behavior Processes. In Chaiken, S. & Trope, Y. (Eds.) *Dual-process Theories in Social Psychology.* New York, Guilford Press.

FIFE-SCHAW, C. & ROWE, G. (2000) Extending the application of the psychometric approach for assessing public perceptions of food risks: Some methodological considerations. *Journal of Risk Research,* **3**, 167–179.

FINUCANE, M. L., ALHAKAMI, A., SLOVIC, P. & JOHNSON, S. M. (2000) The affect heuristic in judgments of risks and benefits. *Journal of Behavioral Decision Making,* 13, 1–17.

FISCHER, A. R. H., DE JONG, A. I. E., DE JONGE, R., FREWER, L. J. & NAUTA, M. J. (2005) Improving food safety in the domestic environment: The need for a transdisciplinary approach, *Risk Analysis*, **25** (3), 503–517.

FISCHHOFF, B., SLOVIC, P. & LICHTENSTEIN, S. (1978) How safe is safe enough? A psychometric study of attitudes towards technological risks and benefits. *Policy Sciences,* **9**, 127–152.

FISCHHOFF, B., BOSTROM, A. & QUADREL, M. J. (1993) Risk perception and communication. *Annual Review of Public Health,* **14**, 183–203.

FLYNN, J., SLOVIC, P. & MERTZ, C. K. (1994) Gender, race, and perception of environmental health risks. *Risk Analysis,* **14**, 1101–1108.

FREWER, L. J. (1999a) Public risk perceptions and risk communication. In Bennet, P. & Calman, K. (Eds.) *Risk Communication and Public Health.* New York, Oxford University Press.

FREWER, L. J. (1999b) Risk perception, social trust, and public participation in strategic decision making: Implications for emerging technologies. *Ambio,* **28**, 569–574.

FREWER, L. J. (2001) Environmental risk, public trust and perceived exclusion from risk management. In Boehm, G. & Nerb, J. (Eds.) *Environmental risks: Perception, evaluation and management. Research in social problems and public policy.* Ukraine, Elsevier Science/JAI Press.

FREWER, L. J., HOWARD, C., HEDDERLEY, D. & SHEPHERD, R. (1996) What determines trust in information about food-related risks? Underlying psychological constructs. *Risk Analysis,* **16**, 473–486.

FREWER, L. J., HOWARD, C., HEDDERLEY, D. & SHEPHERD, R. (1997) The elaboration likelihood model and communication about food risks. *Risk Analysis,* **17**, 759–770.

FREWER, L. J., LASSEN, J., KETTLITZ, B., SCHOLDERER, J., BEEKMANE, V. & BERDALF, K. G. (2004) Societal aspects of genetically modified foods. *Food and Chemical Toxicology,* **42**, 1181–1193.

GRIFFITH, C. J. & WORSFOLD, D. (1994) Application of HACCP to food preparation practices in domestic kitchens. *Food Control,* **5**, 200–204.

GRIFFITH, C. J., WORSFOLD, D. & MITCHELL, R. (1998) Food preparation, risk communication and the consumer. *Food Control,* **9**, 225–232.

HALL, R. L. (1971) Information, confidence, and sanity in the food sciences. *The Flavour Industry,* August, 455–459.

HANSEN, J., HOLM, L., FREWER, L. J., ROBINSON, P. & SANDOE, P. (2003) Beyond the knowledge deficit: Recent research into lay and expert attitudes to food risks. *Appetite,* **41**, 111–121.

HARRIS, P., MIDDLETON, W. & JOINER, R. (2000) The typical student as an in-group member: Eliminating optimistic bias by reducing social distance. *European Journal of Social Psychology,* **30**, 235–253.

HAYWARD, A. C. D. (1997) Consumer perceptions. In Tennant, D. (Ed.) *Food Chemical Risk Analysis.* London, Chapman and Hall.

HOORENS, V. & BUUNK, B. P. (1993) Social comparison of health risks: Locus of control, the person-positivity bias, and unrealistic optimism. *Journal of Applied Social Psychology,* **23**, 291–302.

KRUGLANSKI, A. W. & FREUND, T. (1983) The freezing and unfreezing of lay-inferences: Effects on impressional primacy, ethnic stereotyping, and numerical anchoring. *Journal of Experimental Social Psychology,* **19**, 448–468.

KUNDA, Z.V. (1990) The case for motivated reasoning. *Psychological Bulletin,* **108**, 480–498.

LEE, K. (1989) Food neophobia: Major causes and treatments. *Food Technology,* **43**, 62–**64**, 68–73.

LEK, Y. & BISHOP, G. D. (1995) Perceived vulnerability to illness threats: The role of disease type, risk factor perception and attributions. *Psychology and Health,* **10**, 205–219.

LERNER, J. S. & KELTNER, D. (2001) Fear, anger, and risk. *Journal of Personality and Social Psychology,* **81**, 146–159.

LYNCH, S. & LIN, C. T. J. (1994) Food safety. Meal planners express their concerns. *Food Review,* **17**, 14–18.

McGUIRE, W. J. (1985) Attitudes and attitude change. In Lindzey, G. & Aronson, E. (Eds.) *The Handbook of Social Psychology*, 3rd edn. New York, Random House.

McKENNA, F. P. & MYERS, L. B. (1997) Illusory self-assessments – Can they be reduced? *British Journal of Psychology,* **88**, 39–51.

MEIJNDERS, A. L., MIDDEN, C. J. H. & WILKE, H. A. M. (2001) Role of negative emotion in communication about CO_2 risks. *Risk Analysis,* **21**, 955–966.

MILES, S. & FREWER, L. J. (2001) Investigating specific concerns about different food hazards. *Food Quality and Preference,* **12**, 47–61.

MILES, S. & SCAIFE, V. (2003) Optimistic bias and food. *Nutrition Research Reviews,* **16**, 3–19.

MILES, S., BRENNAN, M., KUZNESOF, S., NESS, M., RITSON, C. & FREWER, L. J. (2004) Public worry about specific food safety issues. *British Food Journal,* **106**, 9–22.

NAUTA, M. J. (2002) Modelling bacterial growth in quantitative microbiological risk assessment: is it possible? *International Journal of Food Microbiology,* **73**, 297–304.

PERLOFF, L. S. & FETZER, B. K. (1986) Self–other judgments and perceived vulnerability to victimization. *Journal of Personality and Social Psychology,* **50**, 502–510.

PETTY, R. E. & CACIOPPO, J. T. (1986) *Communication and Persuasion: Central and peripheral routes to attitude change,* New York, Springer-Verlag.

PETTY, R. E. & WEGENER, D. T. (1999) The elaboration likelihood model: Current status and controversies. In Chaiken, S. & Trope, Y. (Eds.) *Dual-process Theories in Social Psychology.* New York, Guilford Press.

REDMOND, E. C. & GRIFFITH, C. J. (2003a) A comparison and evaluation of research methods used in consumer food safety studies. *International Journal of Consumer Studies,* **27**, 17–33.

REDMOND, E. C. & GRIFFITH, C. J. (2003b) Consumer food handling in the home: A review of food safety studies. *Journal of Food Protection,* **66**, 130–161.

RENN, O., WEBLER, T. & WIDERMANN, P. (1995) *Fairness and Competence in Citizen Participation,* Dordrecht, the Netherlands, Kluwer Academic Publishers.

SPARKS, P. & SHEPHERD, R. (1994) Public perceptions of the potential hazards associated with food production and food consumption: an empirical study. *Risk Analysis,* **14**, 799–806.

WEINSTEIN, N. D. (1980) Unrealistic optimism about future life events. *Journal of Personality and Social Psychology,* **39**, 806–820.

WEINSTEIN, N. D. (1987) Unrealistic optimism about susceptibility to health problems: Conclusions from a community-wide sample. *Journal of Behavioral Medicine,* **10**, 481–500.

WEINSTEIN, N. D. (1989) Optimistic biases about personal risks. *Science,* **246**, 1232–1233.

WELKENHUYSEN, M., EVERS KIEBOOMS, G., DECRUYENAERE, M. & VAN DEN BERGHE, H. (1996) Unrealistic optimism and genetic risk. *Psychology and Health,* **11**, 479–492.

WICKENS, C. D. & HOLLANDS, J. G. (2000) *Engineering Psychology and Human Performance,* Upper Saddle River, NJ, Prentice Hall.

WOOD, W., KALLGREN, C. A. & PREISLER, R. M. (1985) Access to attitude-relevant information in memory as a determinant of persuasion: The role of message attributes. *Journal of Experimental Social Psychology,* **21**, 73–85.

WORSFOLD, D. & GRIFFITH, C. J. (1995) A generic model for evaluating consumer food safety behaviour. *Food Control,* **6**, 357–363.

Part II

Improving design

Introduction

Many food plants were built at a time when the hygienic quality of the food processing environment itself was less important than it is now. The focus was on making the food safe by using the right preservation technique to kill the microorganisms present in or on the food or to prevent their multiplication. Killing was achieved by giving the product a severe heat treatment. Multiplication was prevented by acidification (sometimes by microbial fermentation) and/or the addition of salt or sugar, sometimes in combination with the addition of chemical preservatives. The adverse influence of preservation treatments on the nutritional and sensory quality of food products, and their possible adverse toxicological effects, were either accepted and regarded as unavoidable, or simply not understood.

In more recent decades and in an increasing number of countries, consumers have become more demanding, initially with respect to taste and colour, and more recently with respect to the nutritional value of the products they choose. They have also become more concerned about the use of synthetic preservatives and other chemicals in food. Manufacturers, often with the help of research institutes and universities, have responded by attempting to improve traditional preservation treatments and by developing other ways of preservation. Reducing the severity of heat treatments has been quite successful with the introduction of improved retorts, where heat transfer is dramatically improved by movement of the containers during the heating and cooling stages of the process, and where improving the design of tubular heat-exchangers led to significant improvements in heat transfer. For liquid products, the introduction of plate heat exchangers resulted in some products that, pasteurised, could hardly be distinguished from the untreated reference sample. Steam injection, steam infusion, ohmic and microwave heating all have the potential of reducing the heating time dramatically. Some old approaches to the inactivation of microorganisms, such as high-pressure (invented in the late nineteenth century) and pulsed electric field treatment (invented in the 1960s) have been reassessed and, with improved technology, have reached or almost reached commercial application.

These developments have partly become commercially possible thanks to the improvements in the hygienic design and operation of food factories. Improvements in hygiene have reduced the microbial burden in both plant and raw materials, allowing production run times long enough and clean times short enough to make production economically feasible. In the past few years, the movement has been toward food safety objectives, where a certain minimal reduction in microbial load is no longer the objective of a treatment (for example, for thermal sterilisation a 12 log reduction in bacterial spores), but the final concentration of microorganisms in the product. If a product has a low microbial burden, the treatment required can be much milder to render the product safe. Hygienic design and operation thereby become crucial: if insufficient, the concentration of microorganisms in the product will increase during processing or become contaminated or, after processing, recontaminated. Improving hygiene moreover decreases cleaning time and increases production run time, improving economy.

7

Improving building design

D. J. Graham, Graham Sanitary Design Consulting Limited, USA

7.1 Introduction: sanitation and design

The main premise of this chapter is to show that sanitation and sanitary (hygienic) design are partners in the true sense of the word. Sanitary design deals with details of the hygienic design and construction of the physical structure and the equipment. It is the engineered design of food handling, processing, storage facilities, and equipment to create a sanitary processing environment, and to produce pure, uncontaminated, quality products consistently, reliably and economically. Often it is not the major design criteria that can cause sanitation failure but the smallest details that go with designing a new facility or renovation of an existing facility. For example: Fig. 7.1 shows an expansion joint for a bridge that had been designed to go into a food processing facility. It would have worked very well as an expansion joint but was impossible to clean. Note the 3 inch (75 mm) space in the joint that is under the floor. It becomes a natural accumulator of dirt, water, food particles, etc. Since it is 3 inches wide it would make an excellent passageway for rodents. This is a small item in the overall design of a facility but it turns out to be a very important one when it comes to sanitation.

There are numerous reasons for special consideration of sanitary design when remodeling, changing or building a new grassroots food processing plant. The less processing the product receives the more important the sanitation and sanitary design become. The more microbiologically sensitive the product is, the more sanitation and sanitary design attention is required. Sanitation and sanitary design are true partners.

The one constant for all food processing facilities is change. Plants and facilities are always being expanded, changed, new equipment added, old

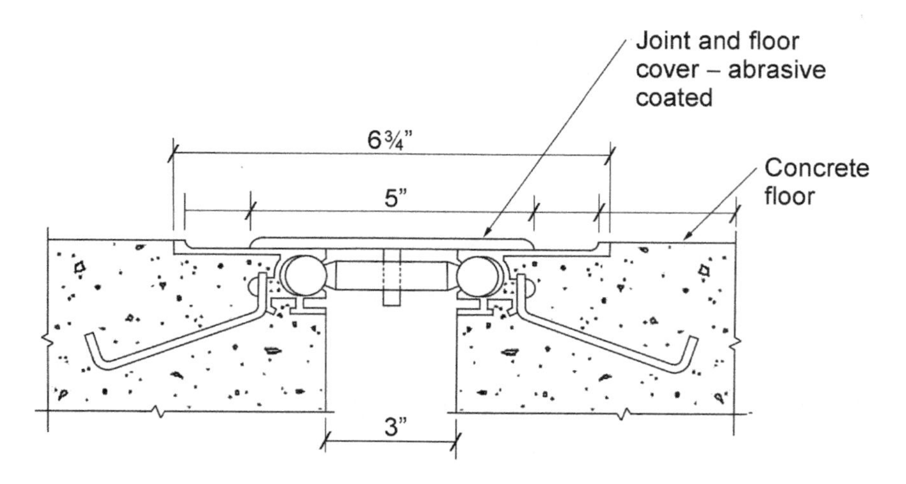

Fig. 7.1 Non-sanitary floor expansion joint (1 inch = 25.4 mm).

equipment removed, new additions added, and new interior walls added or old ones removed. Top consideration to sanitation and sanitary design should be given when plans for a 'change' are being developed. Figure 7.2 shows the effect of early planning and the effect on costs if changes are made as the engineering and construction move past the preliminary planning stage. Sanitation and sanitary design should be given input by any engineering effort, no matter how small, very early in the engineering process. The secret to sanitary design is a process called *mindset.* Sanitary design and sanitation should be one of the main considerations when any physical changes are made in a food processing plant regardless of the type of food processing.

The main purposes of creating top-level sanitary design are to:

- make sanitation programs faster;
- make sanitation programs more efficient;
- make sanitation programs more economical;
- help prevent product adulteration;
- help satisfy regulatory requirements;
- help satisfy consumer/customer audits, demands and requirements.

There are basically three levels of sanitary design for a food and/or a pharmaceutical facility. The three levels are not rocket science. They are simply 'good', 'better,' or 'best.' Each level can be defined as follows:

- **Good** – this level of sanitary design complies with all regulatory requirements only, and is the minimum level that a food processing plant can be designed to meet. Anything less than this is illegal under the regulatory codes in force at the time in the country where the plant is being constructed or renovated.
- **Better** – this level is one step up from the 'good' level and incorporates all the regulatory requirements as well and sanitary design recommendations of

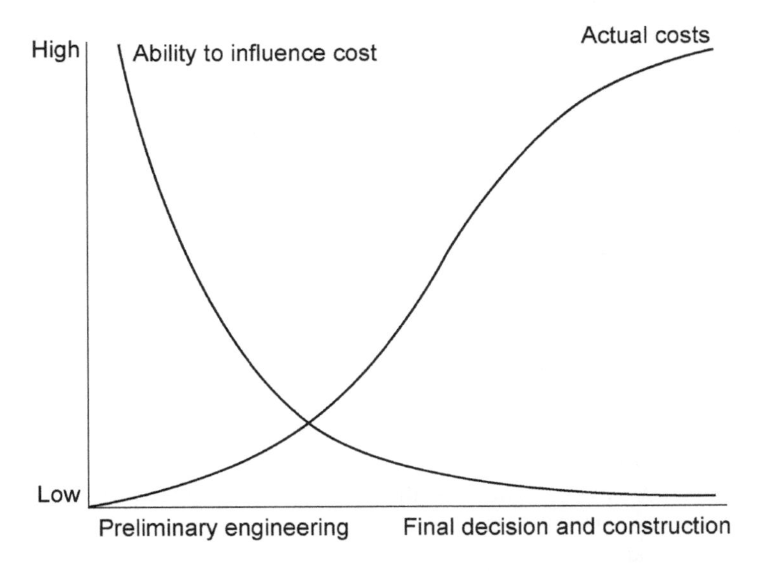

Fig. 7.2 Cost planning curve.

groups affiliated with that section of the food industry. Examples are meat processors heeding design recommendations from groups such as the American Meat Institute, National Food Processors Association, National Sanitation Foundation, and other applicable organizations. Other countries around the world have similar industry-oriented associations that have recommendations that can apply to the food processing industries in their part of the world. These should be consulted. In addition, it behoves any food processor engaging an engineering firm to design and/or build or renovate a facility to question the firm on their knowledge of food processing and especially sanitary design.

- **Best** – the best level is a bit more complicated from the standpoint that this level calls for specialized materials, conditions and knowledge. Designing to the 'best' level is commonly a custom design for that particular product and facility. It usually is confined to a small area of the facility where the product is micro-sensitive and is most likely to become contaminated if the processing/packaging conditions are not clean and the process lacks a kill step either during or after packaging. Special materials for surfaces are often used in the 'best' areas.

7.2 Applying the HACCP concept to building design

Regardless of the level of design, the concepts of HACCP (Hazard Analysis Critical Control Point) should be followed when employing sanitary design. The three hazards (physical, chemical, microbiological) can all be addressed during the design phase. Good sanitary design will go a long way in preventing these hazards from occurring just from the physical facility construction materials or the set-up of the building. Surfaces and construction of floors, walls, ceilings, and equipment are all important in helping to prevent these three hazards from contaminating the products.

7.2.1 Physical hazards

Physical hazards are such things as chipped building materials, dirt from overhead beams, glass from broken light fixtures or from broken windows, chipped paint from overhead painted surfaces, rust from supports made from uncovered mild steel, pieces of insulation, ceiling tiles, things that fall out of employees' pockets or jewelry (such as watches, rings) that is not removed before entering the process areas, and any other physical item one can think of. Consumers, at one time or another have found all of these and more in food products. Most physical contamination comes from employees who are not following good manufacturing practices. This type of contamination also comes from poor planning when the facility is designed and built. Physical contamination (hazards) in food products gives litigious lawyers reason to smile. Physical contaminates are very often considered to be the result of negligence, in its legal sense.

7.2.2 Chemical hazards

Chemical hazards can be as simple as the detergent used to clean the equipment if not rinsed and drained off the food contact surfaces, or out of a tank that has undergone cleaning in place (CIP) and then not been drained and rinsed. Other potential chemical hazards are the sanitizers if they are accidentally added to a product in process or left in tanks or kettles. Pesticides, fungicides and fertilizers are classified as chemical hazards if left on the product or they get mixed with the product.

The most recent additions to the chemical hazard group are allergens, which along with mycotoxins are considered a natural chemical hazard. Allergens can originate as a by-product of a process, be a residue on food contact surface, be in ingredients, and be an inadvertent contaminate from being unable to adequately clean equipment between runs of a product containing an allergen and one that does not. A new hot topic is the design of equipment to allow adequate cleaning of any allergen residue before running non-allergenic products.

Lubricants are considered as chemical hazards as well as poly(tetrafluoroethene) (PTFE, e.g. Teflon) sprays. Any other chemical item that is used for cleaning, lubricating, sanitizing, testing, etc., that is not declared on the label can be considered as a chemical hazard.

7.2.3 Microbiological hazards

Microorganisms are considered to be the biggest hazard for the majority of the food processing industry. Microorganisms are ubiquitous and range from viruses to bacteria to molds, and yeasts. Many of the microorganisms are considered pathogenic; the ones that can give you foodborne illness through formation of toxins or cause infections by ingesting the organisms. Spoilage organisms create short shelf-life in ready-to-eat foods. Organisms such as *Salmonella*, *Listeria* and *Staphylococcus* cause billions of dollars of loss due to foodborne illness around the world each year. An even greater monetary loss is due to microbial spoilage when sanitation practices are not followed owing to poor design of facilities and equipment and subsequent poor sanitation programs. These minute organisms are widespread and are often difficult to control. But controlled they must be in order to produce foods that are safe to consume and have sufficient shelf-life.

Sanitation and sanitary (hygienic) design are partners in conducting an effective sanitation program. This chapter will concentrate on the kinds of design needed in order to have a sanitation program that is effective, efficient, and workable.

7.3 Site selection and plant layout

When designing a new facility, sanitary design starts with site selection. The type of products to be produced should be considered when selecting a site for the new plant. If the products are to have a high fat content (meat, poultry, vegetable oils, etc.) then the site-surrounding areas should be relatively odor-free since fat is a flavor carrier and will pick up odors causing off-flavor very easily if exposed to odiferous conditions. Other conditions that have to be considered are the prevailing winds – are they strong and will they blow contaminates into the plant unless special precautions are taken at the plant site? Ideally the plant will be constructed with receiving/shipping doors on the lee side of the facility so the effect of prevailing winds blowing trash and contaminates into the plant can be minimized.

Is the location near a swamp or wildlife area (lots of insects, rodents, birds, and other potential contaminates) and can they be controlled or prevented from the plant site once the facility is constructed and in operation? Is the site located near an abattoir or a landfill or open agricultural fields where dust from these places can blow into the plant?

Other site considerations include access to major highways, rail, and other infrastructure considerations, depending on the type of products being produced and the logistics necessary for raw material handling and finished product shipping.

Is the plant to be a single story or multistory? There are advantages and disadvantages to each. A single story facility will give 'line of sight' manufacturing. That is, a supervisor can see the entire process line without having to go to another floor. It also provides more flexibility in changing equipment and product flow. From a sanitation standpoint it is more efficient in separating raw materials from product in process and finished goods, thereby reducing the potential for cross-contamination. It allows a simpler straight-line flow and there is less area to cover during sanitation and maintenance. Multistory structures may be required when gravity flow systems are needed. Multistory structures do present unique sanitation and maintenance problems that must be addressed during the design stage.

7.4 Water supply and waste disposal

The site must have adequate supplies of potable water (water suitable for human consumption) available throughout the year, even for future expansion. Incoming water lines must be designed for the necessary volumes and pressure required. If the pressure is not sufficient than a storage tank and booster pumps will be required.

Plant waste disposal (both sanitary sewage and process water) requirements or capabilities will have to be taken into consideration. Waste process water often contains significant amounts of organic material that increases the biological oxygen demand (BOD) and chemical oxygen demand (COD). This can cause problems at off-site or on-site sewage treatment plants. Discharge into rivers is tightly controlled by government environmental agencies so on-site treatment may be required. Solid waste handling (especially in the vegetable, fruit and produce processing industries) requires careful planning so it will not become a sanitation problem. Waste from a food plant always has the potential for being an attractant to rodents, birds, and insects. Never position waste drain lines (process or sanitary) over food products or food contact surfaces. These lines will probably leak one day and contaminate areas below the lines.

All the above items must be taken into consideration when planning the facility and if any are present, the proper design precautions must be taken.

7.5 Landscaping and the surrounding area

Once the new facility is completed or an existing facility is considered there is usually a plan to landscape the grounds. Special considerations must be given when landscaping around a food processing facility. Many existing facilities being considered by a processor for expanding their lines that were not originally designed as a food plant have very attractive landscaping but are not suitable for a correctly designed food processing facility.

Sanitary landscaping will help control rodents by depriving them of places to live (harborage) and keeping dust to a minimum. The landscaping should not include any ponds or streams that attract birds, insects and rodents. The threat of *Salmonella* infestation is a concern when these pests are attracted to a site. The plant itself should have a grass-free strip around the building, where it is not paved to the outside wall of the building. This grass-free strip should be about 30 inches (0.75 m) wide, four inches (100 mm) deep, and lined with a thick poly liner to keep weeds down and then filled with pea gravel. Pea gravel (small rounded gravel) is recommended because it will not bridge if a rodent attempts to burrow through it to get under the plant slab. This strip makes it easier for the plant sanitarian inspecting the building for rodent activity and is a deterrent in itself since rats do not like open areas and will avoid them. It also makes an excellent strip to place bait stations.

Another rodent deterrent is the construction of a horizontal lip fastened to the foundation of the plant. This ledge or lip is located 24 inches (0.6 m) below grade level and extends horizontally 12 inches (0.3 m). This lip can be either concrete or 16 gage metal. It has been shown that rodents burrow at an angle and once they hit that lip they will either burrow along it or will retreat to the surface and try somewhere else.

As part of the landscaping effort bushes should not be placed less than 30 feet (9 m) away from the structure. An alternative is to have them placed far enough away so that when they are full-grown or at the trimmed height that a full-grown person can walk between the bushes and building to inspect for pests. The ground around the bushes should be covered with pea gravel and not mulch or soil. Trees around a plant are not recommended since they provide roosting and nesting spots for birds, so attracting them to the facility. However, if trees are desired or present it is recommended that they be at least 30–40 feet (9–12 m) away from the facility. Select trees that are not attractive to birds for nesting and roosting. Consult a local landscape gardener for a list of such trees that will grow in your location.

All grass areas should be kept mowed or short to prevent harborage for rodents. Driveways, parking lots, and dumpster areas should be paved and sloped to drains which provide sufficient drainage even during storms. The storm sewer inlets should be in accordance with accepted engineering standards for ground water runoff as well as regulatory standards. There should be no standing water anywhere on the premises in order to reduce attraction to birds, insects and rodents or other pests. Dormant water puddles can become a source of foodborne microbiological contamination. The dumpster station should be equipped with a hose so it can be washed down after each removal of full dumpsters.

The perimeter of the grounds should have a chain link type fence to keep out larger animals, people, and children, and to secure the premises. Tall weeds and grass should not be allowed to grow near the fence and it should be kept free of any debris that catches in the fencing. The fence line should be inspected frequently for housekeeping purposes and to make sure it is in good repair. A perimeter fence provides a first line of defense against rodent and is a good location for bait boxes if allowed by local ordinances. These bait boxes should be placed 50 to 75 feet (15–22 m) apart and be secured either to the fence or to the ground.

Rail sidings should be paved with concrete. The contents of rail cars sometimes exhibit damage with subsequent spillage when the car door is opened if the contents were not adequately secured at time of loading. If the rail siding is not paved, the spilled material can fall out of the car and is difficult to clean up. It then becomes an attractant for pests. Concrete is preferred to asphalt as rodents can chew through asphalt to dig tunnels under the plant or siding area and concrete is usually easier to clean than asphalt.

7.6 Roof areas

Roofs are often ignored in sanitary design of a facility. They can be a major source of contaminates, especially if they are constructed of materials that are not cleanable. For example, a tarpaper gravel roof over a vented processing area can become a trap for standing moisture and product spills or bits of product coming from air exhaust vents. The gravel will trap moisture, preventing it from draining away, and thus the roof becomes an open invitation to insects, birds, and even rodents, especially if there is food material mixed in with the moisture. Numerous types of vegetation ranging from bits of grass and weeds to 4 and 5 foot (1.2–1.5 m) trees have been observed growing on gravel-covered roofs. Gravel tarpaper types of roof should be used only over warehouses or other non-process areas. There are numerous single membrane roofs available that are drainable and cleanable. Imholte (1984) recommends avoiding water accumulations that attract vermin; all flat roofs should be designed with downspouts to handle rainwater. Equip the downspout drains with bullet-nose grates that project upward. He also recommends that in regions with a cold climate, downspouts should be located inside the building to prevent freezing. Imholte also recommends all openings through the roof be curbed and flashed. The curb should extend 12 inches (0.3 m) or more above the finished room. Insulation and roofing materials should extend up the outside of the curbed sidewalls of the opening. Do not place insulation on the inside of the curb wall, as it is difficult to clean and frequently becomes infested with insects.

The simplest roof is one that also becomes the interior ceiling. An example of a sanitary design for a roof is the double tee pocket beam construction technique. The old method is to rest the double tee precast roof slabs on pre-cast concrete beams. This method left spaces between the underside of the roof cavity and the top of the beams. A much more satisfactory technique is to install concrete beams with preformed notches and prestressed tendons for structural strength. The double tees are then lowered into the notches, eliminating the cavity created by the older method. The seams are then filled with backer rod (a plastic or foam filler) and caulked with an acrylic-based caulk for elasticity. These roof/wall construction systems have proven to be successful in creating a sanitary condition as well as visual appearance. Other systems call for exposed truss systems under the roof surface. There are numerous types of roofing surfaces ranging from pitch (tar) and gravel to single membrane materials over an insulated concrete roof. Roof penetrations for vents, intakes, oven vents, air-handling systems and some utilities must be sealed to prevent water leakage and subsequent interior air and surface contamination.

7.7 Loading bays

Truck docks and doors are for the purpose of unloading raw material supply trucks or loading out trucks with finished goods or other items. This has to be done while preventing the entrance of pests such as rodents, insects and birds as well as dust and debris. One of the first considerations when designing a facility is to attempt to place dock doors and other frequently used doors on the side of the plant away from the prevailing winds. Then the dock areas must be constructed to be as rodent proof as possible. The older docks that are sunken are not recommended. No matter how good the drain that is supposed to be at the low point, it will become clogged with packing material, debris, and dirt. Water will collect and become a contamination point and an attractant for insects and rodents. Dock driveways should slope away from the building to provide good drainage.

Concrete truck dock walls are easily scaleable by rodents. A simple, inexpensive prevention is to install an 18 inch (0.5 m) wide strip of a very smooth material such as stainless steel under the dock wall overhang. The rodents cannot get a foothold and will be unable to climb the wall to gain access to the facility. If the dock has a dock leveler mechanism and a restraining hook to hold the trailer from rolling during the unloading a rodent can easily jump to the restraining hook, climb up into the dock leveler pit and up through the opening between the plate and the warehouse floor. An inexpensive way to prevent this is to line the dock leveler plate with nylon brushes. Rodents are reluctant to pass or chew through the brushes so they provide an effective barrier to rodent entry.

Some facilities even line the bottom and sides of the overhead truck doors with these brushes to prevent rodent entry through gaps in the door. Rats require 0.5 inches (13 mm) of space to gain entry into a facility while mice require only 0.25 inches (6 mm). Most modern truck docks use dock seals. These seals effectively stop the entry of insects and rodents when the dock door is open and a trailer is backed against the seal. Some docks have a vertical lift dock plate. When the door is opened the dock plate is lowered into place on the truck bed. When finished the dock plate is raised, the truck departs and the door closed. There is no empty space below the plate leading to the outside, which effectively shuts off the access for rodents and insects. The other concern of truck docks is the presence of an overhang or canopy. Many overhangs are constructed with braces and supports that are conducive to bird roosting and nesting. If overhangs are installed they should be designed to be completely smooth underneath to prevent birds from finding a find a perch or a place to build a nest. Older facilities sometimes install mesh netting under the dock canopies of overhangs to prevent birds from gaining access to possible nesting areas. Caution must be exercised if nets are installed. Some birds have been found to peck larger holes in the netting, so gaining access to the protected area where they are trapped and die. This practice, in turn, presents a potential contamination problem and should be avoided in favor of either removing the overhang or changing it to one with a smooth underside.

Rail docks should be wide enough to allow forklift trucks to load and turn as they back out of the rail cars. Bracing and other dunnage should not be stacked against the walls. Rodents can easily build nests behind the materials if they remain more than 24 hours. Rail docks should not be used as storage for extra equipment, product or waste materials. Interior rail docks should be equipped with an overhead door that can be closed while the rail car is being loaded or unloaded. The space beside the rails under the door should be fitted with a compressible rubber plug to keep rodents out of the area when the door is closed. The plug will compress when the rail car flange passed over it and then resume its shape after the wheel has passed over. When the overhead door is closed, the expanded plug will fill the void. These are available from railroad supply houses.

7.8 Entry/exit points and external lighting

Personnel entry and exit doors have the potential to become entry doors for rodents. These doors should have a tight fit with a gap of less than 0.25 inches (6 mm) at the bottom. Personnel doors should not open into the processing areas but into a hallway or directly into the personnel facilities area such as locker rooms. The preferred material for personnel doors is metal with expanded urethane-filled cores. Air curtains should be installed over all personnel entry and exit doors to prevent insect entry when opened.

7.8.1 Dock doors

Vertical lift dock doors are preferred for good sanitary design. If the vertical height for vertical lift is not available then the overhead garage door type can be used. The third choice is a roll-up type door with no housing. Roll-up door housings have been observed harboring insects. Existing roll-up door housing has been cut and hinges installed so it can be opened and cleaned on a routine basis.

7.8.2 External lighting

Dock doors should not have lights positioned above or beside them. Lights can be insect attractants, especially those that emit high levels of ultraviolet. Ultraviolet rays are attractive to flying insects; thus the success of insect electrocutors. If possible, lights should be positioned on standards about 30 feet (9 m) away and shine back on to the doors. High-pressure sodium lights are recommended as they have a whiter light and low ultraviolet emission, and use less energy than others. Low-pressure sodium lights emit an orange light that many find objectionable. Mercury vapor or incandescent lights should be installed at the entrance to the plant site from the street and/or in the parking lots because mercury vapor and incandescent lights emit ultraviolet that attracts insects. Installing these types of lights at the entrance to the plant site or in the parking lots will reduce the insect pressure on the plant, especially the loading docks.

7.8.3 Air curtains

Air curtains, if designed and installed correctly, can help prevent flying insects from entering the facility. There are many homemade versions, consisting of ordinary fans blowing air out the open doors. These do not do the job. A correctly designed air curtain should be positioned above the door on the outside wall. An exception can be made for very large rail or truck dock doors where the air curtain can be installed vertically on each side of the doorway blowing outwards. The units positioned over the doorway must be clear across the door. The air column should be 3 inches (75 mm) thick and have a down and out sweep. Probably the most important criterion is the velocity. The air curtain should have a minimum velocity of 1600 feet (488 m) per minute measured 3 feet (0.9 m) off the floor. For best results the unit should be hard wired to the door opener so when the door starts to open, the air curtain starts and does not shut off until the door closes completely. For the few flying insects that do gain entrance, a well-positioned insect electrocution light trap can be installed. Proper positioning will ensure the unit cannot be seen from the outside. After all, the ultraviolet lights are designed to attract flying insects so they should not be visible from open doors. They are there to attract insects that get inside only. To be the most effective the units should be installed no more than 5–6 feet (1.5–1.8 m) off the floor. Ideally they should be in a corner extending from a few inches off the floor to about the 5 foot (1.5 m) level. The traps work by attracting the insects to the light, which is behind an electric grid. As the fly enters the grid it is electrocuted by the charge on the grid. Some regulators do not want these 'zapper' type units in food processing rooms. Their logic is that the unit blasts the insect into many parts and there is more potential for contamination from the parts than from one intact insect. A number of vendors of these units are now making them more passive so they stun the insect and drop it whole onto a sticky board located in the bottom of the unit.

7.9 Inside the plant

When sanitary design features are incorporated into a new structure or into a renovation or addition plan, they will result in an improved appearance of the structure, and will reduce the time required for sanitation. This will fulfill many of the purposes of sanitary design to make sanitation more effective, faster, and more economical while satisfying regulations and customers' expectations. In this section we will cover the sanitary design recommendations for floors, walls, ceilings, drains, lighting, heating, ventilation and air conditioning (HVAC) systems, personnel facilities as well as miscellaneous items.

7.9.1 Floors

Although there is an entire chapter devoted to floor design, it must be said in this overview chapter that floors are the most abused surface in a food processing plant. Floors must withstand chemical abuse from the use of water, dust, cleaners, sanitizers, acids, and lubricants, and even the abuse from particles and pieces of the food product being produced. They must also withstand the abuse received from mechanical means of dropped equipment and tools, pallets being dragged over the surface, from equipment being moved and holes drilled to fasten it down. Foot traffic and forklift or pallet jack traffic will cause a lot of abuse to floors. The floors can be exposed to temperature swings from clean-up water, spillage of cooking items, hot oil from fryers, cold water from chill tanks, hot and cold water from the sanitation shift, etc. For long-lasting floors, do not try to cut costs by purchasing a cheap covering that will not withstand the use and abuse it will receive. Cutting the capital cost will result in increased maintenance costs down the line. Wood floors are no longer acceptable in food processing facilities. There are many old facilities that still have wooden floors in the dry processing areas for flour, starch, dry grain handling, etc. However, they are not acceptable in wet processing areas.

The most common base material is concrete, which is then covered with a sealer or monolithic coating or a brick/tile material. If concrete is not sealed or otherwise covered, especially in wet processing areas, spalling can occur where the troweled layer wears away or is eaten away and the aggregate is exposed. Water containing high levels of chlorine will rapidly eat away the troweled layer. Acids, food products, and plain water will also attack unsealed or uncovered concrete. Exposed aggregate is a potential home for microbes where they find ideal hiding places and are extremely difficult to remove. Remember also that newly laid concrete floors must have a vapor barrier to prevent migration of moisture from the soil below the floor. Moisture from this source will virtually destroy monolithic floor coatings. Monolithic coatings can be epoxy, urethanes, resins, or combinations of these depending on the type of abuse the floor will receive. Chips in floor coverings in wet areas can lead to water getting under the coating and lifting it off the concrete. As it is doing this, microbiological soup is created under the coating and every time forklift wheels or foot traffic passes over the defect, it is exposed to a loading of microbes from the water expelled through the chip or hole or crack in the coating. This microbial contamination can then be spread wherever the forklift of the foot traffic goes.

For long-lasting floors many companies use acid brick or split pavers or tile. Although the high-end full acid bricks may be more expensive than monolithic coatings, they usually last for many years with minimal maintenance. The main thing to remember is that floors must withstand use, cleaning, and abuse. Monolithic coatings are getting better and better. There are some that bond with the concrete and have approximately the same coefficient of expansion and contraction and are seamless. There are a number of firms that sell these types of floor coating. Degussa Resin Systems (SRS Degadur Corp.) in particular and numerous others in general. Other flooring materials that have been used, but are not recommended in food processing rooms, are vinyl or asphalt tile, wood, metal plates, unless they are stainless, and bituminous/asphalt.

7.9.2 Drains

Wherever there are wet processing conditions drains will be required. Floor drains have proven to be sources of *Listeria* in food processing facilities unless correctly designed, installed, and maintained, and continually cleaned and sanitized. Drainage systems must meet all local and national plumbing codes. The food regulatory agencies are basing their requirements of performance rather than dictate the construction of floor drains. The performance they demand is completely drained floors: no ponding or standing water is allowed on the production floors. The two most common drains are area drains or trench drains.

Area drains must have a p-trap and be spaced at a recommended one 4-inch (0.1 m) drain for each 400 square feet (~40 m^2) of floor space. The floor should be sloped to the drain at a 1–2% slope. Area drains are the most common in meat processing plants and dairies. There are area drains on the market that exhibit sanitary design and are easily cleaned. All drains should be accessible for cleaning and application of sanitizer on a routine basis. Area drains should be a minimum of 4 inches (100 mm) in size and equipped with a removable metal strainer to catch food materials, and to prevent the entry of rodents and some insect pests such as cockroaches. They should also be designed to minimize the reflux of contaminated air that can come when a surge of water enters the drain.

The other most common choice is trench drains or gutter drains. Trench drains should be designed pre-sloped with rounded or coved bottoms. Square bottom drains are no longer recommended because of the difficulty in cleaning them and keeping them clean. Trench drains should be sloped at 1–2% slope for continuous drainage. Trench drains should be cleaned routinely and the grates constructed to withstand forklift traffic and any other wheeled traffic. There are, on the market, preformed trench drains that are easily cleanable and can be quickly installed. Processing or packaging equipment should never be placed over an open area or trench drains. The air from the drains contains aerosols that can contain microbes. These aerosols can contaminate otherwise clean equipment.

7.9.3 Walls

Wall design can be broken into two categories – external and internal. External walls need to be water, rodent, and insect proof. The best material for external walls is concrete, followed by dense concrete block. Medium density block may be available in some areas of the country and will work. Light density or cinder blocks should not be used as they are porous and insects can work their way to the center of the block. Fumigation may be a problem as well since the fumigant will work its way to the center of the block and slowly release into the workspace long after the plant is back in production.

Many facilities use other materials, such as insulated metal panels and corrugated metal especially in pre-engineered buildings. Concrete walls can be cast in place (tilt up), precast, or formed and poured. Many concrete silos are cast in place using slip forms. If you intend to paint or apply epoxy coatings to tilt up walls, remember to match the release agent to the paint or coating material you intend to use. If the two are not matched according to the manufacturer's directions, the coating used may not adhere to the concrete. Precast panels are done at a precaster's location and trucked to the site where a crane of similar piece of equipment hoists them in place onto a poured foundation or footer. The panels can and often are precast with an insulated center surrounded by concrete sealing the insulation into the concrete wall panel. The joints of the precast panels will require caulking with a good caulking compound made with an acrylic base to retain elasticity. There will be some maintenance of these joints required, as the plant gets older. Precasting with the insulation already in the wall has the advantage of not having to attach insulation on the inner wall surface or under an additional surface covering material for refrigerated or otherwise temperature-controlled facilities. Tilt up concrete walls are often used when there is enough space to form and pour the wall panels at the construction site. These are preferred by some construction companies and used with great success. These too can be poured with enclosed insulation.

Pre-engineered metal buildings are not greatly preferred materials or building types. These panels are difficult to keep sealed as they have a high rate of expansion and contraction and can present condensation problems. If the metal panels are sandwich panels they must be equipped with secure and tight end caps to prevent rodent and insect infestation. Rodents can penetrate the insulation and roam freely inside the walls if end caps are not provided. All panel joints should be caulked with a good grade of caulk to prevent insect infestation inside the panels. There are materials on the market that can be sprayed on the interior of the panels to insulate and seal the insulation with a resin material that provides a seamless surface that is easily cleanable and resistant to damage.

When exterior walls are designed rodent proofing can be incorporated into the design. A very simple, inexpensive method is to install a rodent barrier at the base of the wall by installing a barrier of concrete or galvanized metal (anywhere from 16 gauge to 28 gauge) 24 inches (61 cm) down from grade level extending out at least 12 inches (30.5 cm). Rats burrow at an angle and will not try to go around this barrier but will abandon the burrow and go somewhere else. There will be a need for wall penetrations for wiring, plumbing, ventilation, utility pipes, etc. These penetrations should be made, framed, and sealed the same day in order to prevent inner wall infestation by insects and often by rodents. Pipe penetrations require sealing with sheet metal or galvanized hardware cloth or any other long-lasting material that will withstand rodent gnawing. Any penetrations below grade must be protected and sealed. If a number of pipelines enter at one spot then surround them with a galvanized fine mesh screen or hardware cloth to prevent rodent incursion. Cone guards can be used on vertical pipelines as well as flat guards. Whatever type your designer decides on must keep the rodents from gaining access.

7.9.4 Interior walls

Interior walls are constructed from numerous materials, ranging from tile, cement block, concrete, metal, reinforced fiberglass paneling, baked on enamel insulated metal panels, resin materials with built-in antimicrobials as well as dry wall in selected areas. Dry wall should not be used in any area where there is moisture or wash-down cleaning, or any kind of food processing taking place. Whatever type of material is used, there are certain criteria that must be met for it to be considered a sanitary wall for food processing plants. According to Katsuyama (1993), the walls should conform to the following standards of sanitary design and construction:

- The juncture of the roof with the wall should be weather- and rodent-proof. Wall plates should be sealed to prevent insect entry and avoid dust accumulation.
- Double walls of frame construction should have built-in rat stops. The insulation material must be unattractive to rodents for nesting.
- The inside surfaces of the wall should be water-resistant, smooth, washable, and easily cleaned. There should be no ledges to collect dust and debris. All rough or irregular surfaces in concrete walls should be rubbed or ground smooth to reduce dust and dirt accumulation; where grain and flour dust occur, such accumulations can become breeding spots for insects.
- All wall openings should have tightly fitting doors, windows, or screens to exclude rodents, insects, and other pests.
- Flat surfaces, such as horizontal braces, should be sloped at about 45° to prevent their use for storage of personal and miscellaneous items.
- Particular attention should be given to areas that are subject to splash and spray. They should be surfaced to facilitate quick, easy, and frequent cleaning or flushing.
- The floor juncture of framed walls is of primary importance to sanitation. Its proper construction and maintenance are essential to adequate rodent control and general housekeeping. The juncture of the interior wall should be watertight and built on a coved base rising to a height of at least 6 inches (150 mm) above the floor level. Corners should be rounded to facilitate cleaning.

Companies are continually developing wall material that is cleanable and sanitary. Most materials now are white or very light colored. The material needs to be resistant to the cleaning and sanitizing compounds used in the facility.

Resistance to damage is another important criterion since the use of troughs and other mobile equipment can damage and break the surface of the wall, thereby ruining the pest and water proofing of the wall. Protective barriers such as bollards and wall guards are recommended for areas where there is a high potential for damage. If cement block walls are erected in dry areas, the grout lines should be shallow to minimize ledges for dust to collect. Through experience, it has been found that a striking tool the shape of a stainless steel teaspoon works very well in creating a shallow grout line.

In wet areas the tile or block should be constructed using the stacked bond method rather than the running bond construction. Stacked bonding places each cement block or tile directly above the one below it. This yields a vertical grout line so moisture will drain down the grout line to the floor below. A running bond configuration puts the vertical grout line of each course of block directly in the center of the block below and the block above it. Moisture can and will accumulate at each layer where the vertical grout line meets the center of the tile or block below and can create a growth niche for microbes. A word of warning: if the stacked bond method is to be used then the construction structural designers must be notified so the wall can be reinforced by either filling the center of the block with mortar or using reinforcing rods through the center of the blocks, or both if the facility is in a high seismic zone. If block walls are used then the first two courses of block should have the centers filled with mortar. Doing this will not only prevent water or other liquids from seeping under the block to the area adjacent but will also prevent insects from gaining access to the interior of the block wall in case a crack develops at the floor wall junction. Block walls should also be capped to prevent insect and rodent infestation in the center spaces of the block. Walls up to 6 feet (1.8 m) in height should be capped with a concrete cap at an angle of 45° to 60° to prevent tools, clip boards, etc., from being placed on the flat surface. Walls that go all the way to the ceiling or are over 6 feet (1.8 m) can be capped with a flat concrete slab. All interior walls should be constructed so the wall floor juncture has a cove with a radius of 1–3 inches (25–75 mm) to get rid of any crack at the juncture. Joint cracks are very hard to clean and can become harborages for dust, dirt, insects, and bacterial/fungal growth.

If plain concrete walls are used (tilt up, precast or poured in place) and lining or epoxy coating them is not considered as in dry processing areas, warehouses, etc., then a good grade of sealer should be applied to prevent dusting of the concrete. Concrete dust will contaminate open products or settle on packaging material, finished goods and equipment.

Walls in new facilities should be designed without windows, particularly in the raw material storage, preparation, processing, and packaging rooms. Windows require maintenance and are subject to breakage. There should be *no* glass in a food processing facility. Plants with existing windows in the subject areas should replace any glass with a tempered polycarbonate material. The windows should be sealed to prevent opening, which will allow insects, dust, dirt, odors, and anything else present in the air into the facility. Open windows will also destroy required air pressure relationships necessary for good air circulation within the facility. In older facilities where windows already exist, they should be sealed. If they cannot be sealed then they should at least be fitted with screens of 18 × 18 mesh or similar material to keep out insects if opened. Non-opening windows in interior walls are permitted, such as in supervisors' offices, as long as they are non-breakable material. Wire-reinforced glass will break and shatter with small pieces. Plexiglas material will also break into shards if struck so these are not recommended for interior or exterior windows. Windowsills should have a 45° slope to prevent the accumulation of dust and debris and make them easier to clean.

Doors should be tight fitting with less than 0.25 inches (6 mm) clearance at the edges. The doors should be solid, as should the doorframes. Hollow doors and hollow doorframes can and do become harborage areas for insects and rodents. All doorframes should be flush with the wall with no ledges above the door. Any door windows should be of a polycarbonate material and mounted flush to the door on the sensitive product or process side. The other side or exterior of the door should have a sill of 45°. All doors should be self-closing and designed to withstand the use expected. In food processing areas stainless steel doors are always acceptable. There are other materials such as fiberglass and fiberglass resin materials that are acceptable in sanitary areas.

7.9.5 Ceilings

Ceilings should be the easy-to-clean type and should be able to withstand direct impingement of water from hose stations. A ceiling should be a good reflector of light to help make the process area bright and shiny. It must be non-absorbent and above all cleanable. Smooth ceilings will allow better airflow across the ceiling surface and that, in turn, helps to prevent condensate formation. The most sanitary type of ceiling is the walk-on type. This type of ceiling completely seals off the trusses and other structural pieces holding up the roof and connecting the walls. All utilities can be run on the roof side of the ceiling with only vertical drops to the equipment below, thus eliminating horizontal runs of pipe in the process area. There must be access to the above ceiling area, from outside the process room, in order to do maintenance on the lines above the ceiling and for pest control. The space requires ventilation to reduce the possibility of condensate formation. In a well-designed and constructed walk-on ceiling space, the lights in the room below can be changed from above the ceiling. Recessed telescoping sprinkler heads are available, eliminating sprinkler pipes and sprinkler heads in the process room. With this type, ceiling process piping changes, pest control, etc., can all be carried on over the process room without intrusion into the process room envelope.

If at all possible, and the type of processing allows, the interior of the roof becomes the ceiling material. An example is the concrete double tee roof described earlier. The interior surface of the precast concrete double tee can become the ceiling surface. If this type of ceiling is to be used in a high moisture area it is recommended that a good sealer be applied to the interior concrete. This will prevent moisture migration into the double tee concrete and the ultimate rusting of the reinforcement rods in the concrete. It will also eliminate dusting from the exposed concrete and create a smoother surface. Once the reinforcement rod starts rusting, it will cause expansion of the rods and cracking and chipping of the surrounding concrete, creating a potential physical hazard for the products being produced below. This condition can be very expensive to repair and can be prevented by the application of a sealer at the outset. Rusting of the rebar and the cracking of the surrounding concrete can affect the structural integrity of the double tee roof.

Exposed structural steel used in ceiling/roof construction should be encased in concrete or its equivalent. Encasing structural members with concrete sometimes adds a prohibitive load on the roof structure, especially if utilities and equipment are to be hung from the ceiling or air-handling equipment is to be placed on the roof. There are, on the market today, techniques for boxing in overhead beams, steel trusses, and other structural members with foam sheets and then spraying on a fiberglass resin surface. This creates a smooth seamless surface that will withstand moisture and chemicals, and seals the structural members from collecting dust, moisture, and debris, which can fall onto the product below.

Some warehouses use drywall or gypsum ceilings. These may work for dry areas but should not be used in wet environments. Materials such as corrugated metal and metal pan roofing are not recommended for food processing areas owing to the high rate of heat transfer that can cause condensation problems.

Suspended ceilings utilizing 2×4 sq. foot ($0.6 \times 1.2\,m^2$) panels suspended in an aluminum or stainless grid are not recommended for process areas. There are a number of disadvantages to these types of ceiling. The panels are often clipped down or caulked in place when the ceiling is newly installed. Then, whenever any work has to be done on pipelines, etc., that are above the ceiling, a panel has to be displaced. Suspended ceilings usually do not lend themselves to load bearing, so anything above the ceiling has to be accessed through the ceiling itself. When any work is completed, it is virtually impossible to replace a panel and secure it as it was originally. These panels will warp, and any air pressure changes caused by opening or closing doors will create movement in the panels. Often panels get broken or the grid frames bent and this allows air from below to mix and contaminate the air above the ceiling, and the air from above to mix with and contaminate the air below in the process room. It is difficult to treat the above ceiling space for insects since it is not a load-bearing surface. Suspended ceilings, if used, should be confined to office spaces, laboratories or other non-food production areas.

7.9.6 Heating Ventilation Air Conditioning (HVAC) systems

The Food and Drug Administration (FDA, 2003) stated in a presentation: 'Airborne contamination is strongly suspected as the cause of some pathogenic contamination'. Unfortunately this has proven to be true when you look at the Sara Lee hot dog *Listeria* incident (USEPA, 2003). The use of air pressure relationships and filtration becomes more and more important as the sensitivity or risk of products to air contamination increases. Airflow within a facility should flow from clean to dirty. The highest filtered air pressure should be where the product is last in touch with the environment. This is normally the packaging room. In many plants the packaging and final process (before packaging) are located in the same area. The air pressure then flows outward to preprocessing areas such as raw material preparation, raw material receiving and finally to the outside. In the other direction the air flows from processing/packaging to the casing room, warehouse and to shipping. Again the air should flow outward when the dock doors are opened. The pressure differential is not large. It is generally accepted practice to have a pressure differential of 0.01–0.02 inches (0.25–0.5 mm) of water column between rooms. This equates to about 250–300 feet (76–91 m) per minute air velocity from room to room. If the plant operates under negative pressure, that is, air flows into the plant when the doors are opened or comes in through cracks and crevices, then the plant has absolutely no control over the quality of the air coming into the facility. The rule of thumb for plant processing areas is a minimum of 6–12 air turns per hour, i.e. the volume of air in the room is changed 6–12 times per hour.

In areas that operate under refrigeration the cold air can be recycled through filter units. The recommended procedure for doing that is to put an air-handling unit outside the process room (on the roof or beside an outside wall) that has heating and refrigeration capacities. Air is drawn from the room, passed through a filter of 30–50 μ and 40–50% efficient. At this point any make-up air is added from the outside (usually 5–10% of the total volume). The cold air from the process room helps to precondition the outside make-up air. The mix then passes over a refrigeration coil and then through a final filter. The final filter is recommended to be 95% efficient at 5 μm. The 5 μm size has been recommended since most microbes exist in air as passengers (on dust particles), within droplets, and a very few as isolated organisms. Bacteria are on average about 1.5 μm while mold is 2.5–20 μm. Yeast are 4–12 μm. Compared with a human hair at 50–100 μm. A magna-helix gage (or similar airflow gage) should be installed at the final filter stage so when the pressure drop gets too great an alarm will sound in the maintenance office, indicating that the filter requires changing. A clogged air filter can become a positive contamination source. The filtered air is then ducted into the room and flows along the ceiling, down the walls and back to the exhaust vent for recycling. This moving air over the ceiling and wall surfaces will greatly reduce the potential for condensate formation, especially on overhead fixtures.

Air ducts are designed either round or rectangular. Rectangular ducts should be sealed against the ceiling if they are below the ceiling surface. If it is not sealed to the surface the flat surface can become a depository for dust and dirt and make a good runway for rodents. Other ducts are round and present a smaller overhead surface that has to be cleaned. All seams should be welded or interlocked and caulked. Insulation should never be placed inside an air duct. If insulation is required, a sandwich duct construction should be used. The outside surface is hard, the inner surface is hard and the insulation is enclosed between the two surfaces. If insulation is left exposed it will become damp and a substrate for microbial growth. The microbes are then spread throughout the air distribution system. The ducts all require clean-out ports. Ideally they are placed about 5 feet (1.5 m) apart. Some air systems in plants with highly microbial sensitive products are installing CIP systems inside the ducts. The ideal HVAC system:

- cools and heats;
- humidifies and/or dehumidifies;
- filters for clean air;
- keeps ductwork out of the room;
- is not a contamination source;
- pressurizes the room.

In rooms, evaporators are not always the most sanitary choice. The drip pans can become a source of contamination. The drip pans should be sloped to a drain so the condensate can continually drain to a floor drain. It is recommended that a sanitizer block be placed at the low point of the drip pan. It should also be routinely cleaned and sanitized by the sanitation crew. The evaporator fin material should be of a material that will withstand cleaning and sanitizing chemicals without corroding.

Every plant should have access to an air tester to determine the microbial load of the air entering the plant through the air-handling/filter systems. An air tester is even more important for ready-to-eat product facilities. The air should be tested for microbial loads at least once or twice in the spring, summer, fall, and winter. When checking incoming air always look on the roof to make sure exhaust vents are not directing the exhaust air into the intake vents or are upwind of the intake vents. When installing new equipment that requires an exhaust stack direct the installers to make sure there are no intake vents near the exhaust.

Always contact a reputable HVAC engineer to assist in balancing the air within the plant and designing the correct sized air-handling system for the plant's requirements.

7.9.7 Compressed air

Air compressors should be the oilless type. Even if they are rated as oilless the lines should be equipped with coalescing filters and if used for product contact or on food contact surfaces they should be equipped with high-efficiency particulate air (HEPA) filters rated at 99.97% efficiency at 1 μm. Compressed air lines can contain condensate and become a growth medium for microbes. If used to create overrun in selected products or used to open packaging the compressed air can impart microbes, some pathogenic, into or onto food products and food contact surfaces.

7.9.8 Lighting

Interior lighting level requirements have increased considerably over the years. Local ordinances often require a minimum level of lighting. Usually these minimum levels are not adequate for food processing lines. Many facilities are installing lighting that yields upwards of 60–70 foot candles (650–750 lux) at the work surface in the processing areas where inspection is required. Lesser intensities are used in other areas of the facility down to 10-foot candles (100 lux) in areas of shipping and receiving. The light fixtures are all required to have an unbreakable cover to prevent contamination in case of breakage. Regulators will zero in on unprotected light fixtures. Shatterproof bulbs are an acceptable substitution for shields over fluorescent light fixtures. Within the processing areas, lights with low UV emissions should be used to reduce the attraction to flying insects. Metal halide lamps are widely used in the food industry. They can produce much higher intensities than comparable fluorescent lights. They provide better light distribution as they are normally hung below the ceiling surface.

7.9.9 Personnel facilities

In a comprehensive sanitary design program, personnel facilities (rest-rooms, locker rooms, break rooms, hand-washing sinks) are important considerations. Every day one reads accounts of foodborne illness thought to be traced to employees not washing their hands after using the toilet facilities. Vendors of hand-washing sinks and units estimate that foodborne illness could be reduced by at least 25% if adequate hand-washing were accomplished.

Rest-rooms or toilet areas must be designed with sanitation in mind. Special considerations include making sure the toilet areas are vented to the outside of the plant by a fan that is always running or is running when the lights in the room are on. Katsuyama (1993) states that the minimum air flow should be 35 cubic feet (1 m^3) of air per minute for each water closet or urinal. The locker rooms and toilet areas are one of the few areas of a food processing facility that must be under negative air pressure. Air should continually flow into these rooms whenever the doors are opened and through any other openings into the rooms. The preferred entrance/exit to rest-rooms use a maze design so nothing has to be touched going in or out of the room. The facilities should not open directly into a processing area especially if there are open food product or food contact surfaces anywhere near the door. Ideally the rest-rooms open into an anteroom or into a hallway. Floors should be constructed out of moisture impervious materials such as quarry tile or sealed concrete with the floor/wall junction coved for easy cleaning. The floors should be equipped with at least one floor drain and the floor sloped toward the drain. Walls should be of solid construction and extend to the main ceiling and have a smooth, moisture impermeable surface amenable to cleaning with water. Hot and cold water hose bibs should be provided so the floors may be easily cleaned. The individual toilet booth partitions, toilet bowls, urinals, and hand-washing sinks should be ceiling or wall hung to facilitate quick and thorough cleaning. Nothing should be sitting on the floor. Local bylaws and regulations must be considered when designing and constructing rest-rooms and locker rooms.

Hand-washing sinks should be wall hung and the water activated with knee, thigh, or electronic sensors. Foot pedals are no longer recommended because of the perceived cleaning difficulties on the underside of the pedals. The water temperature should be no greater than 110 °F (43 °C). Soap dispensers and paper towels must be within easy reach. A covered trash receptacle with a large open mouth is also necessary and should be close to the paper towel dispenser. Do not install air dryers. These have been shown to recirculate contaminated air and unless they are extremely rapid employees will not spend the necessary time in front of them to dry their hands adequately. Snyder (1999) has shown that the physical action of wiping the hands removes nearly as many microbes as washing them, and it gives more satisfaction to the employee that their hands are clean. Signs reminding employees to wash their hands before leaving the rest-room must be posted in all languages needed for all employees to read and heed. Similar types of hand-washing sinks should be placed in locations on the production floor so employees inspecting or otherwise handling the food products have ample access to hand-washing facilities. Again, there should be no hand controls. Use knee or thigh-operated or electronic sensor-equipped, sinks. If electronic sensor units are used, the ones with heavy-duty transformers should be selected for longer life and to withstand harder use. If employees wear smocks or aprons while on the processing line, a place should be provided to hang these items prior to entering the rest-rooms.

If lockers and locker rooms are provided to the employees, the design of them must also reflect sanitation. Whether half, quarter, or full lockers, the individual lockers (full-sized) or the locker stacks (half or quarter lockers) should be mounted on legs at least 6 inches (150 mm) off the floor so there is ample space to clean under them. Some facilities like to mount lockers on a solid base, but this is not recommended. Insects will penetrate the space between the bottom of the locker and the top of the base. The author has seen numerous instances where cockroaches have thrived in the space between the bottom of the locker and the top of the solid base. Where lockers are stacked the top lockers should not have flat tops. The tops should be pitched at a minimum of 30° to prevent items from being placed on top and forgotten. Locker rooms should be well ventilated and at least part of each locker constructed out of a mesh material so they too are well ventilated. Rules governing the use of lockers such as cleaning them out must be posted and followed. Items left in lockers for a long period of time can become sources of contamination and odor. Management needs to play an active role in keeping locker rooms clean.

Break-rooms/lunch-rooms should be equipped with hard-surfaced chairs and tables for ease in cleaning and preventing absorption of spilled liquids from lunches, cold drinks, etc. If vending machines are present these should be equipped with rollers to easily pull them away from walls for easy access for cleaning. The vending machine operators should be required to clean the inside of the vending machines every time they are serviced or refilled. The insides of vending machines can quickly become a haven for insects including cockroaches.

7.10 Future trends

The increased emphasis on sanitary design of food processing facilities and equipment can be considered a relatively new phenomenon. Up until 10 to 15 years ago it was normally practiced by dairy/cheese related facilities. We can all remember seeing pictures of dairies with tile walls, shiny stainless steel equipment and employees wearing white uniforms with hats over their hair. The food industry outside of dairy was mostly canning and freezing. Meat was delivered in sides or quarters and cut up by the market or store on a per order basis. Over the last few years the world has gotten used to prepackaged foods, ready-to-eat products, and salad bars. This change has added many new stresses to our food processing and distribution systems. Some of these stresses are:

- increased reliance on minimally processed products;
- emergence of new strains of foodborne bacteria;
- centralized growth of large food distributors;
- consumer preferences for ready-to-eat foods;
- growing number of people at high risk for severe or fatal foodborne illness;
- allergens.

All these stresses have produced more and more reliance on and importance of sanitation and its partner, sanitary design. Food processors are demanding processing equipment that not only is efficient and does what it is supposed to but is easily and quickly accessible and cleanable without special keys or tools. The same holds true when they build or renovate the processing facility itself. There must be a minimum number of flat areas, niches, rough surfaces, and cracks and crevices or other hiding places. The processing plants of the future will refine these designs even further. Even today the term 'bright and shiny' is the keyword for designing even the most common processing facility. Microorganisms are ubiquitous and are continually changing through mutation and adaptation to sanitizers, cleaners, and other methods of inactivating them. The plants of tomorrow must exhibit even more dynamics of sanitary design in order to control the numbers of the most serious of the three HACCP hazards – microbial contamination. It is safe to say that we probably will never completely eliminate this hazard but we certainly can control it. New surfaces containing antimicrobial ingredients such as silver ion technology will become more and more commonplace. The vendors are already touting simpler processing lines, fewer overhead fixtures, and tougher floors for more sanitary plants. HVAC systems are undergoing redesign to make them more effective in filtering out organisms that can contaminate food in process. HVAC is an important tool in combating condensation formation in facilities. Condensate is an excellent medium for *Listeria*. Legionnaires' disease bacteria have also been found in some water chillers, in room evaporators, and in other water/air contact chiller units. These units are being designed to be more easily treated with chemicals and sanitizers to control these organisms.

Organizations representing certain special food processing groups are developing their own list of sanitary design criteria for their particular industry. The pendulum of sanitation and sanitary design has already swung past the center point as food processors are developing, marketing, and selling more and more convenience, ready-to-eat and prepackaged food products with the knowledge that sanitation must be designed into the product preparation. It will stay past the center point, slowly approaching the apex of its swing for many years to come. Training of engineers, maintenance, and the general food processing workforce in sanitation and sanitary design will become more and more the norm if a food company is to survive in the years to come. A number of universities that teach Food Science/Technology and Food Engineering are incorporating modules on sanitation and sanitary design or at least exposing their students to it. More pure engineering schools should be incorporating the whys and hows of sanitation and sanitary design in their curriculum for engineers. Even if the graduates do not enter the food processing industry, the training will serve them in everyday life as they dine in restaurants and in home food preparation applications.

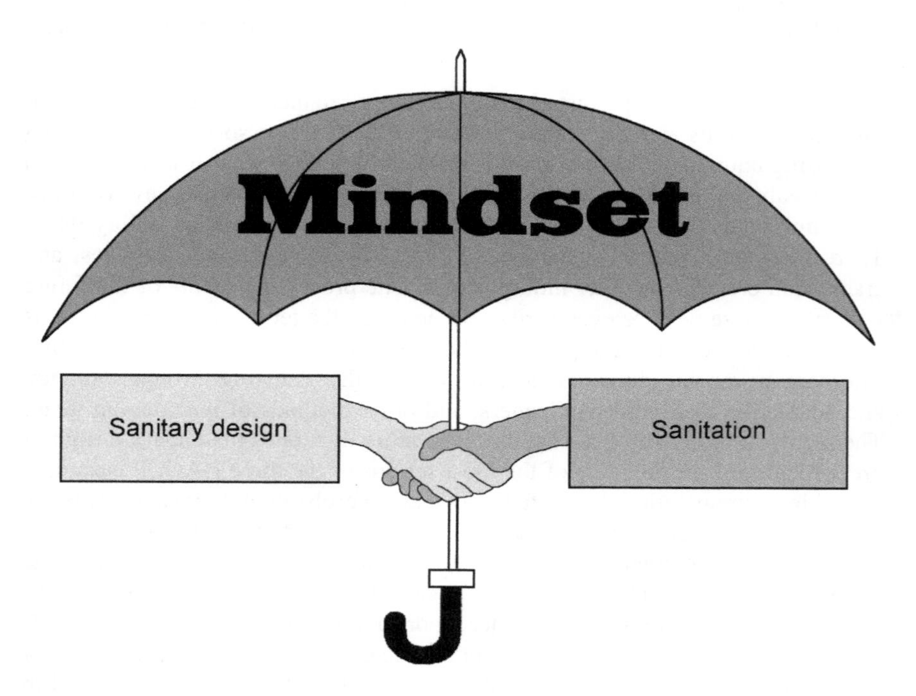

Fig. 7.3 Mindset reminder.

This chapter has not covered everything in sanitary design of facilities, since it is merely an overview. The bibliography will help the reader find more details of the many phases of sanitary design. The key, of course, to continued progress in sanitary design is the development of a *mindset* (Fig. 7.3) by everyone involved with food processing.

7.11 Bibliography

FDA (2003), Federal Register, 13 March, **68** (49), 12188–12189.

GRAHAM, DONALD J. (1991), 'Sanitary Design – A Mind Set (Parts I–VI)', *Dairy, Food and Environmental Sanitation*, **11** (8–12), 388–389, 454–455, 600–601, 669–670, 740–741.

GRAHAM, DONALD J. (1992), 'Sanitary Design – A Mind Set (Parts VII–XII)', *Dairy Food and Environmental Sanitation*, **12** (1–5), 28–29, 82–83, 168–169, 234–235, 296–297.

GRAHAM, DONALD J. (1992), 'Sanitary Design – A Mind Set (Parts XII–XIV)', *Dairy, Food and Environmental Sanitation*, **12** (6–7), 523–524, 578.

GRAHAM, DONALD J. (1992), 'Sanitary Design – A Mind Set (A Checklist Parts 1&2), *Dairy, Food and Environmental Sanitation*, **12** (9, 10 & 12), 636–637, 689–691, 816–817.

GRAHAM, DONALD J. (1993), 'Sanitary Design – A Mind Set (Checklist Parts 2–4)', *Dairy, Food and Environmental Sanitation*, **13** (1–6), 25–26, 91–92, 172–173, 231–232, 291–292, 354–355.

GRAHAM, DONALD J. (2004), 'Using Sanitary Design to Avoid HACCP Hazards and Allergen Contamination', *Food Safety Magazine*, **10** (3), 66–71.

IMHOLTE, T.J. (1984), *Engineering for Food Safety and Sanitation.* Crystal MINN: Technical Institute for Food Safety.

KATSUYAMA, ALAN M. (editor) (1993), *Principles of Food Processing Sanitation.* Washington, DC: The Food Processors Institute.

Personal Communication (2004), SRS Degadur Corp., Piscataway, NJ.

Personal Communication (2004), Arcoplast Corp., Chesterfield, MO.

SHAPTON, DAVID A. and SHAPTON, NORAH F. (eds.) (1991), *Principles and Practices for the Safe Processing of Foods.* Oxford, UK: Butterworth-Heinemann Ltd.

SMYDER, P.O. (1999), *A Safe Hands Handwash Program for Retail Operations: A Technical Review* (Item 15). Florida Environmental Health Association, www.hi-tm.com/documents/handwash-fl99.html.

SPRINGER, RICHARD A. (1991), *Hygiene for Management*, pp.72–79. UK: Highfield Publications, Doncaster.

USEPA (2003), Report of the United States Environmental Protection Agency Region 5, Risk Management Planning Inspection Conducted at Sara Lee Foods, Zeeland Facility, Zeeland, Michigan, 8 October.

8

Improving zoning within food processing plants

J. Holah, Campden and Chorleywood Food Research Association, UK

8.1 Introduction

Factories have always had to be compartmentalised or segregated into specific areas for a number of reasons. These were primarily due to environmental protection (i.e. protecting the product from the wind and rain), segregation of raw materials and finished product, segregation of wet and dry materials, provision of mechanical and electrical services and health and safety issues (e.g. boiler rooms, chemical stores, fire hazards, noise limitation).

More recently, as the nature of food production has changed, particularly with the advent of ready-to-eat products, factories have begun to further segregate or 'zone' production areas for hygiene reasons. A series of higher hygiene, or cleaner, zones have been created to help protect the product from microbiological cross-contamination events after it has been heat treated or decontaminated. In addition, there has also been the recognition that non-microbiological hazards, particularly allergens, have to be controlled by segregating them from other product ingredients.

Finally, label declaration issues such as 'suitable for vegetarians', 'organic', 'does not contain GM materials' or 'Kosher' have all caused food manufacturers to think about how raw materials are handled and processed. This is particularly true if, for example, factories are handling meat, non-organic ingredients, GM ingredients or non-Kosher ingredients. While the presence of, e.g., meat residues in a vegetarian product is not a safety issue, it will be an ingredients declaration issue, which could lead to poor brand perception.

Other than routine food manufacturing issues, access to manufacturing sites by unwanted people, ranging from the media, through petty criminals to bioterrorists, has unfortunately focused attention on site security.

To provide protection from general contamination (physical, chemical and biological hazards) during manufacture, food has historically been protected by a barrier system, made up of up to three barriers (Holah and Thorpe 2000). With the advent of enhanced hygiene control in high hygiene areas, however, this has now been extended to four barriers as shown in Fig. 8.1 (Holah 2003). These encompass the site (1), the factory building (2), a high risk or high hygiene zone (3) and a product enclosure zone (4). In this system the degree of control of the production environment increases such that, finally, fully processed products are manipulated in controlled environments in which contaminants are actively excluded.

With respect to segregation requirements, foods and drinks can be broadly divided into low- and high-risk products dependent on their stability or whether they will be further processed by the food manufacturer or the final consumer. Low-risk products, typically either raw materials or ambient shelf-stable products, include eggs, raw meat and fish, fruit and vegetables, dried goods, canned foods, bakery and baked products, confectionery, snacks, breakfast cereals, oils and fats, food additives/ingredients and beverages. High-risk products, typically short shelf-life ready-to-eat foods, include cooked and smoked meat and fish, prepared vegetables, prepared fruit, milk, cream, cheese, yoghurt, ice cream, sandwiches and ready meals and generally require refrigeration at chill temperatures.

The number of factory barriers required will be dependent on the nature of the food product, the nature of the hazard and the profile of the final consumer, and will be established from the Hazard Analysis Critical Control Point

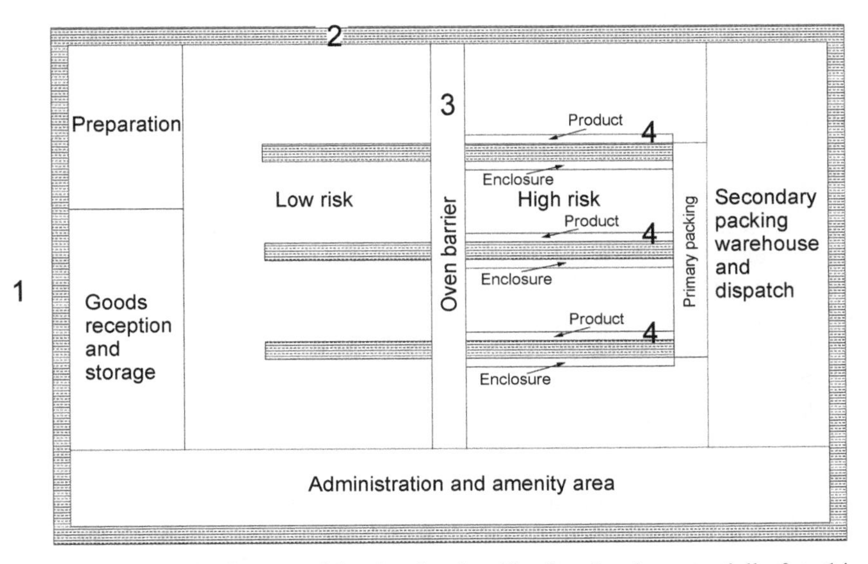

Fig. 8.1 Schematic diagram of the four levels of hygiene barrier potentially found in food factories.

(HACCP) study. For low-risk products, the first two barriers only are likely to be required. For high-risk products, the use of the third barrier is required for microbiological control. The fourth barrier is necessary for aseptic products in which the elimination of external contamination is required, though some fully cooked, ready-to-eat products with extended shelf-life may benefit from the additional controls this barrier affords. Although not absolutely necessary because of hazard control, manufacturers may choose to process food in higher hygiene zones for other reasons. This may be because of local legislation, or they may believe that in the near future their product range will include higher-risk products and it makes financial sense to develop the infrastructure to produce such products at an earlier stage, or simply because they believe it will facilitate brand protection.

8.2 Barrier 1: Site

Attention to the design, construction and maintenance of the site, from the outer fence and the area up to the factory wall, provides an opportunity to set up the first of a series of barriers to protect production operations from contamination. This level provides barriers against environmental conditions, e.g. prevailing wind and surface water run-off, unwanted access by people and avoidance of pest harbourage areas.

At the site level, a number of steps can be taken, including the following:

- The site should be well defined and/or fenced to prevent unauthorised public access and the entrance of domestic/wild animals, etc.
- Measures can be put in place to maintain site security including the use of gate houses, security patrols and maintenance schedules for barrier fencing or other protection measures.
- The factory building may often be placed on the highest point of the site to reduce the chance of ground level contamination from flooding.
- Well-planned and properly maintained landscaping of the grounds can assist in the control of rodents, insects and birds by reducing food supplies and breeding and harbourage sites. In addition, good landscaping of sites can reduce the amount of dust blown into the factory.
- Open waterways can attract birds, insects, vermin, etc., and should be enclosed in culverts if possible.
- Processes likely to create microbial or dust aerosols, e.g. effluent treatment plants, waste disposal units or any preliminary cleaning operations, should be sited such that prevailing winds do not blow them directly into manufacturing areas.
- An area of at least 3 m immediately adjacent to buildings should be kept free of vegetation and covered with a deep layer of gravel, stones, paving or roadway, etc. This practice helps maintain control of the fabric of the factory building.

- Storage of equipment, utensils, pallets, etc., outside should be avoided wherever possible as they present opportunities for pest harbourage. Wooden pallets stacked next to buildings are also a known fire hazard.
- Siting of process steps outside (for example silos, water tanks, packaging stores) should be avoided wherever possible. If not possible, they should be suitably locked off so that people or pests cannot gain unwanted access to food materials.
- Equipment necessary to connect transport devices to outside storage facilities (e.g. discharge tubing and fittings between tankers and silos) should also be locked away when not in use.
- To help prevent flying insects from entering buildings, security lighting should be installed away from factory openings so that insects are attracted away from them.

8.3 Barrier 2: Factory building

The building structure is the second and a major barrier, providing protection for raw materials, processing facilities and manufactured products from contamination or deterioration. Protection is both from the environment, including rain, wind, surface run-off, delivery and dispatch vehicles, dust, odours, pests and uninvited people, and internally from microbiological hazards (e.g. raw material cross-contamination), chemical (e.g. cleaning chemicals, lubricants) and physical hazards (e.g. from plantrooms, engineering workshops). Ideally, the factory buildings should be designed and constructed to suit the operations carried out in them and should not place constraints on the process or the equipment layout.

With respect to the external environment, while it is obvious that the factory cannot be a sealed box, openings to the structure must be controlled. There is also little legislation controlling the siting of food factories and what can be built around them. The responsibility, therefore, rests with the food manufacturer to ensure that any hazards (e.g. microorganisms from landfill sites or sewage works, or particulates from cement works, or smells from chemical works) are excluded via appropriate barriers. The following factors apply:

- The floor of the factory should ideally be at a different level from the ground outside. By preventing direct access into the factory at ground floor level, the entrance of contamination, e.g. soil (which is a source of environmental pathogens such as *Listeria* spp. and *Clostridia* spp.) and foreign bodies, particularly from vehicular traffic (forklift trucks, raw material delivery, etc.) is restricted.
- Openings should be kept to a minimum and exterior doors should not open directly into production areas. External doors should always be shut when not in use and if they have to be opened regularly, should be of a rapid opening and closing design.
- Plastic strips/curtains are acceptable in interior situations only as they are easily affected by weather. Where necessary, internal or external porches can

be provided with one door, usually the external door on an external porch, being solid, and the internal door being a flyscreen door; on an internal porch it would be the opposite configuration. Air jets directed over doorways, designed to maintain temperature differentials when chiller/freezer doors are opened, may have a limited effect on controlling pest access.

- The siting of factory openings should be designed with due consideration for prevailing environmental conditions, particularly wind direction and drainage falls.
- Wherever possible, buildings should be single storey or with varying headroom, featuring mezzanine floors to allow gravity flow of materials, where this is necessary. This prevents any movement of wastes or leaking product moving between floors.
- In addition, drainage systems have been observed to act as air distribution channels, allowing contaminated air movement between rooms. This can typically occur when the drains are little used and the water traps dry out.
- For many food manufacturers and retailers, glass is seen as the second major food hazard after pathogenic microorganisms. For this reason, glass should be avoided as a construction material (windows, inspection mirrors, instrument and clock faces, etc.). If used, e.g. as viewing windows to allow visitor or management observation, a glass register, detailing all types of glass used in the factory, and their location, should be composed.
- Windows should either be glazed with polycarbonate or laminated. Where opening windows are specifically used for ventilation (particularly in tropical areas), these must be screened and the screens be designed to withstand misuse or attempts to remove them. Flyscreens should be constructed of stainless steel mesh and be removable for cleaning.
- If a filtered air supply is required to processing areas and the supply will involve ducting, a minimum level of filtration of >90% of 5 μm particles is required, e.g. G4 or F5 filters (BS EN 779), to provide both suitably clean air and prevent dust accumulation in the ductwork.

Within the internal environment, most factories are segregated into food production areas (raw material storage, processing, final product storage and dispatch) and amenities (reception, offices, canteens, training rooms, engineering workshops, boiler houses, etc.). The prime reason for this is to clearly separate the food production processes from the other activities that the manufacturer must perform. This may be to control microbiological or foreign body hazards arising from the amenity functions, but is always undertaken to foster a 'you are now entering a food processing area' hygienic mentality in food operatives.

Food production areas are typically segregated into raw material intake, raw material storage, processing, packaging and final product warehouse and dispatch. In addition, the flow of ingredients and products is such that, in ideal conditions, raw materials enter at one end of the factory (dirty end) and are dispatched at the opposite end (clean end). Other good basic design principles given by Shapton and Shapton (1991) are:

- the flow of air and drainage should be away from 'clean' areas towards 'dirty' ones;
- the flow of discarded outer packaging materials should not cross, and should run counter to, the flow of either unwrapped ingredients or finished products.

The key differential between segregation barriers at this and the next level (high-care/high-risk areas), is that food operatives are freely able to move between the segregated areas without any personnel hygiene barriers (though hand-washing may be required in order to move between some areas).

While a range of ingredients is brought together for processing, they may need to be stored separately. Storage may be temperature orientated (ambient, chilled or frozen) or ingredient related, and separate stores may be required for fruit and vegetable, meat, fish dairy and dry ingredients. Other food ingredients, such as allergens, and non-ingredients, such as packaging, should also be stored separately. Segregation may also extend into the first stages of food processing, where for example the production of dry intermediate ingredients, e.g. pastry for pies, is separated from the production of the pie fillings. The degree of segregation for storage and processing of ingredients and intermediates is predominantly controlled by the exclusion of water, particularly in how they are cleaned, i.e.:

- Dry cleaning. This applies to areas where no cleaning liquids are used, only vacuum cleaners, brooms, brushes, etc. Although these areas are normally cleaned dry, occasionally they may be fully or partially wet cleaned, when limited amounts of water are used.
- Wet cleaning. This applies to areas where the entire room or zone is always cleaned wet. The contents (equipment, cable trays, ceilings, walls), are wet washed without restrictions on the amount of cleaning liquid used.

In addition to segregating dry areas from a requirement to exclude water, other areas may need to be segregated due to excessive use of water, which can lead to the formation of condensation and the generation of aerosols. Such areas include tray-washer and other cleaning areas.

The control of microorganisms within food processing areas can only adequately be controlled by inclusion of third level (high-care/high-risk) barriers. Other hazards, however, have to be managed at the second barrier level, particularly allergens. This is to prevent the possibility of accidental contamination of products not containing allergens (and particularly those products not labelled as 'may contain allergens') with allergens intended for use in other products. Ideally, manufacturers who manufacture allergenic and non-allergenic products should do so on separate sites such that there is no chance of cross-contamination from different ingredients. This issue has been debated by food manufacturers in both Europe and the USA with the conclusion that it is unlikely to be economically viable to process on separate sites. Segregation of allergenic components will have to be undertaken, therefore, within the same site.

As a preferred alternative to separate factories, it may be possible to segregate the whole process, from goods in through raw material storage and processing to primary packaging, on the same site. If this is not possible, segregation has to be undertaken by time, e.g. by manufacturing products containing non-allergens first and products containing allergens next. Thorough cleaning and disinfection are then undertaken before the manufacture of the products containing non-allergens is then re-commenced. If segregation by time is to be considered, a thorough HACCP study should be undertaken to consider all aspects of how the allergen is to be stored, transported, processed and packed, etc. This would include information on any dispersal of the allergen during processing (e.g. from weighing), the fate of the allergen through the process (will its allergenic attributes remain unchanged?), the degree to which the allergen is removed by cleaning and the effect of any dilution of residues remaining after cleaning in the subsequent product flow.

To a lesser extent and because it is not a safety issue, label declaration issues such as non-organic components in organic foods, genetically modified organism (GMO) components in GMO-free products, vegetarian foods with non-vegetarian components, and 'non-religion' processed components in religious-based foods (e.g. Kosher or Halal) have all caused food manufacturers to think about how raw materials are segregated. As for allergenic materials, segregation is usually by time and by the use of separate ingredient stores. Stores containing key components, e.g. meat in a factory producing vegetarian components, are often locked to prevent inadvertent use of these ingredients when not scheduled, and the locking and unlocking of such stores can be recorded in the quality system.

In the future, as techniques improve with respect to product authenticity testing, there may be the requirement to segregate legally defined components. For example, consider the case of a meat manufacturer producing beef and pork sausages. If he sold pork sausages with, e.g., 50% beef content, something has either gone wrong in the process or he is making false claims. If however, 0.5% beef content was found in his pork sausages, is this 'illegal' or is it that residues from the previous beef sausage run can now be detected in a subsequent pork sausage run? Because such low levels of a component can be detected, does the meat manufacturer now have to have segregated pork and beef sausage lines?

Other than for preventing product contamination, segregation within factories may be required for food operative health and safety reasons. This may be for protection against chemicals, such as the requirement for separate chemical stores, or for the protection from a particular process, e.g. the dosing of chlorine into a product washing system. The requirement for segregation and compartmentation of specific heat processes, e.g. ovens and fryers, or fire hazards such as bulk storage of oils and fats, has long been recognised in the food industry, and these areas are segregated with incombustible materials. Because of fires in chilled food factories, through the use of false ceilings, giving rise to large open spaces above processing areas that allowed the rapid (and unseen) spread of fires, compartmentation of this roof space is strongly recommended. In addition it may be necessary to segregate particularly noisy pieces of equipment; see *Reducing noise exposure in the food and drink industries*, Food Information Sheet No. 32, http://www.hsebooks.co.uk.

Finally, segregation is also now considered as a method of increasing manufacturing flexibility. For example, by splitting down large processing areas into smaller sub-units (e.g. a single 12 line meat slicing hall into three fully segregated sub-units of four slicing lines), cross-contamination between lines can be eliminated. This is particularly the case when some lines need to be shutdown for cleaning or maintenance while the others need to remain in production. Many large, multisite, international food manufacturers are also considering the layout and segregation of new and existing factories such that they are suitable for multiproduct food processing. This allows the manufacturer the flexibility to change the nature of the product produced at the factory within a short time period, to take advantage of ever-changing economic conditions.

8.4 Barrier 3: High-care/risk areas

The third barrier within a factory segregates an area in which food products are further manipulated or processed following a decontamination treatment. It is, therefore, an area into which a food product is moved after its microbiological content has been reduced.

Many names have been adopted for this third level processing area including 'clean room' (or '*salle blanche*' in France) following pharmaceutical terminology, 'high-hygiene', 'high-care' or 'high-risk' area. In some sectors, particularly chilled, ready-to-eat foods, manufacturers have also adopted opposing names to describe second barrier areas such as 'low risk' or 'low care'. Much of this terminology is confusing, particularly the concepts of 'low' areas which can imply to employees and other people that lower overall standards are acceptable in these areas where, for example, operations concerned with raw material reception, storage and initial preparation are undertaken. In practice, all operations concerned with food production should be carried out to the highest standard. Unsatisfactory practices in so-called low-risk areas may, indeed, put greater pressures on the 'barrier system' separating the second and third level processing areas.

To help clear this confusion, the Chilled Food Association in the UK (Anon, 1997) established guidelines to describe the hygiene status of chilled foods (based upon microbiological criteria) and indicate the area status of where they should be processed after any heat treatment. Three levels were described: high-risk area (HRA), high-care area (HCA) and good manufacturing practice (GMP) zones. Their definitions were as follows:

- HRA: an area to process components, *all* of which have been heat treated to $\geq$90 °C for 10 min (for psychrotrophic *Clostridium botulinum* spores) or $\geq$70 °C for 2 min (for vegetative pathogens), and in which there is a risk of contamination between heat treatment and pack sealing that may present a food safety hazard.

- HCA: an area to process components, *some* of which have been heat treated to ≥70 °C for 2 min, and in which there is a risk of contamination between heat treatment and pack sealing that may present a food safety hazard. In practice, the definition of HCA has been extended to include an area to further process components that have undergone a decontamination treatment, e.g. fruit and vegetables after washing in chlorinated water or fish after low temperature smoking and salting.
- GMP: an area to process components, *none* of which have been heat treated to ≥70 °C for 2 min, and in which there is a risk of contamination prior to pack sealing that may present a food safety hazard. In practice, GMP operations are carried out in the second barrier level of processing.

Many of the requirements for the design of HRA and HCA operations are the same, with the emphasis on *preventing* contamination in HRA and *minimising* contamination in HCA operations (Anon, 1997). In considering whether high risk or high care is required and, therefore, what specifications should be met, food manufacturers need to carefully consider their existing and future product ranges, the hazards and risks associated with them and possible developments in the near future. If budgets allow, it is always more economic to build to the highest standards from the onset of construction rather than try to retrofit or refurbish at a later stage.

The requirements for third barrier level high-care/risk segregation for appropriate foodstuffs is now recognised by the major food retailers worldwide and is a requirement in the *BRC Global Food Standard* (Anon, 2003) and the Global Food Safety Initiative, http://www.globalfoodsafety.com.

In general, high-care/risk areas should be as small as possible as their maintenance and control can be very expensive. If there is more than one high-care/risk area in a factory, they should be arranged together or linked as much as possible by closed corridors of the same class. This is to ensure that normal working procedures can be carried out with a minimum of different hygienic procedures applying.

Some food manufacturers design areas between the second 'low-risk' and third barrier 'high-risk' level zones and use these as transition areas. These are often termed 'medium-care' or 'medium-risk' areas. These areas are not separate areas in their own right as they are freely accessed from low risk without the need for the protective clothing and personnel hygiene barriers as required at the low/high-risk area interface. By restricting activities and access to the medium-risk area from low risk, however, these areas can be kept relatively 'clean' and thus restrict the level of microbiological contamination immediately adjacent to the third level barrier.

The building structure, facilities and practices associated with the high-care/risk (referred to simply as high risk in the following text) production and assembly areas provide the third barrier level. This barrier has been under constant development since the late 1980s/early 1990s as part of a three-fold philosophy designed to help reduce the incidence of pathogens, particularly *L.*

monocytogenes, in finished product and, at the same time, control other contamination sources. It was recognised that the major source of pathogens was likely to be the raw materials used in the low-risk area of the factory together with any pathogens that had entered low risk from soil associated with people or vehicular movements. To protect the product being further manipulated in the high-risk area from such pathogens, the philosophy is undertaken to:

- provide as many barriers as possible to prevent the entry of *Listeria* into the high-risk area;
- prevent the growth and spread of any *Listeria* penetrating these barriers during production;
- after production, employ a suitable sanitation system to ensure that all *Listeria* are removed from high risk prior to production recommencing.

Together with the building structure, the third level barrier is built up by the use of combinations of a number of separate components or sub-barriers, to control contamination that could enter high risk from the following routes:

- Structural defects.
- Product entering high risk via a heat process.
- Product entering high risk via a decontamination process. This may include product entering high risk that has been heat processed/decontaminated off-site but whose outer packaging may need decontaminating on entry to high risk.
- Other product transfer.
- Packaging materials.
- Liquid and solid waste materials.
- Food operatives, maintenance and cleaning personnel, etc., entering high risk.
- The air.
- Utensils that may have to be passed between low and high risk.

8.4.1 Structure

Structurally, creating a third barrier level can be described as creating a box within a box. In other words, the high-risk area is sealed on all sides to prevent microbial ingress. While this is an ideal situation, we still need openings to the box to allow access for people, ingredients and packaging and exit for finished product and wastes. Openings should be as few as possible, as small as possible (to better maintain an internal positive pressure) and should be controlled (and shut if possible) at all times. Similarly, the perimeter of the box should be inspected frequently to ensure that all joints are fully sealed.

The design of the high-risk food processing area must allow for the accommodation of five basic requirements, i.e.:

- processed materials and possibly some ingredients;
- processing equipment;

- staff concerned with the operation of such equipment;
- packaging materials;
- finished products.

There is a philosophy with considerable support that states that all other requirements should be considered as secondary to these five basic requirements and, wherever possible, should be kept out of the high-risk processing area. This aids in cleaning and disinfection and thus contamination control. These secondary requirements include:

- structural steel framework of the factory;
- service pipework for water, steam and compressed air; electrical conduits and trunking; artificial lighting units; and ventilation ducts;
- compressors, refrigeration units and pumps;
- maintenance personnel associated with any of these services.

8.4.2 Heat-treated product

Where a product heat treatment forms the barrier between low and high risk (e.g. an oven, fryer or microwave tunnel), the heating device must be designed such that as far as is possible, the device forms a solid, physical barrier between low and high risk. Where it is not physically possible to form a solid barrier, air spaces around the heating equipment should be minimised and the low/high-risk floor junction should be fully sealed to the highest possible height. Other points of particular concern for heating devices include the following:

- Heating devices be designed to load product on the low-risk side and unload in high risk.
- Good seals are required between the heating device surfaces, which cycle through expansion and contraction phases, and the barrier structure that has a different thermal expansion.
- Sealing is particularly critical at the floor level where ovens may sit on an open area or 'sump'. Sumps can collect debris and washing fluids from the oven operation which can facilitate the growth of *Listeria*, and these areas should be routinely cleaned (from low risk).
- Ovens should not drain directly into high risk. In addition, when being cleaned, cleaning should be undertaken in such a way that cleaning solutions do not flow from low to high risk.
- If oven racks of cooked product have to be transferred into high risk for unloading, these racks should be returned to low risk via the ovens, with an appropriate thermal disinfection cycle as appropriate.
- Any ventilation system in the cooking area should be designed so that the area is ventilated from low risk; ventilation from high risk can draw into high risk large quantities of low-risk air.
- Early installations of open cooking vessels (kettles) as barriers between low and high risk, together with (occasional) low level retaining or bund walls to prevent water movement across the floor and barriers at waist height to

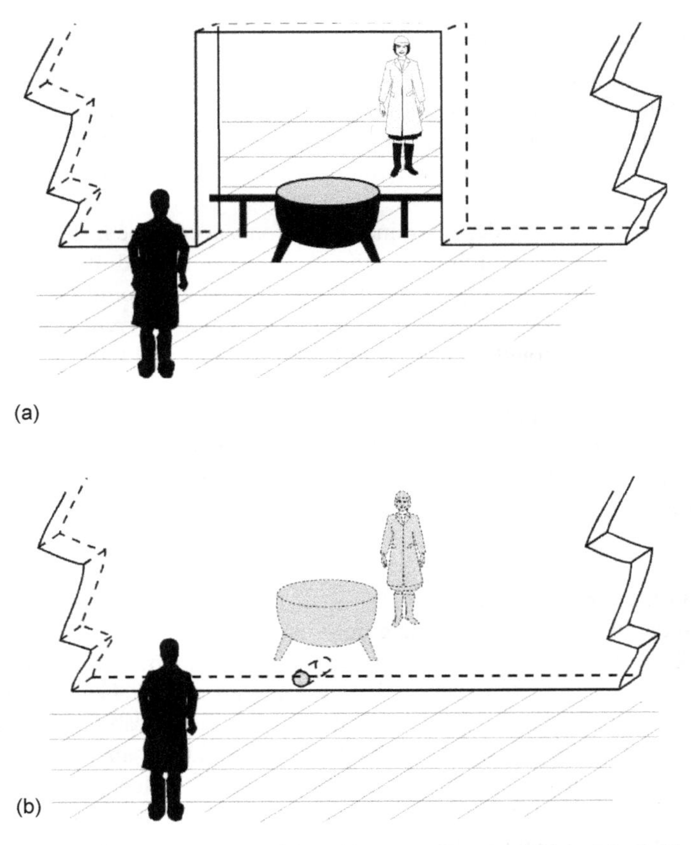

Fig. 8.2 (a) Schematic early low-risk (white-coated worker)/high-risk divide around kettles and (b) more acceptable schematic arrangement in which cooked product is gravity fed or pumped into high risk through pipework. The schematic shows the first approaches in which the kettle exit pipe was too close to the floor. In later, more acceptable, arrangements, the kettles were mounted on mezzanines.

prevent the movement of people, while innovative in their time, are now seen as hygiene hazards (Fig. 8.2a). It is virtually impossible to prevent the transfer of contamination, by people, the air and via cleaning, between low and high risk. It is now possible to install kettles within low risk and transfer cooked product (by pumping, gravity, vacuum, etc.) through into high risk via a pipe in the dividing wall (Fig. 8.2b). The kettles need to be positioned in low risk at a height such that the transfer into high risk is well above ground level (installations have been encountered where receiving vessels have had to be placed onto the floor to accept product transfer). Pipework connections through the walls should be cleaned from high risk such that potentially contaminated low-risk area cleaning fluids do not pass into high risk.

8.4.3 Product decontamination

Fresh produce and the outer packaging of various ingredients may need to be decontaminated on entry into high risk. Decontamination is undertaken using validated and controlled wet systems, using a washing process incorporating a disinfectant (usually a quaternary ammonium compound) or dry systems, using UV light.

As with heat barriers, decontamination systems need to be installed within the low/high-risk barrier to minimise the free space around them. As a very minimum, the gap around the decontamination system should be smaller than the product to be decontaminated. This ensures that all ingredients in high risk must have passed through the decontamination system and thus must have been decontaminated.

For companies that also have ovens with low-risk entrance and high-risk exit doors, it is also possible to transfer product from low to high risk via these ovens using a short steaming cycle that offers surface pasteurisation without 'cooking' the ingredients.

8.4.4 Other product transfer

All ingredients and product packaging must be de-boxed and transferred into high risk in a way that minimises the risk of cross-contamination into high risk. Some ingredients, such as bulk liquids that have been heat-treated or are inherently stable (e.g. oils or pasteurised dairy products), can be pumped across the low/high-risk barrier directly to the point of use. Dry, stable bulk ingredients (e.g. sugar) can also be transferred into high risk via sealed conveyors.

For non-bulk quantities, it is possible to open ingredients at the low/high-risk barrier and decant them through into high risk via a suitable transfer system (e.g. a simple funnel set into the wall), into a receiving container. Transfer systems should, preferably, be closable when not in use and should be designed to be cleaned and disinfected, from the high-risk side, prior to use as appropriate.

8.4.5 Packaging

Packaging materials (film reels, cartons, containers, trays, etc.) are best supplied to site 'double bagged'. When called for in high risk, the packaging material is brought to the low/high-risk barrier, the outer plastic bag removed and the inner bag and packaging enters into high risk through a suitable hatch.

The hatch, as with all openings in the low/high-risk barrier, should be as small as possible and should be closed when not in use. This is to reduce air flow through the hatch and thus reduce the air-flow requirements for the air-handling systems to maintain high-risk positive pressure. For some packaging materials, especially heavy film reels, it may be required to use a conveyor system for moving materials through the hatch. An opening door or, preferably, double door airlock should be used only if the use of a hatch is not technically possible, and suitable precautions must be taken to decontaminate the airlock after use.

8.4.6 Liquid and solid wastes

On no account should low-risk liquid or solid wastes be removed from the factory via high risk and attention is required to the procedures for removing high-risk wastes. The drainage system should flow in the reverse direction of production (i.e. from high to low risk) and whenever possible, backflow from low-risk to high-risk areas should be impossible. This is best achieved by having separate low- and high-risk drains running to a master collection drain with an air-break between each collector and master drain. The high-risk drains should enter the collection drain at a higher point than the low-risk drains, so that if flooding occurs, low-risk areas may flood first. The drainage system should also be designed such that drain access points that can be used for drain cleaning or unblocking (rodding) are outside high-risk areas. Solids must be separated from liquids as soon as possible, by screening, to avoid leaching and subsequent high effluent concentrations. Traps should be easily accessible, frequently emptied and preferably outside the processing area.

Solid wastes in bags should leave high risk in such a way that they minimise any potential cross-contamination with processed product and should, preferably, be routed in the reverse direction to the product. For small quantities of bagged waste, existing hatches should be used, e.g. the wrapped product exit hatches or the packaging materials entrance hatch, as additional hatches increase the risk of external contamination and put extra demands on the air-handling system. For waste collected in bins, it may be necessary to decant the waste through purpose-built, easily cleanable (from high risk), waste chutes that deposit directly into waste skips. Waste bins should be colour coded to differentiate them from other food containers and should be used only for waste.

8.4.7 Personnel

The high-risk changing room provides the only entry and exit point for personnel working in or visiting high risk and is designed and built to both house the necessary activities for personnel hygiene practices and minimise contamination from low risk. In practice, there are some variations in the layout of facilities of high-risk changing rooms. This is influenced by, for example, space availability, product throughput and type of products, which will affect the number of personnel to be accommodated and whether the changing room is a barrier between low- and high-risk operatives or between operatives arriving from outside the factory and high risk. Generally higher construction standards are required for low/high-risk barriers than outside/high-risk barriers because the level of potential contamination in low risk (from raw materials), both on the operatives' hands and in the environment, is likely to be higher.

A generic layout for a changing room should accommodate the following requirements:

- An area at the entrance to store outside or low-risk clothing. Lockers should have sloping tops.

- A barrier to divide low- and high-risk floors. This is a physical barrier such as a small wall (approximately 60 cm high), which allows floors to be cleaned on either side of the barrier without contamination by splashing, etc., between the two.
- Open lockers at the barrier to store low-risk footwear.
- A stand on which captive (remain in high risk), high-risk footwear is displayed/dried. Boot-baths and boot-washers are not recommended as a means of decontaminating footwear between low- and high-risk areas as they are not an effective means of microbial control. Essentially they do not remove all organic material from the treads and any pathogens within the organic material remaining are protected from any subsequent disinfectant action. In addition, boot-baths and boot-washers can both spread contamination via aerosols and water droplets that, in turn, can provide moisture for microbial growth on high-risk floors. The use of boot-washers in high risk should be used only to help control the risk of operatives slipping (if the floors are particularly slippery) by controlling food debris build-up in the treads of the boots.
- An area designed with suitable drainage for boot-washing operations. Research has shown (Taylor *et al.*, 2000) that manual cleaning (preferably during the cleaning shift) and industrial washing machines are satisfactory boot-washing methods.
- Hand-wash basins to service a single, hand-wash. Hand-wash basins must have automatic or knee/foot-operated water supplies, water supplied at a suitable temperature (to encourage hand-washing) and a waste extraction system piped directly to drain. It has been shown that hand-wash basins positioned at the entrance to high risk, which was the original high-risk design concept to allow visual monitoring of hand-wash compliance, may give rise to substantial aerosols of staphylococcal strains that can potentially contaminate the product.
- Suitable hand-drying equipment, e.g. paper towel dispensers or hot air dryers and, for paper towels, suitable towel disposal containers.
- Access for clean factory clothing and storage of soiled clothing. For larger operations this may be via an adjoining laundry room with interconnecting hatches.
- Interlocked doors or turnstiles are possible such that doors/barriers allow entrance to high risk only if a key stage, e.g. hand-washing has been undertaken and detected by a suitable sensor.
- CCT cameras as a potential monitor of hand-wash compliance.
- Alcoholic hand-rub dispensers positioned immediately inside the high-risk production area.

8.4.8 Air

The air is a potential source of pathogens and air intake into the high-risk area, and leakage from it, have to be controlled. Air can enter high risk via a purpose-built air-handling system or can enter into the area from external uncontrolled sources (e.g. low-risk production, packing, outside). For high-risk areas, the goal of the air-handling system is to supply suitably filtered fresh air, at the correct temperature and humidity, at a slight overpressure to prevent the ingress of external air sources, particularly from low-risk operations.

The cost of the air-handling systems is one of the major costs associated with the construction of a high-risk area and specialist advice should always be sought before embarking on an air-handling design and construction project. Following a suitable risk analysis, it may be concluded that the air-handling requirements for high-care areas may be less stringent, especially related to filtration levels and degree of overpressure. Once installed, any changes to the construction of the high-risk area (e.g. the rearrangement of walls, doors or openings) should be carefully considered as they will have a major impact on the air handling system.

Air quality standards for the food industry were reviewed by Brown (2005) and the design of the air-handling system should now consider the following issues:

- Filtration of air is a complex matter and requires a thorough understanding of filter types and installations. The choice of filter will be dictated by the degree of microbial and particle removal required (BS EN 779). For high-care applications, a series of filters is required to provide air to the desired standard and is usually made up of a G4/F5 panel or pocket filter followed by an F7-9 rigid cell filter. For some high-risk operations an H10 or H11 final filter may be desirable.
- The pressure differential between low and high risk should be between 5 and 15 pascals or, through openings, an air flow of $1.5\,m\,s^{-1}$ or greater may be required to ensure one-way flow is maintained. The desired pressure differential will increase as both the number and size of openings, and also the temperature differentials, between low and high risk increases. As a general rule, openings into high-risk areas should be as small and as few as possible. Generally 5–25 air changes per hour are sufficient to remove the heat load imposed by the processing environment (processes and people) and provide operatives with fresh air, though in a high-risk area with large hatches/doors that are frequently opened, up to 40 air changes per hour may be required.
- The requirements for positive pressure in high-care processing areas are less stringent and ceiling-mounted chillers together with additional air make-up may be acceptable.
- As well as recirculating temperature-controlled air, the system may need to be designed to dump air directly to waste during cleaning operations and to recirculate ambient or heated air after cleaning operations to increase environmental drying. With respect to drafts, the maximum air speed close to workers to minimise discomfort through 'wind-chill' should be $0.3 m\ s^{-1}$. This is typically achieved with airsocks, positioned directly over the product lines.
- UK government sponsored work at the Campden & Chorleywood Food Research Association (CCFRA) and the Silsoe Research Institute has

investigated the measurement of both air flows and airborne microbiological levels in actual food factories, from which computational fluid dynamics (CFD) models have been developed to predict air and particle (including microorganism) movements (Anon, 2001). This has allowed the design of air-handling systems that provide directional air. These move particles away from the source of contamination (washrooms, hatches, doors people, etc.), in a direction that does not compromise product safety.
- Relative humidity should typically be 60–70% to restrict microbial growth in the environment, increase the rate of equipment and environment drying after cleaning operations and provide operative comfort. Low humidities can cause drying of the product with associated weight and quality loss, while higher humidities maintain product quality but may give rise to drying and condensation problems that increase the opportunity for microbial survival and growth.
- If the high-risk area is to be chilled, there may be conflict between any national regulations on workroom temperatures and the desire to keep food products cold. To help solve this conflict a document *Guidance on achieving reasonable working temperatures and conditions during production of chilled foods* (Brown, 2000) was published, which extends the information provided in HSE Food Sheet No. 3 (Rev) *Workroom temperatures in places where food is handled*, www.hse.gov.uk/pubns/fis03.pdf.
- Air-handling systems should be installed such that they can be easily serviced and cleaned.

8.4.9 Utensils

Wherever possible, any equipment, utensils and tools, etc., used routinely within high risk, should remain in high risk. This may mean that requirements are made for the provision of storage areas or areas in which utensils can be maintained or cleaned. Typical examples include the following:

- The requirement for ingredient or product transfer containers (trays, bins, etc.) should be minimised, but where these are unavoidable they should remain within high risk and be cleaned and disinfected in a separate wash-room area.
- Similarly, any utensils (e.g. stirrers, spoons, ladles) or other non-fixed equipment (e.g. depositors or hoppers) used for the processing of the product should remain in high risk and be cleaned and disinfected in a separate wash-room area.
- A separate wash-room area should be created in which all within-production wet cleaning operations can be undertaken. The room should preferably be sited on an outside wall that facilitates air extraction and air make-up. An outside wall also allows external bulk storage of cleaning chemicals that can be directly dosed through the wall into the ring main system. The room should have its own drainage system that, in very wet operations, may include barrier drains at the entrance and exit to prevent water spread from

the area. The wash area should consist of a holding area for equipment, etc., awaiting cleaning, a cleaning area for manual or automatic cleaning (e.g. traywash) as appropriate, and a holding/drying area where equipment can be stored prior to use. These areas should be as segregated as possible.

- All cleaning equipment, including hand tools (brushes, squeegees, shovels, etc.) and larger equipment (pressure washers, floor scrubbers and automats, etc.) should remain in high risk and be colour coded to differentiate between high- and low-risk equipment if necessary. Special provision should be made for the storage of such equipment when not in use.
- Cleaning chemicals should preferably be piped into high risk via a ring main (which should be separate from the low-risk ring main). If this is not possible, cleaning chemicals should be stored in a purpose-built area.
- The most commonly used equipment service items and spares, etc., together with the necessary hand tools to undertake the service, should be stored in high risk. For certain operations, e.g. blade sharpening for meat slicers, specific engineering rooms may need to be constructed.
- Provision should be made in high risk for the storage of utensils that are used on an irregular basis but that are too large to pass through the low/high-risk barrier, e.g. stepladders for changing the air distribution socks.

8.5 Barrier 4: Finished product enclosure

The fourth barrier is product enclosure and has the objective of excluding contamination, particularly from microorganisms, from a commercially sterile product. The fourth barrier approach is essential for the production of aseptic foods, but is also being used for the production of some chilled, ready-to-eat foods. Product enclosure can be undertaken by physical segregation (a box within a box within a box) or by the use of highly filtered, directional air currents.

With respect to physical segregation, 'gloveboxes' offer the potential to fully enclose product with the ability to operate to aseptic conditions. Gloveboxes for the food industry work in the same way as gloveboxes for the medical, microbiological or pharmaceutical industry, in which the food is enclosed in a sealed space, totally protected from the outside environment, and manipulated through gloves sealed into an inspection window. They work best if the product is delivered to them in a pasteurised condition, is packed within the box and involves little manual manipulation. The more complicated the product manipulation, the more ingredients to be added, the faster the production line or the shorter the product run, the less flexible gloveboxes become. Operating on a batch basis, pre-disinfected gloveboxes give the potential for a temperature-controlled environment with a modified atmosphere if required (e.g. high CO_2, low O_2 or very high O_2 concentrations), which can be disinfected on-line by gaseous chemicals (e.g. ozone) or UV light.

Gloveboxes may also offer some protection in the future to foodstuffs identified by risk assessments as being particularly prone to bioterrorism.

Gloveboxes are only necessary, of course, if people are involved in the food production line. If robots undertook product manipulation, there would be less microbiological risk and the whole room could be temperature and atmospherically controlled!

Where the use of gloveboxes is impractical, partial enclosure of the product can be achieved by the use of localised, filtered air flows. The high-risk air-handling system provides control of airborne contamination external to high risk but provides only partial control of aerosols, generated from personnel, production and cleaning activities, in high risk. At best, it is possible to design an air-handling system that minimises the spread of contamination generated within high risk from directly moving over product. Localised airflows are thus designed to:

- Provide highly filtered (H11-12) air directly over or surrounding product, and its associated equipment. The air is generated into a box which has a top and sides that direct the air downwards, and a floor that collects the air and wastes or recycles it. In some cases the 'base' of the box may be missing and the air is directed to waste.
- Provide a degree of product isolation ranging from partial enclosure in tunnels to chilled conveyor wells, where the flow of the filtered air provides a barrier that resists the penetration of aerosol particles, some of which would contain viable microorganisms.

By chilling the air, it is possible to keep chilled product cold while operating the high-risk area at ambient conditions. Economically, it is also very expensive to cool the whole of the high-risk area down simply to maintain low product temperatures, thus localised chilling could both cut costs and enhance product safety. Even at the lowest level of product enclosure, localised air conveyor wells (Fig. 8.3), a 1–2 log reduction of microorganisms from the surrounding air can be demonstrated within the protected conveyor zone (Burfoot *et al.*, 2001).

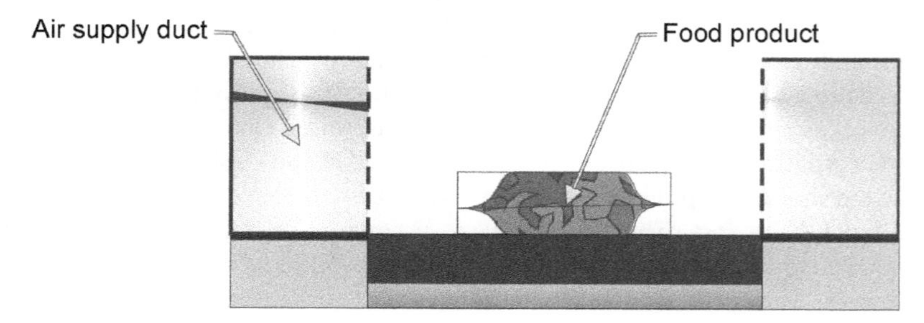

Fig. 8.3 Chilled air is supplied from air ducts on either side of a product conveyor. The chilled air retains the product temperature and its movement, spilling over the duct surfaces, provides a barrier to microorganism penetration.

8.6 References

ANON (1997) *Guidelines for good hygienic practices in the manufacture of chilled foods.* Chilled Food Association, London.

ANON (2001) *Best practice guidelines on air flows in high-care and high-risk areas.* Silsoe Research Institute, Silsoe, Bedford.

ANON (2003) *BRC Global Food Standard.* British Retail Consortium, London, www.brc.org.uk

BS EN 779 (1979) Particulate air filters for general ventilation. Requirements, testing, marking. British Standards Institute, London.

BROWN, K.L. (2000) *Guidance on achieving reasonable working temperatures and conditions during production of chilled foods.* Guideline No. 26, Campden & Chorleywood Food Research Association, Chipping Campden.

BROWN, K. L. (2005) *Guidelines on air quality for the food industry* 2nd edn. Guideline No. 12, Campden & Chorleywood Food Research Association, Chipping Campden.

BURFOOT, D., BROWN, K., REAVELL, S. and XU, Y. (2001) Improving food hygiene through localised air flows. *Proceedings International Congress on Engineering and Food, Volume 2*, April 2000, Puebla, Mexico. Technomic Publishing Co. Inc., Lancaster, Pennsylvania, pp. 1777–1781.

HOLAH, J. T. (2003) *Guidelines for the hygienic design, construction and layout of food processing factories.* Guideline No. 39, Campden & Chorleywood Food Research Association, Chipping Campden.

HOLAH, J.T. and THORPE, R. H. (2000) Hygienic design considerations for chilled food plants. In *Chilled Foods: a comprehensive guide*, 2nd edn. Eds. Mike Stringer & Colin Dennis, Woodhead Publishing Limited, Cambridge, pp. 397–428.

SHAPTON, D. A. and SHAPTON, N. F. (eds) (1991) *Principles and practices for the safe processing of foods.* Butterworth Heinemann, Oxford.

TAYLOR, J.H., HOLAH, J.T., WALKER, H. and KAUR, M. (2000) *Hand and footwear hygiene: an investigation to define best practice.* R&D Report No. 110, Campden & Chorleywood Food Research Association, Chipping Campden.

9

Improving the design of floors

B. Carpentier, Agence Française de Sécurité Sanitaire des Aliments, France

9.1 Introduction

We will here mainly consider the design of floors intended for greasy and/or wet food processing areas where they have to fulfil many requirements to be suitable. Because flooring materials are not food contact surfaces, some may consider that flooring materials are not of paramount importance to obtain the best microbial quality of food product. However, all cleaning systems disperse viable microorganisms in both water droplets and aerosols (Holah *et al.*, 1990), allowing microorganisms to reach food and food contact surfaces. As slipping is one of the main causes of accidents at work, flooring materials need to be rough. Add to these the fact that gravity carries most of the soiling and microorganisms down, and it can be seen why flooring materials are usually more microbiologically contaminated than other inert surfaces of food processing premises. Finally, floors are places where *Listeria monocytogenes* is very likely to be found (Cox *et al.*, 1989; Nelson, 1990). For all these reasons, great attention should be given to the choice and then to the application of a flooring material. The aim of this chapter is to give non-specialists some explanations of what the flooring materials for food processing premises are and to describe what properties are suitable for food processing areas.

9.2 What are floors made of?

9.2.1 The substrate

The material that supports flooring, called the substrate or the floor base, has a great impact on the quality of the flooring material. It is either an existing one, which has to be properly prepared, or a new one, which has to be properly constructed and prepared before applying the flooring in order to allow a good adherence of the latter. It must be dry (a concrete slab must be let to dry for a minimum of 28 days but this time may be far greater if climatic conditions are not optimal) and able to prevent humidity reaching the impervious flooring. It must be capable of withstanding all structural, thermal and mechanical stresses and loads that will occur during service and it must be sloped sufficiently in order for liquids to flow to the drains. This is recommended for resin-based floors as well as for ceramic tiles, even though, traditionally, ceramic tiles are applied on a flat substrate, the slope being given by the screed. Particular attention must be given to joints that are an integral part of the floor system. Nevertheless, it is not the purpose here to detail all the construction rules regarding the substrate. For further information see Timperley (2002).

9.2.2 Flooring

Two families of flooring materials are recommended for food processing areas: ceramic tiles and resin-based floors. Polyvinyl chloride (PVC) sheets are considered unsuitable because they are too easily worn. They can become cracked after the fall of a knife or other sharp object.

Ceramic tiles

Ceramic tiles are made of clay that after shaping is subjected to high temperature. They are manufactured products of constant quality and have been produced for centuries. Vitrified unglazed ceramic tiles are recommended for food processing areas. They are highly resistant to the main constraints that can be encountered in food processing premises, especially to heat shocks. The vitrified tiles can either be pressed (in which case they are usually square or hexagonal) or extruded (they are always rectangular). Dimension tolerances of the pressed tiles are better than those of extruded ones, allowing thinner joints.

Resin-based flooring materials

The first resin-based flooring, the acrylic cementitious systems, appeared in food industry premises during the 1960s and around two decades later synthetic resin flooring was also proposed, with the prospect of achieving a high standard of hygiene because those floorings are seamless. However, a high degree of technical skill is necessary to obtain *in situ* a good final product (only to be applied by a trained operative). As this has not always been respected, there have been many problems with such floors.

Resin-based floors are obtained by application of a mortar made of a mix of one or more organic or inorganic binders, aggregates, fillers and additives, and/or admixture, and can be classified according to the nature of the binder(s) used. There are two families of binders: synthetic resins and the hydraulic binders. The first ones are organic polymers comprising one or more components that react with a hardener at ambient temperature, whereas the hydraulic binders, as cement or lime, need water to harden.

Hydraulic binder

The hydraulic binder used in the construction of flooring material is cement. The main drawback of the use of hydraulic binders is the high porosity due to water evaporation during hardening. The addition of a synthetic polymer reduces porosity, increases the mechanical resistance and reduces cracking risk. Cement can be used in polymer-modified cementitious screed that is defined as a 'screed where the binder is a hydraulic cement and which is modified by the addition of polymer dispersion or re-dispersible powder polymer with a minimum content of dry polymer of 1% by mass of the total composition, excluding aggregate particles larger than 5 mm' (Anon, 2001a). Examples are the acrylic-modified cementitious systems that are the main systems used in the meat industry in France. The main and great advantage of such floorings is that they can be applied onto a damp substrate.

Cement is also used in association with resins, such as epoxy resin and polyurethane. In those cases, resin content is around 5% by mass of the total composition. In such floors, cement is more a filler than a binder, which is why they are considered to belong to the synthetic resins family.

Synthetic resins

Epoxy resins are the more frequently used synthetic binders, followed by polyurethane and methacrylate resins. Polyester resins are seldom used, to the author's knowledge, in the food industry. Characteristics of the different resins change according to the formulation used. The formulation may be changed to adapt to such non-optimal installation conditions as temperature, relative humidity or time available prior to being put into service. Specific formulations proposed may have consequences on the resistance of the final product. It is therefore difficult to give precise rules on curing. The final floor system must be allowed to cure according to the manufacturer's instructions. These generally require 1–3 days at 15–20 °C before trafficking and 3–7 days before washing, before contact with chemicals or before any ponding tests and high traffic loads (Anon, 2001b). The only resin that clearly escapes this general rule is polymethylmethacrylate (PMMA), also called methacrylate or methylmethacrylate. It is characterised by a very short time prior to putting in service: 2 hours. It can be applied at low temperature (−10 °C). However, this resin possesses a strong odour at installation that can irreversibly alter food products present nearby.

The climate above the uncured resin should be maintained at least 3 °C above the dew point. The substrate humidity must also be correct. It must, for instance, be smaller than 3% for an epoxy resin, or 7% for polyurethane–cement flooring.

Aggregates

Aggregates are granular materials that do not contribute to the hardening reaction of the mortar. Roles of aggregate depend on their size and abrasion resistance. Small aggregates that may be called fillers have many roles. Among those are reducing shrinkage and increasing the mechanical resistance. Such aggregates are often made of sand with high silica content (SiO_2 or quartz, 7 on the Mohs scale). Hard aggregates are used to increase resistance to abrasion. Those may be aluminium oxide (Al_2O_3 or corundum, 9 on the Mohs scale), silicon carbide (SiC, commercial name carborundum, 9.5 on the Mohs scale). It may happen that hard aggregates lead to an accelerated wear of shoes and of brushes used for cleaning. Large aggregates are used to increase slip resistance.

Primer
A primer (one or two coats) is most generally used to aid the adhesion of the final flooring and to seal and consolidate the surface of a porous substrate. It consists of a liquid product, which is often a solvent-based epoxy, applied to a substrate.

Coats
Anti-slip resin-based floorings can be obtained by one-coat or multicoat systems. Multicoat systems that are the more frequently proposed are thin flooring (2–5 mm) made of a self-levelling mortar on which large aggregates are sprinkled. One or two coats and then one or two finishing coats can be applied. These finishing coats are very thin, which is why they have a poor durability (1 or 2 years). Such finishing coats can be interesting when they fill the bubble gas holes but their role should not be to maintain large aggregates necessary to slip resistance.

One-coat systems, also called monolithic systems, are made of mortar in which all the aggregates are mixed prior to application. The maximum diameter of the aggregates must be smaller than the third of the flooring thickness (Pollet, 2000). They are thicker (4–12 mm) and the large aggregates necessary to obtain slip resistance are often better maintained.

Gas removal
Gas bubble holes are highly undesirable for hygienic considerations (see below). Surface-active agents can be added to avoid or decrease their formation during polymerisation of resin-based floors. There is also prickle roller, called *hérisson* in French (hedgehog), that is used when the mortar is in the fresh state to release entrapped gas bubbles. In some flooring, such as polyurethane–cement, it is very difficult to remove gas bubbles and to prevent their formation. One important measure is not to apply such flooring when room temperature is increasing.

9.2.3 Jointing

Ceramic tile jointing

A jointing material should completely fill the gap between to ceramic tiles right up to the top edge of the tile as shown on Fig. 9.1. This is often not done

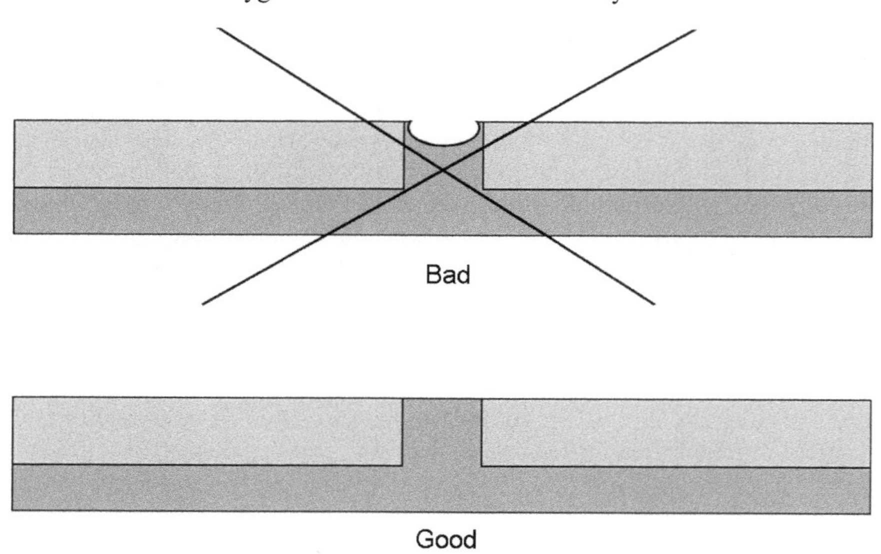

Fig. 9.1 Unhygienic and hygienic jointing.

although it is technically possible. The gap must be as small as possible and applicators must wait (for instance around 1 to 2 hours at 18–20 °C for an epoxy grouting), so that the grout has begun to cure, before the first cleaning of the floor. The jointing material has to absorb dimensional variations of tiles. That is why the better tolerance dimensions of pressed tiles, allow a joint around 5–7 mm wide instead of 6–10 mm for extruded tiles. The smallest joints may be obtained when ceramic tiles are laid in a synthetic resin bed and then subjected to vibration. However, it is not advisable to have joints smaller than 5 mm because it will be impossible to fill them down to the bottom, as tiles for industrial purpose are thick.

A high diversity of grouting products is available (epoxy, vinyl ester, etc.) and descriptions of all of them are not possible here. The choice of system is governed by the chemical stresses expected on the floor surface. Cement grouts are not suitable for food processing areas because they are highly porous, acid sensitive and have a poor durability when subjected to mechanical stresses. In addition, a simple epoxy grouting will not resist the acidic conditions of some food factories such as dairy factories and an anti-acid grouting must be chosen.

Other joints

Among all the joints necessary in the construction of floors are: construction or day joints, expansion joints, movement joints, and isolation joints. All the joints in the subfloor or floor base should be carried through overlay material and filled with a suitable sealant. Joint fillers have to be flexible and are therefore not as capable of withstanding heavy loads or aggressive chemical as the adjacent floor finish. They must be changed when worn.

9.3 Requirements for flooring materials

9.3.1 Slip resistance

In France and the UK, with little variation from year to year, around 20% of all workplace injuries leading to working days lost are caused by falls on the same level (slips, trips, etc.). This cause of injuries ranks second after accidents during manual handling. These injuries are also responsible for 20% of all working days lost, 20% of accidents leading to a permanent incapacity to work and 2% of the fatal injuries (Leclercq and Tissot, 2004). A high risk of slipping exists in the food industry because wet and/or greasy floors are frequent, especially where meat is processed. In slaughterhouses, slipping is the cause of 16% of occupational accidents. In order to decrease slipping accidents, anti-slip floors are necessary. Wearing anti-slip footwear is also necessary but not sufficient and, as for all risks at work, collective measures against accidents must always be taken before individual measures. Unfortunately, anti-slip properties of floors are obtained by increasing surface roughness, while smooth flooring materials, supposed to be the more cleanable ones, are therefore not appropriate. By contrast, efficient cleaning of the floors is necessary to decrease both their slipperiness and their microbial load.

Regulation

According to the European Directive 89/391/EEC employers are responsible for implementing a process of prevention of accidents and other work-related health problems based on nine principles. This process of prevention is based on a hazard assessment. The hazards that cannot be avoided must be evaluated (principle 2), and must be combated at source (principle 3). Collective protective measures must be taken before individual protective measures (principle 8). Appropriate information must be given to employees (principle 9). For instance, an effective cleaning procedure to remove greasy soil and to obtain the correct durability of the floors must be known by employees. Employees must also be instructed not to run.

Surface texture

There is a general awareness that smooth floor surfaces are slippery, especially when wet and/or greasy, and that rougher surfaces are safer, but it is only in the past two decades that scientific research has been conducted on the impact of roughness on underfoot friction (Chang *et al.*, 2001). Grönqvist *et al.* (1990, 1992) proposed three roughness factors that seemed to determine the anti-slip resistance of contaminated floors and footwear: (1) the macroscopic structure (e.g. profile asperities); (2) the microscopic roughness (e.g. R_a, the arithmetic mean roughness) and (3) the microscopic porosity of the floor. Harris and Shaw (1988) from the Health and Safety Executive (UK) proposed the R_z (previously called R_{TM}) which is the average of the single peak-to-valley heights of five adjoining sampling lengths. R_z can be measured by a portable and inexpensive profilometer. For this reason, it is appreciated for measurement of roughness in the field (Chang *et al.*, 2001). More recently, the roughness peak height, also called mean levelling depth, R_{pm}, which is the mean value of the levelling depths of five consecutive sampling lengths, became the preferred roughness parameter of the Health and Safety Executive (Anon, 1999) and of Chang and Matz (2000), cited by Chang *et al.* (2001).

9.3.2 Hygiene

As written in the introduction paragraph, flooring materials may be a reservoir of microorganisms. The pathogen *Listeria monocytogenes*, which is frequently found on floors, can even become persistent in food industry premises (Giovannacci *et al.*, 1999; Miettinen *et al.*, 1999; Chasseignaux *et al.* 2001). Lawrence and Gilmour (1995), using RAPD (random amplified polymorphic DNA) and multilocus enzyme electrophoresis as typing methods, found two coexisting *L. monocytogenes* types widespread on food contact surfaces, floors and drains during an extended period. These bacterial types were also found in the cooked poultry products for at least one year. This highlights the potential for persistent strains to cross-contaminate processed foods.

Regulation

In the European Directive 93/43/EEC, it is stipulated that

> floor surfaces must be maintained in a sound condition and they must be easy to clean and, where necessary, disinfect. This will require the use of impervious, non absorbent, washable and non-toxic materials and require a smooth surface up to a height appropriate for operations unless business operators can satisfy the competent authority that other materials used are appropriate.

European regulations will replace the European Directive 93/43/EEC and many sector-specific Directives on foods of animal origin. They will be applicable as of 1 January 2006. The text concerning floor in the 852/2004 regulation is very similar:

> floor surfaces are to be maintained in a sound condition and they must be easy to clean and, where necessary, disinfect. This will require the use of impervious, non-absorbent, washable and non-toxic materials unless food business operators can satisfy the competent authority that other materials used are appropriate. Where appropriate, floors are to allow adequate surface drainage.

Surface texture

The study by Mettler and Carpentier (1999) on the impact of surface texture on the hygienic properties of flooring materials showed that gas bubble holes, which are frequently found on resin-based flooring materials, are not cleanable (Fig. 9.2). Indeed a smooth polyurethane-based flooring material containing

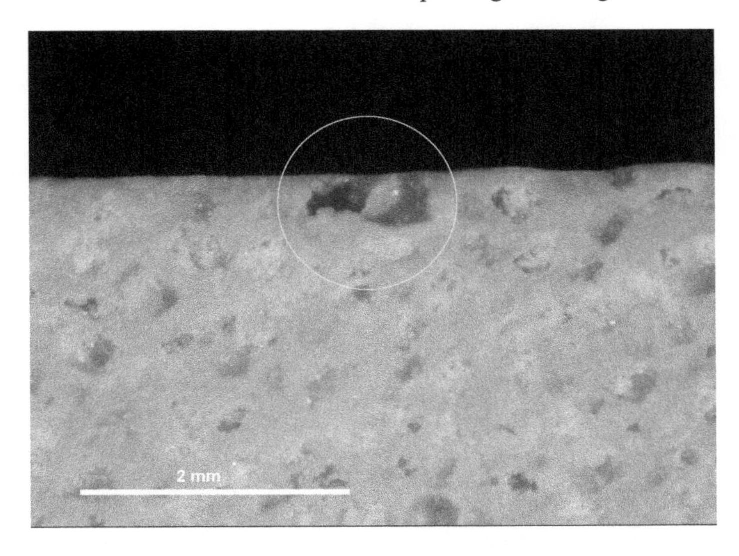

Fig. 9.2 Cross-section of a resin-based flooring material showing gas bubble holes.

many gas bubble holes appears to be, after cleaning, the most contaminated material of those having been inserted in the floor of a cheese-processing site. Furthermore, a significant linear correlation was observed between the number of spherical holes and the cleanability as assessed by a laboratory test. Masurovsky and Jordan (1958) and Holah and Thorpe (1990) have also observed that surfaces which at first glance appeared to have a readily cleanable, smooth surface but were, however, very difficult to clean, were precisely characterised by the presence of small holes when surfaces were examined in more detail. These crevices do not give any slip resistance and are therefore very undesirable. The easiest way to detect such a defect is to observe the material under a stereomicroscope (Fig. 9.3) at ×40 magnification. The observation allows also seeing cracks often found around aggregates, holes left by removed aggregates, 'spongy' aggregates, very deep crevices and other texture defaults. Under the stereomicroscope, it is also interesting to test with a simple metallic point the anchorage of the aggregates. If they are easily removed by manual handling of the metallic point, it means that they will be easily removed when subjected to the 'in-house' mechanical stresses. Such flooring will not maintain their slip resistance and will be difficult to clean. Around 50% of the flooring materials (resin-based or ceramic tiles) received at our laboratory presented such obvious texture defaults visible under the stereomicroscope. Unfortunately, observations are not measurements and in some case, there may be some difficulty in interpretation. In order to reject or accept a flooring material for a food processing area, it is proposed that the observation should be done by at least three trained persons, who should all reach the same conclusion.

Roughness measurement at the microscopic level could be a way to further characterise material cleanability. Mettler and Carpentier (1999) explored

Fig. 9.3 Stereomicroscope for viewing texture defaults.

different roughness parameters at two cut-off wavelengths (2.5 and 0.8 mm) and looked for correlations between those parameters or combinations of parameters and the contaminations that remained after cleaning of six flooring materials inserted for four weeks in a cheese processing site. Asperities taken into account by parameters calculated with the cut-off wavelength of 2.5 mm gave lower correlations, suggesting that there is a threshold value for the diameter of asperities under which the soil is not removed by the mechanical action of the hygiene procedure. This corroborates the finding of Taylor and Holah (1996), who observed that the gross topographic irregularities of floor were not responsible for their cleanability performance. The Mettler and Carpentier study showed that R_{vk}, the reduced valley depth that characterises the depth of the inwardly directed portion of the surface profile, was a better parameter than R_a. As slip resistance is supposed to be linked with other roughness parameters, it should be possible to select cleanable materials with high slip resistance.

Connection between floors and walls
In the past, it was mandatory in France to have a rounded angle between floors and walls. For this purpose, rounded angle pieces were sometime just stuck at the connection between floors and walls. Most of the time humidity and soil were able to reach the back part of such pieces, which rapidly became highly contaminated. Now it is necessary to have cleanable sealed junctions between floors and walls. For resin-based flooring or polymer cementitious systems, it is possible to construct a rounded angle with the same material as the one used for the flooring. In addition, ceramic rounded baseboards may be used; they must be sealed in the wall and in the floor in the same way tiles are sealed, using a correct jointing material.

Cleaning and disinfection of floors
Efficient and frequent cleaning and disinfection operations are necessary to prevent microbial contamination reaching a high level. Between two hygiene operations, the floors should be, if possible, maintained in a dry state. Cold is a common way to decrease growth rate of bacteria, but dryness is also a good measure. All the ways used to decrease relative humidity, water spills, water drop and condensation are good. For instance, flooring must be sloped properly to allow water to drain out; the slopes of the floors must be around 1.5–2% depending on the length of fall. It is also advisable to have cleanable stainless steel drains in the middle of the rooms, so the distance for water to drain is short.

The water used for the cleaning and disinfection operations may be highly contaminated with, for instance, *Pseudomonas* species. Microorganisms transported from inert surfaces of the food processing area may contaminate ends of the ducts used for cleaning (spray nozzle, hose). It has been shown that a *Pseudomonas putida* organism was able to spread by 40 cm within 8 days, i.e. 5 cm per day, upwards in a fixed and straight up duct (Gagnière *et al.*, 2004). To prevent such a contamination the end of the water ducts should be immersed in a disinfectant solution between two hygiene operations.

Floor and jointing material suppliers must give information to the end-users about the compatibility of their products with detergents and disinfectants. For instance, acrylic modified cementitious systems do not tolerate any acidic products. Acidic products attack those floorings so that they return to the colour of the new ones but with formation of crevices that are uncleanable. It is also necessary to have information on the compatibility between chemical products that may accidentally spill on the floor, such as peracetic acid or hydrogen peroxide, which are highly corrosive products.

As flooring materials must be rough to increase their slip resistance they are, of course, not as easy to clean as really smooth materials. An efficient mechanical action is necessary. Using a squeegee cannot be considered as a mechanical action; furthermore, squeegees may be highly contaminated and if they are used to remove excess water, they must be immersed in a disinfecting solution after use. Scrubber brushes or pressure water jets (the use of the latter is no longer recommended because they produce more contaminated aerosols and water droplets than scrubber brushes) are necessary. Of course, brushes must be clean before use and be soft enough not to damage the floors.

9.3.3 Performance requirements

It is necessary to have a durable floor adapted to the effective service characteristics of the food processing area. The main constraints are: mechanical shocks, heavy weight, pressure jet, chemical agents, thermal shocks, wear, shifting and rolling. Falls of heavy objects, knives and other sharp objects may lead to cracks in the floor. Cracks allow water to infiltrate under the floor, which will progressively detach from the substrate. A high thickness of the floor is necessary to reach a good resistance to mechanical shocks and to thermal shocks. The smallest acceptable thickness depends of the size of the aggregates. For a resin-based floor, 3 mm appears to be a minimum but a thickness of 5 mm is strongly recommended. It is of course difficult to know the thickness of a resin-based floor when the application is finished. That is why it is strongly recommended to check whether the right quantity of ingredients has been used by counting the number of bags used at the end of the application. In the case of litigation, core borings may be performed to check the minimum and mean thickness announced. Ceramic tile resistance to thermal and mechanical shocks is higher when the ratio area/thickness is small. The thickness of ceramic tiles adapted for food processing areas ranges from around 8.5 to 20 mm, but for industrial premises with high traffic load a minimum of 12 mm is recommended.

Chemical constraints are essentially linked to the food processed and to the cleaning and disinfecting products used on the flooring materials, but also to the equipment. Sugar, butter, whey and milk products, blood and urine are substrates for microorganisms present on flooring materials. Cleaning and disinfection do not remove or inactivate all microorganisms (Mettler and Carpentier, 1998). Their metabolism leads to the formation of very aggressive acid, e.g. lactic acid formed by lactic acid bacteria. Flooring materials also have to withstand cleaning products such as alkaline, chlorinated-alkaline and acid products (when mineral soil has to be removed) and disinfectants. Peracetic acid and hydrogen peroxide, frequently used to disinfect equipment, are very corrosive and can accidentally spill on the floor.

9.4 Test methods

9.4.1 Slip resistance

'Over 70 machines have been invented to measure slip resistance (Strandberg, 1985), none of them accurately represents the motion of a human foot and at present, there is no generally accepted method of measuring slipperiness' (Chang *et al.*, 2001). The diversity of methods used at present leads to different, sometimes contradictory, floor classifications (Tisserand *et al.*, 1995). This is why it is so difficult to select a method to produce a European standard. It is well recognised that to prevent slipping accidents it is necessary to use two complementary methods for assessment of the slip-resistance of floors in the laboratory and in the field (Leclercq *et al.*, 1994).

Among the numerous laboratory methods to compare new surfaces are the ramp test and tests based on the evaluation of a coefficient of dynamic friction between an oiled surface and an elastomer (as the one chosen by the French National Institute of Research and Safety). The ramp test (German standards) is conducted by two people. They each in turn face downhill with an upright posture, and walk forwards and backwards on the floor surface. During the test, the test person gradually increases the gradient of the floor surface until an (acceptance) angle is reached where the test person either slips or becomes so insecure as to refuse to continue the test. To assess slip resistance in the field where surfaces are often worn and soiled, the French National Institute of Research and Safety uses a portable device developed in Sweden by Ohlsson, called the portable friction tester (PFT). It is based on the continuous evaluation of a coefficient of dynamic friction over a variable distance between the surface to be tested and an elastomer.

9.4.2 Hygiene

The European Hygiene Engineering and Design Group (EHEDG) Tests Method Subgroup is about to produce a guideline document on the testing of the hygienic qualities of flooring materials. Two test methods are proposed: a surface water absorption test and a cleanability test.

The surface water absorption test is based on a test derived from the National Swedish Institute for Materials Testing, Test Regulation CP-BM-2/67-2 (Determination of water transmission under pressure) as described by Taylor and Holah (1996). The method involves sealing a container onto the floor sample and filling it with water to a depth of 10 cm. After 24 h the level of water is examined to see if there has been any absorption into the surface. Ten samples are assessed and all should pass the test (zero absorption) for the surface material to be deemed suitable for use in food factories with respect to water absorption. This test is designed to assess the uptake of large quantities of water (millilitres) into very porous materials. It is not intended to assess the small water absorption of the whole material. This is measured by weighting floor test plates (resin-based) or whole ceramic tiles before and after having been immersed in boiling water for 2 hours.

The second test, a cleanability test, is based on the method used by Mettler and Carpentier (1999). Results of the latter study showed that contamination after cleaning of flooring materials inserted in the floor of a cheese site was linked to their cleanability and not to their disinfectability. That is why a tracer of the microbial soil (a biofilm), spores of *Geobacillus stearothermophilus*, which are not sensitive to alkaline cleaning products, are used to assess the removal of the biofilm. One-day biofilms of *Pseudomonas fluorescens* containing spores of *Geobacillus stearothermophilus* are developed on test

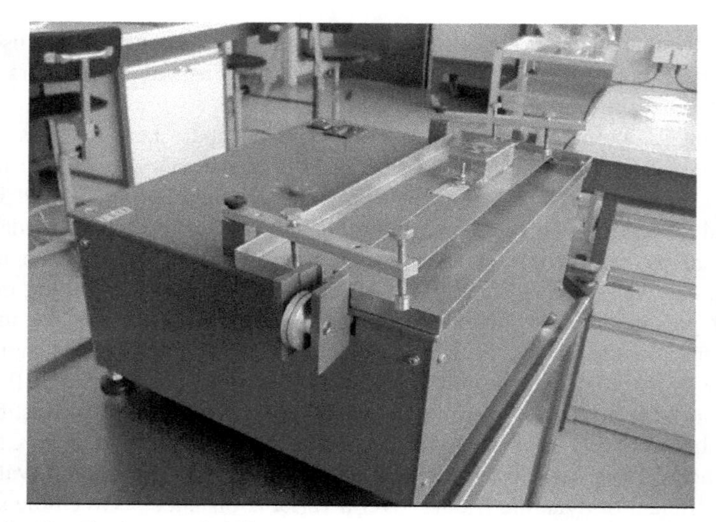

Fig. 9.4 The Gardner washability machine adapted to perform a cleanability test on $4 \times 2.5\,cm^2$ test plates.

plates and subjected to a mild cleaning. The latter consists of a submersion in a 0.01M solution of NaOH followed by a mechanical action provided by 37 reciprocal movements of a scouring pad moved over the plate surface by a Gardner washability machine (Fig. 9.4). After rinsing, residual spores are detached by sonication and counted after growth on Shapton and Hindes agar. Based on the residual spores counts on the test plates and on plates of the control material, flooring materials are classified as more, as or less cleanable than the control material. This test is not intended to accept or reject a flooring material. Only the presence of crevices, cracks or gas bubble holes, etc., observed under a stereomicroscope is considered a criterion to reject a flooring material.

9.4.3 Material resistance

European standardised test methods in accordance with the Construction Products Directive 89/106 EEC have been produced by the CEN committees 'Floor screeds and *in-situ* floorings in buildings' (TC 303) and 'Ceramic tiles' (TC 67) to assess the flooring products or the system's performance.

9.5 Construction of floors

The recommendations needed for the construction of a floor are not provided. For this subject see the sources of further information below. It can be very useful to use questionnaires to check for all the points that need to be examined before choosing a flooring material and before beginning construction. Such questionnaires can be found in the French 'Guide des revêtements de sol' from the Caisse nationale d'assurance maladie des travailleurs salariés (Liot *et al.*, 1998) or in the British 'Guidelines for the design and construction of floors for food production area' (Timperley, 2002).

The time necessary for the construction of a floor encompasses preparation of the subfloor, time for application and time to full cure. End-users often ask for a rapid construction and particularly for a possibility of rapid repair or rapid refurbishment. This decreases the choice of flooring material but, unfortunately, some applicators reduce the construction time at the expense of the quality. In addition, for most resin-based floors, the right relative humidity, temperature and humidity of the substrate are to be respected to obtain a good final product. That is why it is of prime importance to choose a qualified company with trained employees.

9.6 Future trends

Two of the most important future trends should be a better respect of the state of the art when applying flooring materials and the systematic checking to ensure the absence of gas bubble holes and other texture defaults.

The suppliers of resin-based flooring systems are continuously innovating to formulate improved products. However, those new formulations are trade secrets. The AFFAR (the French association of formulators and applicators of resin-based floors) has announced new resin-based floors able to withstand high temperature (more than 100 °C) and others that are able to adhere to wet substrate with short curing times.

Among other possible trends is antimicrobial flooring. Although some few suppliers propose antimicrobial resin-based floorings, no antimicrobial effect has ever been demonstrated in such floorings, to our knowledge. Only PVC floorings, which are not considered suitable for food processing areas (see above), and which all contain an antimicrobial product because PVC is a substrate for fungal microorganisms, have a proven antimicrobial effect (Carpentier, unpublished results). Anyway, if antimicrobial resin-based flooring could ever exist, it would have to be cleaned and disinfected as other floors. Indeed, antimicrobial material may reduce microbial contamination when wet, but cleaning and disinfection allow for further decrease of surface microbial contamination.

The application of ceramic tiles in resin bed is increasing, especially in the dairy industry. It allows withstanding higher variation of temperature, humidity, etc. and smaller joints.

9.7 Sources of further information and advice

9.7.1 Slip-resistance and accidents at work

- European Agency for Safety and Health at Work: http://europe.osha.eu.int/. This website provides statistics of accidents at work in the EU.

- Health and Safety Executive (UK): a paper from Richard Morgan giving the priorities in the food and drink industry may be downloaded from http://www.east-anglian-fishnet.org.uk/docs/oct99.rtf
- An information sheet updates HSE booklet 'Slips and trips – Guidance for the food processing industry' with consideration of the roughness parameter R_{pm} impact on slip resistance of flooring materials: Food sheet no. 22: http://www.hse.gov.uk/pubns/fis22.pdf

9.7.2 Resin-based flooring

- EFNARC (European Federation of producers and applicators of Specialist Products for Structures) was founded in March 1989 as the European Federation of national trade associations representing producers and applicators of specialist building products. Two interesting documents are available at their website: 'Specification and guidelines for synthetic resin flooring' and 'Specification and guidelines for polymer-modified cementitious flooring as wearing surfaces for industrial and commercial use'. A free downloadable pdf copy of those guides is available from http://www.efnarc.org/efnarc/publications.htm
- A French written document of Pollet (2000) from the Scientific and Technical Center for Construction (CSTC) (http://www.cstc.be, website in French and Flemish, English is in preparation): 'Les sols industriels à base de résine réactive' (technical note 216) give some information on resin-based floors, description of tests to assess their performances, recommendations on joints and construction and help in choosing an industrial flooring system.
- AFFAR, the French association of formulators and applicators of resin-based floors has a website (French only) http://www.affar.asso.fr with information on the new technologies, help in choosing the right resin-based flooring, etc.

9.8 References

ANON (1999), 'Preventing slips in the food and drink industries-technical update on floors specifications', Health and Safety Executive information sheet: Food sheet no. 22. HSE Books, Sudbury.

ANON (2001a), 'Specification and guidelines for polymer-modified cementitious flooring as wearing surfaces for industrial and commercial use', EFNARC (European Federation of producers and applicators of Specialist Products for Structures), Farnham.

ANON (2001b), 'Specification and guidelines for synthetic resin flooring', EFNARC (European Federation of producers and applicators of Specialist Products for Structures), Farnham.

CHANG W-R and MATZ S (2000), 'The effect of filtering processes on surface roughness parameters and their correlation with the measured friction, Part I: Quarry tiles', *Safety Sciences*, **36**, 19–33.

CHANG W-R, KIM I-J, MANNING D P and BUNTERNGCHIT Y (2001), 'The role of surface roughness in the measurement of slipperiness', *Ergonomics*, **44**, 1200–1216.

CHASSEIGNAUX E, TOQUIN MT, RAGIMBEAU C, SALVAT G, COLIN P and ERMEL G (2001), 'Molecular epidemiology of *Listeria monocytogenes* isolates collected from the environment, raw meat and raw products in two poultry- and pork-processing plants', *Journal of Applied Microbiology,* **91**, 888–899.

COX LJ, KLEISS T, CORDIER JL, CORDELLANA C, KONKEL P, PEDRAZZINI C, BEURNER R and SIEBENGA A (1989), '*Listeria* spp. in food processing, non food and domestic environments', *Food Microbiology,* **6**, 49–61.

GAGNIÈRE S, AUVRAY F. and CARPENTIER B (2004), 'Vitesse de progression d'un *Pseudomonas putida* Gfp+ dans une conduite contaminée par un aérosol', Poster presented at the National Congress of the French Society for Microbiology, Bordeaux (France), 10–12 May.

GIOVANNACCI I, RAGIMBEAU C, QUEGUINER S, SALVAT G, VENDEUVRE JL, CARLIER V and ERMEL G (1999), '*Listeria monocytogenes* in pork slaughtering and cutting plants use of RAPD, PFGE and PCR-REA for tracing and molecular epidemiology', *International Journal of Food Microbiology,* **53**, 127–140.

GRÖNQVIST, R, ROINE, J, KORHONEN, E and RAHIKAINEN, A (1990), 'Slip resistance versus surface roughness of deck and other underfoot surfaces in ships', *Journal of Occupational Accidents*, **13**, 291–302.

GRÖNQVIST, R, HIRVONEN, M and SKYTT, E (1992), 'Countermeasures against floor slipperiness in the food industry', *Advances in Industrial Ergonomics and Safety,* **IV**, 989–996.

HARRIS, GW and SHAW, SR (1988), 'Slip resistance of floors: users' opinions, Tortus instrument readings and roughness measurement', *Journal of Occupational Accidents*, **9**, 287–298.

HOLAH JT and THORPE RH (1990), 'Cleanability in relation to bacterial retention on unused and abraded domestic sink materials', *Journal of Applied Bacteriology,* **69**, 599–608.

HOLAH JT, TIMPERLEY AW and HOLDER JS (1990), 'The spread of *Listeria* by cleaning systems', Technical memorandum 590. The Campden Food and Drink Research Association, Chipping Campden.

LAWRENCE LM and GILMOUR A (1995), 'Characterization of *Listeria monocytogenes* isolated from poultry products and from the poultry-processing environment by random amplification of polymorphic DNA and multilocus enzyme electrophoresis', *Applied Environmental Microbiology,* **61**, 2139–2144.

LECLERCQ S and TISSOT C (2004), 'Les chutes de plain-pied en situation professionnelle', *INRS – Hygiène et sécurité au travail-Cahiers de notes documentaires,* **194**, 51–66.

LECLERCQ S, TISSERAND M and SAULNIER H (1994), 'Assessment of the slip resistance of floors in the laboratory and in the field: two complementary methods for two applications', *International Journal of Industrial Ergonomics*, **13**, 297–305.

LIOT J-P, CARPENTIER B, LECONTE A.-M, FAU G, VETTER F and SAULNIER H (1998), 'Guide des revêtements de sol répondant aux critères 'Hygiène – Sécurité – Aptitude à l'utilisation' pour les locaux de fabrication de produits alimentaires', Caisse nationale d'assurance maladie des travailleurs salariés (CNAMTS), Direction des risques professionnels, Paris.

MASUROVSKY EB and JORDAN WK (1958), 'Studies on the relative bacterial cleanability of milk-contact surfaces', *Journal of Dairy Science,* **41**, 1342–1358.

METTLER E and CARPENTIER B (1998), 'Variations over time of microbial load and physico-chemical properties of floor materials after cleaning in food industry premises', *Journal of Food Protection*, **61**, 57–65.

METTLER E and CARPENTIER B (1999), 'Hygienic quality of floors in relation to surface texture', *Transaction of the Institute of Chemical Engineers,* **77**, 90–95.

MIETTINEN MK, BJÖRKROTH KJ and KORKEALA HJ (1999), 'Characterization of *Listeria monocytogenes* from an ice cream plant by serotyping and pulsed field gel electrophoresis', *International Journal of Food Microbiology,* **46**, 187–192.

NELSON JH (1990), 'Where are *Listeria* likely to be found in dairy plants?', *Dairy, Food and Environmental Sanitation,* **10**, 344–345.

POLLET V (2000), 'Les sols industriels à base de résine réactive. Note d'information technique 216', Centre scientifique et technique de la construction. Bruxelles (Belgium).

STRANDBERG L (1985), 'The effect of conditions underfoot on falling and overexertion accidents', *Ergonomics*, **28**, 131–147.

TAYLOR JH and HOLAH JT (1996), 'A comparative evaluation with respect to the bacterial cleanability of a range of wall and floor surface materials used in the food industry', *Journal of Applied Bacteriology,* **81**, 262–266.

TIMPERLEY A (2002), Guidelines for the design of floors for food production areas (second edition), Guideline no. 40. Campden and Chorleywood Food Research Association Group, Chipping Campden.

TISSERAND M, LECLERCQ S and SAULNIER H (1995), 'Exigence pour une norme de mesure de la glissance des sols', *Cahiers de notes documentaires,* **159**, 191–198.

10

Improving the design of walls

D. J. Graham, Graham Sanitary Design Consulting Limited, USA

10.1 Introduction

Walls can be considered as the second most abused surface (after floors) in a food processing plant. The Food and Drug Regulations, specifically 21 CFR, Part 110 (the current good manufacturing practice, CGMPs), require that the floors, walls and ceilings in a food plant 'be of such construction as to be adequately cleanable and kept clean and in good repair.' Walls serve a number of purposes in food facilities, depending on whether they are interior or exterior structures. The type of processing that takes place also has a bearing on the type of wall necessary. That is, does it have to withstand hot water wash-down, does the wall enclose a refrigerated or heated space, is it cleanable, and how does it protect the processing space from seasonal changes and extremes in outside temperatures? There are numerous materials that can be used for wall construction, ranging from pre-engineered metal building walls to highly sophisticated gel-coated materials that contain anti-microbial characteristics. The choice depends on climate, plant location, local building codes, operating season, cost and environmental factors. These are a few of the questions that require answers when designing and constructing a sanitary food processing facility. This chapter will deal with external and internal walls, and how to integrate them into the entire facility.

10.2 Exterior walls

Concrete (precast or tilt up) exterior walls are usually load bearing and provide a support for the roof. Exterior walls and the foundation they rest on must provide protection from weather, rodent, insect and water ingress. Exterior walls are constructed of numerous materials such as precast concrete, tilt up concrete, cement block, metal, insulated metal panels or combinations of the materials depending on the size and function of the facility. Precast, or tilt up (poured in place), concrete walls are usually preferred in my experience. They are more durable, withstand more physical abuse and, in many instances, have proven to be about equal in cost to other types of exterior walls.

Whatever type or types of wall material and construction used they should conform to standards of sanitary design and construction according to Katsuyama (1993). The walls should be constructed to be rodent- and weatherproof. This means that the junctures of the roof/wall and the floor/wall must be sealed to prevent insect entry as well as outside weather.

Experience has shown that the best walls for a food processing plant are poured concrete that have been troweled smooth on the inside surface to a standard of no more than 1/8 inch (3.20 mm) diameter hole per square foot (0.1 sq m). Poured concrete walls do not have seams that require caulking found in precast or tilt up construction. Poured concrete is usually more expensive and requires on-site construction of forms and finishing. However, poured concrete in areas where precasting or tilt up construction is not available or feasible may be the only type of concrete wall that can be used.

Precast or tilt up walls have proven to be a rapid and economical way of erecting a food processing plant. Their main disadvantages are the time and expense necessary to adequately caulk all the joints and seams between panels. The caulking must be periodically maintained. A relatively new (since about 1990) innovation using notched beams, notched precast wall panels and double-tee precast roof panels is being used successfully on food processing plants. The technique (called pocket beam construction) entails precasting the wall panels and the roof support beams complete with notches large enough to accommodate the precast double tees of the roof panels. When lifted into place, the double tees fit into the notches rather than resting on top of the beams or walls. By fitting inside the notch, the dust-collecting flat surfaces on top of the beams or wall panels that are usually associated with this type of construction are eliminated. It is then a simple matter to fill and caulk the spaces around the double tees creating a cosmetically attractive and sanitary structure.

A word of caution about precast, tilt up and concrete block should be noted. If a parting agent (sometimes known as a release agent or oil) is used to facilitate the removal of the panel or block from the form, the agent should be tested to make sure it is compatible with any wall covering (epoxy, paint, etc.) *before* it is used. If it is not compatible, peeling will result and, as food processors know well, peeling paints are not welcome in food processing plants.

Rodents like to burrow under building foundations to gain access to the plant through openings in the floor. Rodent-proofing should be incorporated into the initial design of the facility, especially the walls. For example, Graham (1991a,b) reported in *Dairy, Food and Environmental Sanitation* magazine that for a slab floor facilities, the wall footers should be constructed with a rodent flange 24 inches (61 cm) below grade extending 12 inches (30 cm) out at right angles to the foundation. This flange will prevent rats from burrowing under the floor slab and chewing their way through vulnerable places into the plant such as through floor drains or expansion joints. If the building has a basement of cellar, its floor should be tied directly to the solid wall foundation. This will create a solid box that will be an effective pest barrier.

Another rodent deterrent is to construct a clear strip along the outside of an external wall that is about 30 inches (76 cm) wide and 4 inches (10 cm) deep. Line this strip with a heavy duty plastic sheet to prevent weed growth and fill with pea gravel (small diameter rounded stone) that will not bridge when the rodent tries to burrow through it. It will keep collapsing and discourage burrowing.

Other rodent prevention features for exterior walls include the shielding of outside piping and wires to prevent climbing. According to AFIS (1952), rodents can:

- walk along or climb up vertical wires;
- climb the inside of vertical pipes not more than 3 inches (76 mm) in diameter. This would include downspouts or other open drainpipes on the outside of the building;
- climb the inside of vertical pipes not more than 4 inches (102 mm) or less than 1.5 inches (38 mm) in diameter;
- climb the outside of vertical pipes of any size if the pipe is within 3 inches (76 mm) of a wall or other continuous support for a rat;
- jump 26 inches (660 mm) vertically and up to 48 inches (1.2 m) horizontally from a flat surface;
- drop 50 to 80 feet (15–25 m) without being killed;
- burrow vertically in the earth to a depth of 4 feet (1.2 m).

The easiest way to discourage these incursions is to utilize metal shields to prevent the rodents from going around them or install them over wiring to prevent rodents from obtaining a foothold.

If concrete block construction is considered, care should be taken to make sure the hollow cores are blocked to prevent rodents and insects from gaining access and free roaming through the wall interiors. This can be accomplished by filling the block centers with mortar when laid and making sure the tops of the walls are not left open. This caution applies to internal walls as well.

When concrete block is selected for exterior walls then it should be of the high-density type. Volcanic ash or cinder blocks are not acceptable for food processing facilities. They are too porous and will absorb moisture and bacteria and may allow them to penetrate directly to the core of the block where they are virtually impossible to dislodge. Low-density concrete block is not recommended since moisture, bacteria and mold can penetrate the surface and create sanitation problems. However, a good quality sealer can close the pores sufficiently to overcome these disadvantages. Even when sealed in this way, low-density walls still require a good maintenance program to remain effectively sealed.

Another common outside wall material that is used but not recommended for food plants is corrugated metal siding installed over an interior metal framework. Corrugated metal damages easily and is difficult to make rodent and insect proof. If, however, it is used, the outside corrugation must be blocked and caulked at the foundation and at the top, to prevent access by rodents, insects and other pests. Once inside the corrugations, pests can roam up and down the walls at will, finding openings into the plant and so making any pest control program extremely difficult. Using corrugated wall panels often requires separate insulation. If that is the case, do not use fiberglass batting. This material is an attractant for rodents. They like the resin in it and will create nests within the wall insulation. Some suppliers of fiberglass batting will stipulate that their material will not attract pests. If you use fiberglass batting be sure to get this stipulation in writing from the supplier. Otherwise materials such as urethane, foam or foam board are recommended.

Insulated metal panels are often used for exterior and interior walls for plants with refrigerated processing areas. They are also used for constructing freezer and cooler storages. They have the advantage of being interlocking panels, already insulated and can be made rodent and insect proof. These types of walls are usually easily cleanable and have a good sanitary cosmetic appearance. These walls can either be erected against a framework of beams on the inside of the wall, or in some instances they have been observed with the supports mounted on the outside of the plant so the interior is free from any obstructions.

Exterior walls will, at one time or another requires penetrations for access by utilities or for other reasons. These penetrations should be well planned ahead of time and the timing of them coordinated with the utility or other services being taken through the wall. Once the penetration has been made, it should be used and sealed the same day, if at all possible. Leaving it open overnight will probably result in one or more pests invading the wall, which, if it has an exposed, insulated or hollow core, will provide the pests with an excellent home.

10.3 Interior walls

Internal walls should be impervious to moisture, easily cleanable, flat, smooth and resistant to wear, corrosion and impact. In addition, walls in wet processing areas should be resistant to water sprays, cleaning compounds and scrubbing when used. There are a number of acceptable materials that may be used on internal walls. Many plants with wet processing areas or processing areas for microorganism-sensitive products still use ceramic tiles to enhance the cleanability of the walls. Glazed tiles have been used in dairies, breweries and bottling plants as standard for many years. These type tiles are resistant to blood, food, acids, alkalis, cleaning compounds, steam and hot water. These walls are expensive to install but are easily and inexpensively maintained.

New materials for wall construction are appearing constantly. Materials ranging from baked on enamel-insulated panels, to spray-on resins covered with gel coats are on the market. One high-end material is Arcoplast, which consists of sandwich composite panels that are manufactured with different coring materials such as foam, honeycomb, cement or solid glass matrix to suit the user's requirements. The panel cores are reinforced on both sides with multiple layers of glass fibers embedded in a permanent durable polymer resin and finished with a hard, high-gloss gel coat resin. It has incorporated antimicrobial features using silver ion technology. It is highly suitable for highly microbial-sensitive areas in a food or pharmaceutical facility. Another type of application is called Stayflex. It consists of a spray-on foam insulation that is then covered with a resin and gel coat that is extremely hard and damage resistant. This type of application works well for renovation projects with existing walls that are uneven, have rust, peeling paint and hidden niches that can hide dirt and microbes. This spray-on foam will fill the holes and eliminate a hiding place for pests. Other wall materials are reinforced fiberglass board that has been around for a number of years. This material has a smooth non-absorbent surface and is easily cleanable. It is, however, susceptible to damage by forklifts, troughs, carts, etc., banging into it and it must be protected.

Constructing internal walls regardless of the type of material should be considered carefully. Many contractors want to do it the easy way by erecting the wall from floor to ceiling and then pouring a concrete curb wall against the erected wall. This protects the panel material from damage but creates a joint between the concrete and the panel material. In a facility that uses water for processing where there is splashing, moisture will collect between the curb wall and the wall panel. This makes an excellent growth area for mold, bacteria and yeast. In refrigerated rooms this can become a growth area for *Listeria*. A better method is to pour a stub wall with a coved base or install a precast curb wall and set the wall panels on top in a stainless channel that has been filled with caulk. The caulk will prevent moisture from collecting and seal the joint between the bottom of the panel and the channel. If insulation is required in the stub wall it can be incorporated in the precast stub wall. Insulation is required if there is a large temperature differential between the two rooms the wall is separating to prevent condensate formation (see Fig. 10.1).

New ideas and materials for walls and wall coverings in food processing facilities are continually appearing. One only has to attend some of the many trade shows every year to discover new ideas, materials and applications. Sanitation and sanitary (hygienic) design are well up the list of criteria considered when designing a new facility or renovating or adding to an existing facility. The old idea that floors, walls and ceilings are just part of a necessary envelope to hold the equipment is fast disappearing. They have to be considered as an integral part of the sanitation program and must be easy to clean, do not contribute to contamination or adulteration of the product or products being processed/packaged. As said in the beginning of this chapter, walls are the second most abused surface in a food processing plant and this must be taken into consideration.

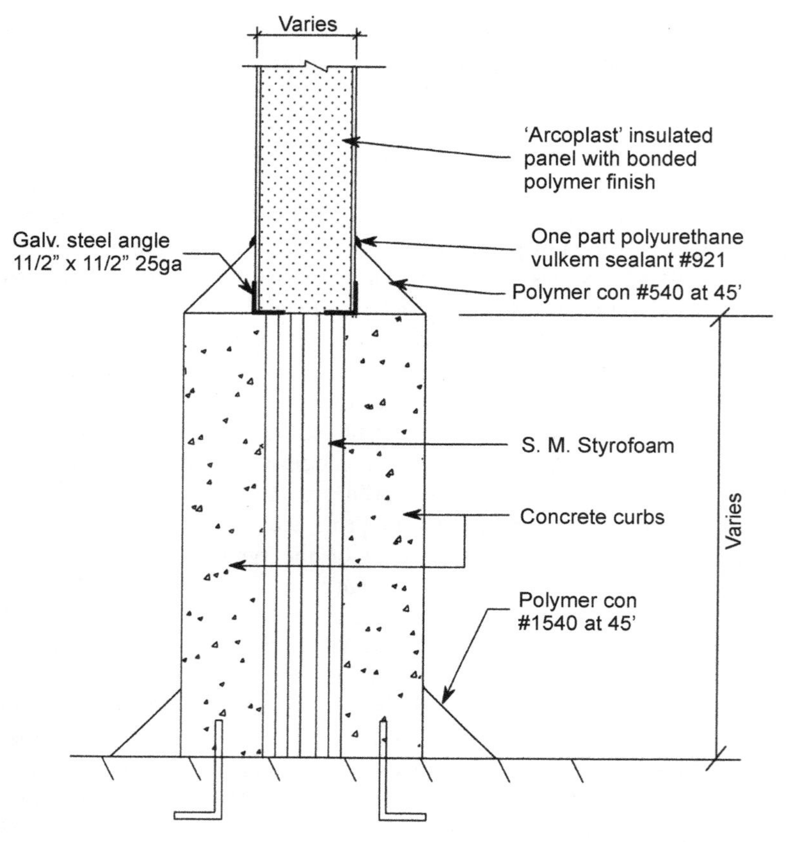

Fig. 10.1 Insulated curb wall drawing (courtesy of Arcoplast Corporation St Charles, MO, www.arcoplast.com).

10.4 Bibliography

AFIS (1952), *Sanitation for the Food-preservation Industries*, New York: McGraw-Hill.

GRAHAM, DONALD J. (1991a) 'Sanitary Design – A Mind Set Part II', *Dairy, Food and Environmental Sanitation*, (August), 454–455.

GRAHAM, DONALD J. (1991b) 'Sanitary Design – A Mind Set Part IV', *Dairy, Food and Environmental Sanitation*, (October), 600–601.

IMHOLTE, T.J. (1984) *Engineering for Food Safety and Sanitation.* Crystal MN: Technical Institute for Food Safety.

KATSUYAMA, ALAN M. (editor) (1993) *Principles of Food Processing Sanitation.* Washington, DC: The Food Processors Institute.

Personal Communication (2004) Arcoplast Corp., Chesterfield, MO. www.arcoplast.com

Personal Communication (2004) Stayflex Systems, Preferred Solutions Inc., Cleveland, Ohio. www.stayflex.com

SHAPTON, DAVID A. and SHAPTON, NORAH F. (eds.) (1991) *Principles and Practices for the Safe Processing of Foods.* Oxford, UK: Butterworth-Heinemann Ltd.

11

Improving the hygienic design of closed equipment

A. Friis and B.B.B. Jensen, Technical University of Denmark

11.1 Introduction: the hygienic performance of closed equipment

Hygienic performance of closed processing equipment for food processing depends on a number of aspects. Some of these aspects are surface material and finish (roughness and topography), gasket material and gasket design, welding quality and welding location, cleaning procedure (time, temperature and detergent) and internal design of the equipment. This chapter focus on the relationship between flow of detergent during cleaning and, as a direct consequence, the design of the equipment. At first glance, flow of detergent might not be the most important aspect in cleaning; nevertheless, it is an aspect that is relatively cheap to optimise through design consideration. Additionally, increased focus on environmental concerns (and taxes) related to energy consumption and use of chemicals has made the obvious choice of increasing the temperature or use of harsher chemicals unattractive from an economical point of view – improved cleaning is obtained more cheaply through proper design of closed equipment.

Maintenance of proper hygiene in closed process equipment is in many ways a complex task. The interaction between the design of the equipment and the nature of fluid flow in the equipment is the main concern. It is already known that dead legs or other types of areas shielded from the main flow can occur and present a hygienic risk (Anon., 1993; Lelieveld *et al.*, 2003). During cleaning the main task of the flow is to bring cleaning agents in the right doses to all parts of the process plant. In turn, the adhesion mechanisms between the soil and the surface of the process equipment must be overcome. Clearly, this is similar to other sanitation procedures; however, in closed processes, validation proves rather difficult, as inspection is often not possible. Hence, a greater basic understanding of the interaction between flow characteristics and soil attached to surfaces can aid validation of hygienic design. Such information can be contained in fluid dynamics theory and models or by rules of thumb. This type of information can be used to assist improvement the design of process equipment with respect to cleaning characteristics and optimisation of cleaning procedures. Prediction of cleaning efficiency in especially complex parts of closed process plants by use of computational fluid dynamics (CFD) has an excellent potential for desktop improvements and computer pre-validation of the hygienic performance of process plants. Hence, the hygienic design of closed equipment related to the movement of detergent can be improved.

11.2 The importance of flow parameters in hygienic performance

The importance of addressing fluid flow in relation to cleaning efficiency in closed processes has been illustrated (e.g. Bénézech *et al.*, 1998; Lelièvre *et al.*, 2002, 2003; Jensen, 2003). In this section, different aspects of the influence of flow on cleaning characteristics are discussed with reference to published studies and new ideas on the subject. Guidelines and legislation on flow conditions in processing equipment to obtain satisfactory cleaning characteristics are discussed and new parameters for validating the cleaning effect of flow are presented. Finally, a brief discussion on methods for visualising flow features is given to underline the need for CFD tools in hygienic design.

11.2.1 Importance of flow in cleaning of closed equipment

From a simplistic point of view, proper hygienic design of closed equipment is an exercise of making detergent (temperature and chemicals) accessible to the soil for a certain period (time) and exposing the soil to a force (mechanical) that is sufficiently large to remove the soil from the surface. Sinner (1960) suggests that for cleaning to take place all four cleaning parameters – temperature, chemicals, time and mechanical action (Fig. 11.1a – Sinner's circle) – should be present. A change in one of the parameters in Sinner's circle must be compensated for by changes in the three other parameters. Additionally, detergent and heat have to be transported to the soil on the surface to be effective and the soil must be removed from the surface and out of the equipment to avoid reattachment. In this section, the importance of flow is illustrated based on Sinner's circle and the effect of flow on the four parts of Sinner's circle is introduced.

Any cleaning procedure can be considered as a process of applying the required energy needed to remove soil from a surface. Sinner (1960) and Holah (2003) divide the energy into four sources: contact time (time), detergent temperature (temp.), detergent strength (chem.) and mechanical action (mech.). The temperature and chemicals weaken the bond between soil and surface as a

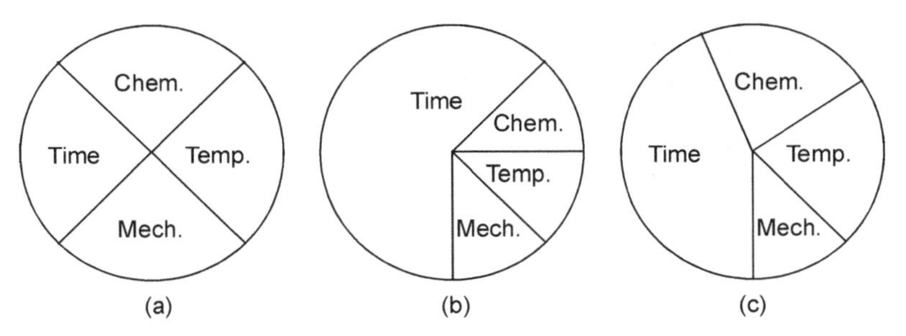

Fig. 11.1 The influence of the transport phenomenon on cleaning time illustrated by Sinner's circle for (a) a fully developed turbulent pipe flow, (b) a recirculation zone in, e.g., a dead-end with poor exchange of detergent and (c) a recirculation zone in, e.g., a valve with a good exchange of detergent.

function of temperature, strength of chemicals and contact time between detergent and soil. The contact time is the time a specified temperature and strength of detergent are present at the interface between soil and detergent. The mechanical effect is a shear force dragging in the soil at the surface. The shear force removes the soil from the surface. For a mild soiling the mechanical effect might be sufficient to remove the soil using only a mild (or even no) detergent solution. In case of a hard soiling, bonds between surface and soil need to be weakened by a stronger detergent to allow the mechanical force to remove the soil from the surface. The mechanical force is applied from liquid moving relatively to the surface (more on this later). Holah (2003) gives details on the combination of the relative energy sources for open equipment and different cleaning techniques.

When cleaning closed equipment, all four of the components of Sinner's circle are influenced, either directly or indirectly, by the flow of detergent inside the equipment. The direct influence is through the mechanical force acting on the soil. The force is generated from motion of the detergent across the surface. The force is also known as the wall shear stress (τ_w). Wall shear stress is a consequence of a velocity gradient occurring because of non-slip conditions at the wall. For fully developed pipe flow the wall shear stress is given by (Shames, 1992):

$$\tau_{w,lam} = -\mu \frac{du}{dr}\bigg|_{r=R} = \frac{4\mu U}{R} \qquad \text{[Pa]} \qquad Re < 2300 \tag{11.1}$$

$$\tau_{w,turb} = 0.03325 \rho U^2 \left(\frac{\nu}{RU}\right)^{0.25} \qquad \text{[Pa]} \qquad 2300 < Re < 3 \times 10^6 \tag{11.2}$$

where μ is the dynamic viscosity (N s/m^2), u is the axial velocity (m/s) at a distance r (m) from the centre of the pipe, R is the radius of the pipe (m), U is the average velocity of the fully developed flow (m/s), ρ is the density (kg/m^3) and ν is the kinematic viscosity (m^2/s).

The indirect influence of the flow on cleaning arises from the fact that the detergent (temperature and chemical) needs to be 'delivered' to the soil by heat and mass transfer. Transport of detergent throughout the entire processing line is necessary. In closed equipment, the obvious and cheapest transport medium is water (Holah, 2003) heated to a specific temperature with the right amount of chemicals dissolved in it.

The influence of flow on contact time is best illustrated by comparing two commonly encountered situations:

1. In a straight pipe detergent flows parallel to the wall and 'fresh' detergent (temp. and chem.) is continuously transported across the soil, resulting in a contact time at ideal cleaning conditions corresponding to the cleaning time (illustrated by Sinner's circle in Fig. 11.1(a) assuming the ideal cleaning conditions are equal amount of 'energy' from all four contributions).
2. In a dead-end or a sudden change of geometry, recirculation zones are present. In such recirculation zones, the detergent is not replaced at the same rate as in the straight pipe (heat and mass transfer are low) and the temperature and strength of the detergent decrease slightly as a function of time. Hence, the resulting contact time at ideal cleaning conditions (here assumed to be the conditions present at the surface of a straight pipe with fully developed turbulent flow) is reduced compared with the contact time in the straight pipe, and the total cleaning time has to be increased to clean the surfaces located in the recirculation zone (illustrated in Fig. 11.1(b) and (c)).

An additional effect of the flow is the transport of detached soil out of the equipment to prevent recontamination.

In Section 11.4 information on local wall shear stress and fluid exchange at the surface is exploited to illustrate good and bad flow patterns in relation to cleaning of closed equipment.

11.2.2 Guidelines on flow conditions during CIP cleaning

A mean velocity of at least 1.5 m/s for cleaning-in-place (CIP) of closed equipment is suggested as a minimum. However, very little, if any, hard evidence has been published stating that 1.5 m/s is a universal value (Timperley and Lawson, 1980), but it should be remembered that 1.5 m/s is used with success for cleaning at present time. In this section, the validity of specifying a minimum mean velocity is discussed with a special emphasis on closed equipment to show that more focus on local flow phenomena is needed to improve the overall hygienic design of processing equipment.

Volume flows corresponding to a mean velocity of 1.5 m/s have been used with success for CIP in many food-processing facilities. At this velocity turbulent flow is guaranteed for straight pipes with an inner diameter above 0.01 m. Turbulent flow is needed to improve cleaning (Majoor, 2003) as it enhances the transport of detergent (mass and heat) from the bulk to the surface compared with laminar flow. Furthermore, the thickness of the so-called viscous (or laminar) sublayer covering the wall reaches an asymptote at a mean velocity of 1.5 m/s, almost independent of the diameter, for pipes with inner diameters above 0.02 m. The thinner the viscous sublayer, the better the cleaning, as a thin viscous sublayer allows a faster (shorter length of diffusion, ∂y) transfer of detergent and heat from the outside of the viscous sublayer to the soil than a thicker viscous sublayer. This is illustrated by Fourier's heat transfer equation:

$$q_s = -k\frac{\partial T}{\partial y}\bigg|_{y=0} \tag{11.3}$$

where q_s is the heat transfer (W/m^2), k is the thermal conductivity (W/(m K)), is the temperature gradient (K) and y is the distance from the soil (m). The effect of velocity and Reynolds number for cleaning in straight pipes is discussed by Timperley (1981). The conclusion of his study is that specifying a velocity of 1.5 m/s in pipes with an inner diameter of 0.038 and 0.076 m is more appropriate than specifying a Reynolds number when evaluating removal of micro-organisms. This is supported by the findings of Bergman and Tragardh (1990) for the removal of clay in a straight duct under turbulent flow conditions. However, recent findings of Lelièvre *et al.* (2002, 2003) show that for more complex equipment the wall shear stress, and thereby the velocity, cannot give a coherent explanation of the results of cleaning tests. Instead, local mass transfer to the surface is shown to be important. In equipment with complex flow patterns, recirculation and separation create fluctuations in the flow and in the boundary layer, creating different levels of mass transfer to the surface and wall shear stress on the surface. In the work of Timperley (1981) and Bergman and Tragardh (1990) the viscous sublayer was hardly affected by the range of velocities and Reynolds numbers (the average velocity was above 1.5 m/s) investigated, which could explain the difference from the conclusions of Lelièvre *et al.* (2002, 2003).

The above considerations presented by Timperley and Lawson (1980) and Timperley (1981) are valid for cleaning of straight pipes. In straight pipes of inner diameters between 0.02 and 0.076 m, with fully developed turbulent flow, a mean velocity of 1.5 m/s produces almost constant wall shear stresses (slightly higher at the smaller diameters). Changing the velocity from 1.5 m/s to, e.g., 0.5 m/s or 2.5 m/s has a large impact on the wall shear stress in the pipe. This is similar to the conditions in all other types of equipment (bends, valves, heat exchangers, etc.) other than straight pipes; local velocities different from the average velocity are encountered at different locations in the equipment.

Much equipment has some areas with velocities higher and some areas with velocity lower than the mean (see Fig. 11.2 for local velocities inside a 90° pipe bend). Furthermore, the velocity at the wall is always zero, so it is impossible to state that a local velocity at the wall should be of a certain magnitude. Hence, specifying a mean velocity as the only indicator for the cleaning effect of fluid flow in a CIP operation is too weak for optimisation purposes. Instead, a combination of wall shear stress and fluid exchange/mass transfer from the bulk to the viscous sublayer should be evaluated to estimate if certain areas of the

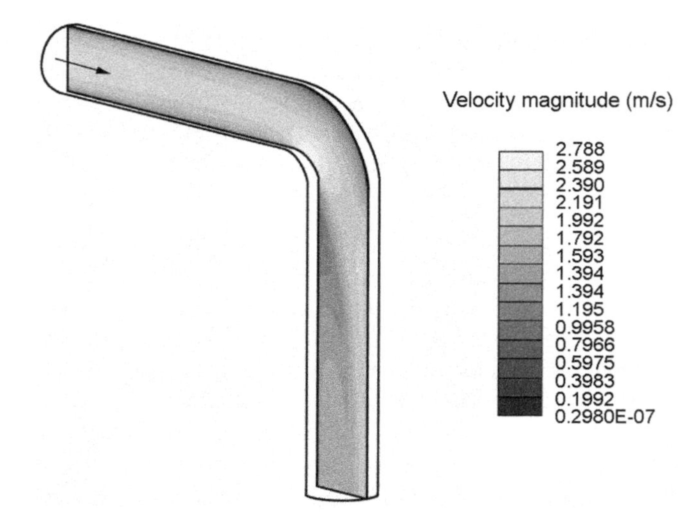

Fig. 11.2 Local velocity estimated by use of CFD calculations. The average velocity is 2 m/s and the liquid is water at 20 °C.

surface are more difficult to clean than other areas (Jensen, 2003). Such considerations would aid designers of closed equipment to identify areas that are difficult to clean in the early stages of the design phase. In the longer term, this should result in an improved understanding of guidelines for hygienic design of closed equipment. Guidelines should cover advantageous flow patterns that promote combinations of wall shear stress and fluid exchange/mass transfer favourable for cleaning. In Section 11.4, examples are given on how CFD has been used for identification of areas potential being a hygienic problem. The CFD method is introduced in the next section.

11.2.3 Flow visualisation methods

Visualisation of flow patterns and parameters is the key parameter for evaluating, and gaining a higher understanding of, hygienic design related to flow of detergents. Visualisation can be done experimentally (EFD – experimental fluid dynamics) or numerically (CFD – computational fluid dynamics). The CFD tools, when validated, have certain advantages compared with EFD. It should be remembered that when using EFD, users have to make certain that they measure what they think they are measuring. The main advantage of using CFD is the fact that data are available in all the control volumes (the same as a huge number of measuring probes). This is hardly possible using EFD techniques such as laser Doppler anemometry, mass transfer techniques and thermal velocity probes. CFD results on the surfaces are of special interest to hygienic design. In order to make a complete evaluation of the hygienic design of a piece of equipment, results must be known over as many portions of the surfaces of the equipment as possible and ideally over the entire surface. Most EFD techniques rely on probes inserted either directly onto the surface, into the flow domain or having a transparent model of the equipment. An additional advantage of having a validated CFD model of flow in a component is that a converged solution contains not only data that were interesting from the outset of the simulations, but also additional data that can be used for further evaluations of, for example, pressure loss and careful processing.

The main reason for suggesting CFD as a tool in the quest of improving the hygienic design of closed equipment is the fact that CFD produces data over the entire domain, not only at discrete points as many experimental methods. Furthermore, experimental methods for measuring wall shear stress and fluid exchange (the two most interesting for evaluating cleanability) are complex to use and modified equipment (transparent or with special measuring probes inserted) is required. Results obtained from CFD simulations should, to some degree, be compared with experimental or analytical data. If the purpose of the CFD simulations is to compare a number of design changes, this is possible without a validation, as the effect of a design change on, e.g., the wall shear stress distribution can then be seen relatively to the results obtained in the related designs.

11.3 Computational fluid dynamics models for optimising hygiene

From the above it is clear that flow is important to the cleaning characteristics of a piece of closed equipment. Hence, to compare the hygienic characteristics of two familiar pieces of closed equipment or the effect of design changes, knowledge of the flow inside the equipment is needed. Flow can be visualised experimentally or by modelling using CFD. Each has advantages and disadvantages as discussed in Section 11.2.3.

In the late 1990s, commercial CFD codes were made available for Windows and Linux platform users. This, combined with increased (and cheap) CPU power and memory capacity of personal computers, has made it feasible to use CFD codes on personal computers. Discretisation of the flow domain (creating the mesh) and the process of creating and setting-up CFD models have been simplified and made more user-friendly over the past 5 to 7 years. Still, a background in fluid mechanics is recommended to make model set-up easier and evaluation of results more straightforward. Some of the most popular commercial codes are CFX (www-waterloo.ansys.com/cfx), Fluent Inc. (www.fluent.com) and Star-CD (www.cd.co.uk), all available for both UNIX and Windows platforms.

Most CFD simulations are based on the same recipe right from the creation of the mesh until a converged solution is obtained.

- The flow domain is divided into small control volumes (called the mesh). Momentum (velocity) and pressure in each control volume, based on the boundary conditions for the control volume, are described by the Navier–Stokes equations (here in incompressible form).

$$\rho \frac{\partial u_i}{\partial t} + \rho \frac{\partial}{\partial x_j}(u_j u_i) = -\frac{\partial p}{\partial x_i} + \frac{\partial}{\partial x_j}(2\mu s_{ij}) \tag{11.4}$$

where t is the time, u_i and x_i the velocity and position vector, p the pressure and s_{ij} the strain-rate tensor. The number of control volumes chosen is a trade-off between accuracy and simulation time. For each parameter solved for (u, v, w and p) the Navier–Stokes equations are set up.

- Physical and empirical models for other phenomena than laminar flow (heat transfer, buoyancy, multiphase, radiation, etc.) are selected. Turbulence in the flow is a special subject treated in details below.
- Boundaries (e.g. inlet, outlet and walls) are defined on appropriate faces of the mesh and the boundary conditions are specified. An inlet velocity (plug flow or an arbitrary profile), total mass flow through the flow domain or a pressure difference between inlet and outlet can be used to generate movement of the liquid. Walls can be specified with different roughness parameters to influence the pressure drop (skin friction) through the flow domain and heat transfer to and from the surface can be estimated.
- Physical properties for the liquid are selected.
- The iteration procedure is initiated and iterations are performed until a specified convergence criterion or divergence is reached.

The influence of turbulence is included in the Navier–Stokes equations through the Reynolds stress tensor (τ_{ij}):

$$\rho \frac{\partial U_i}{\partial t} + \rho U_j \frac{\partial U_i}{\partial x_j} = -\frac{\partial P}{\partial x_i} + \frac{\partial}{\partial x_j}(2\mu S_{ij} - \rho \overline{u'_j u'_i}) \tag{11.5}$$

$$\tau_{ij} = -\rho \overline{u'_j u'_i} = \nu_T S_{ij} \tag{11.6}$$

$$\nu_T = \text{const}\, k^{1/2} l \tag{11.7}$$

where P is the average pressure, S_{ij} the average strain-rate tensor, k the turbulent kinetic energy, l the turbulent length scale and the $'$ denotes fluctuating values. The Reynolds stress tensor is expressed by the average of the product of the fluctuating component, hence, these must be found. This can be done by, for example, direct numerical simulation (DNS), large eddy simulation (LES) or closure models. The closure models solve transport equations for the turbulent kinetic energy and turbulent kinetic energy dissipation (ϵ) in the core flow. In the near-wall layer, production and dissipation of turbulent kinetic energy is estimated by near-wall treatments. The choice of near-wall treatment (see later) prescribes the recommended density of cells in the near-wall layer.

Descriptions of the governing equations for finite volume codes have been extensively described in a number of references (for example Patankar, 1980; Versteeg and Malalasekera, 1995; Ferziger and Peric, 1999).

The mesh (number, size and distribution of control volumes) is probably the single most important part of CFD simulations as this makes the basis for discretisation of the flow domain. The geometry is either imported directly from a commercial computer-aided design (CAD) program or created using an in-the-CFD-code CAD environment. Alternatively, a mesh can be imported from external mesh generators such as ICEM (http://www-berkeley.ansys.com) or GAMBIT (www.fluent.com). Mesh generation can be a very time-consuming process of trial and error. A mesh structure suited for a particular geometry and flow condition might not be appropriate for another, so refinement and coarsening of the mesh is needed (for example flow at different temperatures influences the mesh near the wall – see y^+ in next section).

11.3.1 Near-wall treatment

Chen and Patel (1988) stated the importance of near-wall treatment for the overall success of turbulence models. An important parameter in the success of the different near-wall treatments available is the distance from the wall to the centre point of the first cell, when normalised, called y^+:

$$y^+ = \frac{y_p u_\tau}{\nu} = \frac{y_p \sqrt{\frac{\tau_w}{\rho}}}{\nu} \tag{11.8}$$

where u_τ is the friction velocity (m/s), y_p is the distance from the wall to the centre point of the near-wall cell (m) and ν is the kinematic viscosity (m^2/s). Limits for y^+ depend on the choice of near-wall treatment. The wall function approach has been used with success for modelling flows where near-wall phenomena are less important. Recently, more advanced two-layer models and low *Re* k–ϵ models have been implemented into commercial CFD codes, improving prediction of flow in confined, separating and attaching flows encountered in even slightly complex equipment.

In simulations of flows with adverse pressure gradients and recirculation zones or where flow near the walls, heat transfer, wall shear stress or friction is of special interest, as in the case of hygienic design, Rodi (1991) suggests the two-layer approach to describe flow in the near-wall layer. In contrast to the wall function, the flow in the buffer zone and the logarithmic layer are resolved by a number of cells. A transport equation for turbulent kinetic energy is solved in the near-wall layer and dissipation of turbulent kinetic energy is expressed by an algebraic function. A shift to, for example, the standard k–ϵ turbulence model is done at a distance from the wall where viscous effects become negligible compared with inertia effects. y^+ should be around 3, and approximately 15 points should be placed within the near-wall layer (Anon, 1999). As the flow is modelled all the way to the viscous sublayer, wall shear stress is calculated from the general definition give in equation (11.1)

11.4 Applications of computational fluid dynamics in improved hygienic design

Taking advantage of the possibilities that lie within CFD simulations performed using models and a set-up that is applicable for industrial purposes (reasonable simulation time, time for model set-up and time for meshing) for improving the hygienic design of food equipment is a relatively novel idea. Towards the end of the last century and the beginning of this one, work performed at Campden Chorleywood (Tucker and Hall, 1998; Hall, 1999; Richardson *et al.*, 2001), Cyclone Fluid Dynamics (Hauser and Krüs, 2000) and the Technical University of Denmark (Jensen *et al.*, 2000, 2001) indicates, based on both experimental data and generally accepted mechanisms of cleaning, that data regarding the flow conditions obtained using CFD could in fact be used for explaining why certain areas of different types of equipment were difficult to clean and others were not. The work of the authors of this chapter (Friis and Jensen, 2002; Jensen, 2003; Jensen and Friis, 2005; Jensen *et al.*, 2005) supports these conclusions; it is possible, within certain limits, to predict the outcome of the well-known EHEDG (European Hygienic Engineering and Design Group) cleanability test for assessing the In-place cleanability of food-processing equipment (Anon, 1992).

In this section, the process proposed by Jensen (2003) and Jensen and Friis (2004c) for identifying areas of different levels of cleanability is briefly explained to show how CFD can help in improving hygienic design. This is followed by examples on how CFD has been applied to provide increased understanding of why certain designs are good for cleaning and others are not. These examples cover:

- predicting the outcome of a an EHEDG cleaning test;
- flow in expansions;
- cleaning of a spherical-shaped valve house.

11.4.1 Virtual cleaning test

The outcome of the EHEDG cleaning tests can be predicted from CFD simulations visualising wall shear stress and fluid exchange. The steps needed to make a prediction of the areas with different degrees of cleanability are:

1. a critical wall shear stress under controlled flow conditions is needed for the cleaning test method;
2. wall shear stress and fluid exchange is predicted using a CFD model of the piece of equipment;
3. areas exposed to different levels of wall shear stress in relation to the critical value are identified – a rough estimation of cleanability is possible;
4. areas exposed to different levels of fluid exchange relatively to the fluid exchange in the undisturbed part of the flow are identified;
5. grouping the different areas of wall shear stress and fluid exchange makes the prediction of areas of different cleanability possible.

Jensen (2003) shows this for different types of component commonly found in the food industry. Some of the details in each step are presented below.

Step 1: A critical wall shear stress is required for the cleaning test method investigated. The critical wall shear stress (mechanical) depends on the nature of the soil, the surface of the equipment and the three remaining components of Sinner's circle (cleaning time, detergent temperature and chemical strength). The straightforward approach to obtain a critical wall shear stress is to use the radial flowcell assay (RFC) (Fowler and McKay, 1980) in which removal of an arbitrary soil from a surface can be investigated when exposed to a wall shear stress that gradually decreases from inlet to outlet in the test section. A critical wall shear stress of approximately 3 Pa is applicable for the EHEDG cleaning test using a pickled stainless steel 316L surface with a mean roughness (R_a) of 0.5 μm (Jensen and Friis, 2004a). The critical wall shear stress is found by comparing the cleaning results of EHEDG tests with wall shear stress in the RFC predicted by CFD simulations.

Step 2: Prediction of wall shear stress and fluid exchange should be performed using appropriate mesh and models (for wall shear stress see Jensen and Friis, 2004b, and for fluid exchange see Jensen, 2003 and Jensen and Friis, 2004c). The near-wall region should be resolved and flow simulated using two-layer models or similar approaches. Second order spatial and temporal discretisation should be chosen if possible. Ideally, flow patterns should be validated by use of experimental methods such as laser Doppler anemometry (LDA), laser sheet visualisation (LSV) or particle image velocimetry (PIV). However, such data are seldom available for designers of equipment, hence the CFD simulations must be performed based on best practice (Casey and Wintergerste, 2000) and experience from previous validated simulations (this is an accepted and widely used method in other engineering applications).

Researchers have experimentally shown that different degrees of cleanability in a straight pipe or a duct are linked to the mean wall shear stress (e.g. Duddridge *et al.*, 1982; Bergman and Tragardh, 1990). Others, however, have shown that for more complex flows the mean wall shear stress is not explanatory for the different degrees of cleanability found (Jensen and Friis, 2005); the effect of detergent availability (mass transport) at the soil should also be considered (Lelièvre *et al.*, 2002, 2003; Jensen and Friis, 2004c). Hence, both wall shear stress and fluid exchange have to be known and evaluated for each and every area inside the equipment investigated (Jensen, 2003). One of the main concerns of this approach is the fact that prediction of wall shear stress using CFD is known to be difficult – obtaining data for validation of the wall shear stress and fluid exchange predictions is not trivial. Even though, confidence may be high that the areas predicted as difficult to clean are in fact more difficult to clean than other areas in the component, this may be a problem. All wall shear stresses used (those for obtaining a critical wall shear stress and those for evaluating cleanability in a component) are predicted from CFD simulations. It is assumed that the level of wall shear stress predicted and fluid exchange estimated are comparable.

An example is given below showing the quality of the predictions of cleanability in an upstand geometry tested by the EHEDG test method. This shows that prediction is indeed possible.

11.4.2 Example 1: Prediction of different zones of cleanability – short upstand geometry

Closed processing lines very often contain a number of so-called upstand geometries, e.g. for mounting pressure transducers, thermocouples, sampling equipment or a T-branch for a bypassing pipe. Upstands are known to present a hygiene problem as recirculation zones are present in the dead-ends. Guidelines (Anon, 1993) and legislation (Anon, 1997) states that dead-ends should be avoided and, if unavoidable, the upstand should be as short as possible. Campden Chorleywood published the results of an EHEDG cleaning trail on a short upstand (Richardson *et al.*, 2001) showing that the dead-end itself was uncleaned and, surprisingly, also the pipe surface located just downstream of the dead-end on the side of the main pipe where the upstand was fastened to the main pipe was uncleaned.

From the CFD simulations of the flow in that particular upstand during cleaning, areas of different categories of cleaning level can be identified when comparing wall shear stress and fluid exchange. Figure 11.3 shows the areas exposed to the different cleaning condition types:

- Cleaning condition type 1. Areas exposed to high wall shear stress and very good fluid exchange (good cleaning conditions) in the entire upstream-undisturbed part of the geometry and in the downstream part located on the opposite side of the pipe as the upstand.
- Cleaning condition type 2. Areas exposed to high wall shear stress and intermediate fluid exchange (cleaning conditions not so good) in the downstream region of the upstand on the same side of the pipe as the upstand, but only on the non-horizontal part of the surface.
- Cleaning condition type 3. Areas exposed to intermediate wall shear stress and poor fluid exchange (bad conditions for cleaning) in the downstream region of the upstand in a band running from the upstand and downstream on the horizontal part of the surface.
- Cleaning condition type 4. Areas exposed to low wall shear stress and very poor fluid exchange (very bad for cleaning) in the upstand itself.

Prediction of cleanability in the upstand would be as follows. Cleaning is possible in areas with cleaning condition type 1, areas of cleaning conditions type 2 would be expected to be either cleaned or uncleaned, areas of cleaning conditions types 3 and 4 would be expected to be uncleaned. It is difficult to state whether types 2 and 3 are cleaned or not, as information is lacking to evaluate the levels of fluid exchange – here a critical value is needed. However, the categorisation of cleaning level was performed based on experience from similar investigations on a spherical valve house (Jensen, 2003). Comparing the cleaning

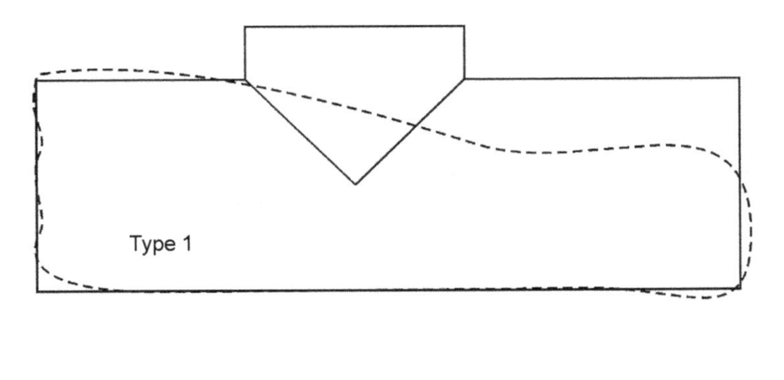

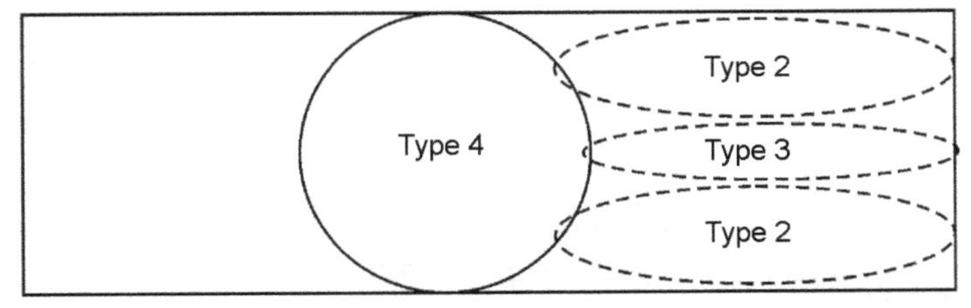

Fig. 11.3 Illustration of the areas in the short upstand exposed to the different types of cleaning condition type defined in Section 11.4.2. Prediction of cleaning characteristics is based on wall shear stress and fluid exchange data available from CFD simulations.

results, wall shear stress levels and relative fluid exchange times for the spherical valve house of a mix-proof valve (shown later in Fig. 11.5) and the similar values for the upstand produce the prediction given above. Hence, a reference of fluid exchange is still needed to perform a sound prediction of cleaning level.

Comparison of the predicted cleanability with the cleaning results shows that a good prediction is possible. The interesting part is that the reason for the cleaning difficulties in the downstream part of the geometry can be identified. Here slow fluid exchange is present, which is a consequence of the disturbance from the short upstand into the main flow because of the geometry. Cleaning is difficult even though wall shear stress is similar to the wall shear stress in regions that are cleaned. From CFD simulations of the short and a long upstand, this disturbance was shown to be non-existent in the long upstand. Investigations are continuing to find a length of the upstand that produces a flow that does not disturb the main flow. The disturbance of the main flow makes the problem out of control, while a longer upstand isolates the problem area to be in the upstand only.

11.4.3 Example 2: Flow in expansions

Expansions are applied in different parts of food production systems from the simple transition from one pipe diameter to another diameter, e.g. to control the velocity of the product (residence time), to the connections between equipment and the pipe system (e.g. from a pump with an off-the-shelf outlet diameter to the pipe diameter of the processing line or from the pipe diameter of the processing line to a unit operation machine that comes from the equipment manufacture with a different inlet or outlet pipe diameter). At first glance, choosing a concentric or an eccentric expansion is a matter of availability, spatial requirements and, in the case of draining against product flow, a case of draining (eccentric is the most drain-friendly of the two – Anon, 1993). However, studying the flow in concentric and eccentric expansions from a small pipe diameter to a larger pipe diameter reveals interesting results in relation to the potential cleaning effect of the flow.

The example given here is expansion from a 1 inch to a 2 inch (25–50 mm) pipe (Fig. 11.4). The expansion angles (8.6° and 16.8°) are based on product

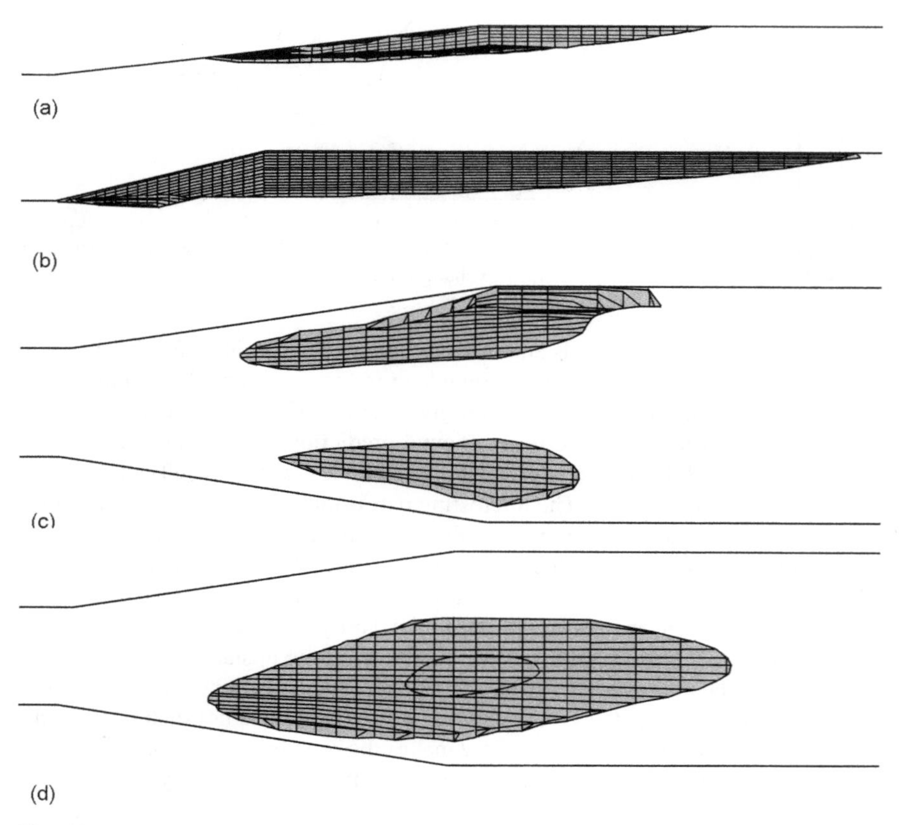

Fig. 11.4 Recirculation zones in eccentric expansions with two different slopes (a) 16.8° and (b) 8.6° and a concentric expansion with a slope of 8.6° at two different times (c) and (d). The solid grey areas illustrate the recirculation zones. Expansions are from 1 inch to 2 inch (25–50 mm) pipe sizes. Length and volume of the recirculation zones are: 331 mm and 71 300 mm^3 for the eccentric expansion with slope 16.8°, 213 mm and 18 200 mm^3 for the eccentric expansion with slope 8.6° and the size is time dependent for the concentric expansion.

catalogue data from a leading equipment manufacturer. Flow at the inlet is fully developed turbulent flow at an average velocity of 1.5 m/s. The simulation results are evaluated with respect to the length and volume of the recirculation zone generated in the expansion part. The measure of the volume provides an indirect measure of the surface area of the expansion swept by the recirculation zone.

Looking at the geometries one would imagine a recirculation zone in the expansion in both the eccentric and the concentric expansions. However, 3D visualisation of the flow using CFD simulations of the flow inside both expansions shows a difference in the size and position of the recirculation zones. In the eccentric expansion a recirculation zone is established in the expansion as expected (in the part of the flow domain just next to the sloped surface). The size of this recirculation zone depends on the slope of the expansion. The higher the slope, the larger the recirculation zone (Fig. 11.4). Hence, for cleaning purposes a very long expansion (low slope) is preferred, as this reduces the slope of the expansion and the size of the recirculation zone.

The concentric expansion, on contrast, does not show a single, large recirculation zone located 360° around the centre axis of the pipe along the walls. Instead, several smaller recirculation zones build, merge or disappear over time. This is a consequence of the Coanda effect (English, 1999). The fact that no steady recirculation zone is observed means that fluid exchange is relatively high in all areas of a concentric expansion. Furthermore, the rotating flow also disturbs the boundary layer, making the diffusion path of heat and mass smaller; hence, mass transfer to the soil is made easier (refer to the beginning of this chapter). In relation to hygienic design, this means that a concentric expansion is preferable to an eccentric one if, and *only if*, it is mounted in a position where it is drainable.

11.4.4 Example 3: Good cleaning of a spherical valve house

Spherical-shaped valve houses have an interesting geometry with respect to flow patterns. The use of these has exploded with the introduction of mix-proof valve types (Fig. 11.5). Cleaning tests have been carried out on spherical-shaped valve houses and many of these have shown that this type of valve house is a very good hygienic design (Jensen 2003). Why is that? The wall shear stress is low in large parts of the valve house because of the relatively large cross-sectional area compared with the inlet and outlet pipes (Jensen and Friis, 2004b) and the wall shear stress can only roughly explain the difference in cleanability shown in cleaning tests (Jensen and Friis, 2005). However, a good fluid exchange in the valve house promotes loosening of the bonds between soil and surface, hence, the wall shear stress needed to remove the soil from the surface is smaller than in areas of poor fluid exchange (Sinner's circle). The reason for the good fluid exchange in large parts of a spherical valve house is discussed below (flow patterns are discussed in detail in Jensen and Friis, 2004b).

As mentioned, the reason for good cleaning is found in good fluid exchange. Looking at construction drawings of a spherical valve house does not provide an

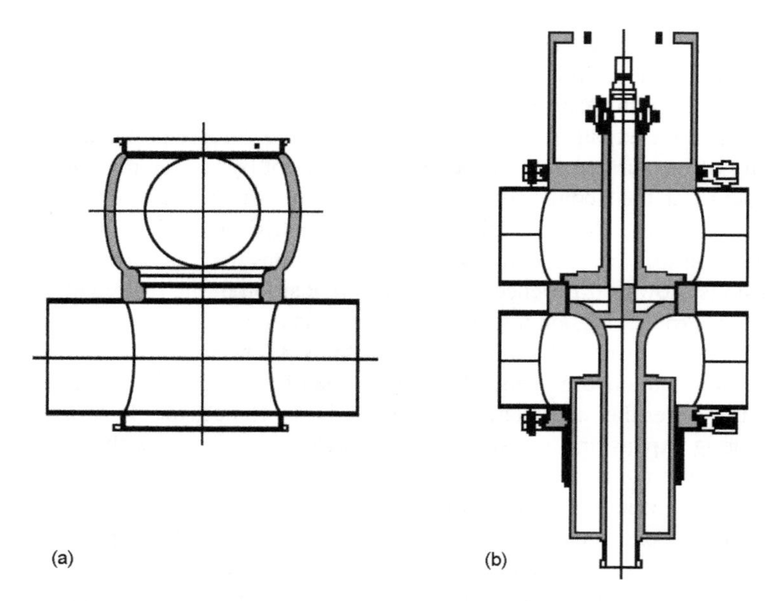

Fig. 11.5 Design of a mix-proof valve: (a) shows the outer surfaces of the valve house only and (b) is a cross-sectional cut down through the valve fully equipped for operation.

explanation. At first glance, recirculation zones will be present in the valve house because of shadow areas. However, performing 3D CFD simulations of the flow in a spherical valve house with the stem in the closed position shows how good fluid exchange is achieved (Jensen and Friis, 2004b). The key to the good fluid exchange is that only a few stationary recirculation zones are present in the valve house. What appear to be shadow areas looking at construction drawings are, in fact, areas where the recirculation zones rotate around axes parallel to the axial axis of the inlet and outlet pipes. In these zones the fluid moves along these axes from just downstream of the inlet to the valve house and towards the outlet of the valve house. Moving recirculation zones are also known as swirl. The advantage of a swirling flow is that the part of the detergent entering through the inlet of the pipe that goes into the swirl is moved downstream in a circling motion. In the case where this twisting motion has one side of the flow path located near a wall, detergent is moved to the wall and the soil is loosened.

In the spherical valve house investigated, two swirl zones in each side of the valve, on top of one other, move down through the valve house generating good fluid exchange in the upper and lower part of the valve house (Fig. 11.6). The part of the wall located between the two swirl zones is exposed to flow conditions of a stagnation zone nature where both fluid exchange and wall shear stress are low. Such a zone is present from the inlet to approximately three-quarters of the way downstream in the valve house. Hence, this is an area of potential hygiene problems. Comparing the flow patterns with data from actual

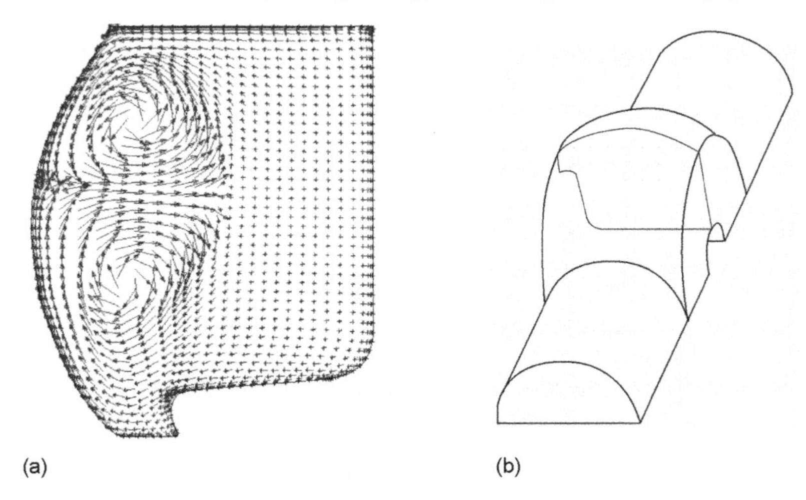

Fig. 11.6 (a) Vector plot showing the flow pattern in a cross-section in the mix-proof valve. (b) The location of the cross-section.

cleaning trails, the conclusion is that the areas that prove most difficult to clean in a standard cleaning test are in fact the areas where fluid exchange and wall shear stress are fairly low.

The swirl in the valve house is generated because of a single design feature (probably not by intention). Certain parts of a pipe mounted on a spherical surface intersect with the valve house slightly before the rest of the pipe intersects with it. This applies especially to part of the inlet pipe located on the equator of the valve house. The same happens to the flow in the pipe. Liquid first enters the valve house at the equator. This liquid experiences empty space to the sides and above and below and it tries to fill the space. Normally this will create a recirculation zone (e.g. flow over a backward-facing step – Durst and Tropea, 1981). This is not the case in the mix-proof valve as the liquid, after turning towards the empty space, hits the spherical-shaped walls and is reflected along the walls until the liquid meets the liquid from the other swirl zone generated simultaneously. Because of the momentum from the inlet pipe, this movement goes along the main flow direction generating the swirl.

This example and the example on expansions illustrate that thinking swirling flow in a design of a piece of closed equipment will promote cleaning of the surfaces located in the swirling zones. Swirl also generates pressure loss and often swirl-generating devices (e.g. winglets and obstacles) present other problems related to cleaning, which are not be discussed here.

11.5 Future trends

It is now possible to manufacture closed processing equipment to optimise the design of equipment with respect to the cleaning effect of fluid flow. This should lead to equipment with a better hygienic design than is seen today. Comparing data from cleaning tests with information on flow patterns obtained from CFD simulations also aids in obtaining a more thorough understanding of certain flow patterns, positive or negative, on the cleaning efficiency. Both increased knowledge and the use of CFD in the design of equipment, could aid in the writing of future guidelines and recommendations on the design of closed processing equipment. This is an excellent opportunity to spread the knowledge obtained using CFD to small manufacturers of equipment to whom CFD tools are too big an investment. Furthermore, data from CFD simulations should be used to visualise areas of potential hygiene problems in closed processing equipment. This would make it clear to non-specialists in hygienic design and fluid dynamics why certain areas are problematic because of unfavourable flow conditions.

The considerations and discussions given in this chapter are based on transient (time-consuming) CFD simulations, as fluid exchange is needed to make a complete prediction of the cleanability of a component. Work is in progress to mirror fluctuations in wall shear stress measured by electrochemical methods using steady-state CFD simulations (Jensen *et al.*, 2005). The fluctuating signal is correlated to the degree of cleaning (Lelièvre *et al.*, 2002), hence, if predictions of these fluctuations are possible by steady-state CFD simulations a relatively fast method is obtained.

Work published shows that making designs that promote the mass and energy transfer to surfaces inside equipment is a path to investigate further. Promotion of mass and energy transfer needs to be done by designs where care has to be taken not to introduce other problems related to cleaning. These problems could be the creation of new shadow zones and areas of very low angles between two meeting surfaces.

It is believed that for future optimisation of design with respect to cleaning characteristics to be possible, taking other process parameters into consideration, CFD is unavoidable. Good hygienic designs exist today, so to improve these either details have to be changed and compared, or totally new design concepts for closed food processing equipment are needed.

11.6 Sources of further information and advice

Further information on the subjects covered in this chapter can be found in the publications given in this chapter. The EHEDG website is a good starting point for advice on hygienic design. Their guidelines present accepted best practice. Research, up-to-date information and experience in the area of hygienic design of closed processing equipment can be found through organisations and companies such as Campden and Chorleywood Food and Research Association in England, TNO in the Netherlands, Cocker Consulting in the Netherlands, Unilever R&D, Insitut National de la Recherche Agronomique (INRA) in France, Technische Universität München in Germany and the Technical University of Denmark.

11.7 References

ANON. (1992), 'A method for assessing the in-place cleanability of food-processing equipment', *Trends Food Sci Tech*, **3**, 325–328.

ANON. (1993), 'Hygienic design of closed equipment for processing of liquid food', *Trends Food Sci Tech*, **4**, 375–379.

ANON. (1997), *European Standard EN 1672-2,* Commission of European Communities.

ANON. (1999), *Computational Dynamics Methodology Manual (on-line) ver. 3.10A,* CD Adapco Group (www.cd-adapco.com)

BÉNÉZECH TH, FAILLE C and LECRIGNY-NOLF S (1998), 'Removal of *Bacillus* spores from closed equipment surfaces under cleaning-in-place conditions', in Wilson D I, Fryer P J and Hasting A P M, *Fouling and cleaning in food processing '98, Jesus College*, Cambridge, 160–167.

BERGMAN B-O and TRAGARDH C (1990), 'An approach to study and model the hydrodynamic cleaning effect', *J Food Process Eng*, **13**, 135–154.

CASEY M and WINTERGERSTE T (2000), *ERCOFTAC Special interest group on quality and trust in industrial CFD – Best practice guide*, ERCOFTAC, US (http://www.ercoftac.org/).

CHEN H C and PATEL V C (1988), 'Near wall turbulence models for complex flows including separation', *AIAA Journal*, **26**, 7–12.

DUDDRIDGE J E, KENT C A and LAWS J F (1982), 'Effect of surface shear stress on the attachment of *Pseudomonas fluorescence* to stainless steel under defined flow conditions', *Biotechnol Bioeng*, **24**, 153–164.

DURST F and TROPEA C (1981), 'Turbulent, backward-facing step flows in two dimensional ducts and channels', *Symposium on turbulent shear flows*, Davis, CA, University of California, 18.1–18.9.

ENGLISH J (1999), 'The coanda effect in maritime technology', *Nav Archit*, April, 18–22.

FERZIGER J H and PERIĆ (1999), *Computational methods for fluid dynamics*, 2nd edn, Berlin, Springer.

FOWLER H W and MCKAY A J (1980), 'The measurement of microbial adhesion', in Berkeley R C W, Lynch J M, Melling J, Rutter P R and Vincent B, *Microbial adhesion to surfaces*, Chichester, Ellis Horwood Ltd, 143–161.

FRIIS A and JENSEN B B B (2002), 'Prediction of hygiene in food processing equipment using flow modeling', *Food Bioprod Process*, **80**, 281–285.

HALL J (1999), 'Computational fluid dynamics: A tool for hygienic design', in Wilson D I, Fryer P J and Hasting A P M, *Fouling and cleaning in food processing*, Luxemburg, European Commission, 144–151.

HAUSER G and KRÜS H (2000), *Hygienegebrechte gestaltung von bauteilen für die lebensmittelherstellung – schwachstellenanalyse durch tests und numerische berechungen*, Waalre, Cyclone Fluid Dynamics (web publication at www.cyclone.nl).

HOLAH J T (2003), 'Cleaning and disinfection', in Lelieveld H L M, Mostert M A, Holah J and White B, *Hygiene in food processing*, Cambridge, Woodhead, 235–278.

JENSEN B B B (2003), *Hygienic design of closed equipment by use of computational fluid dynamics*, Lyngby, Technical University of Denmark.

JENSEN B B B and FRIIS A (2004a), 'Critical wall shear stress for the EHEDG test method', *Chem Eng Process*, **43** (7), 831–840.

JENSEN B B B and FRIIS A (2004b), 'Prediction of flow in mix-proof valve by use of CFD – validation by LDA', *J Food Process Eng*, **27**, 65–85.

JENSEN B B B and FRIIS A (2004c), 'A numerical method for virtual cleaning testing', *International Congress on Engineering and Food (ICEF9)*, Montpellier, France.

JENSEN B B B and FRIIS A (2005), 'Predicting the cleanability of mix-proof valves by use of wall shear stress', *J Food Process Eng*, **28**, 89–106.

JENSEN B B B, ADLER-NISSEN J and FRIIS A (2000), 'Hygienic design of in-line components using CFD', in Holdsworth D, Fryer P J, Grace S, Hasting T, McLeod E and Richardson P, *Food & Drink 2000 – Processing Solutions for Innovative Products*, Rugby, Institution of Chemical Engineers, 26–28.

JENSEN B B B, ADLER-NISSEN J, ANDERSEN J F and FRIIS A (2001), 'Prediction of cleanability in food processing equipment using CFD', in Welti-Chanes J, Barbosa-Canvas G V and Aguilera J M, *Proceedings of the Eighth International Congress on Engineering and Food (ICEF8)*, Lancaster, PA, Technomic Publishing Co, 1859–1863.

JENSEN B B B, FRIIS A, BÉNÉZECH TH, LEGENTILHOMME P and LELIÈVRE C (2005), 'Local wall shear stress variations predicted by computational fluid dynamics for hygienic design', *Transactions of the Institute of Chemical Engineers, Part C Food and Bioprod Process*, **83** (C1), 53–60.

LELIEVELD H L M, MOSTERT M A and CURIEL G J (2003), 'Hygienic equipment design', in Lelieveld H L M, Mostert M A, Holah J and White B, *Hygiene in food processing*, Cambridge, Woodhead, 122–166.

LELIÈVRE C, LEGENTILHOMME P, GAUCHER C, LEGRAND J, FAILLE C and BÉNÉZECH T (2002), 'Cleaning in place: effect of local wall shear stress variation on bacterial removal from stainless steel equipment', *Chem Eng Sci*, **57**, 1287–1297.

LELIÈVRE C, LEGENTILHOMME P, LEGRAND J, FAILLE C and BÉNÉZECH T (2003), 'Hygienic design: influence of the local wall shear stress variations on the cleanability of a three-way valve', *Chem Eng Res Des*, **81** (A9), 1071–1076.

MAJOOR F A (2003), 'Cleaning in place', in Lelieveld H L M, Mostert M A, Holah J and White B, *Hygiene in food processing*, Cambridge, Woodhead, 197–219.

PATANKAR S V (1980), *Numerical heat transfer and fluid flow*, New York, Hemisphere Publishing Corporation.

RICHARDSON P S, GEORGE R M and THORN R D (2001), 'Application of computational fluid dynamics simulation to the modelling of cleanability of food processing equipment', in Welti-Chanes J, Barbosa-Canvas G V and Aguilera J M, *Proceedings of the Eighth International Congress on Engineering and Food (ICEF8)*, Lancaster, PA, Technomic Publishing Co, pp. 1854–1858.

RODI W (1991), *Experience with two-layer model combining k–ϵ model with a one-equation model near the wall,* Report: AIAA-91-0216, Reno, NV, American Institute of Aeronautics and Astronautics, 1–12.

SHAMES I H (1992), *Mechanics of fluids*, Singapore, McGraw-Hill Inc.

SINNER H (1960), *Über das Waschen mit Haushaltwaschmaschinen: in welchem Umfange erleichtern Haushaltwaschmaschinen und -geräte das Wäschehaben im Haushalt?*, Hamburg, Haus + heim verl.

TIMPERLEY D (1981), 'The effect of Reynolds number and mean velocity of on flow on the cleaning in-place of pipelines', in Hallström B, Lund D B and Tragardh C, *Fundamentals and applications of surface phenomena associated with fouling and cleaning in food processing: Proceedings, Tylösand*, Lund, Lund University Reprocentralen.

TIMPERLEY D A and LAWSON G B (1980), 'Test rigs for evaluation of hygiene in plant design', in Jowitt R, *Hygienic design & operation of food plant*, Chichester, John

Wiley & Sons Limited, 79–108.
TUCKER G and HALL J (1998), 'Computational fluid dynamics as an aid to efficient hygienic design of food processing equipment', *Food Review*, June, 10–12.
VERSTEEG H K and MALALASEKERA W (1995), *An introduction to computational fluid dynamics – the infinite volume method*, Essex, Longman Scientific & Technical.

12

Improving the hygienic design of heating equipment

A. P. M. Hasting, Tony Hasting Consulting, UK

12.1 Introduction

Heat transfer is perhaps the most widely used unit operation applied within the food industry and many key processes such as pasteurisation and sterilisation are based around it. Heat transfer can be applied on either a batch or continuous basis and the mechanisms involved can be convection, conduction or radiation, but are usually a combination of these. In addition, heat transfer can take place either through direct contact with the service medium or indirectly across a heat transfer surface.

The most typical practical operations involving heat transfer are:

- heating;
- cooling/chilling;
- freezing;
- evaporation;
- condensation;
- radiation;
- drying.

Heat transfer operations can therefore involve a change of phase in the case of evaporation, freezing and drying. These changes of phase processes are complex operations in their own right and the scope of this chapter is limited to heat transfer equipment for applications with no phase change.

Heat transfer can take place in vessels and tanks but the most common equipment used is the heat exchanger. A wide range of heat exchanger geometries are available in practice and the major ones used for food applications are classified in Table 12.1.

Table 12.1 Classification of main types of food industry heat exchangers

Generic type	Options available	Other potential options
Plate	Standard gap/wide gap Conventional design, fluids A and B flow through alternate channels Dual plate Each 'plate' consists of two plates with an air gap in between Plate in shell Plate pack with alternate plates ungasketed and whole pack installed within a shell. Allows larger volumes of vapour to be handled than conventional design. Liquid flows through gasketed channels	Plates gasketed or welded
Scraped surface	Rotary Blades on rotating shaft scrape internal heat transfer surface. External jacket for heating/cooling medium	Horizontal or vertical orientation
	Linear Scrapers on shaft move in linear, reciprocating motion	Horizontal or vertical orientation, multiple tubes available within shell
Tube (straight)	Monotube A single tube within a tube Multiple tube in tube A single shell within which there are a number of smaller diameter tubes Triple tube Three concentric tubes, fluid A flows through inner tube and outer annulus, product B though inner annulus	Tubes may be smooth or corrugated. Angle of corrugation relative to the vertical can be varied to alter heat transfer characteristics
Tube (coiled)	Monotube A single tube within a tube Triple tube Three concentric tubes, fluid A flows through inner tube and outer annulus, product B through inner annulus	

12.2 Heat exchanger design

On a purely heat transfer basis, the design of any heat exchanger is a balance between achieving the desired thermal duty and the associated capital and running costs. The increasing cost of energy has also led to a far greater implementation of heat recovery and process integration approaches.

Table 12.2 Effect of product and process factors on heat exchanger design

	Impact on exchanger design
Product factors	
• Particles	Heat exchanger geometry
• Specific heat	Thermal load on exchanger and heat transfer area
• Thermal conductivity	Heat transfer area
• Fouling	Heat transfer area
• Rheology/viscosity	Heat transfer, pressure drop, exchanger geometry
Process factors	
• Temperature profile	Heating rate, fouling of exchanger
• Pressure drop	Type of exchanger
• Heating medium	Heat transfer area, fouling
• Heat recovery	Heat transfer area, energy costs, exchanger geometry

In practice the design process is more complex than purely heat transfer and there are a number of product and process factors that can have a significant impact on the design of the exchanger (Tables 12.2 and 12.3). In addition the design of the heat exchanger is often only a small, though important, part of the complete line. Individual heat exchangers are also being required to handle an increasingly broad range of products, which have widely differing characteristics, making optimisation difficult as the heat exchanger will have to be designed to handle the most challenging fluid.

Current food industry guidelines on heat exchanger design, particularly hygiene-related issues, refer more to the overall process within which the equipment is utilised than the individual heat exchanger. The 3 As guidelines have been developed with a strong focus on the dairy industry, whereas the more recent EHEDG guidelines are focused on the principles involved.

Table 12.3 Influence of fluid viscosity on heat transfer performance

Service fluid: Water
Flow regime: Turbulent

Product viscosity ($N\,S\,m^{-2} \times 10^{-3}$)	Heat transfer performance (%)
1	100
2	87
4	74
6	67
8	62
10	59

12.3 Developments in heat exchanger design

Although heat transfer is a very mature technology there have been a number of significant developments within recent years.

12.3.1 Incremental improvements in heat transfer performance

Over recent years the metal thickness of the plates in plate heat exchangers has been reduced from typically 0.8 mm to 0.4–0.5 mm. This has reduced the weight of the material required by 40–50% and hence the cost. There will also be an improvement in heat transfer as the thinner metal results in a reduced thermal resistance. The improvement will be significant only for applications where both fluids are of low viscosity and in turbulent flow. For a typical application with equal thermal resistances on the product and service sides of $10^{-4}\,m^2\,K/W$, a reduction in thickness from 0.8 to 0.5 mm will improve heat transfer performance by 8%.

In tubular heat exchangers, improvements in heat transfer performance have been achieved by using corrugated tubes in place of conventional plain tubes. The corrugations are claimed to enhance heat transfer by disruption of the laminar boundary layer as the fluid flows across it. It is, however, probable that such enhancement would be only minimal for higher-viscosity fluids as the corrugations are unlikely to have a major effect on the fluid dynamics close to the heat transfer surface.

12.3.2 Alternative geometries to address the technical limitations of existing designs

Coiled tubes

Coiled tube designs have been used for a number of years to provide a more compact design of exchanger. Recent work has indicated that the movement of the fluid in a continuous spiral provides an enhanced mixing and heat transfer performance than would be predicted from conventional heat transfer design correlations for linear systems.

Dual plates with air gap

One of the main hygiene concerns with conventional plate heat exchangers is that the two fluid streams are separated from each other by a single, relatively thin metal surface. If this surface becomes damaged through, for example, corrosion or flow-induced vibration, there is potential for cross-contamination to occur between the two streams. If heat recovery is used to heat incoming cold product with hot product, it is possible to contaminate the heat treated (pasteurised/sterilised) product with raw, untreated product.

Current ways of minimising this risk are to maintain a higher pressure on the pasteurised/sterilised side to ensure any flow is from processed to raw product. This does not, however, provide complete assurance as microorganisms can move against a pressure gradient. Another approach is to use a secondary water circuit with a recirculation pump such that direct product/product contact is avoided. This results in the heat transfer area being considerably increased, and careful maintenance of the recirculation circuit is required.

The use of a dual plate with an air gap has been developed to provide additional assurance against cross-contamination. The principle is based on that of the double seat valve, with two plates in place of the valve seats and an air gap to atmosphere between the two plates. Any defect in one of the plates will result in fluid passing into the air gap and out to atmosphere. It is, however, counter-intuitive to deliberately create an air gap within the heat transfer path and in so doing provide an additional heat transfer resistance. Equipment manufacturers have minimised the loss of thermal efficiency by using an air gap of 3–5 μm and thinner plate materials of 3 mm. The loss of performance will be greater the lower the overall heat transfer resistances on the fluid side, for example an application with two low viscosity fluids, rather than where one or more of the fluids is viscous. Table 12.4 shows the effect of the air gap on heat transfer performance for different applications.

Although the reduction in heat transfer performance can be minimised by reducing the air gap, there are a number of potential hygiene issues:

- If a fluid enters the narrow air gap between the two plates due to a defect, surface tension effects may prevent fluid draining out of the system by gravity.
- If a defect does occur, it will be difficult to ensure that any product in the gap can be cleaned effectively due to the minimal flows of cleaning fluid that can be delivered into the gap through the defect.
- Any residual product within the air gap could provide a source for recontamination of product during subsequent production.

Linear scraped surface heat exchanger

Scraped surface heat exchangers are used for processing fluids that other geometries cannot handle, such as large particles. Conventional designs are

Table 12.4 Effect of air gap on heat transfer performance

Air gap (mm)	Duty 1 Dual plates each 0.4 mm thick Product viscosity: $1\,\text{N}\,\text{S}\,\text{m}^{-2} \times 10^{-3}$ Service fluid viscosity: $1\,\text{N}\,\text{S}\,\text{m}^{-2} \times 10^{-3}$ Reduction in performance (%)	Duty 2 Dual plates each 0.4 mm thick Product viscosity: $10\,\text{N}\,\text{S}\,\text{m}^{-2} \times 10^{-3}$ Service fluid viscosity: $1\,\text{N}\,\text{S}\,\text{m}^{-2} \times 10^{-3}$ Reduction in performance (%)
0.000	0.0	0.0
0.001	8.2	5.0
0.002	15.1	9.5
0.003	21.1	13.6
0.004	26.2	17.3
0.005	30.8	20.8

based on a rotating approach whereby scrapers on a rotating shaft continually remove fluid from the heat transfer surface and mix it back into the bulk, while allowing fresh material to reach the surface. A recent development has used linear rather than rotary motion to provide the scraping action, with a number of specially designed baffles attached to the reciprocating shaft. The design of the baffles can be varied to suit the product being processed.

Fluid bed heat exchanger
For severely fouling liquids a fluid bed exchanger has been developed over a number of years for industrial applications in which small particles (1–4 mm diameter) of glass, ceramic or metal are fluidised inside vertical parallel tubes by the upward flow of liquid. The solid particles disrupt the laminar boundary layer to improve heat transfer and in addition the particles have an abrasive effect on the wall of the heat exchanger tubes, helping to minimise the build up of fouling deposits (Klaren 2001). It is claimed such techniques are suitable for food industry applications such as raw juice heating.

Improved working pressure capabilities
In addition to reducing metal thickness, considerable efforts have gone into increasing the range of operating pressures that the plate heat exchanger can operate within. Modified designs are now capable of operating at working pressures of 20 bar, which although lower than tubular systems are still a significant improvement.

Improvements in defect detection
Heat exchanger surfaces can become degraded over time owing to both physical and chemical stresses resulting in cracks or pinholes in the heat transfer surface leading to potential chemical or microbiological contamination of product. Typically plate heat exchangers have to be dismantled before either using a dye penetrant technique or sending the plates away for inspection by the supplier. A technique has been developed (Bowling 1995) whereby an electrolyte is circulated under pressure through the product side of the heat exchanger. On the surface side, water is circulated and the conductivity is monitored. If a defect is present, flow of electrolyte through the defect under the influence of a pressure differential will result in a detectable change in conductivity. It is claimed that this technique has a similar sensitivity to that of traditional dye penetrant methods.

12.4 Future trends

There are number of areas for future improvement in heat exchanger design.

12.4.1 Modified surfaces to reduce fouling/enhance cleaning

Fouling is still one of the major unresolved problems in heat transfer, resulting in reduced performance, the need for regular cleaning and the cost incurred due to loss of production time and cleaning. An increasing amount of work has been carried out on the modification of the heat transfer surface to either reduce the rate of fouling or significantly reduce the time required to clean the exchanger.

A number of potential approaches are currently being researched.

Application of a coating to the surface

It has been known for some time that the adhesion of fouling deposits to surfaces is reduced, the lower the surface energy of the surface. Attempts have been made to coat surfaces with ceramics, poly(tetrafluoroethane) (PTFE) or other non-toxic layers (Zhao *et al.* 2002). However, such coatings must be thin to avoid the loss in heat transfer performance because of their low thermal conductivity. This limits the adhesion between metal surface and coating and hence the ability to withstand mechanical stresses. Recent work using composite coatings of Ni-P-PTFE showed improved mechanical strength and the attachment of thermophilic streptococci could be reduced by more than 99% (Zhao *et al.* 2002).

Modification of the material surface

Novel low-fouling surfaces have been developed by ion implantation, sputtering or electrolytic deposition (Muller-Steinhagen and Zhao 1997). These have the advantage of improved abrasion resistance and strong adhesion. Results for diamond-like carbon (DLC) and sputtered composite coatings (CrN, CrC, Cr_2O_3) showed reductions of 80–99% in thermophilic streptococci (Zhao and Muller-Steinhagen 1999).

12.4.2 Alternative geometries to achieve higher heat transfer area/volume ratios

Plate heat exchangers are generally considered to be the most compact of commercial heat exchanger designs in terms of heat transfer area to volume ratio, 150–350 m^2/m^3. These are due to the fundamental design principle, which uses a narrow gap, 2.5–6.0 mm, between the heat transfer surfaces. An extension of this approach has been the laboratory development of cross-corrugated polymer film heat exchangers with gaps between 0.3 and 1.5 mm, resulting in volumetric heat transfer areas of 500–2500 m^2/m^3 (El-Bourawi and Ramshaw 1999).

12.5 Conclusions

Heat transfer will continue to form a key unit operation within the food industry. Increasing cost pressures and demands for flexibility will continue to challenge the ingenuity of the heat exchanger designers to explore further ways of enhancing the process. It is likely that one of the most promising areas for improvement lies in the modification or coating of surfaces to reduce fouling and enhance cleaning.

12.6 References

BOWLING, M. (1995) Leakage Detection, International Patent Application WO95/16900, 22nd June 1995.

EL BOURAWI, M.S. and RAMSHAW, C. (1999), Fouling mitigation in polymer film compact heat exchangers, in Bott, T.R., Watkinson, A.P. and Panchal, C.P. (eds) *Proceedings of International Conference on 'Mitigation of heat exchanger fouling and its economic and environmental implications'*, Banff, Canada, 169–176.

KLAREN, D.G. (2001) Self-cleaning heat exchangers, in Muller-Steinhagen H. (ed) *Heat exchanger fouling – Mitigation and cleaning technologies*, IChemE, Rugby, UK, 186–198.

MULLER-STEINHAGEN, H. and ZHAO, Q. (1997), *Chemical Engineering Science*, **52**(19), 3321–3332.

ZHAO, Q. and MULLER-STEINHAGEN, H. (1999), Influence of surface properties on heat exchanger fouling, in Bott, T.R., Watkinson, A.P. and Panchal, C.P. (eds) *Proceedings of International Conference on 'Mitigation of heat exchanger fouling and its economic and environmental implications'*, Banff, Canada, 217–228.

ZHAO, Q., LIU, Y. and MULLER-STEINHAGEN, H. (2002), Effects of interaction energy on biofouling adhesion, in Wilson, D.I., Fryer, P.J. and Hasting, A.P.M. (eds) *Fouling, Cleaning and Disinfection in Food Processing*, Jesus College, University of Cambridge, 213–220.

13

Improving the hygienic design of equipment in handling dry materials

K. Mager, Quest International, The Netherlands

13.1 Introduction: principles of hygienic design

Several guidelines for equipment design have been published in order to produce foods in a hygienically acceptable way. The principles are to prevent the contamination of food products by substances that would adversely affect the health of the consumer. In that respect these guidelines describe design principles based on:

- smooth product contact surfaces;
- no dead areas; and
- the avoidance of condensation in the equipment.

These principles of design for equipment in the food industry are originally based on the handling of liquids (EHEDG, 2004). However, food products with other product characteristics may also need to be taken into consideration. This means that the principles of design should also count for dry particulate materials. It is important, therefore, to realise what differences there are in the characteristics of powders as compared with liquids.

13.2 Dry particulate materials and hygienic processing

In this chapter dry particulate materials, more commonly called powders, fall in the size range of less than 10 μm for ultrafine powders up to a several millimetres for agglomerates and granulates. Generally, powders are defined as consisting of individual particles that have a diameter smaller than 150 μm. Larger particulates are often composed of many smaller particles, and substructures can be achieved spontaneously by the natural phenomena of adhesion and electrostatic forces. However, stronger structures are best achieved by forcing particles to bind together using moisture or especially added binders while the particles are being fluidised or mixed.

A definition of a dry product in the sense of microbial stability is not so easily given, since it can change slightly from product to product. As a rule of thumb, one can say that when the water activity is below 60%, little to no microbial growth will occur. Dry materials can be characterised by both their single particle (such as shape and size) and their bulk (such as bulk density and flowability) characteristics. It should be emphasised that the bulk characteristics of industrial dry materials are at least as important as their single particle characteristics, and for each material, the most important characteristics influencing materials handling will vary.

Flowability is an important characteristic for dry material retention in equipment, and generally improves with:

- increase in particle size and particle sphericity;
- decrease of moisture content;
- decrease in fines content;
- decrease in surface stickiness;
- decrease in neutralisation of surface energy/charge.

Hygienic processing also influences the dry material quality properties of:

- aroma;
- chemical, biological or physical activity;
- colour;
- flavour.

In general it can be stated that, based on their characteristics, powders have the tendency to stick to product contact surfaces and are more likely to remain in the process line as compared with liquids. Also, lump formation and hygroscopic properties are important parameters in enhancing this effect. As mentioned earlier, an increased moisture content in the powder (and remaining powder residues!!) can cause serious proliferation of microorganisms. The above-mentioned effects have to be taken into serious consideration when designing equipment processing powders.

13.3 Cleaning regimes

The criteria for hygienic design of equipment and plants for dry materials handling depend upon the moisture content of the dry material and the method of cleaning.

Whether the equipment is cleaned wet or dry has a significant effect on the design criteria. If wet cleaning procedures are applied the design has to fulfil the general requirements for equipment in the liquid area as described in several

EHEDG documents (EHEDG 2001, 2003, 2004). If only dry cleaning procedures are applied, less stringent requirements can be allowed as will be described in the following sections. However, it is important to establish that it is possible to suffice with dry procedures only. Sometimes factories do infrequent wet cleaning and it is known from trend analyses that microbiological contamination occurs after these periods.

If wet cleaning procedures are applied it is extremely important that the equipment is dried immediately, because:

- remaining wet spots can be the cause of lump formation in the subsequent batch;
- the proliferation of microorganisms in the wet spots can contaminate the powders.

Moreover, the combination of powders and water provides an ideal source for microbiological growth! It is particularly important that critical areas such as dead legs, sharp corners are behind seals and gaskets are locations that can be dried within a reasonable time in order to avoid the favourable conditions for microbiological contamination. In this sense it should be emphasised that the need of wet cleaning should be taken into serious consideration. Wet cleaning is a critical hazard in the dry material handling area and dry cleaning procedures are preferred in all respects.

Dry cleaning is applicable for dry food material contact surfaces where:

- dry material remaining in the equipment as loose layers or dust covering does not present any risk of degrading the quality of the dry material subsequently produced;
- possible cross-contamination of dry material during a production change to another material presents no problem to the quality or safety of the dry material subsequently produced;
- dry material remaining in the equipment does not present any risk of microbial growth occurring due to the prevailing moisture content, temperature and humidity conditions;
- dry material is non-hygroscopic and non-sticky.

Dry cleaning procedures include the use of vacuum cleaners, brushes and scrapers. However, procedures can also be applied in which the equipment is rinsed with 'neutral' agents such as salt and starch.

13.4 Design principles

Compared with liquids, dry materials handling must take into account the possibility of material lump formation, creation of dust explosion conditions, high moisture deposit formation in the presence of hot air, and material remaining in the equipment after plant shutdown (even if a degree of self-emptying is achieved). Powders tend to stick in the process equipment more than liquids.

13.4.1 Materials of construction

Construction materials coming in contact with food (including associated adhesives) must be food grade (Food and Drug Administration (FDA) or European Food Safety Authority (EFSA) approved or national equivalent). Selection of construction materials depends greatly upon the dry materials, method of cleaning and cleaning agents to be used. The abrasive characteristics of powders can particularly affect the product contact surfaces.

Metals

Hygienic dry materials handling is best conducted with product contact surfaces of stainless steel. Suitable grades are SS 304, 304L (EN 1.4301/1.4306) and SS 316, 316L (EN 1.4401/1.4404). The 316 grades are more resistant to chloride-containing solutions, especially under wet and hot conditions.

Aluminium and aluminium alloys (coated and non-coated) might also be used as dry material contact surfaces where only dry cleaning is applied. However, the abrasive characteristics of the processed powder shall be considered in this choice. Moreover, if aluminium is specified from an operational or weight aspect, there is a potential corrosion problem when a wet cleaning procedure is applied. Carbon steel can also be considered as a contact surface in components involving dry processing and dry cleaning operations.

Non-metals

Plastics (e.g. polycarbonate, polyetheretherketone (PEEK), polyvinylidene fluoride (PVDF), polyacrylamide (PA) and polytetrafluoroethene (PTFE)) and elastomers (e.g. nitrile butyl rubber (NBR), viton, ethylene propylene diene monomer (EPDM) and silicon rubber) may be used. When in contact with dry materials they must retain their original surface condition and conformational properties when exposed to the processing conditions of temperature and humidity, and also during cleaning operations. It is important to realise that plastics and elastomers in particular are sensitive to abrasive powders and therefore the contact surfaces should be minimised as much as possible.

Glass is a hygienic material, but should not be used because of the risk of breakage and subsequent difficulty in detecting broken glass in dry materials. It is recommended that the glass is replaced with another material, e.g. polycarbonate.

Non-metallic surfaces used in dry materials handling can create electrostatic charges on the material. This can cause surface adhesion by small particles. Electrostatic effects during dry materials handling in pneumatic conveying systems and non-metallic equipment parts, for example, can be problematic, and therefore special attention should be paid to accessibility and cleaning in such systems.

13.4.2 Product contact surfaces

Product contact surfaces should be smooth and resistant against both dry material contact and against liquid chemicals used in wet cleaning. Product contact surfaces therefore should be free of crevices, pitting, pinholes and any hairline cracking that can cause material penetration and cleaning difficulties.

A roughness standard of $R_a \leq 0.8\,\mu m$ is recommended where there is a risk of microbial growth associated with high moisture content in the dry material or wet cleaning. As the surface roughness of cast materials and carbon steels does not meet the recommended figure above, the cleanability of the components made with these materials require further investigation in relation to the actual dry material being handled.

In order to carry out a dry cleaning operation, contact surfaces should be fully accessible for safe manual cleaning and inspection. In order to carry out a hygienic wet cleaning operation, contact surfaces should not be horizontal, but have a slight slope to facilitate drainability of cleaning solutions.

The possibility for product contact on sharp internal corners ($r \leq 6\,mm$) and recesses, etc., where dry material can accumulate, should be avoided. Windows and inspection ports mounted in product contact surfaces should be flush with the surrounding surfaces to minimise dry material build-up. When using non-metallic materials as contact surfaces, the porosity of the materials should be investigated with regard to their ease of cleanability.

13.4.3 Static seals (gaskets) for duct and flange connections

Static seals should be of an elastic material, have a non-porous surface and be cleanable. They should be mounted to create a flush surface without any crevice with the surrounding metallic body (Fig. 13.1a). The seal material should be abrasion-resistant to the dry material being handled. In the case of dry processing and dry cleaning only, closed cell-foamed non-absorbing materials for gaskets or seals can be applied. Open foam material is not allowed. Static seals should be clean before assembly and the possibility for penetration of dry material into the gasket or seal during equipment operation should be avoided.

Misalignment of ducts should be avoided as dry materials can be trapped on the misaligned ridges (Fig. 13.1b). Assembly of seals and gaskets for vessels of large diameter require special attention to prevent operational problems, especially air and liquid (washing) leakage and material dust emissions to atmosphere.

PTFE can be used as a static seal in combination with an elastomer (food grade, FDA or EFSA approved or national equivalent). The PTFE should be of high-density resilient quality. Metal-to-metal contact duct assemblies (Fig. 13.1c) and paper-type gaskets between flanges can be applied where a plant operates at atmospheric pressure and requires no wet cleaning.

13.4.4 Flexible connections

One of the biggest hygienic design concerns in the dry materials handling area is that of the flexible connections in process lines. Flexible connections between duct ends are always liable to cause dry material build-up between the flexible

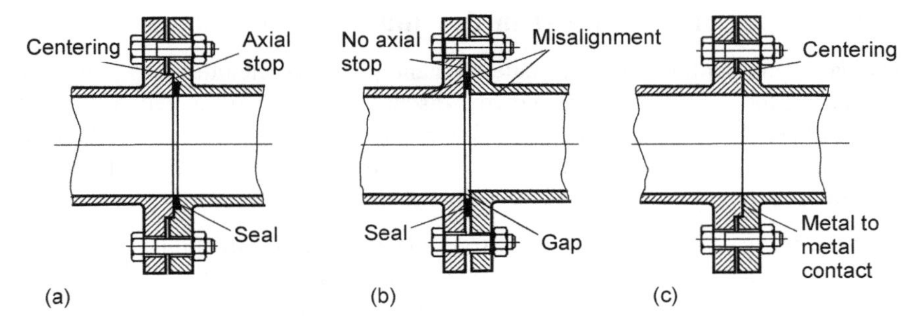

Fig. 13.1 Examples of static flange seals for dry products: (a) hygienically designed seal usable for wet cleaning, (b) seal creating a gap and misalignment, (c) metal-to-metal flange joint (only for dry cleaning).

material and metal duct surface. In smaller diameter devices, duct ends are connected with rubber or plastic sleeves.

Ring clamps for mounting flexible connections should be placed close to or right at the duct end to minimise dead areas for dry material build-up as demonstrated in Fig. 13.2. The plastic sleeve must allow small axial and radial movements without generating axial forces. The flexible material should have a smooth surface that minimises surface build-up of dry material.

Larger diameter duct ends are often connected with rubber type profiles mounted with flanges. These devices are most probably never removed and cleanability is difficult. As such these are critical areas in the process line. This is one of the areas where improvements are needed and present a challenge for the current engineers.

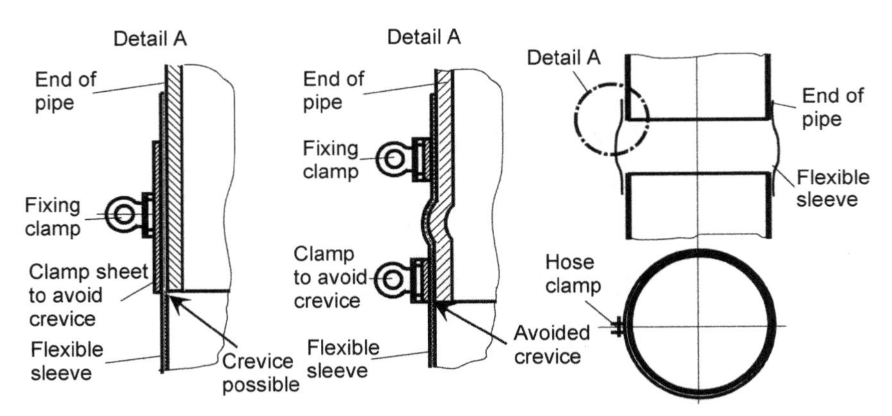

Fig. 13.2 Examples of flexible connection duct ends (right). One ring clamp close to the pipe end used for smaller diameters; crevice not totally avoidable (left). Application of two clamps, one of which is mounted directly at the pipe end to avoid any crevice (middle).

13.5 Types of equipment in dry material handling areas

Typical equipment in the dry material handling area includes, e.g., powder blenders, grinders, rotary valves, powder discharge systems, fluid bed dryers and spray dryers. In principle all the design criteria as mentioned earlier shall be implemented in these equipment. However, typical hazards in all parts of the total process line are considered as well. For example:

- In the *spray driers and fluid bed system* special attention should be paid to the air inlet system. The drying air in fluid bed systems and spray dryers should be filtered to avoid a direct product contamination. EHEDG recommendations are that the filters should be at least of class EU-7 for hot drying air. In the downstream part of the spray dryer the powder has to be transported with cold air and this shall be filtered with an EU-10 filter. The air outlet system can also be a critical area. Dust is collected in these filters and frequently pulsed back into the product. Bag filters in the fluid bed system can be especially contaminated when the cleaning procedures are not carried out according to strict procedures.
- The dust extraction system in the *powder charge cabinets* should be designed in such a way that lumps in the exhaust line cannot fall back into the powder (see Fig. 13.3).
- *Rotary valves* cannot be cleaned in place and therefore special attention has been paid in order to design retractable rotor devices in order to enhance the cleaning procedures.

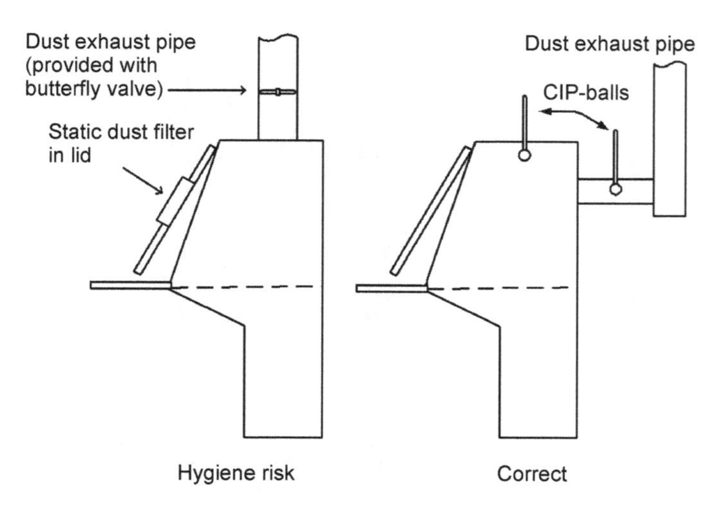

Fig. 13.3 Powder charge cabinet (CIP = cleaning-in-place).

13.6 Conclusions: Improving hygiene in processing powders

It is clear that in the powder-handling area, special design criteria and process procedures are required. A dedicated (pre-) HACCP (Hazard Analysis of Critical Control Point) study should be part of the development process during the design phase of the process lines. Decisions on the required cleaning procedures (wet or dry cleaning) are crucial. In the case that only dry cleaning procedures are to be applied, less stringent design criteria might be possible. On the other hand, the equipment itself in the powder-handling area is often not of a hygienic design when wet cleaning procedures have to be applied. Rotary valves, sieves, bag filters, flexible connections are all examples of equipment and process line components that are still a challenge for equipment manufacturers to improve.

13.7 References

EHEDG (2001) Document No. 22. General Hygienic Design Criteria for the safe processing of dry particulate materials, March. Campden & Chorleywood Food Research Association, Chipping Campden.

EHEDG (2003) Document No. 26. General Hygienic Engineering of plants for the processing of dry particulate materials, November. Campden & Chorleywood Food Research Association, Chipping Campden.

EHEDG (2004) Document No. 8. Hygienic equipment design criteria, April. Campden & Chorleywood Food Research Association, Chipping Campden.

14

Improving the hygienic design of packaging equipment

C. J. de Koning, CFS b.v., The Netherlands

14.1 Introduction

A variety of norms is applicable to the food equipment manufacturing industry. Hygienic engineering and design of packaging equipment should start with a description of the characteristics of the food product that needs to be produced or packed. Furthermore the quality standards of the food producer and the circumstances under which production is performed are important issues. ISO 14159 offers a guideline for applying a systematic approach to equipment design. Other norms are based on the same principles; some are more descriptive and/or dedicated to the application.

Typical for the norm ISO 14159 is the application of a hygienic risk analysis, which will be elaborated in this chapter. ISO 14159 offers a systematic approach to hygienic design of equipment by:

- defining the limits of the machine, its intended use and the products and processes involved;
- applying an analysis on microbiological, chemical and physical hazards;
- applying a risk analysis on food safety aspects of these hazards;
- choosing appropriate materials of construction;
- applying engineering guidelines in order to eliminate possible hazards;
- verifying the hygienic design aspects of equipment;
- documenting the intended use of the equipment for installation, operation, maintenance and cleaning.

14.2 Requirements for hygienic design

Cleaning efficiency is determined by the necessary operational time and the required personnel. In order to reduce the losses on operational time due to cleaning, the equipment needs to be designed for easy accessibility and easy-to-use cleaning methods. In case of cleaning by hand, the reliability of the cleaning result is determined by the skills of the personnel in following the cleaning procedures and the available means and cleaning materials. The design of the equipment is of major importance to the end result.

In the case of closed equipment often cleaning-in-place (CIP) methods are applied. For open equipment these are more difficult to apply and a combination of cleaning by hand with partial dismantling is often applied.

14.3 Application of ISO 14159

14.3.1 Step 1: Definition of the limits of the machine

Define the intended use of the equipment by specifying the following:

- Functional requirements – (i) capacity, (ii) processing conditions and (iii) product to be produced by the equipment.
- Safety requirements – (i) food safety, specifically for the product produced by the machinery, and (ii) operator safety, specifically for the type of equipment.
- Operational requirements – (i) installation, (ii) operation, (iii) cleaning and (iv) maintenance.

14.3.2 Step 2: Hazards

Based on the conditions as specified in step 1, an overview is made of potential microbiological, physical and chemical hazards that are apparent in the packaging process. Table 14.1 shows a typical list.

14.3.3 Step 3: Risk analysis on identified hazards

A quantification of the risks can be made by finding a risk priority number. This is the multiple of Frequency (F) x Exposure (E) x Severity (S) of the hazard as defined in Tables 14.2, 14.3 and 14.4. A limit needs to be defined: an arbitrary value of 60 is chosen in Table 14.5. This value needs to be validated in practice.

With the quantification as stated above, the workflow shown in Fig. 14.1 can be applied during the design process. In this way the designer can keep track of the decision process, which needs to be documented for each step. The decisions will be documented in the construction dossier. In this way design decisions on occupational and food safety issues are recorded. It is particularly useful for evaluating new applications on existing equipment.

After a design decision the risk priority number can be recalculated and documented with the identified hazard (Fig. 14.2).

Table 14.1 Hazard identification

	Hazard existing? Yes	No
Product can be (cross) contaminated via:		
1. Non-processed product (or ingredients)		
2. (Modified air) gas		
3. Processing aids		
4. Packaging materials		
5. Material equipment (product contact surface)		
6. Personnel		
7. Air (environment)		
8. Vermin		
Category M: (Micro-) biological hazards		
Presence (pathogenic) microorganisms		
Cross-contamination (pathogenic) microorganisms		
Growth of (pathogenic) microorganisms		
Microorganisms includes bacteria, yeast and mould, virus		
Category C: Chemical hazards		
Presence of (thermal) stabile microbial toxins		
Presence of (remnants of) (toxical) cleaning and disinfections agents		
Presence/contamination of/with lubrication agents		
Presence/contamination with cooling agents		
Presence of heavy metals (As, Cd, Hg and Pb)		
Presence of dioxins and PCB's		
Presence of pesticides		
Presence of (animal) drugs such as antibiotics, sulphonamides or hormone preparations		
Presence of radioactive agents		
Category P: Physical hazards		
Contamination via (transport) air		
Contamination with burnt (product own) particles		
Contamination with (product own) (dry, hard) rests		
Contamination with glass		
Contamination with paper (rest), stickers, paperclips		
Contamination with wood (particles)		
Contamination with metal (pieces)		
Contamination with plastic (particles)		
Contamination with closing lids		
Contamination with rubber bands		
Contamination with dust, environmental dirt		
Contamination with sand and stones		
Contamination with broken machinery parts		
Contamination with bolt and nuts		
Contamination with (rubber) seals		
Contamination with writing materials		
Contamination with tools, instruments for maintenance		
Contamination by insects, birds, other vermin		
Contamination by leaking packages		
Cross-contamination with return product		

Table 14.2 Frequency of the hazard (F)

Value	Description	Occurrence
10	Probable	Once/batch
8	Realistic	Once/day
4	Feasible	Once/week
3	Improbable	Once/month
2	Most improbable	Less than once/year

Table 14.3 Exposure to the hazard (E)

Value	Exposure
1	Hazard will be neutralised or cannot enter in the product zone
10	Unclear

Table 14.4 Severity of the hazard (S)

Value	Description	Severity
10	Disastrous	Fatal, very serious diseases, great amounts
8	Very serious	Serious disease (admission), a large number of people ill
3	Serious	Doctor's visit, dentist, large number of complaints
2	Less serious	Consumer with internal injury, more complaints
1	Small matter	No health effects, single complaint

Table 14.5 Action

Risk priority value	Next step
Smaller than 60	No precautionary steps necessary
Greater than or equal to 60	Design changes to be considered

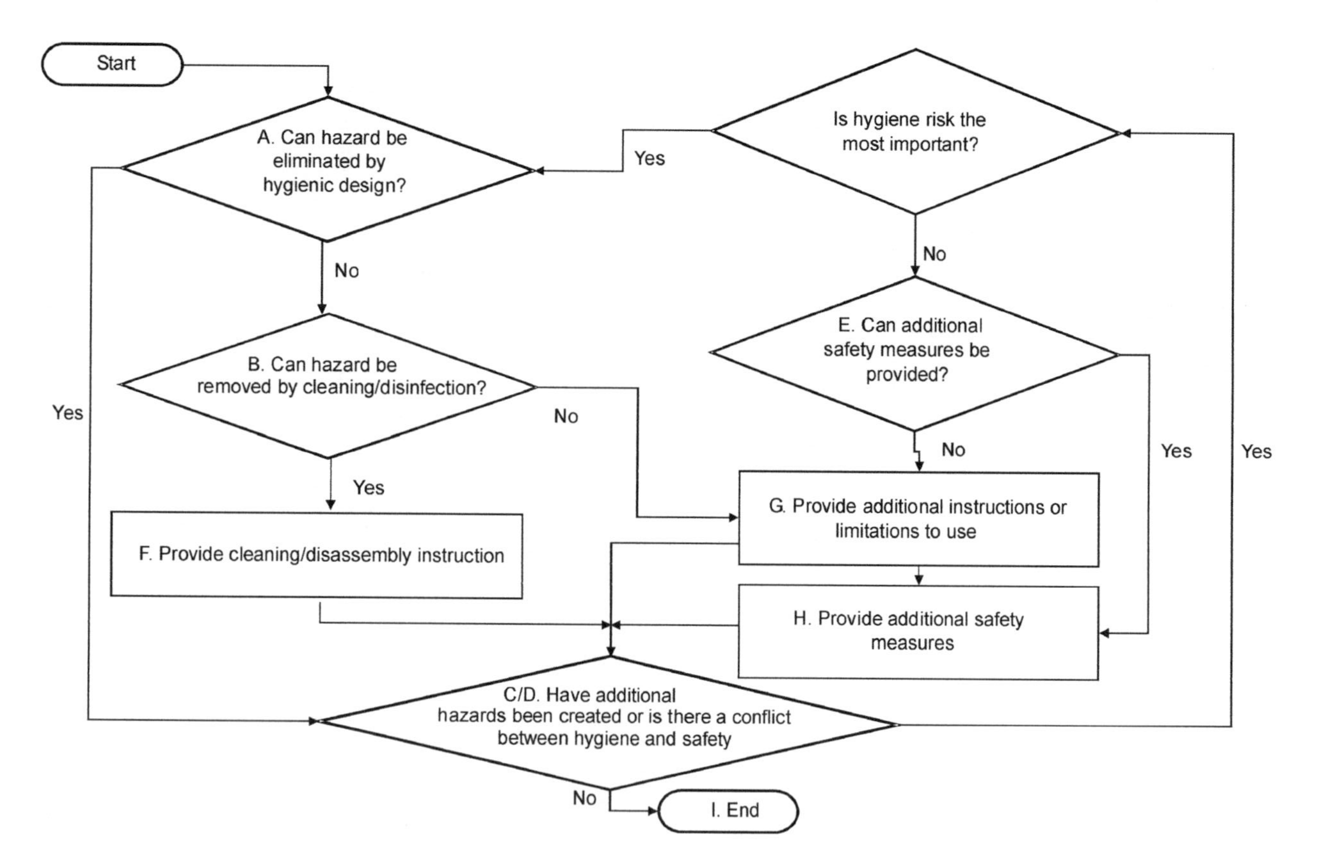

Fig. 14.1 Risk assessment.

Risk assessment according ISO 14159

D. 0083 / 10-03-2003

Machine type: **OptiDipper**

Execution: **Prototype**

Date: **December 2002**

Application:

Process conditions: **stand alone**

	Risk assessment				A: Can hazard be eliminated by hygienic design		B: Can hazard be removed by cleaning / desinfection		C: Have additional safety hazards been created?		D: Have additional hygiene hazards been created?		E: Can additional safety measures be provided?		F: Provide cleaning / disassembly instructions	G: Provide additional instructions or limitations to use.	H: Provide additional safety measures	I: Verify compliance with hygiene requirements as required
					Y	N	Y	N	Y	N	Y	N	Y	N				
Identified hazards	F	E	C	RPN	to C	to B	to F	to G	to E	to I	to A	to I	to H	to A	to I	to I	to D	End
18-Non sterile recipe	10	10	10	1000		X		X								X		O
01-After cleaning, the drainvalve is closed before the machine is completely drained	4	10	3	120		X	X								X			O a
02-While cleaning, pipework and blower are not opened	4	10	3	120		X	X								X			O
03-Dried out splashes of batter falling back onto the products	10	10	1	100		X		X								X		O b
04-Places in the production zone with little circulation	10	10	1	100		X		X								X		O
05-Accessibility/cleanability pivoting point conveyor frames	10	10	2	200		X		X								X		O
06-Crevices between scraper-bar and conveyor frame	10	10	1	100	X					X								V c
07-Crevices between PE-bullit nose and RVS mounting shaft	10	10	1	100		X		X								X		O
08-Crevices between RVS flange and RVS blower	10	10	1	100	X					X								V
09-Cooling fan motor beside productzone	10	10	1	100	X					X								V
10-Two bolts above production zone at take-over from infeed- to main conveyor	4	10	1	40														V
12-No complete draining because of horizontal bottom in level tank	10	10	1	100	X					X								V
13-Threaded parts on belt tensioner close to product area	4	10	1	40														V
14-Valve in bottom-blowerpipe closed during cleaning	4	10	8	320	X					X								V
15-Horizontal part control box, not self draining	4	10	1	40														V
16-Non self draining part in horizontal top-blowerpipe between welded joints	4	10	8	320	X					X								V
17-Complete topside of machine is accessable	2	10	1	20														V
19-Corrosion of non stainless steel components, e.g. motor	2	10	2	40														V
20-Connection tubes for conveyor splice	10	10	1	100	X					X								V
This risk is a customers responsibility and must be emphasized in the manual																		
Risk-Priority-Number (RPN=FxExC) < 60; no preventive measurements are necessary																		
Action to be taken to comply with hygienic requirements																		
Verification done and approved, action (manual) not yet finished																		
Action and verification done and approved																		
Verify compliance with hygienic requirements as required																		
a) hazard 01	4	1	3	12														V
b) hazard 03	10	1	1	10														V
c) hazard 06	3	10	1	30														V

Fig. 14.2 Risk assessments.

14.3.4 Step 4: Materials of construction

An extensive list of construction materials is available for packaging equipment. The specific exclusion of materials is mentioned in the ISO 14159 norm, paragraph 5.2. Under normal conditions the most prominent issues are the use of:

- stainless steel as basic construction material;
- food approved materials, Food and Drug Administration list;
- food grade lubricants.

14.3.5 Step 5: Design and fabrication of equipment

The design process is supported by a number of guidelines that offer the designer examples and design suggestions. The documents published by the European Hygienic Engineering and Design Group (EHEDG) provide very good references. Another very good reference is the so-called 'sanitary design checklist' set up by the American Meat Institute. This checklist offers an overview of the critical issues that need review during the design process. It is based on 10 principles:

1. Cleanable to a microbiological level.
2. Made of compatible materials.
3. Accessible for inspection, maintenance and cleaning/sanitation.
4. No liquid collection.
5. Hollow areas hermetically sealed.
6. No niches.
7. Sanitary operational performance.
8. Hygienic design of maintenance enclosures.
9. Hygienic compatibility with other systems.
10. Validation cleaning and sanitation protocols.

Each equipment manufacturer can make an interpretation of the issues on this list and define the required hygienic design standards suitable for the defined application. Applications could be divided in groups, e.g. convenience products, ready meals, meal components or sliced products.

The characteristics of these products are, e.g., water activity, after-packaging pasteurisation or not, sterilisation or not, distribution through the cool chain or as frozen food. Other relevant characteristics are pH, packaging conditions (modified atmosphere or vacuum), normal lifetime (shelf-life from packaging to retail sales) and open shelf-life (after opening pack for consumption).

The following are the most important hygienic design issues:

- The hygienic requirements of the packing department and the packing process.
- Critical areas for hygienic design of the packing system.
- Performance of the gas supply to the packing machine.
- Packing machine requirements.
- Conveyors with product contact.

Hygienic requirements of the packaging department and the packing process
The packing machine should be installed in an environment appropriate for the handling of hygiene-sensitive products. The following general aspects should be considered:

- The packing machine should be placed such that it is uncluttered and free access is available around the machine.
- Unless mounted such that dust and other foreign matter cannot enter, overhead services (lighting, piping and ducts) should be avoided.
- Clearance under the machine must allow for adequate cleaning and inspection to be carried out effectively.
- Machines should not be positioned over drains if, in doing so, access for inspection and cleaning of the drains is restricted.
- Equipment should be adequately located in position and mounting pads or feet suitably sealed to the floor.
- Services such as air, water and electricity shall be connected in a manner ensuring that proper hygiene of the equipment and area will be maintained.
- The exterior of non-product contact surfaces should be arranged to prevent harbouring of contamination in and on the equipment itself, as well as in its contact with other equipment, floors, walls or hanging supports.

The EHEDG is providing publications on these issues. Consult www.ehedg.org for more information.

Critical areas for hygienic design of the packing system
The product contact surface areas need to be defined for each packaging machine. By doing so, the zones for which specific hygienic design requirements need to be met can be identified.

Performance of the gas supply to the packing machine
The following requirements apply:

- The gas to be introduced for modified atmosphere packaging (MAP) packaging of the solid food should be of high-grade food quality or hospital grade.
- The connections of the gas supply installation to the packing machine should be clean and disinfected.
- All compressed air used for blowing on the product or contact surfaces must be filtered to a minimum of a 0.3 μm level and dried to prevent the formation of moisture in the piping system.

Packing machine requirements
Hygienic food processing equipment should be easy to maintain to ensure it will perform as expected to prevent microbiological problems. Therefore, the equipment must be easy to clean and protect the products from contamination.

Indirect product contact zone areas will be considered as product contact zone areas:

- The equipment shall be installed such that it will not cause contamination of the ingredients, raw foods and end-products.
- Separation between product contact and non-product contact areas prevents cross-contamination during operations. Indirect product contact zone areas must be designed as if they were product contact zone areas.
- Product contact surfaces are designed to prevent build-up of product residue during operations.
- Separation between product contact areas and non-product contact areas has to be determined by a risk analysis.
- All the parts of the equipment should be installed at a distance of 1 m from walls, ceilings and adjacent equipment to allow: transport systems for ingredients and packaging material and for easy access of operating staff (for inspection, cleaning and disinfecting, maintenance and to solve breakdowns).
- Surfaces with direct and indirect product contact are cleanable as measured by <1 colony-forming unit (cfu) per 25 cm^2, <1 cfu per 10 ml when the item is rinsed, acceptable RLU (reflected light unit) (device specific) when measured by residual ATP (adenosine triphosphate), and/or negative for residual protein or carbohydrate when using swabs to detect residual protein or carbohydrate (measured post-installation).

Special attention must be paid to the following aspects:

- Packing machine materials: see ISO 14159, chapter 5: 'Materials of construction'.
- Product contact surfaces and coatings: surfaces of the machine parts that come in direct contact with high-risk ingredients must be made from non-toxic materials and must have finishes that are impervious, non-absorbing, washable, smooth and crack-free to resist microbial settlement and that can be easily cleaned and disinfected. (Note: surface defects on machined components are not acceptable.) For detailed information: see ISO 14159, chapter 7.1: 'Surfaces and geometry' and chapter 7.2: 'Surface finish/surface roughness'.

Conveyors with product contact

Conveyor belts must be made from non-toxic materials and must have finishes that are impervious, non-absorbing, washable, smooth and crack-free so they resist microbial settlement and may easily be cleaned and disinfected. The surface finish must be such that formation of bio-films under the conditions of solid food packing process cannot occur or is limited to a minimum whereby the films can be cleaned.

All belting must be easily removable or the belt tension must be easy to remove without tools so the surfaces underneath can be cleaned. Belt tension is adequate throughout operations to prevent water pooling on belts.

14.3.6 Step 6: Verification of hygiene measures and test methods

The ultimate effect of hygienic design needs to be proven in practice. The basis for the evaluation is the food safety risk assessment, with its hazard analysis documentation and the related design decisions as documented in the construction file. The construction file should also contain a specification of the materials of construction:

- Materials specification.
- Practical and/or functional test (if necessary, e.g. for corrosion).

A design evaluation is made by a visual inspection and/or a practical test. Surface finish, internal angles/corners, etc., can be measured and inspected.

Hygienic operational performance is evaluated by visual inspection and in practice. This should include a test on the microbiology and on practical cleaning performance.

14.3.7 Step 7: Instruction handbook, maintenance and cleaning

The equipment manual should contain all necessary instructions to operate the equipment in a safe, hygienic and efficient way. Specific attention needs to be given to the operational, maintenance and cleaning procedures (see Fig. 14.3). It should be in a format that is easy to understand and easy to follow by operators and technicians. Good examples are available from companies that have developed total productive maintenance.

14.4 Other standards and guidelines

- EN1672-2: On a similar basis as ISO 14159 a European Norm EN 1672-2 (Food processing machinery – Basic concepts – Part 2: Hygiene requirements) is available.
- EHEDG: Many guidelines are available on specific subjects. Very useful practical suggestions can be found in the EHEDG guidelines.
- NSF/ANSI/3A 14159-1-2002: The standardisation organisations in the USA have made a norm that is based on the ISO 14159, however without the risk analysis.
- American Meat Institute (AMI): The USA meat industry organisation AMI has developed guidelines with respect to sanitary design of equipment. In these guidelines 10 principles are defined for hygienic design of equipment. For each of these principles a checklist is created to which a design can be checked and rated. This results in a score that can be used for purchase decisions. Many examples are given, with suggestions for improvements.
- Company-specific requirements: Many food producers have specific guidelines that the equipment producer needs to follow. These guidelines are often additional requirements on the norms as mentioned above and should be taken into account in the design.

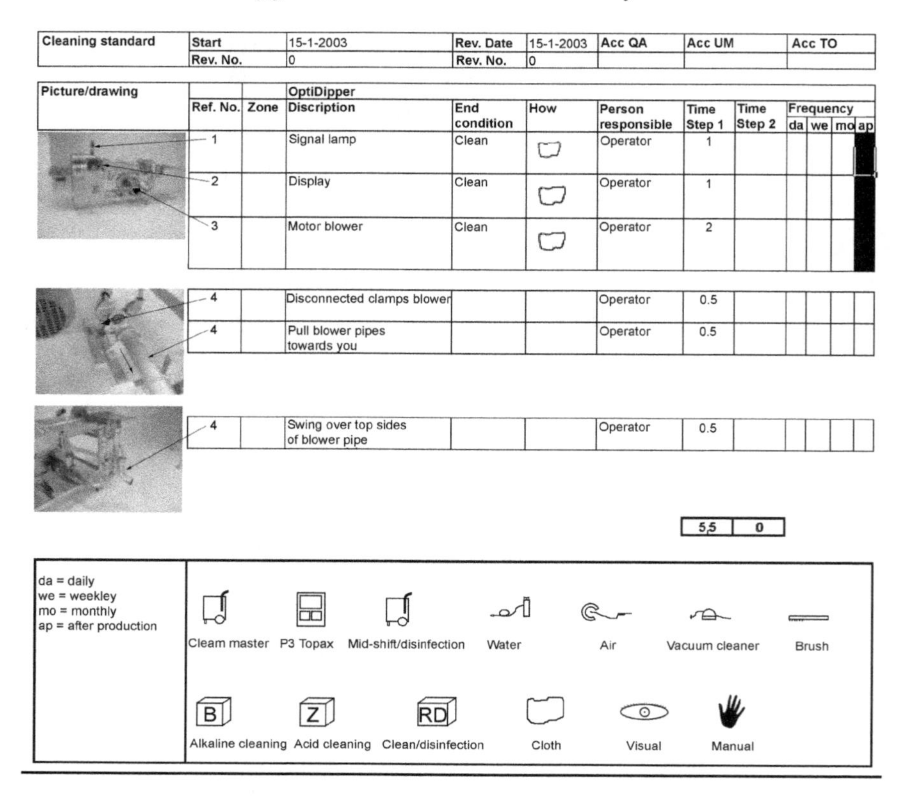

Cleaning standard	Start	15-1-2003	Rev. Date	15-1-2003	Acc QA	Acc UM	Acc TO
	Rev. No.	0	Rev. No.	0			

Picture/drawing	Ref. No.	Zone	OptiDipper Discription	End condition	How	Person responsible	Time Step 1	Time Step 2	Frequency da	we	mo	ap
	1		Signal lamp	Clean		Operator	1					
	2		Display	Clean		Operator	1					
	3		Motor blower	Clean		Operator	2					
	4		Disconnected clamps blower			Operator	0.5					
	4		Pull blower pipes towards you			Operator	0.5					
	4		Swing over top sides of blower pipe			Operator	0.5					
							5,5	0				

da = daily
we = weekley
mo = monthly
ap = after production

Fig. 14.3 Cleaning instructions.

14.5 Conclusion

For hygienic design of equipment the end-product should be taken as the starting point. This means that the hygienic requirements can differ per type of application. Furthermore, the requirements between occupational safety and food safety could be conflicting. The designer needs to find solutions by balancing the risks of occupational and food safety. Furthermore the documentation of the design, the design requirements and the choices made are of utmost importance in the verification of the design. The advantages of hygienic design can be huge; however, strict cost control in the design process is of crucial importance to reach economically viable solutions. Hygienic design of equipment requires a multidisciplinary approach in which food producers and equipment producers work together in finding solutions. A 'quick-fix' is not yet available.

15

Improving the hygienic design of electrical equipment

L. Uiterlinden, GTI Process Solutions BV, The Netherlands, H. M. J. van Eijk, Unilever R&D Vlaardingen, The Netherlands and A. Griffin, Unilever – Port Sunlight, UK

15.1 Introduction

This chapter describes the principles that should be followed when designing and installing power and control/instrumentation field cabling installations in food production facilities that are not in direct contact with product. Based on zoning specifications the level of hygienic installation is defined and described in this chapter for medium and high hygiene areas.

These guidelines should be implemented at an early stage of a project in order to avoid or significantly reduce later costs. Should modifications to a newly installed installation be necessary, in order to achieve its required standard of hygiene, the associated corrective costs will be high and production is likely to be delayed. It is the attention to detail, the skill in procuring the correct materials and fittings, and the close supervision throughout the installation works that differentiate a good installation from an average one. The integrity of the electrical installation should never be, and does not need to be, compromised in order to achieve an installation that is hygienically acceptable. The information in this chapter is based on a wide installation experience in the food industry and on a Unilever guideline for the design of hygienic electrical and control/ instrumentation cabling installations.

15.2 Hygienic zoning

All areas of a plant building where processing takes place should be designated according to one of the following three zones of hygiene classification (Zone B, Zone M, Zone H). This is to ensure suitable equipment selection and the necessary control procedures related to the hygienic hazards of manufacturing, packing and storage of products.

15.2.1 Zone B

Zone B (where B stands for the Basic hygiene zone classification) applies to an area where a basic level of hygienic design requirements suffices. The objective for a Zone B classification is to prevent product contamination by adopting good manufacturing procedures, and to control or reduce the creation of hazardous sources that can affect zones of higher hygienic classification.

There should be no open product handling in a Zone B area. Products need to be in covered and in packed form. This allows products sensitive to contamination risks to be stored in a low-hygiene area. A typical Zone B area is a warehouse. Personnel in this zone area require no special clothing, but clothing must be clean at all times.

15.2.2 Zone M

Zone M (where M stands for the Medium hygiene zone classification) applies to an area where a medium level of hygiene design requirements suffices. The objective for Zone M is to control or reduce the creation of hazardous sources that can affect an associated area of higher zone classification. It is also to protect the interior of food processing equipment from becoming contaminated when exposed to atmosphere. Air entering the area can come from an associated part of the building having either a lower or higher zone classification.

In this area product might be exposed to the environment. Where exposure is likely, e.g. during sampling, additional measures need to be taken to prevent contamination such as placing covers over sampling points. Zone M relates to areas of the plant layout, typically hygienic production areas where:

- contact with the product or with the interior of processing equipment may occur – occasionally for short periods of time (e.g. sampling of products or handling of products with gloved hands, tools etc., to clear blockages or to prevent product build up and further blockage);
- products of low or medium sensitivity to possible atmospheric contamination may be exposed to environmental air for longer periods;
- rooms where closed equipment of a higher hygienic area classification is operating, but where such equipment is likely to be temporarily opened.

Personnel clothing must be clean, with white coats and protective headwear (hairnets) worn. Change of shoes for operators or covered shoes for visitors are not essential.

15.2.3 Zone H

Zone H (where H stands for the High hygiene zone classification) applies to an area where the highest level of hygiene design requirements is essential. The area must be completely contained. This zone classification is typical for open processing, where the highest hygienic design criteria applies to the equipment involved. The objective for this zone classification is to control all product contamination hazards and to protect the interior of food processing equipment from exposure to atmosphere. Filtered air must be supplied to this area.

Zone H relates to areas of the plant layout, typically food production areas where:

- there is continuous exposure to atmosphere of products sensitive to contamination or exposure of food contact surfaces to atmosphere;
- even short exposure to atmosphere of products sensitive to contamination can result in safety or quality hazards.

Zone H areas should be limited in size and contain as simple an equipment layout as possible, allowing enough room for operation, cleaning and maintenance. Therefore, where possible, equipment such as fans, pumps, power supply and ventilation systems should be placed outside the area. Such equipment will then have a lower hygienic status, and a lower degree of cleaning. Protective clothing, headgear (hairnets), change of shoes for operators and shoe covers for visitors should be worn, and put on prior to entering the area.

15.2.4 Wet or dry processing and type of cleaning

The different zones can be further classified according to wet and dry processing areas. In some cases a zone category can also be classified differently if special products are involved requiring even higher hygienic standards, e.g. baby food processing. However, it is necessary to consider what defines the boundary between the zones. Zones may temporarily be classified differently, e.g. for maintenance or cleaning purposes. Release procedures are adopted when the zone reverts to its original classification.

The type of cleaning permitted in a given zone is an essential preventive factor to any area. Three categories should be considered:

- Dry cleaning (d). This applies to areas where no cleaning liquids are used, and cleaning is only by vacuum cleaners, dusting cloths, brooms and brushes. Pressurised air should never be used as a means of blowing down surfaces.
- Controlled wet cleaning (cw). This applies to areas that are defined as dry during processing, although some wet cleaning is permitted. These areas are normally cleaned dry, but occasionally are fully or partially wet cleaned, when limited amounts of water are used. Drying of all surfaces after controlled wet cleaning is essential.
- Wet cleaning (w). This applies to areas where the entire room or zone is always cleaned wet. The contents (equipment, cable trays, ceilings, walls, etc.) are wet washed without restrictions on the amount of cleaning liquid used.

Fig. 15.1 Main cable routing in zone B (outside production zones).

Controlled wet cleaning in zones is often used with dry processing equipment, since full wet cleaning would impose too high a hygienic risk. With controlled wet cleaning, ceilings and walls often stay dry. This may influence the building design and plant layout to ensure these surfaces stay dry.

15.3 Hygienic electrical design principles

This section describes the basic hygienic requirements of equipment as well as the methods of installation for electrical and control systems. It is not practical here to cover all issues and eventualities but give clear objectives: to ensure all installations are electrically and mechanically sound, are easy to clean, avoid the creation of soil traps, and are durable to the environment in which they are installed.

15.3.1 Hygienic design principles

In the food production areas, sound hygienic standards are important to avoid the risk of product contamination by microorganisms. A source of relevant guidelines is the European Hygienic Engineering Design Group (EHEDG). The EHEDG is a consortium of equipment manufacturers, food manufacturers and research institutes.

It is important to keep in mind that edible products or products that are used in direct contact with the human body are being processed. Designers must put as much effort into the hygienic design of the electrical and control systems as in the design of the electrical and control systems themselves.

The following issues should be borne in mind when designing and implementing hygienic installations:

- use of correct materials;
- method of construction;
- cleanability of installation.

All recommendations given in this chapter are also generally applicable for air, water and other kinds of tubing in the field or on a machine. With respect to hygiene, air tubes can, in most cases, be treated in a similar manner as electrical cabling.

15.3.2 Basic requirements

The basic requirements for hygienic design of installations, not in direct contact with product, are as follows:

- The exterior or non-contact material surfaces should be arranged to prevent harbouring, growth and development of soils, microorganisms, pests and vermin in and on the equipment itself as well as in contact with other equipment, floors, walls or hanging supports.
- To design installations such that maintenance and operating personnel do not have to put in unnecessary extra effort to keep the area hygienic because of the access inadequacies in the installation. Instead, the installation should provide a stimulant to help ensure it is kept hygienic.
- In high-hygiene zones all trunkings and conduits for cabling/wiring should be closed and sealed. Their external surfaces should be capable of being cleaned easily. Installing trunking and/or conduit in the horizontal plane within the splash area or contact area should be avoided. Although the use of cable ladder, cable tray and wire tray is widely used in 'normal' areas, these are not suitable for installation within hygienic areas since cleaning is problematic because of the nature of their construction and positioning of cables.
- In medium hygiene zones cable ladders or wire trays and conduits for cabling/wiring should be used. Their external surfaces should be capable of being cleaned easily. Cable ladders should be installed vertically to minimise the space taken in the horizontal plane. Conduits should not be used in dry production areas; small wire trays can be used here.
- Equipment and support structures should be attached to the building (floors, walls, columns, ceiling) in such a way that neither pockets nor gaps exist in which product or soil can accumulate or any gaps should be large enough to be fully cleanable.

15.4 Installation requirements for medium hygiene areas

A significant part of both the electrical and control installations is located within the medium hygiene production area, e.g. power cabling supplying motors or packaged plant/machinery, and control cables connecting sensors via field boxes/cabinets to the plant control system. These cables should be routed, and connections made, in such a manner as to create hygienically acceptable installations.

15.4.1 General requirements for wiring installations

In medium hygiene production areas, the electrical and control installations should be limited to that necessary for the safe and correct operation of the plant. The majority of cabling should be routed outside the production area where less direct hygienic risk is assessed, e.g. mounted on wire trays installed along the pipe bridges not located above the production area.

The preferred installation method for field boxes within the production area, is to hygienically mount field boxes on a convenient wall or its respective machine. In situations where this would prove impracticable, field boxes should be mounted on pedestal type, free-standing brackets. If undertaken properly, these methods of installation avoid potential soil traps being formed. Where the practical solution requires the wall mounting of an enclosure, e.g. a cabinet

Fig. 15.2 Well cleanable and maintainable setup of equipment in Zone M.

having several cables terminated to it, the enclosure should be installed at least 30 mm from the wall, to prevent a soil trap being created at the rear of the enclosure.

It is important that cabling installations in medium hygiene production areas are given special attention and are well laid out, avoiding tangled cable arrangements which may become breeding grounds for vermin and pests. Horizontal surfaces should be minimised. Furthermore, the installation should look neat and be easy to clean. Joints between dissimilar metals (galvanic action) should be avoided. Support equipment and constructions should avoid any dead-ends. The use of conduits reduces the number of supports for cables. Supports are potential places for the build-up of product and soil.

Individual cables or multiple cables of small diameter sharing the same route should be installed in conduits or on wire trays. Individual cables that do not share a common route with other cables, are as hygienic as a single conduit run. However, a cable is usually more difficult to support in a hygienic manner than conduit. A cable is also at more risk of being scuffed (which would result in it being more difficult to clean). Furthermore, should future modifications to the system require the addition of a cable, the installer may be tempted to support this new cable from the previously installed cable. Such a practice leads to an uncleanable and hygienically unacceptable cable bundle. *The use of remote input/output (I/O) and/or bus systems is advisable* as these decrease the number of cables to be installed and, consequently, decrease the number of potential unhygienic situations.

The number of separate connections to a single item or group of items should be minimised as far as is reasonably practicable. For example, most suppliers are able to design their hygienic plant and equipment in such a way that only one electrical and one air connection are required. A particular example is valves and all valve manufacturers should supply valves in such a way that only a single air and electrical connection is required, regardless of the complexity of function of the valve.

15.4.2 Supports

Cable support systems are potential places for the build-up of product and soil. Therefore, support systems and constructions should be totally sealed, i.e. not having open ends where soil can accumulate. The use of wire trays and conduits reduces the number of supports that would otherwise be needed for individual cables.

The material of construction for brackets should be of hygienic material and generally be the same as the equipment being supported (typically, cable trunkings and conduits are preferably manufactured from 304 stainless steel, unless environmental conditions require otherwise, e.g. high temperatures or potential contact with chlorine).

Supports should be constructed in such a way that adequate cleaning is guaranteed by ensuring that:

- sharp edges, recessed corners, uneven surfaces, open hollows, unprotected bolt threads and screws are minimised or avoided;
- inaccessible spaces where product or soil may accumulate are not created;
- supports manufactured from rolled hollow section are totally sealed;
- brackets manufactured from angle or channel are minimised or avoided.

15.4.3 Cable ladders and wire trays

Where practicable and material availability allows, as a general principle it is recommended that for hygienic installations the installer should preferably standardise on 304 stainless steel cable ladders or wire trays and accessories. The layout of the cable routeing installation should be organised in such a manner that allows adequate cleaning. Where cable ladders or wire trays are installed vertically, the cable or cables within should be supported by a suitable means at appropriate intervals in such a manner that the conductor or cable does not suffer damage through its own weight. For shorter vertical distances conduits may be used. In the case of horizontal mounting, special attention should be given to ensure adequate cleaning is practicable because, being horizontal, they offer a flat surface for accumulation of soil. Horizontal cable ways can be installed vertically (on its side) to minimise the horizontal surface.

Where cable ladders or wire trays enter the medium hygiene production area, the opening remaining after the passage of the trunking should be made good with fire-resistant material so as to maintain the degree of fire resistance as well as the hygienic standard of the respective element (e.g. wall, ceiling) (Fig. 15.3). Cable ladders or wire trays should not be mounted above areas where the product is exposed to the environment. This requirement is equally applicable to all elements forming an electrical or control installation.

15.4.4 Conduits

Where practicable and material availability allows, as a general principle it is recommended for hygienic installations that the installer should preferably standardise on 304 stainless steel electrical conduit and accessories. Furthermore, the number of different sized conduits should be minimised so far as reasonably practicable. The conduit's exterior should have a smooth finish and be easy to clean.

It is possible to run more than one cable through a single conduit. When two or more cables partly share a common route, but go to different termination points, the creation of unsealable openings allowing the cable(s) to enter or exit the conduit is possible. For long, vertical wiring routes, cable ladders or wire trays should be used instead of conduit to prevent the conductor or cable suffering damage by its own weight. Conduit systems should provide adequate means of access for drawing in cables. The bending radius of every bend in a wiring system should be such that conductors and cables do not suffer damage. The conduit should be used oversized to allow for wet cleaning. Open conduit

Fig. 15.3 Cables beside each other on a stainless steel cable ladder. Note the closed supports and cleanable conduits. Problem: the horizontal parts and omega rail of the cable ladder!

should not be used in dry production areas, where small wire trays are a better solution, to allow dry cleaning.

15.4.5 Cables and flexible instrument air tubes

Cables and flexible instrument air tubes should have a smooth outer poly(vinylchloride) (PVC) or polypropylene sheath, as applicable, such that they are easy to clean where not protected by cable ladders, wire trays or conduit. When two single cables or tubes are routed together in parallel they should not be tied to each other in order to prevent the build-up of soil and to ease cleaning.

15.4.6 Connections to cabinets and field boxes

To enable manual cleaning and visual inspection of enclosure surface areas on which soil can accumulate, it is necessary to keep these areas as smooth as possible and preferably limit the number of connections to them to a practical minimum. All connections (e.g. cable ladders or wire trays, conduit, cable, air tubes) to cabinets or field boxes should be made via the bottom side of the cabinet or field box. The best way to make the connection between the cabinet or field box and cable is to use a cable gland, complete with earthing tag (if required, mounted inside the enclosure), locknut and shroud.

15.4.7 Cabinets

In general, cabinets should be located outside the medium hygienic production area. For cabinets located in hygienic production areas the requirements detailed below apply. Where practicable and material availability allows, as a general principle it is recommended for hygienic installations that the installer should preferably standardise on 304 stainless steel cabinets. However, coated mild steel or plastic may be acceptable provided its exterior has a smooth finish that is easy to clean.

Cabinets should have a minimum index of protection of IP 55, allowing dust to be removed from the enclosures' exteriors by water cleaning. Horizontal surfaces should be minimised or avoided, by installing a slope roof with a minimum of 30°. The distance between the cabinet base and the floor should be no less than 100 mm (Fig. 15.4). Where cabinets are wall mounted, the enclosure should be installed at least 30 mm from the wall, to prevent a soil trap being created at the rear of the enclosure.

Temperature control in cabinets require special attention. Within hygienic production areas it is not advisable to vent the enclosure or to create an additional airflow since these can lead to microorganisms being transported from one point to another in the area where they can infect the equipment. Also, an uncontrolled airflow from a non-hygienic environment (cabinet internals) into a hygienic area should be avoided. For these reasons the use of ventilators in cabinets should be avoided, especially where the cabinet is located in close proximity (within 2 m) of the product.

The recommended alternative is to make effective use of the self-cooling capabilities of a cabinet by means of creating an internal air circulation and achieving temperature reduction through the cabinet surface. If this does not provide sufficient cooling, then additional cooling could be provided by fixing an air to water type heat exchange to the cabinet.

15.4.8 Field boxes

Field boxes are frequently located in hygienic production areas, and in such circumstances the requirements stated below should be strictly followed. Where practicable and material availability allows, as a general principle it is recommended for hygienic installations that the installer should preferably standardise on 304 stainless steel field boxes. However, plastic may be acceptable provided its exterior has a smooth finish that can be cleaned easily. A

Fig. 15.4 Stainless steel cabinet placed 100 mm from the floor.

field box should have a minimum index of protection of IP, 55 allowing dust to be removed from the enclosure's exterior by water cleaning.

Special attention should be given to the construction of doors of field boxes. In general, there should be a water seal (PVC strip) between the door and the field box. The capillary action should be reduced by providing a folded lip along the top inside edge of the door. For metal enclosures, the door should be provided with a stud-welded earth terminal. Internal to the body of the enclosure, four stud-welded gear plate fixings and a further stud weld for the earth terminal should be provided. In order to prevent condensate dripping from the box into the product, field boxes should not be placed in or above the contact area. Furthermore, field boxes should be located such that easy access for maintenance and cleaning is practicable.

Boxes mounted on the wall or on a frame should be:

- sealed to the wall or frame with a food standard silicon sealant so that soil traps are not created at the rear of the field box, or

- stood-off the wall or frame by a minimum distance of 30 mm to enable the space between the frame and box to be cleaned properly.

Remote I/O blocks and/or valve islands should be installed in cabinets or field boxes. This is because the surfaces of valve islands and remote I/O blocks are not easy to clean.

On/off valves and control valves frequently require one or more instrument air connections, the air being supplied through instrument air tubes. In principle, the installation requirements for these air tubes are similar in nature as those for electrical or control cabling. Valves should be procured such that only a single instrument air connection is required and other air distribution is integral to the valve.

15.4.9 Lighting

Light sources must not be placed above open processes, since should they be damaged, broken fragments will fall into open process equipment. When such a light source cannot be avoided, e.g. above a monitoring point, non-glass tubes or shatter-proof glass should be used. Lighting fixtures and their support are designed to avoid accumulation of dust, especially where cross-contamination risks could arise. Horizontal surfaces need to be minimised.

Light (bulbs or tubes) sources are covered with a protective film or covered by polycarbonate to prevent glass shattering on breakage. These lights should be changed regularly as protective films become brittle in time. Light sources must have watertight enclosures.

Lighting units can be unplugged and removed from the processing area so that bulbs can be changed out of processing area. This facilitates ease of bulb changing.

15.5 Installation requirements for high-hygiene areas

A significant part of both the electrical and control installations is located within the hygienic production area, e.g. power cabling supplying motors or packaged plant/machinery, and control cables connecting sensors via field boxes/cabinets to the plant control system (Fig. 15.5). These cables should be routed, and connections made, in such a manner as to create hygienically acceptable installations.

15.5.1 General requirements for wiring installations

In hygienic production areas, the electrical and control installations should be limited to that absolutely necessary for the safe and correct operation of the plant. Therefore, the majority of cabling should be routed outside the production area where no hygiene classification is required, e.g. mounted on cable trays installed in the service area located above the ceiling of the hygienic production area.

Fig. 15.5 Walkable ceiling with integrated luminaires above the conditioned production zone H and a vertical cable trunk to the machine.

Where it is necessary to install field boxes within the production area, the preferred installation method is to hygienically mount field boxes on a convenient wall or its respective machine. In situations where this would prove impracticable, field boxes should be mounted on pedestal type, free-standing brackets. If undertaken properly, these methods of installation avoid potential soil traps being formed. Where the practical solution requires the wall mounting of an enclosure, e.g. a cabinet having several cables terminated to it, the wall-enclosure outer perimeter surfaces should be sealed with a food standard silicon sealant to prevent a soil trap being created at the rear of the enclosure.

It is important that cabling installations in high-hygiene production areas are given special attention and are well laid out, avoiding tangled cable arrangements which may become breeding grounds for vermin and pests. Horizontal surfaces should be minimised. Furthermore, the installation should look neat and be easy to clean. Joints between dissimilar metals (galvanic action) should be avoided. Support equipment and constructions should avoid any dead-ends. The use of conduits reduces the number of supports for cables. Supports are potential places for the build-up of product and soil.

Where multiple cables drop from the ceiling service area into the hygienic production area, cable trunking should be considered. Individual cables or multiple cables of small diameter sharing the same route should be installed in conduits. If cables are to be installed in rigid conduit, then final connections to plant that is subject to vibration (e.g. motors) should be made via flexible conduit (having a smooth outer surface) or by some other suitable means. Individual cables that do not share a common route with other cables are as hygienic as a single conduit run. However, a cable is usually more difficult to support in a hygienic manner than conduit. A cable is also at more risk of being scuffed (which would result in it being more difficult to clean). Furthermore, should future modifications to the system require the addition of a cable, the installer may be tempted to support this new cable from the previously installed cable. Such a practice leads to an uncleanable and hygienically unacceptable cable bundle. *The use of remote I/O and/or bus systems is advisable* as these decrease the number of cables to be installed and, consequently, decrease the number of potential unhygienic situations.

The number of separate connections to a single item or group of items should be minimised as far as is reasonably practicable. For example, most suppliers are able to design their hygienic plant and equipment in such a way that only one electrical and one air connection are required. A particular example is valves and all valve manufacturers should supply valves in such a way that only a single air and electrical connection is required, regardless of the complexity of function of the valve.

15.5.2 Supports

Cable support systems are potential places for the build-up of product and soil. Therefore, support systems and constructions should be totally sealed, i.e. not having open ends where soil can accumulate. The use of cable trunking and conduits reduces the number of supports that would otherwise be needed for individual cables.

The material of construction for brackets should be of hygienic material and generally be the same as the equipment being supported (typically, cable trunkings and conduits are manufactured from 304 stainless steel, unless environmental conditions require otherwise, e.g. high temperatures or potential contact with chlorine).

Supports should be constructed in such a way that adequate cleaning is guaranteed by ensuring that:

- sharp edges, recessed corners, uneven surfaces, open hollows, unprotected bolt threads and screws are avoided;
- inaccessible spaces where product or soil may accumulate are not created;
- supports manufactured from rolled hollow section are totally sealed;
- brackets manufactured from angle or channel are avoided.

15.5.3 Cable trunking

Where practicable and material availability allows, as a general principle it is recommended that for hygienic installations that the installer should standardise on 304 stainless steel trunking and accessories. Cable trunking should be of a closed type and preferably have no screws. Lids should be fitted with clamps or quick fittings and should preferably have gaskets. Additionally, a proprietary cover plate should be fixed over each lid joint to reduce the ingress of soil. Trunking should be kept out of wet areas since it is difficult to maintain a high degree of protection from water ingress. The exterior should have a smooth finish and be easy to clean.

The layout of the trunking installation should be organised in such a manner that adequate cleaning will be possible. Where trunkings are installed vertically, the cable or cables within should be supported by a suitable means at appropriate intervals in such a manner that the conductor or cable does not suffer damage by its own weight. For shorter vertical distances conduits may be used. In case of horizontal mounting special attention should be given to ensure adequate cleaning is practicable. Horizontal trunking runs should be kept to a minimum because, being horizontal, they offer a flat surface for accumulation of soil. Horizontal trunking mounted hard up to the underside of a flat surface (e.g. a ceiling) and sealed to it is an acceptable method of installation. However, as the trunking lid will be on the underside, cable retainers will have to be employed to retain cables in the trunking with the lid removed.

Where a trunking enters the hygienic production area, the opening remaining after the passage of the trunking should be made good with fire-resistant material so as to maintain the degree of fire resistance as well as the hygienic standard of the respective element (wall, ceiling, etc.). Furthermore, the trunking should be internally sealed so as to maintain the degree of fire resistance of the respective element. Trunking should not be mounted above areas where the product is exposed openly to the environment. This requirement is equally applicable to all elements forming an electrical or control installation.

15.5.4 Conduits

Where practicable and material availability allows, as a general principle, it is recommended for hygienic installations that the installer should standardise on 304 stainless steel electrical conduit and accessories. Furthermore, the number of different sized conduits should be minimised so far as reasonably practicable. The conduit's exterior should have a smooth finish and be easy to clean. The use of pipe rather than conduit is discouraged because of the difficulties in maintaining the integrity of the piping system at cable entries and exits (non-availability of fittings exacerbates this problem). Also, as pipe tends to the thin-walled, less physical/mechanical protection is afforded by pipe. Furthermore, since larger bore pipes are generally selected, these, when used for vertical drops, may lead to unsupported conductors or cables within suffering damage by their own weight. In short, while the use of pipe appears to offer an aesthetically appealing and easy solution it actually creates serious soil traps and can compromise electrical integrity.

Conduits should be suitably sealed at both ends. This should be achieved by using a removable rubber plug at an open end where a cable does not pass through or by a proprietary cable gland/sealing gland where a cable does pass through. In order not to compromise the sealing integrity of the conduit system where cables enter and exit, cable glands should be dedicated to a single cable only. The index of protection for the conduit system should be no less than IP 55. In order to maintain its integrity, it is important that all conduit entries, boxes and couplings are adequately sealed.

It is permitted to run more then one cable through a single conduit. When two or more cables partly share a common route, but go to different termination points, the creation of unsealable openings allowing the cable(s) to enter or exit the conduit should be avoided. For long, vertical wiring routes, trunking should be used instead of conduit to prevent the conductor or cable suffering damage by its own weight.

If cables are installed in rigid conduit, then final connections to equipment that is subject to vibration (e.g. motors) should be made via flexible conduit (having a smooth outer surface) or by some other suitable means. For example, a suitable conduit box could be used to connect the flexible conduit to the ridged conduit system. Since flexible conduit should not be relied upon to provide adequate earth continuity, it is necessary to install a separate protective conductor within the flexible conduit between the conduit box and the equipment. This type of installation is particularly suited to the types of cable that require additional mechanical protection, e.g. PVC insulated single-core cables.

Where practicable, the practice of glanding a cable directly into a conduit coupling should be avoided. Conduit systems should provide adequate means of access for drawing in cables. The bending radius of every bend in a wiring system should be such that conductors and cables do not suffer damage.

15.5.5 Use of cable ladder and cable tray

Owing to their open structure, cable ladders and cable trays require significantly more cleaning effort than that required for trunking and conduit. Consequently, the use of cable ladders and cable trays within the hygienic production area is to be avoided.

15.5.6 Cables and flexible instrument air tubes

Cables and flexible instrument air tubes should have a smooth outer PVC or polypropylene sheath, as applicable, such that they are easy to clean where not protected by trunking or conduit. When two single cables or tubes are routed together in parallel they should not be tied to each other. Instead, they should be separated by a distance of no less than 25 mm to prevent the build-up of soil and to ease cleaning. Stainless steel instrument air tube is easier to route and gives a much better installation and so is preferred over flexible PVC.

It is considered good installation practice to route instrument air tubes and electrical cables in separate trunkings or conduits. Cable glands should have a minimum index of protection of IP 55.

15.5.7 Connections to cabinets and field boxes

To enable manual cleaning and visual inspection of enclosure surface areas on which soil can accumulate, it is necessary to keep these areas as smooth as possible and preferably limit the number of connections to them to a minimum. All connections (e.g. trunking, conduit, cable, air tubes) to cabinets or field boxes should be made via the bottom side of the cabinet or field box. The best way to make the connection between the cabinet or field box and:

- trunking must use a proprietary flare/flange trunking fitting;
- conduit must use a flange (or standard) type coupling and male bush;
- cable must use a cable gland, complete with earthing tag (if required, mounted inside the enclosure), locknut and shroud.

For each of the above connection methods, food standard flexible silicone paste (Permabond adhesives) should be appropriately applied between the fitting, coupling or gland and the enclosure to provide both a hygienically and watertight acceptable connection. Any bolts or set screws used should be of the captive type.

15.5.8 Cabinets

In general, cabinets should be located outside the hygienic production area. For cabinets located in hygienic production areas the requirements detailed below apply. Where practicable and material availability allows, as a general principle it is recommended for hygienic installations that the installer should standardise on 304 stainless steel cabinets. However, coated mild steel or plastic may be acceptable provided its exterior has a smooth finish that is easy to clean. A cabinet should have a minimum index of protection of IP 55, allowing dust to be removed from the enclosure's exterior by water cleaning.

The distance between the cabinet base and the floor should be no less than 100 mm. Where cabinets are wall mounted, the whole of its outer perimeter should be sealed to the wall with a food standard silicon sealant to prevent creating a soil trap at the rear of the field box.

Temperature control in cabinets require special attention. Within hygienic production areas it is not advisable to vent the enclosure or to create an additional airflow since these can lead to microorganisms being transported from one point to another in the area where they can infect the equipment. Also, an uncontrolled airflow from a non-hygienic environment (cabinet internals) into a hygienic area should be avoided. For these reasons the use of ventilators in cabinets should be avoided, especially where the cabinet is located in close proximity (within 2 m) of the product.

The recommended alternative is to make effective use of the self-cooling capabilities of a cabinet by means of creating an internal air circulation and achieve temperature reduction through the cabinet surface. If this does not provide sufficient cooling, then additional cooling could be provided by fixing an air to water type heat exchange to the cabinet.

15.5.9 Field boxes

Field boxes are frequently located in hygienic production areas, and in such circumstances the requirements stated below should be strictly followed. Where practicable and material availability allows, as a general principle it is recommended for hygienic installations that the installer should standardise on 304 stainless steel field boxes. However, plastic may be acceptable provided its exterior has a smooth finish that can be cleaned easily. A field box should have a minimum index of protection of IP 55 allowing dust to be removed from the enclosure's exterior by water cleaning.

Special attention should be given to the construction of doors of field boxes. In general, there should be a water seal (PVC strip) between the door and the field box. The capillary action should be reduced by providing a folded lip along the top inside edge of the door. For metal enclosures, the door should be provided with a stud-welded earth terminal. Internal to the body of the enclosure, four stud-welded gear plate fixings and a further stud weld for the earth terminal should be provided. In order to prevent condensate dripping from the box into the product, field boxes should not be placed in or above the contact area. Furthermore, field boxes should be located such that easy access for maintenance and cleaning is practicable.

Boxes mounted on the wall or on a frame should be:

- sealed to the wall or frame with a food standard silicon sealant so that soil traps are not created at the rear of the field box, or
- stood-off the wall or frame by a minimum distance of 25 mm to enable the space between the frame and box to be cleaned properly.

Remote I/O blocks and/or valve islands should be installed in cabinets or field boxes. This is because the surfaces of valve islands and remote I/O blocks are not easy to clean.

On/off valves and control valves frequently require one or more instrument air connections, the air being supplied through instrument air tubes. In principle, the installation requirements for these air tubes are similar in nature as those for electrical or control cabling. Valves should be procured such that only a single instrument air connection is required and other air distribution is integral to the valve (Fig. 15.6).

15.5.10 Lighting

Commonly, flush-mounted luminaires complete with integral control gear are installed in the ceilings of hygienic production areas for general and emergency

Fig. 15.6 Electrical equipment and valves below the machine in a visual and cleanable compartment 100 mm from the floor.

illumination. For such installations, due consideration should be given to equipment layout, uniformity of light distribution and the proper coordination of the luminaires with the ceiling configuration.

Luminaires should be supported by screwed rod fixed to the building structure so as to avoid the weight of the luminaire being borne by the ceiling (unless the ceiling is of the load-bearing/walk-on type). Final connections from the fixed wiring to the luminaire should be via a 3-pin plug and socket (mounted adjacent to the luminaire) and heat-resisting 3-core flexible cable. Ideally, maintenance of the luminaire (e.g. the replacement of lamps) should be undertaken from within the above-ceiling service area.

15.6 General requirements for construction materials

Materials used for the construction of hygienic production equipment should meet certain specific requirements. Contact materials should, under all operating conditions, be inert to the product as well as to the detergents and chemicals used to free equipment from microorganisms. Furthermore, they should be non-toxic and comply with relevant local legislation (e.g. the Food and Drug Administration). All materials (non-contact and contact) should be:

- corrosion resistant;
- mechanically stable;
- smoothly finished;
- easy to clean;
- such that the original surface finish is unaffected under all conditions of use.

Table 15.1 Non-preferred materials

- Glass – breakable
- Zinc – poison
- Lead – poison
- Cadmium – poison
- Antimony – poison
- Plastics with free phenol – poison
- Plastics with formaldehyde – poison
- Plastics with plasticisers – poison
- Wood – porous
- Copper – may react with cleaning fluids, etc.
- Brass – may react with cleaning fluids, etc.
- Bronze – may react with cleaning fluids, etc.

Certain type of materials should not come into contact with food or food-related materials. These materials include, but are not limited to (in random order), those listed in Table 15.1.

15.6.1 Stainless steel

Frequently, stainless steels are the logical choice for materials of construction for production lines in the food industry. For support and cable infrastructure, stainless steel is preferred to galvanised steel or coated steel because the latter two materials are more susceptible to corrosion for production type environments. It should be borne in mind that corrosion can occur when dissimilar alloys are in electrochemical contact. The types of stainless steel most commonly used are given in Table 15.2.

In summary, the advantages of stainless steel are:

- excellent corrosion resistance;
- maintenance-free in most cases;
- strong construction is achieved from low-gauge materials;
- its attractive finish provides a stimulant for cleaning.

Its disadvantages are:

- difficult to work for both manufacturer and user;
- has high tool wear;
- more costly material than steel.

Table 15.2 Types of stainless steel

AISI 304 (DIN Werkstoff No. 1.4301)	When no chlorides are in the environment or at moderate temperatures (< 60 °C)
AISI 316 (DIN Werkstoff No. 1.4401)	When chlorides are in the environment and higher temperatures (> 60 °C)
AISI 316L (DIN Werkstoff No. 1.4404)	

15.6.2 Plastics

Plastics may, in some cases, have advantages over stainless steel, such as lower cost and weight, as well as better chemical resistance. For cable supports the use of plastics is not recommended due to EMC requirements and potential build-up of static electricity.

The selected plastic should comply with the following regulations:

- USA Code of Federal Regulations of the FDA (CFR 21, parts 170–199 latest edition);
- German list of constituents (BGVV).

In principle, plastics not in direct contact with product and not in the contact area do not require special approval. They should be easy to clean and resistant to the chemicals and temperatures occurring within its immediate installed environment.

When selecting a type of plastic, cleanability is an important factor. The plastics listed in Table 15.3 may be used for construction. The use of poly(tetrafluoroethene) (PTFE) or Teflon requires the particular application to be reviewed carefully since practical experience has shown that PTFE can be very difficult to clean.

Where reasonably practicable, avoid using glass-reinforced plastic (GRP) products as it is known that components of GRP can react with certain wetting agents in detergents. This can be observed by the fact that the material turns black. Of more concern is the risk of small pieces of material becoming dislodged and finding their way into the product.

15.6.3 Elastomers

When the use of rubber is required, e.g. for seals, gaskets and joint rings, the recommended choices are (in random order, the preferences will depend on the area of application) given in Table 15.4. Rubbers that are not in direct contact with product and not in the contact area in principle do not require special approval. However, they should be easy to clean and resistant to the chemicals and temperatures occurring within their immediate installed environment.

Table 15.3 Preferred plastics

- *Polyvinyl chloride unplasticised (PVC)
- *Polycarbonate (PC)
- *Polypropylene (PP)
- Acetal copolymer
- High-density polyethylene (HMWPE)
- Halar (coating) (ECTFE)
- Polyoxymethylene (POM)

* Most frequently used.

Table 15.4 Elastomers

Preferred elastomers	Common trade mark
*Silicon rubber	Silastic
*Nitrile rubber	Perbunan
*Natural rubber	
Nitrile/butyl rubber	NBR/H.NBR
Fluor elastomer	Viton
Polychloroprene	Perbunan C

* Most frequently used.

When using rubber for gasket or seal purposes one should take into account that corrosion problems can occur because of the following reasons:

- Ingress of liquids containing chlorides under gaskets and seals can lead to a high chloride concentration and lead to severe corrosion problems (stainless steel).
- Degradation of the adhesive used to locate the gaskets.
- Degradation of rubber by product or cleaning agents.

15.6.4 Adhesives and sealants

When the use of adhesives is required (e.g. for labelling), the recommended choices are (in random order, the preferences will depend on the area of application) given in Table 15.5. Adhesives that are not in direct contact with product and not in the contact area in principle do not require special approval. They should be easy to clean and resistant to the chemicals and temperatures occurring within their immediate installed environment.

The use of sealants in electrical and control installations is generally to avoid potential soil traps. For example, where the practical solution requires the wall mounting of a cabinet having several cables terminated to it, the wall-cabinet outer perimeter surfaces should be sealed with a food standard silicon sealant to prevent a soil trap being created at the rear of the cabinet.

15.6.5 Surface roughness of construction materials

From a hygiene point of view, the surface roughness or surface finish of construction materials is very important. Tests have shown that the cleaning

Table 15.5 Adhesives

Preferred adhesives	Common trade mark
Single-part epoxies (ESP 108, ESP 110)	PermaBond
Anaerobic adhesive (A 131)	PermaBond
Anaerobic adhesive 518	Loctite
Silicone sealant	Common

time required in order to achieve a set standard of cleaning increases with an increase in the surface roughness or R_a value. The symbol R_a stands for roughness average and is given in micrometres (μm). This term is internationally adopted and is quoted as the main parameter to specify surface roughness (for definition of R_a, see ISO 468).

It is recommended that the surface roughness, R_a, for conduits, trunkings, enclosures and such like for installation in hygienic production areas should not exceed 2.5 μm. For all equipment and materials that are to be used in a hygienic production area, it is important to keep in mind that the higher the *Ra* value, the longer it will take to clean the equipment to achieve the same standard of cleaning.

15.6.6 Use of construction materials

Below are some general dos and don'ts in the use of construction materials.

Do:

- use a food standard silicon sealant to prevent creating soil;
- use nitrile or silicone rubber in environments with edible oils and fats;
- use materials having a R_a value as low as practicable in order to minimise cleaning time.

Don't:

- use nickel plated brass cable glands when there is a chance of direct product contact;
- use GRP products since pieces might become detached (due to wetting agents in detergents) and end up as foreign bodies in the product;
- use ethylene propylene diene monomer (EPDM) rubbers in environments with edible oils and fats.

15.7 Future trends

There will be less electrical wiring/cabling in the future. It is expected that wireless transfer of data between instrumentation and control equipment will replace field buses and control wiring/cabling. Low-energy sensors and actuators will end up battery supplied or energised by the process flow. Energy cabling to motors will still be necessary, but probably projected as an energy ring, with local switching or converting due to further innovation of semiconductors. It sounds futuristic, but when you compare this with the development of the mobile phone. . . .

15.8 Bibliography

Guideline for the design of hygienic electrical and control/instrumentation cabling installations
Revision 0.5 (draft issue) 7 October 1998
Unilever Corporate MAST Group

EHEDG Documents:
No. 13 Hygienic Design of Open Processes, 1996.
No. 26 Hygienic Engineering of Plants for the Processing of Dry Particulate Materials, 2003.

15.9 Appendix: abbreviations

AISI	American International Standards Institute
BgVV	Bundesinstitut für gesundheitlichen Verbraucherschutz und Veterinärmedizin
DIN	Deutsche Industrie Norm
EHEDG	European Hygienic Engineering Design Group
EPDM	Ethylene propylene diene monomer, type of rubber
FDA	Food and Drug Administration (USA federal institute)
GRP	Glass-reinforced plastics
I/O	Input and output
IP	Ingress protection
MCC	Motor control centre
PVC	Polyvinyl chloride, type of plastic
R_a	Symbol used for indication of the surface roughness of materials, see ISO 468 (1982) for definition. The R_a value is given in micrometres (μm).

16

Improving the hygienic design of valves

F. T. Schonrock, 3-A Sanitary Standards Inc., USA

16.1 Introduction

Sanitary valves are ubiquitous throughout processing systems. Without them, modern processing systems would be cumbersome and inefficient. Valves provide operators with the ability to stop, direct, meter, and control the flow of products and ingredients throughout the process. Because of their widespread use, valve design and sanitation can impact on every particle of product passing through the process system.

16.2 Valve types

Valves come in multiple configurations. They are well developed to operate in systems that process fluids, semi-fluids, fluids with particulates, viscous products, and dry products. Some valve designs operate well in more than one of these different environments.

16.2.1 Valves commonly used in fluid product processing systems

The following are descriptions of valves commonly used in processing systems for fluid products, including viscous products and products with semi-solid or solid particulates within the product stream.

- *Plug valve* (Fig. 16.1): a simple design consisting of a tapered plug inserted into a tapered body. The valve can be configured as a shut-off valve or with one or more ports to direct flow to different product streams. The valves are most commonly manually operated but can be affixed with a power actuator.

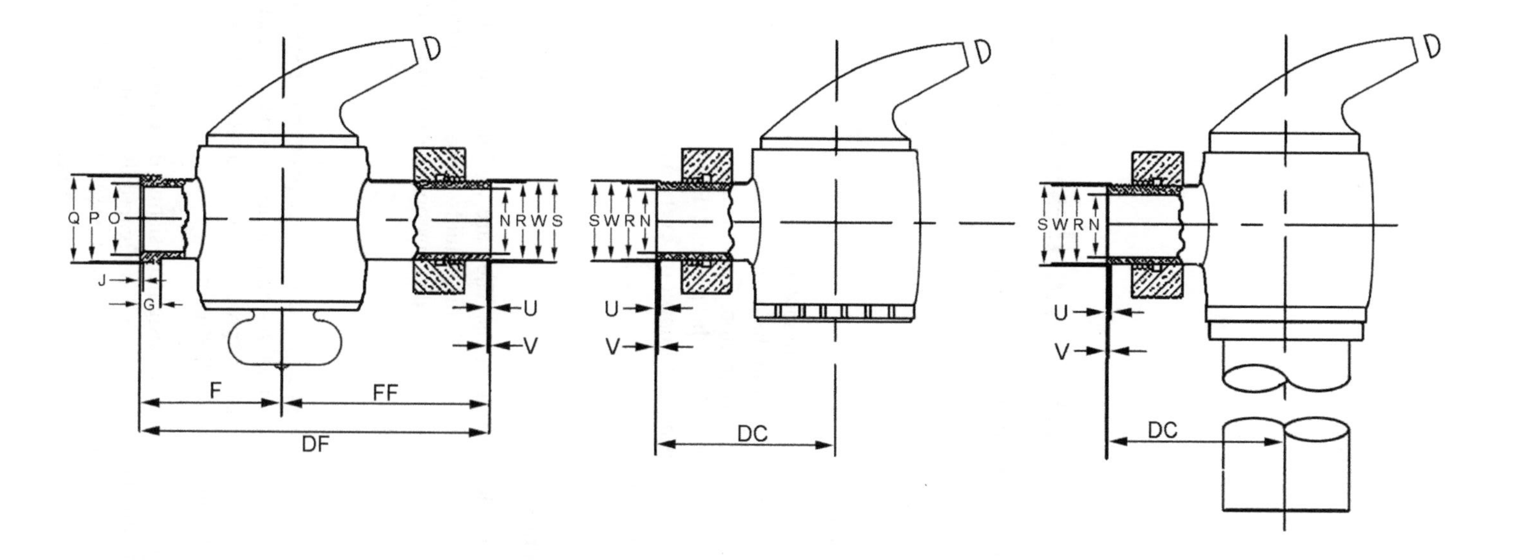

Fig. 16.1 Plug valve.

Plug valves are not generally used in dry product applications. This design is not suitable for mechanical cleaning techniques and requires complete disassembly for manual cleaning.

- *Leak protection valve*: a leak protection valves is a specialized plug-type tank outlet valve that is used on vat pasteurization equipment. The design includes special features that will prevent leakage past the valve by controlling how far the valve may be turned and the inclusion of grooves to provide leak detection. This design is not suitable for mechanical cleaning techniques and requires complete disassembly for manual cleaning.
- *Compression valve* (Fig. 16.2): this design uses a valve seat located on the end of a stem or rod that lifts the movable seat off from a valve seat incorporated into the body of the valve. These valves operate efficiently when located in a variety of positions and have a large, unobstructed valve body to permit optimum flow through the valve. This design is generally suitable for mechanical cleaning methods.
- *Mixproof valve*: a mixproof valve is a specialized compression valve that uses double seats that can be operated independently, separated by a self-draining opening to the atmosphere between the valve seats. The primary design advantage of these valves is to accommodate the separation of two different product streams or product from cleaning fluids during mechanical cleaning.
- *Diaphragm valve* (Fig. 16.3): this design uses a flexible diaphragm to form the seal. The valves are used to shut-off or regulate product flow. They work well with semi-solid and fluid products containing particulates. Diaphragm valves are not commonly used in dry product applications. The valves can be mechanically cleaned provided they are equipped with a power actuator and are installed properly to assure drainage of the valve cavity. Many designs include an orientation mark on the housing to assist with installation.
- *Tank outlet valve*: outlet valves come in a variety of configurations depending on whether they are mounted horizontally or vertically. To eliminate or reduce the amount of product that may be retained in the outlet passage, the design provides for the valve to be as close coupled as possible to the product vessel. They may be manually or mechanically operated and cleaned depending upon their design features.
- *Pressure reducing and regulating valve*: these valves are designed to control product outlet pressure by responding to outlet pressure changes by means of a self-acting actuator. The self-acting actuator raises or lowers the valve seat within the valve body by means of fluid forces within the valve body.
- *Check valve* (Figs 16.4 and 16.5): check valves permit product flow in only one direction. A reversal of product flow or pressure will result in the valve sealing. The actuation of the valve is self-acting through the use of internal balls, valve flaps, or spring-loaded valve seats. Because of their basic design, spring-loaded check valves must be fully disassembled for manual cleaning.
- *Ball valve*: this design uses a ball connected to an actuation shaft. Product flow is directed or stopped through single or multiple passages within the ball. This design incorporates body cavity fillers or encapsulating seals to

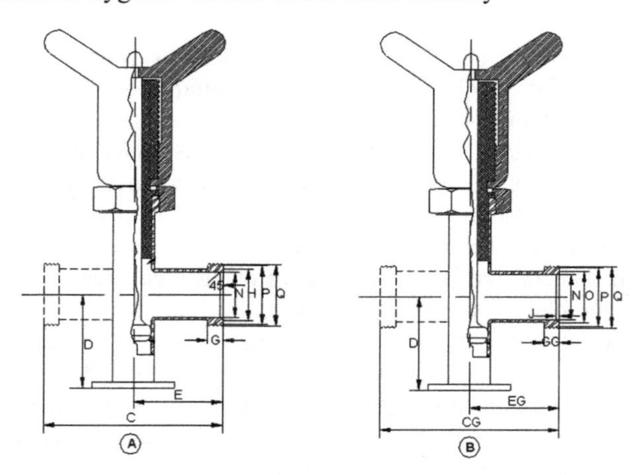

Fig. 16.2 Compression valve.

prevent product flow around the exterior of the ball. This design is not suitable for mechanical cleaning techniques and requires complete disassembly for manual cleaning.

- *Cage ball valve*: this is a variation of the ball valve design. In this design, a solid ball is retained in a movable cage attached to an actuator shaft. Actuation of the valve positions the loosely retained ball so that the product pressure within the valve body causes the ball to block a port of the valve. The action is similar to that which occurs within a ball check valve.

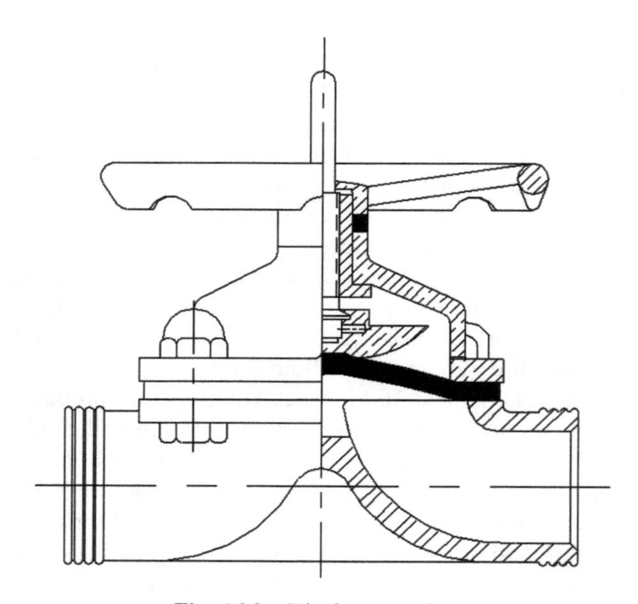

Fig. 16.3 Diaphragm valve.

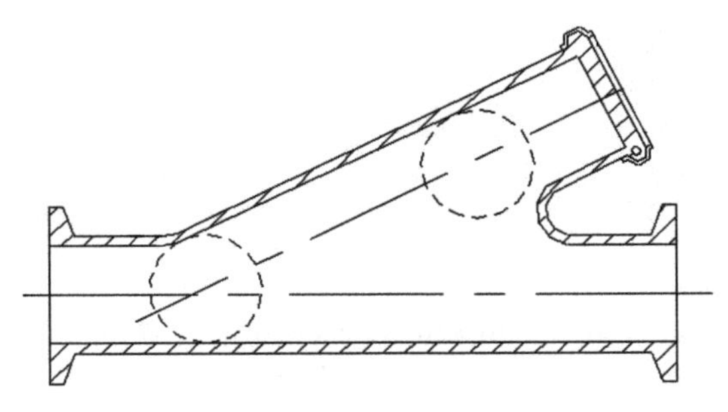

Fig. 16.4 Ball-type check valve.

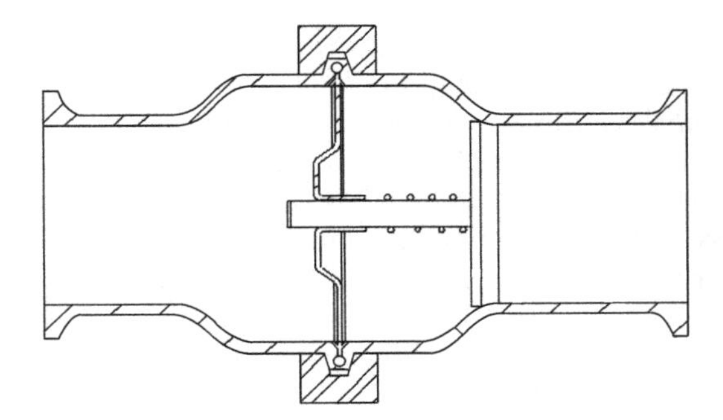

Fig. 16.5 Spring-loaded check valve.

Depending upon the design of a specific model of valve, it may not be suitable for mechanical cleaning.

- *Pinch valve*: a pinch style valve consists of a flexible rubber or polymer liner within a metal tube. The flexible liner is pinched between metal components to restrict or stop the product flow.

16.2.2 Valves used in fluid or dry product processing systems

- *Blender discharge valve*: blender discharge valves come in a variety of shapes and sizes according to the manufacturer of the blender. In most cases, the valves are located on the bottom or one end of the blender. They consist of either sliding gate types or hinged lift gate types. A pneumatic or hydraulic cylinder often powers the valve actuators. These designs require disassembly for cleaning.
- *Butterfly valve*: this design utilizes a more-or-less flat, round disk attached on the edges of the disk to an actuator and support shaft. The shafts pass through a circular rubber or polymer seat that is clamped between two flanges. The valves can be set to regulate or block the flow through the valve. While often cleaned by mechanical means, this design will allow product to migrate along the shafts due to the product pressures in the system and should, therefore, be disassembled for manual cleaning.

- *Rotary valve*: also know as a rotary airlock valve or a star valve, this consists of a multi-chambered shaft that rotates within a cylindrical housing. The valves are particularly efficient for transferring product between zones of differing pressure or vacuum. Some manufacturers have designed models specifically to meet the requirements for mechanical cleaning.

16.2.3 Valves used in dry product processing systems

- *Iris valve*: an iris valve features a fabric or multi-section metal diaphragm that opens or closes by a twisting motion of the valve's outer housing similar to the action of a camera iris.
- *Inflatable seal duct valve*: these valves use a rubber bladder attached to the outer circumference of a valve disk. After the valve is rotated into the closed position, the bladder is inflated to obtain a seal with the duct's interior.
- *Flip-flop valve*: a flip-flop valve has two movable valve disks or flaps, located one above the other, which cycle alternatively by mechanical means. Product is passed into the intermediate chamber between the two valve disks before passing through the valve body. Like rotary valves, flip-flop valves can transfer product between zones of differing pressure or vacuum.
- *Diverter valve*: as its name implies, a diverter valve consists of two or more ports into which the product stream may be directed. Depending on the number of discharge ports provided, the actuating mechanism may be a simple flap valve, a positioning slide valve, or a rotary port plate.

16.3 Hygienic aspects of valve design

The ubiquitous nature of valves within processing systems places them in a position to potentially have an impact on every particle of product passing through the system. Valves can be small or very large; simple or very complex in design. Therefore, their hygienic design is vital to producing a high-quality, safe product.

Valve design plays a major role in whether or not a particular valve or design is suitable for its intended application. Valves that are intended to be mechanically cleaned or cleaned-in-place (CIP) require special design features. The design and operation during cleaning must assure that all components potentially contacted by product will also be contacted by cleaning and sanitizing fluids (including hot water and steam) with sufficient turbulence or flow to thoroughly clean and decontaminate these areas. Generally speaking, any manually operated valve, regardless of its type or design, and valves used in dry product processing systems should not be considered as suitable for mechanical cleaning methods. The inability for automatic cycling of the valve during cleaning prevents cleaning and sanitizing solutions from reaching all areas of the valve seats and other seals. These valves are only suitable for disassembly and manual cleaning.

As with other equipment, hygienic design is based on such features as:

- materials of construction;
- internal surface texture;
- accessibility for cleaning and inspection (including leak detection and seat lifting);
- draining;
- elimination of cracks, crevices, and niches;
- internal angles and corners;
- process and installation concerns.

16.3.1 Materials of construction

Traditionally, hygienic valves for the food industry have been fabricated in stainless steel. The American Iron and Steel Institute, AISI series 300 stainless steels and their equivalent cast grades, have been the materials of choice for the metallic components of the valves. Non-metallic components for seals, valve seats, diaphragms, plunger, and plug encapsulations are to be acceptable to the convening regulatory authority and be non-toxic, relatively inert, non-porous, non-absorbent, and compatible with the environment of intended use, cleaning, and sanitization. Care must be exercised to assure that proper non-metallic materials are selected; with particular emphasis on the fat content and temperature ranges of the products intended to be processed.

16.3.2 Internal surface texture

Product contact surface finishes at least as smooth as an R_a of 0.8 μm (32.0 μinch) on stainless steel free of imperfections such as pits, folds, and crevices in the final fabricated form will clean satisfactorily and are recognized by most hygienic equipment standards. Any deviations rougher than this minimum should be part of the manufacturer's specifications so that the buyer can make an informed decision to use the valve.

16.3.3 Accessibility for cleaning and inspection

The normal cycling of valves during processing and cleaning subject valve components to high stress and wear of movable components. In large fluid lines, these stresses can force product pass seals or removable valve seats. Valves in dry product processes may be subjected to materials that are quite abrasive. Therefore, it is vital that valves be both accessible when installed in a processing system and easy to disassemble so that the wear and cleanliness of components can be periodically inspected. Valves that are supplied with automatic power actuators generally are required to have at least a 25 mm (1.0 inch) space between the actuator and the valve shaft seal that is open to atmosphere so that a failure of the seals is readily observable.

Valves such as diaphragm or pinch valves that have internal components subject to failure, require a drain opening to atmosphere to signal component failure. The drain opening shall be at a point low enough so that leaking product cannot collect within the valve housing.

16.3.4 Draining

Most valve designs have addressed the ability of the valve to be self-draining when properly installed. However, care must be exercised prior to installation, as some manufacturers still require attention to drainability of their designs. Upon installation of any multi-port valve, care must be taken by the installer to assure that dead pockets are not created in the product flow when a port is closed. Some valve designs, such as most diaphragm valves, are inherently non-self-draining. In these cases, the valve should have a clearly identified orientation mark on the valve body to indicate proper installation angles.

Self-draining is not a major consideration in dry product processes. These systems are designed to go for extended periods of time with only dry cleaning. In those rare instances where wet cleaning is necessary because of a contamination or a major maintenance project, the entire system, including any valves, must be disassembled to clean and thoroughly dry the wetted surfaces.

16.3.5 Elimination of cracks, crevices and niches

Some valve designs, even with automatic actuators, are inherently not suitable for CIP or mechanical cleaning methods. These designs tend to have large sliding seal areas as in plug, ball, iris, inflatable seal duct, and diverter type valves. The sliding seal will contain a film of product throughout production. The tolerances of the sliding seals prohibit the transport of the sufficient cleaning and sanitizing fluids to achieve cleaning and decontamination.

A valve's design must take into consideration the proper attachment and compression of elastomeric components. The compression of the elastomeric components must be controlled so that the materials cannot be over-compressed, causing them to extrude into the product flow. Additionally, the compression must be sufficient to assure a tight seal across the full temperature range of processing, cleaning and sterilization.

The attachment of elastomeric components to metallic components shall assure that the intended flexing of the elastomeric components does not create crevices or cracks to open between the components as they are cycled. Lip seal valves can be used only under special conditions where cleaning and sanitation have been validated and documented.

16.3.6 Internal angles and corners

Sharp or decreasing internal angles and corners within a valve must be avoided to assure that hard to clean areas are not created in the design. This is especially important on valves that are subjected to manual cleaning procedures. The internal radii must be sufficiently large to allow for the cleaning implements to reach the surfaces. Generally, internal radii of 3 mm (1/8 inch) are recognized as adequate to accomplish cleaning. Smaller radii are permitted for smaller components within the valve. In these cases where smaller radii are required, they should not be less than 0.8 mm (1/32 inch). The manufacturer of the valve should specify the presence of such radii.

16.3.7 Process and installation

When designing a process system there are additional concerns that must be taken into consideration. Are there legal requirements of the regulatory authority? For example, pasteurization systems, including both batch and continuous systems, have special requirements for the valves used to segregate pasteurized from unpasteurized product. This may require special leak detection and rapid response times for valve actuation. Aseptic systems require valves that can be demonstrated to be bacteria tight.

Generally, valves will be designed for self-closure in the event of a power failure during processing. However, there may be instances where it is more beneficial for the valves to remain open so that the systems can self-drain. Therefore, care must be exercised in the selection of the valves and their placement within the system.

It is quite common to use a 'block and bleed' configuration of valves to assure separation of different streams (Fig. 16.6). Installers should pay extra attention to assure that the bleed lines properly drain and do not retain fluids. The food industry would be greatly helped by the availability of valves that provide the security of a 'block and bleed' system while greatly simplifying draining, avoiding stagnant product. Installers must be careful not to create a common installation error that produces a 'block–block–bleed' configuration.

Care must also be exercised to assure that valves can be easily accessed for periodic maintenance and inspection. Valves commonly include components, which wear and require periodic replacement.

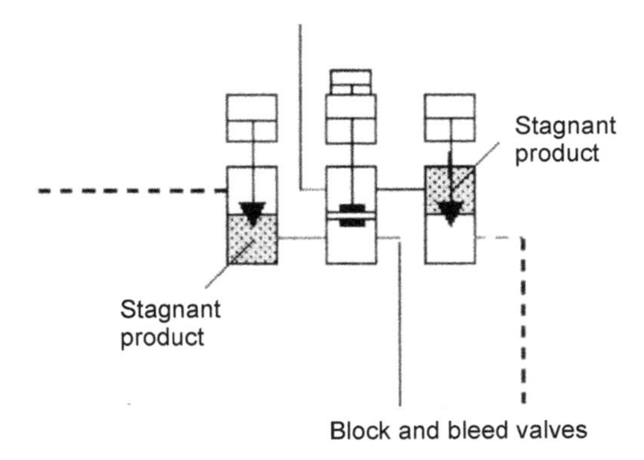

Fig. 16.6

16.4 Current guidelines, standards, and references

3-A SSI, McLean, VA, 3-A Sanitary Standards numbers:
51-01 Plug-Type Valves,
53-02 Compression-Type Valves,
54-02 Diaphragm-Type Valves,
56-00 Inlet and Outlet Leak-Protector Plug-Type Valves,
58-00 Vacuum Breakers and Check Valves,
64-00 Pressure Reducing and Back Pressure Regulating Valves,
66-00 Caged-Ball Valves, and
68-00 Ball-Type Valves.

'Hygienic Requirements of Valves For Food Processing', Doc 14, EHEDG

Dr Roland Cocker, member, EHEDG Valves subgroup, 'Hygienic design and assessment', *New Food*, Volume 7, Issue 1, 2004.

'USDA Guidelines For The Evaluation and Certification of the Sanitary Design and Fabrication of Dairy Processing Equipment', June 2001, US Department of Agriculture, Dairy Programs, Washington, DC.

'Grade A Pasteurized Milk Ordinance', US Department of Health and Human Services, Public Health Service, Food and Drug Administration.

'Milk and Milk Product Equipment; A Guideline For Evaluating Construction', US Department of Health and Human Services, Public Health Service, Food and Drug Administration.

17

Improving the hygienic design of pipes

H. Hoogland, Unilever R&D Vlaardingen, The Netherlands

17.1 Introduction

The main function of piping in a food processing plant is the transport of material. Next to the transport function, piping can be used for processing of material, for example by heating/cooling, mixing, or to provide holding time. This chapter starts by summarising the current requirements for the design of stainless steel piping, and then gives the possible applications of plastic and flexible piping. The chapter also discusses the growth of microorganisms in piping systems and the use of antimicrobial coatings and antifouling coatings. Product recovery from piping systems is briefly discussed, followed by some thoughts on the design of food manufacturing plant.

17.2 Piping design: good practice

A unit of pipe is cheap and easy to clean in comparison with vessels, pumps, extruders and other equipment. To maintain this inherent benefit of simple cleanability some guidelines must be followed when designing pipework:

- Piping material should be food-grade, non-absorbent and resistant to the product and cleaning-in-place (CIP) fluids.[1]
- Internal surfaces should be smooth,[1] also where piping is welded.[2]
- Couplings should be avoided when possible. The first option should be bending followed by welding. When unavoidable, hygienic couplings should be used[3] and a preventative maintenance scheme drawn-up for seal replacement.

- Pipe diameters should not vary too much (to realise sufficient flow during cleaning in all parts of the piping system).
- The pipe system should be drainable,[1] with a pitch of 1 in 100 and supported to prevent sagging.

Although a piping system designed according to the principles as described above should be easy to clean, this could be hampered by complexity. When the piping system contains many branches one has to be very careful when designing the CIP system as usually sequential and repeated switching of CIP valves is needed. Options[4] for the design of piping systems are given in Baumbach and Hoogland.[5]

17.3 Materials of construction

The default choice for piping systems is stainless steel (AISI 304 or AISI 316); however, alternative materials should be considered. The use of plastic piping systems for low-pressure applications could have the following benefits:

- the welding procedures are mainly automatic, requiring a low level of skill;
- they are able to handle highly corrosive products;
- plastic has superior insulating properties – avoid condensation when transporting cold products or reducing heat losses when used for heat treatments (e.g. holding tube);
- the possibility of making lightweight systems in case the plant needs to be mobile.

Next to these benefits there are a number of uncertainties about the use of plastic pipework:

- plastic is more sensitive to wear;
- welding requires special equipment;
- cleanability of plastic piping needs to be verified;
- limited availability of hygienic valves and couplings.

Depending on the local situation, plastic piping might be a good alternative to stainless steel.

Trends such as 'late customisation', 'customisation on demand' and 'distributed manufacturing' often require intermediate products to be stored, transported and coupled to a second process later. A simple way for doing this is the use of containers that can be connected to the processes by flexible tubing. This kind of system is available from a number of suppliers. Selection of a system should be based on the required level of hygiene and the cost of operation. The current trend of healthy/fortified foods might lead to the need for aseptic coupling of containers to processes, for example post-dosing of flavours and micro-nutrients to ultra-heat treated (UHT) sterilised products.

17.4 Product recovery

To minimise product losses during cleaning and or change-over it should be possible to purge the system. The simplest way is by purging the system by water or product. This becomes less effective for very long pipe runs as the mixing zone increases. Also for viscous products purging with a low-viscosity fluid such as water is not effective, as the water will push a hole through the centre of the pipe and the majority of the viscous product will stay behind. The most attractive way of reducing the environmental implications for closed systems is product recovery by 'pigging'. The so-called 'pig' is sent, by means of compressed air or water as the driving force, through the pipeline, and expels most of the product from the system. The pig is often designed from an elastic material such as silicon rubber with a closed surface. Sponge-like pigs are not recommended, as they are difficult to clean. The early pigging systems were operated manually. Today's pigging systems are fully automatic and are cleanable-in-place (CIP), avoiding any chance of contamination by the operator. Also, unless the manual system is carefully designed to prevent the pig from escaping from the line under pressure, it can present a potential danger to the operator.

The drawbacks of these systems are the requirements they put on the design and construction of the piping system, such as:

- all pipes should have the same diameter;
- pipe roundness is important, especially in bends;
- no valves or sensors protruding into the pipeline;
- no welds protruding into the pipeline;
- branches should be constructed in such a way that it is impossible for the pig to go in the wrong direction.

Often these requirements are difficult to meet especially in slightly older existing plant.

For some of the issues mentioned above there are solutions. To avoid protruding sensors some suppliers did develop retractable systems for temperature sensors, pH sensors. Also full-bore ball-valves have been designed which are claimed to be cleanable in place. As standard ball-valves are not 'cleanable-in-place' since the area between the seals cannot be cleaned, this has now been overcome in an inventive way by the use of inflatable seals which are deflated during CIP, allowing the CIP fluid to flow around the ball. The design incorporates two inflatable seals that grip the ball when inflated and released from the ball when deflated. These valves are also available in a three-way version and form the basis for a pigging system as provided by Hygienic Pigging Systems Ltd.

The alternative approach is to make the pig more tolerant for geometrical changes. This has for example been done in a UK-LINK scheme led by the University of Bristol where the pig was made from slush ice. The university has patented this development. It should also be possible to make the pig from food-grade biopolymer.

17.5 Microbial growth in piping systems

Most of the product will remain within the system for a shorter time than that needed for the microorganisms to multiply. Only a very small fraction of the product flows close to the pipe walls and resides sufficiently long to allow growth. Would the growth in this very small volume be able to increase the number of microorganisms in the main flow? To verify this, a simple simulation was made for a viscous Newtonian flow through a straight pipe with an average residence time in the pipe of 5, 8, 16 minutes, much less than the doubling time of 20 minutes. The initial concentration is 100 microorganisms/ml. Figure 17.1 shows the average concentration of microorganisms at the pipe exit. From the simulation we learn that although the volume available for growth is small, the exponential nature will lead to a significant increase in the number of microorganisms in the main flow. This also implies that hygienic equipment will have a limited run length when microorganisms are able to grow within the system.

More and more antimicrobial materials are being introduced to the market and it would be interesting to know if the antimicrobial action also penetrates sufficiently into the product to stop the growth. Examples where this problem

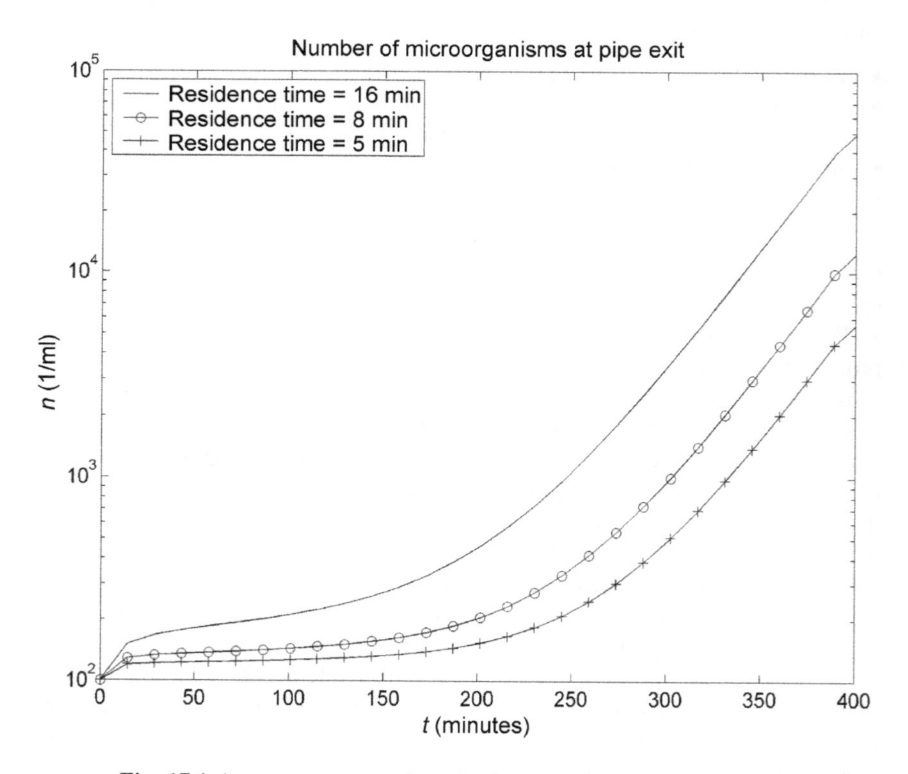

Fig. 17.1 Average concentration of microorganisms at the pipe exit.

could occur are the preheating sections of heat exchangers for pasteurisation of products. In the pre-heating section very favourable growth conditions are found. Although the product is pasteurised afterwards prevention of growth would allow for milder heat treatment or reduces the level/chance of survivors.

When surfaces contain a high level of microorganisms directly from the start of the process, the growth curve will be much steeper. This can be caused by insufficient cleaning, leaving product residues behind that will protect microorganisms against subsequent disinfection. Next to being a source of microorganisms, the residues present a problem in other areas such as heat transfer, source of allergens, etc. Cleaning could be facilitated by the use of non-stick material. Non-stick material is not usually suitable as a material of construction and therefore needs to be applied on the surface of a construction material as a coating. In general the material of construction needs to be resistant to the product and CIP fluids, otherwise any damage to the coating would inevitably lead to local corrosion that continues behind the coating. Often this will lead to stainless steel as the material of construction. From a hygienic point of view it is important that the coating will not blister off the material of construction, even when the colour of the coating matches the colour of the product. As this is very difficult to avoid it could be considered to develop a coating whose integrity could be monitored on-line continuously.

17.6 Plant design

Building on this simplicity the ideal plant should be continuous and consist only of piping. Raw materials could be supplied in containers, which are directly coupled to the piping system and are returned without cleaning to the supplier. A wide range of equipment is already available as pipes. Most unit operations and sensors can be done in piping, static mixers, hose pumps, tube in tube heat exchangers, membrane pressure and temperature sensors, etc. However, it is clear that these devices have their limitations and often need to be replaced with in-line devices that differ in geometry. As long as the devices have a similar cleaning characteristic as a piece of pipe, the principle does not change.

To avoid all problems related to piping such as cleaning, product loss during cleaning and change-over, and residence time distribution, the concept of the pipe-less plant could be applied. Here, small vessels move to the factory, from a dosing station to a processing station, to the filling machine. Such a system requires no piping at all and gives a very accurate control over the residence time. There is still a need to clean the vessels before preparing a new batch. This could be overcome by applying an internal disposable lining, which could be removed easily, for example a thin, plastic foil, shaped to match the vessel. Current developments in automation make this option more and more attractive.

17.7 References

1. CURIEL G J, HAUSER G, PESCHEL P, TIMPERLEY D A, 'Hygienic equipment design criteria', *Trends in Food Science & Technology* (1993), Vol 3(11), p. 277.
2. EASTWOOD C A, WOODALL D L, TIMPERLEY D A, *et al.*, 'Welding stainless steel to meet hygienic requirements', *Trends in Food Science & Technology* (1993), Vol 4(9), pp. 306–310.
3. BAUMBACH, F, DUBOIS, J, GRELL, W, *et al.*, 'Hygienic pipe couplings', *Trends in Food Science & Technology* (1997), Vol 8(3), pp. 88–92.
4. CURIEL, G J, HAUSER, G, PESCHEL, P, *et al.*, 'Hygienic equipment design of closed equipment for the processing of liquid food', *Trends in Food Science & Technology* (1993), Vol 4, pp. 375–379.
5. BAUMBACH F, HOOGLAND H, 'Piping systems, seals and valves'. In: *Hygiene in food processing*. Woodhead Publishing Limited, Cambridge, 2003.

18

Improving the hygienic design of pumps

R. Stahlkopf, Tuchenhagen GmbH, Germany

18.1 Introduction: types of pump used in food processing

In automated food processing technology, pumps are used for the transport and the increase of pressure of low-viscosity, liquid products. To ensure a hygienically and microbiologically perfect condition of the product, high standards are applied to turbo machines and positive displacement pumps with regard to hygiene and cleaning technological requirements. These standards apply to all pumps used in the food processing industry, including centrifugal pumps, piston pumps, rotary pumps, peristaltic pumps, diaphragm pumps, water ring pumps, positive-displacement pumps, screw pumps, gear pumps as well as to homogenisers, dampening devices and finally all valves integrated in the pump body.

Concerning centrifugal pumps, it is common practice to flange the pump directly to the motor. This has the advantage of a more compact pump design. Open couplings as used in the conventional pump design where the motor is mounted on a ground plate are not needed. If motors with special shafts are used, the impeller is plugged on to the motor shaft or the pump shaft is plugged on to the motor shaft of a standard motor (Fig. 18.1). In this case, depending on the pump size and/or axial forces, reinforced bearings, integrated bearing flanges and measures against axial thrust are required.

18.2 Components used in pumps

18.2.1 Impeller

Open impellers without a front cover plate are generally used on centrifugal pumps. The advantage is an easier manufacture of the impeller (Figs 18.2 and 18.3). The impeller blades are mechanically accessible and can be treated to

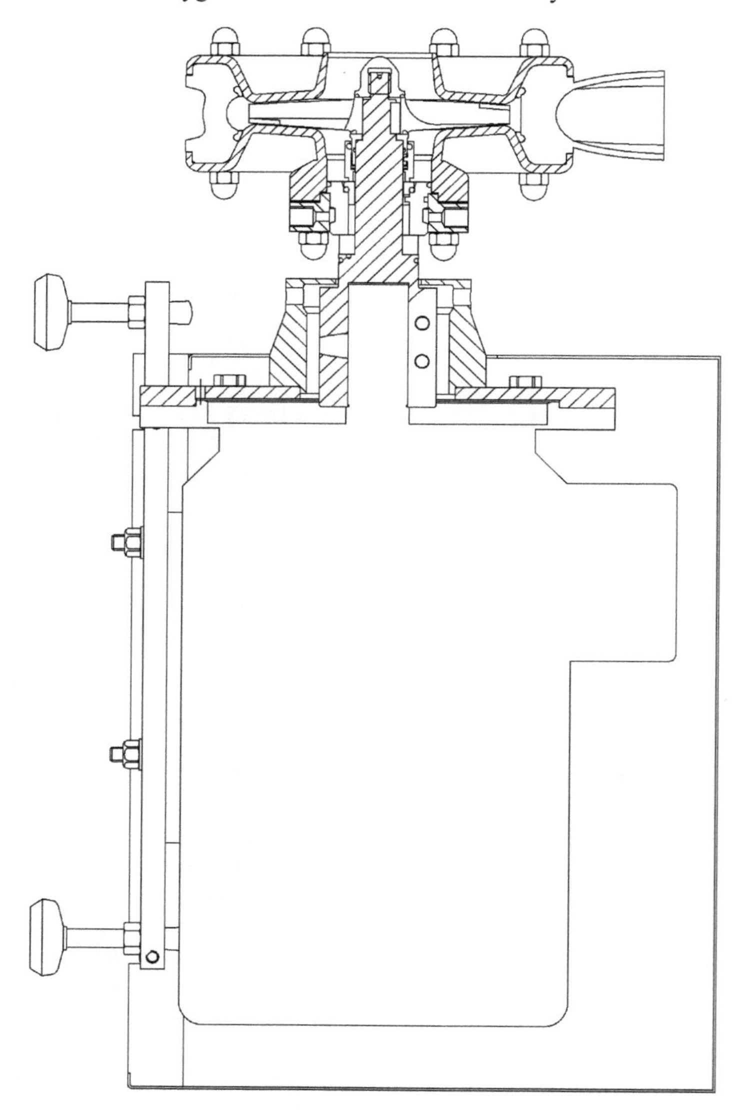

Fig. 18.1 Hygienic centrifugal pump.

achieve the desired surface roughness. Depending on the specific speed n_q, the wheel types are either slow-running radial wheels ($n_q = 10$–30) or medium-running radial wheels ($n_q = 30$–60). The specific speed represents a wheel-type characteristic. For the specific speed n_q consisting of speed n in rpm, volume flow Q in m^3/s and flow head H in m the following numerical equation applies:

$$n_q = n \frac{\sqrt{Q}}{H^{3/4}}$$

Fig. 18.2 Open impeller with compensation bores reducing the axial thrust.

Fig. 18.3 Axial force-free impeller.

18.2.2 Mechanical seals

Mechanical seals have replaced the stuffing box used so far in the food technology. The design of the mechanical seals should provide that the pressing spring for the sliding ring located in the product chamber is encapsulated. Exposed threads (headless screws) or similar should not be located in the product chamber. Garter springs are not permitted on the product side of the seal.

For normal application in breweries and dairies single-acting mechanical seals are used, often in hard/soft pairings (ceramic/carbon). For abrasive media hard/hard pairings are the better choice. For products that crystallise out or paste, quenched mechanical seals are used (mechanical seal and radial shaft seal with flushing). For toxic products or in pharmaceutical applications double-acting mechanical seals are used with pressurised sealing water.

The construction of the shaft and the arrangement of the mechanical seal chamber and lantern should allow for the installation of customer-specific mechanical seals or retrofitting of other mechanical seals for the adaptation to changed media.

18.2.3 Leakage detection

A lantern should be installed between the pump housing and the driving motor. Apart from the geometric connection function between the housing and the motor flange, the lantern is used to connect the pump safely with the motor, in order to make leakage visible in case of damage. In case of leakage, product or cleaning solution must not be allowed to enter the shaft area of the motor where microorganisms could grow unnoticed. The penetration of leakage fluids or splash water from the lantern port into the shaft area of the motor during outside cleaning can be avoided effectively by providing a splash ring on the shaft and a labyrinth seal between the shaft and the lantern housing.

18.2.4 Filling and drainage of the pump housing

For an optimal start of a self-priming pump, the discharge socket should be arranged in a way that air can evacuate autonomously at any operating point (Fig. 18.4b). In this case a radial upwards directed discharge socket is most suitable. With this arrangement, suction and discharge sockets are located on the same axis, which simplifies piping and plant layout.

Non-aqueous products and mainly pharmaceutical applications require the drainage of rinsing water from the pump housing. The European Hygienic Engineering and Design Group (EHEDG) stipulates this as an option for pumps. Drainage valves used in this case must be without dead ends to ensure that the production process is not endangered (Figs 18.4e and 18.4f).

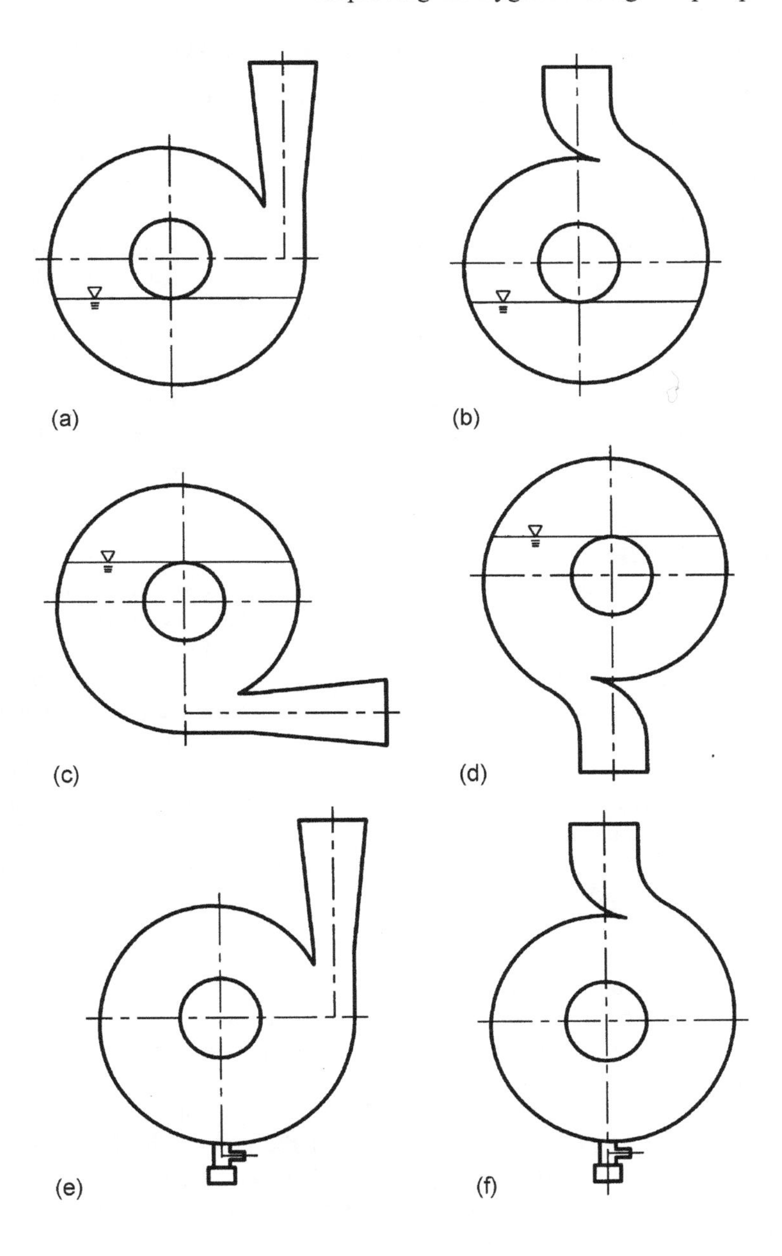

Fig. 18.4 Pump filling and drainage: (a) tangential spiral housing, venting OK, without drainage; (b) radial spiral housing, venting optimal, without drainage; (c) tangential spiral housing, venting poor, with drainage; (d) radial spiral housing, venting poor, with drainage; (e) tangential spiral housing, venting OK, with drainage; (f) radial spiral housing, venting optimal, with drainage.

18.3 Cleanability, surface finish and other requirements

Cleanability of all parts in contact with product in flow (CIP, cleaning-in-place) without the need of disassembly is imperative. The product chamber must show no dead ends and gaps. If the impeller is plugged, the hub of the impeller must be sealed towards the pump shaft and towards the shaft nut. The seal rings must be clamped in a way that they are exposed to and cleaned by the detergent stream. Centring and guiding must be located outside the seals. The $NPSH_{req}$ value of the pump should not exceed 2 m. The reason is that food pumps are often used as CIP pumps with hot CIP media supply (up to $t = 95\,°C$). The chamber of the mechanical seals must have free access for cleaning and cooling the mechanical seals. For all internal corners with an angle $\alpha < 135°$, the radius must be >3.0 mm. All edges must be deburred. Exposed threads, keyways or splines should be avoided. The outer contours of the pump must ensure free drainage of cleaning solutions, condensation water and similar.

18.3.1 Surface finish

The surface finish influences the cleanability of pump parts. The cleaning times decrease with improved surface finishes. Typically high relative velocities and turbulence are created by the rotating elements of pumps during cleaning. Test results have shown that the pump type (geometry) and operation (relative velocities) have a significant influence on cleanability and cleaning of hygienic pumps.

On CIP pumps the surface roughness of the parts in contact with the product should not exceed $R_a = 3.2$ m (see EHEDG, 2004). For customer-specific (aseptic) applications the pump design must allow that the product-contacted surfaces can be brought to a surface roughness (inner surface) of $R_a = 0.8$ m (EHEDG optional, 3A Standard).

18.3.2 Gentle product handling

Gentle handling of the product may be an important aspect with food pumps. The shearing strain acting on the product must not be excessive. The core parameter for this unit is the flow velocity in the impeller channel. The meridian absolute flow velocity should not exceed the calculation point $c_m = 2$ m/s. This is also to the benefit of the $NPSH_{req}$ value of the pump. The flow channels should all be designed in true alignment and without dead ends. The use of spiral housings with a logarithmically shaped spiral, if applicable, would be an advantage.

18.3.3 Additional requirements for aseptic equipment

The equipment shall be designed as to be impermeable to external microorganisms. The design must allow for steam sterilisation at 121 °C for

20 min as a minimum. During steam sterilisation, all product contact surfaces shall reach the required temperatures. The dynamic seals should be double seal arrangements to allow for a secondary barrier of steam or sterile liquids. The static seals may comprise a single seal arrangement, provided that it does not allow any external contamination under any operational conditions. If satisfactory operation cannot be assured with a single seal, then a double seal arrangement with a continuous sterile barrier must be provided. Depending on the application, in particular on the length of the production period, the sterile barrier may consist of a sterile fluid or an antimicrobial fluid (e.g. steam, hot water or chemical solution).

18.4 Materials and motor design

All product contacted parts generally consist of stainless steel (e.g. 1.4301 (X5 CrNi 18-10), 1.4306 (X2 CrNi 19-11), 1.4541 (X6 CrNiTi 18-10), 1.4401 (X5 CrNiMo 17-12-2), 1.4404 (X2 CrNiMo 17-12(13)-2), 1.4571 (X6 CrNiMoTi 17-12-2), 1.4435 (X2 CrNiMo 18-14-3), 1.4581 (GX5 CrNiMoNb 19-11-2), 1.4462 (X2 CrNiMoN 22-5-3), 1.4460 (X3 CrNiMoN 27-5-2),1.4405 (GX4 CrNiMo 16-5-1), 1.3974 (X2CrNiMnMoNNb23-17-6-3), 1.4542 (X5 CrNiCuNb 19-16-1)). Suitable non-metallic materials are carbon, silicon carbide, tungsten carbide, ebonite, nitrile rubbers, ethylene (polyethylenes), propylenes, fluoro rubbers (Viton = DuPont registered trade mark), silicone rubbers, urethane rubbers, natural rubbers, PTFE (poly(tetrafluoroethene)), ETFE (ethylenetetrafluoro-ethene), polyamide, perfluoroelastorners (Kalrez/Chemraz), ceramics and aramides. Products in contact with elastomers should be Food and Drug Administration (FDA), Code of Federal Regulations (CFR) or Bundesinstitut für Risikobewertung (BfR) approved.

Standard three-phase asynchronous motors are preferable to motors with special shafts (block type) because they can be easily replaced throughout the world. Each pump manufacturer sees to it that the modular design of the pump ensures a maximum use of standardised parts for cost reasons and for adapting the pump capacity to any hydraulic requirements.

18.5 Summary

To produce food products in hygienically and microbiologically perfect condition, high standards are given to food pumps with regard to hygienic design and cleaning technological requirements. The main aspects in this respect are hydraulic efficiency, flow characteristics, pocket-free design and low-cost manufacture. This chapter has described the special requirements of the food industry on components, seals, cleanability, material composition and surface quality.

18.6 Bibliography

BOHL, W., *Strömungsmaschinen 2. Berechnung und Konstruktion*. 5 Aufl. Würzburg, Vogel Verlag 1995.

Centrifugal Pump Lexicon, Frankenthal 1989.

EHEDG Guideline No. 17, *Hygienic Design of Pumps, Homogenisers and Dampening Devices*, 2nd edn, September 2004.

PFLEIDERER, C.; PETERMANN, H., *Strömungsmaschinen*, 5th revised edition, Berlin, Heidelberg, New York, London, Paris, Tokyo, Springer-Verlag 1988.

PFLEIDERER, C., *Die Kreiselpumpen*, 5th revised edition, Berlin, Göttingen, Heidelberg, Springer-Verlag 1961.

3-A Sanitary Standards Inc., USA.

TROSKOLANSKI, A.T.; LAZARKIEWICZ, S., *Kreiselpumpen*, 3rd edition, Basel, Stuttgart, Birkhäuser-Verlag 1959.

Tuchenhagen-Variflow, *Centrifugal pumps TP and KN Series*, Büchen 10. 2003.

19

Improving hygienic control by sensors

M. Bücking, Fraunhofer IME Germany, and J. E. Haugen, Matforsk AS, Norway

19.1 Introduction

With an increasing need of better control systems for many food processes, interest has become focused towards on- or in-line control at the production stage. The food processing industry needs a fast, standardised, objective and cost-effective tool to control and improve quality.

Sensor applications range from security controls to landmine detection and from health screening to quality control. In this chapter we want to describe the possible contributions of sensor systems towards hygiene in food processing, i.e. sensors that control the quality of food by giving a response to a certain property of food. The chapter can provide only an overview of highlights in the field of sensors; these examples of the main principles as well as many other systems are described in much more detail in dedicated books (Baltes *et al.*, 1999; Gardner and Bartlett, 1999; Kress-Rogers and Brimelow, 2001; Eggins 2002; Tothill, 2003).

Samples in the food industry are diverse and variable and the food industry has a great need for rapid sensor techniques on the production line. These techniques should be low cost, easy to operate and fast, in the best case on-line for real time quality monitoring of raw material, processing and final product. For Hazard Analysis Critical Control Point (HACCP) purposes, physical sensors as listed in Table 19.1, have found their application in the food industry off- and on-line. They represent the critical parameters of pressure, temperature, relative humidity, water activity and pH, which are some of the factors for controlling the microbial hygienic environment in the production plant. However, there is still a need for rapid sensor techniques for determining the microbial and sensory quality in the production line for the detection of bacterial contamination. Physical sensors will therefore not be discussed, because they are not directly

Table 19.1 Overview of physical sensors used for food industrial purposes

Parameter	Sensor principle/sensor
Temperature	Resistance/Pt-100 sensor
Flow	Volumetric/positive displacement sensors and turbine meters Electromagnetic Mass/Coriolis mass flow meters
Pressure	Resilience/ceramic sensors
pH/redox value/ion concentration	Potentiometric/Ag/AgCl electrode Conductivity/two-electrode sensor with a given cell constant Inductive sensors
Turbidity	Optical/optical sensor
Process status	Ultrasonic/ultrasonic sensor Microwave/propagation of microwaves
Moisture/dew point	Dielectricity/polymer dielectric sensor

related to food quality control, and have been well covered elsewhere (Kress-Rogers and Brimelow, 2001).

Looking at the heart of a sensor system, the sensor element, the task of this apparatus is to respond to a physical stimulus by producing a signal (the sensor is a type of transducer) that provides direct or indirect information about the status quo (Fig. 19.1). The classification of sensor systems results from their material, the operating temperature or their signal/measurement category (e.g. thermal energy, electromagnetic energy, acoustic energy, pressure, magnetism or motion).

As for all analytical instruments, the critical performance requirements for sensor systems are sensitivity, reproducibility and selectivity. Furthermore industrial needs have to be taken into account, i.e. fast measurement times,

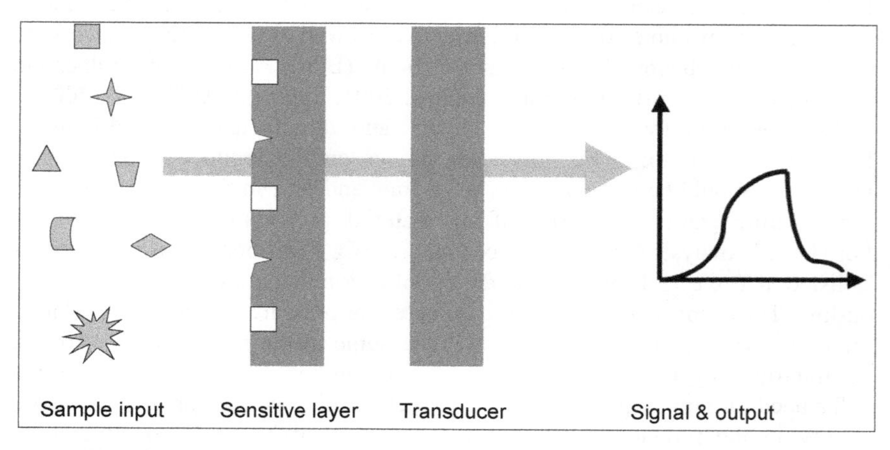

Fig. 19.1 General working principle of chemical sensors and biosensors.

physical robustness (also during common peripheral services), no need for skilled personnel and cost effectiveness (e.g. capital costs, low power consumption). Several companies tried to fulfil these needs by the production of commercial sensor systems. Many of these systems are like standard chemical-analytical devices (e.g. for gas or liquid chromatography), which can be used only in laboratories. Kress-Rogers and Brimelow (2001) give a general overview of these commercial systems and recent papers review commercial systems, e.g. chemical sensors by Stetter and Penrose (2002), or biosensors by Rodriguez-Mozaz *et al.* (2004).

This situation adds one of the most important issues: the ability to use sensor devices as an in- or on-line tool, i.e. direct process monitoring, starting with the incoming goods inspection. In a next step, process monitoring could start directly on farm-side, which would again increase the sensor requirements, especially robustness and ease of handling.

The following section is divided into two parts. The first will concentrate on devices that are capable of controlling by air sampling chemical sensors, while the second one will concentrate on liquid detection, i.e. devices such as electronic tongue devices and biosensor systems.

19.2 Sensor types

19.2.1 Volatile detection – electronic nose

Sensors for volatile (vapour phase analysis) detection are already used in automobiles, medicine, process control and laboratory measurements. The response of these solid-state gas sensors is usually not very specific, for example towards reducing (fuel) gases, oxidising gases or ammonia and molecules with ammonia structure. To overcome the problem of non-specific detection or cross-sensitivity, Persaud and Dodd proposed in 1982 the concept of so-called 'electronic noses', the detection of volatiles by sensor-arrays. Gardner and Bartlett (1993) defined electronic nose 'as an instrument, which comprises an array of electronic chemical sensors (e.g. solid state gas sensors) with partial specificity and an appropriate pattern-recognition system, capable of recognising simple or complex odours'.

To ensure the sensors receive a large amount of information, the systems are equipped with an array of sensors of the same working principle or with sensors of different working principles (so-called hybrid or multi-sensor systems). The amount of raw data from up to 30 single sensors requires complex statistical data processing. Evaluation software of sensor systems often includes functions for these sophisticated statistical approaches such as principal component analysis (PCA), partial least squares (PLS) or artificial neural network (ANN) calculations.

Commercial instruments appeared on the market and in the scientific laboratories, in the early 1990s and since that time hundreds of publications about possible applications have been written. The advantages of these devices

Table 19.2 Overview of advantages/disadvantages and application examples of the different sensor types

	Metal oxide	Acoustic wave	Conducting polymers
Example	Liden *et al.* (2000), On-line measurement	Bender *et al.* (2003), At-line measurement	Gibson *et al.* (1997), Off-line measurement
Advantages	Sensitive (ppm) Fast response Cheap sensors	Low energy consumption High stability of signal	Sensitive Selective
Dis-advantages	Sensor signal shift and sensor drift Poor selectivity High energy consumption	Less sensitive	Reproducibility of fabrication Sensor shift (oxidation)

are relatively low costs for the device itself, its maintenance, as well as its easy handling. The disadvantages are effects such as sensor signal shift and sensor drift (i.e. phenomena that might change sensor signals unpredictably) and its often unspecific detection abilities. However, these issues have partly been solved recently by computational methods (Haugen *et al.*, 1999; Artursson *et al.*, 2000; Tomic *et al.*, 2002, 2004). An overview listing of sensor systems and applications can be found in Table 19.2.

Metal oxide sensors and metal oxide semiconducting field effect transistors

The metal oxide semiconductors (MOS) are the most frequently used sensors for gas sensing and belong to the group of solid state based chemosensors. First developed in the 1960s as detectors for liquid petroleum gases they consist of a metal oxide layer on top of a semiconductor. The gas sensing principle is based on the reaction between adsorbed oxygen on the oxide surface with incoming molecules (Nanto and Stetter, 2002). The output signal is derived from a change in conductivity of the oxide caused by the reaction with the incoming molecule. There are two types of MOS sensors, the n-type (SnO_2, ZnO, Fe_2O_3, WO_3), which respond to reducible gases, and the p-type (CuO, NiO, CoO), which respond to oxidisable gases. The sensors operate at high temperatures between 300 and 450 °C. The metal oxide in the surface layer determines the selectivity of the sensors, an optional added catalyst (i.e. noble metals, mostly platinum) will also influence selectivity as well as their operating temperature. They have a sensitivity range from 5–500 ppm, and are relatively insensitive to water in the humidity range from 30–80% relative humidity (Tomic *et al.*, 2002).

Metal oxide semiconducting field effect transistors (MOSFET) also operate at elevated temperatures (100–200 °C). They consist of three layers: a doped silicon semiconductor, an oxide layer (silica) as insulator and on top a catalytic metal layer (Lundström *et al.*, 1990). This catalytic surface will interact with volatiles by a change of potential – the output signal corresponds to the change of voltage necessary to keep a constant preset drain current, i.e. both, operating temperature as well as metal layer thickness/kind of metal, will influence the sensitivity and selectivity.

A number of feasibility studies have been performed with MOS and/or MOSFET technology on different food applications comprising lipid oxidation in meats, fish and oils, freshness of fish, poultry and meat, or have been compared with standard techniques (e.g. Schaller *et al.*, 1998; Haugen, 2001; Bücking *et al.*, 2002; Miettinen *et al.*, 2002).

Acoustic wave devices

This technique is based upon piezoelectric sensors (vibrating quartz crystals) which are made of quartz, lithium niobate or lithium tantalate (Schaller *et al.*, 1998). These crystals are an inherent part in electronics, electric data processing and high-frequency technology.

The general principle is based upon a physical characteristic of the quartz crystal, the essential part of the system: piezoelectricity. This 'piezo-effect' refers to the production of electrical charges by the imposition of mechanical stress, i.e. these systems apply an oscillating electric field to create a mechanical wave. Their transformation towards a sensor is achieved by coating with a polymer material, e.g. known from chromatographic stationary phases. The coating is achieved by dissolving a chosen polymer in an organic solvent; after this airbrush or spin coating techniques can be used to coat the sensor surface.

The two aspects of this technique, bulk acoustic wave (BAW) and surface acoustic wave (SAW), have different modes of oscillations: the three-dimensional BAW (also called 'quartz micro balances', QMBs) travels at 5–30 MHz through the crystal, while the two-dimensional SAW range at frequencies ranging from 10 MHz to 2 GHz. The sensing layer consists of a polymer, which can interact with a given analyte, i.e. the major measurement parameter is the change in the resonance frequency due to the change in mass, which results from this. Practical consequences of the different layouts are a higher sensitivity of SAWs combined with the need for a more sophisticated control set-up (Ampuero and Bosset, 2003). A sensor array will deliver a 'fingerprint' of the sample if the different polymers attract different volatiles and therefore the variation in mass will induce different frequency shifts in a reproducible way, as long as temperature and humidity are kept constant (Nakamoto *et al.*, 1990; Rapp *et al.*, 1995).

To overcome the disadvantages of relatively insufficient sensitivity pre-concentrating units were combined with these devices, e.g. Tenax tubes (Bender *et al.*, 2003) or solid phase micro-extraction (SPME). This SPME–SAW coupling was used for proof of principle experiments within process control, off-flavour recognition and fruit storage control (Bücking *et al.*, 2005). Other research studies used BAWs as a tool to predict the optimal harvest date of apples (Saevels *et al.*, 2003) or to monitor the ripening process of Emmental cheeses, during which the concentration of 2-heptanone changes characteristically (Bargon *et al.*, 2003).

Conducting polymers
Conducting polymers (CP) have a relatively long history as sensors; Bartlett and Ling-Chung (1989) described the measurement of methanol vapour and so formed the basis of several of the earlier generation of commercial electronic noses (Strike *et al.*, 1999). They operate, unlike MOS and MOSFET, at room temperature. Beside this a wide range of materials can be synthesised, which can respond to a broad range of organic volatiles (Bartlett and Ling-Chung, 1989). Therefore a thin polymer film is deposited between gold-plated electrodes.

The detection principle is based on the change of the electron flow in the system if volatiles are in interaction with the conducting polymer (e.g. pyrroles, anilines or thiophenes); i.e. the change in voltage across the conductive polymer is measured. This change is caused by the disturbance of the conjugated π-electron system, which extends over the whole backbone. Different sensitivities and selectivity are achieved by substitution of side groups of this backbone, the selection of doping ions, the variation of the polymer chain length and condition of the polymerisation.

Although possible applications such as the discrimination of microorganisms were described by Gibson *et al.* (1997) and Craven *et al.* (1996), serious problems such as ageing effects and poisoning should be solved before using these devices in a 'non-lab' environment. Further disadvantages are the poor batch-to-batch reproducibility during production (Schaller *et al.*, 1998), high temperature sensitivity, strong humidity interference and an unpredictable drift due to oxidation processes.

19.2.2 Liquid detection – electronic tongue

In the food industry there are a number of liquid and semi-liquid products (drinks, beverages, dairy products) where liquid sensors may be applied for quality control purposes. Quality properties that may be analysed in the liquid phase may be related to the chemical, microbial or sensory quality of the product.

The term 'electronic tongue' has been used for liquid sensor systems containing an array of unspecific working electrodes in combination with multivariate pattern recognition methods (Winquist *et al.*, 2003). If properly calibrated, the electronic tongue is capable of recognising the quantitative and qualitative overview of important ingredients in multicomponent solutions of different natures, e.g. beverages and foodstuffs.

Different techniques have been used for liquid sensing. The most frequently used are based on potentiometry or voltammetry. Potentiometric devices include ion-selective electrodes (ISE), a reference electrode and a potential measuring unit. A typical ISE is the glass-electrode based pH meter. Recently, ion-selective field effect transistors have been developed. In potentiometry a potential is generated between the reference and working electrode when immersed into a liquid electrolyte solution (Fig. 19.2). The reference electrode is of constant potential and the working electrode responds to target molecules. The output signals correspond to the potential generated across a surface region on the

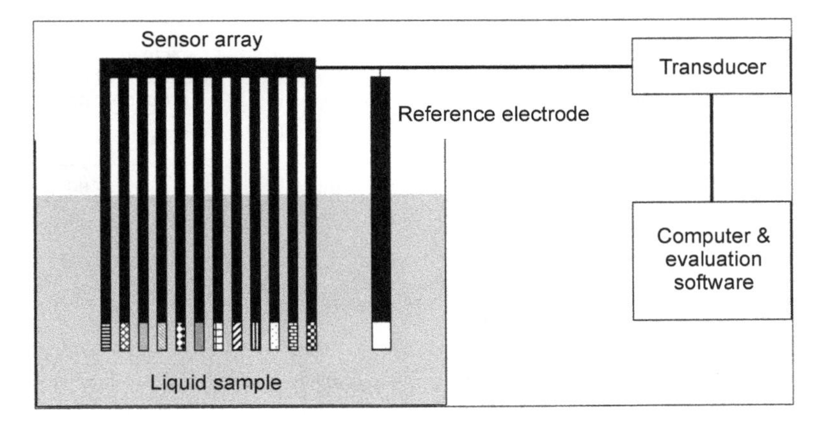

Fig. 19.2 General construction of an electronic tongue.

working electrode relative to the reference electrode. A voltammetric device also consists of one or several working electrodes and a reference electrode, but instead a fixed electrode potential is applied to drive an electron transfer reaction, and the resulting current is measured. The size of the electrode potential determines whether the target molecules will lose or gain electrons. Thus, electroactive chemical species are measured. The technique may be used for measuring ions, redox active compounds, heavy metals, charged particles, conductivity and pH.

Another way of using voltammetry is by large or small amplitude pulsing of the potential (Winquist *et al.*, 2003). These two techniques generate complex multivariate output data and require chemometrics for the data processing and analysis. Piezoelectric sensors such as the SAW devices have also been applied to aqueous phases (Kondoh and Shiokawa, 1994;Yamazaki *et al.*, 2000).

19.2.3 Biosensors

Biosensors, in contrast to the sensors in sections 19.2.1 and 19.2.2, represent

> a self-contained analytical device that incorporates a biologically active material (e.g. enzyme, DNA, antibody, or microorganism) in intimate contact with an appropriate transduction element (e.g. electrochemical, thermal, optical, or acoustic) for the purpose of detecting (reversibly and selectively) the concentration or activity of chemical species in any type of sample (Arnold and Meyerhoff, 1998).

Clark and Lyons developed the first biosensor, an enzyme-based glucose sensor, in 1962. Since then, hundreds of biosensors have been developed in many research laboratories around the world. Over a thousand research papers about biosensors, including reviews and books, have been published since 1995. Their authors classified sensors according to their biological or transduction element, which is used in this chapter (see below; Fig. 19.3) to illustrate these

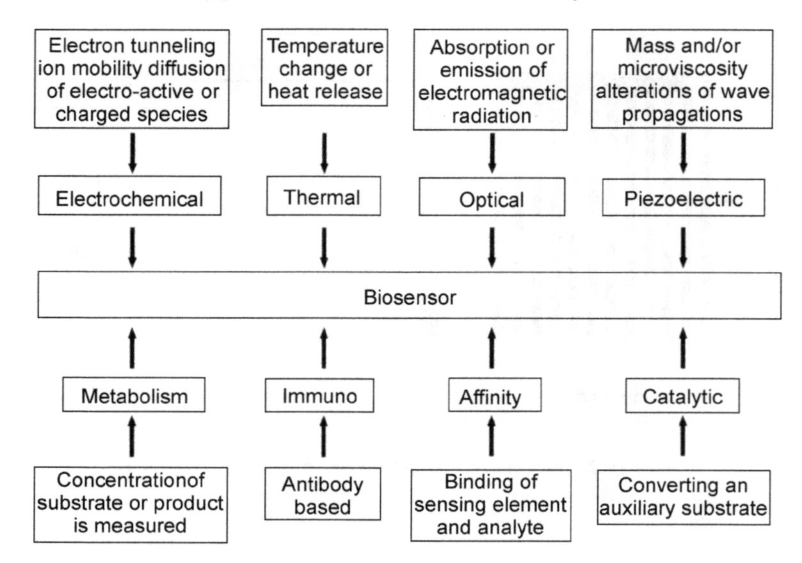

Fig. 19.3 Overview: classification of biosensors by their transduction element (electrochemical, thermal, optical, piezoelectric) or their biological element (metabolism, immuno, affinity, catalytic).

classification options. Further classifications are the 'way of identification' and scaling of 'bioelement integration'.

Advantages of these devices are relatively rapid assay times, low cost, great selectivity, little sample consumption; they can be miniaturised and integrated into one chip, and easy handling. Disadvantages include the dependency towards pH, temperature, limited lifetimes and other general conditions. As a consequence their superior selectivity (compared with chemical sensors) is limited by their biological component which requires mild conditions (O'Connell and Guilbault, 2001a). This also includes the relatively short lifetime of biosensors compared with conventional instrumentation. Because of these specific premises it is important to consider the following aspects:

- Which bioreceptor–transducer combination with which immobilisation method.
- Contamination risk biosensor–sample host system.
- The packaging of the sensor system (e.g. physical robustness).
- Required measurement range and linearity.
- User-friendly maintenance.

Although the biosensor food pathogen testing market was expected to grow to $192 million by 2005 (Alocilja and Radke, 2003), new developments are mainly driven by *in vivo* applications for medical diagnosis. As a consequence, new developments do not find intermediate use in the agricultural and food industry. A list of companies producing biosensors for food control is given by Kress-Rogers and Brimelow (2001). Table 19.3 provides an overview of advantages, disadvantages and application examples of the different biosensor systems.

Table 19.3 Overview of advantages/disadvantages and application examples of the different biosensor types

	Advantages	Disadvantages	Example is given by
Electrochemical	Simple High sensitivity (H_2O_2 electrode) High selectivity (O_2 electrode)	Low sensitivity (O_2 electrode) Low selectivity (H_2O_2 electrode)	Draisci *et al.* (1998), at-line O'Connell & Guilbault (2001b), on-line
Thermal	Versatility Robustness	Large amount of enzyme Expensive	Ramanathan *et al.* (1999), on-line
Optical	Low cost Remote sensing Very good signal/noise ratio	Small concentration range Interference	Biacore AB, mostly off-line
Piezoelectric	Simple Fast response Low cost for readout device	Low sensitivity Unwanted adsorption (e.g. proteins, cell compartments)	Janshoff & Steinem (2001)

Electrochemical biosensors

Electrochemical sensors can be classified according to their transduction principle as amperometric, potentiometric or conductometric sensors (Meadows, 1996). The amperometric sensor for blood glucose concentration based on glucose oxidase is commercially widely used, but systems for food analytes such as ethanol (Yao *et al.*, 2000), ascorbic acid (Akyilmaz and Dinçkaya, 1999), free fatty acids (Schmidt *et al.*, 1996) and different carbohydrates (Bilitewski *et al.*, 1993) have been developed. For this kind of transducer the current produced by an electroactive species is measured and correlated to the concentration of the analyte (Heldman, 2003).

It is also possible to use an amperometric bi-enzyme system as described as an dairy industry application by Scheller *et al.* (1990): a combination of β-galactosidase and glucose oxidase has been used to determinate lactose concentration by the reduction of oxygen in reaction (19.2) or the oxidation of H_2O_2; measurements were performed using a Pt-Ag/AgCl oxygen electrode.

$$\text{lactose} + H_2O \xrightarrow{\beta-\text{galactosidase}} \text{D-galactose} + \beta\text{-D-glucose} \quad (19.1)$$

$$\beta\text{-D-glucose} + O_2 \xrightarrow{\text{glucose oxidase}} \text{D-glucano-}\delta\text{-lactone} + H_2O_2 \quad (19.2)$$

In potentiometry a membrane is measuring a potential resulting from the difference in concentration of H^+ or other positive ions across the membrane. Examples are described by Verma and Singh (2003) for quality control in milk, and by Ercole *et al.* (2003) for *Escherichia coli* detection in vegetable. Conductometric sensors use conductive polymers, which convert the biochemical interaction into an electrical signal. Recently, Mubammad-Tahir and Alocilja (2003) used this technique for the detection of *E. coli* and *Salmonella*.

In contrast to chemical or physical sensors enzyme assays (for electrochemical biosensors and/or in general) are in general unusable for continuous measurements, i.e. enzymatic efficiency is not constant over time (O'Connell and Guilbault, 2001b). To enable constant measurements (in-line/on-line) the flow injection analysis provides a solution by constantly sending samples with recovery phases between each sample. This recovery is used to clean the sensor from sample (residues) and to return the response back to base line.

Thermal biosensors

Thermal biosensors were first developed in the early 1970s and used for continuous measurements and enzyme-reactor control, but they are rarely used for food control. An overview of possibilities is given by Ramanathan *et al.* (1999), including the monitoring of acetaldehyde, ethanol, glucose and penicillin V in industrial fermentation systems. The principle is characterised by immobilised enzymes, which evolve heat during their catalysed reaction, which is in proportion to the amount of substrate in the sample (Harborn *et al.*, 1997). The advantage of this technique is the independence of optical properties, which enables to measure a broad range of bioanalytes with one instrument.

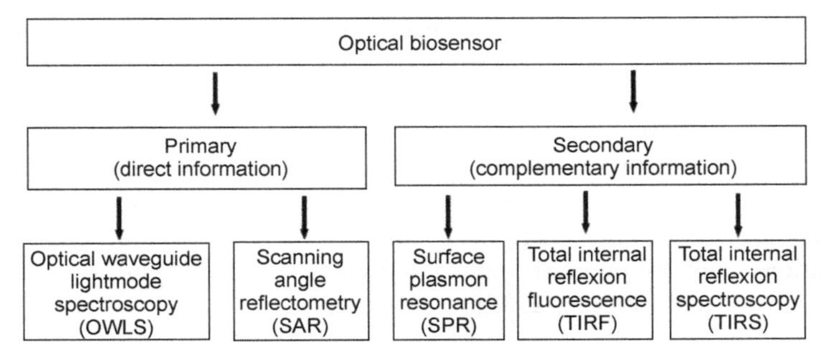

Fig. 19.4 Overview of optical sensor techniques.

Such a device can be fixed in a way that circumvents fouling of the base transducer (Kröger and Danielsson, 1997).

Optical biosensors

Optical biosensors, which can be classified into two groups, depend upon the special properties of light (see Fig. 19.4). In general they offer advantages such as miniaturisation and disposability, and there is no occurrence of electrical interference. Classical optical biosensors consist of fibreglass, which has been coated on one side with an enzyme or antibody. Direct information will be given by such techniques as optical waveguide light-mode spectroscopy (OWLS) or scanning angle reflectometry (SAR). In contrast to this the second group's information is obtained at the expense of a complete description of the adsorbed layer (Ramsden, 1997); their most known examples are surface plasmon resonance (SPR, see Fig. 19.5) and total internal reflexion fluorescence (TIRF) or spectroscopy (TIRS).

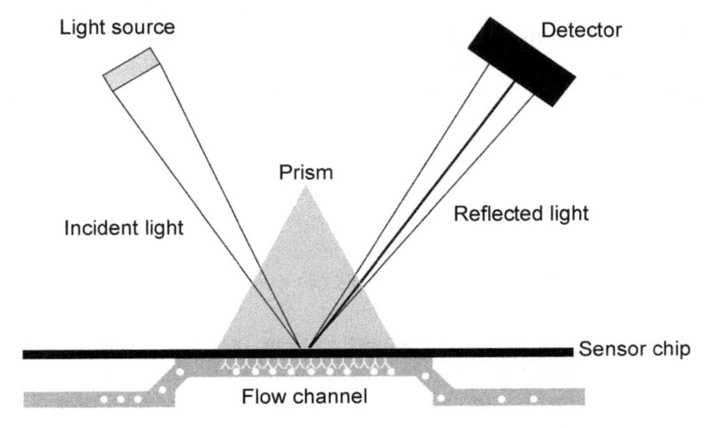

Fig. 19.5 SPR principle: the sample with analyte is flowing over the surface of the sensor chip where the binding partner is located. Interaction with these immobilised molecules can be monitored by a change of reflected light.

More than 4000 articles have been published since 1990, describing numerous applications. Rich and Myszka (2000, 2001, 2002, 2003), Baird and Myszka (2001) and Mehrvar *et al.* (2000) give overviews in their review articles about recent developments.

The SPR technique has been the backbone of a very common commercial biosensor for biomolecular interaction analysis since 1990. Biacore AB (Sweden) developed this instrument, which is now the platform for diverse applications – 90% of published optical biosensor research is performed by these devices (Baird and Myszka, 2001).

Piezoelectric biosensors

Piezoelectric devices are mainly based on the detection of a change of mass density, mostly represented by BAW and SAW, as described in section 19.2.1. Janshoff and Steinem (2001) give an overview about the working principle of these systems in liquids. Owing to high energy losses by measurements in liquids, the sensitivity of these devices is rather low compared with optical devices. Because of their advantages (see Table 19.3) they are still part of some typical bioanalytical research topics.

Bacillus cereus was detected by Vaughan *et al.* (2003) with label-free quartz crystal microbalance (QCM) sensor with a detection of about 10^4 cells/mL. Mannelli *et al.* (2003) developed a DNA piezoelectric sensor for the detection of genetically modified organisms (GMOs). Länge *et al.* (2003) created an immunosensor on the basis of a SAW device, which was able to monitor urease binding at anti-urease-coated SAW devices in real time with good resolution.

19.3 Common industrial applications and future trends

Sensors have a great potential for future use in the food processing industry for process and quality control. The application areas are in the field of pathogen detection related to raw material, processing and quality control of final product. So far, most of the sensors that have been implemented in the food industry on the production line have been for the purpose of environmental monitoring of hazardous gases such as hydrocarbons, ammonia and hydrogen sulphide that may occur during the production process. However, these kinds of sensor, also including physical sensors, provide information on the performance of the process, which in the best case may indirectly contribute to controlling the hygienic quality of the process, but does not provide direct information on the quality of the product being processed. For this purpose, chemical and biosensors are helpful, also with regard to their on- and in-line options. Since most research has been conducted under laboratory conditions these options strongly depend upon the application and the process surroundings.

Two fundamental trends will have the strongest influence on applied research in the field of sensor techniques, which are strongly connected: because of public opinion and resulting political actions food safety issues have become one of the most important topics in daily life. Therefore the industry should be eager to prove food quality. In this connection the most effective way is to verify quality from the production side of raw material via industrial processing to supermarket shelves (tracing 'from farm to fork'). The main requirements for this approach are automation and miniaturisation; i.e. the need for an instrumentation that can work without supervision and that can be implemented in an existing process technology in order not to disturb the ongoing process.

So far, the existing methods cannot serve these needs of regulatory agencies and food producers. Future real-time testing with reliable sensor technology will provide value to food producers through reduced treatment costs and reduced product recalls. As the demands for food safety increase, the request for fast sensing technologies will only increase. A miniaturised total analysis system, as described in Section 19.3.3 for biological and medical applications, could fulfil these requirements.

19.3.1 Chemical gas sensors

Chemical gas sensors have become a useful industrial tool for bioprocess monitoring. They represent a cost-effective tool for rapid assessment of the chemical and microbial status of raw materials, process streams and end-product. Extensive and costly rework or disposal of products that do not fulfil their specifications can be prevented. Most biological processes that can be found in the food and biotechnology industries are probably suited for the application of gas-sensor arrays. This is because they involve significant concentrations of aromatic compounds or volatile secondary metabolites produced by the microorganisms. In a typical bioprocess, cells are grown under sterile conditions in tanks (bioreactors) on liquid media that provide essential nutrients, vitamins, etc. The products from bioprocesses range from enzymes and single cell protein to biopharmaceuticals, which naturally all impose high demands on product quality and safety. Gas sensor array systems have been shown to be very useful for both quantitative and qualitative bioprocess monitoring, which allows real-time determination of cell status, growth rates and product concentration (Liden *et al.*, 2000; Mandenius *et al.*, 1997). Another advantage with this technology is that it can be used to discover bacterial contamination on-line in real time in the bioreactor tank after only a few hours of processing, which is a significant gain compared with traditional microbiological methods. The application of non-invasive on-line monitoring methods such as gas-sensor arrays could therefore certainly contribute to improve the quality of bioprocessed products. This has also been documented by several studies (e.g. Bachinger and Haugen, 2003; Dickert *et al.*, 2003; Pasini *et al.*, 2004).

19.3.2 Biosensors

Although the market is generating a need for pathogen-detecting biosensors, only a few are commercially available or are approaching commercialisation

(Alocilja and Radke, 2003). Up to now medical applications (infection control, etc.) have had the largest market segment.

For the determination of fruit, vegetable, meat and fish freshness, biosensors have been developed for different biogenic amines (histamine, hypoxanthine, xanthine) based on specific oxidase enzymes in combination with amperometric transducers (Draisci *et al.*, 1998) with detection ranges from 10^{-7} to 10^{-3} mol/L.

For rapid detection of bacterial contamination in food, commercial biosensors based on immunochemical assays and DNA hybridisations in combination with different transducer principles have been manufactured. The use of analyte-specific sensors makes it possible to discriminate between different micro-organisms. Several commercial systems exist on the market for detection of specific bacteria in foods. Biacore AB delivers biosensor systems for the detection of food pathogens (e.g. *Salmonella* and *E. coli*) based on immobilised antibodies in combination with SPR technology. Gene-Trak Systems (USA) delivers diagnostic products based on DNA assays for the rapid detection of foodborne pathogens in food. It delivers probes with specificity for respectively *Listeria*, *Salmonella*, *E. coli*, *Campylobacter* and *Staphylococcus aureus*.

The company Applied Biosystems (USA) offer systems based on DNA hybridisation in combination with fluorescence detection for pathogen detection. The Taqman® Detection Systems allow quick determination of the presence of pathogenic bacteria (e.g. *E. coli* and *Salmonella*) and fungi and for further identification of genus and species, the MicroSeq® system is used. The MicroSeq® system is based on comparison of the gene sequence from an unknown sample with genes from known strains. With these pathogen detection biosensors bacterial cells can be detected down to 200–10^3 cells/ml from 10–20 minutes up to 2 hours. However, it should be emphasised that it is the sample clean-up and extraction step of the sample matrix that represents the time-consuming step in these analyses and this may vary from 20 minutes to several hours.

Therefore, these techniques are in principle based on on-line analysis. They are available as portable devices, automated analysers and laboratory instruments. Very few biosensors are used on-line, but in combination with flow-injection systems they could in principle be applied on-line (Tothill and Magan, 2003).

19.3.3 Lab on a chip

In 1990 Manz *et al.* proposed the concept of a 'miniaturised' total analysis system (μ-TAS), a hybrid combining the advantages of a sophisticated analysis system (pre-treatment, separation and detection) with the size of a chemical sensor (Fig. 19.6). Since all these steps are performed in an integrated micro system the more popular term 'lab on a chip' was created. The construction of automated small-sized control instruments requires the combination of classical analytical chemistry with micro-fabrication technology as well as knowledge of biochemistry and biotechnology.

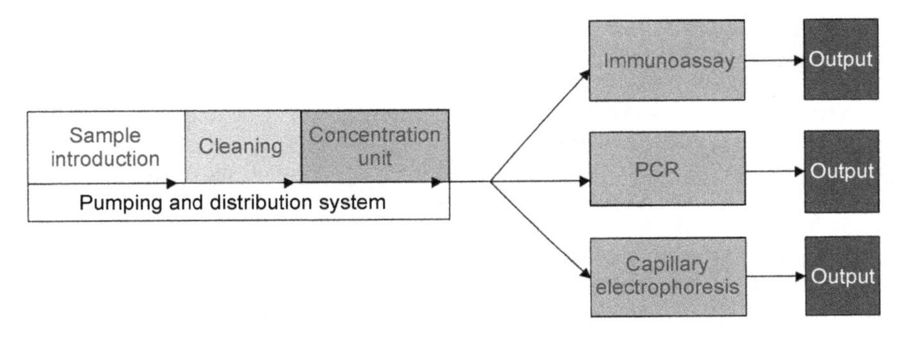

Fig. 19.6 Schematic construction of a 'lab on a chip' with three different miniaturised techniques – the chemical information is transformed into optical and electronic information.

Several reviews (Wang, 2002; Reyes *et al.*, 2002; Auroux *et al.*, 2002; Sheehan *et al.*, 2003; Vilkner *et al.*, 2004) describe status quo, recent developments and advantages of these systems. They illustrate that the advantages of μ-TAS are speed, high performance, versatility, negligible sample/reagent consumption and automation.

The majority of research work in this area has been directed toward biological and medical sciences; examples are (Auroux *et al.*, 2002): cell culture and cell handling, clinical diagnostics, immunoassays, proteins, DNA separation and analysis, polmerase chain reaction (PCR). Direct applications for food hygiene/food safety are not available yet.

19.4 References

AKYILMAZ E, DINÇKAYA E (1999), 'A new enzyme electrode based on ascorbate oxidase immobilized in gelatin for specific determination of L-ascorbic acid', *Talanta*, **50**, 87–93.

ALOCILJA EC, RADKE SM (2003), 'Market analysis of biosensors for food safety', *Biosens. Bioelec.*, **18**, 841–846.

AMPUERO S, BOSSET JO (2003), 'The electronic nose applied to dairy products: a review', *Sens. Actuators*, **B94**, 1–12.

ARNOLD MA, MEYERHOFF ME (1988), 'Recent advances in the development and analytical applications of biosensing probes', *Crit. Rev. Anal. Chem.*, **20**, 149–196.

ARTURSSON T, EKLÖV T, LUNDSTRÖM I, MÅRTENSSON P, SJÖSTRÖM M, HOLMBERG M (2000), 'Drift correction for gas sensors using multivariate methods', *J. Chemom.*, **14**, 711–723.

AUROUX PA, IOSSIFIDIS D, REYES DR, MANZ A (2002), 'Micro total analysis systems. 2. Analytical standard operations and applications', *Anal. Chem.*, **74**, 2637–2652.

BACHINGER, T, HAUGEN JE (2003). 'Process monitoring', in Pearce TC, Schiffman SS, Nagle HT, Gardner JW, *Machine olfaction, Electronic nose technology*, Weinheim, Wiley-VCH, 481–503.

BAIRD CL, MYSZKA DG (2001), 'Current and emerging commercial optical biosensors', *J. Mol. Recogn.*, **14**, 261–268.

BALTES H, GÖPEL W, HESSE J (1999), *Sensors Update*, Vol. 5, Weinheim, Wiley-VCH.

BARGON J, BRASCHOSS S, FLÖRKE J, HERRMANN U, KLEIN L, LOERGEN JW, LOPEZ M, MARIC S, PARHAM H, PIACENZA P, SCHAEFGEN H, SCHALLEY CA, SILVA G, SCHLUPP M, SCHWIERZ H, VöGTLE F, WINDSCHEIF G (2003), 'Determination of the ripening state of Emmental cheese via quartz microbalances', *Sens. Actuators*, **B95**, 6–19.

BARTLETT PN, LING-CHUNG SK (1989), 'Conducting polymer gas sensors part II: response of polypyrrole to methanol vapour', *Sens. Actuators*, **B2**, 141–150.

BENDER F, BARIÉ N, ROMOUDIS G, VOIGT A, RAPP M (2003), 'Development of a preconcentration unit for a SAW sensor micro array and its use for indoor air quality monitoring', *Sens. Actuators*, **B93**, 135–141.

BILITEWSKI U, JÄGER A, RÜGER P, WEISE W (1993), 'Enzyme electrodes for the determination of carbohydrates in food', *Sens. Actuators*, **B15**, 113–118.

BÜCKING M, HAUGEN JE, STEINHART H (2002), 'Classification of peanuts by wet chemical analysis, instrumental methods, electronic nose devices and sensory analysis', in Le Quéré JL and Etiévant PX, *Flavour Research at the Dawn of the Twenty-first Century* (Proceedings of the 10th Weurman Flavour Research Symposium 2001), France, Lavoisier, 528–531.

BÜCKING M, BARIÉ N, RAPP M (2005), 'Surface acoustic wave sensors as a new analytical tool for food quality monitoring', in Schieberle P, Hofmann T, Rothe M, Proceedings 7th Wartburg Symposium on Flavor Chemistry & Biology, Eisenach, Germany, 21–23 April 2004, 322–326.

CLARK LC, LYONS C (1962), 'Electrode systems for continuous monitoring cardiovascular surgery', *Ann. NY Acad. Sci.*, **102**, 29–45.

CRAVEN MA, GARDNER JW, BARTLETT PN (1996), 'Electronic noses – development and future prospects', *TrAC*, **15**, 486–493.

DICKERT FL, LIEBERZEIT P, HAYDEN O (2003), 'Sensor strategies for microorganism detection – from physical principles to imprinting procedures', *Anal. Bioanal. Chem.*, **377**, 540–549.

DRAISCI R, VOLPE G, LUCENTINI L, CECILIA A, FEDERICO R, PALLESCHI G (1998), 'Determination of biogenic amines with an electrochemical biosensor and its application to salted anchovies', *Food Chem.*, **62**, 225–232.

EGGINS BR (2002) *Chemical Sensors and Biosensors*, Chichester, John Wiley & Sons Ltd.

ERCOLE C, DEL GALLO M, MOSIELLO L, BACCELLA S, LEPIDI A (2003), '*Escherichia coli* detection in vegetable food by a potentiometric biosensor', *Sens. Actuators*, **B91**, 163–168.

GARDNER JW, BARTLETT PN (1993), 'A brief history of electronic noses', *Sens. Actuators*, **B18**, 211–220.

GARDNER J W, BARTLETT P N (1999), *Electronic Noses: principles and applications*, Oxford, Oxford University Press.

GIBSON TD, PROSSER O, HULBERT JN, MARSHALL RW, CORCORAN P, LOWERY P, RUCK-KEENE EA, HERON S (1997), 'Detection and simultaneous identification of microorganisms from headspace samples using an electronic nose', *Sens. Actuators*, **B44**, 413–422.

HARBORN U, XIE B, VENKATESH R, DANIELSSON B (1997), 'Evaluation of a miniaturized thermal biosensor for the determination of glucose in whole blood', *Clin. Chim. Acta*, **267**, 225–237.

HAUGEN, JE (2001), 'Electronic noses in food analysis', in Rouseff R and Cadwallader C (eds.), *Headspace Analysis of Foods and Flavours: Theory and practice*. New

York, Kluwer Academic/Plenum Publishers, 43–56.

HAUGEN JE, TOMIC O, KVAAL K (1999) 'A calibration method for handling the temporal drift of solid state gas-sensors', *Anal. Chim. Acta*, **407**, 23–39.

HELDMAN DR (ed.) (2003), *Encyclopedia of Agricultural, Food, and Biological Engineering*, Marcel Dekker, New York.

JANSHOFF A, STEINEM C (2001), 'Quartz crystal microbalance for bioanalytical applications', *Sensor Update*, **9**, 313–354.

KONDOH J, SHIOKAWA S (1994), 'New application of shear horizontal surface acoustic wave sensors to identify fruit juices', *Japan Journal of Applied Physics*, **K 33**, part I, 3095–3099.

KRESS-ROGERS E, BRIMELOW CJB (2001), *Instrumentation and Sensors for Food*, Cambridge, Woodhead Publishing Limited.

KRÖGER S, DANIELSSON B (1997), 'Calorimetric biosensors', in Kress-Rodgers E, *Handbook of Biosensors and Electronic Nose: Medicine, Food and the Environment*, New York, CRC Press, 279–298.

LÄNGE K, BENDER F, VOIGT A, GAO H, RAPP M (2003), 'A surface acoustic wave biosensor concept with low flow cell volumes for label-free detection', *Anal. Chem.*, **75**, 5561–5566.

LIDEN H, BACHINGER T, GORTON L, MANDENIUS CF (2000), 'On-line determination of non-volatile or low-concentration metabolites in a yeast cultivation using an electronic nose', *Analyst*, **125**, 1123–1128.

LUNDSTRÖM I, SPETZ A, WINQUIST F, ACKELID U, SUNDGREN H (1990), 'Catalytic metals and field-effect devices – a useful combination', *Sens. Actuators*, **B1**, 115–120.

MANDENIUS CF, EKLOV T, LUNDSTRÖM I (1997), 'Sensor fusion with on-line gas emission multisensor arrays and standard process measuring devices in baker's yeast manufacturing process', *Biotech. Bioeng.*, **55**, 427–438.

MANNELLI I, MINUNNI M, TOMBELLI S, MASCINI M (2003), 'Quartz crystal microbalance (QCM) affinity biosensor for genetically modified organisms (GMOs) detection', *Biosens. Bioelec.*, **18**, 129–140.

MANZ A, GRABER N, WIDMER HM (1990), 'Miniaturized total chemical analysis systems: A novel concept for chemical sensing', *Sens. Actuators*, **B1**, 244–248.

MEADOWS D (1996), 'Recent developments with biosensing technology and applications in the pharmaceutical industry', *Adv. Drug Deliv. Rev*, **21**, 179–189.

MEHRVAR M, BIS C, SCHARER JM, MOO-YOUNG M, LUONG JH (2000), 'Fiber-optic biosensors – trends and advances', *Analytical Sci.*, **16**, 677–692.

MIETTINEN SM, TUORILA H, PIIRONEN V, VEHKALAHTI K, HYVONEN L (2002), 'Effect of emulsion characteristics on the release of aroma as detected by sensory evaluation, static headspace gas chromatography, and electronic nose', *J. Agric. Food Chem.*, **50**, 4232–4239.

MUBAMMAD-TAHIR Z, ALOCILJA EC (2003), 'A conductometric biosensor for biosecurity', *Biosens. Bioelec.*, **18**, 813–819.

NAKAMOTO T, FUKUNISHI K, MORIIZUMI T (1990), 'Identification capability of odor senor using quartz-resonator array and neural-network pattern recognition', *Sens. Actuators*, **B1**, 473–476.

NANTO H, STETTER JR (2002), 'Introduction to chemosensors', in Pearce TC, Schiffman SS, Nagle HT and Gardner JW (eds), *Handbook of Machine Olfaction*, Weinheim, Wiley-VCH, 79–104.

O'CONNELL PJ, GUILBAULT GG (2001a), 'Sensors and Food Quality', *Sensor Update*, **9**, 255–282.

O'CONNELL JP, GUILBAULT GG (2001b), 'Future trends in Biosensor research', *Anal. Lett.*, **34**, 1063–1078.

PASINI P, POWAR N, GUTIERREZ-OSUNA R, DAUNERT S, RODA A (2004), 'Use of a gas-sensor array for detecting volatile organic compounds (VOC) in chemically induced cells', *Anal. Bioanal. Chem.*, **78**, 76–83.

PERSAUD K, DOOD K (1982), 'Analysis of discrimination mechanisms in the mammalian olfactory system using a model nose', *Nature*, **299**, 352–355.

RAMANATHAN K, RANK M, SVITEL J, DZGOEV A, DANIELSSON B (1999), 'The development and applications of thermal biosensors for bioprocess monitoring', *Trends Biotechnol.*, **12**, 499–505.

RAMSDEN JJ (1997), 'Optical biosensors', *J. Mol. Recogn.*, **10**, 109–120.

RAPP M, BÖß B, VOIGT A, GEMMEKE H, ACHE HJ (1995), 'Development of an analytical microsystem for organic gas detection based on SAW resonators', *Fresenius J. Anal. Chem.*, **352**, 699–704.

REYES DR, IOSSIFIDIS D, AUROUX PA, MANZ A (2002), Micro total analysis systems. 1. Introduction, theory, and technology', *Anal. Chem.*, **74**, 2623–2636.

RICH RL, MYSZKA DG (2000), 'Survey of the 1999 surface plasmon resonance biosensor literature', *J. Mol. Recogn.*, **13**, 388–407.

RICH RL, MYSZKA DG (2001), 'Survey of the year 2000 commercial optical biosensor literature', *J. Mol. Recogn.*, **14**, 273–294.

RICH RL, MYSZKA DG (2002), 'Survey of the year 2001 commercial optical biosensor literature', *J. Mol. Recogn.*, **15**, 352–376.

RICH RL, MYSZKA DG (2003), 'Survey of the year 2002 commercial optical biosensor literature', *J. Mol. Recogn.*, **16**, 351–382.

RODRIGUEZ-MOZAZ S, MARCO M P, DE ALDA M J L, BARCELO D (2004), 'Biosensors for environmental applications: future development trends', *Pure Appl. Chem*, **76**, 723–752.

SAEVELS S, LAMMERTYN J, BERNA AZ (2003), 'Electronic nose as a non-destructive tool to evaluate the optimal harvest date of apples', *Postharvest Biol. Tec.*, **30**, 3–14.

SCHALLER E, BOSSET JO, ESCHER F (1998), ' "Electronic noses" and their application to food', *Lebensm.-Wiss. u.-Technol.*, **31**, 305–316.

SCHELLER WF, RALIS EV, MAKOWER A, PFEIFFER D (1990), 'Amperometric bi-enzyme based biosensors for the detection of lactose – characterisation and application', *J. Chem. Tech. Biotechnol.*, **49**, 255–256.

SCHMIDT A, STANDFUSS-GABISCH C, BILITEWSKI U (1996), 'Microbial biosensor for free fatty acids using an oxygen electrode based on thick film technology', *Biosens. Bioelec.*, **11**, 1139–1145.

SHEEHAN AD, QUINN J, DALY S, DILLON P, O'KENNEDY R (2003), 'The development of novel miniaturized immuno-sensing devices: a review of a small technology with a large future', *Anal. Lett.*, **36**, 511–537.

STETTER J R, PENROSE W R (2002), 'Understanding chemical sensors and chemical sensor arrays (electronic noses): past, present, and future', *Sensors Update*, **10**, 189–229.

STRIKE DJ, MEIJERINK MGH, KOUDELKA-HEP M (1999), 'Electronic noses – a mini-review', *Fresenius J. Anal. Chem.*, **364**, 499–505.

TOMIC O, ULMER H, HAUGEN JE (2002), 'Standardisation methods for handling instrument related signal shift in gas-sensor array measurement data', *Anal. Chim. Acta*, **472**, 99–111.

TOMIC O, EKLÖV T, KVAAL K, HAUGEN JE (2004), 'Recalibration of a gas-sensor array system related to sensor replacement', *Anal. Chim. Acta*, **512**, 199–206.

TOTHILL, IE (2003), *Rapid and On-line Instrumentation for Food Quality Assurance*, Cambridge, Woodhead Publishing Limited.

TOTHILL, IE, MAGAN N (2003), 'Rapid detection methods for microbial contamination', in Tothill IE (ed.), *Rapid and On-line Instrumentation for Food Quality Assurance*, Cambridge, Woodhead Publishing Limited.

VAUGHAN RD, CARTER RM, O'SULLIVAN CK, GUILBAULT GG (2003), 'A quartz crystal microbalance (QCM) sensor for the detection of *Bacillus cereus*', *Anal. Lett.*, **36**, 731–747.

VERMA N, SINGH MA (2003), 'Disposable microbial based biosensor for quality control in milk', *Biosens. Bioelec.*, **18**, 1219–1224.

VILKNER T, JANASEK D, MANZ A (2004), 'Micro total analysis systems. Recent developments', *Anal. Chem.*, **76**, 3373–3385.

WANG J (2002), 'On-chip enzymatic assays', *Electrophoresis*, **23**, 713–718.

WINQUIST F, KRANYZ-RÜLCKER C, LUNDSTRÖM I (2003), 'Electronic tongues and combinations of artificial senses', in Pearce TC, Schiffman SS, Nagle HT, Gardner JW, *Machine Olfaction, Electronic Nose Technology*, Weinheim, Wiley-VCH, 267–291.

YAMAZAKI T, KONDOH J, MATSUI Y, SHIOKAWA, S (2000), 'Estimation of components in mixture solutions of electrolytes using a liquid flow system with SH-SAW sensor', *Sens. Actuators*, **B83**, 34–39.

YAO Q, YABUKI S, MIZUTANI F (2000), 'Preparation of a carbon paste/alcohol dehydrogenase electrode using polyethylene glycol-modified enzyme and oil-soluble mediator', *Sens. Actuators*, **B65**, 147–149.

Part III

Improving hygiene management and methods

Introduction

Hygiene GMPs, such as personal hygiene, good housekeeping and cleaning and disinfection, are established to ensure that contamination of foodstuffs from the processing environment (e.g. environmental surfaces, food processing equipment, people, condensation, cleaning fluids, the air, etc.) is minimised and controlled. Monitoring and verifying that these GMPs are effective is thus essential for food product safety and quality.

It is fundamental, of course, to construct a manufacturing environment, select appropriate food processing equipment and environmental materials and adopt personal hygiene practices that can be intrinsically controlled by hygiene measures. Taking cleaning and disinfection as an example, food processing equipment and environmental surfaces must inherently be capable of being cleaned to an acceptable level. In addition, it is necessary to have some understanding as to how deficiencies in equipment or surface design, together with changes due to production practices and wear, will affect such cleanability. Equally, the selection of appropriate cleaning equipment and chemicals along with the design of a suitable cleaning and disinfection programme is an essential prerequisite.

The control of hygiene GMPs should follow the same principles as adopted by HACCP plans for the food product. Again taking cleaning and disinfection as an example, 'critical control points' (CCPs) could be identified as the number of cleaners needed, detergent and disinfectant concentrations, water temperature and pressures and contact times. Such CCPs can be monitored before or during the cleaning and disinfection programme, and if out of specification, would negate the requirement for further assessment as the programme would not have been undertaken optimally and would need to be rectified.

The ongoing operation of hygiene GMPs can be monitored firstly by visual audits. These may be undertaken within a time frame such that if faults are found, control measures can be implemented immediately. For the hygiene of surfaces, rapid methods are also available that can detect the presence of soiling (food residues and microorganisms), beyond visible cleanliness, in a time frame (approximately <10 min) allowing process control. Such methods include the detection of specific chemical markers of soiling, e.g. adenosine triphosphate (ATP) or physical parameters such as pressure drops or heat transfer rates.

Verification of the success of hygiene GMPs is usually undertaken by microbiological sampling. Samples can be analysed for markers of overall hygiene (total viable count), indicators of poor hygiene practice (Enterobacteriaceae or coliforms) or specific pathogens. Results are available within an extended time frame (24–48 hours or longer) such that process control is not possible. Verification results are thus used to assess the overall performance of the GMP, and can be trend analysed over time periods to ensure that the GMP is effective and/or can be improved.

Finally, if problems do occur, troubleshooting exercises can be undertaken to identify any failings in the system. Modern microbiological approaches using genetic fingerprinting techniques now allow a much more in-depth assessment of the microbial ecology of food manufacturing plants than was hitherto possible and are leading to enhanced detection and control of environmental microbial niches.

20

Risk assessment in hygiene management

I. H. Huisman, Nutricia, The Netherlands and E. Espada Aventín, Unilever R&D Vlaardingen, The Netherlands

20.1 Introduction

If one searches the Internet for 'risk assessment', the result is a long list with a large variety of web pages. These pages relate to a wide range of disciplines such as medicine, safety at the workplace, finance, insurance, fire-fighting, and on-line gambling. Risk assessment is a tool that has been used since the earliest history and is used in almost any area, also in daily life.

Central in every risk assessment is a 'hazard': an event with a possibly negative impact that is poorly predictable or has a random character. In a risk assessment one evaluates the probability of occurrence of the hazard and the adverse effect of the hazard. As a daily-life example of a risk assessment, we can think of a commuter who takes the bus to work every morning to arrive at 8:20 at the office. A couple of times a year the bus has a delay that causes him to arrive at 8:35. One morning, this commuter has an important meeting at 8:30. He decides to take an earlier bus to make sure he arrives at the office in time. Of course the probability of having a delay is no different from any other day, but the negative impact of the delay will be much larger on the day of the meeting. The 'risk' for the commuter, which is a function of probability and adverse effect, is thus unacceptably high on the day of the 8:30 meeting, hence his decision to take an earlier bus.

As a risk assessment is an evaluation of the probability of occurrence and the adverse effect of a hazard, every risk assessment can be thought to consist of four parts: (1) identify the hazard, (2) determine the probability of occurrence, (3) determine the adverse effect, and (4) combine the above to estimate the risk. Not every risk assessment scheme mentions these four parts explicitly, yet the microbiological risk assessment according to the Codex Alimentarius

Commission (see below), uses a scheme very similar to this general four-step model.

Because part of a risk assessment is the estimation of probabilities, it relies heavily on statistics. Quantitative risk assessment in almost every field needs large amounts of data and/or statistical model calculations. The development in the past decades of powerful computers that analyse statistical data and perform model calculations has boosted the use of risk assessment studies in all fields.

20.1.1 Historical evolution of risk assessment in food hygiene

Although risk assessment has been used in various areas since the earliest history, its application to ensure hygienic food production is relatively new. This is mainly because knowledge on food hygiene was very limited until the first half of the 20th century (Notermans *et al.*, 2002).

Even long before the underlying causes of foodborne diseases were known, there were rational approaches to the control of food safety. Experience and trial and error resulted in rules for food hygiene that were often expressed as religious taboos. A well-known example is the ban on eating pork in the Jewish and Muslim religions (Tannahill, 1973). In the absence of knowledge about the causes of foodborne diseases, legislators used an approach similar to religious taboos. This approach is also known as the prohibition principle: it was prohibited to produce or consume certain types of food after it was realised that a specific food could cause illness or death. As an example, in the Middle Ages, in the Swiss cities of Zurich and Basel, it was forbidden to sell fish that had been left over from the day before (Notermans *et al.*, 2002). Even nowadays, although knowledge on food safety has increased dramatically, legislators sometimes use an approach comparable to the prohibition principle, now called the precautionary principle. This approach implies that policy should err on the side of caution when scientific evidence on the food safety of new technologies is incomplete, so that the risk to the consumer is minimised. The precautionary principle was for example invoked in the Cartagena Protocol on Biosafety (Anon, 2000). The document states that even when there is a lack of scientific evidence that products produced through biotechnology are likely to cause harm, a country can ban the import of those products.

The first half of the 20th century saw a vast increase in knowledge on how heat treatments can be used to destroy microorganisms in food. With this increase in understanding a new, more sophisticated, way emerged to guarantee food safety. This relied on the use of certain process conditions (minimum time and temperature of heating) that would guarantee the destruction of a target microorganism in the food under consideration. A well-known example is the development of process conditions for the sterilisation of low-acid canned foods by Esty and Meyer (1922). They derived minimum time–temperature values for the destruction of spores of *Clostridium botulinum*, in order to minimise the risk of botulism.

During most of the 20th century food safety was guaranteed by process conditions as described above, combined with hygienic manufacturing practices, and microbiological testing of the final product to verify the process. Originally there were attempts to routinely test prepared foods for all pathogenic organisms and toxins of concern, but that approach was impractical owing to the large number of possible organisms and toxins in most foods. In the 1950s and 1960s the approach thus shifted towards the testing for 'indicator organisms': organisms that are not pathogenic but that would indicate under-processing or recontamination of the food. *Escherichia coli* was often used as an indicator organism. 'Total aerobic count' was another frequently used indicator for under-processing or recontamination. It was, however, realised that microbiological testing of final products had its limitations. If one wants to ensure that a target pathogen is absent from a certain batch of food, uneconomically large numbers of samples must be taken. This led to the understanding that food safety management should not be based on control of the final product, but on control of the food production process. This invoked the introduction of quality management systems such as GMP (good manufacturing practices) and HACCP (Hazard Analysis Critical Control Point) in the 1970s and onwards.

Hygienic manufacturing practices, such as the use of cleanable equipment and personal hygiene of the operators, have been used for many years. Their use, however, had been non-systematic and non-verifiable. The quality management system GMP aims to place these manufacturing practices, which are the results of long practical experience, in a more systematic framework. Still, the GMP system is largely subjective and qualitative in its benefits. It has therefore been extended by the introduction of HACCP. HACCP is a systematic approach to the identification, assessment and control of hazards at a food processing plant. HACCP was originally developed around 1970 for the NASA space program, to guarantee 100% food safety. In the 1970s it became mandatory for canned food production in the USA, but only in the late 1980s its application was broadened to various categories of canned and non-canned foods inside and outside the USA (Corlett, 1998).

An important step in the implementation of HACCP is the systematic assessment of possible hazards in foods and their associated risks. Increased use of HACCP in the 1990s thus led to an increased need for knowledge on systematic risk assessment in food processing. At the same time, the development of risk assessment was strongly stimulated when in 1995 the World Trade Organization (WTO) was established and a free trade in safe food was agreed. This was formalised in the Agreement on the Application of Sanitary and Phytosanitary Measures, the SPS agreement (Anon, 1995). This agreement requires that food safety legislation be scientifically based and that the process of risk assessment be applied, for example when using microbiological criteria for controlling imported foods. The SPS agreement also requires that countries should take into account the risk assessment technique developed by relevant international organisations. As a result of this the Codex Alimentarius Commission (CAC) of the United Nations Food and Agriculture Organisation

(FAO) and the World Health Organization (WHO) began to provide member countries with principles, guidelines and examples of food safety risk assessments. Although these documents are aimed to support governments in developing food safety legislation, the principles of risk assessment, as summarised in Section 20.2 below, are equally valid for the individual food processor. Moreover, the examples of risk assessment provided by the CAC contain many details that are of interest for the individual food processor, although the extension of these example assessments is well beyond the scope of the risk assessments typically carried out by the food industry.

20.2 Quality management and risk assessment

20.2.1 Risk management: HACCP and its validation

HACCP is the most important food safety management programme around the world. It is recommended by the Codex Alimentarius Commission in the General Principles of Food Hygiene (CAC, 1997), recognised by many governments and regulatory authorities and encouraged by food processors and suppliers as a means to enhance food safety. HACCP is a systematic and scientifically based protocol that focuses on prevention of problems occurring. It identifies specific hazards and measures for their control to ensure the safety of food. HACCP is based on the application of seven principles. There are, in addition, a number of prerequisites that need to be accomplished before the application of the HACCP principles. These prerequisites include assembling a team that will assume the responsibility to develop the HACCP plan, and obtaining knowledge about the product, its intended uses and all the processes involved from primary production until consumption. The seven principles are summarised below:

- *Principle 1: Conduct a hazard analysis*. The HACCP team should list all the hazards that may be reasonably expected to occur at each step from primary production until the point of consumption. Hazard is here defined as 'a biological, chemical, or physical agent in, or condition of, food with the potential to cause an adverse health effect' (CAC, 1997). Moreover, the HACCP team should conduct a hazard analysis to identify which hazards are of such a nature that their elimination or reduction to acceptable levels is essential to the production of safe food. The evaluation of hazards should include the following:
  - the likely occurrence of hazards and severity of their adverse health effects;
  - the qualitative and/or quantitative evaluation of the presence of hazards;
  - survival or multiplication of microorganisms of concern;
  - production or persistence in foods of toxins, chemicals or physical agents; and
  - conditions leading to the above.

The HACCP team must then consider what measures can be applied to control each hazard.

- *Principle 2: Determine the Critical Control Points (CCPs).* The HACCP team must identify the CCPs or steps in the production process where control can be applied and is essential to prevent or eliminate a food safety hazard or reduce it to an acceptable level.
- *Principle 3: Establish critical limit(s).* A critical limit is defined as a criterion that separates acceptability from unacceptability. The critical limit must be specified and validated if possible for each CCP.
- *Principle 4: Establish a system to monitor control of the CCP.* Monitoring is the scheduled measurement or observation of a CCP relative to its critical limits. The monitoring procedures must be able to detect loss of control at the CCP. Further, monitoring should ideally provide this information in time to make adjustments to ensure control of the process to prevent violating the critical limits.
- *Principle 5: Establish the corrective action to be taken when monitoring indicates that a particular CCP is not under control.* Specific corrective action must be developed for each CCP to deal with deviations from the critical limits. The actions must ensure that the CCP has been brought under control. Actions taken must also include proper disposition of the affected product.
- *Principle 6: Establish procedures for verification to confirm that the HACCP system is working effectively.* Examples of verification activities include review of the HACCP system and its records, review of deviations and product dispositions and confirmation that CCPs are kept under control.
- *Principle 7: Establish documentation concerning all procedures and records appropriate to these principles and their application.* Documentation and records are essential to demonstrate safe product manufacture and that appropriate action has been taken for any deviations from the critical limits.

Although the use of HACCP is widespread, some of its details are relatively underdeveloped. For example, until recently the *validation* of critical limits, mentioned under Principle 3, had received very little attention. To fill this gap and provide guidance on how one should validate food safety control measures, draft guidelines were prepared by the Codex Alimentarius Committee (CAC, 2004). In these guidelines 'validation' is defined as 'the obtaining of evidence that the food hygiene control measure or measures selected to control a hazard in a food are capable of consistently controlling the hazard to the level specified by the performance objective'. Thus validation of control measures requires that effectiveness is measured against an expected outcome (for example reduction of the level of *Salmonella* by 99.999%).

The authors of the draft guidelines note that in the current environment of flexibility with the selection of hygiene control measures, the concept of validation acquires increased importance. For many years the use of mandated processing conditions was the main way to guarantee food safety, for example the pasteurisation of milk must deliver a lethality equivalent to or more than 72 °C for 15 seconds. However, the emergence of new pathogens, increased knowledge of the survival capacity of pathogens, the development of novel processing technologies and the marketing of minimally processed foods create a greater need for process validation.

The draft guidelines set out a number of approaches that may be used to validate food hygiene control measures. One may make reference to previous validation studies or scientific knowledge (for example for the heat treatment for pasteurisation of milk), or conduct scientifically valid experimental trials on laboratory or pilot plant scale (for example to document log reduction of pathogens by a thermal process). Other approaches to validation are the collection of biological, chemical and physical contaminant data, both during process establishment/commissioning and during normal operating conditions and the use of statistically designed surveys (for example to validate the effect of hygienic equipment design or personal hygiene). Mathematical modelling is also mentioned as an approach (for example to estimate the combined performance of a combination of already validated control measures).

20.2.2 Risk assessment

The globalisation and liberalisation of world food trade, while offering many benefits and opportunities, also presents new risks. Because of the global nature of food production, manufacturing and marketing, infectious agents can be disseminated from the original point of processing to any place in the world. To assist governmental bodies to achieve an appropriate level of protection derived from the free international trade of food, the Codex Alimentarius Commission has published guidelines with generic principles of risk assessment (CAC, 1999). According to the CAC, risk assessment consists of the following steps: (1) hazard identification, (2) exposure assessment, (3) hazard characterisation and (4) risk characterisation. In addition to these steps and prior to the beginning of a particular risk assessment, the specific purpose of the risk assessment should be clearly stated and the output form and possible output alternatives should be defined. Other important principles are that the risk assessment should be based upon science, should be transparent and its risk estimation should not be influenced by preferential application of particular control measures of the risk.

Hazard identification consists of the identification of biological, chemical and physical agents capable of causing an adverse health effect, which may be present in a particular group of foods. Information on hazards can be obtained for example from scientific literature and studies of government agencies or international organisations.

Exposure assessment is the evaluation of the likely intake of biological, chemical and physical agents via food as well as exposures from other sources if relevant. For microbiological agents, exposure assessment determines the likelihood of consumption and the likely dose of the pathogen or its toxins to which the consumers may be exposed in a food. Exposure assessment is one of the most uncertain aspects of microbial risk assessment. Hence it is normally based on predictive microbiology modelling and simulations. A number of relevant factors have to be considered to determine the frequency and level of contamination of a pathogenic agent in food. These include the initial contamination of the raw material, the processing, packaging, distribution and storage methods of the foods and preparation steps prior to consumption. Also consumption patterns may be considered which are related to socio-economic and cultural backgrounds, seasonality, age, etc. The level of microbiological pathogens may also increase considerably in case of exposure to abusive conditions. Therefore the exposure assessment may predict the range of possible exposures depending on the effects of process conditions such as hygienic design, cleaning and disinfection, time/temperature conditions, food handling and consumption patterns.

Hazard characterisation is the evaluation of the nature of the adverse health effects associated with the hazard. In the case that data are available, a dose–response assessment should be performed. For microbiological hazards it relates to the ingestion of microorganisms or their toxins. The factors to be considered can be divided into three groups: factors related to the pathogen, the food or the human host. The factors related to the pathogen are its virulence and infectivity, which may depend on its interaction with the host, and its capacity of resistance, adaptation, replication and transmission in specific conditions. The food matrix and food process conditions can affect the pathogenicity of a microorganism. High fat content may protect the organism from stomach acid and hence increase the chances of survival and consequent infection. Factors to consider related to the host are the variation of susceptibility depending on age, health or immune status.

Risk characterisation is the final step of the risk assessment. It is the process of estimation of the probability of occurrence and severity of known or potential adverse health effects in a given population based on hazard identification, hazard characterisation and exposure assessment.

Risk assessments may vary considerably in depth of assessment and structure, depending on the purpose of the exercise. Some of them are focused on one aspect, e.g. hazard characterisation, or do not use the Codex Alimentarius structure and definitions. Assessments conducted by national or international food safety organisations (USDA, FAO/WHO) normally contain extensive literature reviews and in-depth dose–response assessments. However, as these risk assessments cover whole sectors within the food industry (for example: ready-to-eat foods) and cover the total production chain, these assessments cannot focus on the details of the individual food processor. Risk assessments conducted by individual food processors are less detailed on public health issues, but focus on the prevalence and concentration of a recognised pathogen in their food ingredients and finished product. Risk assessments conducted by food equipment manufacturers focus on the hygienic design (e.g. selection of construction materials, cleanability specifications) to minimise the risk of transferring hazards to a food product during manufacturing (Holah, 1998).

Risk assessment consists of the evaluation of the probability of occurrence of the hazard and the adverse effect of the hazard. Such an assessment can contain valuable information for the implementation of a management programme, such as HACCP, which is the procedure that ultimately will set the means to eliminate, control or reduce the hazard to acceptable levels. Risk assessment assists during the implementation of the HACCP to identify the hazards associated with a product. In addition risk assessments collect information about the influence of process or storage conditions that may affect the viability of foodborne pathogens. This will help HACCP to identify the steps in the food production that are critical to food safety and at which control actions would produce a great reduction in risk. Hence, it is of potential use for CCP identification. Finally risk assessments provide a valuable scientific foundation to determine the critical limits related with each CCP.

20.3 Examples of risk assessments

20.3.1 *Listeria monocytogenes* in ready-to-eat foods

An assessment of the risk to public health from *Listeria monocytogenes* in selected ready-to-eat foods was conducted by the United States Department of Agriculture (USDA, 2003a, 2003b). This risk assessment estimated the potential level of exposure of three age-based groups to *L. monocytogenes* from 23 food categories and related this to public health consequences. The three age-based groups were perinatals, elderly and the intermediate-age group. Predictions from the risk assessment were given in two forms: the estimated rate of fatal infection on an individual serving basis and the estimated number of fatal infections per year in the US ('per annum risk'). For each food category and each age-based group the per annum risk is derived from the per serving risk by multiplying the latter by the yearly number of servings of the food category by that specific age-based group.

The risk assessment was based on the four steps according to the CAC (1999). In the *hazard identification* step, the health effects associated with *L. monocytogenes* were identified. The main part of the study is the *exposure assessment* which estimates how often consumers eat food contaminated with *L. monocytogenes* and estimates the number of bacteria likely to be in that food. The contamination data used were from published and unpublished studies that mainly concerned food samples collected at retail. Mathematical models were then used to calculate the changes in contamination levels during refrigerated storage and reheating in the home, thus obtaining the contamination level at the moment of consumption. These models considered factors such as refrigerator temperatures during storage, the specifics of the various food categories, and the estimated length of time that food is stored. Finally, the number of servings, necessary to calculate the 'per annum risk', was estimated for each food category and for each age-based group using data from two large nationwide US food consumption surveys conducted in the 1990s.

In the *hazard characterisation* step, a dose–response relationship was described, as a percentage of each age-based group that will become seriously ill or die after being exposed to a particular level of *L. monocytogenes*. The dose–response relationship was estimated from literature data on listeriosis outbreaks and from studies with animals. In the *risk characterisation* step, the results of the exposure assessment and the hazard characterisation were combined to estimate the likelihood of adverse health effects. This was done by running Monte Carlo simulations with statistical data resulting from the exposure assessment and the hazard characterisation.

Results from the risk assessment showed that the risk of listeriosis on both per serving and per annum basis varies greatly among the various categories. The category 'deli meats' ranked as highest risk on both per serving and per annum basis, with estimated risks being 7.7×10^{-8} cases per serving or 1599 cases per year. The categories 'pasteurised fluid milk' and 'high fat and other dairy products' ranked moderate risk on a per serving basis (about 2×10^{-9} cases per serving) but ranked high risk on a per annum basis (about 70 cases/year), because of the high yearly consumption of these products. Hard cheese ranked low risk on both per serving and per annum basis, with estimated risks being 4.5×10^{-15} cases per serving or <0.1 cases per year. The risk assessment also showed that susceptible subpopulations (such as the elderly, perinatal and individuals with chronic illnesses) are much more likely to contract listeriosis than the general population. This suggests that strategies targeted to these susceptible groups would result in the greatest reduction in public health impact.

Finally, the risk assessment model was used to calculate some 'what if?' scenarios to estimate the likely impact of control strategies for foodborne listeriosis. It was, for example, calculated what the impact would be if home refrigerators could not operate above either 45 °F (7.2 °C) or 41 °F (5 °C). These scenarios predicted that the total number of cases of listeriosis would be reduced from 2105 to 656 cases per year by ensuring that all home refrigerators operated at 45 °F or less. The number of cases would reduce to 28 per year (a 98% reduction) if home refrigerators would always operate below 41 °F. Similarly it was shown that reduction of storage time (shelf-life) for deli meat from 28 days to 14 days reduces the number of cases of listeriosis in the elderly population by 13.6%. In another set of 'what if?' scenarios it was predicted that a 1 log reduction in contamination at retail would reduce the number of predicted cases in the elderly population by 50%; a 2 log reduction would result in 74% less cases.

20.3.2 Production line for pasteurised milk

Most food processors have carried out risk assessments of some kind on their production lines, often as part of a HACCP plan. Although such risk assessments are widely performed, very few of them are published owing to the confidential information contained in such assessments. The example given below is

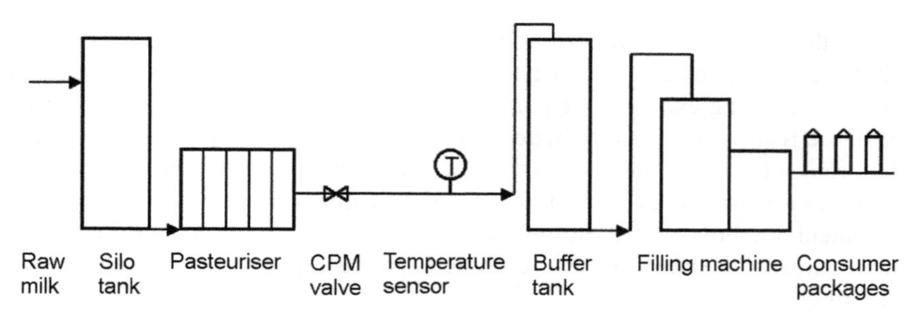

Fig. 20.1 Production line for pasteurised milk.

therefore a risk assessment of an imaginary production line of pasteurised milk. It is based on the experience of the present authors and on the work of Eneroth (1999). The mathematics used in the assessment is of an approximate nature and is meant to give rough estimates only. Throughout the example contamination levels are expressed as colony-forming units (cfu) either per mL or per L; 0.1 cfu/L should be read as 1 cfu/10 L.

The production line to be assessed is schematically shown in Fig. 20.1. This plant produces 10^8 litres of pasteurised milk per year, which corresponds to 200 000 L per batch. It consists of silo tanks for raw milk, a pasteuriser, buffer tanks and a filling machine. Numerous valves and sensors are present on the production line. This example-assessment considers a constant pressure valve (CPM valve) present downstream of the pasteuriser and a temperature sensor placed on a T-piece downstream of this valve. The filling machine fills gable-top cartons ('tetras') of 1 L and is in open contact with the surrounding air. Water is used for chilling the fillers, for lubricating the conveyer chains, and for rinsing of milk residues.

The risk assessment is based on the four steps according to the CAC (1999). In the *hazard identification* step, the microbiological hazards associated with pasteurised milk are identified. From literature reviews and past experience it is clear that the main organisms of interest can be divided in two groups: (i) Gram-negative psychrotrophs (such as *Pseudomonas*, *Enterobactereaceae* and *Aeromonas*) and (ii) Gram-positive spore formers (mainly *Bacillus cereus*). The Gram-negative psychrotrophs (GNPs) are all killed by a suitably performing pasteurisation step, but they may be present in the final consumer packages due to recontamination before sealing the gable-top package. *B. cereus* spores survive pasteurisation, and are typically present in pasteurised milk.

The *hazard characterisation* step discusses the effect that above-mentioned microbiological species have on quality and safety of pasteurised milk. This step is totally based on literature reviews and is not dependent on the details of the production line. The group of GNPs includes well-known pathogens, such as *E. coli* O157:H7 and *Aeromonas* species. Moreover, many species of this group can cause product off-flavours, i.e. spoilage. It is therefore generally understood that the level of these bacteria in the final product must be below 10^6 to 10^7 cfu/mL at the moment of consumption. *B. cereus* can cause food poisoning if present at high levels, and can produce quality issues in milk such as sweet curdling. For these reasons most European countries have set limit values of 10^4 or 10^5 cfu/mL for *B. cereus* at the time of consumption. Pasteurised milk normally contains about 100 cfu/L of *B. cereus*. At a storage temperature of 7 °C these will multiply in 8 to 10 days to 10^4 or 10^5 cfu/mL. Pasteurised milk may contain 0 cfu/L of GNP, but often contains a few cfu/L GNP. At 7 °C these will multiply in 8 to 10 days to 10^6 or 10^7 cfu/mL. It is thus clear that after a storage time of about 9 days both the level of *B. cereus* and the level of GNP may be at its maximum and spoilage can occur by either of these bacteria groups.

The *exposure assessment* step is the part of the risk assessment that relates to the specifics of the production line. It identifies the sources of (re)contamination by the two groups of bacteria and evaluates the possible level of each group present in the finished product. A detailed on-site inspection of the production line indicated eight possible sources of (re)contamination: (1) raw milk, (2) poorly cleanable heat exchangers/gaskets in the pasteuriser, (3) constant pressure valve, (4) poorly cleanable temperature sensor on T-piece, (5) direct air contact in buffer tanks, (6) water used at filling machine, (7) air near filling machine, and (8) packaging material (gable-tops). The contribution of each of these sources of contamination was investigated by microbiological testing. Three kinds of samples were collected: (i) milk samples of 1 L each from various sample points along the production line during production, (ii) swab samples, taken after a production run and cleaning-in-place (CIP), after dismantling some parts of the production line, and (iii) environmental samples, such as air samples and water samples. Each sample was analysed for GNPs and for *B. cereus*.

Based on the results (given in Table 20.1), the effect of contamination from the eight identified sources can be evaluated. These results indicate that recontamination of milk with *B. cereus* takes place in the pasteuriser: upon passing through the pasteuriser the level of *B. cereus* increases from 100 to 200 cfu/L. The most probable cause of this recontamination is a fouling layer which leaks *B. cereus*. The occurrence of difficult to clean layers containing milk soil and *B. cereus* spores on the plate heat exchangers and gaskets of pasteurisers is well known. The results in Table 20.1 also show that the pasteurisation step works well and effectively kills off the GNPs, thus reducing the risk associated with GNPs in raw milk to a minimum.

When during the inspection the constant pressure valve was dismantled, it was seen that the diaphragm was not perfectly in place and that it contained crazes. The diaphragm was swabbed and this swab showed 50 000 cfu of GNPs. If these were to be released over one batch of milk (200 000 L), it would result in a contamination of 0.25 cfu/L. It is generally believed that swabs estimate only approximately 10% of the microflora sampled, i.e. the real number of GNPs in the diaphragm would be about 500 000 cfu. It is, however, very unlikely that all these bacteria would end up in one batch of produced milk. For the process line considered in this example assessment, it can be estimated that the total number

Table 20.1 Microbiological test results for a risk assessment of a pasteurised milk production line (each number is a rounded average of a large number of samples taken, or gives an approximate range)

Sample location of liquid milk	Sample location swab	*B. cereus*		Gram-negative psychrotrophs	
		cfu/L	cfu/swab	cfu/L	cfu/swab
Silo tank		100		5 million	
Directly down-stream of pasteuriser		200		< 1	
	Diaphragm of constant pressure valve		< 10		50,000
	T-piece of temperature sensor		< 10		10,000
Buffer tank		200		0 ... 3	
	Package material before usage		< 10		< 10
Filled and sealed consumer package		200		0 ... 4	
Estimated number of GNPs in air near buffer tank		n.d.		100 cfu/m^3	
Estimated number of GNPs in air near filling machine		n.d.		1000 cfu/m^3	
Water samples at filling machine		n.d.		100 cfu/m^3	

of GNPs that can be detached by one batch of milk is roughly similar to the number of bacteria that can be detected by a swab sample: 50 000 cfu. The poorly cleanable annular gap around a temperature sensor placed in a T-piece was found to contain 10 000 cfu of GNPs, which would result in a contamination of 0.05 cfu/L if these 10 000 GNPs were to be released over one batch.

An empty buffer tank contains air with aerosols contaminated with about 100 cfu/m^3 GNPs (Table 20.1). These bacteria move towards the walls and bottom of the tank by sedimentation and diffusion and then stick to the wet walls. Upon filling the buffer tank with milk, these bacteria will be released into the milk, thus contaminating the pasteurised milk with 100 cfu/m^3, i.e. 0.1 cfu/L. During filling, the milk can be recontaminated by bacteria originating from the air, from the rinsing water or from the packaging material. The level of recontamination is more difficult to estimate directly from the data in Table 20.1. Based on literature data (e.g. Eneroth, 1999), rough estimations can be made: each of these three sources around the filler will contribute about 0.2 cfu/L GNPs.

In the *risk characterisation* step the results of exposure assessment and hazard characterisation are integrated and compared. As explained above,

Table 20.2 Contamination contribution of various sources of (re)contamination

Source	*B. cereus* contribution (cfu/L)	Gram-negative psychrotrophs contribution (cfu/L)
Raw milk	100	0
Pasteuriser	100	0
Constant pressure valve		0.25
Dead leg around temperature sensor		0.05
Air in buffer tank		0.1
Air around filler		0.2
Water around filler		0.2
Packaging material		0.2
Total	200	1

spoilage of pasteurised milk can occur by either of the bacterial groups after a storage time of about 9 days. If the contamination level of one of the two bacterial groups is increased during processing, that will probably be the limiting factor that reduces shelf-life. As stated above, during the 9-day shelf-life of the milk, the level of *B. cereus* increases by a factor of 1000 (from about 100 to 10^5), whereas the level of GNPs increases by a factor of a million (from 1 to 10^6). An increase in initial contamination level by a factor of two will thus have more impact for *B. cereus* than for GNPs (two out of a thousand is more than two out of a million, relatively speaking). It can be calculated that an increase in initial *B. cereus* level by a factor of two will reduce the shelf-life at 7 °C by 10 hours; the same increase in initial GNP contamination will reduce shelf-life by only 5 hours (Eneroth, 1999).

Table 20.2 shows an overview of contamination levels attributable to the different sources. It is clear that the recontamination with *B. cereus* in the pasteuriser is the largest recontamination source in this plant. It is, however, not necessarily true that this is also the highest risk. To evaluate the risk, a link is made with shelf-life, as done by Notermans *et al.* (1997). The extra contamination in the pasteuriser increases *B. cereus* levels by a factor of two and thus reduces shelf-life by 10 hours. All the other contamination sources increase the level of GNPs by amounts adding up to 1 cfu/L. Assuming that 1 cfu/L is a normal contamination level that is associated with a 9-day shelf-life, an additional contamination by 1 cfu/L can be seen as a twofold increase in GNP level, which corresponds to a reduction of shelf-life by only 5 hours. It is thus clear that the risk associated with the *B. cereus* contamination in the pasteuriser is the highest, as it reduces shelf-life most.

20.4 Future trends

The development of risk assessment in food hygiene over the past decade has been very rapid. The two main drivers for this fast development were the SPS agreement that required food hygiene legislation to be based on risk assessment, and the increased use of HACCP and other assessment-based quality systems in the food industry. The Codex Alimentarius and other organisations have published a number of large risk assessments recently. These risk assessments tend to have a bias towards the developed countries of North America and western Europe. It is to be expected that in future risk assessments will also be carried out on foods and production methods used in developing countries and non-western cultures.

In the past decade much effort has been put in the development of methods for risk assessment, and much knowledge has been gained from a number of published risk assessments. It is expected that in the coming years these methods and this knowledge will be used by food industry in hygiene management, for example by improving their HACCP systems.

A further trend may be towards using quantitative risk assessment approaches to assess the effect of hygiene control measures that normally do not fall under HACCP, but are included in GMP, such as hand-washing or the use of hygienically designed equipment. Moreover, computer programs for risk assessment and risk calculations will be further developed in the next years and may well be linked to extensive databases.

20.5 Sources of further information and advice

Books

ALLI I., *Food quality assurance – Principles and practices*, CRC Press, Boca Raton, 2004.

CORLETT D.A., *HACCP user's manual*, Aspen Publishers, Gaithersburg, 1998.

FORSYTHE S.J., *The microbiological risk assessment of food*, Blackwell Science, Oxford, 2002.

Internet pages

Codex Alimentarius Commission (www.codexalimentarius.net)
FDA Bad Bug Book (http://www.cfsan.fda.gov/~mow/intro.html)
Food Safety Risk Analysis Clearing House (http://www.foodriskclearinghouse.umd.edu/)
World Health Organisation (http://www.who.int/foodsafety/micro/riskassessment/en/) (about risk assessment)

20.6 References

ANON (1995), Agreement on the Application of Sanitary and Phytosanitary Measures, World Trade Organisation, Geneva, http://www.wto.org/english/docs_e/legal_e/15-sps.pdf

ANON (2000), The Cartagena protocol on biosafety http://www.biodiv.org/biosafety/articles.asp

CAC (1997), Recommended International Code of Practice – General Principles of Food Hygiene. CAC/RCP 1-1969. Codex Alimentarius Commission, Rome.

CAC (1999), Principles and guidelines for the conduct of microbiological risk assessment, CAC/GL-30. Codex Alimentarius Commission, Rome.

CAC (2004), Proposed draft guidelines for the validation of food hygiene control measures, Codex Alimentarius Commision, Rome.

CORLETT D.A. (1998), *HACCP user's manual*, Aspen Publishers, Gaithersburg.

ENEROTH Å. (1999), Bacteriological hygiene in the production of pasteurised milk, PhD Thesis, Lund University, Sweden.

ESTY J.R., MEYER K.F. (1922), The heat resistance of spores of *Bacillus botulinus* and allied anaerobes, *XI J. Inf Dis* **31**, 650–663.

HOLAH J.T. (1998), Hygienic design of food equipment, *Food Australia* **50** (7), 336–339.

NOTERMANS S., DUFRENNE J., TEUNIS P., BEUMER R., TE GIFFEL M., PEETERS WEEM P. (1997), A risk assessment study of *Bacillus cereus* present in pasteurised milk, *Food Microbiology* **30**, 157–173.

NOTERMANS S., BARENDSZ A.W., ROMBOUTS F. (2002), The evolution of microbiological risk assessment in food production, in: Brown M and Stringer M, *Microbiological risk assessment in food processing*, Woodhead Publishing, Cambridge.

TANNAHILL R. (1973), *Food in history*, Stein and Day Publishers, New York.

USDA (2003a), Interpretive summary: quantitative assessment of the relative risk to public health from foodborne *Listeria monocytogenes* among selected categories of ready-to-eat foods, http://www.cfsan.fda.gov

USDA (2003b), Technical document: quantitative assessment of the relative risk to public health from foodborne *Listeria monocytogenes* among selected categories of ready-to-eat foods, http://www.cfsan.fda.gov

21

Good manufacturing practice (GMP) in the food industry*

J. R. Blanchfield, Consultant, UK

21.1 Introduction

What is 'good manufacturing practice' (GMP) in relation to food? In the USA the term GMP has a legal status – it is codified in Food and Drug Administration (FDA) Current Good Manufacturing Practice (CGMP) Regulations for foods – including a general regulation covering all foods[1] and specific regulations for specific food categories.[2–5] It is very clear that these are wholly concerned with hygiene requirements for food safety. It is interesting to note that among the definitions provided, there is no definition of 'good manufacturing practice' – in the absence of which it is assumed to consist of the sum total of the stated regulatory requirements. In May 2004 the FDA announced its plans and processes for modernising the GMPs used for foods, which were last revised in 1986.[6]

Outside the USA, GMP in its wider aspects has been authoritatively defined and detailed since 1986 by the four successive editions of the Institute of Food Science & Technology's 'Food and Drink – Good Manufacturing Practice: A Guide to its Responsible Management©',[7] familiarly known as 'the IFST GMP Guide'. It is a tribute to the IFST that it is impossible in this chapter to avoid reflecting the nature, structure or subject matter of what has been so effectively expounded at length in the IFST GMP Guide. Although developed in the UK/EU context, when taken in conjunction with local food regulations in different countries, it has wide international application. Within the USA the approach described here as GMP has been taken in a combination of other designated systems, one of which is standard operating procedures (SOPs) (see Chapter 22).

Food and drink industry managers, food scientists and food technologists recognise the special responsibilities of the food and drink industries to make available products that are 'safe' and consistently offer maximum value and desirability to consumers and therefore to their retailer customers. However, what is meant by 'safe'? There is no such thing as absolute assurance of absolute safety. Hazards abound in every aspect of life, and food is no different. The target is to achieve absence of *unacceptable* risk or, to use a term borrowed from the World Trade Organization, 'an appropriate level of protection' (ALOP).[8] What constitutes ALOP is determined by legislators in the form of laws and regulations. Moreover, regulatory requirements, apart from country-to-country variation, are not immutable and may themselves be changed. There is, of course, nothing to prevent a manufacturer seeking to achieve a higher level of protection that the ALOP of the country for which the food is intended.

In recent times, the Codex Alimentarius Food Hygiene Committee has adopted the approach of science-based risk analysis based on risk profiling related to achieving ALOP.[9,10]

> In brief, a Risk Profile describes: 1) the pathogen and food commodity combination(s) of interest; 2) the public health problem; 3) the current state of knowledge regarding relevant food production, processing, distribution, and consumption practices; 4) risk assessment needs and risk management questions; 5) available information and knowledge gaps; and 6) a recommendation for work and the type of Committee documents required (e.g., risk management control guidance document, code of hygienic practice, code of practice, amendments to existing Codex hygiene texts, establishment of microbiological criteria or food safety objectives). It is expected that the material provided within the Risk Profile will prove useful in further considering microbiological risks during the subsequent development of the proposed microbiological risk management guidance document.

For practical purposes, references to 'safety' and 'safe' in this chapter should be interpreted as meaning 'achieving ALOP'. It follows that the food or drink manufacturer needs to ensure that the manufacturing process delivers a product that achieves ALOP while providing as effective freedom from contamination as it is realistically possible to achieve, and while complying with relevant regulatory requirements and also uniformly fulfilling the quality parameters (e.g. colour, flavour, texture, mouthfeel, freedom from defects and foreign matter) built-in to the product specification. It is in this wider sense that GMP will be discussed in this chapter. This is the approach taken by successive editions of the IFST 'Food and Drink – Good Manufacturing Practice: A Guide to its Responsible Management'.[7] This approach is also taken in USA but under other terminologies.

The key is the word 'management'. The components of GMP are summarised in this chapter but some are described much more comprehensively in other chapters. GMP is more than the sum total of all these components. *GMP is first and foremost an integrated management system* and requires, therefore, that there is full and detailed specification of the product and of everything that goes into making, storing and distributing it; and management of the materials, resources, measures and precautions to ensure that the specification requirements are fulfilled. This implies that GMP is the prime concern of company chairmen, presidents and chief executives. Though not directly concerned in designing or implementing GMP measures, it is they who are responsible for establishing GMP policies and for providing authority, facilities and resources to the functional managers and staff (and that includes provision of competent managers and staff with appropriate skills).

GMP is the concern and responsibility of every kind of manager in industry. These are included in Table 21.1. However, there are others who need to be 'GMP-literate', including those in legislation and regulatory enforcement, and likewise those responsible for teaching the next generation of all those managerial functions. Academics cannot legitimately take the line 'GMP is for industry and does not concern academia'.

It is evident that GMP has two complementary, and interacting components; the manufacturing operations themselves, and a quality control/quality assurance system which can go by many designations but which the IFST has designated 'food control'.

Both of these components must be well designed and effectively implemented. The same complementary nature and interaction must apply to the respective managements of these two functions, with the authority and responsibilities of each clearly defined, agreed and mutually recognised. In this context, the management of the manufacturing function also embodies the important other key functions contributing direct services or advice to the manufacturing function.

Table 21.1 Who needs GMP?

- Company chairmen, presidents, chief executives
- General managers
- Plant managers
- Technical managers
- Quality assurance managers
- Food hygiene managers
- Production managers
- Plant engineers
- Warehouse managers
- Distribution managers
- Work study managers
- Everyone involved in training
- All involved in food legislation and food law enforcement
- All involved in teaching the next generation of all the above

Finally, there is no section headed 'Food Hygiene' in this chapter because GMP does not treat food hygiene as a compartmentalised topic but as a continuous thread underlying every aspect.

21.2 Effective manufacturing operations and food control

Every aspect of the manufacturing operation must be properly specified in advance and systematically reviewed regularly, including all resources and facilities, premises and space, equipment, trained people, raw materials, packaging materials, storage, transport, operating procedures, cleaning procedures, establishment of critical control points, management and supervision, and services (see Table 21.2). And, critically important, these must be provided in the right quantities, at the right times and places; and the thoroughness and effectiveness of their use must be regularly monitored.

To ensure that manufacturing operations go according to plan it is essential that:

- relevant written procedures are provided to operators, in instructional form and in clear and unambiguous language, and are specifically applicable to the facilities provided;
- operators are trained to carry out the procedures correctly;
- records are made (whether manually or by recording instruments or both) during all stages of manufacture, which demonstrate that all the steps required by the defined procedures were in fact carried out, and that the quantity and quality produced were those expected;
- a traceability system is in place to identify and track all inputs into, and outputs from, the manufacturing process;
- records are made and retained in legible and accessible form to enable the history of the manufacture and distribution of a batch to be traced;

Table 21.2 Effective manufacturing operations

Every aspect of manufacture fully specified in advance; all resources and facilities specified, namely:

- measures and precautions at Critical Control Points based on hazard analysis
- adequate premises and space
- correct and adequately maintained equipment
- appropriately trained people
- correct raw materials and packaging materials
- appropriate storage and transport facilities
- written operational procedures and cleaning schedules
- appropriate management and supervision
- adequate technical, administrative and maintenance services

- a system is available to deal with unexpected emergencies, including terrorism, sabotage, or the need to withdraw or recall from sale or supply any batch of product, if that should become necessary.

21.2.1 Effective food control

The other and complementary major component of GMP is effective food control. Effectiveness requires:

- well-qualified and appropriately experienced food control management participating in the drawing up of specifications;
- adequate staff and facilities to do all the relevant monitoring of suppliers, inspection, sampling and testing of materials, and monitoring of process conditions and relevant aspects of the production environment (including all aspects of hygiene);
- rapid feedback of information (accompanied where necessary by advice) to manufacturing personnel, thereby enabling prompt adjustment or corrective action to be taken, and enabling processed material either to be passed as fit for further processing or for sale as the case may be, or to be segregated for decision as to appropriate treatment or disposal.

There are two schools of thought on the relationship between food safety and food quality. One school holds that the two are separate and should be controlled separately; and in particular that the preventive Hazard Analysis Critical Control Point (HACCP) approach should be applied only to hazards (some even aver only microbiological hazards) to food safety. The other school holds that if 'quality' in this context resides in consistently fulfilling the specification embodying the requirements of the law and the marketplace, then food safety is a critically essential part of quality and not a thing apart. Likewise if a hazard is defined as 'an intrinsic property of a system, operation, material or situation that could in certain circumstances lead to an adverse consequence' (i.e. giving rise to a risk), then the preventive approach of HACCP can and should equally be applied to preventing the risk of unacceptable departures from other defined quality parameters. It follows that rather than two separate food control systems, one for food safety and another for other quality factors, a single integrated food control system is desirable. This, of course, does not preclude, but on the contrary demands, within the food control system, specialists such as food microbiologists, food chemists, sensory scientists, regulatory specialists, as well as process control technologists.

This chapter firmly adopts the latter school of thought. It is also the approach adopted since 1998 by the British Retail Consortium's Technical Standard, now further developed and entitled 'BRC Global Standard – Food'.[8] Originally introduced to eliminate multiple audit by retailer technical and third party technical representatives of food manufacturers supplying UK retailers with own brand products, it has gained much wider application and significant international recognition for its content, format and supporting system. Its objective is to specify food safety and quality criteria required to be in place within a manufacturer's organisation to supply product to UK retailers. The format and content of the Standard is designed to allow an assessment of the supplier's premises and operational systems and procedures by a competent third party, thus standardising food safety criteria and monitoring procedures. It is essentially a GMP-based standard.

21.3 Personnel and training

Compatible with the size and type of the business there should be sufficient personnel at all levels with the ability, training, experience and, where necessary, the professional and technical qualifications appropriate to the tasks assigned to them. Their duties and responsibilities should be clearly explained and recorded as job descriptions or by other suitable means.

Potential employees should be appropriately qualified for the requirements of the task. Due regard should be given to the potential suitability of a candidate for the task in hand. For example, persons who have a clean and tidy appearance may be more likely to appreciate the principles required in the hygienic handling of food than those who do not.

Where employment is terminated for any reason, consideration should be given to any possible risks to safety or quality arising through disaffection or simple lack of continuing interest and commitment from the employees concerned, and appropriate precautions taken as necessary.

All production and food control personnel should be trained in the principles of GMP, in all relevant aspects of food hygiene and in the practice and underlying principles of the tasks assigned to them. Similarly, all other personnel (e.g. those concerned with maintenance or services or cleaning) whose duties take them into manufacturing areas or bear on manufacturing activities, should receive appropriate training. Records should be kept of the training of each individual.

Training should be in accordance with programmes approved by the respective functional managers. Instruction should not be regarded as an adequate substitute for training, which provides not only information on what should be done, but also an understanding of why it is important. Training should embody the requirements of personal hygiene and the reasons why they are important. Adequate facilities must be provided to enable personnel to comply fully with those requirements. Training should be given at recruitment and repeated, augmented and revised as necessary. Both in training itself, and with regard to the need for personnel to be able to understand and follow written instructions and procedures, notices, etc., particular attention should be given to overcoming language or reading difficulties.

Periodic assessments of the effectiveness of training programmes should be made, and checks should be carried out to confirm that designated procedures are being followed.

21.4 Documentation

Effective documentation is an essential and integral part of GMP and, in particular, one of the essential features of a properly operated HACCP system. The purposes of documentation are to define the policies, materials, operations, activities, control measures and products; to record and communicate information needed before, during or after manufacture; to reduce the risk of error arising from oral communication; and to form a vital part of the audit trail necessary for tracing the history of the components used to produce the final product. The system of documentation should ensure that, as far as is practicable, the history of each batch of product, including utilisation and disposal of raw materials, intermediates and bulk or finished products, can be ascertained.

Documents fall into four main classes, namely, those setting out policies; those setting out requirements, instructions and procedures; those setting out programmes to be carried out in particular periods; and those requiring data to be entered before, during and after manufacture to provide a record of what has happened. Failure to maintain appropriate documentation and records will nullify many of the benefits of GMP.

21.5 Premises, equipment, product and process design

Buildings should be located, designed, constructed, adapted and maintained to suit the operations carried out in them and to facilitate the protection of materials and products from deterioration, contamination or cross-contamination. Equipment should be designed, constructed, adapted, located and maintained to suit the processes and products for which it is used; to facilitate protection of the materials handled or the resulting products from contamination, cross-contamination or deterioration; and to facilitate effective cleaning.

GMP relates not only to products and processes that already exist. The development of a new product or a new process, or a significant change to an existing product or process, provides the opportunity to optimise in respect of GMP. Products and processes should be so designed as to ensure that the end-product meets consumer expectation within the intended and anticipated duration and circumstances of use, and to ensure that product design and performance have been fully evaluated for the required function in respect of microbiological safety, chemical safety, physical safety and sensory quality. To this end, the value of the multidisciplinary hazard analysis technique usually known as Hazard Analysis Operability Studies (HAZOP) is emphasised, preferably applied from the earliest stages of product and process development with a view to eliminating or minimising potential hazards wherever possible and incorporating effective control parameters into the product and process design.

Similar considerations apply where changes are made that would affect integrity, safety or stability of a product. Such changes could include those made to ingredients, formulation or recipe, operations, machinery, processes or process parameters, packaging, storage, distribution, and consumer use. The list is not exhaustive, but is based on experience of situations where failure to take such changes into account has resulted in serious human or commercial consequences or both.

21.6 Manufacturing and operating procedures

The operations and processes used in manufacture should, with the premises, equipment, materials, personnel and services provided, be capable of consistently yielding finished products that conform to their specifications and are suitably protected against contamination or deterioration. Definition and documentation of manufacturing procedures, including associated activities and precautions, are necessary to ensure that all concerned understand what has to be done, how it is to be done and who is responsible, and avoidance of mistakes that could compromise safety or quality. For each product, this should be provided in comprehensive master manufacturing instructions (MMI).

Before the introduction of MMI for a product, trials should be carried out to establish whether the formulation, methods and procedures specified therein are suitable for factory production, and are capable of consistently yielding products within the finished product specification. If necessary, amendments and further trials should be made until these conditions are satisfied. Similar evaluation should be carried out in connection with any significant proposed change of raw material, plant or method; and should be carried out periodically, to check that the MMI are being followed, that they still represent an effective and acceptable way of achieving the specified product and that they are still capable of consistently doing so.

There should be provided premises, equipment, materials, suitably trained personnel, services, information and documentation, in each case of appropriate quantity and quality, to enable the requisite quantity and quality of finished products to be produced. Production should be carried out in full compliance with the MMI, from which no departure should be permitted except by written instructions from the managers responsible respectively for production and for food control, indicating the nature and duration of the departure, and agreed and signed by them.

Incentive bonus schemes for production operators can create potential hazards and, viewed from the standpoint of safety and quality, are best avoided. If, however, the provision of an incentive bonus scheme is company policy, it should be so designed as to discourage operators from taking unauthorised 'short-cuts', for example by building into the formula for bonus calculation a 'quality factor' and/or penalty for observed deviation. In general, prevention of unauthorised short-cuts is primarily a task for management and supervision. Where operators have ideas for process simplification or improvement, they should be encouraged to raise them (for example through suggestions schemes) so that they may be properly evaluated.

As part of the preparation of operating instructions for production operators it is highly advantageous to consult the relevant operators, thereby both gaining the benefit of their practical experience and giving them a sense of 'stakeholder ownership' of the instructions. Operating instructions should be written in clear, unambiguous instructional form, and should form a key part of operator training. Due regard should be given to reading or language difficulties of some operators.

Particular attention should be paid to problems that may arise in the event of stoppages, breakdowns or emergencies, and written instructions should be provided for action to be taken in each case.

21.7 Ingredients and packaging materials

Each ingredient should comply with its specification. Each delivery or batch should be given a reference code to identify it in storage and processing, and the documentation should be such that, if necessary, any batch of finished product can be correlated with the deliveries of the respective raw materials used in its manufacture and with the corresponding laboratory records. Deliveries should be stored and marked in such a way that their identities do not become lost.

Deliveries of raw materials should be quarantined until inspected, sampled and tested and released for use only on authority of the manager responsible for food control, taking account of any certificate of analysis or conformity accompanying a delivery. It may not necessarily be enough to assume that the description of a consignment of a raw material on the packages or on the corresponding invoice is accurate. Where the identity is not absolutely obvious beyond question, the identity of each consignment of raw material should be checked to verify that it is what it purports to be.

Particular care should be taken where a delivery of containers appears from markings to include more than one batch of the supplier's production, or where the delivery is of containers repacked by a merchant or broker from a bulk supply. Where appropriate, immediate checks should be carried out for off-flavours, off-odours or taints, and, particularly in the case of additives, testing should include test of identity, i.e. establishing that the substance is what it is purported to be. (NB. In a multi-container delivery it is impracticable to check the identity of the contents of every container on arrival, but operators should be trained and encouraged to report immediately anything unusual about the contents when a fresh container is brought into use.)

Temporarily quarantined material should be located and/or marked in such a way as to avoid risk of its being accidentally used. Material found to require pre-treatment before being acceptable for use should be suitably marked and remain quarantined until pre-treatment. Material found totally unfit for use should be suitably marked and physically segregated pending appropriate disposal. In the case of a bulk delivery by tanker, preliminary quality assessment should be made before discharge into storage is permitted.

All raw materials should be stored under hygienic conditions, and in specific conditions (e.g. of temperature, relative humidity) indicated in their specifications, and having regard to any regulatory requirements for storage of hazardous raw materials. Stocks of raw materials in store should be inspected regularly and sampled/tested where appropriate, to ensure that they remain in acceptable condition.

Authorised procedure and documentation should be established and followed for the issue of raw materials from store. Correct stock rotation should normally be observed, unless otherwise authorised or specified by the food control manager. When a raw material has been issued but not used as planned (e.g. because of a plant stoppage), Food Control should advise as to its disposition.

Depending on the product being manufactured, the ingredients involved, and the nature of the process and equipment, the dispensing of the required quantities of ingredients could take various forms, including manual dispensing by weight or volume, automatic dispensing of batch quantities by weight or volume, or continuous metering by volume; the form(s) actually taken will be stated within the MMI. In each case, the weighing and/or measuring equipment should have the capacity, accuracy and precision appropriate to the purpose, and the accuracy should be regularly checked.

Where batch quantities of an ingredient have to be dispensed manually into containers in advance, this should be done in a segregated area. Where manual pre-dispensing of relatively small and accurate quantities (for example of additives) is required, this should be done by, or under direct supervision of, laboratory staff.

Records should be kept to enable the quantities of materials issued to be checked against the quantity or number of batches of product manufactured. Where an operator controls the addition of batch quantities of one or more ingredients to a batch, the addition of each ingredient to a batch should be recorded at the time on a batch manufacturing record, to minimise risk of accidental omission or double addition.

Each packaging material should comply with its specification (including any legal requirements). The specification should be such as to ensure that:

- the product is adequately protected during its expected life under normally expected conditions (with a safety margin for adverse storage);
- in the instance of packaging coming into immediate contact with the product, there is no significant adverse interaction between product and packaging material;
- where the packaged product undergoes subsequent treatment, whether by the manufacturer, caterer or consumer, the packaging will adequately stand up to the processing conditions and no adverse packaging/product interaction occurs;
- the packaging is capable of providing the necessary characteristics and integrity where the preservation of the product depends on the pack;

- the finished pack will carry the statutory and other specified information in the required form and location.

Where packaging material carries information required by law (e.g. labels, printed packages, lithographed cans), the food control manager should ensure that the specification is updated as required to comply with new regulatory provisions, and that stocks of packaging materials that no longer comply are quarantined for modifications (if possible and desired) or destruction.

Each delivery or batch should be given a reference code to identify it in storage and processing, and the documentation should be such that, if necessary, any batch of finished product can be correlated with the deliveries of the respective packaging materials used in its manufacture and with the corresponding laboratory records. Deliveries should be stored and marked in such a way that their identities do not become lost.

Deliveries of packaging material should be quarantined until inspected, sampled and tested, and released for use only on authority of the food control manager. Operators should be trained and encouraged to report immediately anything unusual about the appearance, colour, odour or behaviour of packaging materials issued.

Temporarily quarantined packaging material should be located and/or marked in such a way as to avoid risk of its being accidentally used before release. Material found totally unfit for use in packaging operations should be suitably marked and physically segregated pending appropriate disposal.

All packaging materials should be stored in hygienic conditions, and as indicated in their respective specifications. Stocks of packaging materials in store should be inspected regularly to ensure that they remain in acceptable condition.

Authorised procedure and documentation should be established and followed for the issue of packaging materials from store. In issuing packaging material from store for production use, stock rotation should normally be observed, unless otherwise authorised or specified by the food control manager.

Where a company manufactures more than one product, or more than one version of a single product, the greatest care should be taken to check that the correct packaging is issued for the product to be manufactured, and that no incorrect packaging materials left over from a previous production run of a different product or a different version are left in the production area where they might accidentally be used. In no circumstances should primary food packaging be used for other than its intended purpose.

Where packaging is reference-coded and date-marked in advance for use, care should be taken to ensure that only material carrying the correct date is used. Surplus material left from earlier production and bearing a no longer valid reference or date should not be left in the production area. Where the reference and/or date are applied during the manufacturing operation, care should be taken to ensure that the marking machine is set for the correct reference and date.

21.8 Managing production operations: intermediate and finished products

Where a company manufactures more than one product or more than one version of a product, and there is more than one production line, production layout should be such that confusion and possible cross-contamination are avoided.

Whether in single-line or multiple-line production, particular care should be taken, in terms of production layout and practices, to avoid cross-contamination of one product by another. This is especially important when handling, unavoidably in the same building, ingredients or manufactured products containing one or more major food allergens, and those that do not. On a production line, the name and appropriate reference to the product being processed/packaged should be clearly displayed.

Before production begins, checks should be carried out to ensure that the production area is clean, and free from any products, product residues, waste material, raw materials, packaging materials or documents not relevant to the production to be undertaken; and that the correct materials and documents have been issued and the correct machine settings have been made. All plant and equipment should be checked as clean and ready for use.

Processing should be strictly in accordance with the MMI subject to any variations approved, and by detailed procedures set out for operators in the Plant Operating Instructions. Process conditions should be monitored and process control carried out by suitable means including, as appropriate, sensory, instrumental and laboratory testing, and on-line checking of correct packaging and date-marking. Where continuous recorders or recorder/controllers are in use, the charts should subsequently be checked by food control.

There should be regular and recorded checks, by appropriate personnel, on the accuracy of all instruments used for monitoring processes (e.g. thermometers, temperature gauges, pressure gauges, flowmeters, checkweighers, colour measurement devices, metal detectors, X-ray machines). The frequency of checks should be established to ensure that instruments are always correctly calibrated, with an accuracy related to national standards.

Effective cleaning of production premises and equipment must be carried out. All persons working in or visiting the production area must comply with the requirements of personal hygiene, and adequate facilities must be provided.

General 'good housekeeping' should be practised, including prompt removal of waste material, precautions to minimise spillage or breakage, prompt removal and clean-up of any spillage or broken packaging occurring, and the removal of any articles that might enter the product as foreign matter. Where appropriate, foreign matter detectors should be used.

21.8.1 Intermediate products

After its preparation, an intermediate product should be quarantined until checked and approved by the food control manager for compliance with its specification. If required to be stored before further processing, it should be stored as designated in that specification, and suitably reference marked and documented so that it can be correlated with the lots of raw materials from which it was made and the batch(es) of finished product in which it is subsequently incorporated.

A batch of intermediate product found to be defective should remain quarantined pending reworking or recovery of material or outright rejection as the case may be.

21.8.2 Finished products

Packed finished products should be quarantined until checked and approved by food control for compliance with the appropriate finished products specification. An approved batch of finished product should be suitably flagged to identify it, and it should be stored under the appropriate conditions (e.g. of temperature or relative humidity) stated in the finished product specification.

Where a batch of finished product fails to meet the specification, the reasons for failure should be thoroughly investigated. Defective finished product should remain quarantined pending reworking or recovery of materials or disposal as the case may be.

21.9 Storage and movement of product

In addition to specific references in earlier sections to storage practices for raw materials, packaging materials, intermediate products and finished products prior to distribution, the following considerations should be met:

- Access to material and product storage areas should be restricted to those working in those areas and other authorised persons.
- Materials and products should be stored under conditions specified in their respective specifications. There should be effective protection of materials and products from contamination. Particular attention should be paid to the avoidance of microbiological cross-contamination and tainting. Where special conditions are required, they should be regularly checked for compliance.
- Materials and products should be stored in such a way that cleaning, the use of pest control materials without risk of contamination, inspection and sampling, retention of delivery identity or batch identity, and effective stock rotation can be easily carried out.
- Effective cleaning of storage premises and equipment must be carried out with the frequency and using the methods and materials specified in well-designed cleaning schedules and instructions.
- Products that have been recalled or returned, and batches that have been rejected for re-working or recovery of materials or disposal, should be so marked and physically segregated, preferably in an entirely separate storage facility.
- Material deliveries and product batches temporarily quarantined pending the results of testing, should be so marked, suitably segregated, and effective

organisational measures implemented to safeguard against unauthorised or accidental use of those materials or despatch of those products.
- If a batch of finished product has to be temporarily stored unlabelled, to be labelled at a later date, the greatest possible care should be exercised in maintaining its exact identity.
- Storage areas should be regularly inspected for cleanliness and good housekeeping, and for batches of products which have exceeded their shelf-life or, in the case of date-marked products, leave insufficient time for retail display.

21.9.1 Internal transport

Materials or products should be transported within the factory premises in such a way that their identities are not lost; that there is no mixing of materials or products approved for use or despatch with those that are quarantined; that by-products, particularly those not intended for human food use, do not lead to contamination of other materials; that no spillage is caused and no breakage or other physical damage is caused to the goods being transported; and that goods being transported are not left in adverse conditions or otherwise allowed to deteriorate.

21.9.2 Warehousing and external transport

Effective warehousing operations should be designed to ensure that all products are easily acceptable for load assembly as required; to ensure that aisles and assembly areas are planned so that unimpeded movement is possible to and from all parts of the warehouse; to facilitate proper stock rotation, particularly important in relation to short-life and date-marked foods; and to obtain maximum utilisation of available space, consistent with the foregoing requirement.

Storage and transportation of finished products should be under conditions that will prevent contamination, including the development of pathogenic or toxigenic microorganisms, will protect against undesirable deterioration of the products and the containers, and ensure the delivery of safe, clean and wholesome foods to consumers. This deterioration includes, but is not limited to, contamination from insects, rodents and other vermin, toxic chemicals, pesticides and sources of flavour and odour taint.

The buildings, grounds, fixtures and equipment of food warehouse and vehicles should be designed, constructed, adapted and maintained to facilitate the operations carried out in them and to prevent damage.

21.10 Special requirements for certain foods

The GMP requirements described elsewhere in this chapter apply to all foods. However, certain classes of foods pose additional problems and have additional requirements.

21.10.1 Heat-treated foods

Heat-treated foods that are designed to be microbiologically stable must be heat treated to an extent that will prevent the outgrowth of microorganisms under the conditions, including packaging, and during the period, in which the foods are intended to be stored. All low-acid foods having in any part of them a pH value of 4.5 or above and intended for storage under non-refrigerated conditions must be subjected to the minimum botulinum process, i.e. one that will reduce the probability of survival of *Clostridium botulinum* spores by at least 12 decimal reductions unless the formulation or water activity or both, of the food is such that it can be demonstrated that growth of strains or forms of the organism cannot occur.

21.10.2 Chilled foods

Chilled foods, which offer high risk of potential growth of pathogenic organisms, need the most stringent hygiene precautions. Product safety must be determined by the proper consideration of ingredients' hygienic quality, product formulation/characteristics, processing parameters, intended use of product, storage and distribution conditions, manufacturing hygiene and intended shelf-life.

21.10.3 Frozen foods

Frozen foods are critically dependent for their safe preservation on stringent hygiene and careful maintenance of the cold chain, i.e. an organised system governing the conditions under which frozen foods are stored and handled by the producer, distributor and retailer, and which ensure that temperatures maintained during storage, distribution and sale are those consistent with the maintenance of quality and safety.

21.10.4 Dry foods

Dry foods and processes involving dry materials have problems associated with dust, particularly those of cleaning, the possible creation of an explosive dusty atmosphere and the risks of cross-contamination by dust particles. It is important, therefore, to contain dust as far as possible in an enclosed system and, with the aid of dust removal and extraction systems, to maintain a high standard of cleanliness. The general environment of the plant and equipment including ledges and girders, etc. should be regularly cleaned and an effective air extraction system should be installed. Such a system should discharge through a filter and at a point situated so as to minimise the risks of the discharge being able to contaminate other plant or products. Dust extraction systems should be properly maintained, cleaned and serviced; they become heavily coated inside duct-work, and cleaning and filter changing can create a very dusty atmosphere. Dusty atmospheres should be considered as potentially dangerous explosion hazards. It may be desirable therefore to use flameproof motors and switches or ensure that they are situated in a relatively clean environment. Adequate protective clothing and other equipment should be provided for those involved physically in the cleaning operations, and during production if necessary.

21.10.5 Compositionally preserved foods

Compositionally preserved foods are those that depend for their preservation and/or specific properties and maintenance of their quality during their expected life on the achievement of a particular quantitative composition (for example the attainment of requisite – and, in the UK and the other EU member states, also legally required – refractometric solids in jams with no added preservative, or of a preservation index of 3.6% acetic acid in the volatile constituents in non-pasteurised pickles and sauces). In products where a quantitative compositional factor is critical, the training of production supervisors, operators and quality control staff should emphasise the critical nature of such compositional factors. Production methods and control procedures should be such as to ensure that the required composition is consistently achieved.

21.10.6 Novel foods and foods produced by novel processes

Care should be taken in the use of novel food or food ingredients produced from raw material that has not hitherto been used (or has been used only to a small extent) for human consumption in the area of the world in question, or that is produced by a new or extensively modified process not previously used in the production of food (and would thus include genetically modified materials). This must include attention to food safety considerations, compliance with the relevant regulations of the country for which the food is intended, and provision of label information to enable the purchaser or consumer to make an informed choice.

21.10.7 Genetically modified foods

Genetically modified foods are subject, in the EU, to additional special requirements as to labelling and traceability.[11,12]

21.10.8 Foods manufactured for food service operations

Foods manufactured for food service operations additionally should have regard to the special requirements that relate to the intended use. 'Manufactured' in this context applies not only to food products made by a food manufacturing company and sold to a food service operator, but also to food prepared in a central production unit by factory-style processing, by a food service organisation for use in its own food service outlets, as distinct from preparation in 'cook-serve' form. Particular regard should be given to the circumstances and conditions of use, the probable expertise of the food service operator and staff, and the interactions likely to occur between the product and its subsequent environment. The manufacturer should be prepared to offer technical advice to users on the suitability of products for the uses intended and on any appropriate precautions to be observed.

21.10.9 Foods for use in vending equipment

Foods for use in vending equipment should take account of possible interaction effects between product, environment and equipment. In the manufacture of products for these purposes, the manufacturer should ensure awareness of such potential hazards as within-machine environment, hygiene and cleaning needs, product flow properties, variability of throw or dispensation, as well as product and machine interactions and interactions with other products or ingredients. The requirements for vending operations may call for particular product performance standards, e.g. dispersion at sub-scalding temperatures or interchangeability with competitors' products in an identical vending situation.

21.11 Rejection of product and complaints handling

Inevitably rejection of product will be necessary from time to time, and proper means of disposal should be considered and agreed with the Food Control manager, the Production manager, and any other interested parties such as the Purchasing manager or Sales/Marketing Departments. In determining disposal, due regard should be paid to protecting the public, and complying with appropriate legislative or local authority requirements, protecting the company or brand name and the needs of securing cost recovery.

Material may be recovered or reworked or reprocessed by an appropriate and authorised method, provided that the material is safe and suitable for such treatment, that the resulting product complies with the relevant specification and that the related documentation accurately records what has occurred.

Because there are so many different circumstances that can arise with different kinds of food products and processes, it is not possible to be specific here about each of them. The matters referred to here, however, may be classified under three main groups, namely systematic, semi-systematic and 'occasional'. However, in all circumstances, appropriate precautions must be taken to avoid microbiological contamination, introduction of undeclared ingredients, cross-contamination with allergens or the introduction of foreign matter.

Meat or poultry material left over from the previous day should never be carried forward. The possible carrying forward of other perishable material should be subject to a most stringent hazard analysis by the Food Control manager and, where approved, should be subject to precautions against material in which it was incorporated being carried forward again. Where a quantitatively known product residue from previous production is systematically utilised as one of the starting materials for the same or another product (e.g. dough trimmings in biscuit manufacture), that should be written into the MMI, and the rate or conditions of use there specified should not be departed from other than through the established procedure for varying MMI.

'Semi-systematic' applies to instances where a variable quantity of intrinsically satisfactory but extrinsically unacceptable product occurs and can be reused (e.g. misshapen or short-weight moulded chocolate bars); or to instances where a usable starting material can be extracted from wholesome but defective product (e.g. recovery of sugar as a syrup, from misshapen or erroneously formulated sugar confectionery). In such circumstances, provision for such recovery should be made in the MMI, specifying a maximum limit to the rate of incorporation.

'Occasional' instances are all those other than referred to in the foregoing paragraphs. They should in all cases be subject to hazard analysis and critical assessment by the food control manager before any decision as to disposal.

In any re-labelling of packs, any identifying marks carried by the original labels should be carried by the new labels; and where the pack carries a durability indication on the label, the new label should carry a date no later than the original durability indication.

21.11.1 Complaints procedure

The full significance of a quality complaint may only be appreciated with the knowledge of other related complaints. A procedure must therefore be provided for appropriate channelling of all quality complaint reports. The system for dealing with complaints should follow written instructions, which indicate the responsible person through whom the complaints must be channelled. If the responsible person is not the Food Control manager, the latter should be fully informed and closely consulted. The responsible person should have the appropriate knowledge and experience, and the necessary authority to decide the action to be taken.

Where possible, product quality complaints should be thoroughly investigated by the food control manager, with the cooperation of all relevant personnel, and a report prepared as a basis for action and for the records. Action should include responding to the complainant, and must include responding to any regulatory enforcement authority involved. Where the complaint is justified, steps to remove or overcome the cause and thus prevent recurrence should be taken; and the defective material that the complaint sample might represent should be dealt with, including possibly a product recall.

Complaints reports should be regularly analysed, summarised and reviewed for any trends or indication of a need for a product recall or of any specific problem requiring attention. Appropriate summaries including comparative data should be regularly distributed to the company Board and Senior Management.

21.12 Product recall and other emergency procedures

A product defect coming to the manufacturer's attention, whether through a complaint or otherwise, may lead to the need for a product withdrawal from the retail distribution system, or a public product recall also involving return of products by members of the public. In such situations, product identification and traceability of products, their ingredients and their packaging materials are crucially important.

There should be a predetermined written plan, clearly understood by all concerned, for the recall of a product or a known batch or batches of product known or suspected to be hazardous or otherwise unfit, or the withdrawal or recall of wholesome but substandard product which the manufacturer wishes to withdraw or recall. A crisis procedure and management team should be established. A responsible person, with appropriate named deputies, should be designated to initiate and coordinate all withdrawal and recall activities, to liaise with retailers, and to be the point of any contact with the regulatory authorities on recall matters.

The design of manufacturing records systems and distribution records systems, and the marking of outer cartons and of individual packs should be such as to facilitate effective traceability, withdrawal or recall if necessary. A good system of lot or batch marking will pinpoint the suspect material and help avoid excessive withdrawal or recall.

There should be written withdrawal and recall procedures, and they should be capable of being put into operation at short notice, at any time, inside or outside working hours. The procedures should be shown to be practicable and operable within a reasonable time by carrying out suitable testing. They should be reviewed regularly to check whether there is need for revision in the light of changes in circumstances or of the responsible person.

Product withdrawals or recalls may arise in a variety of circumstances, which, however, fall into three main categories:

1. where the regulatory authorities become aware of a hazard or suspected hazard, and information and cooperation from the manufacturer or importer is necessitated;
2. where the manufacturer, importer, distributor, retailer or caterer becomes aware of a hazard or suspected hazard;
3. where there is no hazard or suspected hazard involved, but there is some circumstance (e.g. substandard quality, mislabelling) which has come to light and which prompts the manufacturer, importer or retailer to decide to withdraw or recall the affected product.

In case (3), the company will itself have to organise the withdrawal or recall operation. In cases (1) and (2), consideration may be given to issuing a public Food Hazard Warning. Generally this would be done in consultation among the manufacturer or importer, the distributor or retailer, and any relevant enforcement authority interest. Normally any arrangements for recall would be discussed so that the most appropriate methods could be effected or endorsed by the authorities, and would also take into account any requirements for or arising from the information indicated below.

Although a defect or a suspected defect leading to withdrawal or recall may have come to light in respect of a particular batch or batches or a particular period of production, urgent consideration should be given to whether other batches or periods may also have been affected (e.g. through use of a faulty material, or a plant or processing defect), and whether these should also be included in the recall.

The procedures should lay down precise methods for notifying and implementing a withdrawal or recall from all distributive channels, retailers and of goods in transit, i.e. wherever the affected product might be. It should also include a procedure to prevent any further distribution of affected goods. The recall procedure should also provide for method of public notification.

Notification of recall should include the following information:

- name, pack size and adequate description of the product;
- identifying marks of the batch(es) concerned;
- the nature of the defect;
- action required, with an indication of the degree of urgency involved.

Recalled material should be quarantined, pending decision as to appropriate treatment or disposal.

Emergency procedures should be planned to meet possibility of real or threatened hazard arising from deliberate sabotage or contamination or poisoning of product or ingredients by bioterrorists, extremists or disaffected staff. The first intimation of a specific problem in this area could come from a variety of sources, e.g. complaint from a consumer retailer, the media, the police, the regulatory authorities or employees, or by telephone, e-mail, post or personal contact with any company location or any employee at any time. It is therefore essential that any personnel engaged in manufacture should be aware of company procedures to be followed in dealing with such threats both within and outside normal working hours, and that suitable arrangements exist for calling in key personnel out of hours in such an emergency. The extent to which any such emergency procedures may override normal lines of management should be explicitly stated. Faced with an emergency situation, the recall procedures described above will apply, while the expertise of those involved in food control and other relevant functions should be put at the disposal of the crisis management team responsible for handling the emergency.

The possibility of such sabotage and even site invasion may indicate a need for particular security precautions in vulnerable areas, locked rooms, use of seals, etc. Any emergency or recall situation is likely to involve retailers, wholesalers or food service operators, and a smooth and effective interface with their procedures should be achieved as early as possible during the crisis.

21.13 'Own label' and other contract manufacture

Where complete or part manufacture is carried out as an own-label, private-label, distributor's-own-brand, contract packing or similar operation, except where responsibility is specifically excluded by mutual agreement between the Contract Giver and the Contract Acceptor the obligation is on the Contract Acceptor (the actual manufacturer) to ensure that production is carried out in accordance with GMP in the same way that would be expected were he manufacturing for distribution and sale on his own account. This may be facilitated if the Contract Acceptor is certified under the BRC Global Standard – Food.

The Contract Acceptor should ensure that the terms of the contract are clearly stated in writing and that raw materials and end-products are covered by adequately full specifications (as outlined earlier in this chapter). Any special GMP requirements should be clearly emphasised, and quality control, record transfer, coding, rejection, dispute and complaint procedures should be identified and agreed. Items of possible confidentiality should be identified and any appropriate safeguards be mutually agreed.

It is normal practice for Contract Givers to impose contractual conditions that ensure quality standards and GMP. This is desirably achieved, at least in the first instance, by a visit to the manufacturing unit, by the Contract Giver's Food Control manager. The visit should include the following objectives:

- to ensure that, within the manufacturing environment, the food can be produced safely;
- to agree a detailed product specification covering all aspects of product, process, pack and delivery, embracing parameters to be used for acceptance or rejection, and any legal requirements relating thereto;
- to agree levels of sampling of finished products by the customer and sample plans to be used in case of dispute;
- to evaluate the adequacy of the control resources, systems, methods and records of the manufacturer;
- to agree, wherever possible, objective methods of examination; while subjective measurements should conform to recognised and accepted standards if possible.

Agreement in all five areas is essential for any manufacturer/customer trading relationship and should benefit both parties.

21.14 Good control laboratory practice (GLP)

A control laboratory should be designed, equipped, maintained and of sufficient space to suit the operations to be performed in it, and should have appropriate premises, facilities and staff, and be so organised, as to enable it to provide an effective service at all relevant times necessary to fulfil GMP requirements. This should include provision for writing and recording and the storage of documents and samples, and refrigerated storage for samples, as required. The resources required will depend on the nature of the materials to be tested. It is essential that facilities are appropriate to the needs of the tests, whether chemical, physical, biological or microbiological. Staff should be properly trained, well motivated and well managed. Standards should be set at the highest level and maintained by careful attention to approved and agreed methods and method checks using, where appropriate, reliable outside expertise. Methods should be chosen with care to fulfil the needs of the analyses. For food control purposes, the chosen method should be that most efficacious for the accuracy and speed of results needed, and the skill of the staff concerned. When possible, methods acceptable to any enforcing authority, or which are internationally acceptable, should be used. In all cases method checks need to be incorporated into any analytical scheme to ensure reproducibility, repeatability and operator independence.

Chemical, biological and microbiological laboratories should be separated from each other and from manufacturing areas. Separate rooms may be necessary to protect sensitive instruments from vibration, electrical interference, humidity, etc. Care should be taken to avoid contamination in either direction between laboratories (particularly microbiological laboratories, where access and exit controls should be strictly followed) and manufacturing areas, and reagents or materials that could cause taint should ideally be kept in a separate building. Provision should be made for the safe storage of waste materials awaiting disposal. There should be careful compliance with regulations related to control of hazardous substances. For example in the UK the Control of Substances Hazardous to Health Regulations 1994, SI 1994 No. 3246 as amended by SI 1996 No. 3138 affect the choice of safe laboratory working methods. All methods written up should include an assessment of the hazard of each of the chemicals used in the analysis and appropriate instructions to contain any hazard. If necessary, monitoring of the exposure to hazardous chemicals should be carried out.

Control laboratory equipment and instrumentation should be appropriate to the testing procedures undertaken. Equipment and instruments should be serviced and calibrated at suitable specified intervals by an assigned competent person, persons or organisation. Approved laboratories working to traceable national standards should in turn calibrate measuring equipment and test pieces used in the calibration process. Records of the calibration procedure and results should be maintained for each instrument or item of equipment. These records should specify the date when the next calibration or service is due.

The EU has adopted acceptance of the Organisation for Economic Cooperation and Development (OECD) principles of GLP. EU Directive 2004/9/EC concerns the inspection and verification of GLP.[13]

21.15 Future trends

The factors most likely to impact on GMP in the future are: changes in manufacturing methods; changes in requirements for, and facilities for carrying out, traceability; changes in various countries of the acceptable level of protection (ALOP) demanded by legislators on behalf of society; and further developments in the power of major retailers to make stringent quality demands on their suppliers.

One possible future change of manufacturing method is heralded by a research project currently funded by the UK Department for the Environment, Food and Rural Affairs (DEFRA), that of the 'food factory in a pipe'.[14] The one-year Link Bridge project brings together the Centre for Robotics & Automation at Salford University under Professor John Gray, machine systems experts AMTRI, and Marks & Spencer and Heinz. Their aim is to see if it is possible to put together flexible automated food processing lines using a series of sealed units or modules which link to each other through standard interfaces. Each module would contain its own processing equipment (such as robots) in its own sealed-in processing environment (such as high or low temperature, high or low pressure, modified atmosphere, aseptic conditions, and so on). The modules would link to each other through barriers that prevent contamination. Using standard interfaces, they could then be quickly and easily changed around to reconfigure the 'factory in a pipe' to suit different processing needs. If the concept proves feasible, the researchers hope to invite other industrial partners to take the work forward into a full DEFRA Link project and build a demonstrator plant.

If this concept of the 'factory in a pipe' proves feasible, then the benefits could be enormous – far less factory floor space, fewer people, energy bills slashed, better hygiene control, and the potential for extending shelf-life and for developing innovative new products by processing food in environments hostile to humans. For example the preparation of chilled ready meals in such a system would greatly facilitate hygiene control while chilling just the closed environment instead of a whole factory building and the people in it.

Future traceability systems may involve the effective use of newer techniques, such as radio frequency identification (RFID), already being tried in retailing and likely to expand in use as the cost comes down. However, the problem remains of the use which manufacturers and retailers can make of traceability 'documentation' and this depends in turn on improved data management systems, possibly with use of the Internet.

Traceability has always been an essential component of GMP, but is likely to assume a greater importance in legislators' concepts of what constitutes ALOP. Already in the EU and its member states stringent mandatory traceability requirements have been established for genetically modified food and feed products and ingredients,[11,12] and as from 1 January 2005 such requirements have been extended right across the board for all foods and feeds.[15]

Current and future research on food safety issues (such as *trans* fatty acids, formation of acrylamide in foods, formation of furan in foods) may lead to other changes in legislators' concepts of their ALOP requirements, in turn impacting on aspects of GMP.

21.16 References

1. Title 21 Code of Federal Regulations, Part 110 (21 CFR 110).
2. Quality control procedures for assuring the nutrient content of infant formulas (21 CFR 106).
3. CGMP regulations for thermally processed low-acid foods in hermetically-sealed (air-tight) containers (21 CFR 113), and for acidified foods (21 CFR 114).
4. CGMP regulations for bottled water (21 CFR 129).
5. 21 CFR, Part 110.110 allows the FDA to set maximum levels of natural/unavoidable defects in food for human use that present no inherent health hazard.
6. FDA announcement of modernisations of food GMPs, 6 May 2004; http://www.fda.gov:80/bbs/topics/news/2004/NEW01063.html
7. Blanchfield, J.R., Ed. (1998) *Food and Drink – Good Manufacturing Practice: A Guide to its Responsible Management*, 4th Edition. IFST, London.
8. British Retail Consortium (2003) BRC Global Standard – Food; http://www.brc.org.uk/TechMaster.asp?id=81&sStd
9. World Trade Organization: Agreement on the Application of Sanitary and Phytosanitary Measures, Article 5: Assessment of Risk and Determination of the Appropriate Level of Sanitary or Phytosanitary Protection.
10. Codex Alimentarius: Codex Committee on Food Hygiene, Paper CX/FH 04/5 'Work in Microbiological Risk Assessment/Risk Management' http://www.fsis.usda.gov/OA/codex/fh04_05e.pdf
11. Regulation (EC) 1829/2003 of the European Parliament and of the Council of 22 September 2003 on genetically modified food and feed; http://europa.eu.int/eur-lex/pri/en/oj/dat/2003/l_268/l_26820031018en00010023.pdf
12. Regulation (EC) 1830/2003 of the European Parliament and of the Council of 22 September 2003 concerning the traceability and labelling of genetically modified organisms and the traceability of food and feed products produced from genetically modified organisms and amending Directive 2001/18/EC; http://europa.eu.int/eur-lex/pri/en/oj/dat/2003/l_268/l_26820031018en00240028.pdf
13. EU Directive 2004/10/EC of 11 February 2004 (OJ L50/44) on application and verification of principles of good laboratory practice; http://europa.eu.int/eur-lex/pri/en/oj/dat/2004/l_050/l_05020040220en00440059.pdf
14. Anon (2004), 'The food factory of the future could be in a pipe', *Food Manufacture*, March, 37.
15. Regulation (EC) No. 178/2002 of the European Parliament and of the Council of 28 January 2002 laying down the general principles and requirements of food law, establishing the European Food Safety Authority and laying down procedures in matters of food safety (OJ L 031, 01/02/2002 p. 0001–0024). Article 11, Traceability; http://europa.eu.int/eur-lex/pri/en/oj/dat/2002/I_031/I_03120020201en00010024.pdf

22

The use of standard operating procedures (SOPs)

R. H. Schmidt, University of Florida, USA and P. D. Pierce, Jr, US Army Veterinary Corps

22.1 Introduction: defining standard operating procedures (SOPs)

In addition to following good manufacturing practices (GMPs) as described in Chapter 21, current hygienic and sanitary techniques in a food processing and handling facility also should include developing and implementing standard operating procedures (SOPs). The primary difference between SOPs and GMPs is specificity. While GMPs are general practices applied throughout the facility or system, SOPs are very specific and focus on routine or repetitive activities, tasks, or functions within the facility or system. They provide individuals with very specific and directed information and instructions to effectively perform these functions.

The development and use of SOPs for key functions are an integral part of the overall food handling or processing operation. They may describe technical and/or administrative operational functions of the organization. Effective SOPs promote consistency in implementing processes or procedures (even when there are personnel changes) and may increase efficiency through reduced employee work-load. They also provide a framework for personnel training and minimize the potential for misunderstanding and miscommunication. An additional advantage of well-written SOPs is improved data comparability, credibility, and legal defensibility.

In the US, federal regulatory agencies require specific SOPs for a variety of functions throughout the industrial sector. Certain food industry personnel may already be familiar with the laboratory SOPs required by the Environmental

Protection Agency (EPA)[1] or those under Occupational Safety and Health Administration (OSHA)[2] regarding worker safety.

If the facility is operating under the Hazard Analysis Critical Control Point (HACCP) system, SOPs are usually included as precursory, prerequisite, or foundation programs. Valid SOPs are required under HACCP regulations (see below), or as part of third party HACCP audits or other customer audits. International Organization for Standardization (ISO)[3] accreditation also requires the use of SOPs, but addresses them as work instructions. For example, to receive accreditation under ISO 9000, a company must establish a quality management system that includes developing and implementing SOPs.

The US Army Veterinary Corps uses SOPs as a means to evaluate a food establishment that is producing food destined for the US Department of Defense.[4–6] Under this program, military auditors require and evaluate SOPs for process controls, food sanitation, control of food safety hazards, and quality systems against the Military Standard for compliance. While auditing food processing facilities for compliance with Military Standard 3006A, SOPs are essential to ensure that a facility is working within its outlined parameters as part of their quality management system.

To be effective, SOPs should be carefully written such that they are usable and are followed. Obviously, very general and poorly written SOPs are of limited use. In addition, well-written SOPs are not effective if they are not followed. As a well-implemented SOP program provides an outline of compliance with GMPs as well as other practices, careful planning is needed to design a documentation and tracking scheme into SOPs. In addition, SOPs should be clearly identified and readily accessible for reference in the specific work area to those individuals performing the tasks.

With regard to food hygiene and sanitation, the most usable SOPs are those that detail the work processes and provide documentation framework for those activities necessary to the production of safe food products. SOPs which specifically address sanitation conditions are often referred to as Sanitation SOPs (SSOPs). A partial list of SOP categories that may be appropriate to food sanitation and hygiene are presented in Table 22.1.

22.2 The key components of SOPs and SOP programs

Valid SOPs are important to ensure that sanitation programs in a food handling facility are effective. The use of SOPs is an excellent way to demonstrate proficiency of tasks or functions such as a process controls (e.g. testing and calibration of a pasteurizer or cooker, thermometer calibration), daily sanitation (e.g. cleaning and sanitizing of equipment, hand-washing) or sanitation system (e.g. sanitary design and construction of equipment and buildings). Well-documented SOPs (written and implemented) are a very good indicator to an auditor that an establishment has control over their processes. However, the key

Table 22.1 A partial listing of general categories where standard operating procedures (SOPs) may be used in food sanitation and hygiene programs

1. Facilities	Design, maintenance, and construction Cleaning and sanitation
2. Equipment	Design, maintenance, and construction Cleaning and sanitizing food contact surfaces Cleaning and sanitizing non-product contact surfaces
3. Food	Receiving and storage Ingredients and use Processing and production Distribution
4. Allergen control	
5. Chemical use, storage, and control	
6. Pest management	
7. Extraneous matter control	
8. Consumer complaints	
9. Recall and traceability	
10. Security and tampering prevention	
11. Labeling compliance	
12. Training	

components of these SOPs may vary slightly with the individual requirements of third party auditors and/or military and regulatory agencies.

SOPs should be written as a group or team project with input from all affected individuals with sufficient knowledge and experience regarding the procedures and processes involved. There should be ample time given to analysis and evaluation of the processes and system. Consideration should be given to the number of SOPs required, a description of the main tasks, and the general order or steps. Consideration should also be given to training requirements and needs.

Before discussing the key components of SOPs, it is important to note that each SOP is written as part of an overall sanitation program for specific sanitation conditions and practices. This program also includes:

- monitoring of important SOP conditions and practices;
- correction of those SOP conditions and practices not in conformance; and
- records to document the monitoring and corrections.

22.2.1 General SOP components

The general key components of an SOP include the following.

Identification

SOPs should be clearly identified as follows:

- *Title*. The title should be descriptive and clearly define the activity or procedures. Facilities with several SOPs for different activities should adopt an identification (ID) numbering system for adequate retrieval and reference.

- *Date of issue*. The date of issue, as well as the date of any revision, should be clearly listed.
- *Identification of responsible individual(s)*. The signature requirements for SOPs may vary, but should include the signatures (and signature dates) for the individual(s) who prepared and approved the SOP, as well as for those responsible for daily implementation and their supervisors. A space for regulatory signatures may also be appropriate for those SOPs required by regulation.

Table of contents

A table of contents is optional and may not be necessary for each SOP, depending upon its complexity. If SOPs are collectively kept in a single document, however, a table of contents should be used and is helpful as a quick reference for locating information.

Purpose

Other terminology may be used for this section (e.g. goal, mission, policy). This should be a concise statement of the specific purpose of the SOP.

Scope

A simple statement of the scope of the SOP should be given. What specifically is covered by the SOP?

Regulatory requirements and implications

If the SOP is a part of a prerequisite program required under a HACCP regulation, this should be so noted.

Definitions

If appropriate, a listing of definitions of specialized terms, abbreviations, acronyms, etc. may be included for each SOP. If SOPs are collectively kept in a single document, the definitions could all be combined into one section of the overall document.

Special precautions

If appropriate, SOPs should include a clearly written warning of any risk of personal injury involved with performing the tasks.

Procedures

The procedures section of an SOP is the framework and fabric of the document. Since wordy and lengthy procedural statements and SOP documents are difficult to implement, the format should be concise, step-by-step, and easy-to-read. The following should be considered in writing the procedures section:

- *Individual steps or procedural items*. Each step should define one action or task. Combining steps should be avoided.

- *Number of steps*. As employees generally tend to ignore long SOPs, it is often recommended that SOPs contain not more than 10–12 procedural items or steps. If an SOP goes beyond this number of steps, it may be appropriate to break it up into sub-SOPs or into two SOPs.
- *Language or writing style*. SOPs should be written in the language appropriate to the ethnic background of the workforce. For example, an SOP written in English cannot be effectively followed by employees who do not read English. The most effective SOPs are those written in active (rather than passive) voice and present (rather than future) verb tense. Active voice/present verb tense usually provides stronger directive statements to the person performing the task while using few words. Thus, the SOP is easier to follow. Comparative examples are shown in Table 22.2 for the steps involved in sanitizing food equipment.
- *Sufficient detail*. Procedural steps should be written with sufficient detail such that someone with limited experience, but with basic understanding, of the procedures can adequately follow the procedures without supervision. However, care should be taken when providing highly specific data (e.g. precise temperatures, concentrations) in procedural steps, unless it is absolutely necessary. The provision of precise data implies that these values will be precisely monitored and documented. For example, providing a precise temperature for the detergent solution used to clean a piece of equipment implies that this temperature is precisely monitored and is maintained at that temperature. Depending upon the procedure or condition and how it is to be monitored, it may be advisable to provide a minimum value, a maximum value, or a range of values. The use of general descriptive terminology (e.g. warm, cold, dilute) may be used, but only when precise monitoring is not appropriate or necessary. However, if monitoring is

Table 22.2 Examples of SOP procedural statements written in passive *vs.* active voice and present *vs.* future verb tense

Passive voice/future tense	Active voice/present tense
Equipment will be rinsed with warm water (x to y °F) to remove cleaning solution.	Rinse equipment with warm water (x to y °F) to remove cleaning solution.
A sanitizer solution will be prepared by dissolving x oz of concentrated sanitizer per gal of water.	Prepare a sanitizer solution by dissolving x oz of concentrated sanitizer per gal of water.
The sanitizer solution strength will be checked to ensure proper level (between x and y ppm).	Check the sanitizer strength to ensure proper level (between x and y ppm).
The equipment will be sanitized by rinsing with the sanitizer solution for a minimum contact time of 30 seconds.	Rinse with the sanitizer solution for a minimum contact time of 30 seconds.

necessary to determine the effectiveness of the procedure, these descriptive terms should not be used, as they do not provide enough information for consistent day-to-day monitoring operations. Use of phrases such as *see manufacturer's recommendations* or *according to manufacturer's recommendations* should be avoided in SOPs. An SOP should be a stand-alone document to be adequately followed by employees. Manufacturer's recommended procedures may or may not be readily available to the employee performing the tasks. Thus, it is more appropriate to list these recommended procedures as steps in the SOP.

- *Revisions of SOPs*. It is important that SOPs are kept current and up to date. Whenever significant changes in the processes and/or practices occur, the SOPs should be revised accordingly.

22.2.2 SOP monitoring and record keeping

The old adage, 'if you did not write it down, you did not do it', definitely applies to SOPs. This is especially true where SOPs are required by commercial or regulatory entities. The monitoring and record-keeping program should be formatted in such a way that it is easily implemented. When developing an SOP monitoring program the following must be determined:

- What is to be monitored and how?
- How frequent will monitoring be performed?
- Who will do the monitoring?
- How will it be recorded?

A wide variety of monitoring procedures are used. Monitoring of SOPs is usually subjective and periodic, but could involve continuous monitoring using instrumental devices. For example, typical monitoring procedures used to evaluate an SOP for cleaning and sanitizing of a food contact surface may include: visual observation that the tasks are completed, visual inspection that the surface is clean, chemical testing (e.g. test strips or kits) of solutions used, or verification check (e.g. swab tests).

As much as is practical and practicable, an SOP monitoring program should be designed with the specific SOP and the procedures and practices listed therein in mind. However, it is not necessary and may be redundant to repeat all the SOP steps on the monitoring record. The individual responsible for monitoring may, in some instances, be the same individual as performed the tasks, the supervisor, or both. The monitoring frequency also varies with the requirements of the SOP. For example, an SOP established for the condition and maintenance of food equipment may require monthly monitoring, while monitoring for an SOP established for cleaning and sanitizing equipment would be daily or more frequently.

As with the SOPs themselves, SOP monitoring records should be formatted in a concise, simple, and easy to use style. The following information should be included on an SOP monitoring record:

- minimum expectations or goals of the tasks being monitored (e.g. chlorine sanitizer concentration, clean food contact surface);
- monitoring information (e.g. monitoring method, frequency, action limits);
- roles and responsibilities of employees involved;
- definition of corrective actions to be taken;
- statement of corrective actions, if taken;
- date and time of monitoring;
- signature or initials of person(s) performing the monitoring; and
- signature or initials of supervisor.

A commonly used format for an SOP monitoring record is a simple check-list (or questionnaire) where an employee responds to a series of questions by placing a check mark (or initials) in the appropriate box as yes or no. Another format is to have the individual responsible for monitoring put their initials and date in a box of the table indicating that certain tasks have been completed. It should be noted that SOP monitoring forms or records are not inspection forms or auditing forms. Thus, it is usually not appropriate to ask individuals for subjective judgment. For example, it may not be appropriate to use a ranking or scoring scale (e.g. rating of 1 to 5) or the use of judgment terms (e.g. unsatisfactory, satisfactory) in an SOP record.

SOP records provide documentation of continuous day-to-day practices and should be up to date with actual practices being conducted. Thus, SOP monitoring forms may be revised from time to time as deemed appropriate to provide the most accurate reflection of current practices. Revision of the monitoring forms may also be necessary when the SOPs themselves are revised. Whenever the monitoring form or an SOP is revised, an archive record should be kept on file with a notation of the reason(s) that revisions were made.

The length of time that SOP records should be kept on file varies with the type of product being manufactured, as well as with the requirements of regulatory agencies or customers. In general, SOP records should be kept for a minimum time period which is at least as long as the expected shelf-life of the product.

22.2.3 SOP verification and review

The purpose of verification of an SOP program is to establish that the SOPs and the procedural steps accurately depict what is required for the sanitation condition and practices, and that these steps are being performed as prescribed. If an SOP is being used as a control measure for a food hazard, it should also be validated through appropriate evaluation procedures to demonstrate that it is effective in controlling the hazard.

Verification and review should be done by individual(s) with appropriate training and experience. The ideal situation is that verification be done by someone other than the individual(s) who wrote the SOP. Verification should be done initially or prior to finalizing the SOP and whenever significant changes in the processes and practices warrant revision of the SOP. As SOPs need to be kept up to date, an annual verification (or more frequent) should be done. A record of verification activities should be kept on file as described above for SOP monitoring records.

22.3 SOP requirements under regulatory HACCP programs

SSOPs are included as prerequisite programs as part of US regulatory HACCP programs under the US Department of Health and Human Services (DHHS)/ Food and Drug Administration (FDA) and US Department of Agriculture (USDA)/Food Safety and Inspection Service (FSIS). While the regulations required by these agencies are slightly different, their basic intent in regard to SSOPs is similar.

22.3.1 FDA HACCP regulations

The FDA has promulgated HACCP regulations for seafood and for fruit and vegetable juice. A voluntary HACCP program has also been implemented under the National Conference on Interstate Milk Shipments (NCIMS) program for Grade A fluid milk and fluid milk products. With respect to SSOPs and prerequisite programs, these HACCP programs have similarities, but there are certain subtle differences, which are discussed in the following.

FDA Seafood HACCP regulations

HACCP regulations for fish and fishery products[7,8] were promulgated by FDA in 1995. In addition to requiring implementation of the HACCP system for domestic and imported fish, fishery products and molluscan shellfish, these regulations emphasize compliance with prerequisite programs including:

- current good manfacturing practices (cGMPs);[9] and
- sanitation control procedures

These regulations are summarized below:

- *Sanitation SOP*. Each processor *should* develop and implement written SSOPs.
- *Sanitation monitoring*. Each processor *shall* monitor the conditions and practices during processing with sufficient frequency to ensure conformance with cGMPs that are appropriate to the plant and food being processed, and relate to the following eight sanitation conditions or practices:
  - safety of water that comes in contact with food or food contact surfaces, or is used in the manufacture of ice;
  - conditions and cleanliness of food contact surfaces;
  - prevention of cross-contamination from unsanitary objects to food, food packaging materials and other food contact surfaces;
  - maintenance of hand-washing, hand-sanitizing, and toilet facilities;

- protection of food, food packaging material, and food contact surfaces from adulteration with potentially hazardous chemical, physical and biological agents;
- proper labeling, storage, and use of toxic compounds;
- control of employee health conditions; and
- exclusion of pests.

- *Sanitation control records*. Each processor *shall* maintain sanitation control records, in English, that, at a minimum, document the monitoring and corrections prescribed in the sanitation monitoring program above. These records shall be retained at the processing facility or at the importers place of business in the US for at least 1 year from the date of manufacture for perishable or refrigerated products, and for at least 2 years or the shelf-life of the product (which ever is greater) for frozen, preserved, or shelf-stable products.

FDA Juice HACCP regulations
More recent FDA HACCP regulations, promulgated in 2001, for domestic and imported fruit and vegetable juice and juice products[10,11] are very similar to those for fish and fishery products. A notable exception is that written SSOPs are required rather than recommended (should *vs.* shall) as is the case in the seafood HACCP regulation. These SSOPS are required for the eight key sanitation conditions and practices (listed above) before, during and after processing. Monitoring, record keeping, and record retention requirements are similar to those described above for the seafood HACCP regulation.

Under the Juice HACCP regulations, the FDA recommends implementing *rigorous SSOP* controls to avoid cross-contamination of milk allergens into juice if shared equipment is used for both milk and juice.[12] These SSOPs should include adequate logging and documentation to ensure that such equipment has been cleaned properly before it is used to process juice.

NCIMS voluntary fluid milk HACCP
A voluntary HACCP program has been phased in under the NCIMS Program for fluid milk and fluid milk products.[13] Under this cooperative state/federal program, milk processors have the choice of being inspected and listed under the traditional NCIMS rating system, or they can be listed as a HACCP facility, whereby they must develop and implement the HACCP system, and are audited accordingly by the regulatory agencies. To comply with the HACCP program, each milk plant, receiving station or transfer station *shall*:

- develop and implement prerequisite programs (PPs) (e.g. SSOPs) that address the eight sanitation conditions (listed above). Each PP shall have a brief written description or checklist that the programs can be audited against to ensure compliance;
- monitor the conditions and practices of all required PPs with sufficient frequency to ensure conformance with those conditions that are appropriate both to the plant and the safety of the food being processed;

- correct those conditions and practices that are not in conformance; and
- maintain records that document the monitoring and corrections, and retain these records on file (as described above for seafood and juice).

22.3.2 FSIS meat and poultry HACCP regulations

The HACCP regulations promulgated in 1996 by the FSIS[14,15] modified existing sanitation regulations for meats and poultry. The SSOP requirements under these regulations are summarized below:

- *SSOP development.* SSOPs shall
  - describe all procedures an official establishment will conduct daily, before and during operations, sufficient to prevent direct contamination or adulteration of product(s);
  - be signed and dated by the individual with overall authority upon initial implementation and upon any modification;
  - address, at a minimum, the cleaning of food contact surfaces of facilities, equipment, and utensils; and
  - specify the frequency with which each procedure is to be conducted and identify the individual responsible for implementation and maintenance of the SSOPs.
- *SSOP implementation.* Each official establishment shall
  - conduct the preoperational procedures in the SSOPs before the start of operation;
  - conduct all other procedures in the SSOPs at the frequencies specified; and
  - monitor daily implementation of procedures in the SSOPs.
- *SSOP maintenance.* Each official establishment shall
  - routinely evaluate the effectiveness of the SSOPs and the procedures therein in preventing direct contamination or adulteration of products; and
  - revise the SSOPs, as necessary, to keep them effective and current with respect to changes to facilities, equipment, utensils, operations, or personnel.
- *SSOP corrective actions.* Corrective actions shall be taken
  - when either the establishment or FSIS determines that the SSOPs, or the implementation or maintenance of the SSOPs, may have failed to prevent direct contamination or adulteration of product(s); and
  - that include procedures to ensure appropriate disposition of product(s) that may be contaminated, restore sanitary conditions, and prevent recurrence of direct contamination or adulteration of product(s), appropriate reevaluation and modifications of the SSOPs or appropriate improvements in the execution of SSOPs.
- *SSOP record-keeping requirements.* Daily records shall
  - document the implementation and monitoring of the SSOPs and any corrective actions taken;

  - be authenticated by the initials of the responsible employee for implementation and monitoring of the procedures specified in the SSOPs and dated; and
  - be maintained for at least 6 months and made accessible and available to the FSIS.
- *Agency verification of SSOPs.* The FSIS shall verify the adequacy and effectiveness of the SSOPs including:
  - reviewing the SSOPs;
  - reviewing the SSOP records;
  - direct observation of the implementation of the SSOPs and any corrective actions taken; and
  - direct observation or testing to assess the sanitary conditions in the establishment.

22.4 Common problems in implementing SOPs effectively

22.4.1 Personnel and management issues

Based upon the experience of the authors, the most common issues or problems with regard to implementing SOPs effectively are related to personnel and management issues. Some of the common issues which can lead to problems in implementing SOPs are as follows.

Inadequate team building

Too often, writing SOPs falls upon the shoulders of one individual (usually the quality control or quality assurance manager). The same individual is also usually the person who monitors the SOP, and conducts pre-operational inspections and audits. This creates a 'we *vs.* they' situation between this individual and those expected to perform the tasks and erosion of trust, which hinders effective implementation. It is imperative that a team approach is used and that buy-in is attained of all employees involved. This includes involvement of those actually doing the work. An SOP written by an individual who has never done the task has limited effectiveness in implementation.

Inadequate support by upper management

Without support of upper management, an SOP implementation program cannot be effectively implemented. Management must create a work atmosphere in which every employee from supervisor to line worker has buy-in and empowerment in implementing SOPs with appropriate personnel management and reward programs. A very top-down, directive management system can lead to ineffective SOP implementation.

Inadequate employee time

Another personnel-related problem, which is very common, is that many facilities are under-staffed and are highly production driven. Thus, sanitation and SOP implementation may not be deemed important or, at least, not of as high importance as it should be. Developing and implementing SOPs requires a major time commitment to be effective. Without adequate encouragement and time allotment, the SOP program will fail.

Inadequate training
An effective training program is imperative to effective SOP implementation. Well-written and conceived SOPs can actually be used to facilitate employee training. If insufficient emphasis is given to training, SOP implementation will not be effective.

22.4.2 SOP functional issues

Inadequate monitoring and record-keeping programs
A functional problem, which often impedes effective implementation of SOPs, is that the monitoring and record-keeping program may be cumbersome, ill-conceived, or ineffective. It should be remembered that SOPs are not exciting to all employees and there are those individuals that may not religiously follow the monitoring frequency and simply fill out the forms at the end of the day or less frequently than prescribed. Supervision and appropriate chain of responsibility checks are imperative to effective implementation.

Improper details
As described above, a major challenge in writing SOPs is determining the appropriate amount of detail. Too much wordiness and detail results in SOPs that are not usable in day-to-day operations. Conversely, SOPs that have insufficient detail are not effective and of limited use. It is imperative that the detail level is evaluated during verification and review.

Impact of regulatory and third party audit requirements
While the overall intent and impact of HACCP and SOP regulations and customer audit requirements has been positive, there may be certain negative aspects to these programs. The effectiveness is directly related to how strong the partnership is between the auditor and the food handling entity. One mistake, which is born out of lack of trust or partnership, is to purposely write very sketchy SOPs that contain little general descriptive terms rather than providing data for parameters (e.g. temperatures, concentrations). This is done out of fear that a third party auditor or regulatory official may invalidate the SOP if the monitoring and record-keeping program failed to demonstrate compliance with restrictive prescribed details. The incorrect and short-sighted philosophy at play here is that if you do not give the specific data, the person doing the audit has less to investigate and is less likely to fail you. It is important to determine through discussion with the team, as well as with auditors, where such detail is needed within the goal set for the SOP.

Finally, the very prescriptive nature of regulatory requirements may decrease uniqueness and individual creativity. Excellent regulatory guidance documents and model SOPs are available for a variety of applications. However, these are not necessarily universally applicable. An easy way out approach by industry personnel might be to focus more on the minimum regulatory requirements and these model documents than on taking sufficient effort to develop personalized SOPs for their own operations. As a result, the personnel do not have the necessary growth and team-building processes that are associated with developing specialized SOPs, which can be effectively implemented in their specific operation.

22.5 Sources of further information

There is a burgeoning amount of information available regarding SOPs and SSOPs and their development and implementation. The following is a partial list of publications and internet sources of information:

(a) Environmental Protection Agency
- *Guidance for Preparing Standard Operating Procedures (G-6),* March 2001, EPA/240/B-01/004 (http://www.epa.gov/quality/qs-docs/g6-final.pdf)

(b) Food and Drug Administration (www.cfsan.fda.gov)
- *Fish and Fisheries Products Hazards and Control Guidance – Third Edition, June 2001*
- *Juice HACCP Hazards and Controls Guidance – First Edition, March 2001*
- *NCIMS HACCP Pilot Program Phase II Expansion*
- *USDA/FDA HACCP Training Programs and Resources Database*

(c) USDA/Food Safety and Inspection Service (FSIS) (www.fsis.usda.gov).
- *HACCP Guidance*
- *Key Facts: Sanitation Standard Operating Procedures*
- *Supervisory Guideline for the Pathogen Reduction/HACCP Regulatory Requirements*
- *Guidebook for the Preparation of HACCP Plans*
- *Sanitation Performance Standards*

(d) Seafood HACCP Alliance (http://seafood.ucdavis.edu/haccp/training/training.htm)
- *HACCP: Hazard Analysis Critical Control Point Training Curriculum*

(e) Meat and Poultry HACCP Alliance (http://haccpalliance.org/index.html)
- *Sanitation Standard Operating Procedures* (http://haccpalliance.org/alliance/fsisSept28.pdf)

(f) State Regulatory Agencies – some state agencies have published training materials and documents. For example:

- *Sanitation Standard Operating Procedures (SSOPs)*. Michigan Department of Agriculture, October 1998. (http://www.michigan.gov/documents/mda_CiderMill_8297_7.pdf)

(g) University Cooperative Extension Programs – a wide array of information, documents and materials are available through university cooperative extension programs. For example:
- *Fresh Juice Processing GMPs*, University of Florida IFAS Extension (http://edis.ifas.ufl.edu/FS078)
- *Standard Operating Procedures/Prerequisite Program Checklists,* Iowa State University Coop. Extension (http://www.iowahaccp.iastate.edu)
- *Food Safety Toolkit: Equipment, Facilities & Premises Cleaning Checklists*. Purdue University Coop. Extension (http://www.cfs.purdue.edu/RHIT/foodsafety/Checklists_Cleaning.htm)
- *Food Safety Inspections: Basic Compliance Checklists for GMPs, GAPs, SSOPs and HACCP*. Clemson University Coop. Extension (http://www.clemson.edu/psapublishing/Pages/Foodsc/EC708.pdf)

(h) International Organization for Standardization (*http://www.iso.org/iso/en/iso9000-14000/index.html*)

(i) Trade Association publications
- *Guidelines for Developing Good Manufacturing Practices (GMPs), Standard Operating Procedures (SOPs) and Environmental Sampling/Testing Recommendations (ESTRs) – Ready to Eat Products*. In cooperation with North Am. Meat Processors; Central States Meat Association; South Eastern Meat Association; Southwest Meat Association; Food Marketing Institute; National Meat Association; and American Association of Meat Processors (http://www.nmaonline.org/files/guifinal.pdf)

22.6 References

1. Environmental Protection Agency, *Guidance for Preparing Standard Operating Procedures (G-6)*, March 2001, EPA/240/B-01/004, http://www.epa.gov/quality/qs-docs/g6-final.pdf.
2. US Dept. of Labor, Occupational Safety and Health Administration, www.osha.gov
3. International Organization for Standardization, http://www.iso.org/iso/en/iso9000-14000/index.html
4. United States Medical Command Regulation 40-28; Medical Services *Veterinary Services Standardization and Policies,* 25 August 2003.
5. Army Regulation 40-657, Medical Services *Veterinary Medical and Laboratory Services*, 21 February 2005.
6. Department of Defense Standard Practice, Military Standard 3006A, *Sanitation Requirements for Food Establishments,* 7 June 2002.

7. US Dept. of Health and Human Serv., Food and Drug Admin., Center for Food Safety and Appl. Nutr., Procedures for the Safety and Sanitary Processing and Importing of Fish and Fishery Product; Final Rule, 18 December 1995, *Federal Register,* **60**: 65095–65202.
8. Food and Drug Admin., Center for Food Safety and Appl. Nutr., Fish and Fishery Products. *Code of Federal Regulations* Title 21 Part 123.
9. Food and Drug Admin., Center for Food Safety and Appl. Nutr., Current Good Manufacturing Practices, *Code of Federal Regulations* Title 21 Part 110.
10. Food and Drug Admin., Center for Food Safety and Appl. Nutr., Hazard Analysis and Critical Control Point (HAACP) (sic); Procedures for the Safe and Sanitary Processing and Importing of Juice; Final Rule, 19 Jan. 2001, *Federal Register* **66**: 6137–6202.
11. Food and Drug Admin., Center for Food Safety and Appl. Nutr., Hazard Analysis and Critical Control Point (HACCP) Systems, *Code of Federal Regulations* Title 21 Part 120.
12. Food and Drug Admin., Center for Food Safety and Appl. Nutr., *Juice HACCP Hazards and Controls Guidance* – First Edition, March 2001, http://www.cfsan.fda.gov/~dms/juicgu10.html
13. Food and Drug Admin., Center for Food Safety and Appl. Nutr., NCIMS HACCP Pilot Program Phase II Expansion, May 2001, http://www.cfsan.fda.gov/~comm/daipilo2.html
14. US Dept. of Agriculture, Food Safety and Insp. Serv., Pathogen Reduction; Hazard Analysis and Critical Control Point (HACCP) Systems, Final Rule, 25 July 1996, *Federal Register* **61**: 38856–38906.
15. Food Safety and Insp. Serv., Hazard Analysis and Critical Control Point (HACCP) Systems, *Code of Federal Regulations* Title 9 Part 417.

23

Managing risks from allergenic residues

R. W. R. Crevel, Unilever Colworth, UK

23.1 Introduction

Food allergy has been long recognised as a clinical phenomenon, with numerous reports in the 20th century medical literature.[1–3] However, while it was known that patients could suffer extremely severe and sometimes fatal reactions following ingestion of minute amounts of the offending food, food allergy was perceived as a problem for the individual sufferers. Since 1985, however, this perception has changed, and food allergy is now recognised as an important public health problem. A major factor in this increased concern is probably the rise in the prevalence of atopic disease,[4] of which it can be considered a manifestation. The prevalence and incidence of food allergy and the number of severe reactions[5] may be increasing, although the lack of sound baseline epidemiological data precludes firm conclusions. The new perception of food allergy has been accompanied by the recognition that the solution to the problem lies with collaboration between all the stakeholders, including patients and those who look after them, clinicians, public authorities and the food industry.

The ultimate aim for all stakeholders is to avoid food allergy sufferers reacting to the allergens to which they are sensitised. This can be achieved in two ways. One is to ensure accurate allergen declaration through labelling, the other is to ensure that where a specific allergen is not declared, the product does not contain it in an amount that would pose a risk and food allergy sufferers can assume it is safe for them. Both these requirements can only be fulfilled by detailed knowledge of the composition of products. Food manufacturing processes are extremely complex. This complexity derives from several factors, including material sourcing, processing, efficient use of equipment and other resources, and product formulation. Managing allergen risks requires an integrated approach, which takes into account all these factors throughout the supply chain, from ingredient suppliers through to retailers, and ultimately the consumer.

Total avoidance of cross-contact and therefore absence of specific allergens from products where they are not part of the formulation is often not practicable. Such circumstances require an analysis of the risk arising from residual allergen, and subsequently a quantitative risk assessment. Although knowledge of minimum provoking doses for many allergens is inadequate, knowing how much allergen is present in a product is a key element in this assessment, and the subsequent management of the allergen risk. Allergen detection methods can play an important role at several points in the analysis of this risk. These include the initial analysis phase, in which the current risk is determined, and the validation of specific risk management procedures, such as line cleaning. Upstream of the food manufacturer, analytical methods can be used as part of the supplier audit process. Detection of allergen also plays a role in investigations of incidents, and compliance with process standards. This chapter will discuss the role of allergen detection methods, but will not address details of individual methods, except where these could have a bearing on the specific use of such methods.

23.2 Food allergy and product safety

Before considering the role of detection methods in ensuring the safety of food allergic consumers, it is useful to take a broader view of food allergy as a safety issue. A first consideration with food allergy is that the nature of the hazard differs from that of other toxicants. Indeed schemes for the classification of adverse effects of food usually distinguish it from toxic reactions, which potentially affect anyone who eats the food.[6] Specifically, food allergy affects only a defined section of the population, and food allergens present no risk to non-allergic persons, irrespective of the level of intake. However, while risk assessment in food allergy must therefore focus on the specific population at risk, it is still valuable to analyse it through the accepted framework of hazard identification, hazard characterisation, exposure assessment and risk characterisation.

The first step, hazard identification, is encompassed in the definition of the problem, namely food allergy. The hazard under consideration is any reaction to a food mediated by the immune system, although for practical purposes it extends only to those responses in which antibodies of the IgE class are implicated. Intrinsic in the accepted definition of allergy is the concept of clinical reactivity. It thus excludes situations where people are only sensitised, as revealed by skin prick testing or measurement of specific IgE, but do not otherwise react to contact with the food.

Risk characterisation consists of establishing the relationship between the dose of a material and the response it produces. In conventional toxicology, this is usually achieved by experiments in suitable animal species. These provide information about dose–response relationships and result in the definition of a no-observed adverse effect level (NOAEL) from which a safe dose for people is calculated. Hazard characterisation in food allergy differs from this situation, firstly because animal experimentation is irrelevant, and secondly because it can be viewed at both an individual level and a population one. The population dimension is probably the most relevant from the public health point of view, and consequently for the food manufacturer, while the individual dimension is most relevant to the clinician advising a patient on management of the condition.

For ethical reasons it is extremely difficult to obtain information about individual responsiveness to different doses of allergen. Full characterisation, up to the point of the most severe reaction can only occur inadvertently, where the dose increment used proves to be too large. Characterisation of the population response to food allergens is, however, feasible and ethical, in the context of helping patients manage their condition. This is achieved through studies using double blind placebo-controlled food challenge (DBPCFC) with increasing doses of allergen up to the lowest dose producing an objective response. Such experiments provide information on the frequency of response to particular doses in the population tested. However, such studies have a number of limitations as tools to identify a precise NOAEL. Firstly, because of their logistics, they can be performed only in relatively small numbers of allergic individuals, which limits their statistical power. In a typical clinical study, based on binomial distribution, 29 or 58 individuals would be challenged, and it can be shown that, statistically, such numbers give 95% confidence that fewer than 10% or 5% respectively of the population from which they are drawn would react to the dose identified as the NOAEL.[7] A further limitation is that individuals who have experienced a severe reaction to the allergen of interest are excluded from the studies by some clinicians. A modelling approach to overcome such problems has recently been proposed.[8] Although this approach is proving promising, much validation work is still required before it can be used in risk management.

Once the hazard has been characterised, and a NOAEL defined, exposure assessment is required. For individuals with a food allergy, the risk arises from acute ingestion of often quite small amounts of allergen, rather than ingestion over a period of time. The usual measure to consider for purposes of risk assessment is the amount of allergen that could be present in a portion of food. An important unresolved issue exists, however, with respect to the period over which the intake of allergen should be summed, as well as the possible effect on the provoking dose of exposure to small amounts, unable by themselves to stimulate an allergic reaction. Indeed some publications have attributed increased reactivity to exposure to such doses via other routes than the oral one.[9]

23.3 Management of food allergy risks

23.3.1 Management of food allergens: the aims

The first aim of food allergen management must clearly be to protect food-allergic consumers, while not limiting their food choices unnecessarily. This implies a risk assessment, as described above, as the alternative would be to use precautionary labelling for almost all products in some instances. Indeed the role of detection methods for allergenic residues is predicated upon a risk-based approach to management of food allergens in food manufacturing. If allergens were declared under all circumstances, irrespective of the risk they pose, there would be no need for methods to detect residual allergen. The allergic consumer would simply be informed of the presence of particular allergens, and left to manage the risk individually. However, this approach is viewed by allergic consumers as an abdication of responsibility, and much disliked.[10,11] Furthermore, rather than safeguarding allergic consumers, it can actually place them at increased risk by leading them to erroneous conclusions about the safety of products. From the manufacturer's point of view, managing allergens on a risk basis means effectively taking a view as to what proportion of the allergic population it is feasible to protect, based on knowledge of NOAEL for individual allergens, coupled with an assessment of achievable residual allergen content for particular products. Defining the aims of allergen management is important not only in setting process control objectives, but also in providing a basis for clear communication with stakeholders, such as allergic consumers, and health practitioners who they consult for advice. For instance, decisions must be made about whether the policy aims to avoid all reactions in allergic individuals or just severe ones. The implications of achieving this aim must also be evaluated in a wider socio-economic context. For instance, more thorough cleaning procedures may result in undesirable environmental consequences, or the introduction of water into dry systems may introduce a microbiological hazard. The range of responsiveness of allergic people is extremely wide and some react to extremely low doses. Protection of such consumers may only be achieved by advising them not to consume manufactured foods. If so, what does that mean for assays?

23.3.2 Integrated approach

Current approaches to the management of the allergen risk in the food industry recognise that it has to be integrated into the whole product life cycle from its design right through to the point at which the consumer eats the product. It is within that integrated approach that the role of allergen detection methods must fit. Major food manufacturers have devised specific corporate policies for the handling of allergens, supplemented by guidelines which provide practical advice to individual manufacturing units. These methods ensure that a high minimum standard exists for the handling of allergens throughout the company. For instance, Unilever has a policy for dealing with allergens which states that it shall declare the presence in its products of any allergen that is a common cause of allergic reactions. At a minimum, any allergen required by local regulations will be declared. However, beyond that, the allergenic risk from foods not commonly known to be allergenic may be assessed if clinical or epidemiological data indicate the need. If then classed as a common cause of allergic reactions, this food component would be declared on labels and included in HACCP plans. Unilever also undertakes to inform any consumer on request about the presence of uncommon allergens in specific products.

Allergen management guidelines need to ensure that allergens are correctly and intelligibly declared in products, but also to make sure that allergen is not present inadvertently at levels likely to cause adverse effects on health. Such guidelines specifically need to address all stages in the product life cycle, from its design, through the sourcing of ingredients to manufacture, labelling and distribution. Specifically, it needs to deal with the following:

- *Innovation*: is the use of the allergenic ingredient necessary for the functionality of the product or could an equivalent non-allergenic ingredient serve as well?
- *Supply chain*: control of allergens in the supply chain requires a close relationship with suppliers, so that they understand our needs and can meet our requirements. Typically, the starting point of the supplier assessment will be a questionnaire about allergens handled and precautions in place to avoid cross-contact, including the existence of a HACCP plan. This is backed up by periodic audits of the suppliers' facilities. Additionally, suppliers are required to seek agreement to any change in the formulation of the ingredient they supply.
- *Manufacturing protocols* are another critical element. Main considerations are the inclusion of common allergens in HACCP plans, production scheduling to minimise cross-contact, validated cleaning procedures and clear labelling and separation of specific allergenic ingredients within the factory. Procedures need to cover rework, where sound product is not packaged but 'recycled'. Staff training to understand the importance of allergen control procedures is vital and improves support for what can be additional procedures in the production process. Finally, the same degree of attention is needed whether the company's own manufacturing facility is concerned or that of co-packers.
- *Packaging, promotion and advertising*: packaging carries the label and therefore the allergen information. Care is required to ensure that information remains with the product until it reaches the consumer. Other considerations include warnings if the formulation has changed to include an allergenic ingredient previously not present.
- *Retailers*: generally, the manufacturer's allergen information will be sufficient. However, situations such as in-store promotions require care to ensure that the consumer is fully informed. Sound product, which fails to meet all standards for general sale, may be repackaged and sold on in specialised outlets or even in a different market. The manufacturer needs to

ensure that appropriate allergen information is retained and available to the ultimate consumer.

- *Food professionals*: most allergic reactions to foods occur outside the home, in conditions where the product is often not labelled and even when asked, food professionals fail to provide correct information. Where pre-prepared food is provided to that sector, the manufacturer has a responsibility to ensure that accurate allergen information is provided and conveyed to the consumer.

23.3.3 Role of allergen detection in the integrated approach

Consideration of the various phases of the product life cycle reveals a number of points where the detection of allergenic residues has a role. These will include all the points at which there is uncertainty about the presence of allergens or their levels. By definition detection applies only to allergens that could be present inadvertently and will therefore apply to the manufacturing stages and any upstream of those stages. Detection of allergenic residues will have a relatively limited role at the innovation stage, although it may be useful in making a choice between suppliers of an ingredient. As discussed, assessing the risk arising from inadvertent allergen presence begins with ingredient suppliers, and clearly measurement of residual allergen levels can provide valuable information.

As mentioned earlier, the use of detection methodology in allergen risk management must be guided by the objectives of the policy. Total elimination of the allergen risk, in the sense of a guarantee that no allergic individual may be affected, irrespective of their sensitivity or the severity of reactions they experience, is rarely, if at all, possible, unless specific allergens are excluded from manufacturing facilities. A key element in deciding what needs to be achieved in order to afford a specified level of protection to the allergic population is the minimum dose which provokes a reaction in such individuals (threshold). Data on such doses are unfortunately still scarce, and subject to much debate (see, e.g., ref. 9), even in the case of the most common food allergens. They can be difficult to use confidently in management of risk allergens, particularly since the uncertainties surrounding their derivation are difficult to quantify. Recently attempts have been made to investigate the distribution of such doses in the population, and to estimate by mathematical modelling the level below which residual allergenic protein in a food must be kept in order to protect a specified proportion of the allergic population.[8] As work progresses on defining such levels, they will provide more effective ways of monitoring the success of risk management measures. Methods for the detection of residues will thus increase in importance in such roles as confirming that products and ingredients meet set specifications, and the validation of risk measures such as cleaning. They will also provide, of course, the basis for assessment of compliance by authorities.

23.4 Role of allergen detection and other considerations

23.4.1 Why do we need detection methods?

Methods for the detection of allergenic residues can be deployed for a variety of uses. In industry, these will include what is effectively the exposure assessment part of risk assessment. Typical activities would be assessment of the extent of cross-contact at different points, as part of a HACCP study and subsequently validation of the measures put in place to control the extent of low level homogeneous cross-contact. Extending up the supply chain, such methods could also be used to confirm suppliers' statements about their ingredients, as part of the audit of their processes, while downstream, product analysis could be envisaged where incidents have occurred, or there is a suspicion that allergenic residues may exceed specification. Similarly, confirmation that residues are present, and in what amount would be an important starting point for investigation of incidents. However, industry is not the only potential user of detection methods. Public authorities need to provide evidence to support compliance activities, and demonstration of the presence of residual allergen in products which are not supposed to contain them can form a strong part of such evidence. Allergic consumers may also be potential users of such methods, although none is currently suitable for this type of application.

Different users probably require different methods with different characteristics, with respect to detection limits, quantitation, robustness and ease of use. Risk assessment activities imply quantitative evaluations, and require methods that measure accurately and reliably the residues of interest, even in complex matrices. In HACCP studies, although desirable, it is probably not essential for the method to be easy to use. In contrast, enforcement authorities will only in practice be interested in quantitation if the relevant regulations specify an action level. If no level is specified, it would presumably be sufficient for a method to have an adequate detection limit and to be known not to produce false positives. For potential allergic users, a key requirement is no false negatives, as well as an adequate limit of detection.

23.4.2 What should assays for allergenic residues detect?

The allergenic activity of a food usually depends on a range of proteins, and it has been shown many times that the pattern of response of allergic people to the different proteins can differ considerably.[12] It has also been shown more recently both in allergic patients[13] and in experimental animals[14] that the overall response to an allergenic food is a summation of the responses to the individual proteins. The implication of those observations is that immunoassays for food allergens should essentially be considered as means of quantitating the relevant protein(s), rather than measuring allergenic activity in the food, which will differ for each allergic patient. Another implication is that quantitation of single allergenic proteins may be valuable if one is trying to monitor the effect of processing on such proteins, but may give highly misleading results if used in an assay directed at other purposes, such as the estimation of the extent of cross-contact, or establishing whether a product contains more allergenic material than a set limit. The key consideration with respect to assay development should therefore be the purpose of the assay. The allergenic protein, although the obvious candidate, may not always be the optimal choice. However, detecting the protein is probably the most common approach and it is therefore appropriate to discuss the options available.

In developing an assay based on protein detection, two main choices of methodology exist: monoclonal and polyclonal antibody technology. Both have advantages and drawbacks. As monoclonal antibodies recognise single epitopes on proteins, this technology usually results in a highly specific assay, with a relatively low incidence of cross-reactivity, even with closely related proteins, provided the antibodies have been correctly screened. Theoretically there is also an endless supply of antibody with exactly the same performance characteristics as the original antibody. However, the narrow specificity of the monoclonal antibody can also be its Achilles' heel, as detection of the protein of interest will take place only if the antibody binding site remains intact and accessible in the various food matrices in which the protein may be present. Polyclonal antibody technology relies on the production of a range of antibodies, either to a single protein of interest, or to all the proteins within the food, depending on the preparation used. This provides a detection system that is less likely to fail completely to identify the presence of proteins of interest, although quantitation may remain problematical, if processing alters the relative proportions of the different immunochemically active proteins in a food. However, the main drawbacks of polyclonal technology lie in the need for extensive purification procedures that may need to be applied to the protein(s) of interest, as well as to the resulting antiserum to ensure specificity and absence of unwanted cross-reactivity, together with the need to develop procedures to ensure batch to batch reproducibility.

Once the technology itself has been selected, there remain several possibilities in developing immunoassays based on protein. As discussed, monoclonal technology results in a highly specific detection system, but it can nevertheless be broadened by using a combination of detection antibodies against different epitopes of the same protein, and/or different proteins. However, beyond a few proteins, this becomes complex to optimise. Polyclonal technology leaves open the choice of the material against which the antiserum can be raised. Thus it can be as general as a protein extract of the whole food, or as specific as a highly purified protein.

Detection of the protein, or proteins, may not be necessary or even the method of choice in all instances where detection of allergenic residues is sought. Instead a marker molecule, for which a robust and sensitive analytical method exists, and which is always found in a known ratio to the proteins, can be used. An example would be lactose in milk which could be used as a tracer for estimating the amount of milk protein left on a line by cross-contact. Similarly, a marker compound could be used in supplier audits. However, only measurement of the protein could help an allergic individual decide whether a particular product is safe to eat. Similarly, compliance activities by food safety authorities are likely to be required to demonstrate directly the presence of the offending allergen, rather than a marker material.

The reverse transcriptase polymerase chain reaction (RT-PCR) assay has recently become the subject of considerable interest, and is described elsewhere in this book. It takes advantage of amplification of any relevant (species) DNA present in the food of interest to the point where it can be detected. It can also be used in a mode which can be considered semi-quantitative. However, in relation to allergens, it relies on an implicit assumption that the presence of DNA in a processed food denotes the presence of (allergenic) protein, a contention that needs experimental justification.

23.4.3 Limit of detection

The limit of detection of any assay is an important parameter, but it needs to be considered together with the other parameters that make up the assay. Again, the purpose of the assay should dictate this factor, as it does the others. For most purposes, such as monitoring the effectiveness of allergen control measures, or verifying compliance with set limits, it would be reasonable for the detection limit to be such that the assay could detect allergenic material in an amount in a portion of food that was close to the lowest amount shown to provoke some reaction under controlled clinical conditions. This, of course, begs the question as to what those levels are, and to what extent any uncertainty in determining minimum provoking doses should also be allowed for. Minimum provoking doses vary considerably among individuals[15,16] and Bindslev-Jensen *et al.*[8] have recently described their cumulative frequency versus log-normal dose in the population of allergic patients as a sigmoidal log-normal plot. The implication of this distribution is that there is a small proportion of individuals who will respond to very small amounts of allergen. Assays could be developed to detect such small amounts, but, even leaving aside possible technical issues of signal to noise ratio, it is questionable whether such sensitivity is actually required, except for forensic purposes. A recent suggestion, based on clinical findings,[17] proposed that measures to minimise the inadvertent presence of allergen in industrial food manufacture should aim to protect 95% of the allergic population. On that basis, the lowest levels that residual allergen should not exceed were of the order of 5 ppm protein, a requirement that most current commercially available assays meet, assuming minimal losses during extraction. It should also be borne in mind that the concept of a valid lower detection limit is only meaningful in the context of an assay developed to detect a representative range of proteins in a food product. Assays developed to detect a single protein could be very difficult to interpret in these circumstances, for reasons already discussed.

23.4.4 Characteristics of the ideal allergen detection assay

Having considered various parameters of assays for the detection of allergenic residues, it is appropriate to examine what an ideal assay might look like. However, simply in formulating such a question, one is inevitably drawn to question whether there can be a single 'ideal assay' equally suitable for all purposes. The 'ideal assay' concept will therefore be examined in some of the different contexts in which it might be used.

Assays for monitoring effectiveness of allergen risk management measures
A typical application of such an assay would be to assess effectively different line cleaning measures, for instance. A primary requirement would be that it is sensitive enough to detect relevant levels of allergen, and that it possesses adequate accuracy, providing a true measure of the analyte. A high degree of precision is not as important as accuracy, however. High specificity is also a requirement, as food products will often contain a number of proteins from different sources. Detection of irrelevant proteins could lead to inappropriate decisions with regard to allergen management, implementing more stringent measures than necessary, or leading to defensive labelling, if the analysis suggests the problem cannot be overcome. However, in addition to the appropriate technical parameters, the assay also needs to be designed for those who will be the primary users. In the factory environment, these will not necessarily be people with specialised laboratory training. The assay design must therefore be sufficiently robust for use by non-laboratory personnel, and formats such as calibrated test strips, for instance, are worth investigating. Similarly, experience of the development of in-home diagnostic assays used for clinical monitoring may hold valuable lessons.

Assays for measuring residual allergen in finished products
Although measuring residual allergen in finished products is not a common application of assays, there are occasions when it is necessary. These may be, for instance, where a process failure leads to the conclusion that there is a high probability of residual allergen being present at a level likely to endanger allergic individuals, or where there have been reports of reactions to the product. While this type of application requires of an assay many of the same properties as described for monitoring assays, the assay is likely to be performed by laboratory personnel, rather than in a factory, and therefore 'user-friendliness', while always desirable, is of a lower priority.

Assays for investigating compliance
Where limits have been set on the residual allergen amounts in food products, the assay must meet more stringent requirements in some respects than in the previous instances. Sensitivity and specificity are key elements as before, while precision is more important than for other assays, given the potential legal implications of the test detecting allergen. In particular, the assay should have a low incidence of false positives.

Assays for measuring single allergenic proteins
Many assays have been developed for single allergenic proteins. While these can sometimes serve for the purposes described above, they have many

shortcomings in this regard, as discussed earlier. However, they could have a role in evaluating the effects of processing, for instance, on the protein of interest. Such assays will essentially be highly specific, and will probably also be sensitive. However, they will be used almost exclusively in the laboratory, and therefore simplicity of operation will not be a primary design consideration.

23.4.5 Common limitations

Most of the assays discussed above have a number of limitations, some of which are inherent in the methodology, while others result from particular combinations of methodology and the substrate in which the analyte is sought. The most significant limitations relate to extraction of the analyte from the food for analysis, interference by other components of the matrix, which cannot be readily separated, and changes in the analyte itself, which reduce the ability of the method to detect it.

Variability in extracting the analyte from the food

Most common methods, including many of those for total protein analysis (e.g. Bradford, bicinchoninic acid (BCA)) and enzyme-linked immunosorbent assay (ELISA) methods, operate in an aqueous environment and require extraction of the protein prior to analysis. The efficiency of this extraction will depend on the solubility of the protein(s) of interest in the aqueous buffer used for extraction. Many foods and food products include lipids as part of their formulation, and many proteins, including allergens, are associated in the food or product with the lipid component. In experiments to measure the total residual protein content of edible oils, we consistently recovered only 50% by extraction into phosphate-buffered saline, based on a comparison with the content measured by excited nitrogen analysis, which does not require extraction. This effect can be difficult to detect. In the example quoted, recovery of protein spiked into oils was virtually quantitative, presumably because of differences in their physico-chemical properties, compared with the proteins remaining in oils after refining. In a different context, Keck-Gassenmeier *et al.*[18] found very low recoveries (2–3%) of peanut protein added to chocolate products, but were able to improve this to near-quantitative recovery by the addition of fish gelatine to the extraction buffer. These experiences indicate the need for a thorough knowledge of the physicochemical characteristics of both the matrix and the protein(s) of interest in order to obtain reliable results.

Matrix interference

As well as interfering with the recovery of the analyte(s) of interest, the food matrix, or some of its components, may actually interfere with the subsequent assay, if those components are co-extracted in sufficient amounts. For instance, we have found that on occasions, solutions with very high sugar content (although within the range used in several foods) reduced considerably the recovery of β-lactoglobulin (unpublished results). Other materials commonly used in foods, such as colours could obviously also interfere with the performance of assays based on colorimetric end-points, depending on their fate during extraction.

Changes to proteins due to processing
Food processing probably poses the greatest challenges with respect to allergen detection, particularly for the most common type of assay, namely immunoassays. Processing can alter either the allergenicity of a protein, or its ability to be detected in the food matrix, either because of changes in immunoreactivity or in the interactions between the protein and the matrix, or indeed both. Thus, fermentation of milk with certain strains of lactobacilli reduces the IgE-binding capacity of the product compared with native milk, suggesting a reduction in its ability to provoke reactions. Under these circumstances, detection of lower amounts of milk protein would be a true reflection of a reduction in hazard to the allergic individual. A similar situation occurs in the case of the apple allergen, which is known to be heat-labile.[19] However, assays can also significantly underestimate the content of heat-treated milk proteins, as a result of what must be assumed to be altered recognition of the protein analyte, since a total protein assay yielded total recovery (unpublished results). Similar findings were reported by Koch *et al.*[20] for roasted peanut proteins. Clinical data on reactivity to heated milk and peanut proteins suggest that under those circumstances the apparent reduced protein content does not reflect a reduction in hazard to the allergic person. These examples illustrate the need for a thorough understanding of the pitfalls of an assay before it is used to generate data that will be used in risk assessment.

23.5 Future trends

The need for detection of allergenic residues has now been established as the importance of food allergy as a public health problem has become acknowledged. Several current trends are likely to influence the development and application of allergen detection. One is the developing legal framework, which will ultimately lead to defined action levels. Another is the determination of NOAELs for many of the main allergenic foods. A third may be the need for ways of monitoring allergenicity of particular foods, as they are modified to reduce their allergenicity. Finally, while it not currently feasible, pressure from allergic people and their support groups for the means of monitoring foods for the presence of cross-contact allergens may lead to development of some rapid assays. The likely influence of each of these trends will be examined separately.

The legal framework with respect to food allergens is developing fast, with Switzerland, Japan, Australia/New Zealand and the European Union bringing in legislation specifying which allergenic foods must be declared. The lists are usually based on the list of allergens in the Codex General Standard on Labelling, but extend it to cover allergens of regional importance such as celery and mustard in the European Union. This legislation can be anticipated to drive food manufacturers to use test kits much more extensively to demonstrate for legal purposes that their allergen risk management procedures are effective. Although allergen testing has not proved to be the primary mechanism of enforcement in some legislatures where it has a longer history, such as the USA, enforcement authorities will undoubtedly seek to use them to support other evidence. Except for Switzerland, current legislation does not address the issue of allergen presence through cross-contact, and action levels have not yet been set. However, as allergen test kits become used to a greater extent, pressure is likely to grow, particularly from manufacturers, for defined action levels, below which the presence of the allergen would not constitute an infringement of the law. If action levels are not set by the agencies or the legislators, they will probably be defined by case law, which is probably not an ideal mechanism for this type of issue.

Determination of NOAELs and their use will provide manufacturers with defined targets for their allergen management policies, in terms of what amounts constitute a risk to what proportion of food allergy sufferers. They will also provide manufacturers with information for improved control of allergen hazards. Such control will, however, require that they know what level of allergenic residues are present in their products. Measurement of allergenic residues at appropriate points during the manufacturing process will be one way to obtain this information and could therefore increase considerably from its current relatively limited use.

Monitoring the allergenicity of certain foods or food products is another area where detection of allergenic residues could play an increasing role, as manufacturers seek to provide foods with reduced allergenicity. However, this area is likely to have a lower impact than the previous two, as looking for residual allergenicity by protein quantification is only one of several steps in defining reduced allergenicity.

Food allergy significantly impairs quality of life for sufferers.[21,22] Greater control over their condition by food allergy sufferers would undoubtedly help restore some of this quality. Demand for means to do so could spur an extension of the measurement of allergenic residues to this totally new area. This prospect is probably still quite distant, inasmuch as it requires methods that are simple to use and robust. A critical question will be the extent of the test manufacturer's legal liability in the event of an allergen not being detected and producing a reaction in a sufferer.

23.6 References

1. PRAUSNITZ, C., KÜSTNER, H. (1921) Studien über die Ueberempfindlichkeit. *Centralbl Bakteriol Abt Orig* **86**, 160–169.
2. LOVELESS, M.H. (1950) Milk allergy: a survey of its incidence: experiments with masked ingestion test. *J Allergy* **21**, 489–499.

3. TUFT, L., BLUMSTEIN, G.I. (1942) Studies in food allergy II. Sensitization to fresh fruits: clinical and experimental observations. *J Allergy* **13**, 574–581.
4. LEWIS, S., BUTLAND, B., STRACHAN, D., BYNNER, J., RICHARDS, D., BUTLER, N., BRITTON, J. (1996) Study of the aetiology of wheezing illness at age 16 in two national British birth cohorts. *Thorax* **51**, 670–676.
5. SHEIKH, A., ALVES, B. (2000) Hospital admissions for acute anaphylaxis: time trend study. *BMJ* **320**(7247), 1441.
6. BRUIJNZEEL-KOOMEN, C., ORTOLANI, C., AAS, K., BINDSLEV-JENSEN, C., BJORKSTEN, B., MONERET-VAUTRIN, D., WUTHRICH, B. (1995) EAACI position paper. Adverse reactions to food. *Allergy* **50**, 623–635.
7. AMERICAN ACADEMY OF PEDIATRICS, COMMITTEE ON NUTRITION (2000) Hypoallergenic infant formulas, *Pediatrics* **106**(2), 346–349.
8. BINDSLEV-JENSEN, C., BRIGGS, D., OSTERBALLE, M. (2002) Can we determine a threshold level for allergenic foods by statistical analysis of published data in the literature? *Allergy* **57**, 741–746.
9. STEENSMA, D.P. (2003) The kiss of death: a severe allergic reaction to a shellfish induced by a good-night kiss. *Mayo Clin. Proc.* **78**, 221–222.
10. SAID, M., WEINER, J.M. (2004) 'May contain traces of . . .': hidden food allergens in Australia. *MJA* **181**, 183–184.
11. KOSA, K.M., CATES, S.C., POST, R.C., CANAVAN, J. (2004) Consumers' attitudes toward labeling food products with possible allergens. *Food Prot Trends* **24**, 605–611.
12. DE JONG, E.C., VAN ZIJVERDEN, M., SPANHAAK, S., KOPPELMAN, S.J., PELLEGROM, H., PENNINKS, A.H. (1998) Identification and partial characterization of multiple major allergens in peanut proteins. *Clin Exp Allergy* **28**(6), 743–751.
13. LEWIS, S.A., WARNER, J.O., HOURIHANE, J. (2003) Promiscuity of IgE binding to peanut allergens correlates with clinical reactivity to whole peanut during double-blind placebo controlled challenge. Poster 136, presented at the XXIInd Congress of the European Academy of Allergy and Clinical Immunology, Paris.
14. KNIPPELS, L.M.J., PENNINKS, A.H., BANNON, G.A. (2003) The sensitizing potential of peanut proteins in four different mice strains. Poster 681, presented at the XXIInd Congress of the European Academy of Allergy and Clinical Immunology, Paris.
15. TAYLOR, S.L., HEFLE, S.L., BINDSLEV-JENSEN, C., BOCK, S.A., BURKS, A.W., CHRISTIE, L., HILL, D.J., HOST, A., HOURIHANE, J.O., LACK, G., METCALFE, D.D., MONERET-VAUTRIN, D.A., VADAS, P.A., RANCÉ, F., SKRYPEC, D.J., TRAUMAN, T.A., MALMHEDEN YMAN, I., ZEIGER, R.S. (2002). Factors affecting the determination of threshold doses for allergenic foods: how much is too much? *J Allergy Clin Immunol* **109**, 24–30.
16. CREVEL, R.W.R., KERKHOFF, M.A.T., KONING, M.M.G. (2000a) Allergenicity of refined vegetable oils. *Food Chem Toxicol* **38**, 385–393.
17. MORISSET, M., MONERET-VAUTRIN, D.A., KANNY, G., GUÉNARD, L., BEAUDOUIN, E., FLABBÉE, J., HATAHET, R. (2003) Thresholds of clinical reactivity to milk, egg, peanut and sesame in immunoglobulin E-dependent allergies: evaluation by double-blind or single-blind placebo-controlled oral challenges. *Clin Exp Allergy* **33**, 1046–1051.
18. KECK-GASSENMEIER, B., BENET, S., ROSA, C., HISCHENHUBER, C. (1999) Determination of peanut traces in food by a commercially-available ELISA test. *Food Agric Immunol* **11**, 243–250.
19. VIETHS, S., AULEPP, H., BECKER, W.-M., BUSCHMANN, L. (1996) Characterization of labile and stable allergens in foods of plant origin. In *Food allergies and intolerances: proceedings of a symposium, Bonn, May 1995*. Weinheim: VCH Publishers, 130–149.

20. KOCH, P., SCHAPPI, G., POMS, R.E., WUTHRICH, B., ANKLAM, E., BATTAGLIA, R. (2003) Comparison of commercially avilable ELISA kits with human sera-based detection methods for peanut allergens in foods. *Food Add Contam* **20**, 797–803.
21. AVERY, N.J., KING, R.M., KNIGHT, S., HOURIHANE, J.O. (2003) Assessment of quality of life in children with peanut allergy. *Pediatr Allergy Immunol* **14**, 378–382.
22. SICHERER, S.H., NOONE, S.A., MUNOZ-FURLONG, A. (2001) The impact of childhood food allergy on quality of life. *Ann Allergy Asthma Immunol* **87**(6), 461–464.

24

Managing contamination risks from food packaging materials

L. Raaska, VTT Biotechnology, Finland

24.1 Introduction

The pulp, paper and board industry supplies basic raw materials for packaging and hygiene applications. The total output of paper and paperboard in 2002 was approximately 90 million tonnes in Europe, which is one-third of 2002's world production. Fibre-based food contact materials produced were about 10 million tonnes, which is about 11% of the world production.

Packaging serves as a major defence against external hazards. However, undesirable interactions between packaging material and food can give rise to potential problems, e.g. migration of packaging components or penetration of microorganisms, insects and rodents through packages. The importance of hygiene in the paper and packaging industry has increased considerably as a result of more specific demands in legislation, tighter international competition and increasing customer requirements. This chapter will concentrate on clarifying potential microbiological and hygienic risks related to packaging materials and especially on how the extremely high hygienic level of food packaging materials is achieved in the paper and packaging industry.

Traditionally paper and packaging manufacturers have not necessarily considered safety and hygiene requirements as such an important factor as is considered in the food industry, but as more emphasis is placed on food safety and hygiene, an awareness of the critical role of packaging is increasing among the paper and packaging manufacturers. Manufacturers of packaging materials for food have become increasingly conscious of customer demands relating to concerns of food safety. As important raw material suppliers for the food industry, manufacturers of packaging materials are expected to bring their standard hygiene in line with the expectations of the food industry.

The importance of hygiene and microbial management has increased considerably as a result of extending the role of package and changes in the manufacturing process of fibre-based packaging materials. The contact time between package and food has lengthened owing to longer selling periods and extended transport time from production plant to market. Furthermore, the package is no longer used only as a cover of food product during transportation and storing which is the traditional function of package, but often also as a tray in which food can be warmed up and served as well (Gerding *et al.*, 1996). Simultaneously today consumers prefer minimally processed food with no preservatives, low fat, sugar and salt content and a long time of use, making food more susceptible to microbial contamination. Important microbial growth-enhancing changes such as increased use of recycled fibre as a raw material and decreased consumption of water (closing of water circulation system) has also taken place in paper and board manufacturing, which also stress the importance of hygiene and efficient microbial management.

Today's legislative demands include in particular consumer protection and environmental concern. The Framework Directive 89/109/EEC applies to all materials and articles for food-contact use and the Food Hygiene Directive 93/43/EEC set the minimum demands for the safety and hygiene of food and materials used in food applications (EC Council Directive, 1989, 1993). These Directives have been implemented into national laws of EC member states, e.g in Finland (Ministry of Trade and Industry, 1993), Germany (BGVV, 2000) and the Netherlands (VgB, 1999). National regulations exist also in USA (FDA, 2000). The basic idea of Framework Directive 93/43/EEC is formulated in Article 2:

> Materials and articles must be manufactured in compliance with the good manufacturing practices so that they do not transfer their constituents to foodstuffs in quantities which could endanger human health, bring about unacceptable change in the composition of the foodstuffs or a deterioration in the organoleptic characteristics thereof.

To date, a legal requirement exists for chemical quality of raw materials and food packages, but microbiological requirements are usually missing. The legislation on food contact materials does not give any specifications for the microbiological quality of the materials. The Council of Europe (2002) gave a Policy Statement concerning paper and board materials and articles intended to come into contact with food. It states that materials intended to come in contact with food should be of suitable microbiological quality, taking into account the intended end use, and that for materials intended to come in contact with aqueous and/or fatty foodstuffs particular attention should be paid to pathogens. When completed the Council of Europe Resolution will form the basis for a forthcoming directive. If the resolution is accepted as such into the forthcoming directive it will be a clear tightening of microbiological demands compared with the present Framework Directive 89/109/EEC.

While the legal requirements for microbiological quality are missing, the paper and packaging industry follows a few available recommendations for microbiological quality of food-contact materials. For dairy products, the so-called 'Dairyman's standard' states the limit for microbiological contaminants in liquid packaging material (250 cfu g^{-1}). The Food and Drug Adminstration (FDA) standard (FDA, 1991) also states that microbial numbers should not exceed 1 cfu cm^{-2} or 250 cfu g^{-1}. The total numbers of yeasts, moulds and bacteria accepted must be low, and no pathogenic bacteria, including enterobacteria and *Escherichia coli*, should be detected (May, 1994). A German DIN-10082 standard (1996) sets a limit of six yeasts and two moulds dm^{-2} for greaseproof paper.

Legal aspects are important, but equally important are the demands and requirements set by the customers, the food industry, which most often are clearly tighter than the legal ones and presuppose an active Hazard Analysis Critical Control Point (HACCP) programme to confirm the safety of paper products used in food applications. In addition the hygiene and safety demands set by the company itself to give the impression of a company that considers safety aspects seriously have become an important marketing argument, e.g. in the pulp industry.

24.2 Potential microbiological problems with packaging

Production of fibre-based food contact materials include several phases in which microbiological problems can occur. However it must be stressed that the threat of microbial contamination from packaging materials is more theoretical, while food itself is always the most prominent source of microbial contamination. There are no published cases where microorganisms originating from packaging material have migrated into the food, multiplied there and caused illness of a human. However there are cases when microorganisms originating from packaging materials have spoiled the food product. Paper and board machines offer microorganisms a favourable environment to grow and multiply, and hence their numbers may be quite high without causing significant problems for runnability or end-product safety. Therefore, it is more important to determine the nature of the microorganisms (e.g. pathogen, slime producer, toxin former, producer of off-odours) than their exact numbers.

Microbiological problems in the paper industry can be grouped as follows:

- Microorganisms causing spoilage of raw materials, since many microorganisms break down cellulose fibres, starch, casein and rosin sizings, e.g. *Pantoea agglomerans* and *Bacillus subtilis* spoil starch.
- Microorganisms causing problems in the process: producers of slime and deposits, e.g. *Burkholderia cepacia*, *Deinococcus* spp. and *Bacillus* spp.
- Microorganisms threatening process safety: microorganisms harmful to human health in raw materials, process environment and end-products, e.g. *Bacillus cereus, Klebsiella pneumoniae* and *Staphylococcus aureus.*
- Microorganisms reducing the quality of end-product: microorganisms harmful to human health and their metabolic products (toxins),

microorganisms that can cause colour defects in products and microorganisms harmful to the hygiene of the end-product, e.g. *Bacillus cereus*, *Staphylococcus aureus*, coliforms and moulds.
- Microorganisms that can cause smell or taste defects, such as the volatile, foul-smelling compounds produced during microbial metabolism, e.g. *Clostridium* spp., *Desulfovibrio* spp. and actinobacteria.

Important potential contaminants in the paper and packaging industry and their characteristics are summarised in Table 24.1.

There are two ways in which microorganisms may contaminate the end-product: during the manufacturing process when microorganisms are found within the products (mainly due to the microbial quality of the fibre, water, aerosols and pulp additives), or at a later state during recolonisation, when microorganisms are found on the surface of the product (airborne micro-organisms, aerosols, handling by personnel and storage conditions). Micro-organisms can enter the production facilities via raw materials, e.g fresh water, raw materials and additives such as fillers, pigments, starches and coatings. Furthermore harmful microorganisms can enter via packaging materials, transport equipment, open windows, doors and openings, air conditioning, insects and pests and personnel and settle into the production facilitities. Contamination risk depends on the source of contaminant, number of micro-organisms, quality of microorganisms, conditions in production premises (temperature, moisture, pH), availability of nutrients and efficiency of control methods (Fig. 24.1). The characteristics of some important groups of contaminants are described in the following sections.

24.2.1 Spore-forming bacteria in end-products

Heat and steam in the dryer normally kill bacterial vegetative cells. However, certain species of bacteria form endospores that are highly resistant to heat, desiccation, disinfectants, chemical biocides, ultraviolet light and ionising radiation. The surviving free spores can remain dormant until external conditions are favourable for germination and development into vegetative cells. Heat-stable enzymes and toxins can also be problematic if migration into delicate foodstuffs occurs (Pirttijärvi, 2000). This is most likely with products having a high grammage, where the thickness of the product protects the microorganisms from heat during the drying stage. Microorganisms also withstand heat well in end-products containing minerals by adhering to the mineral crystals. From the point of view of end-products safety, the most important microorganisms include spore-forming bacteria (*Bacillus* spp., *Clostridium* spp.), bacteria that are injurious to health, certain moulds, actinobacteria and anaerobic bacteria.

Aerobic endospore-forming bacteria are widely distributed in nature. The spores are resistant to heat, desiccation, disinfectants, ionising radiation and ultraviolet light (Russell, 1990). Spore-forming bacteria are common in mill water circulations and have also been isolated from board samples. Studies have shown that the main contaminants in food-packaging paper and board are

Table 24.1 Important potential contaminants in the paper and packaging industry and their characteristics

Characteristics	*Bacillus* spp.	*Clostridium* spp.	Sulphate-reducing bacteria (SRB)
Growth requirements	Aerobic, facultative anaerobic Temperature optimum 25–45 °C, also psychrophilic and thermophilic strains Wide pH range	Strictly anaerobic, some microaerophilic species Temperature optimum 30–37 °C or 40–45 °C, also thermophilic species that can grow even at 80 °C pH optimum pH 6.5–7.0 for most species	Strictly anaerobic species Temperature optimum 30–40 °C, also thermophilic species that can grow even at 80 °C
Formation of spores	Formation of endospores which are resistant to, e.g., heat, desiccation, disinfectants, radiation and ultraviolet light	Formation of endospores	Members of *Desulfotomaculum* genus produce endospores
Pathogenesis	Same species are pathogenic, e.g. *Bacillus cereus* causes food poisoning	Several pathogenic species, e.g., *Clostridium perfringens, C. botulinum, C. tetani, C. difficile, C. paraputricum*	Non-pathogenic species
Other problems	Often involved in biofilm and slime formation Some produce starch degrading enzymes causing, e.g., spoilage of raw materials used in paper manufacturing	Production of hydrogen (H_2) and volatile organic fatty acids, e.g. butyric, propionic and valeric acids: Offensive odours and explosive gases	Reduction of sulphur compounds to hydrogen sulphide (H_2S) Cause odour problems and corrosion of stainless steels
Habitat, living environment	Widely distributed in the environment especially soil and water Common in conifer trees used in paper manufacturing Often detected in vegetable and cereal products	Common in soil, anaerobic fresh water sediments, intestinal tracks of human and domestic animals Often detected in vegetable and animal product	Common in anaerobic water sediments, intestinal tracks of human and animals
Appearance in paper industry environment	Detected in circulating waters, mechanical pulp, papermaking chemicals, broke, slime, end-product	Detected in circulating waters, broke, slime, end-product	Detected in circulating waters, broke, kaolin, starch, slime

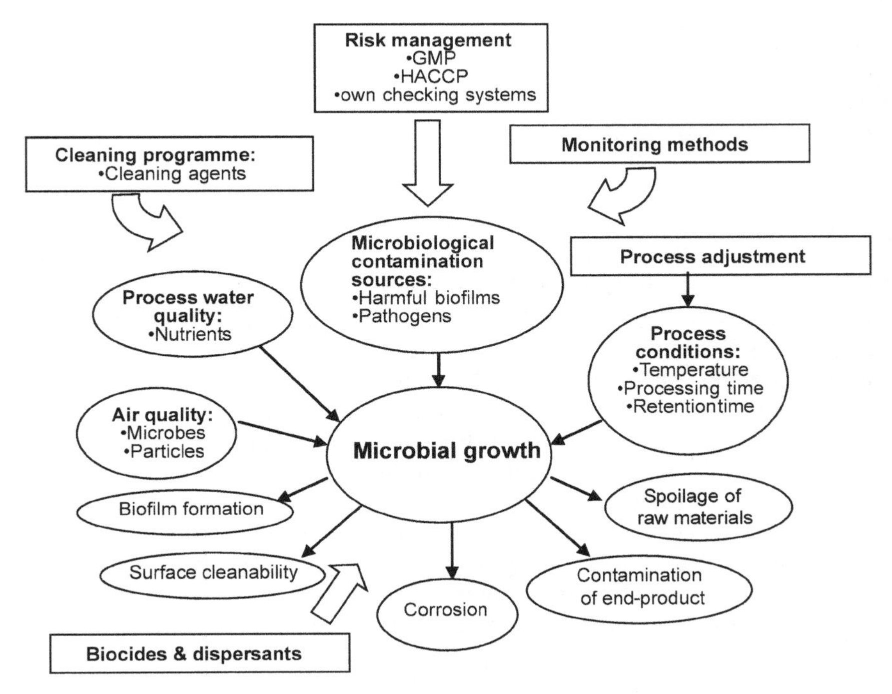

Fig. 24.1 Sources, problems and control of microbial contaminants in the paper and packaging industry.

aerobic spore-forming microbes belonging to the genera *Bacillus* and *Paenibacillus* (Väisänen *et al.*, 1989, 1991; Pirttijärvi *et al.*, 1996; Suihko *et al.*, 2004). The dominance of bacilli in paper and board products may be explained by the following factors: (1) the recycling of broke (components rejected during manufacturing that are reused later in the process), which contains large numbers of spore-formers, constantly contaminating the stock in the headbox; (2) the thermoresistant spores survive the heat during drying of the paper and board; (3) most bacilli are amylolytic with some cellulolytic activity and therefore grow well in environments rich in starch and cellulose; and (4) bacilli are very resistant to many slime-destroying agents used in paper and board machines (Väisänen *et al.*, 1991).

The genus *Bacillus* comprises a diverse collection of saprophytic bacteria widely distributed in the environment, especially soil and wood. Some, e.g. *B. subtilis,* spoil raw materials, whereas some produce slime, which disturbs the process and causes defects in the end-products. The species *B. cereus* is pathogenic, and causes two types of foodborne diseases – diarrhoea and emesis. Apart from the emetic toxin (cereulide), three different enterotoxins are produced by *B. cereus* (Granum, 1997). There are strains of *B. cereus* which have been shown to cause food poisoning at a very low infectious dose (10^3–10^4 bacteria/g) (Andersson *et al.*, 1995). *Bacillus cereus* group cells, at least *B.*

cereus, *B. thuringiensis*, *B. anthracis* and *B. mycoides,* share a high degree of chromosomal sequence similarity, and all of them have been shown to be potentially toxigenic and cause foodborne illness outbreaks and opportunistic infections (Jackson *et al.*, 1995; Damgaard *et al.*, 1996; Beattie and Williams, 1999; Hsieh *et al.*, 1999; Hansen and Hendriksen, 2001).

Microbial numbers in food-grade paper and board have usually been below the limit set by the FDA. The spectrum of microorganisms isolated from food-packaging boards and blanks has also been reported to be relatively narrow (Väisänen *et al.*, 1991; Kneifel and Kaser, 1994; Pirttijärvi *et al.*, 1996). Pirttijärvi *et al.* (1996) studied liquid packaging boards and characterised over 200 isolates. Contaminants found were aerobic spore-forming bacteria, of which *B. cereus* group was often present. Approximately 50% of the *B. cereus* group strains were positive in enterotoxin tests. Later it was concluded that the numbers of *B. cereus* were low in paper and board and only one of nine strains (11%) isolated was producing enterotoxin (Suihko *et al.*, 2004).

The anaerobic *Clostridium* species are a very heterogenic and scarcely studied group of potential end-product contaminants. Most clostridia produce gases as fermentation products such as butyric, acetic, propionic and valeric acids, and as a consequence can cause offensive odour problems in end-products. This spore-forming anaerobic group of bacteria have also several pathogenic species. *C. sporogenes* is quite common in papermaking systems. This spore-forming bacterium can be found in soil, water, broke, additives, recycled fibre and recycled sludge (Robichaud, 1991). Another *Clostridium* species, *C. perfringens,* can cause food poisoning. *C. perfringens* is ubiquitous in nature. Anaerobic *C. perfringens* produces spores that are highly resistant to environmental stresses such as radiation, desiccation and heat (McClane, 1997). It can be isolated, e.g. from soil and dust or from intestinal tracts of humans and domestic animals. *C. perfringens* causes two different types of human disease that can be transmitted by food, i.e. *C. perfringens* type A food poisoning and necrotic enteritis. *C. perfringens* type A food poisoning is most common in industrialised societies.

The genetic diversity of anaerobic bacteria present in paper mill environments has been studied recently (Suihko *et al.*, 2005). A total of 177 anaerobically grown isolates were screened for aerotolerance, from which 67 obligate anaerobes were characterised by automated ribotyping and one isolate from each ribogroup was further identified by partial 16S rDNA sequencing. The mesophilic isolates indicated 11 different taxa (species) within the genus *Clostridium* and the thermophilic isolates four taxa within the genus *Thermoanaerobacterium* (formerly *Clostridium*). The most frequently occurring clostridia were closest related to *C. magnum*, *C. peptidivorans*, *C. puniceum* and *C. thiosulfatireducens*. The phylogenetic positions of the mill isolates indicated that most of them (69%) are potential members of new species (partial 16S rDNA sequence similarity <99%) or even of new genera (similarity <95%). Slime and clay were observed to be good reservoirs for these species in paper mills, but their spores tolerated the heat treatment of the process and they were also present in end-products.

24.2.2 Fungi and actinobacteria in end-products

Fungi may present a potential food spoilage risk because of their ability to degrade, ferment or convert substrates as well as to produce odorous metabolites (off-flavours) and mycotoxins. The growth of fungi on paper and paperboard is quite common, especially under conditions of high moisture and low illumination (Piluso, 1987). The presence of fungi is primarily due to postcontamination via air and water during and after product manufacture. *Aspergillus*, *Cladosporium*, *Fusarium*, *Mucor* and *Penicillium* species have been isolated from fibreboard cartons (Tindale *et al.*, 1989). Suihko and Hoekstra (1999) isolated yeasts and moulds from recycled fibre pulps and paperboards. They observed that the drying stages of the paper and boardmaking processes drastically eliminated the fungi present in the recycled fibre pulps and process water, but not during the heat dispersal. Moulds were present in all paperboard samples; however, the numbers were close to the detection limit (10 cfu g^{-1}). The dominant species occurring in all the recycled fibreboards was *Gilmaniella humicola*; *Paecilomyces variotii* was as well also found (Suihko and Hoekstra, 1999). The mill environment (air) or the boardmaking chemicals were considered to be the likely contamination sources. The low level of contamination by microorganisms of food-grade paper surfaces indicates that the risk for food contamination is low. However, when pasteurised and sterile foods are packed the demands for the packaging materials are high. Fungal ascospores can contaminate products of low pH during long-term storage if migration through pinholes occurs. The spoilage of fruit juice by *Penicillium* species originating from carton material have been reported (Narciso and Parish, 1997).

Actinobacteria are a large group of morphologically and physiologically diverse, Gram-positive bacteria, having in common DNA with an unusually high GC content (Ensign, 1992). The morphological diversity ranges from the micrococci to the branched filament-forming species. Actinobacteria include, e.g., the genera *Actinomyces, Mycobacterium, Frankia, Nocardia, Micrococcus* and *Streptomyces.* Actinobacteria are the most prevalent producers of organic odours. The mycelial actinobacteria are able to cause problems in paper and board production and odour problems in the final product by producing volatile metabolites, mainly geosmins. Furthermore, many streptomycetes have been reported to produce antibiotics and some, like *S. anulatus,* may produce genotoxins. Thermophilic actinobacteria have been isolated up to 10^2 cfu g^{-1} dry weight from board samples on which recycled fibre pulps had been used (Suihko and Skyttä, 1997).

24.2.3 Migration of microorganisms into foods from the packaging material

Most packaging materials have proven to be completely impervious to microorganisms. Moreover, the microbial load of fresh food products is incomparably large relative to the amounts of microorganisms that could permeate through a package, so this phenomenon can usually be neglected (Höutmann, 1999).

However, even a low number of bacteria in the packaging material could be of concern for aseptic foods if bacteria migrate across the package (Suominen *et al.*, 1997).

The routes of contamination from the packaging material to food include the surface, cutting dust or direct contact with the raw edge of the paperboard. To eliminate the raw edge, cartons intended for liquid foods with long shelf-life and for aseptic packaging are usually skived (a carton configuration which eliminates raw edges) (Pirttijärvi, 2000). The dust generated during the skiving process may be scattered on the inner surface of the carton. During extended storage it has been found to be an indication of more frequent spoilage of skived chemi-thermomechanical pulp (CTMP) cartons compared with non-skived (NordFood, 1999; Pirttijärvi, 2000).

With scanning confocal laser microscopy (SCLM) Suominen *et al.* (1997) studied the microbiological barrier properties of food-packaging paperboards, coated with polyethylene, mineral pigment or a biodegradable polymer and of high-density paper. They reported that spatial distribution of microscopically observable bacterial cells is uneven inside the paperboard. The numbers of bacteria at the interface between the polyethylene coating and the cellulose fibres were 100–200 times higher than inside the cellulose matrix. The bacteria in the interface and in the mineral coating layer grew in response to access of nutrients and water, whereas no growth was observed inside the fibre web, even after extended exposure for up to 90 days. The factor limiting growth and migration of bacteria inside the fibre web was probably the limited access to free water, even under conditions of excessive moisture. According to the authors, the microbes residing between the paperboard and its polymer coating facing the food could potentially contaminate the food. Mineral-coating pigments were reported to be a source of contaminating microbes. Therefore the use of high hygienic-quality surface-sizing chemicals is essential (Suominen *et al.*, 1997).

24.3 Improving hygienic production and management

Microbial management is divided into two main strategies: to prevent the entrance of microorganisms to production premises and to eliminate micro-organisms from production facilities if they have entered the premises. As sterile conditions are impossible the preventive actions should be focused so that the restriction of microorganisms is economically and operationally reasonable. Focus should be on microorganisms that cause harm, damage or danger to product, process, worker, environment or customer. The ultimate goal of all these activities carried out in microbial risk management is prevention.

The control of microbial growth in a paper machine environment is traditionally based on the use of biocides (Fig. 24.1). However, the amount of microorganisms is not always proportional to the severity of microbial problems. The most important factors in the use of biocides are selection of suitable agents and the optimal feeding strategy. Owing to variability and complexity of the microbial flora in paper machines there is no single biocide that would alone be effective in microbial control. Resistance of microbes is a commonly faced problem. Furthermore for environmental reasons new methods are constantly researched. Today the focus is more on the combined use of several control methods. The extreme importance of good housekeeping in successful microbial management has been emphasised. Important housekeeping-related precautions include good control of the entire process and plant operation, minimising unnecessary delays for easily spoiling materials, and cleaning tanks, containers and machinery sufficiently and on a regular basis. The process parameters are generally not changed in order to control the microbial growth. However, especially in the tanks of microbiologically sensitive raw materials, the rise of temperature from 45 °C to 70 °C substantially reduces the microbial activity.

24.3.1 Hygiene and safety management system

Most of the paper and packaging plants have a quality system which is based on ISO 9000 series of standards. Good manufacturing practice (GMP) is a fundamental part of quality control and product safety assurance. GMP identifies the basic requirements and sets the basic rules of operation. Preventive microbial risk management in the paper and packaging industry is performed mostly through GMP and by building up a hygiene and safety management system (own-checking system or HACCP programme) which is recommended to be based on HACCP concept. The HACCP principles have been internationally accepted and approved and details of the approach have been published by, e.g., the Codex Alimentarius Commission (CAC, 1997; NACMCF, 1998; Ewartt, 2000). The prerequisites based on GMP form the foundation upon which a hygiene and safety management system can be built (Fig. 24.2).

Control of hazards that are generic, those that can have an influence at any stage of the process and those that are not associated with specific process points are examples of hazards that are controlled under GMP. Conversely, the Critical Control Points (CCPs) identified in HACCP evaluation are strictly associated with specific process steps and a monitoring system is implemented to ensure that the threshold level set by the HACCP team is met. The HACCP system is compatible with the implementation of total quality management (TQM) system based on the ISO 9000 series of standards. Integration of the HACCP programme and the ISO 9000 quality management system offers the best basis for optimum business operations (Unnevehr and Jensen, 1999). The HACCP programme is often also certified according to standards, e.g. Danish Standard (1998) or British Standard (BRC 2001). In recent years some published cases of applying a HACCP approach on the paper and packaging industry have appeared (Bovee *et al.*, 1997; Blakistone, 1996; Galeano *et al.*, 1999; Sjöberg *et al.*, 2002).

Process phases determined as CCPs that need systematic control depend on the nature of the process. Conditions in paper and board mills are favourable for microbial growth as the process includes many wet process phases, and several

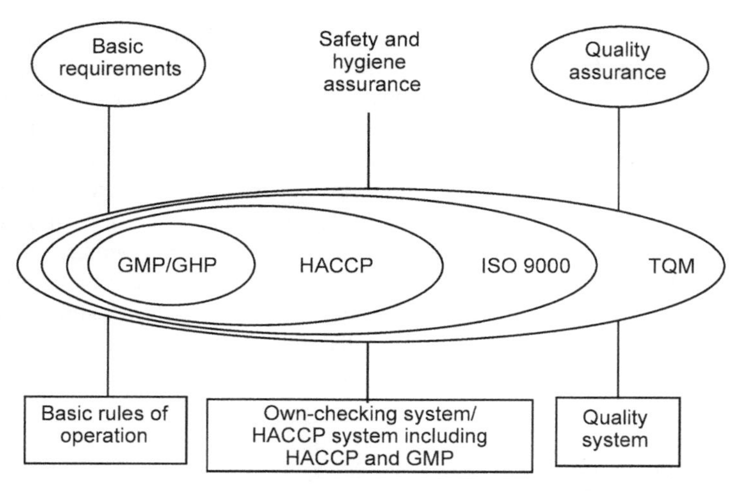

Fig. 24.2 Dependency of quality and safety management systems (TQM = total quality management, GHP = good hygiene practice).

raw materials, such as starch, kaolin, calcium, carbonate and carboxy-methyl cellulose, susceptible to microbial contamination are used. As a consequence microbiological risks clearly associated to specific process steps are often considered to be the most significant ones from the product safety point of view. Thus more emphasis is placed on building up an efficient and cost-effective HACCP programme.

However, conditions in factories further converting paper or board are usually dry and microbiological risks are mainly related to the use of starch-based glues and water-soluble lacquers (Raaska *et al.*, 2002). On the other hand hygiene problems arising through insufficient pest control and cleaning programme, personnel working practices or through air are considered to be important risks to product safety. Thus the implementation of GMP is crucial in factories further converting paper or board. Some examples of important process phases in hygiene and safety management in the paper and board industry and in factories further converting paper and board are presented in Table 24.2.

The driving forces in carrying out an effective hygiene and safety management system in the food industry are especially the legal demands and restrictions, while the demands set by the customers are more emphasised in the paper and packaging industry. The image of a company considering safety and hygiene aspects important is becoming more and more important in the paper and packaging industry, as it is in the food industry. Employees' attitudes towards HACCP-related activities mostly performed by line workers and their knowledge on food hygiene and safety aspects are clearly still superior in the food industry compared to those in the paper and packaging industry. GMP measures have been implemented in the food and packaging industry, but in the paper industry more effort has been placed on building up an efficient HACCP programme (Table 24.3).

Table 24.2 Examples of important process phases in safety and hygiene management

Process phase	Identification of hazard
(a) Paper and board industry	
Cellulosic raw materials	Physical risk, e.g. sand, stones, pieces of metal, dirt
Circulating process waters	Microbiological risk, e.g. problems in runnability, high content of microbes in end-product
Storing and use of broke	Microbiological risk, e.g. bad odours in factory and in end-product caused by anaerobic bacteria
Surface sizing agents and coating compounds	Microbiological risk, e.g. problems in runnability, high content of microbes in end-product caused by microbial growth in starch, CMC, $CaCO_3$
Factory environment	Biological risk, e.g. insects contaminating the product Physical risk, e.g. cleave of paint and cement rendering, glass particles, condition of building
(b) Industry further converting paper and board	
Storing of raw materials, e.g. paper sheets and rolls, and end-products	Biological risk, e.g. insects, rodents or birds and their faeces contaminating the product
Refining equipment	Chemical risk, e.g. oil and other lubricants contaminating the product
Lacquers and glues	Microbiological risk, e.g. high content of microbes in end-product, bad odours in end product caused by microbial metabolites
Printing	Chemical risk, e.g. bad odours in end-product caused by the use of wrong printing colours
Packing	Microbiological risk, e.g. microbial contamination of end-product caused by wrong working practices, insufficient ventilation or old and dirty pallets Physical risk, e.g. contamination of product by foreign bodies; glass, hair, nails, metal pieces
Factory environment	Physical risk, e.g. contamination of product by dust

Although HACCP has been internationally accepted and approved, and it has been used as a tool in food safety management, there is open acknowledgement that companies have had problems with its practical application and implementation. If product safety aspects are confused with quality, the evaluation can result in too many and/or overlapping CCPs, which make the programme unmanageable; thus people fail to see the point of carrying out the control measures determined for every CCP. However, quality aspects are sometimes very difficult to distinguish clearly from safety aspects, especially in the paper and packaging industry. Commitment of management as well as employees is crucial for the success of implementation and application of hygiene and safety management systems. Written documentation behind the development of specific

Table 24.3 Comparison of driving forces, employee attitudes and knowledge on hygiene and safety management aspects in food and paper and packaging industries

	Food	Paper	Packaging
Legal demands	**xxx**	xx	xx
Customer demands	xx	**xxx**	**xxx**
Company image	xxx	xxx	xxx
Employee knowledge on hygiene and end-product safety aspects	**xxx**	x	x(x)
Employee attitudes towards hygiene and safety management activities	xx(x)	x(x)	xx(x)
Establishment of GMP	**xxx**	x	**xxx**
Implementation of HACCP	xx	**xxx**	xx
Documentation	x(x)	xx	x(x)

xxx = high, very positive ; xx = medium, neutral ; x = low, negative
Bold crosses = most important characteristics
xx(x), x(x) = depending on size of company, attitude may be very positive or medium, neutral

HACCP programmes, which would ease validation and reassessment, has also been difficult to produce, particularly in small and medium-sized enterprises or where no previous quality management system exists and thus record-keeping is not part of the company culture (Mortimore, 2001; Panisello and Quantick, 2001). An important problem in risk assessment at the manufacturing level is that a more quantitative systematic approach should be used, and that a risk assessment procedure should be based on scientific knowledge and performed by a team that has the knowledge and experience needed to perform reliable evaluation of risks. Such hazard analysis comprises four elements: (1) hazard identification (what may be present), (2) hazard assessment (which of the identified hazards need control), (3) control measure assessment (which control measures are effective and to what extent taking into consideration also the cost effect) and (4) combination of control measures. Often this kind of expertise is missing from industry, which does not always have the resources needed.

Emphasis on more powerful tools for assessment and control of hazards is being stressed in a new ISO standard (Food Safety Management System, CD ISO 22000, ISO TC34/WG 8). To help in overcoming at least some of these problems, a practical and easy-to-use risk assessment tool for establishing GMP and HACCP principles in the pulp, paper and packaging industry is under development. PaperHYGRAM® is a risk assessment model based on the concepts of HACCP and GMP for performing and maintaining risk assessment of process hygiene and safety aspects in the pulp, paper and packaging industry. It is a Windows (Windows NT/XP/2000 is required) based program that does not require special computer skills. The tool, called HYGRAM® (hygiene risk assessment model), has been especially developed for assessing risks in the food industry and has been used as a basis for building up a separate model for the paper industry (Tuominen *et al.*, 2003).

The model offers software-assisted guidance through the procedure and includes a databank on risk assessment and problematic microorganisms in the pulp, paper and packaging industry. PaperHYGRAM® includes a background information sheet of the factory, process and products to be evaluated, a checklist for meeting the requirements of GMP, an analysis tool for identifying, describing and assessing the risk groups of a HACCP programme (physical, chemical, biological and microbiological risks) and additional risk groups chosen by the user. All necessary information can be documented in the model, and it is easily updated when changes occur in the process. Both HYGRAM® and PaperHYGRAM® currently exist in Finnish, but translation into English will be performed in the near future.

24.4 Future trends

Based on research results obtained hitherto the numbers of potential pathogens are very low, most of the potential pathogens have been shown not to be virulent and the migration of microorganisms from packaging materials into food is very unlikely. Furthermore the implementation and application of GMP and HACCP principles in the paper and packaging industry, as well the easy and practical tools under development to help industry in assessing and maintaining GMP and HACCP programmes will ensure the hygiene and safety of packaging materials in the future.

Competence in microbiology is required, however, to perform reliable risk assessment and to focus the risk assessment measures efficiently and cost-effectively. Furthermore microbiological risk assessment and risk management activities at industry level require a significant amount of microbiological analysis performed by using sufficiently sensitive and rapid detection methods. Traditional cultivation methods are far too slow compared to the turnover of fibre-based products and in many cases are not sensitive enough to detect pathogens that may cause illness at very low concentration, and there is thus a clear need for novel, rapid and sensitive detection as well as identification methods.

Over the past decade many improvements have been seen in both conventional and modern methods for the detection and identification of bacteria in industrial environments (Maukonen *et al.*, 2003). Phenotypic analyses – what an organism looks like, what it does, what enzymatic activities it has – have traditionally played an important role in bacterial identification and classification. However, bacteria are small, contain relatively few structural clues and the physiological properties may be changed. Thus, the genotypic analyses have been found to be very useful and accurate in the identification and classification of these bacteria. Effective use of the typing methods, however, requires that good identification libraries are created. Isolates are usually well adapted to their specific environments and do not possess the typical characteristics of any species hitherto described. In many cases the results of physiological, chemotaxonomical (FAME (fatty acid methyl ester analysis), SDS PAGE (sodium dodecyl sulphate polyacrylamide gel electrophoresis)) and biomolecular (partial 16S rDNA sequencing, ribotyping) tests are not in good agreement, and the isolates have been proposed to represent new species (Raaska *et al.*, 2002).

Bacillus cereus has been traditionally detected from paper and board by culturing and verified by additional biochemical tests. This is slow and complicated. Recently a real-time PCR-based rapid and semi-quantitative method was published to screen and detect *B. cereus* group bacteria directly from various paper and packaging samples (Priha *et al.*, 2004). Quantitative results can be obtained in one day, whereas plate count and API results take a minimum of one week to obtain. The detection method enables industry to obtain accurate information on contamination routes and the occurrence of *B. cereus* group bacteria in end-products, as well as to respond quickly to the changes detected in the microbiology of manufacturing process.

Another example of advanced detection of important contaminant in paper industry is the detection of sulphate-reducing bacteria (SRB) with molecular techniques. So far the detection of SRB has been hampered by the inability of cultivation to detect the number and nature of these bacteria correctly. According to Maukonen *et al.* (2005) *Desulfovibrionaceae*-related bacteria can be detected without prior enrichment directly from paper industry samples rapidly and reliably.

The emergence of detection and characterisation methods based on molecular biology are linked to the need for a better assessement and management of the microbiology in industrial raw materials, processes and products. The development of molecular methods that are valid in practical industrial use is crucial for the better management of microbiology in industrial processes.

24.5 Sources of further information and advice

- HYGRAM® hygiene risk assessment model: http://hygram.vtt.fi
- Policy statement concerning paper and board materials and articles intended to come in contact with foodstuffs: www.coe.int/soc-sp
- Risk World provides links to risk analysis-related sites: http://www.riskworld.com

24.6 References

ANDERSSON A, RÖNNER U and GRANUM P E (1995), What problems does the food industry have with the spore-forming pathogens *Bacillus cereus* and *Clostridium perfringens*? *Int J Food Microbiol,* **28**, 145–155.

BEATTIE S H and WILLIAMS A G (1999), Detection of toxigenic strains of *Bacillus cereus* and other *Bacillus* spp. with an improved cytotoxicity assay. *Lett Appl Microbiol,* **28**, 221–225.

BGVV (2000), Empfehlungen des Bundesinstitutes für Gesundheitlichen Verbraucherschutz. In: *Kunstoffe im Lebensmittelverkehr* XXXVI, Papiere, Kartons und Pappen. Carl Heyman Verlag KG, Berlin, pp. 115–122 d/13.

BLAKISTONE B A (1996), Targeting food packaging materials for HACCP. *The 211th Annual Meeting of the American Chemical Society*, New Orleans, 26 March 1996, pp. 115–123.

BOVEE E H G, DE KRUIJF N, JETTEN J and BARENDSZ A W (1997), HACCP approach to ensure the safety and quality of food packaging. *Food Additives Contaminants*, **14**, 721–735.

BRC (BRITISH RETAIL CONSORTIUM) (2001), Technical standard and protocol for companies manufacturing and supplying for packaging materials for retailer branded products. The Stationery Office, London.

CAC (1997), *Hazard Analysis and Critical Control Point (HACCP) System and Guidelines for its Application*. Codex Alimentarius Commission CAC/RPC 1-1969, Rev. 3, Annex, Rome.

COUNCIL OF EUROPE (2002), Partial Agreement in the Social and Public Health Field. Public Health Committee. Committee of Experts on Materials Coming into Contact with Food. Policy statement concerning paper and board materials and articles intended to come into contact with foodstuffs. Version 1, p. 67.

DAMGAARD P H, LARSEN H D, HANSEN B W, BRESCIANI J and JORGENSEN K (1996), Enterotoxin-producing strains of *Bacillus thuringiensis* isolated from food. *Lett Appl Microbiol,* **23**, 146–150.

DANISH STANDARD (1998), DS 3027 E. Food safety according to HACCP (Hazard Analysis and Critical Control Points) – Requirements to be met by food producing companies and their subcontractors, p. 7.

DIN STANDARD 10082 (1996), Packmittel. Buttereinwickler. Technische Lieferbedingungen. DIN Deutsches Institut für Normung. e.V., Berlin.

EC COUNCIL DIRECTIVE (1989) 89/109/EEC. 28 Nov 1989. Council Directive on the approximation of the laws of the Member States relating to materials and articles intended to come into contact with foodstuffs. *Official J European Communities*, **L347**, 37–44.

EC 93/43/EEC (1993) EC Council Directive 93/43/EEC (1993), *Official J European Communities*, **L175**/2.

ENSIGN J C (1992), Introduction to the actinomycetes. In Balows A, Trüper HG, Dworkin M, Harder W. and Schleifer K-H (Eds.). *The Prokaryotes,* 2nd ed., Vol. 1. New York: Springer-Verlag, pp. 811–815.

EWARTT M (2000), Hygiene and HACCP in the manufacture of food packaging. *Food Packaging Bull*, **9**, 5–7.

FDA (1991), Food and Drug Administration Standard, Fabrication of single-service containers and closures for milk and milk products. US Department of Health and Human Services, Washington, DC, Government Printing Office.

FDA (2000), Code of Federal Regulations, Title 21, Food and Drugs, part 176 – Indirect Food Additives: paper and paperboard components, Food and Drug Administration Washington, DC, US Government Printing Office.

GALEANO S F, GARNER B A, RETZKE, J A and KERSNICK C R (1999), Enhancing food packaging safety procedures – the HACCP opportunity. *Tappi J*, **82**, 68–73.

GERDING T K, RIJK A A H, JETTEN J, VAN DER BERG F and DE KRUIJF N (1996), Trends in food packaging: arising opportunities and shifting demands. *Packag Technol Sci,* **9**, 153–165.

GRANUM P E (1997), *Bacillus cereus*. In: Doyle M P, Beuchat L R and Montville T J (eds), *Food Microbiology. Fundamentals and Frontiers*. ASM Press, Washington, DC, pp. 327–336.

HANSEN B M and HENDRIKSEN N B (2001), Detection of enterotoxic *Bacillus cereus* and *Bacillus thuringiensis* strains by PCR analysis. *Appl Environ Microbiol*, **67**, 185–189.

HÖUTMANN U (1999), Germ load on packaging paper and board: transfer of micro-organisms. *Tappi J*, **82**, 74–77.

HSIEH Y M, SHEU S J, CHEN Y L and TSEN H Y (1999), Enterotoxigenic profiles and polymerase chain reaction detection of *Bacillus cereus* group cells and *B. cereus* strains from foods and food-borne outbreaks. *J Appl Microbiol*, **87**, 481–490.

JACKSON S G, GOODBRAND R B, AHMED R and KASATIYA S (1995), *Bacillus cereus* and *Bacillus thuringiensis* isolated in a gastroenteritis outbreak investigation. *Lett Appl Microbiol*, **21**, 103–105.

KNEIFEL W and KASER A (1994), Microbiological quality parameters of packaging materials used in dairy industry. *Archiv für Lebensmittelhygiene*, **45**, 38–43.

MAUKONEN J, MÄTTÖ J, WIRTANEN G, RAASKA L, MATTILA-SANDHOLM T and SAARELA M (2003), Methodologies for the characterization of microbes in industrial environments: a review. *J Ind Microbiol Biotechnol*, **30**, 327–356.

MAUKONEN J, SAARELA M and RAASKA L (2005), Detection and characterization of desulfovibrionaceae-related groups from paper mill environment with molecular techniques and culture. *J Ind Microbiol Biotechnol*, submitted.

MAY O W (1994), Development of microbiological guidelines for food-grade paperboard: a historical perspective. *Tappi J*, **77** (12), 41–43.

McCLANE B A (1997), *Clostridium perfringens*. In: Doyle M P, Beuchat L R and Montville T J (eds), *Food Microbiology. Fundamentals and Frontiers*. ASM Press, Washington, DC, pp. 305–326.

MINISTRY OF TRADE AND INDUSTRY (1993), Decision 143/93 of the Ministry of Trade and Industry on Paper and Board intended to come into contact with Food. Finland.

MORTIMORE S (2001), How to make HACCP really work in practice. *Food Control*, **12**, 209–216.

NACMCF (NATIONAL ADVISORY COMMITTEE ON MICROBIOLOGICAL CRITERIA FOR FOODS) (1998), Hazard analysis and critical control point principles and application guidelines. *J Food Prot*, **61**, 762–775.

NARCISO J A and PARISH M E (1997), Endogenous mycoflora of gaple-top carton paperboard used for packaging fruit juice. *J Food Sci*, **62** (6), 1223–1239.

NORDFOOD (1999), Hygienic milk packages, Summary report, project 93118, In: *NordFood* 3/1999, Nordisk Industrial Fond, Oslo.

PANISELLO P J and QUANTICK P C (2001), Technical barriers to Hazard Analysis Critical Control Point (HACCP). *Food Control*, **12**, 165–173.

PILUSO A J (1987), Fungi proliferation leads to several operating problems in the paper mill. *Pulp Paper*, **61** (8), 98–101.

PIRTTIJÄRVI T (2000), Contaminant aerobic sporeforming bacteria in the manufacturing processes of food packaging board and food. PhD Thesis. Department of Applied Chemistry and Microbiology, University of Helsinki, Helsinki.

PIRTTIJÄRVI T S M, GRAEFFE T H and SALKINOJA-SALONEN M S (1996), Bacterial contaminants in liquid packaging boards: assessment of potential food spoilage. *J Appl Bacteriol*, **81**, 445–458.

PRIHA O, HALLAMAA K, SAARELA M and RAASKA L (2004), Detection of *Bacillus cereus* group bacteria from cardboard and paper with real-time PCR. *J Ind Microbiol Biotechnol*, **31**, 161–168.

RAASKA L, SILLANPÄÄ J, SJÖBERG A-M and SUIHKO M-L (2002), Potential microbiological hazards in the production of refined paper products for food applications. *J Ind Microbiol Biotechnol*, **28**, 225–231.

ROBICHAUD W T (1991), Controlling anaerobic bacteria to improve product quality and mill safety. *Tappi J*, **74** (2), 149–152.

RUSSELL A D (1990), Bacterial spores and chemical sporicidal agents. *Clin Microbiol Rev*, **3** (2), 99–199.

SJÖBERG A-M, SILLANPÄÄ J, SIPILÄINEN-MALM T, WEBER A and RAASKA L (2002), An implementation of the HACCP system in the production of food-packaging material. *J Ind Microbiol Biotechnol*, **28**, 213–218.

SUIHKO M-L and HOEKSTRA E S (1999), Fungi present in some recycled fibre pulps and paperboards. *Nordic Pulp Paper Res J*, **14** (3), 199–203.

SUIHKO M-L and SKYTTÄ E (1997), A study of the microflora of some recycled fibre pulps, boards and kitchen rolls. *J Appl Microbiol*, **83**, 199–207.

SUIHKO M-L, SINKKO H, PARTANEN L, MATTILA-SANDHOLM T, SALKINOJA-SALONEN M and RAASKA L (2004), Description of heterotrophic bacteria occurring in paper mills and paper products. *J Appl Microbiol*, **97**, 1228–1235.

SUIHKO M-L, PARTANEN L, MATTILA-SANDHOLM T and RAASKA L (2005), Molecular characterization of mesophilic and thermophilic clostridia isolated from paper mills. *Systematic and Applied Microbiology*, in press.

SUOMINEN I, SUIHKO M-L and SALKINOJA-SALONEN M (1997), Microscopic study of migration of microbes in food-packaging paper and board. *J Ind Microbiol Biotechnol*, **19**, 104–113.

TINDALE C R, WHITFIELD F B, LEVINGSTON S D and NGUYEN T H L (1989), Fungi isolated from packaking materials: their role in the production of 2,4,6-trichloroanisole. *J Sci Food Agric*, **49**, 437–447.

TUOMINEN P, HIELM S, AARNISALO K, RAASKA L and MAIJALA R (2003), Trapping the food safety performance of a small or medium-sized food company using a risk-based model. The HYGRAM® system. *Food Control*, **14**, 573–578.

UNNEVEHR L J and JENSEN H H (1999), The economic implications of using HACCP as a food safety regulatory standard. *Food Policy*, **24**, 625–635.

VÄISÄNEN O M, ELO S, MARMO S and SALKINOJA-SALONEN M (1989), Enzymatic characterization of bacilli from food packaging paper and board machines. *J Ind Microbiol*, **4**, 419–428.

VäISäNEN O, MENTU J and SALKINOJA-SALONEN M (1991), Bacteria in food packaging paper and board. *J Appl Bacteriol*, **71**, 130–133.

VGB (VERPAKKINGEN EN GEBRUIKSARTIKELENBESLUIT) (1999), Regling Verpakingen en gebruiksartikelen (Warenwet) Hoofdsuk II – Papier en karton. Koninklijke Vermande, Lelystad, Netherlands.

25

Improving hygiene in food transportation

E. U. Thoden van Velzen and L. J. S. Lukasse, Wageningen University and Research Centre, The Netherlands

25.1 Introduction

The hygiene aspect of food transport has become an issue for European transport operators. This development started roughly in 1990, when national governments urged transport operators to act on food safety. However, nowadays retailers and food producers are demanding more hygiene measures from transport operators. Transport operators need to fulfil an increasing number of hygiene schemes to be allowed to participate in transport tenders of the large retailers and food producers.

In every food supply chain there are at least a few but usually dozens of transport steps. These transport steps can differ greatly in duration, conditioning and type of food product. Consequently, all these transports will have their own food safety profile. Since the majority of food transports have a short duration (1–4 hours) in comparison with the duration of the complete supply chain (at least 3–7 days for fresh foods) the hygiene risks are usually limited in comparison to the risks associated with the longer supply chain steps (food production, retail and storage at the consumer). Obviously, there are also several transports in food supply chains that are more delicate, such as live animals, hung carcasses, etc.

25.2 Legislation

The most relevant European legislation on transport and hygiene is summarised in Table 25.1. The Food Hygiene Directive demands that foodstuffs should be

Table 25.1 The most relevant European laws and directives dealing with transport and hygiene, in sequence of importance

Number	Name
93/43 EEC	Food Hygiene Directive
179/2002 EC	General food law
96/3/Euratom/ESCS/EC	Derogation of certain provisions of 93/43 EEC as regards the transport of bulk liquid oils and fats by sea
98/28/EC	Derogation of certain provisions of 93/43/EEC as regards the transport by sea of bulk raw sugar
89/397/EEC	Official control on Foodstuffs Directive
91/628/EC	Animal Transport Directive
2002/99/EC	Animal Health Directive

transported hygienically. This text is not very specific and gives general instructions. The General Food Law makes retailers and food producers responsible for the safety of foodstuffs they sell on the market. Consequently, it forces retailers and food producers to demand from their suppliers (including transport operators) that relevant hygiene procedures are enforced.

However, from 1 January 2006 the European Union will enforce new legislation in the field of food safety, with possibly also new rules and implications for transport operators. The total of 16 current Directives dealing with hygiene and food safety has been gradually developed from 1964 to regulate the internal market and to protect the consumer. The EU is now also convinced that these 16 Directives are too complex and intermingled and is striving to create a more straightforward, transparent legislation. The discussions and negotiations on this new legislation (often named the Hygiene Package) have been going on for a number of years and are now being finalised. The 16 Directives will be replaced by three laws: 852/2004, 853/2004 and 854/2004. The first law will deal with food hygiene, the second with hygiene of foods from animal origin and the third will describe how the official inspections for foods of animal origin should be organised. For the moment relatively few changes are expected for the food transport sector; however, the laws are not yet final and amendments can still be made.

Additionally, the Codex Alimentarius Commission of the FAO/WHO has developed global standards for handling and transporting food commodities. Standards have been established for many food products, which advise on transport temperatures, hygiene, packaging and labelling.[1]

25.3 Implementation of the current legislation

The Dutch government implemented the food hygiene directive in the 'Warenwetregeling Hygiëne van Levensmiddelen' on 14 December 1995. Paragraph 10 is completely devoted to transport hygiene. In summary it states that all units of transport equipment[2] used for foods:

- should be designed and constructed such that they can easily be cleaned and disinfected;
- should be kept clean and well maintained;
- are not to be used for other materials than foods that could contaminate food transports;[3]
- should be marked 'for food only';
- should separate foods (and other loads) sufficiently to avoid cross-contamination during combined transport;
- should be cleaned sufficiently in between different loads, to avoid cross-contamination.

This law does not contain any specific instructions on temperature and hygiene management. It does oblige transport operators to establish a certified Hazard Analysis Critical Control Point (HACCP) scheme to minimise the food safety risks, or to adopt the hygiene code of the Dutch transport board (Transport en Logistiek Nederland).[4] The majority adopted the hygiene code. The group of transport operators with their own HACCP schemes grows steadily. The hygiene code is a large document that covers all types of transportation and foods and is consequently not clear and precise. It does contain lists of required and recommended transport temperatures for foodstuffs, the preferred stacking arrangements for pallets in trucks without interfering with the circulation of chilled air, drafts for cleaning plans, etc. The HACCP schemes of individual transport operators contain clear daily working routines that state how often units of transport equilibrium are cleaned, how they are cleaned, how the effectiveness of the cleaning operation is monitored, at which temperatures the foods are transported and how these temperatures are monitored and controlled.

Both the hygiene code and the HACCP schemes of individual transport companies are not publicly available as literature. Consequently, it is not possible for interested third parties to verify if a certain transport company really puts effort into hygiene. Retailers and food companies that order transportation generally have full insight in the hygiene schemes that the operator follows. Consumers and independent researchers do not have access to these schemes. They are informed that a transport operator is ISO certified or follows the BRC standard, but this does not give information on how clean the truck actually is. In order to gain more confidence from the general public by showing that transport operators really act on food safety, it would be wise to be more open on these hygiene issues. For instance, as a first step transport companies could describe on their websites how often and well trucks are being cleaned.[5]

25.4 Examples

De Greef Woudenberg BV, a transport company in the middle of The Netherlands, is putting much effort in maintaining a high level of hygiene in its transportation activities. The majority of the products transported are complete round Dutch cheeses and packed fresh meat produce. Occasionally hung carcasses are transported. The company has developed its own HACCP scheme and is audited by Netherlands Controlling Authority for Milk and Milk Products (COKZ) twice a year. This HACCP scheme precisely describes the set-points of the temperatures and how this temperature is monitored, and forbids mixed transports.

Furthermore, the scheme also describes the cleaning procedure. The 20 self-owned trucks are cleaned every month with a high-pressure water spray gun and scrubbed with an aqueous solution of a biodegradable detergent. Stains on the interior are sanded away, the complete interior of the truck is treated with a steam spray gun and finally it is treated with a dedicated and approved disinfectant (see Fig. 25.1). This procedure has been developed over five years. Its effectiveness is tested twice every month by microbial surface sampling and testing. Samples are taken at ten spots in the interior and exterior of the truck. The average score should not exceed the limit of 1.4 cfu/cm^2 after cleaning. This cleaning procedure hardly ever fails. When it does, the truck needs to be cleaned again. Besides having its own trucks, this company also serves as distribution centre for transport partners from Central Europe (Czech Republic, Romania). All trucks from Central Europe are cleaned following the same procedure every time they stop at Woudenberg (about 100 per month), not only to maintain the hygiene of the truck interior, but also to clean the often highly polluted exterior. In this transport company one employee is continuously busy cleaning and maintaining trucks. Hence, maintaining a high level of hygiene requires serious attention and adds significantly to the overall costs.

As mentioned previously, Greef-Woudenberg BV is an advanced example in the meat and cheese sector, where temperature control and avoiding cross-contamination are important. It is not representative for all transport operators. On the other end of the hygiene spectrum, there are the fruit and vegetable transporters. They use a mix of closed and open vehicles. Since cross-contamination is not an issue in the transport of fruits and vegetables (most loads are heavily contaminated with *Botrytis*, *Fusarium* and *Rhizopus*), the cleaning of the vehicles receives little attention. However, this could change in the future, especially for very perishable fruits such as strawberries. In order to increase the shelf-life of the fruits, there are several initiatives on-going to reduce the initial *Rhizopus* load on strawberries. Once successful these will need to be transported in clean trucks to keep the load low and the shelf-life longer.

To summarise, Europe has a complicated legislation regarding the hygiene of food transport, but it is expected to be simplified in a few years time. Transport operators have dealt with this legislation and have established procedures to manage the temperature and avoid cross-contamination. In essence, this works fine. The only downside, perhaps, is the poor transparency for outsiders. There are several pending issues.

25.5 Temperature management

The refrigeration units and insulation of refrigerated trucks have been designed to be able to maintain the temperature of the chilled load even under hot summer

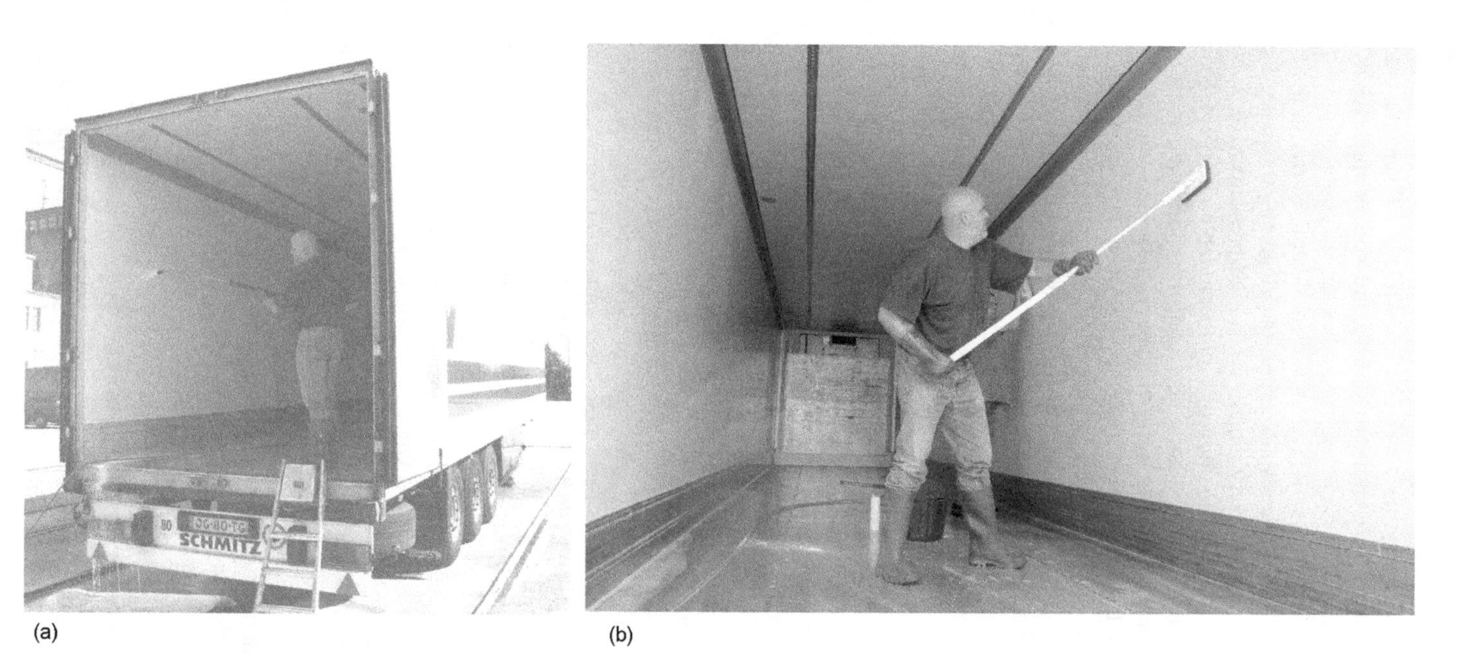

Fig. 25.1 Photographs showing a typical cleaning operation for trucks. The first step is cleaning with a high pressure water gun (left), the second step is scrubbing and sanding the interior (right).

conditions outside the truck. Usually they are not able to cool the load down during transport. The European Agreement on the Transport of Perishables (ATP)[6] developed test standards to certify insulation materials and refrigeration units for trucks. Improvements are still regularly being developed, such as improved insulation materials that are thinner and lighter and refrigeration units that are more energy efficient.[7] The performance of insulation foams had temporarily been diminished when chlorofluorocarbon (CFC) blowing agents were forbidden. However, the foam industry is continuously developing improved new foams without CFCs and with improved insulation properties. These improvements are expected to continue in the coming years and will bring the costs for refrigerated transport down slightly.

In those cases that the temperature in a supply chain was insufficiently managed and the load was subjected to too high temperatures, the problem most often did not arise from poorly refrigerating vehicles. A few exceptions have been reported in the past, such as: too dense a stacking of pallets in front of the chilled air outlet and drivers that turned the noisy refrigeration units off for the night in order to have a good night's sleep. The pallet stacking is usually no longer a problem, as long as the pallets are hollow and there is sufficient head space over the load (20–30 cm is enough) and between the load and the back doors. Such an arrangement will allow the chilled air to circulate from the engine at the top front over the load to the back doors and through the hollow pallets back to the front. Nowadays, both problems hardly occur. This is partly due to the hygiene code, educating the loaders and drivers and to the increasing use of dataloggers (time–temperature recorders). Dataloggers are usually placed in the rear of the vehicle and effectively detect temperature abuse. The first datalog systems could only detect an abuse situation long after the transport had ceased, when the truck was returned and the dataloggers were dismounted from the truck and the data were loaded on the computer system of the operator. Modern datalog systems alert the driver of temperature abuse situations during transport and some can even automatically alert the transport operator's headquarters via cell phone systems. These modern systems allow the driver to act and avoid the loss of a full cargo due to temperature abuse.

Temperature abuse is more likely to occur at loading and unloading of vehicles and whenever the supplier of the load did not bring the temperature of the load down sufficiently prior to transport. Cooling loads can take much time, up to a day for carcasses, and hence there is a clear motive for hasty suppliers in the hectic food supply chains to dash off the pre-cooling. Since refrigerated trucks are not designed to cool a cargo during transport, the consequence can be major. In order to avoid claims, modern transport operators will measure the temperature of the load themselves prior to loading, or use a modern datalog system, which would register temperature abuse from the start of the trip.

Within the sector of fruit and vegetable transport, temperature control is currently playing an important role. Traditionally, the fruit and vegetable sector was divided in two temperature regimes (10–12 °C and below 7 °C). Since retailers and transport operators are forced to reduce costs, there is a tendency to transport all fruit and vegetables at about 7 °C. This is bad for the food quality (not necessarily bad for the food safety) of tropical fruits and convenience products. Hence, the frequency of chilling injury of bananas and other tropical fruits has increased and the shelf-life of pre-cut vegetables has decreased. The strong rise in sales of convenience vegetable products combined with the trend to transport these products at too high temperatures is resulting in a growing visibility of the shelf-life problem.

Temperature abuse at unloading occurs reasonably frequent with transports to shops. At small shops the goods reception is not chilled. The driver leaves the load at reception and sometimes no shop personnel are directly available to attend. We have occasionally recorded time periods of ½ to 2 hours for loads left uncooled with our own datalogger tests of Dutch supply chains. Again a (modern) datalog system will assist the transport operator whenever claims are filed as a result of temperature abuse at unloading. In order to avoid these situations, some supermarkets allow drivers to bring the load to chilled cabinets or only allow the driver to unload after the supermarket personnel are present to receive the chilled load.

Food transportation with frequent stops and many door openings forms a special challenge in maintaining the temperature of a conditioned transport, especially for foodstuffs with low specific heat capacities, such as bakery products. Lamellae and air curtains at the doors of the vehicle can reduce the heat leakage during (un)loading significantly.

In our modern supply chains an increasing amount of refrigerated shipments is air-freighted. Refrigerated air shipments have their own unique problems. Since normal refrigeration units are forbidden in aeroplanes, freights are either unchilled and insulated or shipped in special containers chilled with dry ice. Most refrigerated loads are well pre-cooled and insulated for the flight. A very limited number of shipments make use of the specially designed Envirotainer.[8] This is an air-cargo sized container where dry ice (solid carbon dioxide) can be introduced in a separate cupboard. The carbon dioxide evaporates and chills the load. A fan forces the air to circulate in the container. The relative high rent prices of these Envirotainers currently limit their use. Consequently, most loads are pre-cooled and insulated. Temperature abuse can happen as a result of slow loading or unloading of planes. When food or flowers are flown over borders (to other continents) the veterinary or phytosanitary inspection service first has to clear the load. Sometimes it takes them a few hours to attend and to inspect at airports. The insulated container will wait all that time outside the plane – often in full sunshine – until the inspectors approve it. A few hours of full sunshine at a tropical destination can have disastrous effects on the temperature of the load. Improved air cargo containers can be expected in the near future, with, for instance, electric Peltier coolers and improved insulation materials.

Conditioned sea transports (reefer boats and refrigerated containers) usually perform outstandingly well in comparison with conditioned truck transport. In north west Europe the Coolboxx initiative (intermodal reefer containers for short sea, rail and road) has improved the logistics of fresh produce during the last decade. It allows transport to be fast and accurately conditioned. It is expected to expand further into central and south Europe in the coming years, also embracing inland shipping and controlled atmosphere storage technologies.[9] Reports of temperature abuse on sea-going vessels are related to unconditioned transports. For instance, temperatures of 80–90°C have been reported for unconditioned containers that are on top position on the boat and sail through the tropics. These extremely high temperatures have detrimental effects on the foodstuffs present in the hot container.

25.6 Avoiding cross-contamination

Cross-contamination can be avoided by taking the correct precautions. Some loads should not be transported simultaneously or should be well separated by a barrier, such as pallet sleeves or pallet wraps. Alternatively, transport units could be cleaned prior to and after transport with a high risk for contamination: liquids, bulk goods and hung carcasses. In the past cross-contamination problems have arisen with return freights and wooden pallets.

Return logistics such as returnable poly(ethyleneterephthalate) (PET) bottles form a potential contamination hazard in retail trucks. These trucks serve several shops from one distribution centre and deliver food stuffs to the shops and pick-up returnable PET bottles. These used bottles contain sugar-rich residues and are perfect growth media for microorganisms. A limited amount of bottles are misused by consumers and contain residues of motor oil, pesticides, urine, etc. Since these bottles do not all have their screw caps tightened, a pallet rack of these returned bottles forms a dripping stack. The fluids will drip on the floor of the truck and cause cross-contamination to other loads. Some incidents have been reported with returned bottles in the past. Nowadays, these bottles are collected in big bags or closable crates in the supermarkets. These bags or crates are closed, mounted on pallet trolleys and transported without hygiene problems.

Another example of a troublesome return freight is spoilt meat. Fresh meat that has reached its best-before date in the shop needs to be collected separately and destroyed at an approved facility. Some supermarket chains also use the normal transport trucks to return this load from the shops to the distribution centres. In cases where the spoilt meat was packed under modified atmospheres, there is no problem; otherwise the spoilt meat should be packed air-tight before return trucking. Empty plastic crates form another large return load in retail transports. This return load is often non-cooled and can cause temperature control problems when combined with loads of chilled food products.

The hygiene of wooden pallets has been heavily debated, as a result of the microbial flora found in other wooden articles, such as cutting boards.[10,11] Based on the differences in structure and cleanability it appears to be likely that plastic pallets will be more hygienic than wooden pallets. Hence, on the internet several pallet suppliers promote hygienic plastic pallets that are more cleanable than wooden pallets.[12] However, from the available scientific literature it cannot be concluded that wooden pallets are indeed less hygienic than plastic or metal pallets. Studies have revealed some differences in microbial flora between plastic and wooden pallets,[13] but as long as pallets are not stored outdoors or in very moist locations both types of pallets are safe to be used.[14] Nevertheless, the reality is that several countries (e.g. Australia) have forbidden import on wooden pallets to reduce the chance of spreading plant pests. Other countries (e.g. USA) allow import on wooden pallets only if they are heat-treated or fumigated. Furthermore, many individual food companies will not allow loads on wooden pallets to enter their facility. Moreover, from 1 March 2005 the European Union has forbidden import on wooden pallets (all wooden packages, including crates, etc.) that have not been adequately treated.[15] Allowed treatments are heat treatments and methyl bromide fumigation treatments according to the ISPM 15 standards.[16] The treated wooden pallets need to be marked with the International Plant Protection Convention (IPPC) logo. Therefore, the business of plastic and metal pallets is flourishing.

25.7 Future trends

The new contracts that transport operators will negotiate with food companies and retailers will contain financial claims in case temperature abuse is noticed at the destination. In order to avoid unjust claims, transport operators are implementing datalog systems. These systems record the actual temperature in the unit of transport equipment and are useful to refute unjust claims. The modern variants of these systems contain many added functionalities such as alerts to the driver when temperature abuse is becoming apparent. These temperature control systems will become more advanced in the near future and are expected to become fully fledged elements in advanced supply chain management solutions. A global positioning system (GPS) will continuously monitor the location of the vehicle, dataloggers will continuously determine the actual temperature in the truck, radiofrequency identification (RFID)-tags on pallets, crates and individual packages will describe the precise content of the load. All this information will periodically be exchanged via cell phones with the central computer of the transport operators and of their clients. This data avalanche will be processed by powerful computers that will continuously describe the real-time stock levels, its relative quality (based on the experienced temperature history) and location. These systems will continuously guard the temperature of the loads and give alerts in case of temperature abuse and in case forbidden mixed loads are present inside one transport unit. The incoming supply chain data will be further processed to optimise the allocation and prices of perishables based on actual market information. Currently, much value of perishables is lost in the supply chains, owing to the lack of information of stock levels, quality of loads and prices. The shrinkage for perishables is usually between 5 and 10% in the developed world and up to 40–50% in the developing world, implying a financial loss of at least 25 billion euros per annum. Advanced quality-differentiated stock management systems for perishables that will help to reduce the shrinkage to a few per cent, will improve the financial results of retailers. Additionally, these systems will reduce the environmental impact of food supply chains significantly. These systems might appear to be fuzzy science-fiction; however, it may be expected that the resulting savings will be an incentive for implementation in the coming years.

25.8 Acknowledgements

Many people contributed to this chapter, mostly via personal communications. Because hygiene is a relatively sensitive subject, many informers would not like to have their names and company names mentioned. We respect that and thank them for their sharing their knowledge and experiences with us.

25.9 References and notes

1. The complete list of standards is available via: http://www.codexalimentarius.net/standard_list.asp
2. The term 'units of transport equipment' might sound cumbersome, but it is the technical term used by the ATP (European Agreement on the Transport of Perishables) to refer to all types of transportation: boats, planes, trains, trucks, etc.
3. In agreement with European legislation, two exceptions are described: bulk transport of liquid oils and fats and crude sugar by sea vessels. When properly cleaned, these sea vessels are allowed to transport oils, fats and sugars and non-food cargo.
4. KEIJSERS, W.G.P. 'Hygiënecode, code voor hygiënisch werken bij transport, opslag en distributie van levensmiddelen', version 2/2001, published by EVO, Koninklijk Nederlands Vervoer and Transport en Logistiek Nederland, Zoetermeer 2001, in Dutch.
5. Transport operators regard precise cleaning instructions as competitive information. Nevertheless, it should not be a problem to publish generalised hygiene information on the web, which will still be interesting for a broad public.
6. For more information on the ATP see website: www.unece.org/trans/main/wp11/atp.html
7. PANOZZO, G.; MINOTTO, G.; BARIZZA, A. 'Transport et distribution de produits alimentaires: situation actuelle et tendances futures', *International Journal of Refrigeration* 1999, **22**, 625–639, in French.
8. See for instance: http://www.envirotainer.com/
9. References to Coolboxx can be found out: http://www.vdgp.nl/referentie_detail.asp?ProjectID=799 (Dutch), http://www.dunelmpr.co.uk/Geest-CoolboxxDeliveries-PR.htm and www.agro.nl/innovatienetwerk/doc/Coolboxx.pdf (Dutch report).
10. AK, N.O.; CLIVER, D.O.; KASPAR, C.W. 'Cutting boards of plastic and wood contaminated experimentally with bacteria', *Journal of Food Protection* 1994, **57**(1), 16–22.
11. RODEL, W.; HECHELMANN, H.; DRESEL, J. 'The hygiene aspects of wooden and plastic cutting boards', *Fleischwirtschaft* 1995, **75**(3), 277–280.

12. See for instance http://www.vicfam.com/hygiene.html, http://www.goplasticpallet.com
13. KLAASSEN, R.K.W.M.; VAN AKEN, J.A. 'Vergelijking van het hygiënisch gedrag van houten en kunststof pallets', SHR report number 95.175, Wageningen, 18 April 1996, in Dutch.
14. BEYER, G.; GUDBJÖRNSDOTTIR, B. 'Project P99095, Wood in the food industry – guidelines for handling wooden pallets and packaging', Nordic Industryfond, Herning January 2002. Copies available via: http://www.treteknisk.no/Tema/naeringsmiddel/publikasjoner/Report-8.pdf
15. Commission Directive 2004/102/EC, see for the English version: http://europa.eu.int/eur-lex/pri/en/oj/dat/2004/l_309/l_30920041006en00090025.pdf
16. See for instance: https://www.ippc.int/IPP/En/default.htm

26

Improving the control of insects in food processing

E. Shaaya, The Volcani Center, Israel, R. Maller, Pepsi Co., USA M. Kostyukovsky, The Volcani Center, Israel, and L. Maller, United States Department of Agriculture

26.1 Introduction

In the past, the main management efforts were focused on preharvest agriculture, and only recently has there also been emphasis on post-harvest management. This chapter will provide a basic overview on the hygiene management of dry commodities where the moisture content of the products is less than 17%, such as grain, dry fruits and vegetables, spices, flour products and other sorts of dry food.

At least seven thousand years ago when humans started to settle and cultivate the land, they began to store food. There are two important reasons for food storage. First, to enable the farmer to keep the commodity for the time that it can be sold at a good price. Second for security, countries import food for local consumption; they have to keep stocks of grains and other commodities for emergency use. Today, grain is the main source of food in the world. About 80% of the food used by people originates from plants such as wheat, rice, millet, sorghum, rye, barley, legumes, soy, cassava, sweet potato, coconuts and banana. In most developing countries the yearly consumption of grain per capita is 180 kg = (1/2 kg/day) and most of the proteins come from plant origin, compared with developed countries where most of the proteins for human requirement originating from animal sources.

When considering the distribution of grain: in the Eastern part of the world where rice is the main grain culture, Thailand is the biggest producer followed by China, Japan and Indonesia, whereas in the West, wheat is the main grain culture in the USA, Australia and Europe, and in South America corn. In general the total production of grain in the world is around a billion tonnes. The damage during storage of grain in the developing countries is between 5% and 40% depending on the period of storage. Besides spoilage of the commodities because of the metabolism of insects and microflora, oxygen is consumed and heat, water and carbon dioxide are released. This biological process causes the formation of hot spots in the grain bulk, with temperature that rises up to 60 °C. The continuation of the biological heating is the chemical heating that is caused by the oxidation of unsaturated fats. In this process the heat produced can reach up to 300 °C and the commodities ignite. It should also be mentioned that some insects are able to survive with moisture content of the grain as low as 1.0%. Therefore, the hygiene management of dry stored products is of utmost importance for safe storage.

26.2 The grain bulk as an ecosystem

The grain bulk is regarded as a dynamic system. Seeds are living matter but in a state of dormancy and their activity is very low. This situation is the best for storage: the deeper the dormancy of the seeds, the better. To understand the ecology of storage one should consider a grain bulk as an ecosystem (Fig. 26.1). The grain bulk ecosystem consists of biotic and abiotic factors, and the environmental factors of temperature, atmospheric gases and moisture.

26.2.1 Biotic factors

Biotic factors are the live components, and consist of:

- grain;
- insects and mites, damaging components;
- microflora = (fungi, bacteria and yeast), also damaging component.

These will be dealt with later.

26.2.2 Abiotic factors

Abiotic factors are inanimate, and consist of:

- dust and foreign material;
- intergranular air;
- water vapour;
- storage structure.

Dust, dockage and foreign material

These are good substrates for insect and fungi development.

Intergranular air (space)

The intergranular air is the air between the seeds or other commodities. It occupies 20–50% of the volume of the bulk grain. The intergranular air is important for the following reasons:

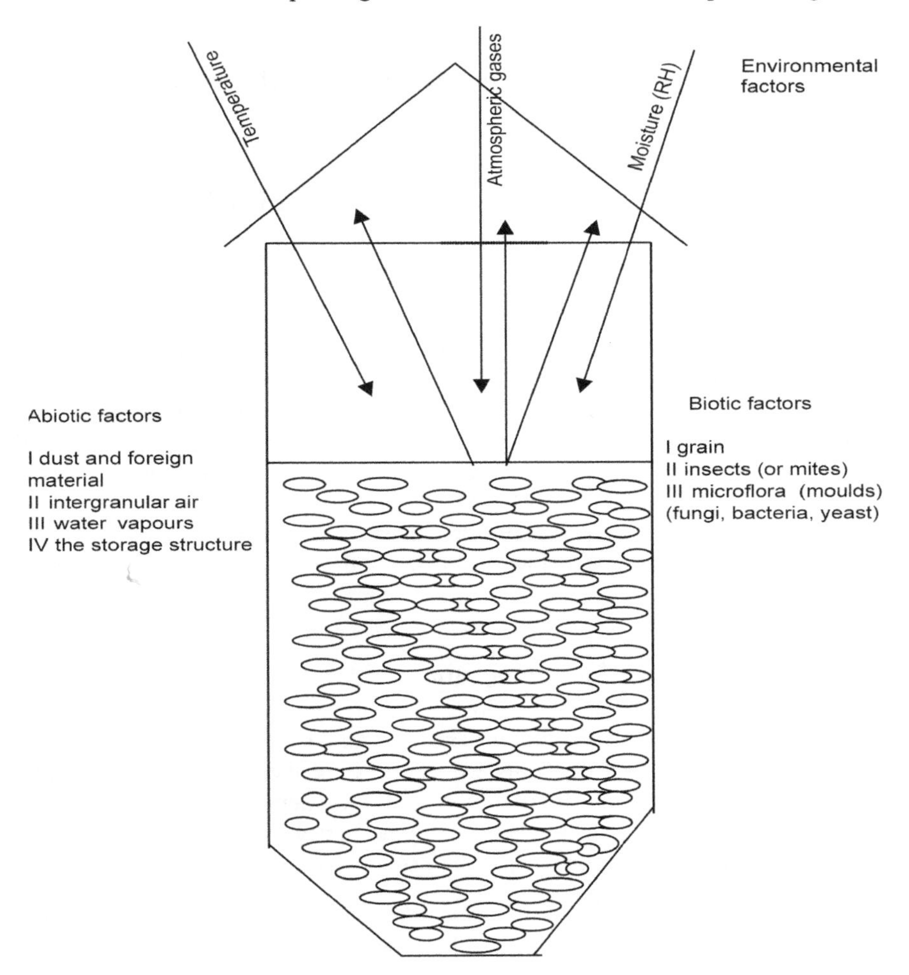

Fig. 26.1 The grain bulk as an ecosystem.

- According to the percentage of the air in the grain, it is possible to calculate the weight of the seeds that can be stored in 1 m^3. For example, for wheat about 700 kg/m^3, and for flour 500 kg/m^3.
- In the intergranular space, air currents (= convection currents) can move from one place to the other and carry water vapour.
- The intergranular space facilitates fumigation treatments for insect control.
- By changing the concentration of the atmospheric gases in the intergranular air it is possible to control or depress insects and fungal development.

Water vapour
The water vapour is located in the intergranular air. If there is a temperature gradient in the commodity, the water vapour is carried from the hot to the cold

area by convection currents and condensates. The water increases the humidity of the commodity, causes fungal development and finally the spoilage of the commodity.

Storage structure
Storage structure is important as it facilitates aeration, fumigation and the transport of the commodity from one place to other.

26.2.3 Temperature

Stored product insects are mainly of tropical and sub-tropical origin. Because insects cannot control their body temperature, their rates of development and reproduction increase with rising temperature and become inactive at low temperatures. For each species there is an optimum temperature for growth and development, which is between 28 and 34 °C for most major stored product insects. There are also minimum (17–22 °C) and maximum temperatures (around 40 °C) between which they are able to develop. Temperature 5–10 °C above optimum will be lethal and below will cause a drop in their reproduction and feeding activity.

26.2.4 The role of moisture (humidity)

Moisture is the most important factor in storage. We have to remember that water is connected with life and there is no life without water. Therefore, the less water in the commodities, the better we can preserve them. For example, seeds are in deep dormancy, and water and oxygen wake them up. This is a negative process that is not reversible and helps the fungi to develop that finally spoil the commodity. Humidity can be divided into two parts:

1. *Water inside the seeds or the commodity*. The percentage of H_2O in the commodity plays an important role in the storability of the commodity. High humidity results in high activity of biotic factors such as microflora and causes the spoilage of the commodity. H_2O is an integral part of the grain and its products and it has the same commercial value as the commodity.
2. *Water in the outside air or relative humidity of the air (RH)*. Cereal grains are hygroscopic and gain or lose moisture to achieve equilibrium with the ambient air. If we store dry grain or other commodity in humid air of 80–90% RH, the grain or other commodity will absorb H_2O until it arrives at equilibrium with the moisture of the ambient air. In dry air of 50–40% RH the opposite will happen: wet grain will lose H_2O. The dynamic equilibrium between the moisture of the ambient air and of the commodity moisture is called the equilibrium moisture content (EMC). This is one of the most important factors in predicting the storability of the grain. EMC values for grain vary by differences in grain variety, maturity and chemical composition, (mainly oil content). EMC increases, with increasing relative humidity and decreasing temperature.

Another important factor is the critical moisture content (CMC): at this moisture the grain is not resistant to storage and facilitates the development of microflora. Each type of seed has its CMC. For example the CMC of wheat is around 15%, which is in equilibrium with 70–75% RH of the ambient air. Usually seeds with high oil content have much lower CMC than seeds with low oil content.

For safe storage, farmers use a safe moisture content, which is usually about 2% lower than CMC, and secure safe storage. For example wheat 13% for a period of 1 year storage, 11–12% for 5 years.

Methods for measuring grain humidity

- *Direct methods – oven drying.* These are usually accurate methods. The common method is oven drying procedures. Whole grains or milled products are dried in one step according to the temperature and time specified for each grain type or its products. The moisture content can be calculated on a wet or dry basis. Wet weight is calculated on the total weight of the product tested. Dry weight is the dry weight of the product tested after the water has been removed.
- *Indirect methods – electronic moisture meters.* These are based on electrical conductance or capacitance of grains, which are largely dependent upon the moisture content of the grain. This is a practical method for determining moisture content of the grain or other commodities. Electronic meters are reliable, fast, simple to operate and of low cost per test. Various types of such moisture meters are available in the market for different uses.

26.2.5 Atmospheric gases

By changing the percentage of the various gases in the intergranular air, increasing the concentration of CO_2 or decreasing the concentration of O_2, it is possible to control or depress insects and microflora development, for example by using hermetic storage or modified atmosphere.

26.3 Moisture migration in the grain bulk

This is also called movement of moisture as a result of non-uniform temperatures in the grain bulk. In the intergranular space there are convection currents, which also carry water vapour from the hot area to the cold area of the grain bulk. In the cold area the water condenses and the humidity of the seeds increases, which accelerates microflora development and hot-spot formation, which cause biological deterioration of the commodities. In metal silos this phenomenon is very common. Moisture migration is a limiting factor for the use of hermetic storage and in this type of storage it is important to store dry commodities. To prevent moisture migration, it is recommended to aerate so as to cool the grain and also to isolate the parts of the storage area that receive direct sun. In storage with sacks, usually we see spoiled grain or other commodities in the bottom part of the storage area. This is because the floor is cold and the moisture migrates from the upper hot places to the bottom. To overcome this problem, the sacks should be stored on wooden beams.

26.4 Dry- and wet-grain heating

During the metabolism of insects and microflora, O_2 is consumed and heat, water and CO_2 are released. The very poor thermal conductivity of the grain prevents the heat from dissipating, and hot-spots are formed with temperatures that rise up to 40 °C. This process is called dry-grain heating and is caused by insects. High temperatures are unfavourable for insect development and the active forms move away from hot spots to form new ones. In addition the metabolic water in the form of water vapour is carried upward through the bulk by convection currents and is deposited by condensation at the cool surface. Consequently, dry-grain heating caused by insects is often followed by wet-grain heating caused by microflora as moisture is transferred from the hot to the cold part of the grain bulk (Fig. 26.2).

26.4.1 The process of heat production by insects and microflora

Heat is produced by insects and microflora by aerobic or anaerobic processes. Examples of the aerobic process are:

- grain with high carbohydrate content, 1 mole glucose produce:

$$180\,\text{g glucose} + 134\,\text{L}\ O_2 \longrightarrow 134\,\text{L}\ CO_2 + 108\,\text{g}\ H_2O + 677\,\text{kcal}$$

- Grain with high oil content, 1 mole tripalmitin produce:

$$807\,\text{g tripalmitin} + 1624\,\text{L}\ O_2 \longrightarrow 142\,\text{L}\ CO_2 + 883\,\text{g}\ H_2O + 7617\,\text{kcal}$$

Examples of the anaerobic process are:

$$1\,\text{M Glucose} \begin{array}{l} \nearrow \text{Lactic acid} + 22.5\,\text{kcal} \\ \longrightarrow \text{Ethyl alcohol} + 22\,\text{kcal} \\ \searrow \text{Acetic acid} + 15\,\text{kcal} \end{array}$$

The above equations reveal that equivalents of O_2 consumed and CO_2 produced vary with the type of substrate oxidized. Also, the energy released per unit of substrate in the aerobic process is much higher than the anaerobic process.

We differentiate between biological and chemical heating. Biological heating is caused by insects and microflora occurs at a maximum temperature of 50–60 °C. It is caused by both dry-grain heating and wet-grain heating.

Dry-grain heating is caused by insects and is characterized by:

- the moisture content of the grain or other commodity less than the critical moisture;

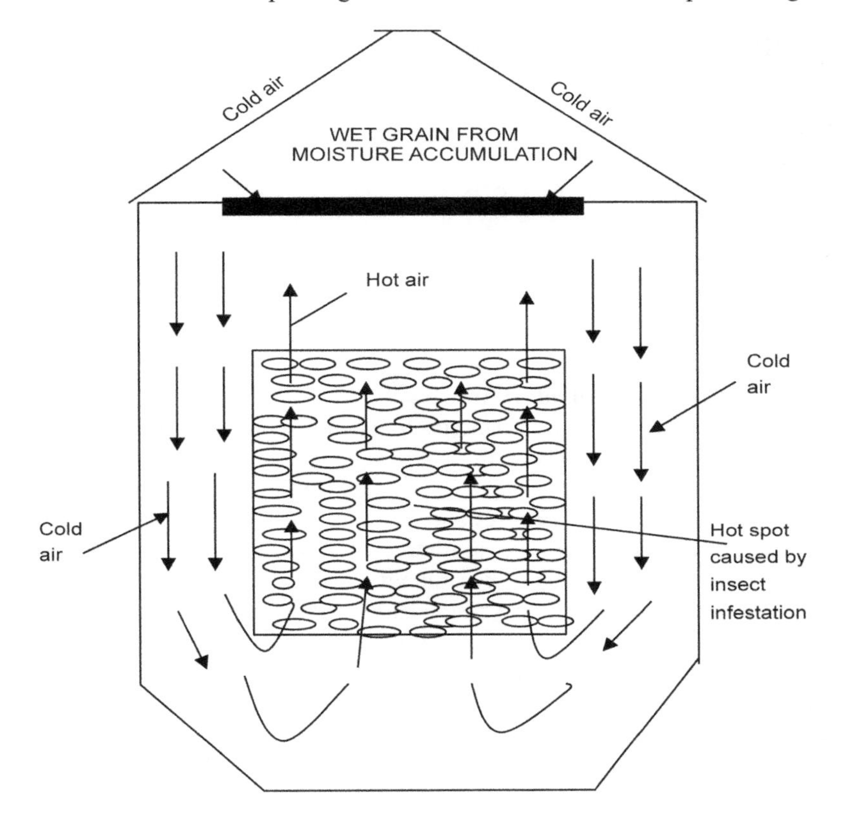

Fig. 26.2 Moisture migration in stored grain as a result of hot-spot formation caused by insects.

- the temperature of the grain around 40 °C;
- insects are present.

In this case, the control measures are chemical, such as fumigation.

Wet-grain heating is caused by microflora and is characterized by:

- the moisture content of grain higher than the critical moisture;
- the temperature is higher than 40 °C and it can reach to 60 °C;
- no insects are present.

In this case aeration is recommended, to lower the humidity and temperature of the commodity.

Chemical heating is the continuation of the biological heating and is caused by the oxidation of the unsaturated fats. In this process, the heat produced can reach over 100 °C, in some cases up to 300 °C, which makes it possible that the commodity will ignite.

26.5 Insects in stored products

Stored product insects are of tropical or subtropical origin; they live and reproduce best under warm, humid conditions. They are small in size and capable of hiding in cracks and crevices. They feed on foods of low moisture content. Their reproduction rate is rapid, and most stored product insects complete their development within a month in optimum temperatures. For example, red flour beetle populations can increase 70-fold per month. It should be mentioned that stored product insects are not known to transfer pathogenic microorganisms that cause human diseases.

26.5.1 Primary and secondary insects

In the warmer climates of North America a good example of a primary infesting insect is the Rice Weevil (*Sitophilus oryzae*) and the angoumois grain moth (*Sitotroga cerealella*), which attack the crops in the field just before harvest. In fact, it is this harvest primary infestation that has been the original infestation from which the grain stores, the silos, truck transport, ship holds, and warehouses of modern commerce also became infested. Primary insects are uniquely adapted both anatomically and physiologically to attack the specific food supply they have evolved with over time. The primary insect's larvae and adults feed entirely within the kernels of grain or chew through the outer coat and devour the inside of the grain or cereal. The common primary insects are: *Sitophilus oryzae*, *Sitophilus granaries* (*L*), *Sitophilus zeamais Motschulsky*, *Rhizophertha dominica* (*F*), *Lasioderma serricorne* (*F*), *Trigoderma granarium Everts*, Bruchidae sp., *Sitotroga cerealella*.

Secondary insects are opportunistic insects that feed on the grain only after the seed coat has been broken either mechanically or by a primary insect. Secondary insects typically feed on materials that are out of condition and damp, and have some mould growth present. Some secondary insects feed on the mould rather than the food product. Typically one does not find secondary insects creating the initial primary damage in the field; they are insects that infest and attack the food supply in storage. Secondary infesters are not necessarily uniquely adapted to attack the food supply in the field, although they have become adapted to attack food supply already damaged by primary field insects. A good example of this would be the saw toothed grain beetle (*Oryzaephilus surinamensis*), which is typically found as a secondary infester of cereal grain, flour and dry fruits and vegetables. Other common secondary insects are: *Tribolium castaneum* (*Herbst*), *Tribolium confusum Jacquelin du Val*, *Cryptolestes* sp., *Plodia interpunctella* (*Huebner*) and *Ephestia cautella* (*Walker*).

26.5.2 Insect damage

Very few people outside the pest control and hygienic design fields realize the value insects extort from people in their efforts to store and distribute food and other consumable commodities. It is estimated that the annual loss sustained by insect infestation in cereal and grain storage is approximately 10% worldwide. In developing countries the damage may account for over 40%, depending on the period of storage. If we could eliminate this loss, or reduce it significantly, there would be significant positive economic savings and there would be an improved chance to avert famine by utilizing these cereals and grains to feed those in developing and overpopulated countries.

Insects usually come into the warehouse with infested grain or cereal and come into the home with infested packaged foods. Under favourable conditions, insects greatly increase their numbers and may then crawl or fly into uninfested food. They can enter through extremely small cracks and in some cases can chew or bore directly through the wrapper or container.

Insects not only consume the commodity, they also contaminate the food with faeces, webbing, insect fragments, ill-smelling and metabolic product. One must remember that foods that are infested after processing and packaging are considered adulterated and can no longer be saved through fumigation or cleaning. Once infested, the material must be destroyed in most cases. This creates millions of dollars of loss for food processors and manufacturers. A good example of unseen adulteration can be illustrated by the metabolic by-products of the grain weevil (*S. oryzae*). In many developing countries corn is the stable food for the poorer people. However, the uric acid produced by the weevils can render the corn unpalatable before its storage shelf-life. This seriously reduces the value of the food as a nutritional source and has obvious economic impact.

In dry food the standard is not more than five insect fragments in 50g. In some countries, keeping this standard can limit the export of the commodities. There are attempts to develop immunological tests based on the concentration of insect-specific proteins such as myosin and others to standardize the insect fragment method.

26.5.3 Ecology of stored product pests

Climate has a significant impact on insect ecology because it can affect insects' competitors by favouring or retarding them. The most significant effect is that climate can govern the geographical distribution of insects. Some insects thrive in a temperate climate, while others thrive in a tropical climate. For most insects there is an optimum climate in which they grow and increase most rapidly.

Effects of humidity, temperature, food, light, atmospheric gases, and insect density

Humidity

Primary infesters, such as the rice and grain weevils, cannot breed in grain with a moisture content of 8.0% or less and they will die if restricted to such grain for food. The flour beetle and the saw toothed grain beetles are examples of insects that can reproduce at low moisture content. The flour beetle and the saw toothed grain beetle can survive with moisture contents down to 6.0% and 1.0% respectively.

Temperature

Temperatures above 40 °C are not favourable for the development of most stored product insect pests, while temperatures below 21 °C will retard insect development. Food temperatures of 18 °C or below significantly inhibit reproduction of primary grain infesters since they do not lay eggs at these temperatures.

Food

Store product insects feed on plant material except for insects of the Dermestidae family which feed on animal skins. However, the type of food that insects consume directly affects their ability to increase their populations. A good example of this is the cocoa moth: apart from the proteins and carbohydrates the moth needs to build body parts and supply energy, this moth cannot develop, thrive or compete without vitamin B. Insects that bite and chew their food may drink water; however, most of the dried storage pests obtain their necessary moisture from the solid food they consume. Some insects that infest very dry food material obtain their water supply through condensation due to temperature changes in the product.

Light

The effects of changes in light intensity or photoperiodicity on insect development has shown its importance in relation to the maturation of the reproductive organs in stored product insects.

Atmospheric gases

Atmospheric gases such as oxygen and carbon dioxide have a direct effect on an insect's ability to survive. In fact, insects cannot survive in the absence of oxygen. Grain insects sealed up in air-tight vessels, with or without wheat, die as soon as the oxygen has been used up, and corresponding amount of carbon dioxide is produced by live wheat. There is also an optimum moisture content in the wheat that will result in maximum production of carbon dioxide. Wheat stored in airtight containers, at the optimum moisture level, will render itself immune to insect attack. High levels of carbon dioxide are less deadly than the effect of diminishing oxygen, but it has an intoxicating effect on the insects. Insects are able to survive for a long period in the presence of high concentrations of carbon dioxide particularly in low temperatures. Immature stages of many species show higher levels of tolerance than adults.

Insect density

Competition between insects may take the form of competition for food and competition with predators and parasites. Competition for food is most severe between species that have the same nutritional requirements and the same temperature and humidity preferences. Sometimes insects may enter into severe competition if the temperature is suitable to both; however, a drop in temperature or change in moisture content can be detrimental to one species, while the competing species may not be affected. This will typically favour the species that is adapted to the change in temperature and moisture. In addition, as an insect infestation and population become established and of high number it can become susceptible to parasites and predators. A predator insect is one that seizes and carries off and devours its victims. An insect that lives attached to or within its victim or host is called a parasite. Parasites do not always kill their host and one well-known effect of a parasite on its host is to produce sterility. Bacterial and protozoan agents of insects have been known to destroy up to 90% of a population's larvae stage. The larger and denser the population of an infesting insect, the more susceptible the population is to damage via decreasing supply of nutrients, changes to the environment due to metabolic waste products, and the spread within the population of bacterial and protozoan parasites.

26.5.4 Methods of insect detection

There are a number of methods of detection at the disposal of the entomologist and sanitarian: one of the most effective ways is thorough inspections. Inspections need to encompass the facility where dry products are stored and manufactured along with the equipment that processes and packages the product and the frequency of inspection needs to be within the life cycle of the target insect(s). This can be accomplished by inspecting interior surfaces of the processing equipment, especially in dead end areas or dead spots in the pneumatic piping used to transport products. In addition, dust control systems are to be inspected at the same frequency because this is an area that will definitely show signs of activity if either ingredients or processing equipment are infested.

An infestation of stored product pests may be recognized not only through the presence of insects themselves, larvae and adult, but also by the holes in the packaging, webbing in the food material, insect feces, and odours that are by-products of insect metabolism. All of the insects mentioned above, along with many other infesters of dried products not mentioned, create millions of dollars of damage in grain mills and warehouse operations. It is also critical that incoming raw materials are not bringing insects into the plant and these can be effectively inspected through proper sifting with inspection of sifting tailings for adults and larva.

26.6 Measures of control

The prevention and control of infestation is the ultimate aim of the dried product industry. The effective control of pests is very complex and requires the use of many tools and disciplines. The process starts with good hygienic design of equipment and facilities, insect-free ingredients and packaging materials, high level of knowledge of insect behaviour and ecology, high level of inspection, integrated sanitation procedures and the use of proven pest control methods and systems. Control measures can be divided into three major groups: preventative measures; chemical control; and physical measures.

26.6.1 Preventative measures

Design of storage facilities, good housekeeping and sanitation

Good housekeeping and sanitation are two of the most important elements of insect control. The first level of prevention is designing storage facilities that do not have false ceilings, porous insulation, false walls, and numerous cracks and crevices where dust can collect and insects can be harboured. Every aspect of a dried product handling facility needs to be designed to keep dust and insects out. Adequate dust control systems need to be installed to keep all product dust isolated to a sealed dust collection system that has easy access and is easy to clean. All equipment must be installed to facilitate inspection, cleaning, maintenance, and insect control.

Pneumatic conveying systems and dried product-processing equipment must be designed to be free of deadends and other dead accumulation points. Access doors and ports need to be strategically designed so that interior process surfaces can be inspected and cleaned within the life cycle of the insect.

The plant or warehouse must have all attractants to insects eliminated. The entire facility must be free of weeds or tall grass, all litter picked up promptly, no unused equipment, lumber, pipes stored outside unprotected and grounds should slope away from the wall and not contain low spots or areas that permit water to stand. Refuse must be stored in insect-proof containers and such containers must be stored on a smooth concrete slab equipped with a hose and cleaning station, with the entire slab discharged to a sanitary drain. All openings to the outside must be equipped with tightly closed doors, all windows containing screens with no damage, and all cracks to the outside completely sealed off. All vents, ducts, exhaust and air inlets must be examined to be certain they are effectively sealed, screened, or closed off to the ingress of insects. All storage within the plant and warehouse must be at least 0.45 m from walls to allow for inspection and to apply insect monitoring and control devices. Columns and doors that intrude into the space must be taken into account and at least 0.45 m left for access past a column or door.

Sanitation is critical to the control of insects and to minimize infestation. Periodic cleaning to prevent the accumulation of food materials will reduce the possibilities of infestation. In addition, equipment and plant production storage facilities need to be inspected and cleaned at a frequency that is within the life cycle of the target insect. A master sanitation schedule needs to be developed to assure that all areas associated with processing equipment and facilities get cleaned. It is essential that loose food or dust be removed from ceilings, overhead pipes, in cracks and crevices in walls and floors and any other areas that will trap dust and harbour insects. A dried product operation without a detailed sanitation program will become infested and has no hope of controlling the infestation. Fumigation and other means of control can have a significant impact on reducing insect activity; however, without a good sanitation programme one will never eliminate the source of the infestation.

Inspection and monitoring

Direct methods

As mentioned before it is essential that large stock of stored ingredients and products, particularly if they are not used or moved at frequent intervals, be inspected regularly (within the life cycle of the insect) for signs of infestation, so that control measures can be applied before the infestation becomes serious. One of the most effective ways of detecting insects in raw ingredients is to install and inspect sifting equipment. The first and most reliable method passes ingredients over and through the sifter screen and trails the insects over to reject. The finer the ingredient particle size, the finer the mesh of the sifting screen must be, and the more efficient it is in removing all forms of insect life, including eggs. Stored-product insects are most prominent in grains/grain-based food products and dehydrated fruits and vegetables. To remove adult stored-product pests, a minimum 30-mesh screen with a (0.4 mm) mesh opening needed. The 30-mesh will remove most of the insects in the larva stage also. A 10XX screen is needed (0.13 mm) to remove eggs. The important detection method is frequently examining the tailings for evidence of insect infestation. This is very effective at minimizing large infestations that are brought into the storage and processing system. For bulk loads of products coming in the sifter, tailings need to be inspected at the beginning, middle, and end of the unloading process. For in-process inspections, tailing should be inspected once per shift for evidence of insect infestation.

Another means of detection is the use of UV light Flintrol units. The UV wavelength emitted from these units is attractive to many flying stored product adult insects and when they fly into the unit, the insects are electrocuted and they fall into a collection pan. Weekly inspection of the collection pans can provide the first sign of a problem with an ingredient or finished product stored in a warehouse or process area.

Glueboard zone monitors are similar to rodent glueboards; however, they are made to be attachable to walls, columns or pallets and are designed to monitor for insect activity. Several types of zone monitors utilize a food attractant to increase its effectiveness.

Pheromone traps are specialized monitoring devices that utilize a sex pheromone attractant to lure a particular insect to the trap. This device is particularly useful because it can monitor activity over a large area of the facility. Care must be used when utilizing this trap because one does not want to lure specific insects into a facility that are also indigenous to the area of the storage/processing facility. As a result, they must be utilized and monitored with great vigilance.

Indirect methods

Nearly all stored product insects cause an increase in temperature. Therefore, it is extremely important to be aware of temperature changes in the grain bulk by taking temperature readings on a continuous basis. The readings can be taken using either portable temperature probes or permanent installations of thermocaples, capable of providing readings from all depths of the grain bulk.

Moisture measurements are a very important criterion for grain storability. The measurements are carried out using special moisture meters and also by laboratory methods.

26.6.2 Chemical and physical control measures

Although there are many tools and methods available to help prevent insect infestation, it is almost impossible to stop the entry of insect eggs into grain storage areas and in dried product processing and packaging plants. As a result there are a number of chemical and physical control measures available as a supplement to preventive measures.

Chemical controls

It must be considered that most pest control procedures usually involve the application of chemicals. These include fumigants, aerosols and residual insecticides.

Fumigation is the most effective method for insect control. Under controlled temperatures the fumigant can penetrate into packaging materials and bulk grain to kill all stages of insect development. The components are highly toxic and need to be handled with great care. However, their toxicity and penetrating ability make them highly effective. In fumigation a toxic substance in the form of a gas is added to an enclosure (fumigant chamber or silo) or structure (building/processing plant or warehouse) or under gas-proof sheets, which can be kept sufficiently gas-tight for this purpose. The required concentration of gas must be maintained for a specific period of time, with the container or building ventilated after meeting minimum times and concentrations. The amount of fumigant is dependent on the target insect, temperature, and level of container or building tightness.

Two major fumigants are very common for the control of stored product insects, methyl bromide and phosphine (PH_3). Methyl bromide is classified as a serious ozone-depleting substance and its use will be phased out in the near future in all developed countries. Studies also revealed that insects are developing resistance to phosphine. Besides, PH_3 is corrosive and slow acting, with an exposure time of between 4 and 28 days, which depends on temperature and the commodity. PH_3 is available in a number of formulations: in bags, plates, pellets, tablets, sachets and also cylinder formulation, phosfume (a pre-mixed phosphine in 98% CO_2). Three types of solid phosphine formulations have been developed and can be used for various purposes: aluminum phosphide for hot ambient; magnesium phosphide for cold ambient and methyl phosphide against resistant insects. Among other fumigants, sulfuryl fluoride (Vikane ProFume) is used as a space fumigant primarily for controlling dry wood insects, termites and other pests of structure. Ethyl formate and methyl iodide are not licensed yet.

Fumigants are volatile chemicals, their vapours entering the insect by inhalation or through the body surface. Fumigation is used in dried foodstuffs, buildings and ships. They are typically applied as a gas; however, PH_3 is used in a solid state and when contacted by atmospheric moisture, hydrogen phosphide gas (PH_3) is released to kill its target insect. Fumigants are highly effective and have no residual properties after ventilation.

Aerosols are effective against exposed adults in processing plants and warehouses. Aerosols are effective because they take the insecticide in water or oil-based emulsions and the aerosol generator creates a mist or fog of fine particles of liquid. The fine particles can remain suspended in air for several hours, until they meet an insect: the particle will then stick on the insect, eventually penetrating its waxy outer surface, then enter the body and poison the insect. The benefit of this treatment is that you can treat large areas through the ULV (ultra low volume) concept and utilize natural air currents in a building or space to carry the insecticide droplets to the insects. Typically droplet sizes of 1 to 15 μm are most effective; larger particle sizes fall out of the atmosphere too quickly and are less effective. This treatment is effective against flying insects and relatively ineffective against larval stages in food material or crawling insects in cracks and crevices. The disadvantage with aerosols is that they treat the infestation activity on the surface and need to be reapplied frequently to effect control. Large facilities today have installed aerosol systems that are designed to deliver specific droplet size insecticides through strategically placed application nozzles that guarantee complete coverage of the space being protected.

Residual insecticides utilize an insecticide applied in such a way that it remains active for a considerable period of time. Dusts, sprays, and lacquers are used to deliver long-lasting insecticides with which an insect may make contact with over time. A number of contact insecticides, which are active against a wide spectrum of stored product, are in common use in grain and other commodities. Methoprene is a juvenoid, with a recommended dose of 2–4 mg/kg. Pirimiphos-methyl (Actellic) organophosphores dosage: 4–10 mg/kg; Deltamethrin, a synthetic pyrethroid, 0.5–2 mg/kg; against *R. dominica* and other stored product insects a mixture of pirimiphos-methyl 4–8 mg/kg and deltamethrin 0.3–0.5 mg/kg is recommended. The chemicals can be applied to the grain either by pump sprayers or by gravity techniques during the transfer of the grain to the bin.

Physical measures

These are environmentally safe measures which should be used as a substitute to the chemical methods if possible.

Entoleter

This is a device that has been around for a long time. The entoleter is used in flourmills and flour receiving plants as a mechanical means of destroying insects. Flour that is transferred to the entoleter is blown into the machine and the flour is thrown by centrifugal force between two flat steel discs or plates that revolve on a central shaft at very high speed. Small, round, hardened steel posts are closely spaced in two concentric rings between the two discs. The impact of the flour against the revolving discs and posts and against the housing of the machine means that all stages of insects and mites, including eggs, that may be in the flour are killed. The treated flour passes out through a spout at the base of the machine.

Heat treatment

Heat treatment is often used in the sterilization of grains and dried fruit to free them of insects. Many of the dried product infesters can be killed if the products are exposed to temperatures of 55 and 60 °C, for 1 hour depending on the insect and stage. However, difficulties in attaining these temperatures in insulated grain mass limit the use of this technique.

Freezing

With certain types of infesters freezing for long periods of time can also help kill all stages of insect development. With certain commodities (very low moisture) holding the product at −18 °C or below for 14 days will kill eggs, larvae, and adults. It is critical that −18 °C be achieved and that the material be held for the full 14 days.

Aeration and refrigeration

Aeration is the process of cooling the commodities with ambient air. Refrigeration involves passing ambient air through refrigeration system to lower its temperature, before it is blown through the grain. The purpose is to prevent moisture migration, retard the development of insect infestation and prevent biological deterioration of the commodities.

Ionizing radiation

This provides chemical residue free process to disinfest commodities of live insects. The sensitivity of the various live organisms to ionizing radiation is varied. For example, insects are about a hundred times more resistant to radiation than humans. The sensitivity of insects to gamma rays varies according to the species and stage of development. The egg stage is more sensitive than larvae and adults. There is a dosage limitation of 10 kGy of gamma radiation in food. The common use of radiation is to prevent sprouting of potatoes and onions and the control of microbial infections of spices and meat.

The disadvantages of this method are:

- economically, it is an expensive method;
- the difficulties in transferring the commodity to the radiation area;
- proof that no chemical changes occurred in the commodity after radiation;
- the acceptance of the radiated commodity by the consumer.

Hermetic storage
In hermetic storage the grain or other commodities are placed in an airtight enclosure. The natural respiratory activities of the grain and the associated insects and fungi reduce the concentration of oxygen and increase the concentration of carbon dioxide to a lethal level to insects and fungi.

Controlled atmospheres
This type of storage involves the use of externally generated gas to displace the intergranular gases. The aim is to reduce the concentration of O_2 to about 1% or to raise the carbon dioxide to about 80%. By doing so, insects and fungi infestation are retarded or controlled.

Inert dusts
Inert dusts are made from silica aerogels, various clays, diatomaceous earth and other silicates. The dusts function as insect killer by injuring the cuticle of the insects and causing them to bleed which result in losing water and death by desiccation. High price, cleaning and health hazard to the workers limit the use of this control method.

26.7 Future trends

Methyl bromide and phosphine are the most widely used fumigants for controlling pest infestation in grain cereal, dry food products and quarantine insects. Now we are facing the problem that methyl bromide will be phased out in developed countries in the near future and certain stored product insects developed resistance against phosphine. Methyl bromide is known to have wide spectrum effectiveness and it is assumed that only a different combination of alternative fumigants can replace its use. This situation will have serious impacts on industries and consumers, who rely on this fumigant for structural fumigation and post-harvest commodity until suitable alternatives will be developed.

This situation has led to the emergence of other fumigants. Among them are sulfuryl fluoride, for space fumigation and for the control of dry wood termites and other pest structures. Also, carbonyl sulfide was rediscovered as an alternative to methyl bromide. The combination of heat, carbon dioxide and phosphine, which has been introduced recently, was found to be very effective.

The future trend is to reduce the use of toxic chemicals which are hazardous to people and the environment:

- The use of integrated pest management (IPM). In this approach emphasis on routine inspection and cleaning to detect the problems before an outbreak occurs. The use of various traps (pheromone, light, glue board) is very useful.
- The use of phytochemicals: substantial work is underway to evaluate the potential of essential oils from aromatic plants for the control of stored product insects.

- The use of insect growth regulators (IGRs). Methoprene is very effective as residual insecticide. Also biopesticide, such as *Bacillus thuringiensis* and baculovirus, may be used.
- Physical control methods. These are environmentally safe measures that should be used as a substitute to the toxic chemical methods if possible. These include heat treatment, cold storage, radio frequency waves, irradiation. The various control methods should be tailored to different commodities that need treatment.

26.8 Acknowledgement.

We like to thank Ms Esther Ngumbi for her great help in preparing this manuscript.

26.9 Bibliography

BAUR J F (1992), *Insect Management for Food Storage and Processing*, St Paul, MN, American Association of Cereal Chemists.

BEIRNE B P (1967), *Pest Management,* London, Leonard Hill Books.

BELL H C, PRICE N and CHAKNABARI B (1996), *The Methyl Bromide Issue*, New York, John Wiley and Sons.

IMHOLTE T J (1984), *Engineering for Food Safety and Sanitation*, Crystal, MN, The Technical Institute of Food Safety.

MUELLER K (1998), *Stored Product Protection, a Period of Transition*, Indianapolis, IN, Insect Limited.

SAVER D B (1992), *Storage of Cereal Grains and their Products*, St Paul, MN, American Association of Cereal Chemists.

SHAAYA E and KOSTYUKOVSKY M (1998), Efficiency of phyto-oils as contact insecticides and fumigants for the control of stored product insects. In Degheele D and I Ishaaya (eds), *Insecticides with Novel Modes of Action: Mechanism and Application*, Berlin, Springer-Verlag.

SMITH E H and WHITMAN R C (1992), *NPCA Field Guide To Structural Pests,* Dunn Loring, VA, National Pest Control Association.

SUBRAMANYAM B and HAGSTRUM W (1996), *Integrated Management of Insects in Stored Products*, New York, Marcel Dekker.

TRUMAN L C, BENNETT G W and BUTTS W L (1976), *Scientific Guide to Pest Control Operations*, 3rd Edition, Cleveland, OH, Harvest Publishing Co.

27

Improving cleaning-in-place (CIP)

K. Lorenzen, Tuchenhagen GmbH, Germany

27.1 Introduction: limitations in current CIP systems

A cleaning-in-place (CIP) system can be only as effective as its design, construction, installation and operation. There are a number of opportunities to minimize or eliminate the limitations in today's CIP systems through improved hygienic design. Design personnel need to be familiar with good manufacturing practice (GMP), Hazard Analysis Critical Control Point (HACCP) methods, and the hygienic design criteria and principles laid down by the European Hygienic Engineering and Design Group (EHEDG, Documents 9, 10, 13, 14, 16, 17, 18, 20 and 25). All components that come into contact with product and/or the CIP system should be designed according to the Hygienic Equipment Design Criteria (EHEDG Document 8).

Engineering, installation and process parameters are directly related to hygienic design, and influence the hygienic and functional integration of all elements of the CIP system. If the design and installation of the plant result in the presence of pits, crevices, gaps, sharp edges, threads and/or dead ends, cleaning time will be greatly increased and there will be a risk of contamination.

Inadequate or inaccurate installation and the presence of porous or rough surfaces may allow bacterial colonies to proliferate, or permit the build-up of biofilm (substances excreted by bacteria to improve their ability to adhere to surfaces) and plaque. Cleaning and disinfecting fluids can only attack the top surface of the biofilm. Bacteria in a biofilm cannot be killed by saturated steam at 121 °C for 30 minutes, and thus, under normal operating conditions, these sections cannot be sterilized and there is a serious risk of contamination of the product. The T-joint is one example where there is a major risk of insufficient

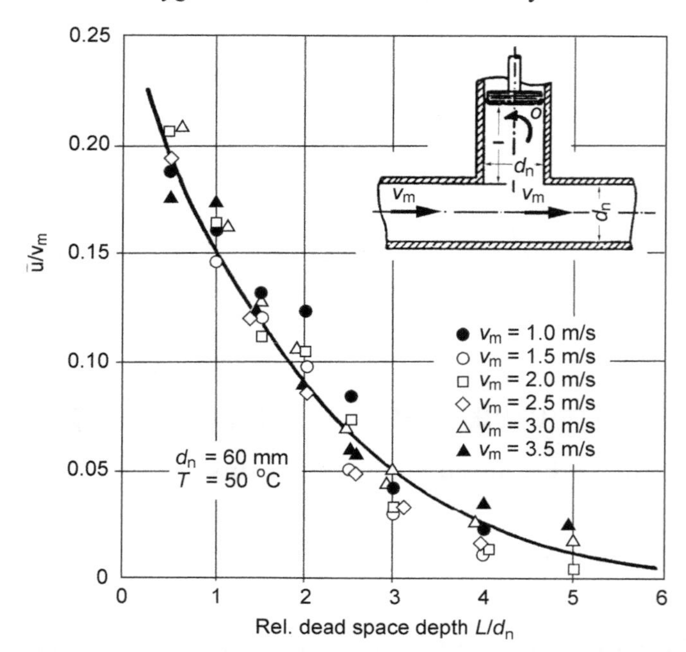

Fig. 27.1 Fluid motion in a dead space. (Federal Institute of Milk Research, Kiel, Germany.)

cleaning and therefore formation of a biofilm. Assuming the cleaning detergent in a process line has a velocity of 2 m/s, a T-section with a length equal to the diameter of the main pipe line will receive only 15% of this velocity (2 m/s), which is 0.3 m/s (see Fig. 27.1). Modern cleaning detergents are often used as a last resort to minimize cross-contamination, but a better solution would be to address the hygienic design issues that led to the contamination in the first place.

The composition of the detergents used in CIP systems is important. For optimum cleaning, the detergents used and their effectiveness in relation to the cleaning parameters must be adjusted to one another. Finally, lack of education and training for system operating personnel may be a limitation. Personnel who are better trained and educated in the use and methods of the system will be better able to respond to problems and to suggest improvements.

27.2 Cleaning and disinfection parameters

27.2.1 Cleaning parameters

The four cleaning parameters that are independent of the installation are shown in the Sinner's circle (see Fig. 27.2): detergent (R in the diagram), cleaning time (Z), fluid mechanics (F) and temperature (T).

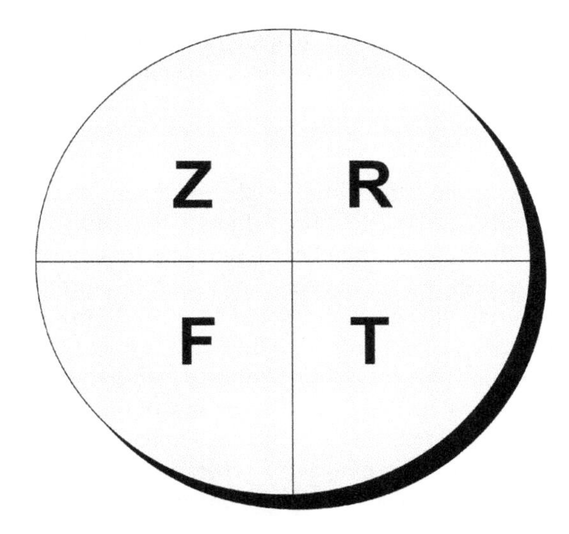

Fig. 27.2 Four cleaning parameters.

Detergents
The effectiveness of detergents is related to their chemical activity, composition, concentration, surface tension properties and dispersing power. The detergent must dissolve rapidly and completely in water, and must act quickly on the soil or deposits it is expected to remove (such as protein, fat, beer scale, hop resins, burnt yeast, trub, pulp).

The detergent should also have a high soil-carrying capacity and be easy to flush off with water. In addition, the detergent should not foam, and should be compatible with the other materials used in the plant. Cleaning detergents must have a conductivity well above that of water, and demonstrate a linear relation between concentration and conductivity. Modern cleaning detergents can reduce the cleaning parameter R through improving one or more of these factors.

Cleaning time
Cleaning time relates to the total duration of the cleaning process, and must be sufficient to allow for the dissolution of soil, swelling of soil, saponification, dispersion and final rinsing. The last residual layer of the deposit on a surface can only be attacked by a detergent if it is allowed long enough to react, with the reaction time being the time the detergent is in contact with the deposit at the right concentration and the right temperature.

The duration of cleaning is determined by the time it takes to complete the different phases of the cleaning operation. These phases are the pre-rinse, where all loose components are flushed out, the main wash, which must be checked at critical points (e.g. heat exchanger plates, pipe bends, tank, internals), the intermediate rinse, where the solution being flushed out is checked for neutrality, and the final rinse, where the solution being flushed out is checked to make sure there are no traces of the characteristic components of the previous solutions. The actual cleaning steps and times depend on the production process to be cleaned.

Fluid mechanics
Fluid mechanics influence the flow of the detergent through the system. Turbulent flow results from friction where fluid is in contact with surfaces, and is characterized by the Reynolds number (in practice *Re* is approximately 10 000 for smooth, round pipes) at the boundary layer with the surface.

Flow velocity during pipe cleaning should be around 2 m/s (minimum 1.5 m/s). If the flow velocity is too high, this will cause a pressure drop in the pipe system. Systems should not incorporate pipes of several different diameters, because the resulting flow velocities will be too low or too high. If different pipe diameters cannot be avoided, they should be limited to a maximum of two different sizes. It is also important to avoid separating the pipe path into parallel streams, as this causes undefined flow and contamination.

Cleaning temperature
Temperature influences the effectiveness of the detergent, and the optimum temperature for CIP is determined by the cleaning task and the detergents used. Pre-rinse steps above 40 °C are not recommended, since soil containing protein or starch will undergo chemical changes at these temperatures and thus impede the subsequent washing steps. Typical temperature ranges for different detergents are given in Table 27.1.

27.2.2 Disinfection parameters

Disinfection must be a separate cleaning step, and must only be carried out after cleaning, because soil residues can protect microorganisms against direct

Table 27.1 Typical temperature ranges for detergents

Detergent type	Temperature (°C)	Cleaning task
Acid (HNO_3)	60–65	Tanks, pipes, milk pasteurizers, etc.
Caustic (NaOH)	Cold up to 40	Fermentation tanks, storage tanks, bright beer tanks, fillers. Higher temperatures are desirable, but problematic from a technical point of view.
	50–80	Milk collection tankers, milk tanks, cream tanks, quarg and yoghurt tanks, filling machines.
	70–90	Brewhouses, lauter tubs, mash tuns, wort coolers, milk pasteurizers, pipes.
	90–130	UHT-plants, sterilizers for puddings and desserts, etc.

contact with the disinfectant. The strength of the disinfectant used can be affected by decomposition of soil, and bacteria that survive owing to insufficient disinfection can multiply in soil residues and contaminate the product. Lower soil concentrations permit the use of combined cleaning and disinfection agents.

Thermal disinfection
Thermal disinfection is effective only if all plant components are subjected to the required temperature and time. Thermal disinfection kills bacteria, but does not remove them from the system – 'sterile soil' is formed.

Methods of thermal disinfection include hot water (preferably with direct flow through equipment and pipes), steam at a temperature between 70 °C and 80 °C, maintained for 15 minutes (for production plants with a large volume, e.g. tanks), and hot pressurized water. Spores of some bacteriophages and some mould fungi are killed only if kept at temperatures of 130–140 °C for at least 20 minutes. It is important to prevent the intake of contaminated air during the cooling phase after thermal disinfection.

Chemical disinfection
The effectiveness of the various different chemical disinfectants depends on concentration, temperature, reaction time and soil load. *Activated chlorine* decomposes rapidly with increasing temperature, and therefore can only be used cold and without collection. It cannot be used for standing disinfection because of its corrosive effect on stainless steel. *Peroxide compounds*, such as hydrogen peroxide and peracetic acid, are also suitable as additives to acid cleaning solutions. Peracetic acid can be used as a cold disinfectant at low temperatures. *Iodophors* have good bactericidal and sporicidal effects, but there are problems with flushing them out and a risk of reaction with plastic parts or corrosion effects. They are not suitable for temperatures above 40 °C. *Quaternary ammonia compounds* have good wetting properties and surface activity. Disadvantages include residue formation and a strong tendency to foaming. They are rarely used in CIP processes. *Halo acids* have comprehensive biological effects and are used in combination with acid carriers. They are completely free of foam and have good flushing properties.

The top diagram in Fig. 27.3 shows a pocket-free design for process pipelines and all integrated components in a functional system, described as a hygienic design. By comparison with the standard installation (the lower diagram in Fig. 27.3), the hygienic design reduces cleaning time significantly, which means that production time can be extended.

27.3 Factors determining the effectiveness of a CIP system

27.3.1 Type of CIP system

The cleaning parameters can be optimized by selecting the right CIP system. The four most commonly used CIP systems are described below.

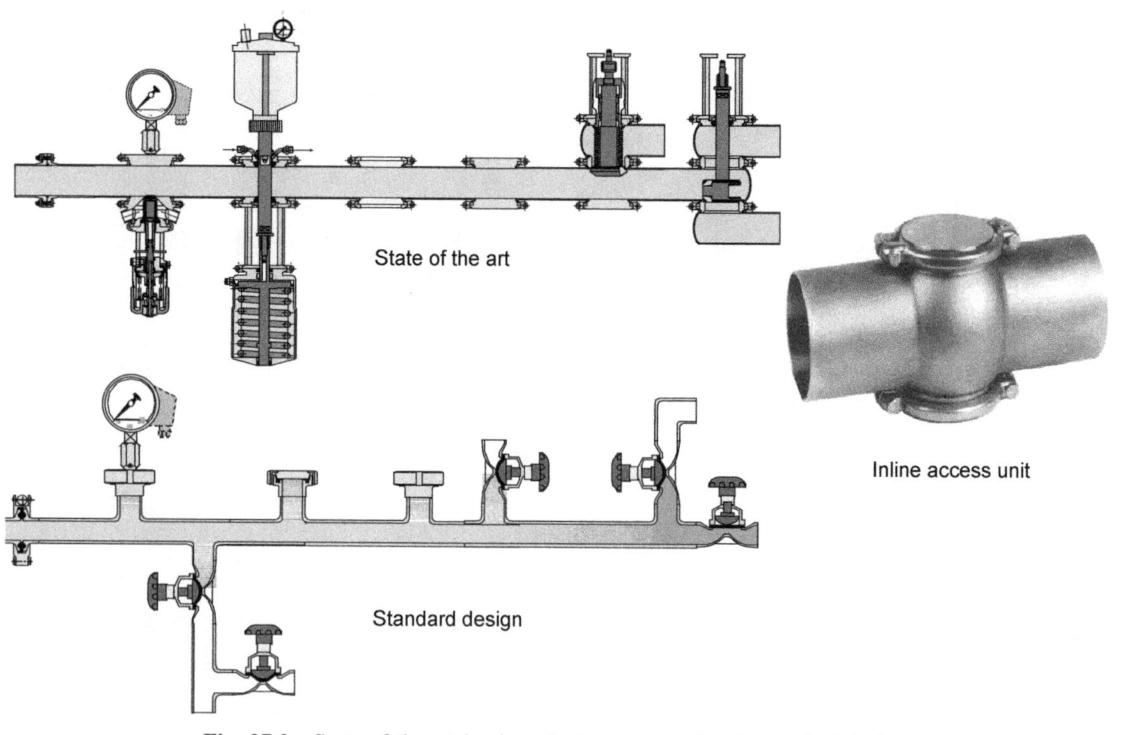

Fig. 27.3 State of the art hygiene design compared with standard design.

Single-path CIP system
In the single-path CIP system, a freshly made-up cleaning solution is supplied to the plant and then drained, and there is no residual circulation in the cleaning system. The single path system is used when there is a high degree of soiling. It is also appropriate when it is critical to prevent cross-contamination. This system is commonly used in the pharmaceutical industry.

Single-use CIP system
A freshly made-up cleaning solution is supplied to the plant, circulated, and then drained. This system is used when there is a high degree of soiling or when cross-contamination must be avoided, and the volume of cleaning solution in the system is low.

Recovery CIP system
This system allows for multiple use of the concentrated cleaning solution, and can encompass several cleaning tasks carried out simultaneously. The cleaning solution is stored after circulation. There are large circulation volumes in the cleaning system, and the cleaning plant is installed centrally.

Satellite CIP system
This method combines recoverable and single-use cleaning. The cleaning solution is supplied from a main CIP station to different single-use CIP systems in different sections of the plant, independently circulated in the different plant sections, and then drained or returned to the main CIP station. The cleaning solution can be used more than once, and several cleaning tasks can be carried out simultaneously and at different temperatures if necessary. The circulation volumes in the cleaning system are low to average, and the main installation is centralized, with satellite systems set up where needed around the plant.

27.3.2 Dosing methods for cleaning detergents

Dosing can be done directly or in-line. A disadvantage of direct dosing is that the concentration of the cleaning solution must be adjusted to the piece of equipment that requires the highest concentration for cleaning, so it will be higher than necessary for the rest of the equipment.

In-line dosing, directly into the CIP circuit, has the advantage that the concentration of the cleaning solution can be adjusted to the individual requirements of the equipment to be cleaned. This requires including the correct dosing technology in the CIP system, as shown in Fig. 27.4.

27.3.3 Use of conductivity probes

Conductivity probes consist of electrodes, and are used to control dosing and to make sure all cleaning fluids have been removed from the system once cleaning is complete. They do this by measuring the conductivity of the fluids in the

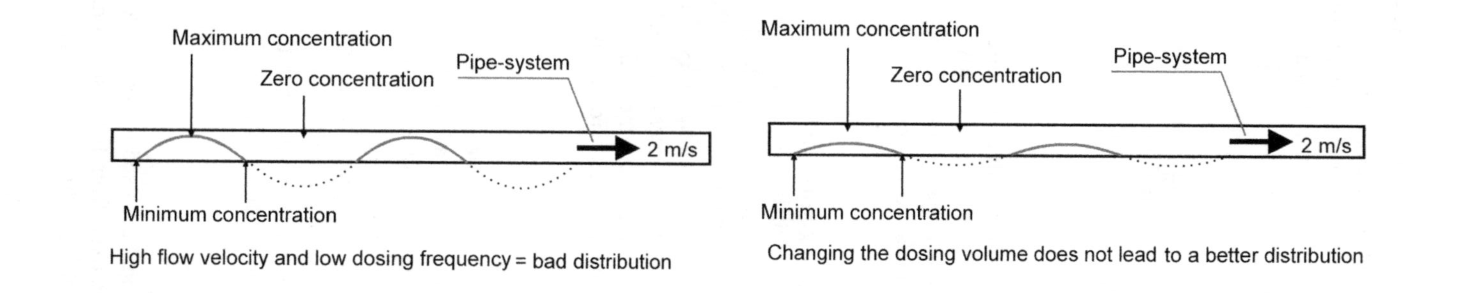

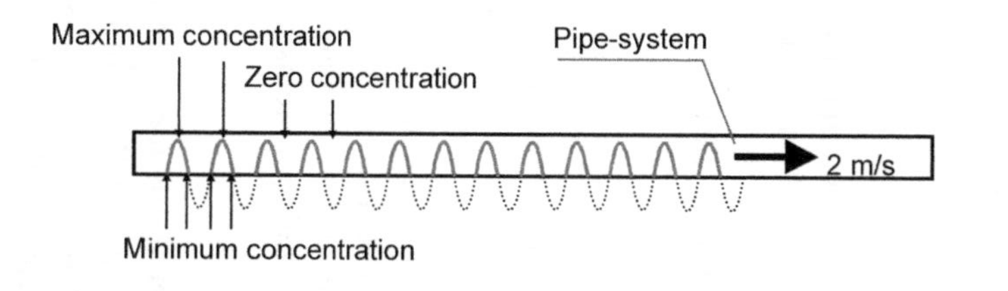

Fig. 27.4 In-line dosing requires the use of the right dosing technology.

system, which gives a measure of concentration. Conductivity probes indicate that rinsing has been successful once the conductivity of the rinse water falls below a specified threshold.

The measuring electrode must be temperature-compensated to allow for the effects of temperature changes on concentration, and must be resistant to 'ageing', deterioration caused by chemical changes over time. The electrodes must be checked regularly to ensure that they are free of soil deposits. They should not be installed where there is a heavy dirt load, as this tends to increase salt concentrations and therefore conductivity, which will be assumed to indicate high detergent concentration. The measuring electrode must also be installed where there is sufficient turbulence, because low turbulence will result in lower measurements.

Figure 27.5 gives the conductivities of various cleaning system fluids, and shows the influence of temperature on conductivity.

27.3.4 Components used for cleaning inside tanks

Spray balls

Spray balls come in several different types. They must generate 30–50 litres of liquid per minute per metre of the tank circumference to maintain a continuous liquid film on the tank wall.

Rotating jet cleaner

Rotating jet cleaners use different sizes of fan-shaped nozzles, and rotate slowly inside the tank. The liquid film runs down the tank wall in spirals, which has the effect of a pulse–pause operation. Rotating jets can generate high specific liquid impact on the tank wall at a low total consumption.

Orbital cleaner

Different types of orbital cleaners operate at high, medium or low pressures, and come with or without an external motor. They provide narrow planetary scanning of interior surfaces of tanks and large containers.

27.3.5 Tanks for collecting cleaning solutions

Tanks for cleaning solutions and disinfectants

These tanks store cleaning solutions for all CIP circuits that operate simultaneously.

Fresh-water tank

The fresh-water tank separates the CIP system physically from the water mains. Fresh-water tanks must have the capacity to supply all CIP circuits that might be running at the same time, with their water levels falling to not less than one-third full. The fresh water in the tank must be used as soon as possible to reduce the risk of contamination.

Specific conductivities at 20 °C

Solution	S/cm	mS/cm	µS/cm
Purest water			0.04
Tap water, soft		0.18	180
Tap water, hard		0.46	460
Tap water, saline		0.75	750
Brackish water		2	2,000
Soda solution at 1%	0.012	12	12,000
NaOH solution at 1%	0.0475	47.5	
NaOH solution at 2%	0.09	90.0	
NaOH solution at 3%	0.127	127.0	

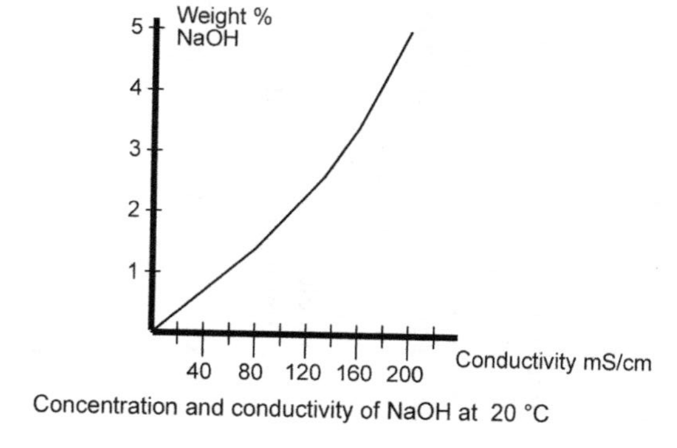

Concentration and conductivity of NaOH at 20 °C

Solution	20 °C	70 °C	85 °C
NaOH at 1%	47 mS	93 mS	107 mS
NaOH at 2%	90 mS	177 mS	203 mS
NaOH at 3%	127 mS	254 mS	293 mS

Influence of temperature on conductivity

Fig. 27.5 Conductivities of various cleaning system fluids.

Return water tank
This tank collects the water from the intermediate and final rinsing steps, holding it ready for the next pre-rinse. The tank must be large enough to cope with the volume of water being used in the rinsing steps that supply it.

27.3.6 Tanks for storing detergent concentrates

Tanks used for storing detergent concentrates must be type-tested, and must be installed and maintained only by an approved expert company. The tanks must be closed, should not be operated under pressure, and should be protected against mechanical damage and freezing. Storage tanks must be installed in a collecting pan that can hold the entire contents of the tank. They should have separate filling pipes for each product, and when concentrates are supplied from a road tanker, it must be ensured that the hose connections cannot be mixed up.

Tanks should be equipped with level indicators and devices to protect against overfilling, including automatic interruption of the filling process if there is a risk of overfilling. The internal diameter of the venting and overfilling pipes must be at least 20% larger than the filling pipe. The filling pipe must have a shut-off valve and a non-return valve, and venting pipes must be connected to the atmosphere – gas and steam exhaust in the storage area must be avoided. It is also important to ensure that no unwanted chemical reactions occur in grouped venting pipes.

Components for heating the tank contents must be protected against dry running and should have temperature limit switches. All the relevant laws and regulations for installing, filling and using these storage tanks must be taken into account.

27.3.7 Non-self-priming centrifugal supply pumps

The pressure and the flow rate in a pipe system are not produced by the pump, but by the resistance in the system itself, so the pump must have the capacity to supply the entire system and cope with all cleaning tasks. In practice, a single pump is generally used to supply several different CIP circuits, which all have their own system characteristics. The pumps must therefore be rated for the maximum required flow rate and delivery pressure, and must then be adjusted to the other circuits by means of throttling in the pipe system or speed adjustments via frequency control.

27.3.8 Self-priming centrifugal pumps for CIP return, used for tank cleaning

These pumps are gas-conveying periphery pumps, and have very good suction capacity. In contrast to non-self-priming pumps, centrifugal pumps continue operating even if the liquid being sucked out of the tank contains air or other gaseous media. A non-self-priming pump can be throttled back by the drop in pressure until it stops conveying, and therefore power consumption decreases. A self-priming pump cannot be throttled back in this way, or can be throttled only very slightly, and so power consumption increases.

Owing to the small clearance between impeller and housing, the periphery pump is very sensitive to soiling. If dirt ingress is likely, dirt traps should be provided on the suction side.

27.3.9 Design considerations for CIP systems

A number of factors must be taken into account when designing CIP systems to ensure that they are both technically and commercially suitable. A major factor is the production process that the cleaning system will be associated with. Plant downtimes are important, because they indicate when and for how long cleaning can take place. The number of shifts running also affects timing of cleaning operations, as does identifying which areas can be cleaned simultaneously. When the system is highly complex, a production and equipment allocation plan is useful. Customer requirements and specifications must also be considered.

CIP circuit groups

CIP circuits can be grouped where one or more of the following factors are similar in the areas to be cleaned:

- production conditions;
- type and degree of soiling;
- conditions required for cleaning solution (composition, concentration, temperature and reaction time);
- flow rates (avoid extreme changes in nominal widths); and
- time allowed for cleaning.

Matching flow velocity in grouped CIP circuits is important. Figure 27.6 shows the influence of different nominal widths on flow velocity and pressure drop.

Volume flow rate for a flow velocity of 2 m/s			Flow velocity at a volume flow rate of 50 m^3/h		
DN (mm)	Q (m^3/h)	Δp_{100m} (bar)	DN (mm)	w_m (m/s)	Δp_{100m} (bar)
50	14.1	0.90	50	7.1	10.0
65	24.6	0.65	65	4.1	2.4
80	37.1	0.50	80	2.7	1.0
100	56.6	0.40	100	1.8	0.3
125	88.4	0.29	125	1.1	0.1

DN = the nominal diameter of the pipeline; Q = the flowrate;
Δp_{100m} = the pressure drop of 100m of pipeline

Fig. 27.6 Influence of different nominal widths on flow velocity and pressure drop.

Flush-out processes
Flush-out processes can be controlled in a number of ways. Setting a flush-out time is the least favoured solution, because of the dependence on supply pressure. Controlling flush-out by conductivity readings is the optimum method for transition zones between cleaning solutions and water, and controlling via volume readings is a good solution for optimizing water flush-out phases.

It is important when designing flush-out processes to consider not only the medium being used to do the flushing out, but also that which is being flushed out. For instance, flushing out cleaning solution with water has different characteristics from flushing out water with cleaning solution. This is most obvious if the same times are specified for carrying out both processes.

Tank cleaning
When tanks are cleaned, all elements installed in them, such as measuring instruments, safety equipment and agitators, must also be cleaned. Therefore it is important to avoid spray shadows behind installed equipment, and consider all aspects of tank geometry.

If spray balls are used for tank cleaning, the recommended flow of liquid is 30–50 litres per minute per metre of tank circumference. So, for instance, a heavily soiled tank with a diameter of 4.7 m (circumference approximately 15 m) needs a flow rate of cleaning solution of approximately 45 000 litres per hour. The flow rate for rotating jet cleaners is lower, at about 50–35% of this amount.

When designing a system to clean large tank outlet pipes, it is important to remember that the flow rates of supply and return pumps only match in a few cases, and that trapped air and low flow in the outlet pipe can lead to unreliable cleaning. Large outlet pipes are generally cleaned in a pulse–pause manner, with rinsing pulses from the bottom of the tank to ensure reliable and complete wetting of the tank outlet zone. It is also important to ensure that there is a sufficient volume of cleaning solution in the CIP-buffer tank.

Tank phases
For hot cleaning, in particular when alternating hot and cold spraying processes in tanks, the effects of temperature on gas volume must be taken into account. For instance, when changing from cold pre-rinse to hot detergent rinse, the gas volume in the tank strongly expands. Generally, the result of this expansion is not that the pressure exceeds the permitted level, but that the cleaning solution violently bursts out when the safety and vacuum relief valves are open.

When changing from hot cleaning to a cold final rinse, the gas in the tank cools down rapidly and its volume decreases accordingly. This drop in gas volume must be compensated for via a vacuum relief valve, otherwise the pressure will fall below the permitted vacuum level and the tank will be destroyed.

Typical problems during cleaning system operations

- Spray head partly clogged: leads to formation of deposits at certain points due to spray shadows, followed by pressure build-up and atomization of the cleaning solution.

- Supply pump defects (impeller, motor, leaks): results from the formation of deposits due to insufficient pressure and flow velocity.
- Return pump defective or insufficient feed: results in the formation of deposits at the bottom of the tank because too much liquid collects at the bottom and there is insufficient turbulence to prevent deposit build-up.
- Turbulence at the tank outlet, return pump draws air: similar problem to the above. With non-self-priming return pumps, recirculation is interrupted.
- Leaks at pipe connections: pump delivery collapses, foam forms and flow rate is reduced.

Corrosion

Materials for the production and CIP systems must be carefully selected to allow for the corrosive effects of the products, cleaning agents and disinfectants. The chemical behaviour of industrial water with respect to corrosion under operating conditions must be examined. The corrosive effects of detergents and disinfectants should be specified by the manufacturers.

27.4 Improving CIP systems

27.4.1 Cold tank and hot pipe cleaning

Figure 27.7 is a schematic diagram of an optimized CIP system for cleaning tanks and production systems. Single-use cleaning tanks are recommended for cold tank and hot pipe cleaning in the brewery and beverage industry, in combination with recoverable cleaning for the pipe and production systems (satellite cleaning). Recoverable cleaning has been demonstrated to have clear advantages over single-use cleaning with regard to chemical consumption and effluent rates. However, single-use cleaning is more suitable for cold tank cleaning because it avoids the risk of microbiological contamination that is associated with the use of recovered cleaning solutions.

Single-use system

The single-use CIP system should be located close to the tank group to be cleaned, in order to minimize consumption of cleaning solutions and reduce effluent rates. Locating the cleaning system near the tanks reduces circulation volumes because the inlet and outlet paths for the cleaning media are short, and losses from intermediate rinsing steps and flush-outs are reduced.

The CIP circulation pump and the CIP return pump must be tuned to each other at the supply pump by controlling their speeds (frequency adjustment). If the flow rates are not coordinated, the buffer tank between the pumps will be overfilled and/or emptied continuously.

The level of the sump produced by internal cleaning of the tank should be kept as low as possible and should pulsate (fluctuate) during the cleaning operation. The sump inside the tank should at least cover the tank outlet and prevent eddying at the outlet. Changing the height of the pulses prevents a layer

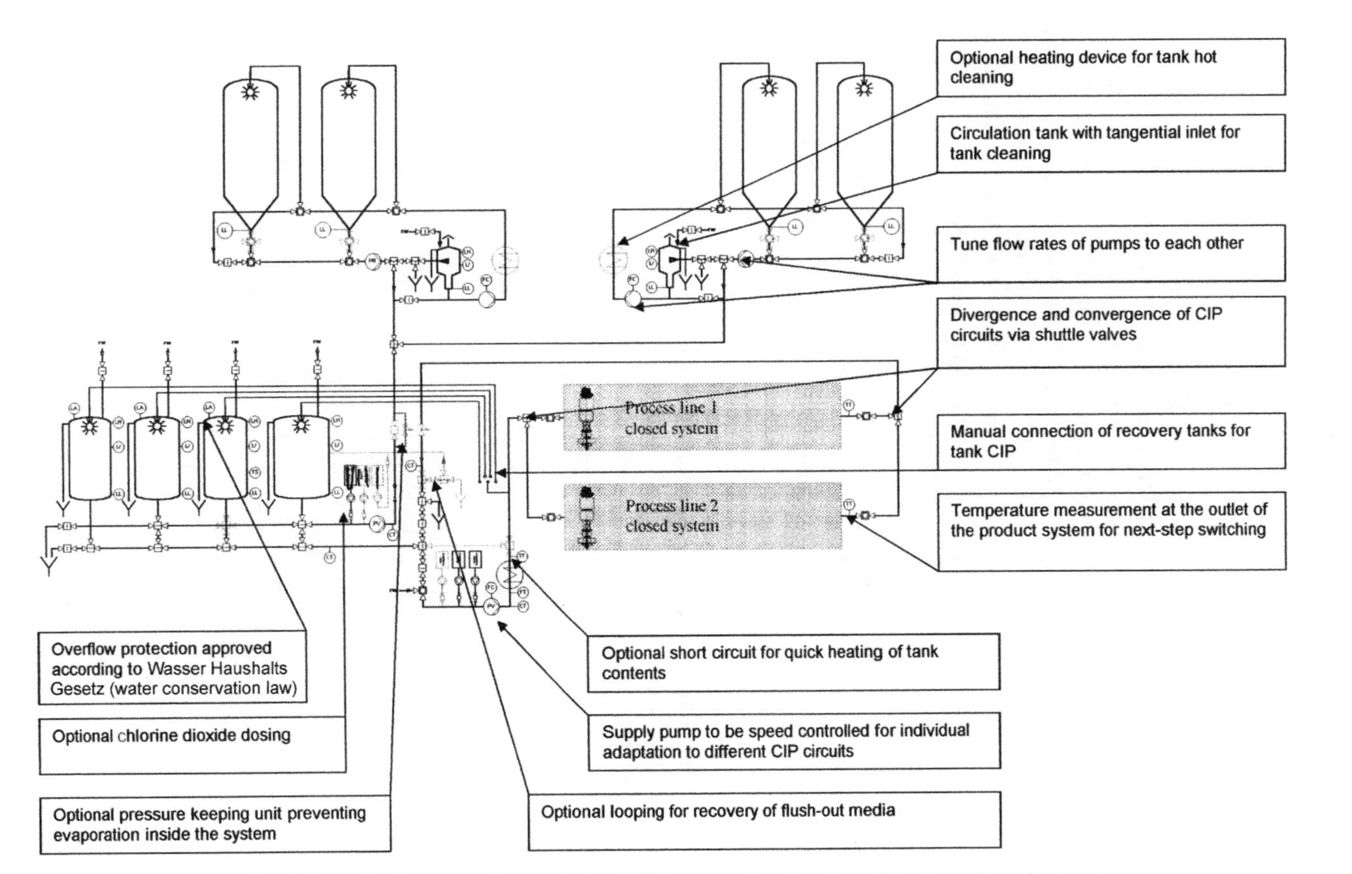

Fig. 27.7 Schematic diagram of an optimised CIP system for cleaning of tanks and product systems.

of soil forming at the upper rim of the sump. The volume of the CIP circulation tank is calculated according to:

Total volume = supply volume + pulsating volume.

It is essential to take the shape of the tank to be cleaned into account, since the pulsating volume depends on tank shape, particularly at the bottom of the tank. Otherwise, losses of cleaning solution will rise, because the circulation tank is constantly overfilled and requires new cleaning solution as a result.

The CIP return pipe should be inserted tangentially into the circulation tank. This causes the liquid in the tank to rotate, resulting in a quicker evacuation of air out of the tank. The CIP return from the tank to be cleaned should be designed so that air cannot accumulate in the return system and/or back gassing into the tank cannot take place. Air that accumulates in the return system forms air cushions, which hinder proper cleaning of the pipes and reduce flow. If the design does not permit automatic back gassing into the tank, then the gas must be forced back to the tank by inversion of the flow direction. The optimal solution is a vertical tank outlet complete with the pipe matrix system, 'ECO-MATRIX'.[1]

Recovery system

The recovery CIP system for cleaning pipes and production lines should be located centrally and as near as possible to the object to be cleaned. Pipes for supply and return of CIP media should be as short as possible and of the appropriate diameter. Pipes that are long or too large increase the losses during rinsing and flush-out operations. When calculating appropriate pipe diameters, it is important to remember that smaller diameters and longer pipes lead to greater pressure drops, which must be compensated for by the use of larger supply pumps and booster pumps.

The circulation of cleaning solutions in the system should bypass the CIP recovery tanks. If CIP recovery tanks are included in the circulation, there is a risk that the recovery tanks and production line could be contaminated with microorganisms. Thus a recoverable CIP system should be designed so that highly polluted cleaning solutions from the initial cleaning phase are drained directly and/or conveyed to the neutralization plant, to prevent pollution and microorganisms being distributed throughout the system.

Flow measurements (e.g. inductive flowmeter) should be used to optimize the cleaning media supply to the conditions of the production system to be cleaned. Recoverable CIP systems are generally used for cleaning several different production systems, so it must be possible to adjust the cleaning system to the specific cleaning conditions of each CIP circuit, and this is achieved by flow rate control. To do this, the CIP supply pump must have sufficient capacity to accommodate the total consumption of the system. If this is not the case, then the CIP circuits must be supplied by several pumps, or booster pumps must be installed at suitable locations in the pipe system.

The volume of fluid used for the intermediate rinsing steps should be kept to a minimum. Intermediate rinsing has no real cleaning task and merely serves as a separator between the cleaning steps (e.g. from caustic solution to acid), so sending an appropriately sized water barrier through the pipe system is sufficient. This barrier can be produced as a defined volume using a volumeter. Intermediate rinsing phases that fill the entire pipe system significantly increase fresh-water losses and effluent rates, and this effect is particularly extreme in systems of long and/or wide pipes.

The volume of media used for flush-outs during a phase change-over should be calculated using appropriate measuring methods. In this instance, conductivity and volume measurements are adequate. Controlling by timing is not recommended because this method requires a constant flow rate, which cannot be guaranteed because of fluctuating pressure from the water mains.

Cleaning system parts should not be arranged in parallel with the production system to be cleaned. Parallel pipe systems divide flow and lead to undefined flow in the associated partial systems, they also increase transition zones during flush-out and ultimately contribute to detergent losses.

Measuring units used for next-step switching of the cleaning steps should be installed at optimal points in the pipe system. Positioning measuring points carefully avoids unnecessary waste of time. For instance, temperature measurements should be taken at the outlet of the production system, to ensure that the required temperature has been reached. Downstream CIP returns do not need to be heated before next-step switching can take place.

CIP supply systems should be equipped with shuttle valves that distribute CIP media to the individual CIP circuits. For CIP return, shuttle valves are used to combine CIP media from the individual CIP circuits. The shuttle valves prevent contamination or mixing of the media from different parts of the system.

27.4.2 Other operations within the CIP system

Refilling the recovery tanks with fresh water

This procedure takes place only when the CIP circuits associated with the recovery tank concerned are not operating, as otherwise the tank would overflow. The recovery tanks for detergent, caustic and acid must be refilled automatically with fresh water when the volume falls below filling level HL, to lower the primary concentration. During the subsequent circulation, the required concentration is re-established in the circuit by in-line dosing. After the contents of the circuit are flushed out into the recovery tank, the primary concentration increases again, and may reach the maximum concentration for use.

The recovery tank for return water needs no additional refilling because more return water is produced than needed, so this tank tends to overflow rather than empty. In the unusual event that the level in the recovery tank for return water drops too low, it is possible to switch over to fresh-water supply. When the level in the return water recovery tank reaches the higher level, LH, it must be switched to drain to prevent over-filling. The tank over-flow provided for emergencies should not be used.

Cleaning the recovery tanks

The recovery tanks must be cleaned occasionally to remove any micro-organisms. This is done by emptying the tank and connecting it to the cleaning system. Freshly made-up cleaning solutions at the required concentrations are used, supplied in small amounts via the dosing unit.

Fresh water supply

Fresh water should be supplied from adequate ring water mains. The connections between the system and the mains should be leak-proof, using mixproof valves (double-seat valves) or a low-volume buffer tank. If the capacity of the water mains is insufficient, the rinsing operations of the CIP system should be designed to be supplied from a water collection vessel between the system and the mains. Fresh water must not introduce microbiological contamination to the system, and a ring main is safer in this respect, because continuous water exchange takes place.

Fresh water is often stored over long periods of time in large-volume buffer tanks, and prevention of microbiological contamination under these circumstances requires expensive measures such as aeration via sterile filters or dosing with chlorine dioxide.

Insulation

The vessel used to recover hot cleaning media must be sufficiently insulated, at least around the bottom and the shell. Hot cleaning media should be circulated once a day for cleaning and to prevent excessive cooling of the CIP recovery tank. Using a suitable insulation and circulating hot cleaning solution at least once a day reduces the heat energy input because only the plant elements need to be heated. Heat loss from the circulating cleaning agents can be minimized by insulating the pipes.

Return water tanks

Return water tanks must have the correct capacity for the entire cleaning system. If the return tank is too small, it will overflow continuously, increasing wastage rates and mains water usage.

27.4.3 Design of CIP plants

The requirements for hygiene placed on a cleaning system are the same as those for a production system. Contamination from microorganisms from an inadequately designed CIP plant will enter the production system and contaminate the product.

Preventing water hammer

Water hammer occurs when liquid is suspended in the pipe system, resulting in evaporation and the condensation of steam. This phenomenon occurs typically if the inlet is closed but the outlet is still open, when a bigger height of fall exists and the medium has already reached a higher temperature (water of 20 °C will evaporate in a suspended liquid column of approximately 10 m). If the cleaning system and plant parts to be cleaned are arranged on different levels, appropriate measures must be taken to prevent the suspension of liquid columns, for instance by using overflow valves in descending pipes.

In-line dosing
In-line dosing systems must be equipped with dosing pumps that have adjustable stroke volume and stroke frequency to ensure optimal distribution of cleaning solutions in the flow. The distribution of cleaning solution in the flow becomes more homogeneous if the required volume is distributed throughout the cleaning cycle in a large number of small doses rather than a smaller number of large doses. Larger doses can result in wave-like accumulations of high and low concentrations.

Double-seat valves
During cleaning, the valve disks in the double-seat valves installed in the system must be lifted so that they are cleaned properly. Alternatively, the system should be cleaned by looping so that the valves can open fully and the valve seats be cleaned from below. Seat lifting should occur periodically during each cleaning phase. The duration of the lifting pulses and the intervals between them depend on the level of soiling, and are generally between 10 and 60 seconds in duration, with 3–5 minutes between pulses. To minimize the consumption of cleaning agents, Varivent double-seat valves are equipped with the LEFF module.[2] Using the LEFF function, the detergent rate can be reduced to a minimum during the lifting procedure, while the cleaning effect remains unchanged.

The maximum number of double-seat valves that can be combined in one lifting group is five. If too many valves are lifted at the same time, the pressure in the downstream pipe and valve system drops. The lifting function always requires pressure of at least 1 bar above atmospheric pressure, otherwise the valve seat will not be cleaned and air will be drawn into the system instead. This must be taken into account when designing valve systems where seat-lifting will be required for cleaning, particularly in the case of tank return.

Flushing out with mains water should never be time controlled. The flushing out medium must always be at constant pressure and in practice this is never the case with mains water supplies. Therefore controlling flushing out by timing would result in varying volumes of water being introduced into the system. Flushing out should be controlled via volume measurements, or direct measurement of a parameter such as conductivity or pH value.

27.5 Future trends

Hygienic design and certification are important elements in the improvement of CIP systems, and will be used to develop improved components, process models and process installations. Testing and certification are based on the EHEDG-developed test procedure (EHEDG Document 2) and hygienic design criteria (EHEDG Document 8).

Detergent compositions and their applications will continue to be refined. Processes will become fully automated to guarantee replicable, high quality and safe cleaning operations. Education in hygienic practices will be improved at all levels, through training courses, e-learning, lectures, work shops at secondary school, university, seminars and symposia.

27.6 References and further reading

1. ECO-MATRIX (GEA Tuchenhagen Dairy Systems GmbH, Am Industriepark 2-10, D-21514 Büchen, Germany)
2. LEFF Module (GEA Tuchenhagen, Tuchenhagen GmbH, Flow Components, Berliner Strasse 25, D-21514 Büchen, Germany (http://www.tuchenhagen.com))

EHEDG guidelines
EHEDG training material
Information databank: http://www.hygienic-design.de

28

Improving cleaning-out-of-place (COP)

L. Keener, International Product Safety Consultants, USA

28.1 Introduction

It is a given that cleaning and sanitizing food processing equipment are fundamental to protecting public health. This has not always been the case. These fundamental truths about food safety have not always been understood by sanitarians and food processors. Upton Sinclair's seminal work, *The Jungle* (Sinclair, 1906) provides graphic accounts of putrid materials held in overtly soiled containers, prior to their being processed for human food on equally soiled processing equipment. This 1906 depiction of the sanitary conditions within food processing resulted in a massive intervention by the US government. Both the Pure Food Act and the US Federal Meat Inspection Act have their genesis in *The Jungle*. However, if the conditions described by Mr Sinclair, and not to justify them for one moment, are put into context of the time, a different picture emerges as to why such pernicious conditions would have existed. The industry of that era simply did not understand that their lack of hygiene in food processing jeopardized the health of the nation. Consider that Joseph Lister's theories on antisepsis were not universally accepted until 1876, and in Germany, by 1878, Robert Koch was demonstrating the usefulness of steam for sterilizing surgical instruments and dressings. Pasteur's germ theory was not finally consolidated and accepted until 1880. The collective efforts of Lister, Koch and Pasteur in illuminating the mechanisms of microbial contamination and infection provided important insights into developing strategies for their prevention. Lister's use of carbolic acid for wound treatment and equipment disinfection may represent the first modern case in which a chemical intervention was used for preventing the spread of infectious agents.

It is unrealistic, then, to assume that the food processing industry of Upton Sinclair's day would have had a meaningful understanding of these important scientific developments. Mr Sinclair's writings do suggest, however, a fundamental understanding by some within the industry of that era, of the effectiveness of heat in minimizing the outward manifestations of putrefaction. He describes the use of 'white hot pokers' for treating the overt manifestations of spoilage in de-boned hams. Heat and or hot water would certainly have been the media of choice during that epoch, for both cleaning and disinfecting food processing equipment. It is also clear from the literature that the preponderance of cleaning activities, if not all, would have been carried out on a clean-out-of-place (COP) basis, i.e. removed from its normal place of use in a food processing operation specifically for the purposes of cleaning.

Since 1906 there have been many great innovations in microbiology and food processing. These technological advances have led the development of a scientific basis for cleaning and disinfecting food processing equipment. And while much has changed in the food processing industry since 1906, it remains a fact that hot water and heat are still widely used cleaning and disinfecting agents within the industry. Moreover, COP cleaning remains a leading method for wet-cleaning, food processing equipment. Sinks, dishwashers, tote-washers, and similar devices are widely used for cleaning pipes, pump parts, valves, utensils, and the other portable appurtenances used in the mass production of foods intended for human consumption.

This chapter will provide a review and discussion of the common methods and types of COP cleaning currently in use by the food industry. The chapter will also focus on current best practices and methods for developing and validating an effective science and risk-based COP cleaning process.

28.2 Best practices in developing an effective COP process

The simple act of extracting and delivering machine parts, portable utensils, or other implements into an agitated tank filled with hot water and a cleaning compound will not ensure that they are properly cleaned. Successful COP cleaning requires knowledge of all the factors that may limit or otherwise compromise the effectiveness of the applied cleaning process.

A process, by definition, is a series of planned activities that have been carefully constructed to achieve specific objectives or outcomes. In the case of a COP process, the objectives are the removal of objectionable odors, food residues, other extraneous debris and undesirable microorganisms from the contact surfaces of food processing equipment. On the face of it, COP cleaning is a very straightforward process consisting of five basic unit operations: pre-rinse; wash; post-cleaning rinse; sanitization (disinfection in Europe); and final rinse. These basic procedures constitute the fundamental elements of all wet-cleaning processes (cleaning by hand, dishwashing machines, CIP and COP). It is noteworthy that, while the elements of the process are identical, the outcomes of applying them may yield highly variable results. Variability in the outcomes of applying a COP process, in most instances, can be attributed to the following factors: type of soil that is targeted for removal; cleaning water chemistry; detergent type; processing temperatures; exposure times; type of sanitizing agent; and time of exposure to the sanitizing compound. It is broadly accepted that process variability arises from either 'common cause' or 'special cause' circumstances (Deming, 1993). 'Common cause' variation is the normal oscillations of a stable process within the boundaries of its control limits. By contrast, 'special cause' variation is indicative of an unstable process. Processes that exhibit 'special cause' variation are unreliable. The most profound distinction between a 'common cause process' and a 'special cause process' is predictability. In the case of a COP wet-cleaning process, a 'special cause' variation might involve high microbial counts (100 colony-forming units, cfu)/unit area when the upper control limit specifies less than 10 cfu/per unit area examined. Properly defining the capabilities and therefore expectations of a COP process is critical to ensuring its effectiveness.

28.3 Defining the process

The first step in developing a COP process is the determination of an appropriate cleaning objective. The cleaning objective, or more precisely the process objective, is the aim or intent of the applied process. In other words it is the specification delineating what the process is expected to deliver. Cleaning objectives can be constructed around two fundamental concepts; soil removal and reduction in microbiological burden. Cleaning objectives are customarily defined by four major criteria:

1. *Cleaning* – refers to the complete removal of food soil, other residues and their associated odors using appropriate cleaning chemicals in accordance with recommended protocol.
2. *Disinfection* – refers to inanimate objects and the action of inactivating all vegetative microorganisms (not spores).
3. *Sanitize* – is a US term and denotes a process whereby microorganisms are reduced to safe levels as defined by public health authorities. The official US definition (Association of Official Analytical Chemists) of sanitizing for product contact surfaces is a procedure that reduces the contamination level by 99.999% (5 logs) in 30 seconds. The procedure as defined for non-contact surfaces requires a 99.9% (3 log) reduction in contaminants. The standard test organisms are *Staphylococcus aureus* and *Escherichia coli*.
4. *Sterilization* – refers to the statistical destruction and elimination of all living organisms.

Establishing cleaning objectives is a risk-based exercise which cannot or should not be conducted in isolation. Risk assessment is a logical, systemic activity that provides identification, measurement, quantification, and evaluation of the hazards associated with certain man-made activities (Haimes, 2004). The risks of an ineffectual cleaning process are the potential proliferation of both spoilage bacteria and human pathogens. The consequences of such a process failure are economic spoilage or human illness and suffering. The likelihood of either or both events occurring must be understood in order to properly define a cleaning objective.

The first consideration in establishing the objective is a critical assessment of the fate of the product in which the cleaned equipment will ultimately be used. The cleaning objective must also consider the intended use of the finished product. For example, the cleaning objectives established to support production of ready-to-eat meats will assuredly differ from those required for thermally processed, low-acid, canned foods.

Establishing an effective cleaning objective requires broad consideration of its impact and implications for the entirety of the manufacturing process. A cleaning objective defined solely as a function of the unit operation where the cleaning activity occurs is both myopic and dangerous. The cleaning objective must, therefore, be linked with other elements of the sanitation program, as well as with elements of the integrated food safety system. Ultimately, the cleaning objectives must be defined in the context of achieving broader regulatory and company food safety objectives.

28.4 Elaboration of process parameters

After setting the cleaning objective, the next step in the process is the elaboration of the discrete procedures that will be used for achieving the desired outcomes of applying the COP process. Setting the process requires a closer examination of the individual steps that define the process. As noted previously the basic constituents of a COP process are pre-rinse; wash; rinse; sanitization; and the post-sanitization rinse. Each process step contains activities that are critical to achieving the process objective. These critical-to-cleaning activities must be identified and elaborated on a stepwise basis across the process. The following is a summary of the critical-to-cleaning activities for each step of a COP process.

28.4.1 Pre-rinse

Pre-rinse is the process step in which water is used to remove coarse debris and other food residues from the surfaces of the component that is intended for cleaning. Soil reduction is an important aspect of cleaning since the carryover of excessive organic matter will adversely impact the effectiveness of the wash step. The limiting factors for the pre-rinse are soil type, applied force (mechanical action) and rinse water temperature.

Soil type is essential for defining the requisite rigor of the pre-rinse step. The soil's composition and other relevant physical and chemical characteristics are crucial for determining how it can best be removed. Moreover, understanding soil type is a requirement for properly specifying the quantity of soil that must be removed in order to achieve the cleaning objectives. For example, if the soil type under consideration is known to be a good microbiological growth medium, then it would be prudent, at this step, to reduce its residuals to the absolute minimum.

Rinse water temperature and pressure are also important to achieving soil removal. The physical and chemical nature of the targeted residues will dictate the force (water pressure or agitation) and temperature of the rinse water stream. Excessively hot or cold water, depending on soil type, may give rise to film formation or other conditions that may complicate the cleaning process. Force and temperature requirements for removing emulsified fats and oils will differ radically from those required to remove discrete pieces of vegetable matter. Moreover, when defining the pre-rinse parameters it is important to specify, qualitatively, the amount of residual matter that may remain attached to the food contact surface. For example, the specification might simply read 'no visible residual matter'. An effective pre-rinse is critical to achieving the specified cleaning objective. Failure to deliver an effective pre-rinse may limit or compromise the performance of the entire COP process.

28.4.2 Washing

Washing is the systematic detergent-aided scouring, scrubbing, or other procedures that are intended to render the element of food processing equipment free from offensive odors, residual food and other extraneous materials. Arguably, washing is the most important step of a COP wet-cleaning process. It is also the most complex. Its complexity arises from the number of critical-to-cleaning factors that are associated with this step in the process. The critical-to-cleaning factors include the following: soil type; water chemistry; detergent type and concentration; surface finish; temperature; agitation; and exposure time.

Soil type

The issue of soil type was introduced and discussed briefly in the preceding section. Because of its importance to the overall process, a more robust and in-depth discussion of the subject is provided here. Food soils vary considerably in their composition and complexity. They are the unwanted residues of food processing that adhere to the surfaces of the processing machinery, its appurtenances and the other implements of production. Soil may be visible or invisible. The residues may be moist and malleable or dry and brittle. The residues may exist as discrete particles or as a fine film with imbedded bacteria. For purposes of discussion it is convenient to assign food soils, broadly, to one of four categories: fat-based; carbohydrate-based; protein-based; and mineral-based.

Fat-based residues are usually present in food processing operations as emulsions. Edible oils, lard, and other hydrogenated nut and vegetable oils are excellent examples of fat-based food soils. The equipment used on the churn deck of a margarine production plant would likely be soiled with fat-based residues. Likewise, the implements involved with handling meat emulsions used in sausage manufacturing would certainly be fouled with fat residues.

Carbohydrate-based residues include the simple sugars, starch, and starch-containing compounds. Carbohydrate residues are associated with the production of confectionary products, baked goods, jelly, and juice products. The moguls used in molded candy manufacturing are notorious for harboring both starch and sugar residues. Fruit and vegetable processing equipment is also likely to be contaminated with carbohydrate residues.

Protein-based soils arise in food processing from a diverse range of foodstuffs including, meat, milk, nuts, soy, shellfish, and various grains. When heated, protein residues are easily denaturated (coagulation). Denaturation may also promote or contribute to film formation. Protein-based films are very difficult to remove. Milk (casein) and egg products (albumin) stabilized by heating readily form protein-based film. It is also noteworthy that protein residues, on food processing equipment, are of great concern to the food industry and public health officials because of their associated allergenic potential.

Mineral salts, under certain processing conditions typically involving heat and low pH, form insoluble complexes that precipitate onto the surface of food equipment. Typical mineral salts include milk stone (limestone), calcium phosphate, and calcium oxalate (spinach). Other difficult mineral deposits include iron and manganese.

Understanding the chemical and physical properties of the various soil types, as discussed above, is a necessary prerequisite for developing a coherent cleaning process. For example, food soils exhibit a wide range of miscibility and solubility properties. With a perspective toward developing a COP wet-cleaning process, soil solubility properties can be a valuable indicator of the likelihood that a cleaning process will be successful. If the residue under consideration is not soluble in a proposed cleaning solvent, it is reasonable to conclude that a process based on that solvent will not be effective. Solubility characteristics are frequently used as a criterion for grouping food soils into four categories:

1. soluble in water (sugars, some starches, and salts);
2. soluble in acid (milk stone, mineral deposits, and calcium oxalate);
3. soluble in alkali (protein, fat emulsions);
4. soluble in water, alkali, or acid (carbohydrates, minerals, and fats).

Solubility may be influenced by the physical condition of the residues. Burned-on, desiccated and hardened food films will exhibit different solubility characteristics from their moist counterparts. It is also true that most processed foods are formulated with a number of food ingredients, all of which, independently, may exhibit differing solubility properties; thereby, further complicating the design and implementation of a COP cleaning process. Soil classification with the corresponding removal characteristics are summarized in Table 28.1.

Table 28.1 Food soil removal characteristics (after Guthrie, 1972)

Soil type	Solubility properties	Removal characteristics	Reactions to heat
Sugar	Water	Easy	Caramelization, more difficult to clean
Fat	Alkali	Difficult	Film formation, more difficult to clean
Protein	Alkali	Very difficult	Denaturation, more difficult to clean
Starch	Water and alkali	Easy–moderate	Glue-like formation
Monovalent salts	Water/acid	Moderate	Not significant
Polyvalent salts	Acid	Difficult	Reactions

Water quality and chemistry

The first and foremost consideration in evaluating the suitability of a water source for use in any food processing application, including cleaning, is its potability. A water source that contains no harmful elements and is deemed suitable for drinking is generally considered potable. While the nations of the world differ in their interpretation of harmful elements, it is safe to say that their primary focus is on precluding pathogenic microorganisms and other toxic substances from waters that are intended for human consumption. Again, and it is worth emphasizing, all water used in food processing establishments should be derived from a potable source.

Water is the basic solvent used in a majority of wet-cleaning applications. Its primary functions are to contain and deliver the cleaning agent to the target substrate. Water is also frequently used as the applied motive force, essential for removing food film and residues. The important functions of water in a COP process may be negatively influenced by its chemistry.

Aside from potability, perhaps the most important characteristic of a cleaning water source is its hardness. Originally, water hardness was understood to be a measure of the capacity of water to precipitate soap (American Public Health Association, 1985). Soap precipitated in water is chiefly caused by the presence of calcium and magnesium ions. Other polyvalent cations also may precipitate soap, but they are often in complex forms, frequently with organic constituents, and their role in water hardness may be minimal and difficult to define. Water hardness is frequently discussed as either carbonate or non-carbonate hardness (Bakka, 1997). The former is also referred to as temporary hardness and is removed by heat. Non-carbonate hardness or permanent hardness is not affected by heat. Consistent with current practice, total hardness is defined as the sum of the calcium and magnesium concentrations, expressed as calcium carbonate, in milligrams per liter of water. The hardness may range from zero to hundreds of milligrams per liter, depending on the source and treatment to which the water has been subjected (American Public Health Association, 1985). Total water hardness is calculated using the following formula:

Table 28.2 US Geological Survey water hardness data

Hardness	Concentration
Soft water	0–51.3 mg/L (1.0–3.5 grains/gallon)
Moderate hard water	51.3–119.7 mg/L (3.5–7.0 grains/gallon)
Hard water	119.7–179.5 mg/L (7.0–10.5 grains/gallon)
Very hard water	Greater than 179.5 mg/L ($>$ 10.5 grains/gallon)

$$\text{Mg equivalent } CaCO_3/L = 2.4797\ [Ca,\ mg/L] + 4.118\ [Mg,\ mg/L]$$

The US Geological Survey's Water Supply Paper Number 658, a frequently cited standard in the US (American Public Health Association, 1985), classifies water hardness according to parameters presented in Table 28.2. The relative effectiveness of the various classes of cleaning compounds in achieving the various functions of cleaning is summarized in Table 28.3.

As noted previously, water primarily serves as the carrier of the cleaning compound. Water hardness may impact its ability to retain the cleaning agent in solution while simultaneously maintaining its optimal state of activity. Moreover, the minerals in hard water will react with certain soaps and detergents to form sticky, insoluble films. This film will deposit onto the cleaning tank's walls and also on the surface of the cleaned equipment. Additionally, hard water can promote calcium deposits and the formation of rust. An effective COP cleaning process must have an adequate supply of soft, hot water. In those instances where soft water is not readily available, it is recommended that the supply be conditioned. Water conditioning, rendering hard water soft, is simply the removal of the calcium and magnesium ions from the water source. Characterizing the quality and chemistry of the water source

Table 28.3 Activity and function of cleaning compounds

Function in cleaning	Activity
Corrosion	Rust prevention
Rinsing	Reduction of the surface tension of water thereby permitting clean equipment to drain dry
Dispersion	Reduction of clumps of suspended matter into small particles
Emulsification	Reduction in fat globule size and their dispersion in the cleaning solution
Peptizing	Formation of colloidal solutions from partially soluble materials
Wetting	Action of reducing the surface tension of water
Saponification	Action of alkali on fats to form soap
Chelating	Reaction of organic compounds with metal ions dissolved in water (i.e. calcium, magnesium, manganese, and iron) to prevent film formation

is a necessary first step in the process of selecting an appropriate cleaning agent or detergent.

Detergent type
Cleaning compounds are routinely assigned one of five classifications: basic alkalis; complex phosphates; surfactants; chelating agents; and acids (Bakka, 1997). Occasionally detergents are formulated using enzymes that catalytically react with specific food soil components to cause their degradation. However, the vast majority of cleaning compounds are formulated to perform the cleaning related functions summarized in Table 28.4. Detergents and cleaning compounds are most often a mixture of ingredients that are formulated to react with food soil by either a physical or chemical means.

Physical acting cleaning compounds
Physical acting detergents alter the properties of the targeted food soil. For example, compounds of this class may alter the solubility or colloidal stability of the soil. Surfactants, or surface acting compounds, are examples of agents that work by way of a physical mechanism. The most efficient compounds of this class have a chain length of between 13 and 15 carbon atoms (Elliot, 1980; Katsuyama, 1980). These organic compounds have functional properties where a portion of their molecular structure is hydrophilic (water-loving) and a portion is hydrophobic (repulsed by water). Surfactants function in cleaning compounds by promoting the physical dispersion, emulsification, penetrations, foaming or wetting of the target soil. Surface acting agents are either ionic or non-ionic (Elliott, 1980). Ionic surfactants are divided further into three sub-classes: anionic (negatively charged in water), cationic (positively charged in water) and amphoteric (charge is pH dependent). Amphoteric surfactants are cationic under acid conditions and anionic under alkaline conditions.

Table 28.4 Cleaning compound effectiveness by class and cleaning function (after Guthrie, 1972)

Cleaning function	Strong alkalis	Mild alkalis	Poly-phosphates	Weak acids	Strong acids	Surfactants
Chelating	0	1	4	0	0	0
Saponification	4	3	3	3	3	1
Wetting	1	2	1	1	0	4
Peptizing	4	3	1	2	3	0
Emulsification	1	2	2	0	0	4
Dispersion	2	3	1	3	0	3
Rinsing	3	3	2	1	0	4
Corrosion	4	2	0	2	4	0

Key: 4 = excellent; 3 = very good; 2 = moderate; 1 = low; 0 = none.

Anionic wetting agents are essentially pH neutral and usually compatible with acid or alkaline cleaners and certain classes of soap. They frequently provide a synergistic effect when combined with sodium sulfate, phosphates, sodium carboxymethylcellulose, and some natural gums (Elliot, 1980; Katsuyama, 1980). The hundreds of compounds with this designation are typically placed into five categories (Elliot, 1980; Katsuyama, 1980): sulfated alcohols; sulfated hydrocarbons; aryl alkyl polyether sulfates; sulfonated amides; and alkyl aryl sulfonates. Anionic wetting agents, across all classes, exhibit good to excellent detergency with moderate to high foam production.

Cationic wetting agents are limited to the quaternary ammonium compounds (QACs). As wetting agents, QACs are not as efficient as anionic or non-ionic compounds (Elliot, 1980; Katsuyama, 1980). Cationic wetting agents also tend to react unfavorably with minerals and food soils. They have low levels of detergency and are therefore of little use in cleaning applications. By contrast, QACs do have good antibacterial properties (Elliott, 1980) which make them of interest as disinfectants or sanitizing agents. Amphoteric wetting agents are compatible with cationic, anionic and non-ionic wetting agents as well as with some soap (Elliott, 1980). They work by loosening and softening protein and carbohydrate soil types. Amphoteric-type detergents are widely used for cleaning food contact surfaces.

The remaining major class of wetting agents comprises the non-ionic compounds. The materials in this class do not dissociate when dissolved in water and have the broadest range of cleaning properties. They are better detergents for oil than the anionic or cationic species. Non-ionic wetting agents are only marginally affected by water hardness (Elliot, 1980; Katsuyama, 1980). They also vary considerably with regard to their foaming properties. Foam-producing capacity is a function of the ratio of the hydrophilic/hydrophobic balance. This balance is influenced by the temperature of the cleaning solution. For example, with an increase in the temperature of the cleaning solution the hydrophobic property and solubility of a non-ionic wetting agent decreases. At the point of minimum solubility (cloud point), non-ionic surfactants generally act as defoamers. At conditions below the cloud point they vary considerably with regard to their foam producing properties (Arizona Department of Health Services, 2003).

Chemically active cleaning compounds

Compounds that react chemically with soils to affect cleaning include both acid and alkaline types of detergents. Acid cleaners are compounds that are formulated with either organic or inorganic acids. Organic acids are also called weak acids. Common organic acids include acetic, butyric, citric, and lactic acid. By contrast the inorganic acids, also called mineral or strong acids, include hydrochloric (HCl), sulfuric (H_2SO_4), nitric (HNO_3) and phosphoric acids (H_3PO_4).

The hydrogen ion (H^+) is the active ingredient in both strong and weak acids. The reaction between the food soil and acid causes the soil to break-down and dissolve into the cleaning solution. Acid cleaners are especially effective in removing milk stone, mineral deposits, and water stains caused by hard water. Unfortunately many acid-based detergents are corrosive to metals, particularly galvanized iron and stainless steel. HCl, in particular, is especially corrosive to stainless steel. The strong acids are considerably more corrosive than their weak counterparts. For this reason, strong acids are used sparingly and most often only in special cases.

Alkaline cleaners combine with fat to form soap and with proteins to form soluble compounds that are easily removed by water. Alkalis are the principal constituent of most cleaning formulations. The active alkalinity of a chemical is the measure of its value as a detergent. Alkaline solutions with a pH of less than 8.3 have little or no cleaning effect (Elliot, 1980; Katsuyama, 1980). Alkaline cleaners are of three types: strong alkalis; mild alkalis; and soaps.

Sodium hydroxide (NaOH), also called lye and caustic soda, is the strongest of the strong alkaline cleaners. It has excellent detergency. Sodium hydroxide is also highly corrosive to aluminum, galvanized metal, and tin. It has poor rinsing and wetting characteristic in the absence of functional additives. Sodium hydroxide, like other strong alkalis, causes precipitates to form in hard water.

Mild alkalis have greater dissolving power and are far less corrosive than strong alkalis. As mentioned previously, alkalis are the major active ingredient of many industrial cleaning compounds. The bulk of these are mild alkali. The mild alkali sodium carbonate is a common constituent of many cleaning compounds.

Soaps are cleaning agents made from the reaction of alkali and fats (saponification) or fatty acids. Soaps, in general, are not very soluble in cold water. They form precipitates and difficult to remove scum in hard water. Moreover, in solution they break down to fatty acids and lose their detergency. Soap is seldom used in food processing operations, except for hand-washing.

Other important considerations

In the context of defining the parameters of the wash cycle there are important factors other than those previously discussed that require consideration. These factors include the following: surface finish or surface characteristics of the item intended for cleaning; wash cycle temperature and temperature control during wash; exposure time or duration of the wash; and the applied force or the amount of agitation required for cleaning.

Surface finish

The point has been made that a number of the acid and alkaline cleaners are highly corrosive to a range of materials that are commonly used as the materials of construction for food processing equipment. In addition to susceptibility to corrosion, some surface finishes are vulnerable to film formation; both mineral and biological. Special consideration must be given to equipment that is fabricated from materials that do not have smooth, polished or impervious surfaces. Rough-surfaced machine parts, hand tools, and utensils may harbor food soil and shield harmful microorganisms against the rigor of the applied cleaning process. In addition to stainless steel, plastic, and rubber, a host of other polymers are used widely for constructing valve components, pump parts and other components used in a modern food processing operations. The polymers must be reviewed on a case by case basis to determine their compatibility with chemicals and temperatures used in the cleaning process.

Temperature and temperature control

The temperature of the washing solution is extremely important in achieving the cleaning objective. Most cleaning compounds are designed to be active over a relatively narrow range of temperatures. Exceeding the limits of that range, on either the high or the low side, may impair the effectiveness of the cleaning process. There are a host of other adverse effects that might be expected to occur as a result of using inappropriate washing temperatures. For example, elevated temperatures may adversely affect the detergency of a selected cleaning compound by altering its foam-producing capacity as a result of changing the compound's ratio of hydrophobic/hydrophilic balance. Inappropriate washing temperatures may also impact the physical or chemical stability of the target soil and result in the formation of precipitates or films. Recall that many proteins are denatured at temperatures above 185 °F (85 °C) causing them to form difficult to clean films. Not only is it important to select the correct wash cycle temperature, it is imperative that the selected temperature is controlled and monitored during the entirety of the specified wash cycle.

Exposure time

Determining the duration of the wash cycle or the length of time required to achieve the complete removal of the undesirable soils is an essential for setting an effective process. It is also complex. Exposure time must consider soil type, quantity, and its physical form. Water hardness, detergent type, and washing temperature will also impact setting of exposure time. Another major consideration in this regard has to do with the application of force during the cleaning process. A static, force-free process, will typically require a greater exposure time than will a force-aided cleaning process.

Force or agitation

The applied mechanical energy in the form of shear forces created by turbulence, scrubbing, acoustic action or other forms of agitation is an important element of a COP cleaning process. The systematic application of force during cleaning will greatly expedite the process. For example, the use of power ultrasound during cleaning has been shown to radically reduce cleaning cycle time. It is more commonplace, with COP cleaning, to employ agitation during the wash cycle. Agitation within a standard COP tank usually results from the use of a high-speed recirculation pump and specially designed delivery nozzles.

28.4.3 Post-wash rinse

The post-wash rinse is frequently omitted from the literature discussing the basic steps in a COP cleaning process. This is in fact a significant omission because the post-wash rinse is critical to the success of the sanitization or disinfection step that it precedes. The post-wash rinse is normally carried out using soft water at a temperature in the range of 50–75 °C (120–170 °F). Rinsing is responsible for removing and or preventing the redeposit of food soils onto the cleaned surfaces of the processing equipment. Carry-over of food film and residues to the disinfection step are likely to significantly limit the effectiveness of the sanitization process. A post-wash rinse step is an important and necessary element of a sound COP cleaning process.

28.4.4 Sanitization

Discussions of sanitizing agents, disinfectants, and sterilants are necessarily a discussion of microbiology. It is a discussion of the mechanics and kinetics of microbial inactivation. The primary objective in sanitizing, disinfecting, and sterilization, in the context of cleaning food processing equipment, is the reduction of undesirable microorganisms, on the food contact surfaces of equipment, to specified and acceptable levels. Logically then, it follows that any discussion related to sanitizing and disinfection must first focus on identifying and characterizing a target microorganism. Owing to the complexities of this process the skills of a specialist are frequently required. Food processors, small processor in particular, often rely on cleaning chemical vendors for support in this activity.

Identifying a target microorganism is an exercise in risk assessment. The procedure must give consideration to the impact and implications of spreading undesirable microorganisms across the expanse of a manufacturing supply chain. The consequences of this action, which may include economic spoilage, loss in product shelf-life, or life-threatening foodborne illness outbreaks, warrants close examination.

Identifying the target organism requires exquisite and detailed knowledge of the foods that are being manufactured and also of the manufacturing processes. This information is necessary for projecting the outgrowth potential of the various microorganisms likely to be associated with the food and its manufacturing processes. Simply stated, certain environmental factors confer selective advantage on various microorganisms within a given population. Food products and or their residues are capable of conferring this kind of advantage. Therefore, it is probable that the organism in a population best suited to exploit this advantage will be relevant in terms of defining the appropriate control measures. For example *Schizosaccharomyces pombe* (spoilage organism) may be important to fermented steak sauce production, but have little or no relevance to a meat rendering plant. Likewise, *Listeria monocytogenes* (human pathogen) is likely to be important to a fresh lettuce packer but of no importance to a producer of thermally processed vegetables. Properly identifying the target microorganism and establishing risk are fundamental to developing coherent science-based sanitization strategies.

It would be ideal if the yeast, molds, viruses, and bacteria, associated with food processing, responded equally to the various methods of sanitization and disinfection. The fact is, however, that they do not. Among the bacteria, for example, there is great disparity in susceptibility to chemical and heat treatments. As a rule, Gram-positive bacteria tend to be more resistant to heat and chemicals than are their Gram-negative counterparts. This disparity is evident even within species: consider that *Salmonella enterica* serovar *Senftenberg* ($D_{72} = 0.09$ min) is several times more resistant to heat than is *Salmonella enterica* serovar *typhimurium* ($D_{72} = 0.003$ min) in the identical heating medium (Adams and Moss, 1995). This inequality must be taken into account before making a decision regarding sanitization methods. Thermal and chemical sanitizing are the preferred methods of the food processing industry.

Thermal sanitization methods

Thermal sanitizing with hot water is commonplace in COP cleaning operations. Hot water is a very effective broad spectrum sanitizer, disinfectant, or sterilant. It is effective against yeast, molds, viruses, bacteria, and bacterial spores. Achieving the desired lethality or level of inactivation is dependent on temperature, pH, and exposure time. Hot water is easily applied and has great penetrating ability. It is minimally corrosive and therefore compatible with the wide range of metals and plastics used in food processing. Hot water sanitizing is a slow process as it must allow for both come-up and cool down times. Hot water sanitizing also has the disadvantages of forming or contributing to film formation, and also shortening the life of certain equipment. As a general recommendation, hot water sanitizing requires a water temperature of 180 °F (82 °C) with a minimum exposure time of 20 min.

Chemical sanitization methods

It is often written of the ideal chemical sanitizer that it should be approved by regulatory officials for food contact surface application; it should have a wide range or scope of activity and rapidly destroy microorganisms. The ideal sanitizer must also be stable at conditions of use and storage. It should be soluble and possess some detergency while having low toxicity and corrosivity. It will become abundantly clear after an evaluation of the chemicals discussed in this section that there is no single compound for the job of sanitization.

Chemical sanitization is somewhat more complicated than thermal methods. As previously noted there is great disparity in the response and susceptibility of microorganisms to chemical treatments. The effectiveness of chemical sanitizing is dependent on concentration as delivered and exposure time. Other factors that may limit the effectiveness of chemical sanitizing include temperature and the presences of food or detergent films. Surfaces that contain biofilms cannot be effectively sanitized.

The chemicals approved for use in food processing operations vary widely from country to country. In general, approved sanitizing agents are of the following classifications: chlorine-releasing agents; iodophors; QACs; acid–anionic surfactants; fatty acid sanitizers; and peroxides.

Chlorine compounds

Chlorine compounds have broad spectrum antimicrobial properties. They are the most widely used sanitizer in US food processing operation. Chlorine-based sanitizers form hypochlorous acid (HOCl) in solution. HOCl exerts its germicidal effect by attacking microbial membranes and by inhibiting vital metabolic processes. The effectiveness of chlorine is affected by pH, temperature, and organic load. The major disadvantages to chlorine compounds are corrosiveness and worker health and safety issues.

Iodine compounds

Iodophors are the class of sanitizers that contain the halogen iodine. The compounds exist in many forms but are most often formulated with a surfactant as a carrier. Their antimicrobial activity is achieved by way of direct halogenation of cellular proteins and through damage to the cell wall. Iodophors are active against bacteria, yeast, molds, protozoa, and viruses. The major disadvantages to using iodophors are temperature and staining. Iodine evaporates at temperature in excess of 120 °F (49 °C). Thus iodophors are limited to relatively low-temperature applications. Staining of some surfaces, especially plastic, have been reported in applications where this iodophors have been used.

Quaternary ammonium compounds

QACs are surface-active compounds and it is generally accepted that their germicidal activity is related to interactions with the microbial cell wall. The germicidal activities of these compounds are formula dependent and it is therefore difficult to speak in general terms about their overall effectiveness. QACs can be formulated to be effective against both Gram-positive and Gram-negative bacteria. They also have activity against yeast and molds. QACs are also active and stable over a broad range of temperatures. The major disadvantages to using this class of compounds include excessive foam production in automated systems, incompatibility with wetting agents, and low hard water tolerance.

Acid–anionic sanitizers

Acid–anionic sanitizers are surface-active compounds. They are formulated with an inorganic acid and a surfactant. Acid–anionic compounds achieve germicidal activity by way of interaction and subsequent damage to the microbial cell wall. These compounds are effective against vegetative bacteria. The major disadvantage of using acid–anionic sanitizers has to do with their low activity against yeast and molds and also to the fact that they are effective over a very narrow pH range (pH 2 to 3). Because they are surfactants they may also result in excess foam production when used in automated cleaning systems.

Fatty acid sanitizers

Fatty acid sanitizers are formulated with carboxylic acid and other mineral or organic acids. They have a broad range of activity against vegetative bacteria. They are not very active against yeasts and molds. Fatty acid sanitizers also have low germicidal activity at a pH above 4.0. They also lose activity at temperatures below 10 °C (50 °F).

Peroxides

Peroxides are strong oxidizing agents (see Table 28.5) that achieve germicidal activity by disrupting structural and functional properties of the cell. Hydrogen peroxide (H_2O_2) and peroxyacetic acid (PAA) are widely used in food processing operations. Hydrogen peroxide is an excellent broad-spectrum antimicrobial with activity against yeast, molds, viruses, and vegetative bacteria. At concentration 33% or higher and temperatures above 180^0F (82 °C), H_2O_2 is also effective against bacterial spores. Peroxyacetic acid is effective against both Gram-positive and Gram-negative bacteria. Its mode of action has not been fully elucidated; however, it is generally accepted that it too derives its germicidal power from disrupting cell wall integrity. Peroxyacetic acid has also been shown to be effective in the removal of biofilms.

Emerging chemical and mechanical sanitization methods

There are a number of chemical and mechanical methods that are currently used or show promise for use as sanitizers or disinfectants. Notable among these are ozone (O_3), power ultrasound, and cold plasma.

Table 28.5 Activity comparison of common sanitizers

Activities/ limitations	Chlorine-based	Iodine	QACs	Acid–anionic	Fatty acid	Peroxides
Sporicide	Yes	Low	Low	Low	No	Yes
Gram+ bactericide	Yes	Yes	Yes	Yes	Yes	Yes
Gram− bactericide	Yes	Yes	Low	Low	Yes	Yes
Fungicide	Yes	Yes	Yes	Low	Poor	Yes
Neutral pH effective	Yes	+/-	Yes	No	No	Yes
Acid pH effective	Yes	Yes	+/-	Yes (<3.5)	Yes (<4.0)	Yes
Alkaline pH effective	Yes	No	Yes	No	No	Low
Stability (150 °F)	No	No	Yes	Yes	Yes	No
Foam	No	Low	Yes	Yes	Low	No
Hard water	No	Low	Yes	Low	Low	Low
Corrosive	Yes	Low	No	Low	Low	Low

Ozone

Ozone, or more properly tri-atomic oxygen (O_3), has been investigated and shown to be an effective method for sanitizing pre-cleaned nonporous surfaces. Testing conducted by Air Liquide and Del Industries Inc., (witnessed and reported by the National Sanitation Foundation, July 2001) (Boisrobert, 2002) show a 5 log reduction in *E. coli* within 30 s of exposure to ozone at concentrations that ranged from 2.1 to 4.2 ppm. Ozone activity has also been reported, by these same workers (Boisrobert, 2002), against *Staph. aureus*, *Salmonella choleraesuis*, *Listeria*, *Camplyobacter*, *Aspergillus*, and other fungi. Tri-atomic oxygen is produced by reacting two oxygen atoms (O) with two oxygen molecules (O_2) to form the very unstable tri-atomic species:

$$2O + 2O_2 \longrightarrow 2O_3$$

The germicidal effect of ozone is derived by way of oxidation (see Fig. 28.1). Specifically the compound attacks and disrupts bacterial cell walls and also oxidizes DNA. Ozone is partially soluble in water (more so than oxygen) and its germicidal activity is not adversely affected by hard water.

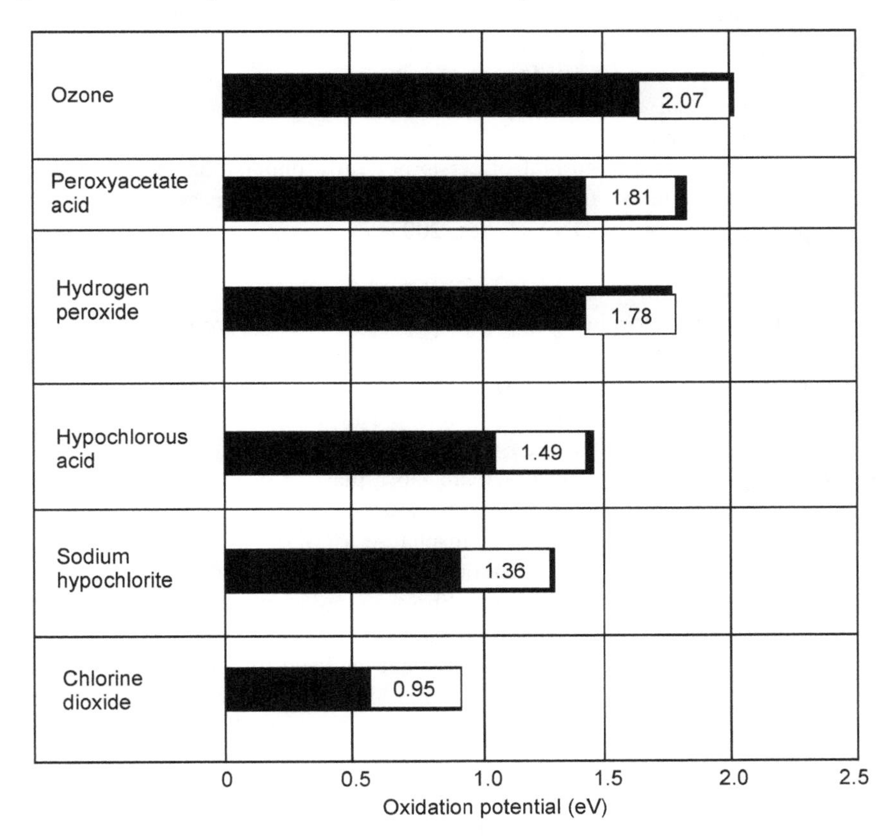

Fig. 28.1 Antimicrobial strength comparison (after Del Ozone, 2004).

Power ultrasound

Ultrasonic cleaning has been around for the last 40–50 years in industry utilizing the concept of welding low power (20 000–40 000 kHz) ultrasonic transducers to the bottom of a steel COP tank. The technology was shown to be highly effective for batch cleaning of engineering components in combination with chemical detergents. However, because of low power and energy efficiency conversion, the technology has always been limited to small-scale batch applications and very long residence times.

Recent advances in low-frequency, high-power ultrasound (20 kHz to 1 MHz) make this technology attractive for a large number of food processing applications including cleaning and sanitizing food contact surfaces. Ultrasound causes tiny bubbles, naturally present in a liquid, to expand (rarefaction) and contract extremely rapidly to the point where they collapse. This occurs thousands of times per second (e.g. 40 000 times per second at 40 kHz). The temperature in the induced cavitation bubble reaches about 5000 °K and the pressure about 2000 atmospheres (Bates, 2004), so a very large amount of energy is transferred into the liquid system, which can be harnessed to produce high shear energy waves and micro-streaming (high-velocity liquid). The high shear energy waves are responsible for producing the ultrasonic cleaning effect. A synergistic effect is reported when using ultrasound in conjunction with an accepted germicidal agent (Bates, 2004). There are also reports in the literature of ultrasonic activity against protozoa, bacteria, yeast, and molds. The mechanism of the antimicrobial effect is not fully understood; however, it is believed to be due to the mechanical disruption of the cell wall and a corresponding leakage of DNA (Bates, 2003).

Non-equilibrium atmospheric pressure plasma

Non-equilibrium atmospheric pressure plasma, frequently referred to as cold plasma, comprises partly ionized gases, which are generated in high-voltage electric fields. So-called cold plasma is formed at or about room temperature. Scientists have known of low-temperature plasma since at least the end of the 19th century; only within the past several years have techniques emerged to make cold plasma generation practical. Cold plasma at atmospheric pressure can be generated by various methods, among these, are dielectric barrier discharge (DBD), resistive barrier discharge (RBD) and also atmospheric pressure plasma jet technology (Laroussi, 1996). These methods can generate relatively large volumes of non-equilibrium, low-temperature plasma at or near atmospheric pressure. Cold plasma can be created using a number of different gases. Air or other gas mixtures can be used to generate the plasma. The gases are sources of UV and visible light and IR radiation, and free radicals such as atomic oxygen and hydroxyl groups, all of which play an important role in microbial inactivation. Mounir Laroussi has devised an apparatus that creates mini-plasma inside a Plexiglas cube by passing an electric current through helium gas via specially calibrated electrodes. Laroussi's process is scalable; cold-plasma containers of virtually any size are feasible. No vacuum pumps are required, since the plasma is generated at normal atmospheric pressure. These developments will give rise to new and broader food processing related applications of the technology including sanitization of food contact surfaces.

Cold plasma gases have been shown to have antimicrobial activity against a wide range of microorganisms. Roth (1998) has reported activity of cold plasma against vegetative bacteria, bacterial spores, yeast, mold and viruses. Inactivation times for many of these microorganisms, on a variety of surfaces, were at least the equivalent of those reported for conventional thermal and chemical methods. The mechanisms of inactivation are not fully understood. Cold plasma technology is still in development. If successful, this innovative technology will offer the food industry a non-chemical, low-temperature alternative means of disinfection.

28.4.5 Post-sanitizing rinse

The final step in developing a COP wet-cleaning process involves the removal and or inactivation of the sanitizing or disinfecting chemical. Because many of the chemicals used as sanitizing agents have toxic properties, it is necessary to provide measures that will either remove or reduce the compounds to safe and acceptable levels. Because of their toxicity, many world governments have promulgated regulations that specify allowable levels or maximum residue levels (MRLs) for sanitizing chemicals and disinfectants. In the USA, for example, this regulatory activity is administered jointly by the US Environmental Protection Agency and the US Food and Drug Administration. It is also noteworthy that the residues of certain sanitizers may cause taints, off-odors, and other unacceptable organoleptic manifestations in food products. These product quality-related changes are also justification for rinsing the sanitizer from the food contact surfaces of processing equipment.

The rinse step in COP cleaning is accomplished most often by flooding or immersing the sanitized equipment in potable water. The rigor required of the post-sanitizing rinse is dependent on the toxicity of the compound and the requirements for its removal from the food contact surface of the sanitized substrate. The rinse cycle must ensure that carry-over of sanitizer into food production is consistent with public health and other regulatory requirements.

28.5 Validation

The COP process is based on a certain set of assumptions related to the manufacturing process. The assumptions underpinning the COP process require validation. That is, the assumptions must be shown to be both valid and relevant to setting the parameters of the process. A COP process developed around faulty assumptions will not be reliable. An effective process, COP or otherwise, must yield predictable results. Validation requires documented evidence that a process consistently conforms to requirements. Validation is intended to demonstrate that the process is stable.

There are a number of good statistical tools that can be used as part of the validation process. These include control charts, designed experiments, capability studies, sampling plans, and failure modes and effects analysis. Control charts, sampling plans, and capability studies are commonplace within the food industry. These methods are well suited to validating a COP process. The selected methods must include conformation of the critical-to-cleaning criteria established at each step in the process. For example, preventing film formation during the washing step is a critical-to-cleaning activity. Therefore, the validation protocol must confirm that film formation is not occurring and will not occur, provided that the process is properly executed. Likewise, inactivation of the target microorganism to levels specified by the process is also critical to the sanitization step of the process and must, therefore, be confirmed. The validation methods must expose and illuminate the causes of process variation. It is imperative that common cause variation is distinguished from special cause variation. When the validation process reveals special cause variation, it is important that root cause analysis is conducted and that aggressive measures are provided to eliminate the source of variation.

Another important point related to the validation process is its impact on related activities. Validating the COP process may well impact other aspects of manufacturing. It is entirely reasonable to conceive validation affecting the sanitation, quality control, GMP, and HACCP programs. For example, validating the elements of the COP process may highlight weaknesses in the master sanitation schedule or confirm the validity of a critical limit specified in the HACCP plan. Validation, in other words, should confirm that the COP cleaning of equipment is not an exercise completed in isolation. The implication and risk associated with improperly cleaned equipment will not remain confined to the COP tank, pipe washer or automated tote washer.

28.6 Records and process documentation

If there is not a record, it did not happen. Good record keeping is fundamental to food safety and it is also vital to process verification. Records confirming that the critical-to-cleaning activities specified for each step in the COP process are compulsory. The record-keeping program must include the data that will confirm that the particulars of the process have been achieved. For example, if the process specifies a wash temperature of 180 °F (82 °C) for the duration of the wash cycle, then there must be a compelling and unambiguous document to demonstrate that the temperature was achieved and maintained for the entirety of the wash cycle. Likewise, when a MRL has been established for a sanitizing chemical or disinfectant, then your records must provide proof that the tolerance was not exceeded. Moreover, and in addition to daily process records, there must be a body of documentation that provides a historical record of how the COP process was originally developed. These records should identify all key personnel involved with the development process. The record must also include evidence of the risk assessment process as well as information supporting or justifying the critical-to-cleaning criterion established at each step of the process. Perhaps the most important aspect of record keeping that is frequently overlooked is the need to document change to the process. When the process is changed, or the nature of the object to be cleaned is changed, these events must be included in the COP process records. Moreover the actions taken in response to these changes must also become part of the record. Records are an integral and necessary part of the COP process.

28.7 Summary

COP cleaning of food processing equipment is considerably more involved than scrubbing the component with a brush and then placing it into a tank containing hot water. While these were the accepted methods when Upton Sinclair's *The Jungle* was first published, it is now, however, a generally accepted fact that COP cleaning is a science and risk-based activity. The formal COP process is composed of five discrete steps: pre-wash rinse; cleaning wash; post-wash rinse; sanitization; and a post-sanitization rinse. Moreover, each step of the process contains discrete activities that are critical to the success of the entire COP process. The delineation of the critical-to-cleaning criteria, on a stepwise basis, is essential for developing a stable COP process. Critical-to-cleaning criteria include selecting a detergent based on the physical and chemical characteristic of a specific soil type. They also involve temperature and the amount of force applied during the post-wash rinse. Critical-to-cleaning criteria (temperature, concentration, and exposure time) for the sanitizing step of the process, must be based on resistance data of the target organism. Using a sanitizer that is capable of a 5*D* reduction in the target organism when 6*D* is required will produce a process failure. Understanding the capability of the process is vital to predicting the outcomes of applying that process. As stated previously, COP cleaning is widely used within the food processing industry. Pipe washer, tote and tub washers, batch and continuous dishwasher, COP tank-type washer and sinks are used for this application. It is noteworthy, with regard to the discussion of stable processes, that manual COP cleaning applications are inherently unstable. This type of COP process relies exclusively on human interventions for its execution. The facts are that each willing worker's approach to the task, though well intended, will be different. This difference is a source of special cause variation and will therefore make it impossible to predict the outcomes of their actions in applying a COP process. This is an important consideration in terms of risk management.

The capability of the process must also be validated. That is, measures must be provided that will enable confirmation of the assumptions that underpin the process. Validation is a critical but often overlooked aspect of COP process development. Validating the efficacy of the entire COP process is important for understanding how the identified risk, associated with improperly cleaned equipment, are managed. Moreover, the validation process will likely raise questions about other aspects of the sanitation and hygiene programs. Validating the COP process is not an exercise that can be conducted in isolation. The consequences of a COP process failure are much greater than dirty or contaminated equipment. The consequences are likely to also include loss of product shelf-life, economic spoilage, or even life-threatening foodborne illness outbreaks. In any event it is certain that the outcome of repeated COP process failures will be a negative impact on the company's bottom line.

This chapter opened with a discussion of the dark days of food processing and of primitive cleaning and hygienic practices. The scientific and technological advances in cleaning food processing equipment, since 1906, have also been noted. Perhaps the most important and profound advances to date have been in the area of microbiology. Sanitarians and food processors now understand, absolutely, the necessity of protecting their businesses against the pernicious impact of undesirable microorganisms within the food processing environment. There have been advances in other areas related to cleaning as well. The cleaning power of water containing suspended nanoscale particles such as detergent surfactants has been appreciated for many years. However, until recently scientists have not fully understood the details of how this process worked. Research in the area of nanofluid flow has greatly increased our knowledge of the mechanics of how particle containing fluids spread out and lift soils off a surface. The growth in the understanding of nanofluid flow (Wasan and Nikolov, 2003) and its impact on soil removal, portend great opportunity for innovation in the area of detergent chemistry and the design of cleaning compounds for the future. These scientific and technological breakthroughs will, again, provide the food processing industry with the additional tools necessary for supporting the safe mass production of foods for human use and consumption.

28.8 Bibliography

ADAMS, M.R. and MOSS, M.O. 1995, *Food Microbiology*, The Royal Society of Chemistry, Cambridge, England.

AMERICAN PUBLIC HEALTH ASSOCIATION 1985. *Standard Methods for the Examination of Water and Wastewater*, 16th edition. American Public Health Association, Washington DC, USA.

ARIZONA DEPARTMENT OF HEALTH SERVICES 2003. Food Equipment Cleaning and Sanitizing (www.hs.state.az.us/phs/oeh/fses/fecs_weq.htm).

BAKKA, R.L. 1997. *Making the Right Choice – Cleaners*. Ecolab, Food and Beverage Div., St. Paul, MN, USA.

BATES, D. 2003. *Characteristics of High Power Ultrasound*, Chicago, IL, USA.

BATES, D. 2004. Ultrasound in food processing operation. *IFT/IFC Conference*, Sydney, Australia.

BOISROBERT, C. 2002. Ozone as a surface sanitizer. *Food Safety Watch*, Vol. 12, No. 3, Illinois Institute of Technology, Chicago, IL, USA.

BOUFFORD, T. 2003. *Making the Right Choice – Sanitizers*. Ecolab, Food and Beverage Div., St. Paul, MN, USA.

CLARK, J.P. 2004. Ozone – cure for some sanitation problems. *Food Technology*, Vol. 58, No. 4, 75–81.

DEMING, W.E. 1993. *The New Economics for Industry, Government and Education*. Massachusetts Institute of Technology, Cambridge, MA, USA.

DE VRIES, J. 1997. *Food Safety and Toxicity*. CRC Press, Boca Raton, FL, USA.

DEL OZONE 2004. Surface sanitation (www.delozone.com).

ELLIOTT, R.P. 1980. *Principles of Food Processing Sanitation*, Chapter 4. The Food Processors Institute, Washington DC, USA.

FROBISHER, M. and MICHENER, H.D. 1974. *Fundamentals of Microbiology* 6th edition. W.B. Saunders Company, Washington DC, USA.

GUTHRIE, R.K. 1972. *Food Sanitation*. AVI Publishing Company, Inc., Westport, CT, USA.

HAIMES, Y.Y. 2004. *Risk Modeling, Assessment, and Management*. John Wiley-Interscience, Hoboken, NJ, USA.

IMHOLTE, T. J. 1984. *Engineering for Food Safety and Sanitation*. Technical Institute of Food Safety, Minneapolis, MN, USA.

KATSUYAMA, A.M. 1980. *Principles of Food Processing Sanitation*. The Food Processors Institute, Washington DC, USA.

LAROUSSI, M. 1996. *Plasma-based Sterilization*. Old Dominion University, Norfolk, VA, USA.

MIKEL, H. and SCHROEDER, R. 2000. *Six Sigma*. Doubleday, Minneapolis, MN, USA.

ROTH, J.R. 1998. IEEE Trans *Plasma Science*, Vol. 20.

SCHMIDT, R.H. 1997. *Basic Elements of Equipment Cleaning and Sanitizing in Food Processing and Handling Operations*. University of Florida, Tampa, FL, USA.

SINCLAIR, U. 1906. *The Jungle*. National Penguin Press Inc., New York, USA.

WASAN, D.T. and NIKOLOV, A.D. 2003. Spreading of nanofluids on solids. *Nature*, Vol. 423, pp. 156–159.

29

Improving the cleaning of heat exchangers

P. J. Fryer and G. K. Christian, University of Birmingham, UK

29.1 Introduction

29.1.1 Fouling

Fouling can be described as the unwanted build-up of deposits on a surface. This is a major problem in the food industry, particularly during thermal treatment. The build-up of deposit increases pressure drop, owing to the increase in surface roughness and the decrease in cross-sectional area of the flow channels, and reduces heat transfer efficiency. Increased costs are therefore incurred to operate the plant. Fouling can also compromise product quality, by cross-contamination or microbial growth on the deposit. It is necessary to stop production to clean the process plant, often daily. The overall productivity of the process plant is therefore reduced, and failure to clean could compromise product quality or sterility.

Often the deposit has a very different chemical composition from the process fluid. The fouling process and the ways to minimise it have been investigated for many years (for detailed examples, see the series of conferences including Fryer *et al.*, 1996 and Wilson *et al.*, 1999, 2002). There is now focus on understanding how the deposit is removed in order to improve the efficiency of cleaning processes. Milk and other dairy fluids have been the subject of much of the research in this field, but other food materials, such as starches, cause processing problems.

Fouling occurs in many different situations and can arise from different mechanisms. The fouling process generally involves a number of steps (Epstein, 1983): initiation, transport, attachment, build-up and ageing. Table 29.1 summarises mechanisms, one or more of which may be involved in a particular fouling process. Fouling in food processing is common: for example, Bird

Table 29.1 Summary of the fouling mechanisms and the underlying process involved (Bott, 1995).

Fouling mechanism	Underlying process
Crystallisation or scaling	Formation of crystals on the surface. Deposits formed from solutions of dissolved substances on to heat transfer surface. Cooled surfaces are subject to fouling from normally soluble salts and fats and waxes. Inversely soluble salts, e.g. calcium carbonate deposits onto heated surfaces.
Particulate	Small suspended particles such as clay, silt or iron oxide deposit on heat transfer surfaces. Where settling by gravity is the determining factor this is then called *sedimentation fouling*.
Chemical reaction	The deposit formed on the surface (particularly heat transfer surfaces) is not the initial reactant (e.g. in petroleum refining, polymer production, dairy plants).
Corrosion	The material of the heat transfer surface is involved in reactions with components of the fluid to form corrosion products on the surface, i.e. a specific type of chemical reaction fouling.
Biological	Microbial fouling – deposition and growth of organic films consisting of microorganisms and their products. Macrobial fouling – attachment and growth of macroorganisms, such as barnacles or mussels.
Freezing	Deposit formed from a frozen layer of the process fluid.

(1992) reported papers on the cleaning of a number of foods, such as: chocolate desserts (Bird, 1992; Rene *et al.*, 1988), coffee solutions, corn syrup, meat products, soya oil protein (Wilkinson, 1982), starch (Maruyama *et al.*, 1991) and tomato soft solids (Cheow and Jackson, 1982a,b). The diverse nature of food fluids led to a number of components being deposited on heat transfer surfaces (see Table 29.2); the fouling mechanism is often a combination of chemical reaction (for example, of proteins) and crystallisation (of mineral salts).

Table 29.2 The nature of food deposits deposit during production, the effect of heating and the solubility of the deposit (Grasshoff, 1997)

Component deposited	Solubility	Ease of removal	Change upon heating
Sugar	Water soluble	Easy	Caramelisation: more difficult to clean
Fat	Water and alkali soluble	Difficult	Polymerisation: more difficult to clean
Protein	Water insoluble, alkali soluble, slightly acid soluble	Very difficult	Denaturation: denatured protein deposits are very difficult to clean
Mineral salts	Water solubility is variable but most are acid soluble	Easy to difficult	Interactions with other constituents: generally easier to clean

29.1.2 Cleaning

Cleaning can be described as the removal of foreign bodies to return a system to its original state before fouling occurred. The need to clean and the consequences of an ineffective cleaning regime have been discussed above. The cleaning requirement of a system varies on the process. Different classifications for the extent of cleaning are possible:

- *atomically*: clean on a nano-scale;
- *physically*: no physical measurement of the deposit is possible and none can be optically detected;
- *chemically*: absence of substances that may interfere with product processing;
- *biologically* (or sterile): free of microorganisms.

Most cleaning research in food processing has studied milk because of the ubiquity of the problem. Three phases are involved in the removal of proteinaceous milk foul: the heat transfer surface, the deposit and the alkali cleaning solution. Any of the following reactions may be involved in cleaning of food deposits (Plett, 1985): melting, mechanical break-up, wetting, swelling, desorption, emulsification, hydrolysis, saponification and dispersion. Removal may be governed by a combination of mass transfer, diffusion and reaction (Bird and Fryer, 1991), any of which may be the controlling factor. Jeurnink and Brinkman (1994) concluded that the process cannot be diffusion controlled since the diffusion coefficient is small and cleaning would take hours by diffusion; it takes only a matter of minutes, however.

The extent of fouling and cleaning is often monitored by pressure drop and heat transfer changes throughout the plant. Historically, process equipment was opened and cleaned individually. Large-scale production led to the need for cleaning-in-place (CIP) systems, in which chemicals are circulated to remove the deposit; alternatively surfaces can be subjected to sprays or jets of the cleaning chemicals. Rinsing stages, where water is used, are often involved. These processes have become highly developed and automated, but are rarely, if ever, optimised. Cleaning regimes generally involve a number of cycles:

- *Pre-rinse* – circulation of water to remove loosely bound substances from the surface.
- *Detergent cycle* – action of the cleaning chemical (acid or alkali) to release the deposit from the surface. The resulting components are held in solution and removed with the fluid flow. The majority of cleaning takes place during this cycle.
- *Post-rinse* – all traces of deposit and cleaning chemical are removed from the system by circulation of water.
- *Sanitisation* – disinfection and surface conditioning.
- *Final rinse* – circulation of water prior to product processing.

Currently processes are carried out under empirical conditions, either of fixed time or cleaning solution volume known to give repeatable clean results. Much research has been carried out to optimise the operation of milk processing plants through understanding of fouling (for example Georgiadis *et al.*, 1998; Petermeier *et al.*, 2002; Grijspeert *et al.*, 2004). This approach might be combined with cleaning schedule optimisation to optimise overall plant operation. Smaili *et al.* (1999) considered CIP of sugar process plants, and investigated the scope for reducing the cost of cleaning by optimising the fouling and cleaning cycle, including scheduling and minimising the length of the cleaning period (Smaili *et al.*, 2001, 2002a,b). Starling and Nicol (2001) investigated CIP systems to determine the optimal layout of different plants.

29.1.3 Cost of fouling and cleaning

Fouling and subsequent cleaning of food production plants has both economic and environmental impact. The direct costs of fouling and cleaning have been categorised as the following (Pritchard, 1988):

- *Loss of production*: reduced process efficiency and the need to shut down to clean.
- *Maintenance costs*: due to the necessity to install complex cleaning processes.
- *Fuel costs*: increased heating and pumping power to maintain process conditions.
- *Capital expenditure*: overestimating heat exchanger area and installation of extra pump capacity to allow for fouling.

Environmental effects are increasingly important. There are increasing global concerns to reduce the amount of waste from food production. Solutions used for dairy cleaning must be neutralised to a pH range between 6.5 and 10 before being released, so the products of this neutralisation step are of mainly environmental concern (Grasshoff, 1997). Products include sodium carbonates, sodium hydrocarbonates and sodium salts of nitric and phosphoric acid. Phosphates (added to cleaning agents to reduce water hardness) are of known ecological risk due to excess water phytilication/eutrophication. Complexing agents added to cleaning solutions may also remobilise heavy metals (Grasshoff, 1997). Other possible additives (e.g. active oxygen and active chlorine) are also of concern, despite the increase in cleaning rates on their addition. Active oxygen can cause splitting of water and molecular oxygen on addition to hot cleaning solutions (Grasshoff, 1989). Addition of such chemicals to increase cleaning rates should be kept to a minimum until more is known of their environmental effect. Recovery and reuse of CIP chemicals are used in some cases to reduce the load on sewerage, although in all cases the effluent must at some point be released. This can take place by: (1) processing of the waste on site and direct disposal to an outfall ditch, or (2) release to a local sewerage plant for processing, i.e. indirect disposal.

29.1.4 Understanding cleaning

If fouling did not occur there would not be a need to clean; however, extensive research has not yet found a prevention method and so cleaning must still take place. Understanding cleaning would require knowledge of the following aspects:

- *Deposit removal*; understanding of how the deposit is removed will allow the process to be improved.
- *Cleaning process parameters*; these include temperature, chemical type, chemical concentration, flow rate (turbulence, etc. is also associated with plant design). Determination and use of optimum conditions will reduce the time to clean.
- *Process plant design*; i.e. reduce the 'dead legs' within the system and constitution of the plant with material that is easily cleaned and maintained (using guidelines such as those of the European Hygienic Engineering and Design Group (EHEDG), described elsewhere in this book).
- *Cleaning regime*; such as the order and duration of circulation of cleaning chemicals and rinse waters. Knowledge of a regime to ensure cleaning in the shortest time and with the lowest use of chemicals and water is desirable.
- *Monitoring the extent of cleaning*; all of the above factors could be optimised if the state of the level of cleaning throughout the plant is sensitively and accurately known.

Better understanding of the processes involved with fouling and cleaning would essentially lead to reduced expenditure and environmental impact through:

- reduced process conditions (i.e. temperature, flow rate, chemicals and process water), and
- shorter cleaning cycles and hence reduced down-time.

29.2 Processing effects on fouling and cleaning

29.2.1 Fouling from milk

Fouling and cleaning from milk and other dairy products have been extensively studied and will be used as an example. Milk is a complex fluid with a number of thermally unstable components. The composition of the deposit formed during thermal treatment depends on the process temperature and differs greatly from the composition of the fluid. Two major types of deposits have been classified (Burton, 1968); the composition and appearance of the two deposit types (Type A and Type B) are given in Table 29.3. The protein content of Type A deposit is mainly β-lactoglobulin (β-lg, a whey protein); at least 50% of the protein content of deposit formed at 70–80 °C is β-lg (Lalande *et al.*, 1985; Tissier and Lalande, 1986) although it is only *ca.* 10% of the total protein in milk (Walstra and Jenness, 1984). In the native state this protein is a dimer; above 50 °C it dissociates and between 60 and 70 °C the monomers unfold (i.e. denature). Upon unfolding β-lg free –SH groups are exposed, which may react with disulphide bonds on other β-lg molecules (or with other proteins) in a polymerisation chain reaction (Roefs and de Kruif, 1994). This is not reversible; the aggregates that result are water insoluble. However, de la Fuente *et al.*

Table 29.3 Composition of deposits formed by milk at different temperatures. The balance of the percentage composition is made up by other components such as carbohydrate (Burton, 1968)

	Processing conditions	Temperature	Composition	Appearance
Type A	Pasteurisation	up to 100 °C	Protein: 50–60% Mineral: 30–35% Fat: 4–8%	Soft voluminous, curd-like material, white or cream in colour
Type B	UHT	100–140 °C	Protein: 15–20% Mineral: 70% Fat: 4–8%	Brittle, gritty and grey in colour

(2002) state that the unfolded molecules, oligomers, dimers and monomers that dissociate from larger aggregates by treatment with SDS would not be predicted by the Roefs and de Kruif model.

The difficulty of storage and cost of using raw milk in research led to the use of whey protein concentrate (WPC) solutions by a number of authors (such as Gillham, 1997; Robbins *et al.*, 1999). Use of WPC avoids the natural variations associated with milk and enables reproducible fouling behaviour to be obtained. Heat treatment of WPC solution causes a deposit similar to Type A from milk; heavy protein fouling of this type occurs at all temperatures in the range 85–140 °C (Robbins *et al.*, 1999). However, in contrast to WPC processing, milk does not produce a deposit between 100 and 120 °C. Also, unlike milk, WPC does not produce a mineral scale at UHT temperatures. The extent of fouling from WPC solution increases with increasing temperature.

Hege and Kessler (1986b) noted that after an hour, where Type A deposit is initially formed, a mineral-rich layer (mainly calcium and phosphorus) developed next to the heat transfer surface. The protein remained as an outer layer, thus forming two distinct layers within the deposit. Tissier and Lalande (1986) suggested this was due to diffusion of minerals through the protein layer and crystallisation of calcium phosphate at the heat transfer surface. Type B deposit is hard and granular rather than soft and spongy like Type A (Lyster, 1965). A mineral-rich layer (calcium, phosphorus and magnesium) forms at the heat exchanger surface (Foster and Green, 1990). The same authors found that, despite there not being distinct layers within the Type B deposit, protein was concentrated near the outside of the deposit. Lalande *et al.* (1985) found that the calcium phosphate deposited on surfaces within a plate heat exchanger (PHE) is a mixture of dicalcium phosphate dihydrate (DCPD) ($CaHPO_4 \cdot 2H_2O$) and octacalcium phosphate (OCP) ($Ca_8H_2(PO_4)_6 \cdot 5H_2O$).

Fouling from milk processing is thought to be due to a combination of a number of different fouling mechanisms. Often there is a lag (induction) period, during which no change in heat transfer coefficient of pressure drop occurs, before fouling commences. Belmar-Beiny and Fryer (1992, 1993) studied the early stage of fouling from whey protein solutions on stainless steel pipes at pasteurisation temperatures. From scanning electron microscopy (SEM) images they determined that the metal surface first became covered with a layer of material (120 seconds), to which aggregates adhere or grow. The results (from X-ray elemental mapping, X-ray photoelectron spectroscopy (XPS)) for deposits formed over 60 minutes show:

- the initial film formed has a composition close to the theoretical composition for pure β-lg;
- after 150 seconds: calcium and sulphur but not phosphorus are detected – the deposit has a non-uniform nature;
- after 60 minutes: both calcium and phosphorus are detected – a high concentration of these minerals was found at the deposit–stainless steel interface.

These results suggest a change in deposit structure with time; with protein being deposited first, then a mineral layer forming at the surface. Visser *et al.* (1997) proposed three fouling mechanisms for solutions such as milk or whey protein solutions with milk salts:

1. Denatured whey proteins aggregate with themselves or other proteins and adhere to the surface.
2. Inversely soluble calcium phosphate and calcium citrate precipitate upon heating. Crystallisation at the surface can occur and form a mineral scale.
3. Inclusion of product fluid within the growing fouling layers allows the formation of different layers, which may have different compositions. This mechanism is dependent on the process and the product conditions.

De Jong *et al.* (2002) describes the use of models based on β-lg kinetics to predict fouling; this type of model has been used by authors such as Grijspeerdt *et al.* (2004) and Sahoo *et al.* (2005) to study fouling in a variety of situations. The simulations suggest that holding times at intermediate temperatures reduce fouling through reactions taking place in the bulk rather than on the surface. This type of holding section has been used empirically for many years; the models allow rational selection of process times and conditions.

29.2.2 Factors affecting fouling

The factors affecting fouling have been summarised by a number of authors (such as Changani, 2000; Tuladhar, 2001) and include the following:

- *Natural variation:*
  - composition and mineral content including seasonal variation and cattle diet (Burton, 1966, 1968; Fryer, 1985);
  - pH (Skudder *et al.*, 1986; Hege and Kessler, 1986a);
  - air content (Burton, 1968; Fryer, 1985).
- *Engineering factors:*
  - age or holding, i.e. keeping milk prior to treatment (Burton, 1968; Jeurnink, 1991);

- pre-treatment (heating the milk to 40 or 50 °C prior to pasteurisation) (Burton, 1968; Visser and Jeurnink, 1997);
- flow rate (Gordon *et al.*, 1968; Fryer, 1985; Belmar-Beiny *et al.*, 1993);
- bulk and wall temperature (Lalande *et al.*, 1985; Paterson and Fryer, 1988; Belmar-Beiny *et al.*, 1993);
- processing time (Fryer, 1985; Hege and Kessler, 1986a).

Surface treatments have been investigated for a number for years; for example, Britten *et al.* (1988) studied the effect of coatings such as polymethylacrylate, nylon and cellulose acetate and Yoon and Lund (1994) investigated electropolished stainless steel, Teflon- and polysiloxane-coated plates. Once the initial layer of deposit has formed, subsequent deposition is unaffected, however, the adhesion strength is altered. Ion implantation magnetron sputtering, plasma-enhanced vapour deposition, dynamic mixing, and auto-catalytic Ni–P–PTFE (poly(tetrafluoroethene)) are among the more recent surface modifications investigated to reduce fouling (Zhao *et al.*, 2002). Although little advance has been found in reducing fouling by dairy products using these modification techniques, results show (Beuf *et al.*, 2003) that cleaning efficiency of NaOH is considerably increased with Ni–P–PTFE-treated surfaces. Rosaninho *et al.* (2003) found that surface with low electron donor values (such as Ni–P–PTFE-treated surfaces) are less susceptible to fouling from calcium phosphate solutions. The effect of modifying surface energy on cleaning will be discussed later.

29.2.3 Cleaning of dairy deposits

There are a number of issues to be considered when looking at cleaning: (1) process parameters (e.g. cleaning solution concentration, temperature and flow rate), (2) the circulation regime (i.e. order and circulation time of each stage), (3) monitoring of the extent of fouling and hence cleaning and (4) plant hygienic design. Extensive laboratory-scale research has been carried out to determine whether optimal or minimal cleaning conditions exist for dairy deposit removal. Such information could lower the cost of the cleaning process and reduce the time the plant is out of production, which would also save money. Laboratory-scale experiments allow controlled process conditions, including the use of reproducibly fouled samples, to investigate the kinetics and mechanisms of cleaning. However, the complex inter-relationship between temperature, viscosity and flow rate makes it difficult to determine the specific effect of each parameter on cleaning, even under controlled conditions. Larger-scale experiments (such as in pilot-scale plate heat exchangers) aid in the overall evaluation of the cleaning process and determination of the required duration and order of cleaning solution circulation.

Industrial CIP involves the circulation of hot cleaning fluids through closed systems of pipes, tanks and heat exchangers; this avoids the need to dismantle equipment. Traditionally dairy CIP involved circulation of alkali and acid solutions, although now single stage cleaners are often used, as they involve only one chemical rinse and result in a cleaner surface (Timperley and Smeulders, 1987). Where two-stage cleaners are used the order of circulation of acid and alkali steps is subject to the nature of the deposit and varies for each plant. It would be advantageous to know the exact required length of each step in order to reduce the overall process time.

Cleaning mechanism

An understanding of how the deposit is removed from the heat transfer surface and the controlling processes involved would aid in optimising the overall cleaning process. Until the behaviour of deposit removal is understood under all conditions, the plant may not be cleaned in the shortest time and waste from the cleaning process will not be minimised.

Water alone will not remove milk deposit; cold water runs off without wetting the surface, fats within the deposit prevent contact. Hot water is able to melt the fats, allowing better contact with the deposit. A cleaning fluid must be selected in which the deposit is soluble. Proteins are water insoluble, alkali soluble and slightly acid soluble. Reaction of NaOH with the protein deposit involves hydrolysis of the peptide bonds, which link the amino acids in the protein structure. The product is more water-soluble than the native deposit. The water solubility of minerals is variable; solubility of calcium phosphate decreases with increasing temperature, hence the scale forms at increased temperatures. Most minerals are acid soluble (Grasshoff, 1997) and therefore removed more efficiently by acid than alkali.

Protein deposits swell when they adsorb water, alkali solutions can reinforce this swelling and also dissolve the protein (Jeurnink and Brinkman, 1994); cracks can then form, which increase penetration of the cleaning solution into the deposit layer. Plett (1985) gave a detailed description of the possible steps involved in the overall removal of milk soil:

- Bulk reaction between components of the cleaning chemical and the bulk fluid, e.g. hard water ions. Much of the chemical may not even be involved in the cleaning process.
- Transport to the surface of the cleaning chemical. The transport process through the boundary layer is affected by temperature, concentration and flow conditions.
- Transport into the deposit layer: penetration of the chemical components into the deposit is dependent on the deposit structure. Reaction zones may appear at the deposit/fluid interface in thick deposits. Surface-active agents can increase penetration owing to their wetting ability.
- Reaction between deposit and cleaning chemical including melting, mechanical break-up, wetting, swelling, desorption, emulsification, hydrolysis, saponification and dispersion.
- Transport to the interface: reaction products diffuse out of the deposit.
- Transport to the bulk: concentration gradients and hydrodynamic conditions allow the transport of the reaction products into the bulk.

Visualisation of the cleaning of milk and whey protein concentrate deposits (such as Christian, 2004; Bird and Bartlett, 1995; Grasshoff, 1989) have shown the non-uniform nature of deposit removal, i.e. the deposit is removed in discrete islands. The process has three stages (Fig. 29.1(a)):

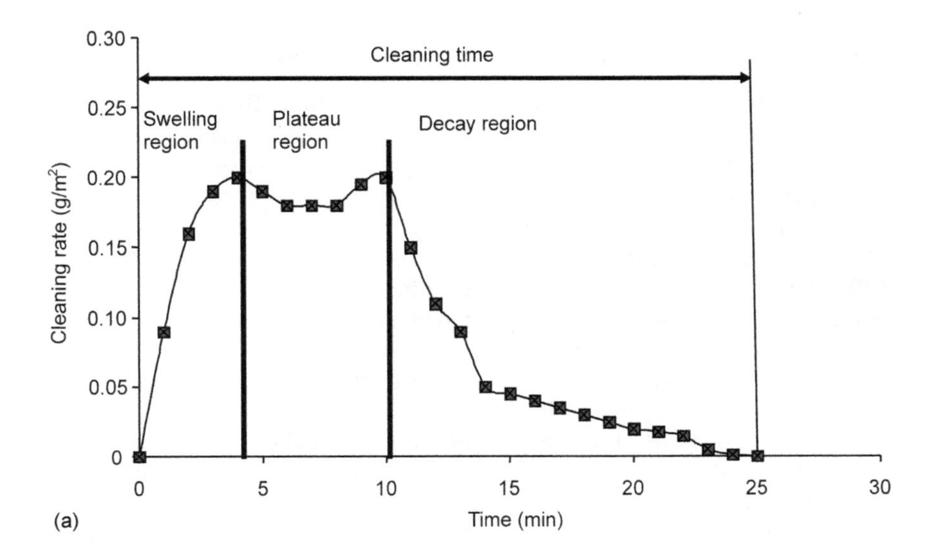

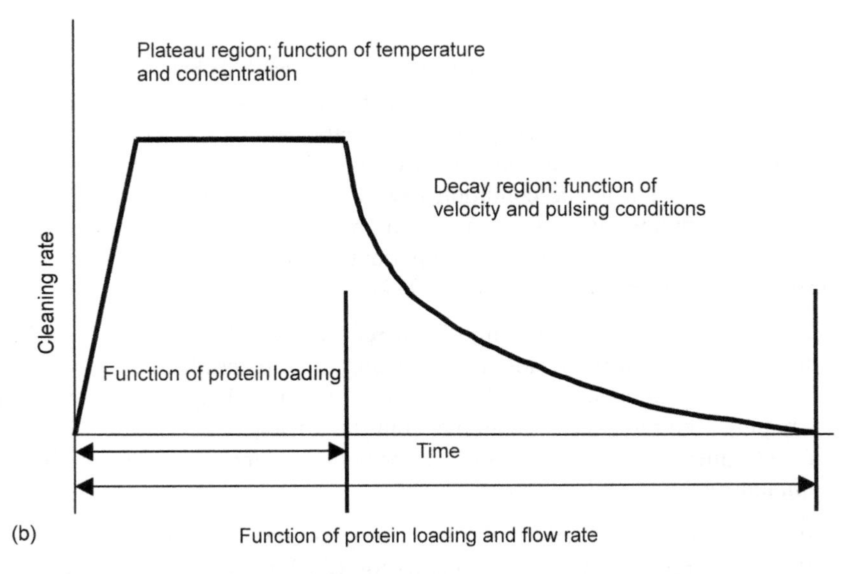

Fig. 29.1 Schematic diagrams of typical cleaning curves (a) shape of removal curves (b) effect of process variables on each stage of cleaning (both from Gillham, 1997).

- *Swelling* – alkali solution contacts the deposit and causes swelling, forming a protein matrix of high void fraction.
- *Erosion* – uniform removal of deposit by shear stress forces and diffusion; there may be a *plateau region* of constant cleaning rate, but this depends on the balance between swelling and removal.
- *Decay* – the swollen deposit is thin and no longer uniform, removal of isolated islands occurs by shear stress and mass transport.

This behaviour has been seen by many workers, such as Bird (1992), Gillham (1997), Grasshoff (1997) and Tuladhar (2001), using visual methods, heat transfer monitoring and fluid dynamic gauging. SEM analysis showed that a more open structure is formed when whey proteins come in contact with alkali solution (Bird, 1992; Gillham, 1997). At higher concentrations (e.g. above 2 wt% NaOH) the structure can become less susceptible to shear forces and more difficult to remove. Protein deposits are often removed in large chunks, and hence may remove minerals from different layers with them. Figure 29.1(b) shows the proposed dependency on the cleaning process parameters, which suggests that the plateau region is a function of temperature and the decay region is dependent on the flow velocity.

Christian (2004), Bird and Fryer (1991) and Grasshoff (1997) carried out mechanistic studies to determine the effect of process parameters (temperature, flow rate and chemical concentration) on each stage of the cleaning process, discussed below. They conclude that diffusion and reaction will be important in the initial stages as the deposit swells; however, their investigation also showed that a key step is removal of swollen deposit in lumps by shear.

Monitoring cleaning

It is first important to assess what is meant as 'clean'. What is sufficient for some cases may not be for others; for example in the pharmaceutical industry, where a single microbe is potentially catastrophic, systems must be microbiologically clean. In food processing, 'clean' is often considered as returning the system to the condition it was in before fouling occurred; after cleaning, the fouling rate is the same as in previous runs – although this might miss small amounts of fouling. It is important to monitor fouling and cleaning. Types of monitors include (Karlsson, 1999):

- *continuous* – allows monitoring of soil removal, or *discontinuous* – enables the amount of soil remaining after cleaning to be determined;
- *in-situ* – measurements carried-out in the experimental equipment, or *ex-situ* – requires the removal of a sample from the set-up;
- *direct* – measurements at the surface, or *indirect* – monitors cleaning by the amount removed.

Ideally monitors should tell both when a process run should stop and when a plant is clean. Fouling increases pressure drop (ΔP) and decreases heat transfer coefficient (HTC); these values can be monitored, together with (for example) turbidity monitoring to confirm when rinsing is complete and the fluid flowing down a pipe is clear. Some monitors, such as the 'Monitex' program (Müller-Steinhagen, 2000), analyse information, to allow scheduling of cleaning. Key parameters are given below. The pressure drop across a system (ΔP) is defined as:

$$\Delta P = P_{\mathrm{I}} - P_{\mathrm{O}} \tag{29.1}$$

where P_{I} and P_{O} are the inlet and outlet pressure, and P_{O} increases with time during fouling. The change in HTC during fouling can be accounted for by including a fouling resistance, R_{f}, in the equation relating the initial clean HTC, (U_0) to that at time t, (U_t):

$$\frac{1}{U_t} = \frac{1}{U_0} + R_{\mathrm{f}} \tag{29.2}$$

The extent of fouling may be expressed by a Biot number (Bi), which accounts for deposit thickness (x) and thermal conductivity (λ): $Bi = R_{\mathrm{f}} U_0$ where $R_{\mathrm{f}} = x/\lambda$ for the deposit. Both heat transfer and pressure drop measurements can be insensitive to small amounts of fouling and do not give good information about the level of deposit left on the heat transfer surface. Tuladhar (2001) defines a deposit resistance during cleaning (R_{d}):

$$R_{\mathrm{d}} = \frac{1}{U_t} - \frac{1}{U_{\mathrm{c}}} \tag{29.3}$$

where U_t and U_{c} are the HTC at time t and the HTC of the final clean system. A combination of this approach with in-line turbidity sensors that show when a line is clean of product, and thus the end of rinse stages, can give reasonable monitoring of processes.

29.3 Investigations into cleaning process parameters

29.3.1 Temperature effect

In general, as temperature increases the cleaning rate increases (for example, Alfa Laval AB, 1995; Hankinson and Carver, 1968; De Goederen *et al.*, 1989; Fryer and Bird, 1994). Timperley and Smeulders (1988) investigated the effect of temperature (60–90 °C) cleaning of a PHE, using a single-stage cleaner. They observed a 60% decrease in cleaning time when the temperature of the cleaning fluid was increased from 60 to 90 °C, 40% of which occurred between 60 and 75 °C. The increase or decrease in the rate of a process (e.g. of cleaning) for each 10 °C increase in temperature has been reported for cleaning of milk deposits using sodium hydroxide solutions at temperatures up to *ca.* 90 °C; values of between 1.6 and 2 have been reported by a number of authors (Jennings, 1959; Gallot-Lavallee *et al.*, 1984; Grasshoff, 1989).

Optimal temperatures have been reported. Hankinson and Carver (1968) found an optimum temperature of 55 °C for cleaning milk deposits by water alone, for temperatures between 35 and 90 °C. Nisbet and Langdon (1997) reported that milk films heated to 80 °C became more resistant to desorption. Above the optimum temperature heat denaturation effects may cause the deposit to be more tenaciously held and hence not removed. Experiments on the removal of whey protein deposits from stainless steel pipes show a strong dependence on temperature (Gillham *et al.*, 1999). The initial swelling phase does not seem to be a strong function of temperature; however, the uniform phase (even removal of the deposit) and decay phase (random removal of large chunks of deposit, Section 29.2.3) show a strong sensitivity to temperature. The length of the decay phase significantly decreased when the deposit–liquid interface temperature exceeds 50 °C, above this temperature little further effect on the length of the decay phase is seen.

29.3.2 Chemical effect

Although water alone fails to remove milk and other dairy deposits, it generally constitutes 95% of typical cleaning fluids (Grasshoff, 1997). Alkali solutions remove the highly proteinaceous deposit whereas acidic solutions are more efficient in removing deposits of high mineral content. Commercial cleaning chemicals now often consist of an alkaline detergent-based cleaning solution with other chemicals added to improve the removal of the deposit (Table 29.4).

Increasing cleaning solution concentration decreases the time to clean (for example, Gallot-Lavallee *et al.*, 1984). Plett (1985) reported a near linear increase in cleaning rate with increasing detergent concentration, but that a maximum cleaning rate occurs. Some authors have reported optimal chemical concentrations which minimise cleaning time (including De Goederen *et al.*, 1989; Jeurnink and Brinkman, 1994; Fryer and Bird, 1994). This action of the cleaning solution on the deposit structure has been studied using SEM (Tissier and Lalande, 1986; Belmar-Beiny and Fryer, 1993). After exposure to a 2 wt% NaOH solution, images showed a less open structure than at 0.5 wt%, that may not be as susceptible to fluid shear, and is therefore more difficult to remove than more open structures. Other authors have explained this effect by: (i) the swelling of the deposit inhibiting transportation at higher concentrations (Plett, 1985), (ii) formation of a 'glassy' deposit (Bird and Fryer, 1991) or (iii) a rubber-like deposit (Jeurnink and Brinkman, 1994). Figure 29.2(a), from Christian (2004) shows cleaning data for a set of different flows and sodium hydroxide concentration.

Optima reported are difficult to compare as deposits, cleaning solutions and equipment vary. Bird (1992) reports an optimum concentration of 0.5 wt% NaOH for removal of both milk and whey protein deposits from bench-scale equipment. Tuladhar (2001) quotes an optimum at 0.5–1.0 wt% NaOH for bench-scale removal of WPC deposit, at temperatures of 50 °C and below. Changani (2000), studying removal of similar deposits from a pilot-scale PHE, did not observe an optimum in the range of 0.1 and 1.5 wt% NaOH. However, a value of 0.3 to 0.7 wt% NaOH was suggested for most efficient deposit removal, above this value there is little benefit of increasing concentration.

Table 29.4 Major classes of components in alkaline detergents, and their function, with examples of common substances (based on Kane and Middlemiss, 1985, and Karlsson, 1999)

Component	Function
Alkalis Sodium hydroxide Sodium silicates Trisodium phosphate	Soil dissolution, protein hydrolysis and solubilisation, fat saponification
Sequestering agents Sodium polyphosphate EDTA NTA	Formation of complexes with calcium and magnesium ions to avoid precipitation of salts
Surfactants Alkyl aryl sulphonates (anionic) Alkyl sulphonates (anionic) Alkylphenol ethoxylates (non-ionic)	Wetting, soil removal, solubilisation, emulsification
Oxidizers, e.g. hydrogen peroxide and sodium hypochlorite	For intensifying the cleaning effect
Minor components Dispersing agents Antifoaming agents Anticorrosion agents Stabilisers	Promoting dispersion, preventing foam formation, preventing alkaline erosion, prolonging storage

The balance between cleaning from water and hydroxide was shown clearly by Christian (2004); a pilot-scale PHE was cleaned by hydroxide and water in rotation. Figure 29.2(b) shows typical results for pressure drop as a function of time when the flow is switched. Essentially, cleaning stops when sodium hydroxide is replaced by water.

Two-stage cleaning involves the circulation of an alkali solution (commonly NaOH) and an acid solution (nitric or phosphoric), usually separated by a water rinse step (Romney, 1990). The order or circulation depends on the type of deposit; a protein-rich soil is removed more effectively by initial circulation of the alkali solution, but this may be reversed where a deposit with high mineral content is present. Minerals present within the protein matrix are removed with the protein during circulation of the alkali solution. The acid step that follows removes minerals remaining on the surface. The overall process may be time consuming, uses large amounts of water and may not give sufficiently clean results (Timperley and Smeulders, 1987). Rinsing steps may be omitted to save time and water, although this results in ineffective usage of the chemicals.

In contrast, single-stage cleaning uses a complex blend of chemicals, containing formulated detergents, surface-active agents (including wetting agents) and chelating compounds (Kane and Middlemiss, 1985; Romney, 1990). They usually contain sodium hydroxide or other sodium salts. Single-stage cleaning

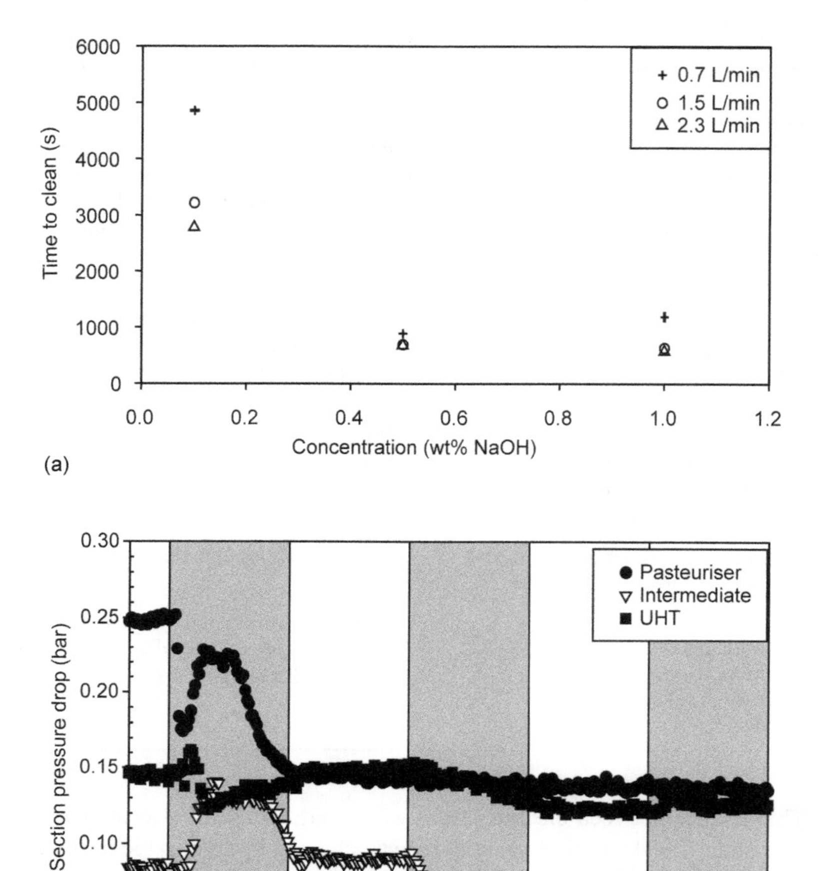

Fig. 29.2 (a) Variation in cleaning time for dairy deposit as a function of NaOH concentration and flow rate, showing minimum cleaning times for 0.5%, and (b) the effect of changing from sodium hydroxide (grey) to water (clear) on cleaning of the three stages of a plate heat exchanger; no change in pressure drop occurs when water replaces hydroxide (both from Christian, 2004).

typically requires only three steps: (rinse, clean, rinse), compared with a minimum of five needed for two-stage cleaning: (rinse, alkali clean, rinse, acid clean, rinse) (Timperley and Smeulders, 1987). The alkali step generally takes place before the acid, although this may be reversed if the deposit is of high mineral content. Trials to compare single and two-stage cleaning regimes showed that although chemicals used for single-stage cleaning are more expensive (Timperley and Smeulders, 1987; De Goederen *et al.*, 1989), overall cost can be reduced.

29.3.3 Mechanical effect

In CIP processes shear force is provided by fluid flow. It is important to find the minimal effective conditions as the energy expenditure for pumping increases as the square of the flow velocity. The flow may be characterised in terms of mass flow, Reynolds number (*Re*), surface shear stress (τ) or flow velocity (v). A given flow rate will have different velocities, *Re* and surface shear stress depending on the system. The flow rate of the cleaning solution also affects mass transfer of both the chemical and the reacted deposit. The individual nature of each system makes it difficult to suggest a minimum or optimum flow rate. For example, Alfa Laval AB (1995) state that flow rates between 1.5 and 3.0 m/s should give good scouring effect on the surface of equipment.

Increased flow rates induce greater surface shear on the deposit. However, this reduces the contact time; cleaning solution must remain in contact with the soil for long enough for it to take effect. The underlying effect of the cleaning solution flow rate on cleaning is difficult to determine as both mass transfer and surface shear stress are related to flow. Since the thickness of the boundary layer is also affected by fluid flow rate, it has been suggested that boundary layer thickness may control cleaning, i.e. as the boundary layer becomes thinner more deposit is protruding into the turbulent flow and is removable (De Goederen *et al.*, 1989). This takes into account the effect of the deposit on the boundary layer. Turbulence (in terms of *Re*) has been found to be the most important factor in the cleaning process, although the significance becomes less as chemical efficiency increases (Jennings *et al.*, 1957). Some workers have found a threshold below which the mechanical effect of flow is negligible (Jennings *et al.*, 1957; Schlüssler, 1976, in Jackson and Ming Low, 1982) on tomato deposits, although Bird and Fryer (1991) noted that there was no significant change in cleaning rate when moving from laminar to turbulent flow, and Bird (1992) found no minimum flow velocity. Timperley and Smeulders (1988) found that the cleaning time of a PHE decreased with increasing flow rate, the greatest reduction occurring upon increasing the flow velocity from 0.2 to 0.5 m/s. In general, it is clear that the higher the flow rate, the shorter the cleaning time – but the cost of pumping the cleaning fluid may become excessive.

29.3.4 Materials properties of deposits

The above has shown the complexity of cleaning, which is complicated as the materials properties of deposits are not well known, so that the effect of process variables will differ between different systems. We have tried to gain empirical understanding of the materials properties, using micromanipulation probes that

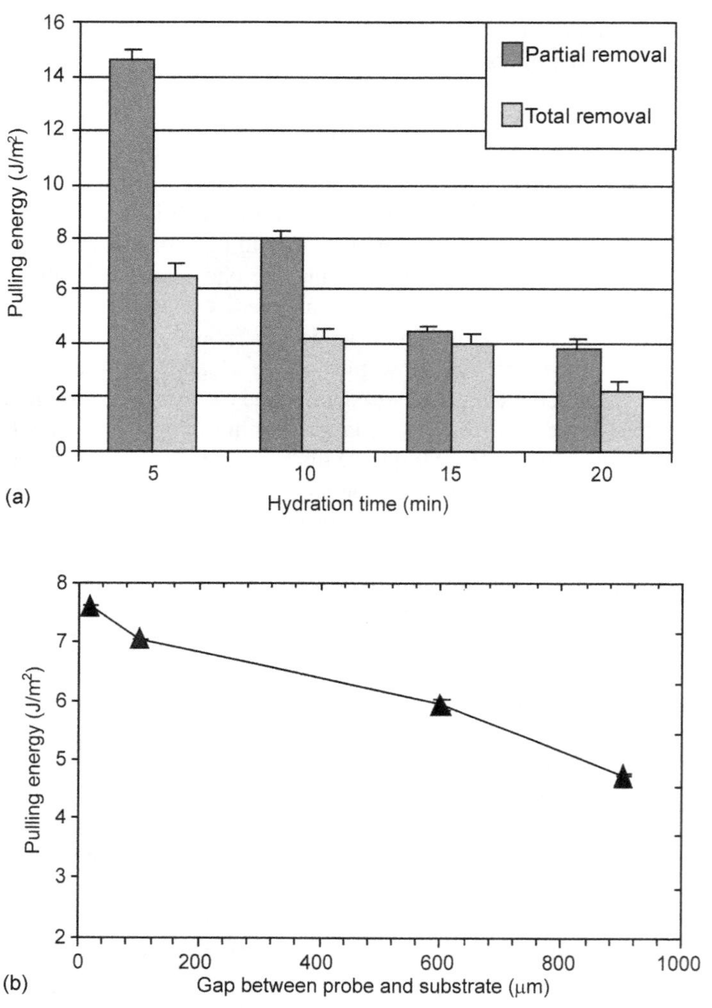

Fig. 29.3 Comparison of the energy required for partial and total removal of deposits of (a) baked tomato paste; (b) milk proteins (Liu *et al.*, 2005).

have been developed to remove layers of deposit at millimetre level (Liu *et al.*, 2002). The force required to disrupt a surface film can be determined by drawing a probe across the film, and the effectiveness of removal followed both by filming the process and examining the surface afterward. The method was developed to study biofilms (Chen *et al.*, 1998), but has been extended to study foods.

The force required to remove the deposit is measured by drawing the micromanipulation probe across the surface of the deposit. The apparent adhesive strength of a fouling sample, σ (J/m^2), defined as the work required to remove the sample per unit area, is given by

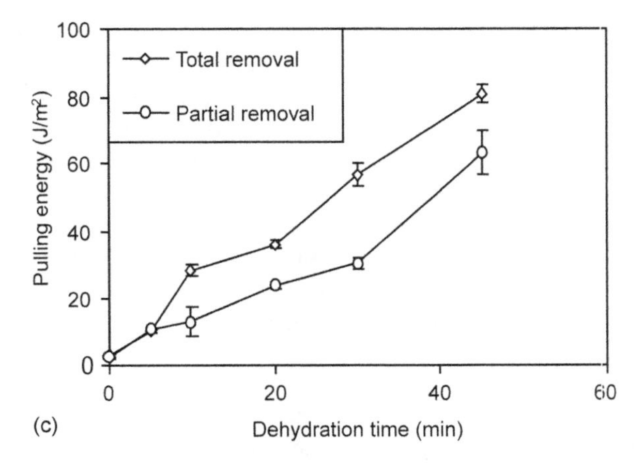

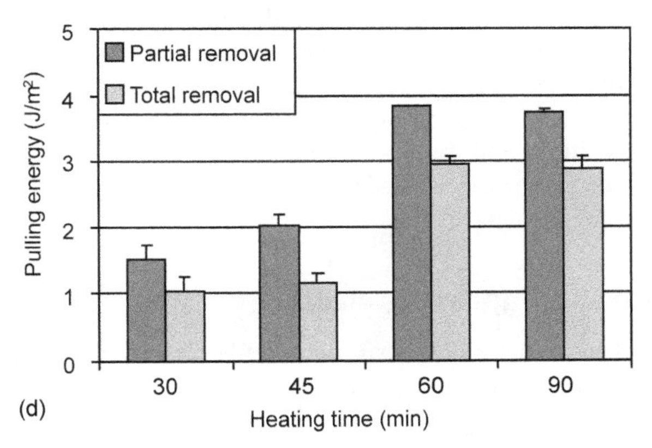

Fig. 29.3 (continued) Comparison of the energy required for partial and total removal of deposits of (c) bread dough, and (d) egg albumin showing differences in cohesive and adhesive behaviour (Liu *et al.*, 2005).

$$\sigma = \frac{W}{\alpha A} \tag{29.4}$$

where A (m^2) is the disc surface area, and α is the fraction covered by the sample measured by image analysis as described above. The relationship between σ and the actual adhesive strength between the surface and the deposit is not clear, as the measured force not only removes the deposit from the surface but also deforms it.

Two types of measurement are possible: (i) total removal, in which both adhesive and cohesive forces are overcome, and (ii) partial removal of deposit to measure cohesive strength. This has been done for four deposits. Figure 29.3 compares the results for full removal (adhesive) and partial removal (cohesive) for four types of deposit. For tomato paste (Fig. 29.3a), the force required for partial removal of the deposit *exceeds* that for the total removal, showing that cohesive forces between the deposits exceed those of adhesion between surface and deposit, the gap between the probe and substrate is 10 μm for total removal and 50 μm for partial removal. For milk (Fig. 29.3b), the initial thickness of the deposit layer was around 1300 μm. Force measurements were taken after leaving the gap between the probe and substrate to 900, 600, 100 μm (partial removal) and 20 μm (total removal) respectively. Here, the cohesive forces between elements of the deposit are weaker than those of adhesion. This is opposite to the behaviour of tomato starch, in which it is easier to remove the whole of the deposit than it is to remove a surface layer. For bread dough (Fig. 29.3c), the cohesive force was measured by leaving a 800 μm gap between the probe and the substrate during removal. The result shows that the adhesive strength is the same as the cohesive strength at the first 5 min of dehydration, but, beyond this time, the adhesive strength exceeds the cohesive strength and the difference between the adhesive and cohesive strengths are virtually constant while the dehydration time increases. For albumin (Fig. 29.3d), the cohesive strength was measured by leaving a 800 μm gap between the probe and the substrate during removal. The result shows that the cohesive strength is greater than the adhesive strength. These measurements show clearly that different materials have different balances between deposition and removal. This may enable us to select different cleaning protocols for different deposits.

29.4 Ways of improving cleaning

Conventional cleaning methods simply circulate chemicals. A number of non-standard methods have been tried at laboratory or larger scales, and are discussed here.

29.4.1 Non-uniform flow

The effect of flow rate has been discussed above. Non-uniform flows have also been investigated to determine whether they could have increased cleaning behaviour. Cleaning of milk soils using entrained air gave limited improvements (Trägårdh, 1981). Compression waves increased the removal of brittle boiler scales (Hanjalic and Smajevic, 1994), whereas water hammer gave little improvement (Hankinson and Carver, 1968) in removal of dried milk deposits. Cleaning of milking systems has typically been carried out by single phase flow of the cleaning chemical through the equipment. Using two phase flow (air/water) reduces the water required for circulation, increases flow velocities and enhances mechanical cleaning action (Reinemann, 1996). The use of ultrasonics has been investigated (Grasshoff, 1997; Mott *et al.*, 1995) to improve removal; the penetration depth of ultrasonic waves limits application to real dairy plants.

Pulsing of fluid flow at low frequency and high amplitude enhances heat and mass transfer (Keil and Baird, 1971). There is some conflict in the literature: the influence of pulsing in tubular systems on heat transfer can show: an increase (Karamercan and Gainer, 1979), no enhancement (Martinelli *et al.*, 1943) or a decrease (Lemlich, 1961) in the heat transfer coefficient. Low-frequency/large-amplitude waves generated by a solenoid valve arrangement enhanced removal of whey protein deposits (Farries and Patel, 1993). Gillham *et al.* (2000) concluded that pulsing had negligible effect on the initial 'swelling' phase, but had significant effect on both subsequent phases. Cleaning behaviour in the 'uniform' stage was more sensitive to pulse amplitude than frequency. Enhancement of cleaning is evidently controlled by a combination of surface shear stress (amplitude) and frequency effects, it does not solely depend on the shear stress imposed on the surface. Flow reversal seems to be especially useful in speeding cleaning – but this might be difficult to do in practice.

29.4.2 Reduction of environmental impact: novel technologies

Initial issues with respect to the use of enzymes to clean dairy equipment included high costs and low cleaning efficiency (Grasshoff, 1997). However, with increasing environmental concern, enzymatic cleaners are a promising alternative to traditional chemicals (Grasshoff, 2002). The textile industry has employed such methods, resulting in reduction in the chemicals required and reduced heating, hence energy saving. Enzymes have been successively used for the cleaning of cold milk processing equipment (Potthoff *et al.*, 1997) and membrane cleaning. A number of investigations on the use of enzymes to clean milk heaters have been reported (Grasshoff, 1997).

The use of ozonated cold water (10 °C) prior to cleaning of a dairy soil (reconstituted non-fat dry milk) from stainless steel plates has been investigated (Guzel-Seydim *et al.*, 2000). Results showed that the ozonated water pre-treatment removed 33% more of the deposit than pre-treatment with warm water at 40 °C. Trials on a novel 'pigging' system, which uses a mixture of water and ice combined with a freezing point depressant, to remove deposits such as jam, margarine, salad cream and toothpaste have been reported (Quarini, 2002). The advantages of this technique include low environmental impact and the ability to separate and recover products. Results show a significant improvement on cleaning behaviour compared with water at 20 °C; however, the results are not compared with cleaning with chemicals. It is also not clear yet to what extent pigging removes very thin layers of deposit, for example – but the method may significantly reduce rinsing times.

29.4.3 Effect of surface modification

The physical properties of the deposit and the surface will affect removal. Fouling deposits form as a result of adhesion of species to the surface and cohesion between elements of the material. The forces between elements of the deposit depend on the nature of the material; deposits may be covalently bonded (for example, reacted egg or milk proteins) or held together physically (such as gelled biopolymers). Understanding of the interaction between deposits and surfaces is clearly critical in cleaning. On a nanoscale, atomic force microscopy (AFM) has been used to characterise surfaces and fouling (such as Parbhu *et al.*, 2002; Weiss *et al.*, 2002). Low-adhesion coatings (Müller-Steinhagen and Zhao, 1997) reduce fouling in some situations such as mineral scales and work is underway to study the application of modified surfaces in food cleaning (Santos *et al.*, 2004, for example).

Zhao *et al.* (2004) demonstrate that biofouling can be reduced by changing surface energy, and link this to adhesive energy between surface and deposits; there is evidence for the minimum in terms of foulant attachment when the surface free energies cover the range 20–40 (mN/m). Theory gives that a *minimum* adhesion energy between deposit and surface exists, given by:

$$\sqrt{\gamma^{LW}_{surface}} = \left(\frac{1}{2}\right)\left(\sqrt{\gamma^{LW}_{foulant}} + \sqrt{\gamma^{LW}_{fluid}}\right) \quad (29.5)$$

where $\gamma^{LW}_{surface}$, $\gamma^{LW}_{foulant}$ and γ^{LW}_{fluid} are the Lifshitz-van der Waals (LW) surface free energy of the surface, foulant and fluid (e.g. water), and can be determined experimentally from contact angle measurements (Zhao *et al.*, 2004). In general, fouling formed on processing equipment may consist of various types of foulants (such as the mixtures of proteins and minerals deposited from milk) and $\gamma^{LW}_{foulant}$ in equation (29.5) is the average LW surface energy of the fouling deposit, which can be determined by measuring contact angles. If the surface energy of the stainless steel surface is reduced to the fouling-resistant value, foulant adhesion force to the surface could be decreased significantly, and the fouling deposit could be removed more easily. The effectiveness of this approach has been demonstrated by Zhao *et al.* (2004) for microbial adhesion.

Liu *et al.* (2005) discuss adhesion to a range of stainless steel disks modified by Ni–P–PTFE composite coatings. Contact angles of the Ni–P–PTFE composite coatings and tomato paste (baked and unbaked) were obtained and surface energies calculated (van Oss *et al.*, 1986) as

- $\gamma^{LW}_{surface} = 26$ mN/m, for which the adhesion of baked tomato paste is minimal;
- $\gamma^{LW}_{surface} = 25$ mN/m, for which the adhesion of unbaked tomato paste is minimal.

The effect of surface treatment can be quantified by the probe. Experiments have studied the removal of tomato paste from surfaces with surface energies ranging from 15 to 40 mN/m. Details of the surfaces are given in Zhao *et al.* (2002, 2004): the topography and roughnesses can be significantly different.

Figure 29.4(a) shows data for baked and unbaked pastes; in both cases there is a minimum adhesive strength between the surface energies of 20 and 25 mN/m, with an increase in the adhesive strength on either side. The data are scattered, but the minimum is clear in both cases; it is more obvious for the unbaked material in which the change in adhesive strength with free energy is smoother. The minimum is in the region predicted by the theory (see above); at either side of this minimum the force increases, an effect especially marked for

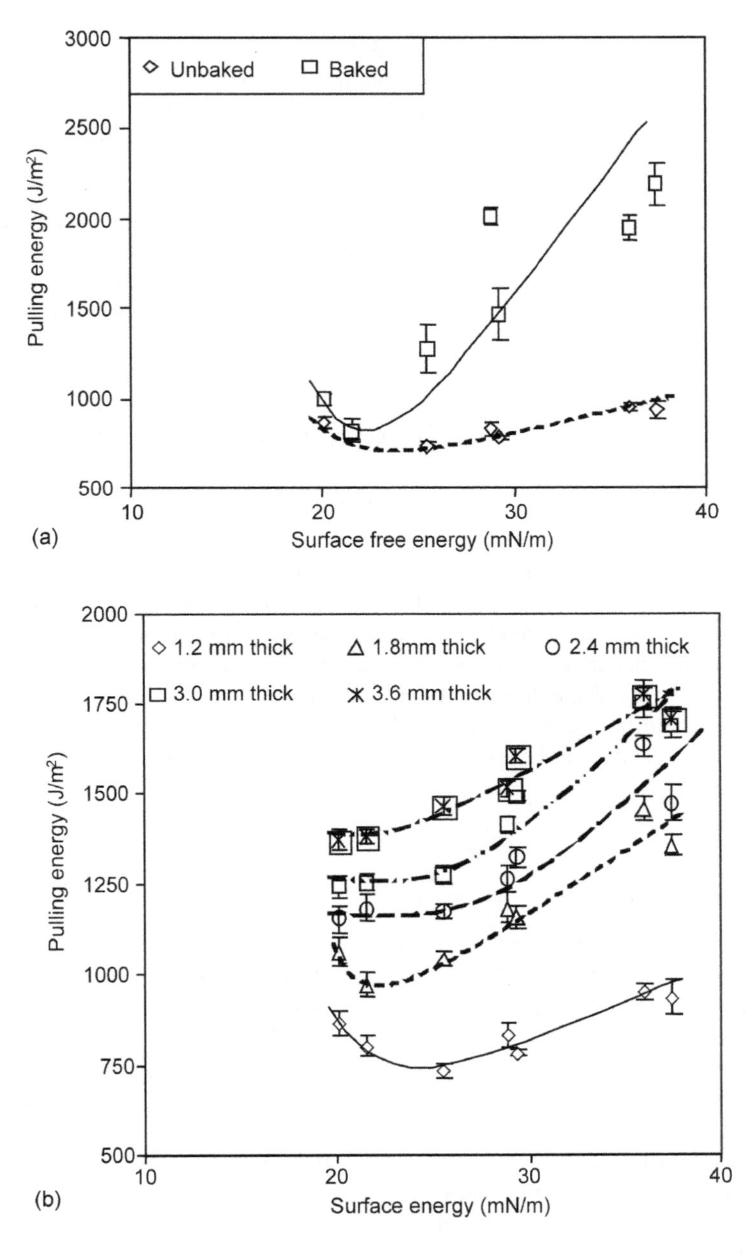

Fig. 29.4 The effect of changing surface energy on removal of (a) baked and unbaked deposits, showing a minimum forces of adhesion, and (b) different thicknesses for tomato pastes, showing that the effect of the surface decreases with increasing thickness (Liu *et al.*, 2005).

the baked-on sample, as would be expected, as that has been cooked onto the surface and would be expected to bond more strongly.

The force measured by the micromanipulation probe is a composite of the cohesive forces between deposit elements and the adhesion to the surface. This is shown explicitly by Fig. 29.4(b), which shows data for different thicknesses of unbaked paste:

- As the thickness increases, the total force required to remove the deposit increases. The increase reflects the need to overcome the cohesive forces between elements of the deposit and to force the deposit to break and flow with the probe away from the sample surface; as the thickness increases, so does this force.
- A similar minimum to that found in Fig. 29.4(a), in similar regions of surface energy. However, the minimum becomes more difficult to identify as the thickness of deposit increases, reflecting the decreased contribution of the surface forces to the whole. At the highest thicknesses used, there is no measurable minimum; the curve simply flattens out at the lowest surface energies.

These results demonstrate that modifying the surface energy can affect the energy required to clean the surface. The problem in practice will be in producing surfaces that are cheap (and safe) enough for use in the food industry, and ensuring that they do not get broken down in operation; the surfaces must stay effective over the lifetime of a plant. It might be that a more effective solution would be the addition of a chemical during the last stages of cleaning which changes the surface properties to resist fouling and/or aid subsequent cleaning.

29.5 Conclusions

The problem of fouling, and the associated problems of cleaning, have been reviewed, with particular reference to milk fluids, as they have been most thoroughly studied. Fouling from milk results from protein and mineral deposition, each of which results in different problems for cleaning. Cleaning time is a function of a number of variables; both chemical (such as the cleaning chemical type and concentration, and the temperature) and physical (such as the flow rate, which affects the fluid shear on the surface of the deposit). In milk cleaning, protein deposit is first swollen by action of hydroxide and then removed by shear. The cleaning rate increases with increasing temperature and surface shear stress, but there can be an optimal concentration of hydroxide, above which the deposit becomes difficult to remove. A number of methods have been tried to increase the cleaning rate, including pulsed flows, enzyme cleaners and ice pigs. Surface modification has been tried by a number of workers.

It is possible to separate adhesive and cohesive effects in cleaning: some deposits (such as tomato paste) are essentially cohesive, while protein deposits (such as milk) are adhesive; cleaning involves reaction of the deposit into a viscoelastic gel that adheres to the surface. This classification is useful as it allows decisions to be made about selection of cleaning protocols. Use of modified surfaces shows that (i) the theoretical minimum adhesion condition can be found in practice (at least for tomato deposits) but that (ii) the effect of the surface decreases with increasing deposit thickness, as would be expected.

Work is currently underway at Birmingham to quantify the relationship between the measured apparent adhesive strength and theoretical adhesive and cohesive strength. This could enable understanding in more detail the relationship between adhesion and deposit chemistry, and how to clean most effectively.

29.6 Acknowledgements

We wish to acknowledge financial support from the UK Biotechnology and Biological Sciences Research Council (BBSRC), DEFRA UK and the Food Processing Faraday.

29.7 References

ALFA LAVAL AB (1995). *Dairy Handbook*. Alfa-Laval, Lund, Sweden.

BELMAR-BEINY, M. T. and FRYER, P. J. (1992). Bulk and surface effects on the initial stages of whey fouling. *Trans IChemE* **70**(C) 193–199.

BELMAR-BEINY, M. T. and FRYER, P. J. (1993). Preliminary stages of fouling from whey protein solutions. *J Dairy Res* **60** 467–483.

BELMAR-BEINY, M. T., GOTHAM, S. M., PATERSON, W. R., and FRYER, P. J. (1993). The effect of Reynolds number and fluid temperature in whey fouling. *Journal of Food Engineering* **19** 119–139.

BEUF, M., RIZZO, G., LEULIET, J. C., MÜLLER-STEINHAGEN, H., YIANTSIOS, S., KARABELAS, A., and BENEZECH, T. (2003). Potency of stainless steel modifications in reducing fouling and increasing cleaning of plate heat exchangers processing dairy products. In: *Heat Exchanger Fouling and Cleaning Fundamentals and Applications*, Santa Fe, New Mexico, USA.

BIRD, M. R. (1992). Cleaning of food process plant. PhD Thesis, University of Cambridge, UK.

BIRD, M. R. and BARTLETT, M. (1995). CIP optimisation for the food industry: relationships between detergent concentration, temperature and cleaning time. *Trans IChemE* **73**(C) 63–70.

BIRD, M. R. and FRYER, P. J. (1991). An experimental study of the cleaning of surfaces fouled by whey proteins. *Trans IChemE* **69**(C) 13–21.

BOTT, T. R. (1995). *Fouling of Heat Exchangers*. Chemical Engineering Monographs no. 26, Elsevier, Oxford.

BRITTEN, M., GREEN, M. L., BOULET, M., and PAQUIN, P. (1988). Deposit formation on heated surfaces: effect of interface energetics. *J Dairy Res* **55** 551–562.

BURTON, H. (1966). A comparison between a hot-wire laboratory apparatus and a plate heat exchanger for determining the sensitivity of milk to deposit formation. *J Dairy Res* **33** 317–324.

BURTON, H. (1968). Reviews of the progress of dairy science. *J Dairy Res* **35** 317–330.

CHANGANI, S. D. (2000). An investigation into the fouling and cleaning behaviour of dairy deposits. PhD Thesis, University of Birmingham, UK.

CHEN, M. J., ZHANG, Z., and BOTT, T. R. (1998). Direct measurement of the adhesive strength of biofilms in pipes by micromanipulation. *Biotechnology Techniques* **12** 875–880.

CHEOW, C. S. and JACKSON, A. T. (1982a). Circulation cleaning of a plate heat exchanger fouled by tomato juice I. Cleaning with water. *J Food Technol* **17** 417–430.

CHEOW, C. S. and JACKSON, A. T. (1982b). Circulation cleaning of a plate heat exchanger fouled by tomato juice II. Cleaning with caustic soda solution. *J Food Technol* **17** 431–440.

CHRISTIAN, G. K. (2004). Cleaning of carbohydrate and dairy protein deposits. PhD thesis, University of Birmingham, UK.

DE GOEDEREN, G., PRITCHARD, N. J., and HASTINGS, A. P. M. (1989). Improved cleaning processes for the food industry. In: *Fouling and Cleaning in Food Processing*, Prien, Germany, Kesslar, H. G. and Lund D. B., eds. 115–130.

DE JONG, P., DE GIFFEL, M. C., STRAATSMA, H., and VISSERS, M. M. M. (2002). Reduction of fouling and contamination by predictive kinetic models, *Int Dairy J*, **12** 285–292.

DE LA FUENTE, M., SINGH, H., and HEMAR, Y. (2002). Recent advances in characterisation of heat-induced aggregates and intermediates of whey proteins. *Trends Food Sci. Technol* **13** 262–274.

EPSTEIN, N. (1983). Fouling in heat exchangers. In: *Heat Exchangers Theory and Practice*, Taborek, J. Hewitt G. F. and Afgan N., Eds., 795–815, Hemisphere Publishing, New York.

FARRIES, R. and PATEL, H. (1993). The effect of flow reversal and pulsing on surface cleaning. Part II project report, University of Cambridge, UK.

FOSTER, C. L. and GREEN, M. L. (1990). A model heat-exchange apparatus for the investigation of fouling of stainless steel surfaces by milk II. Deposition of fouling material at 140 °C, its adhesion and depth profouling. *J Dairy Res* **57** 339–348.

FRYER, P. J. (1985). Modelling of heat exchanger fouling. PhD Thesis, University of Cambridge, UK.

FRYER, P. J. and BIRD, M. R. (1994). Factors which affect the kinetics of cleaning dairy soils. *Food Sci Technol Today* **8**(1) 36–42.

FRYER, P. J., HASTING, A. P. M. and JEURNINK, TH. J. M. (eds) (1996). *EUR 16894 – Fouling and Cleaning in Food Processing,* Office for Official Publication of the EC, Luxembourg.

GALLOT-LAVALLEE, T., LALANDE, M., and CORRIEU, G. (1984). Cleaning kinetics modelling of holding tubes fouled during milk pasteurisation. *J Food Process Eng* **7** 123–142.

GEORGIADIS, M. C., ROSTEIN, G. E., and MACCHIETTO, S. (1998). Optimal design and operation of heat exchangers under milk fouling. *AIChE J* **44**(9) 2099–2111.

GILLHAM, C. R. (1997). Enhanced cleaning of surfaces fouled by whey protein. PhD Thesis, University of Cambridge, UK.

GILLHAM, C. R., FRYER, P. J., HASTING, A. M. P., and WILSON, D. I. (1999). Cleaning-in-place of whey protein fouling deposits: mechanisms controlling cleaning. *Trans IChemE* **77**(C) 127–135.

GILLHAM, C. R., FRYER, P. J., HASTING, A. M. P., and WILSON, D. I. (2000). Enhanced cleaning of whey proteins soils using pulsed flows. *J Food Eng* **46** 199–209.

GORDON, K. P., HANKINSON, D. J., and CARVER, C. E. (1968). Deposition of milk solids on heated surfaces. *J Dairy Sci* **51** 520–526.

GRASSHOFF, A. (1989). Environmental impact of the use of alkaline cleaning solutions. In: *Fouling and Cleaning in Food Processing*, Kesslar, H. G. and Lund D. B., eds. Prien, Germany, 107–114.

GRASSHOFF, A. (1997). *Fouling and Cleaning in Heat Exchangers*, H. Visser, ed., IDF Bulletin No. 328.

GRASSHOFF, A. (2002). Enzymatic cleaning of milk pasteurizers. *Trans IChemE Part C: Food Bioproducts Processing* **80**(4) 247–252.

GRIJSPEERDT, K., MORTIER, L., DE BLOCK, J., and VAN RENTERGHEM, R. (2004). Applications of modelling to optimize ultra high temperature milk heat exchangers with respect to fouling, *Food Control*, **15**, 117–130.

GUZEL-SEYDIM, Z. B., WYFFELS, J. T., GREENE, A. K., and BODINE, A. B. (2000). Removal of dairy soil from heated stainless steel surfaces: use of ozonated water as a prerinse. *J Dairy Sci* **83**(8) 1887–1891.

HANJALIC, K. and SMAJEVIC, I. (1994). Detonation wave technique for local deposit removal from surfaces exposed to fouling. Part I: Experimental investigation and development of the method and Part II: Full scale application. *Eng Gas Turbines Power J* **116** 223–236.

HANKINSON, D. J. and CARVER, C. E. (1968). Fluid dynamic relationships involved in circulation cleaning. *J Dairy Sci* **51**(11) 1761–1767.

HEGE, W. U. and KESSLER, H-G. (1986a). Deposit formation of protein-containing dairy liquids. *Milchwissenschaft* **41** 356–360.

HEGE, W. U. and KESSLER, H-G. (1986b). Die Ansatzbildung beim Erhitzen von Milch und Molke. *Internationale Zeitschrift für Lebensmittel-Technologie und Verfahrens-technik* **37** 92–100.

JACKSON, A. T. and MING LOW, W. (1982). Circulation cleaning of a plate heat exchanger fouled by tomato juice III. The effect of fluid flow rate on cleaning efficiency. *J Food Technol* **17** 745–752.

JENNINGS, W. G. (1959). Circulation cleaning. III. The kinetics of simple detergent system. *J Dairy Sci* **42** 1763–1771.

JENNINGS, W. G., McKILLOP, A. A., and LUICK, J. R. (1957). Circulation cleaning. *J of Dairy Science* **40** 1471–1479.

JEURNINK, TH. J. M. (1991). Effect of proteolysis in milk on foiling in heat exchangers. *Netherlands Milk Dairy J* **45** 23–32.

JEURNINK, TH. J. M. and BRINKMAN, D. W. (1994). The cleaning of heat exchangers and evaporators after processing milk or whey. *Int Dairy J* **4** 347–368.

KANE, D. R. and MIDDLEMISS, N. E. (1985). Cleaning chemicals – state of the knowledge in 1985. In: *Fouling and Cleaning in Food Processing*, eds Lund, D. B., Plett, E. and Sandu, C., University of Wisconsin, Madison, WI, USA, 312–355.

KARAMERCAN, O. E. and GAINER, J. L. (1979). The effect of pulsations on heat transfer. *Ind. Eng Chem Fundam,* **18**(1) 11–15.

KARLSSON, A-C. (1999). *Fouling and Cleaning of Solid Surface: The influence of surface characteristics and operating conditions*. Lund University, Sweden.

KEIL, R. H. and BAIRD, M. H. I. (1971). Enhancement of heat transfer by flow pulsation. *Ind Eng Chem Process Des Develop.* **10**(4) 473–478.

LALANDE, M., TISSIER, J. P., and CORRIEU, G. (1985). Fouling of heat transfer surfaces related to β-lactoglobulin denaturation during heat processing of milk. *Biotechnology Progess* **1**(2) 131–139.

LEMLICH, R. (1961). Vibration and pulsation boost heat transfer. *Chem Eng* **68**(10) 171–176.

LIU, W., CHRISTIAN, G. K., ZHANG, Z., and FRYER, P. J. (2002). Development and use of a micromanipulation technique for measuring the force required to disrupt and remove fouling deposits. *Trans IChemE Part C: Food Bioproducts Processing* **80**(4) 286–291.

LIU, W., FRYER, P. J., AB AZIZ, N., BRIDSON, M. C., ZHANG, Z., ZHAO, Q., and LIU, Y. (2005). Cohesion and Adhesion in the Cleaning of Food Fouling Deposits, World Congress of Chemical Engineering, Glasgow, on CD-Rom.

LYSTER, R. L. J. (1965). The composition of milk deposits in an ultra-high-temperature plant. *J Dairy Res* **32** 203–210.

MARTINELLI, R. C., BOELTER, L. M. K., WEINBERG, E. B., and YAKAL, S. (1943). Heat transfer to a fluid flowing periodically at low frequencies in a vertical tube. *Trans ASME* **65** 789.

MARUYAMA, Y., SHIBATA, Y., and WANTANABE, K. (1991). A proposed design method for predicting parameters of the bio-plant. In: *4th World Congress on Chem. Engineering*, Karlsruhe, DECHEMA, ed., Frankfurt.

MOTT, I. E. C., STICKLER, D. J., COAKLEY, W. T., and BOTT, T. R. (1995). Ultrasound in the control of biofilms. In: *Fouling Mitigation of Industrial Heat-exchange Equipment*, Panchal, C. B., ed., Begell House, New York.

MÜLLER-STEINHAGEN, H. (2000). *Heat Exchanger Fouling – Mitigation and Cleaning Technologies*, IChemE, Rugby.

MÜLLER-STEINHAGEN, H. and ZHAO, Q. (1997). Investigation of low fouling surface alloys made by ion implantation technology,*Chem Eng Sci* **52** 3321–3332.

NISBET, T. J. and LANGDON, A. G. (1997). Milk protein interactions at stainless steel/aqueous interfaces. *New Zealand Dairy Sci Technol* **12** 83–87.

PARBHU, A. N., LEE, A. N., THOMSEN, S. J., and SIEW, D. C. W. (2002). Atomic force microscopy applied to monitoring initial stages of milk fouling on stainless steel. In *Fouling, Cleaning and Disinfection in Food Processing*, Wilson, D. I., Fryer, P. J., and Hasting, A. P. M. eds, Dept of Chemical Engineering, University of Cambridge, Cambridge, 33–40.

PATERSON, W. R. and FRYER, P. J. (1988). A reaction engineering approach to the analysis of fouling. *Chem Eng Sci* **43** 1714–1717.

PETERMEIER, H., BENNING, R., DELGADO, A., KULOZIK, U., HINRICHS, J., and BECKER, T. (2002). Hybrid model of the fouling process in tubular heat exchangers for the dairy industry. *J Food Eng* **55** 9–17.

PLETT, E. A. (1985). Cleaning of fouled surfaces. In: *Fouling and Cleaning in Food Processing*, Lund, D. B., Plett E., and Sandu C., eds, University of Wisconsin, Madison, WI, USA, 286–311.

POTTHOFF, A., SERVE, W., and MACHARIS, P. (1997). The cleaning revolution. *Dairy Industries Int* **62** 25–29.

PRITCHARD, A. M. (1998). The Economics of Fouling. In: *Fouling Science and Technology*, NATO ASI Series E 145, Melo, L. F., Bott, T. R. and Bernardo, C., eds, Kluwer, Dordrecht, 31–45.

QUARINI, J. (2002). Ice-pigging to reduce and remove fouling and to achieve clean-in-place. *Appl Thermal Eng* **22** 747–753.

REINEMANN, D. J. (1996). Technical design and assessment of tube equipment using two-phase flow for cleaning and disinfection. *Int J Hygiene Environmental Med* **199**(2–4) 355–365.

RENE, F., LEULIET, J. C., GOLDBERG, M., and LALANDE, M. (1988). Fouling and cleaning study of a plate heat exchanger used for chocolate desserts processing. *Le Lait* **68**(1) 85–102.

ROBBINS, P. T., ELLIOT, B. L., FRYER, P. J., BELMAR, M. T., and HASTING, A. P. M. (1999). A comparison of milk and whey fouling in a pilot scale plate heat exchanger: implications for modeling and mechanistic studies. *TransIChemE* **77**(C) 97–106.

ROEFS, P. F. and DE KRUIF, K. G. (1994). A model for the denaturation of β-lactoglobulin. *European J Biochem* **226** 883–889.

ROMNEY, A. (1990). *CIP: Cleaning In Place*, 2nd Edition, Society of Dairy Technology, London.

ROSANINHO, R., RIZZO, G., MÜLLER-STEINHAGEN, H., and MELO, L. F. (2003). The influence of bulk properties and surface characteristics on the deposition process of calcium phosphate on stainless steel. In: *Heat Exchanger Fouling and Cleaning Fundamentals and Applications*, Santa Fe, New Mexico, USA (http://services.bepress.com/eci/heatexchanger/).

SAHOO, P.K., ANSARI, I. A., and DATTA, A.K. (2005). Milk fouling simulation in helical triple tube heat exchanger. *J Food Eng* **69** (2) 235–244.

SANTOS, O., NYLANDER, T., ROSMANINHO, R., RIZZO, G., YIANTSIOS, S., ANDRITSOS, N., KARABELAS, A., MÜLLER-STEINHAGEN, H., MELO, L., BOULANGÉ-PETERMANN, L., GABET, C., BRAEM, A., TRÄGÅRDH, C., and PAULSSON, M. (2004). Modified stainless steel surfaces targeted to reduce fouling – surface characterization, *J Food Eng* **64** 63–79.

SCHLÜSSLER, H. J. (1976). Zur von Reinigungsvorangen an Festen oberflachen Brauwissenschaft. *Brauwissenschaft* **29** 263–268.

SKUDDER, P. J., BROOKER, B. E., ANDREW, D. B., and ALZAREZ-GUERRERO, N. R. (1986). Effects of pH on the formation of deposit from milk on heated surfaces during ultra high temperature processing. *J Dairy Res* **53** 75–87.

SMAILI, F., ANGADI, D. K., HATCH, C. M., HERBERT, O., VASSILIADIS, V. S., and WILSON, D. I. (1999). Optimisation of scheduling of cleaning in heat exchanger networks subject to fouling: sugar industry case study. *Trans IChemE* **77**(C) 159–164.

SMAILI, F., VASSILIADIS, V. S., and WILSON, D. I. (2001). Mitigation of fouling in refinery heat exchanger networks by optimal management of cleaning. *Energy Fuels* **15**(5) 1038–1056.

SMAILI, F., VASSILIADIS, V. S., and WILSON, D. I. (2002a). Long-term scheduling of cleaning of heat exchanger networks – comparison of out approximation-based solutions with a backtracking threshold accepting algorithm. *Chem Eng Res Design* **80**(A6) 561–578.

SMAILI, F., VASSILIADIS, V. S., and WILSON, D. I. (2002b). Optimization of cleaning schedules in heat exchanger networks subject to fouling. *Chem Eng Commun* **189**(11) 1517–1549.

STARLING, S. E. and NICOL, R. S. (2001). Cleaning system design tools. In: *6th World Congress of Chemical Engineering, September, 2001*, Melbourne, Australia.

TIMPERLEY, D. A. and SMEULDERS, C. N. M. (1987). Cleaning of dairy HTST plate heat exchangers: comparison of single- and two-stage procedures. *J Soc Dairy Technol* **40**(1) 4–7.

TIMPERLEY, D. A. and SMEULDERS, C. N. M. (1988). Cleaning of dairy HTST plate heat exchangers: optimisation of the single-stage procedure. *J Soc Dairy Technol* **41**(1) 4–7.

TISSIER, J. P. and LALANDE, M. (1986). Experimental device and methods for studying milk deposit formation on heat exchange surfaces. *Biotechnol Progess* **2**(4) 218–229.

TRÄGÅRDH, CH. (1981). Cleaning in air water flow. In: *Fundamentals and Applications of Surface Phenomena Associated with Fouling and Cleaning in Food Processing*, Hallström, B., Lund D. B., and Trägårdh Ch., eds., University of Lund, Tylösand,

Sweden, 424–429.

TULADHAR, T. R. (2001). Development of a novel sensor for cleaning studies. PhD Thesis, University of Cambridge, UK.

VAN OSS, C. J., GOOD, R. J., and CHAUDHURY, M. K. (1986). The role of van der Waals forces and hydrogen-bonds in hydrophobic interactions between bio-polymers and low energy surfaces. *J Colloid Interface Sci* **111**(2) 378–390.

VISSER, H. and JEURNINK, TH. J. M. (1997). Fouling of heat exchangers in the dairy industry. *Exp Thermal Fluid Sci* **14**(4) 407–424.

VISSER, H., JEURNINK, J. M., SCHRAML, J. E., FRYER, P. J., and DELPLACE, F. (1997). Fouling of heat treatment equipment. In: *Fouling and Cleaning of Heat Treatment Equipment*, Visser, H., ed., IDF Bulletin No. 328.

WALSTRA, P. and JENNESS, R. (1984). *Dairy Chemistry and Physics*, John Wiley and Sons Inc., New York.

WEISS, A., BIRD, M. R., CHUKWUEMEKA, J. K. and PRICE, R. (2002). In: Fouling, Cleaning and Disinfection in Food Processing, Wilson, D. I., Fryer, P. J., and Hasting, A. P. M., eds, Dept of Chemical Engineering, University of Cambridge, Cambridge, 151–164.

WILKINSON, P. J. (1982). An investigation of protein fouling and its removal in stainless steel piping. PhD Thesis, University College, London, UK.

WILSON, D. I., FRYER, P. J., and HASTING, A. P. M. (eds) (1999). *Fouling and Cleaning in Food Processing '98,* EUR 18804, Office for Official Publications for the European Communities, Luxembourg.

WILSON, D. I., FRYER, P. J., and HASTING, A. P. M. (eds) (2002). *Fouling, Cleaning and Disinfection in Food Processing,* Dept. of Chemical Engineering, University of Cambridge, Cambridge.

YOON, J. and LUND, D. B. (1994). Magnetic treatment of milk and surface treatment of plate heat exchangers: effects on milk fouling. *J Food Sci* **59**(5) 964–980.

ZHAO, Q., LIU, Y., and MÜLLER-STEINHAGEN, H. (2002). Graded Ni–P–PTFE coatings and their potential applications. *Surfaces Coatings Technol* **155** 279–284.

ZHAO, Q., WANG, S. and MÜLLER-STEINHAGEN, H. (2004). Tailored surface free energy of membrane diffusers to minimize microbial adhesion. *Appl Surface Sci* **230** 371–378.

30

Improving the cleaning of tanks

S. Salo, VTT Biotechnology, Finland, A. Friis, Technical University of Denmark and G. Wirtanen, VTT Biotechnology, Finland

30.1 Introduction to cleaning tanks

The hygienic state of food production surfaces has a crucial effect on the quality of food products. Therefore, the hygienic requirements of the cleaning procedure must be included in the process design and integration. Hygiene is important in all processes, because the production cannot be run if it is possible that microbes will infiltrate the process through surfaces and equipment that are used in association with the tank. Contamination on tank surfaces can be fatal to the product quality due to long processing times with nutritious raw materials, where microbial growth leads to discarding large product batches. It has been proved that specific hydrodynamic parameters control cleaning in closed process systems. The fluid flows are important both in production and in cleaning. Computational fluid dynamics (CFD) is a tool for improving the hygienic design of equipment components and their integration into the process line.

Tanks are crucial for the operation of food production plants. Major applications are the storage of raw materials and end-products, buffers for intermediate products, fermentation, mixing, heating and cooling. Big tanks are needed in brewery, dairy and fruit juice processes (Birus, 2003). The sizes of the tanks used in dairies vary from 100 L to 230 000 L. The biggest tanks in dairies are silo tanks that are used for collection and reception of milk. The holding capacities of other storage tanks in dairies normally vary from 1000 L to 50 000 L. There are also mixing and processing tanks in different sizes (Bylund, 1995). For fermentation purposes the equipment, including tanks, must be clean, in some cases even sterile, at the starting point, because fermentation processes are prone to contamination. Microbes in nutrient-rich environments grow well in fermentation (Storgårds, 2000). Hence a few unclean niches harbouring harmful microbes can initiate fatal contamination of the fermentation. In severe cases, poor cleaning can cause a biofilm formation on equipment surfaces, which can both contaminate the process and also cause corrosion and even health problems (Wirtanen, 1995; Geesey and Bryers, 2000).

The hygienic state of tanks is important in order to avoid contamination of the end product and, therefore, tank cleaning is receiving increasing attention. The specific cases studied comprise dairy- and brewery-process tanks. These tanks are made of high-grade stainless steel but for financial reasons the outer shell can be made of a lower-quality steel coated with anti-corrosion paint. The bottom of a tank slopes downwards with an incline of about 6% towards an outlet in order to make complete drainage easy (Bylund, 1995). Improper cleaning of tank surfaces in dairies leads to re-contamination of pasteurised milk with either psychrotrophic or thermotolerant bacteria (Geesey and Bryers, 2000; Wirtanen *et al.*, 2002). It has also been reported that thermotolerant contaminants extend the production time of fermentation milk products and cause undesired growth in closed processes (Austin and Bergeron, 1995; Wirtanen *et al.*, 2002).

The hygiene of process surfaces is also crucial in beer production because the fermentation of nutritious wort into beer including maturation in the fermenting tank may take several weeks. This long production time lead to changes in restricting factors for growth of contaminants. The common sizes of tanks in breweries vary from 10 000 L to 900 000 L. For instance the size of main fermentation tanks in three biggest breweries in Finland varies from 250 000 L to 510 000 L. Minibreweries can have batch sizes as small as 60 L. The consequences of contamination in the fermentation can, therefore, be disastrous to the quality of the beer. Hence it is very important to design the production lines so that they are compatible with the cleaning procedure used (Storgårds, 2000). The fermentation tanks in breweries mostly have conical bottoms with an inner angle of 70–80°. They are advantageous because they allow simple emptying of the yeast aseptically, effective temperature control, easy cleaning with cleaning-in-place (CIP) systems, easy collection of carbon dioxide and small losses of beer compared with tanks with flat and inclined bottoms. The tanks used for storage and stabilisation are normally tanks with flat and inclined bottoms (Enari and Mäkinen, 1993). The whole fermenting process can take place in the same tank, which is called a combi tank. These combi tanks have also conical bottoms.

30.2 Factors affecting cleaning efficacy

It is generally understood that the most successful way of preventing microbial contamination is to ensure that the cleaning is properly performed. In the food industry there is a trend towards longer production runs with short intervals for sanitation (Lelieveld, 1985). Factors affecting the cleaning process are based on mechanical and chemical impact, holding time and temperature (Wirtanen,

1995). There is still the challenge of designing and optimising the cleaning systems in terms of efficiency and economy. This implies choosing the right spray ball for a given tank in such a way that the hygienic state required is obtained all over the tank. In designing the cleaning procedure, attention should also be paid to the quality of the process water, steam and other additives used. Aspects such as surface topography and material properties, the fluid dynamics as well as the specific microbial flora and other soil components affect the cleaning efficiency (Wirtanen, 1995; Storgårds, 2000). Several studies have, however, shown that there are only weak relationships between the surface roughness and the cleaning efficiency (Verran *et al.*, 2001). To be able to minimise costs, the consumption of water and cleaning chemicals and also the length of the cleaning time must be minimised.

The initial hygienic design of the process equipment has a big impact on reducing the risks of food becoming contaminated during production, which also means that the shelf-life of products is improved. Hygienic requirements should be adopted at the initial stage in developing process equipment and components. On the basis of EU Machinery Directive 98/37/EEC equipment is granted a CE mark, which in theory should ensure safe food, but unfortunately this directive clearly falls short of hygiene requirements. The European Hygienic Engineering & Design Group (EHEDG) has developed an evaluation and certification programme which ensures that there is hygienically designed process equipment available on the market (Kastelein and Wirtanen, 2003). The guidelines, which are important in evaluating the hygienic design of process equipment and components in closed systems, are 'Hygienic equipment design criteria', Document No. 8 (1993a), 'Hygienic design of closed equipment for the processing of liquid food', Document No. 10 (1993b) and 'A method for the assessment of in-place cleanability of food processing equipment', Document No. 2 (Timperley *et al.*, 2000).

In closed cleaning systems pre-rinsing with cold water is carried out to remove loose soil, and the CIP treatment is normally performed using a hot cleaning solution, but a cold solution can also be used in handling fat-free products (Wirtanen, 1995). The chlorine-based agents loosen the biofilm from the surface, break it and finally eliminate the microbes. Chelating agents in the cleaning solution enhance the breakage and removal of biofilms. The cleaning agent manufacturers can deliver effective cleaning programmes applicable for most practical situations. Mechanical grinding, chemical bathing or passivation is used as post-treatment of the surface to obtain stainless properties on the tank surfaces. The passivation can be performed using an oxidising acid, e.g. $< 20\%$ nitric acid at 0–60 °C for 3–5 min (Taulavuori *et al.*, 2004).

The CIP treatment in dairies is normally performed using hot cleaning solutions, but cold solutions can be used in the processing of fat-free products, e.g. in breweries (Chisti and Moo-Young, 1994; Wirtanen, 1995; Storgårds, 2000). In dairy CIP cleaning the warm alkaline cleaning solution, normally of 1–1.5% sodium hydroxide (NaOH), is heated to 75–80 °C and the cleaning time is 15–20 min. The equipment is rinsed for 3 min with cold water before the acid treatment (0.5–1%) is performed at approximately 60–70 °C for 5 min. The effect of chlorine-based agents can be divided into three phases: loosening the biofilm from the surface, breakage of the biofilm and the disinfective effect of the active chlorine (Costerton *et al.*, 1985; Wirtanen, 1995). The cleaning solutions should not be reused in processes aiming at total sterility, because the reused cleaning solution can contaminate the equipment (Chisti and Moo-Young, 1994). Single-phase cleaning agents for CIP treatment are commonly used because the processing industry wants to save time (Husmark, 1994). In single-phase cleaning procedures the time for one cleaning process, normally the acid treatment, and a rinsing step can be saved. If the disinfection is performed using hot water, the water should be heated to 90–95 °C and the duration should be 5 min (Bylund, 1995).

The main detergent products used in breweries are based on sodium hydroxide. These agents are highly effective in removing organic soil. The main disadvantage in using these agents in brewery tanks is that they react with carbon dioxide to form a less effective carbonate salt while removing precious carbon dioxide from the process vessel (Sedgwick, 1999). One cubic metre of carbon dioxide at one atmosphere and 20 °C will neutralise 2 kg of sodium hydroxide, which is equivalent to 100 L of a 2% w/v solution (Gingell, 1999). Acid-based sanitisers are more efficient and cost effective in removing and preventing beerstone and hard water deposits (Gingell, 1999). In the tank cleaning in breweries a CIP technique comprising of an acidic step and a disinfecting step represents an advanced, more effective cleaning procedure under carbon dioxide pressure at low temperatures (Sedgwick, 1999). A typical classical acid-cleaning regime mainly used in breweries for lightly soiled tanks consists of a 5–7 min pre-rinse using recovered water, followed by 40–60 min of acid cleaning using 1.5–2.0% v/v detergent, a 5–7 min intermediate rinse with water, 25–30 min of acid sanitation and beerstone prevention using a 0.5–0.6% v/v circulation acid sanitiser and ending with a 10 min post-rinse with microbiological pure water (Gingell, 1999).

In breweries the cleaning procedure should be based on the oxygen content left in the tank after cleaning and sanitising. Dirksen and Duca (2000) compared acid and alkaline CIP cleaning procedures. The procedures were a 10 min water rinse, 20 min acidic detergent wash, 10 min water rinse and 10 min peracetic acid sanitiser as well as 15 min water rinse, 25 min alkaline detergent wash, 10 min water rinse and 10 min peracetic acid sanitiser respectively. They found that the amount of oxygen introduced into the tank by caustic cleaning is substantially higher than with acid cleaning. High oxygen levels cause oxidation flavouring of the beer. Furthermore, solutions of peracetic acid-based sanitisers make only a small oxygen contribution to beer, if regular concentration and drainage time are observed (Vasconcelos *et al.*, 2003). A study was made of the efficacy of chlorine dioxide solutions (5–7.5 ppm) as an alternative to hot water (85 °C) in the disinfection step of the brewery CIP treatment of blending tanks, finishing tanks and process lines under carbon dioxide pressure, and the results showed that chlorine dioxide could successfully replace hot water disinfection.

The advantages of chlorine dioxide disinfection include a shorter duration and reduced energy and water consumption. Experience with optimised new sanitation procedure has shown no major problems in beer quality. Furthermore, interruptions into the production have been reduced and equipment is subjected to less thermal stress (Agius *et al.*, 2004). Han *et al.* (1999) showed that 10 mg/L of chlorine dioxide gas were effective in reducing spoilage organisms in fruit juice production in epoxy-lined storage tanks when the exposure was 30 min at 9–28 °C with a high relative humidity (>90%) or at 25–28 °C with the relative humidity > 69%.

The design and operation of spray balls have been improved to enhance cleaning and today many suppliers can produce customised systems with an excellent cleaning efficiency. Spray balls are hollow and usually spherical, with diameters from 35 to 150 mm with an inlet tube of about 25–50 mm. A number of holes, typically with a diameter from 1.6 mm to 2.4 mm, are drilled in the body in positions that give the desired spray distribution (Morison and Thorpe, 2002). The traditional static spray ball provides a simple means of distributing the cleaning fluid onto the tank walls. Rotating spray heads spray the walls in a uniform pattern and this combination of cleaning pattern and mechanical impact removes residues in a shorter time and with less water than a static spray ball. The rotating nozzle head is particularly suitable for tanks that have complicated internals because a single rotating jet hits the various tank components from both sides because of its horizontal rotation (Müller, 2001). In general, the process of soil removal involves the initial wetting and subsequent softening or dissolution of the fouled material, followed by complete removal by impingement or irrigation. The cleaning by a jet with a stable diameter of less than 4 mm was not performed by direct impingement alone. The effectiveness of a spray ball depends on the wetting area around and below the point of impingement. According to the studies by Morison and Thorpe (2002), the temperature of the water and the addition of a small quantity of detergent were two of the most significant variables to affect the wetting rate. Increasing the pressure of the jet also enhances the surface wetting.

30.3 Hygienic design test methods

Hygienic design test methods can be used to find the challenging critical spots in cleaning equipment. These tests are intended as basic screening tests for the hygienic design of equipment and are not indicative of performance in industrial cleaning situations. These methods are based on standardised procedures. They are very sensitive, so that it is possible to indicate even very small poorly designed or defective areas, such as crevices arising from, e.g., the wrong design of a gasket (EHEDG, 1993b). The hygienic design test methods always have three steps. Firstly the equipment has to be soiled with a standardised soil containing a tracer substance. The soil is dried to the equipment surface and it is cleaned afterwards with an adapted cleaning procedure that should leave some soil residue on the surfaces tested. The last step is the detection of the residual soil in the equipment. To give a statement on the cleanability of the equipment it is important to have a reference for which the cleanability level is known. Subsequently the assessment of the cleanability of the test equipment could be compared with the reference. Most of the published test methods use microbes as tracers in the soil. To be able to assess the cleanability of large-scale equipment or whole process lines in the industry an organic or inorganic non-toxic tracer should be used.

At BioCentrum-DTU a 80 L pilot scale stainless steel tank with an inner diameter of 400 mm and height of 800 mm and a conical bottom was tested with a spray ball mounted in the tank lid. The length of the shaft of the spray ball was 145 mm and the rotating tip measured 50 mm. The position of the spray ball was 60 mm from the centre of the tank. In the first situation the tank was soiled with beer containing yeast cells. In the second case the tank was soiled with sour milk containing *Bacillus stearothermophilus* var. *calidolactis* (NIZO C953) spores and subsequently dried at room temperature for 3 hours. The cleaning was performed using a pilot plant CIP cleaning system consisting of pre-rinsing with cold water for 1 min, cleaning with 60 °C detergent solution (1% EHEDG Testcleaner) for 10 min and a final rinsing with cold water for 1 min. The mean flow velocity in the pipe before the spray ball in the tank lid was adjusted to 1.5 m/s. The hygienic state of the tank was studied using contact agar methods (Hygicult TPC® and Petrifilm AC®) and ATP-assay. Samples were taken from the tank lid, the spray ball and selected positions on the tank walls (upper, middle and lower parts). On the basis of the detection methods used, the smooth parts of the tank did not contain any microbial contaminants after the cleaning procedure. The tests showed, however, that some residual microbes were present on the rougher parts of the tank. These places were the screws in the lid, the gasket between the tank and the lid and some unhygienic scratches on the tank wall.

30.4 Detecting the cleanliness of tanks

In dealing with closed processes, one problem is always the validation of the cleaning procedure. In some cases visual inspection can be utilised, which is the case for large tanks. Different tests for evaluating cleaning efficiency by means of fluorescent substances are commercially available (Wirtanen, 1995; Storgårds, 2000). By using UV light, fouling not visible with normal lamp light or daylight becomes visible (Kold and Skræ, 2004). Despite this, it was occasionally difficult to determine if shining was due to the surface finish, the type of steel or actually because of fouling. The fouling appeared as a dim white and occasionally pink coating. A thick layer of fouling was on some occasions found in pipes connected to cleaning nozzles, on areas around lids and on gaskets. Jacob and Brandl (2002) also noticed limitations in the visual inspection when they monitored the efficiency of the cleaning and disinfection procedures in tanks in breweries. They have also used UV illumination to detect deposits, Fuchsin staining to detect proteins and Lugol's iodine solution to detect starch. According to Paez *et al.* (2003) a commercial ATP-bioluminescence system can be used to evaluate the cleanliness of milking equipment, bulk tanks and milk transport tankers. The outlet of the plate cooler, the outlet pipe of the bulk tank and the internal surface of the manhole lid in the milk transport tanks were the most critical points. Results indicated that bioluminescence results were not reliable for testing rinse water, and therefore swabbing was needed to complete the hygienic assessment. Odebrecht *et al.* (2000) studied the applicability of bioluminescence methods for testing hygiene in beer fermentation and maturation tanks and yeast tanks in breweries. Results indicated that the bioluminescence method was not suitable for the detection of microbial contamination, as the results did not correlate with those obtained by conventional microbial culturing techniques. It may, however, be useful as a method for monitoring the efficiency of cleaning and disinfection of tanks, pipelines and other brewery equipment.

30.5 Using computational fluid dynamics (CFD) to assess cleanability of closed process lines

CFD is used in many applications to model the bulk parameters of fluid flows. Recently model developments have made it possible to resolve what happens in specific positions on and near walls, which is of interest in the study of cleaning processes (Jensen, 2003). CFD models of tanks exist for purposes of optimising the operation of processes such as mixing, heating and cooling. The CFD model developed by Tress *et al.* (2004) predicts mixing patterns in silo tanks. It provides information about the behaviour of two-phase flows stirred with two pitched blade turbines placed differently. The results of the studies are applied to reduce the energy requirements for mixing and optimise mixing efficiency. The extension applies to combining the models for the prediction of the hygienic design of valves, pipes, etc. made by Jensen (2003) with conventional tank flow models and the establishment of more information on the connections between surface characteristics and the cleaning effect of fluid flow. This is believed to result in a tool suitable for the evaluation of the efficiency of cleaning procedures in tanks. The hypothesis is that both validation of the cleaning procedure and the design of a proper cleaning system can be supported and improved using CFD models. The aim is to evaluate the suitability of CFD simulations for estimating and improving tank cleaning and to perform cleaning tests in a pilot plant in order to establish a correlation with the results of CFD simulations. Preliminary results of the CFD simulations yield information about wall shear stresses in the tank and the flow rates in different parts of the system. Some studies have been carried out on the flow simulation suitable for the accurate evaluation of the hygienic state of tanks during and after spray-ball cleaning.

More fundamental studies are needed concerning the effect of hydrodynamics in tank cleaning processes. CFD has been shown to be a useful tool for the optimisation of the hygienic design of plain, closed process equipment (Jensen, 2003). A combination of wall shear stress, fluid exchange and turbulence conditions can be used to predict areas that are not properly cleaned in both simple and complex flow systems (Friis and Jensen, 2002). The surface roughness is difficult to implement in flow models, but the fact that the above-mentioned results on surface roughness are not a significant factor in cleaning is encouraging the use of CFD simulations as tools for optimising the cleaning procedure. In the CFD modelling tool the flow phenomena, e.g. wall shear stress and fluid exchange, are used for predicting the cleanability in the flow systems. The combination of surface topography, fluid dynamics and surface microbiology provides a good basis for studying and solving issues relating to hygiene in closed systems.

30.6 Future trends

The results of the preliminary experiments studying a simple case gave each other good support in the simple case study performed. This has provided evidence for the hypothesis that a combination of knowledge in fluid dynamics and microbiology also provides an excellent basis for the hygienic design of integrated tank and CIP cleaning systems. The main focus has been on the experimental part since a sound basic knowledge was required in order to establish a proper CFD model that can be used to simulate this type of process system. The CFD model should be extended to cover different tanks and spray balls and validated with the use of modified microbial test methods. Development of these test methods is needed in the future studies. Chism and Smith (2004) have patented a method for refurbishing beverage storage tanks. The interior surface of the beverage storage tank is sprayed with a fluid jet to remove lining and surface contaminants. The interior surface of the beverage storage tank is abrasive blasted and a beverage-safe coating material is applied to the prepared interior surface of the beverage storage tank prior to reuse. A future trend will also be to use improved coating materials for the tank surfaces in order to facilitate the cleaning procedures.

30.7 References

AGIUS G, BURKEEN S and MYNATT J (2004) Benefits of using chlorine dioxide as an alternative to hot-water sanitation, *Mast Brew Ass Amer Tech Quart*, **41**, 42–44.

AUSTIN JW and BERGERON G (1995) Development of bacterial biofilms in dairy processing lines, *J Dairy Res*, **62**, 509–519.

BIRUS T (2003) Use of stainless steel tanks in the fruit juice industry, *Fluessiges-Obst,* **70**, 658–660.

BYLUND G (1995) *Dairy processing handbook*, Lund, Tetra Pak Processing Systems AB.

CHISM SD and SMITH EC (2004) Refurbished beverage storage tank, US 2004/0159657 A1 (US2004/0159657A1), US; United States Patent Application Publication.

CHISTI Y and MOO-YOUNG M (1994) Cleaning-in-place systems for industrial bioreactors: design, validation and operation, *J Ind Microbiol*, **13**, 201–207.

COSTERTON JW, MARRIE TJ and CHENG K-J (1985) Phenomena of bacterial adhesion, in Savage DC and Fletcher M *Bacterial adhesion*, New York, Plenum Press, 3–43.

DIRKSEN J and DUCA J (2000) Minimizing oxygen content in bright beer tanks by use of acid cleaning, *Mast Brew Ass Amer Tech Quart*, **37**, 435–436.

EHEDG (1993a) Document No. 8, Hygienic equipment design criteria, *Trends Food Sci Technol*, **4**, 225–229.

EHEDG (1993b) Document No. 10, Hygienic design of closed equipment for the processing of liquid food, *Trends Food Sci Technol*, **4**, 375–379.

ENARI T-M and MÄKINEN V (1993) *Panimotekniikka*, Porvoo, Kirjapaino t-t.

FRIIS A and JENSEN BBB (2002) Prediction of hygiene and flow modelling in food processing equipment, *Trans IChemE*, **80**, Part C, 281–285.

GEESEY G and BRYERS J (2000) Biofouling of engineered materials and systems, in Bryers J *Biofilms II Process analysis and application*, New York, John Wiley-Liss Inc., 237–279.

GINGELL K (1999) Latest developments in acid cleaning and related sanitation in breweries, *Brewer*, **85**, 186–192.

HAN Y, GUENTERT AM, SMITH RS, LINTON RH and NELSON PE (1999) Efficacy of chlorine dioxide gas as a sanitizer for tanks used for aseptic juice storage, *Food Microbiol*, **16**, 53–61.

HUSMARK U (1994) *Sanitering I mejeri: Raport 1 'En sammanställing över disk och desinfektionsrutiner i svensk mejeriindustri'*, Göteborg, SIK.

JACOB F and BRANDL A (2002) Auffinden von Schwachstellen in Reinigungssystemen mit UV-Licht, *Brauwelt*, **142**, 1162–1163.

JENSEN BBB (2003) *Hygienic design of closed processing equipment by use of computational fluid dynamics,* Lyngby, BioCentrum-DTU, Technical University of Denmark.

KASTELEIN J and WIRTANEN G (2003) EHEDG procedures for evaluating, testing and certification of process equipment, *EHEDG Newsletter*, **5**, 5–6.

KOLD J and SKRÆ AM (2004) A method for assessing the cleanability of open processing equipment, in Wirtanen G and Salo S *DairyNET – Hygiene control in Nordic dairies*, Espoo, Otamedia, 213–216.

LELIEVELD HLM (1985) Hygienic design and test methods, *J Soc Dairy Technol*, **38**, 14–16.

MORISON KR and THORPE RJ (2002) Liquid distribution from cleaning-in-place sprayballs, *Trans IchemE*, **80**, Part C, 270–275.

MÜLLER O (2001) Efficient tank cleaning – untapped potential, *Brauwelt Int*, **2**, 141–143.

ODEBRECHT E, SCHMIDT HJ and FRANCO BGM (2000) Studies on applicability of bioluminescence in the brewery. Comparative studies and critical evaluation, *Brauwelt*, **140**, 1904–1905, 1908–1915.

PAEZ R, TAVERNA M, CHARLON V, CUATRIN A, ETCHEVERRY F and COSTA LH DA (2003) Application of ATP-bioluminescence technique for assessing cleanliness of milking equipment, bulk tank and milk transport tankers, *Food Protection Trends*, **23**, 308–314.

SEDGWICK A (1999) Innovative acid cleaning, *Brewer*, **85**, 193–195.

STORGÅRDS E (2000) *Process hygiene control in beer production and dispensing, VTT Publications 410,* Espoo, VTT Offsetpaino.

TAULAVUORI T, KYRÖLÄINEN A and TARKIAINEN R (2004) *Ruostumattomat teräkset*, Tampere, Tammer-Paino Oy.

TIMPERLEY AW, BOURION F, BENEZECH T, CARPENTIER B, CURIEL GJ, HAUGEN K, HOFMAN J, KASTELEIN J, RONNER U, TRAGARDH C and WIRTANEN G (2000) A method for the assessment of in-place cleanability of food processing equipment, EHEDG Document No. 2, Chipping Campden, Campden & Chorleywood Food Research Association.

TRESS AJG, STUBBE P, JENSEN BBB and FRIIS A (2004) Prediction and optimization of asymmetric mixing patterns in tanks, *International Conference on Engineering and Food*, Montpellier, France.

VASCONCELOS ICF, JORGE K, NOTHAFT A, MARINHO SV, MIRANDA GHP, SCHMIDT B and GRAB L (2003) The impact of a novel mixed peracid-based sanitizer on beer – a sensorial and colloidal evaluation, *Mast Brew Ass Amer Tech Quart,* **40**, 199–203.

VERRAN J, BOYD RD and HALL K (2001) The effect of wear on fouling and cleanability of hygienic food contact surfaces, in *Proceedings of the 32nd R^3 Nordic Symposium*, Enköping, ReklamProducenterna, 172–184.

WIRTANEN G (1995) *Biofilm formation and its elimination from food processing equipment. VTT Publications 251,* Espoo, VTT Offsetpaino.

WIRTANEN G, LANGSRUD S, SALO S, OLOFSON U, ALNÅS H, NEUMAN M, HOMEID JP and MATTILA-SANDHOLM T (2002) *Evaluation of sanitation procedures for use in dairies*. VTT Publications 481, Espoo, Otamedia Oy.

31

Ozone decontamination in hygiene management

L. Fielding and R. Bailey, University of Wales Institute Cardiff, UK

31.1 Introduction

Discovered in 1840 by German chemist Christian Schönbein, ozone is an allotrope of oxygen, naturally present as a colourless gas with a distinctive odour. It is produced in the upper atmosphere by the action of ultraviolet radiation on oxygen molecules and at ground level as a by-product of photochemical reactions involving oxygen, nitrogen and hydrocarbons (Graham 1997).

Ozone is an effective antimicrobial agent due to its high oxidizing potential (+0.27 volts) compared with oxidizing agents such as hypochlorous acid (HOCl), the active chlorine species in aqueous solution, which has an oxidation potential of +1.49 volts. The only substances with a higher oxidation potential are fluorine, fluorine dioxide and the oxygen radical (Gurley 1985).

31.1.1 Chemical characteristics and production of ozone

Because of its highly reactive nature, ozone autodecomposes over a relatively short time to produce oxygen. This means that ozone does not persist in the environment and will only be present for a short period of time after its application. It cannot, therefore, be stored and must be produced *in situ*. The concentration of ozone is quoted as either parts per million (ppm) or milligrams per litre ($mg\,L^{-1}$). 1 ppm is equivalent to $0.00196\,mg\,L^{-1}$ (the equivalent of $1.96 \times 10^{-6}\,mg\,m^{-3}$).

Ozone can be used as an antimicrobial agent in two forms, either in the gaseous state or dissolved in purified water to produce ozonated water. Gaseous ozone is generated by a number of methods depending on the concentration required. Low concentrations (0.03 ppm) can be produced by the exposure of air to radiation with a wavelength of 185 nm from UV lamps. Higher concentrations can be produced by the corona discharge method. A high voltage alternating current is applied across a discharge gap in the presence of air or oxygen. This causes electron excitation, which induces splitting of oxygen molecules. Ozone is then formed by the recombination of the free oxygen atoms.

The concentration of ozone produced depends on many factors such as voltage, current frequency, discharge gap, dielectric material and thickness. Oxygen or an oxygen concentrator can be used instead of air to increase the yield but if air is used, it should be dried to prevent the formation of nitric acid (Kim *et al.* 1999).

Ozonated water can also be produced by a number of different methods. Ozone is 12 times more soluble in water than oxygen but it has a lower partial pressure, making it difficult to obtain high concentrations. Under normal conditions of temperature and pressure, no more than a few milligrams per litre can be dissolved. To obtain the maximum oxidative effect, conditions must be created where there is maximum transfer across the gas/water interface. Normal methods of dissolving oxygen in water are unsuitable for ozone because of its high rate of decomposition. Alternative methods include allowing a large number of small bubbles containing ozone, produced by a fine porous diffuser, to rise through a column of water. The rate of rise of the bubbles and the volume of the liquid column are adjusted to maximize absorption in relation to the rate of decomposition of ozone. Injecting ozone into water through a series of water jets or violently mixing ozone and water in emulsion turbines can also achieve ozonation. Sonication can be employed in conjunction with these methods as it forms extremely fine bubbles, which aids dissolution of ozone in water.

31.1.2 Mode of action

It is thought that ozone initially targets the bacterial membrane glycoproteins, glycolipids or amino acids, causing cell death through a change in cell permeability and cell lysis. It can also react with the sulphydryl groups of some enzymes causing disruption of normal cellular activity. Ozone will also affect the nucleic acid components of the cell by modifying the pyrimidine and purine bases.

When ozone is dissolved in water it decomposes to form molecular oxygen and other highly reactive free radicals ($\bullet HO_2$, $\bullet OH$ and $\bullet H$). The hydroxyl radicals, which are formed as a product of ozone decomposition, can have a powerful biocidal effect (Cho *et al.* 2003). The sites of action of ozone dissolved in water are similar to those of gaseous ozone. Gurley (1985) highlights the primary sources of attack as the cell membrane, possibly through ozonolysis of the carbon–carbon double bonds in the membrane lipids, leading to cell lysis. Other possible sites of action are the amino acids and nuclides of the cell membrane or enzymes. The actual method of microbial inactivation is probably a combination of both these processes.

31.2 Historical uses of ozone

31.2.1 Water treatment

Some of the earliest recorded industrial uses of ozone are in the treatment of water supplies. There are many municipal drinking water facilities in France that have employed ozone as the primary disinfectant since 1906 (Gurley 1985). The ability of ozone to retain its high oxidation potential when dissolved in water is necessary for water disinfection as ozone disinfects by oxidizing microbial cells.

The ozone molecule and other species present will react with any organic molecules and will ultimately be oxidized to carbon dioxide and water. Different sources of water will contain varying amounts of organic matter and the water to be disinfected will, therefore, have a specific ozone demand. This is the amount of ozone consumed by a given volume of water in oxidizing the organic matter and must be satisfied before there is any antimicrobial effect. The ozone demand and the concentration of dissolved ozone are the two major factors that determine the degree of disinfection. There must be a residual concentration of ozone to act as an antimicrobial agent.

A study by Broadwater *et al.* (1973) indicated that ozone disinfects by an 'all or nothing' action in water. If the applied dosage is high enough to oxidize the organic matter and the microorganisms, all the microorganisms will be inactivated. When the dosage of ozone is too low, it is quickly exhausted by oxidation of the organic matter and a significant number of microorganisms will survive.

Ozone is also used to disinfect water containing natural organic matter (Cho *et al.* 2003) and has proved effective against a number of organisms, including *Cryptosporidium parvum* oocysts and *Clostridium perfringens* spores, which are relatively resistant to chlorine (Finch *et al.* 1993; Venczel *et al.* 1997).

31.2.2 The food industry

The use of ozone as an antimicrobial agent in the food processing industry has excited interest due to a number of factors. Ozone was approved for use by the US Food and Drug Administration on 26 June 2001 for use as an antimicrobial agent for the treatment, storage and processing of foods in the gaseous and aqueous phases.

It has been investigated for use as a decontaminant of actual food products but also has a major role to play in the decontamination of equipment and the food processing environment. Ozone gas has been found to be effective at reducing the levels of viable organisms attached to stainless surfaces (Moore *et al.* 2000). As it leaves no residue on the surface, it may have potential for use as a terminal sanitizer for food contact surfaces. It must, however, be employed following cleaning, as ozone is highly reactive and its microbicidal effects are lost upon contact with any other organic material, such as food debris. It is not, therefore, a replacement for good hygiene practices such as cleaning but can enhance the efficiency of a good cleaning policy if used correctly. Case studies concerning the effects of ozone as a terminal sanitizer in the food industry, for both the environment and equipment, are presented in section 31.5.1.

31.3 The effect of ozone on microorganisms

31.3.1 Vegetative bacteria

Vegetative bacterial cells in water tend to be very susceptible to ozone. Amounts ranging from 10^6 to 10^7 cfu/ml can be reduced to undetectable levels after 5 min when exposed to ozone concentrations ranging from 0.01 to 1.0 mg L^{-1} (Haufele and von Sprockhoff 1973; Venosa 1972).

Moore *et al.* (2000) found gaseous ozone to be effective against a range of Gram-positive and Gram-negative organisms at levels of 0.05 to 2 ppm. The effects were also proven to be time dependent. This study also investigated the effects of organic soil on ozone efficacy and concluded that its presence reduces the antimicrobial effects. The study confirms that ozone is only an effective sanitizer when the surfaces are clean and free from organic soil.

31.3.2 Bacterial endospores

Bacterial spores are much more resistant to ozone, owing to the presence of the spore coat and the extra protection this provides to the protoplast. Broadwater *et al.* (1973) found that spores of *Bacillus* spp. were up to 15 times more resistant than their vegetative cells. A study by Ishizaki *et al.* (1986) found that increasing the relative humidity increased the effectiveness of ozone as a sporicidal agent and that no significant inactivation of spores occurred at humidities of less than 50%.

31.3.3 Fungi and fungal spores

Moore *et al.* (2000) found that 2 ppm ozone significantly reduced the number of fungi attached to stainless steel coupons while Foarde *et al.* (1997) determined that levels in excess of 6 ppm were required (9 ppm when attached to a surface).

31.3.4 Viruses

Viruses have an extremely low tolerance to ozone in low ozone demand waters. The ozone inactivation of viruses in water is limited by the rate of diffusion through the protein coat into the nucleic acid core, resulting in damage to the viral RNA. Komanapalli and Lau (1998) found that bacteriophage λ was completely inactivated by 600 ppm ozone in phosphate buffer solution after 10 minutes.

31.4 Undesirable effects of ozone

31.4.1 Construction materials

The reactive nature of ozone means that it reacts with many materials including textiles, organic dyes, metals, many plastics, paints and natural rubber. It can also react explosively with grease and oils (Anon 1998a). The rate of degradation of natural rubber can be reduced by the addition of anti-ozonants. This can, however, significantly increase the cost (10–25%) and the volume of the rubber component (34%) (Anon 1993). Greene *et al.* (1994) tested the resistance of gaskets made from a number of materials commonly found in the food processing environment, to ozone and chlorine. The results were that the use of ozone did not affect most of the gaskets more than chlorine, with the exception of polytetrafluoroethylene (PTFE). Latex, most silicones, viton, polycarbonate and rigid polyvinyl chloride are resistant to the effects of ozone (Dusseau *et al.* 2004). A thorough audit must, therefore, be conducted prior to using ozone to replace these materials with alternatives that will not be degraded by ozone.

31.4.2 Health and safety considerations

As ozone is a highly reactive substance, it reacts with any organic substance, including the human body. The health effects are mainly at the sites of initial contact, normally the respiratory tract, lungs and eyes, and there is a degree of variation among individuals in sensitivity to ozone. The principal effects are irritation and damage to the small airways of the lung, and exposure to high levels (1 ppm and above) may lead to more serious health effects including lung damage (Anon 1998a). If this exposure exceeds 24 hours, then the damage may be irreversible (Dusseau *et al.* 2004).

The Occupational Exposure Standard (OES) in the UK for ozone is 0.2 ppm averaged over a 15 minute reference period (Anon 1998a). The US has a threshold limit value–long-term exposure limit (TLV-LTEL) of 0.1 ppm for a normal 8 hour day (40 hours per week) and a TVL-LTEL of 0.3 ppm for 15 minutes (Xu 1999).

31.5 Practical applications of ozone

31.5.1 Case studies of the use of ozone in the food industry

Gaseous ozone

The use of ozone for process plant decontamination was implemented into a small cheese-making factory that had problems with contamination of the processing environment. The heterotrophic aerobic plate count and levels of Enterobacteriaceae on both food contact and environmental surfaces were reduced by treating the whole plant with 2 ppm ozone overnight for a period of two months. When ozone use was discontinued, the heterotrophic plate counts and levels of Enterobacteriaceae increased. It is important to note, however, that these reductions were observed after cleaning had taken place. Ozonation must be used as a terminal disinfection step and is not a replacement for cleaning or good hygiene practices.

Ozonated water

It is well reported that cells within a biofilm are more resistant to antimicrobial compounds such as commercially available cleaning solutions (Lewis 2001). In addition to this, many of the cells within the biofilm are physically protected from the effects of cleaning agents by the structure of the biofilm and so are able to carry on growing normally after cleaning (Holah *et al.* 1994, Lewis 2001). One of the major problems is that chemical-based cleaning regimes have little effect on the structure of the biofilm. A cleaning protocol that destroys microorganisms and removes the biofilm would, therefore, be beneficial.

The effectiveness of ozonated water and traditional chemical beer line cleaner was investigated against biofilm in beer lines. The ozonated water was produced by a WL2 Advanced Oxidant Generator (AOG, Ozone Manufacturing/ IMI Cornelius). The investigation revealed that both systems were effective at reducing biofilm from initial levels by 99.98% (ozone) and 99.85% (chemical). The ozone gave a statistically significantly lower number than the chemical cleaner for tubing samples in a model beer line system. These findings were supported by the direct observation of the biofilm using epifluorescent microscopy, which showed that the ozone removed large portions of the biofilm from the glass slide whereas the chemical cleaner merely killed a proportion of the cells.

31.5.2 Monitoring ozone levels in air and water

Kim *et al.* (1998) grouped the analytical methods for the detection and measurement of ozone into three categories. These are physical methods, physicochemical methods and chemical methods. The physical methods are based on measuring the intensity of absorption of light in the UV, IR or visible regions of the electromagnetic spectrum. Examples are the Ozone Photometer, which can measure ozone (gaseous or aqueous) and the API Model 450 Ozone Monitor, which can measure gaseous ozone in the region of 1–1000 ppm (Anon 1999).

Physicochemical methods rely on measuring the physical effects produced when ozone reacts with suitable reagents. The effects measured include chemiluminescence or heat of reaction. Chemical methods rely on measuring the products formed when ozone undergoes a chemical reaction with an appropriate reagent, or the reduction in molecular weight of a polymer when it reacts with ozone. An example of a chemical detection method is the Indigo colorimetric method. Ozone adds across the carbon–carbon double bond of sulphonated indigo dye and it decolourizes. The amount of ozone present is proportional to the amount of decoloration, which can be determined spectrophotometrically. Dräger tubes consist of a tube and a pump, which draws air into the tube. The ozone reacts with the indigo in the tube filling. These devices measure 0.05–0.7 ppm and 20–300 ppm, which excludes the range commonly used in food processing (Anon 1998b).

31.5.3 Dispersal of ozone

The dispersal of ozone is mediated by its reaction with organic compounds. Ozone dispersal is more rapid in water than in air, initiated by chain reaction involving •OH (Bott 1991). Staehelin and Hoigne (1985) determined that the rate of decomposition of ozone in water is affected by the concentration and type of organic solutes. Compounds may initiate ozone decomposition (e.g. glycoxylic acid), promote decomposition (e.g. formic acid, primary alcohols, benzeze and humic acid) or inhibit decomposition (e.g. *tert*-butyl alcohol, bicarbonate). Decomposition is also more rapid at alkaline pH values (Hoigne and Bader 1976).

For ozone dispersal in air, use of ventilation can rapidly reduce the residual concentration (Masaoka *et al.* 1982). If using ozone gas as a terminal sanitizer in an unoccupied space, an evaluation must be made of the rate of decomposition to ensure that there is no risk to the health of returning personnel.

31.6 Future potential

The low efficacy of ozone in certain circumstances has led to research into the use of ozone in combination with other treatments. These include heat treatment, negative air ions and natural oils. It has been shown that there may be synergy between ozone and these other treatments, which would lead to an enhanced effect, compared to the individual treatments.

31.7 Conclusion

Ozone has been proved to be an effective antimicrobial for a number of applications. It is not, however, a cure-all and there are significant health and safety implications and detrimental effects to food and the environment. It is also difficult to advise as to the exact concentration of ozone, either gaseous or aqueous, needed for a certain application owing to many factors such as the amount of organic material present, the desired effect and the operating conditions of the system. It does, however, have the potential to be used as a terminal sanitizer for food contact equipment and the environment, provided that personnel can be excluded during its application and degradation. Levels of ozone of 2 ppm have been shown to be effective against a range of microorganisms in the laboratory (aerosolized and surface attached) and in industry. Ozonated water can also be used for decontamination of surfaces and equipment, and has the potential for use in cleaning-in-place systems. It also has fewer health and safety considerations than the use of ozone gas. A thorough investigation must, however, precede any use of ozone or ozonated water to ensure product, equipment and personnel safety.

31.8 Sources of further information and advice

International Ozone Association, IOA-EA3G Secretariat, Bât. ESIP - 40 av. du Recteur Pineau, 86022 POITIERS cedex, France (http://www.ioa-ea3g.org/homepage.asp).

31.9 References

ANON (1993) *Ozone in the United Kingdom: Third Report*. United Kingdom Photochemical Oxidants Review Group, London, pp. 83–107.

ANON (1998a) *Ozone: health hazards and precautionary measures*. Guidance Note EH38 (revised). Health and Safety Executive, Bootle, UK.

ANON (1998b) *Dräger-Tube Handbook*, 11th Ed. Dräger Sicherheitstechnik GmbH: Lübeck, Germany.

ANON (1999) *Instruction Manual, Model 450 Ozone Monitor*. Teledyne Instruments, Advanced Pollution Instruments Division, San Diego, CA.

BOTT, T.R. (1991) Ozone as a disinfectant in process plant. *Food Control*, January, 44–49.

BROADWATER, W.T., HOEHN, R.C. and KING, P.H. (1973) Sensitivity of three selected bacterial species to ozone. *Applied Microbiology*, **26**, 391–393.

CHO, M., CHUNG, H.M. and YOON, J. (2003) Disinfection of water containing natural organic matter by using ozone-initiated radical reactions. *Applied and Environmental Microbiology*, **69** (4), 2284–2291.

DUSSEAU, J., DUROSELLE, P. and FRENEY, J. (2004) Sterilisation: gaseous sterilisation. In *Disinfection, Preservation and Sterilisation* 4th ed. (eds. Fraise, A., Lambert, P. and Maillard, J.) Blackwell Publishing Ltd, Oxford, pp. 401–435.

FINCH, G.R., BLACK, E.K., GYUREK, L. and BELOSEVIC, M. (1993) Ozone inactivation of *Cryptosporidium parvum* in demand-free phosphate buffer determined by *in vitro* excystation and animal infectivity. *Applied and Environmental Microbiology*, **59** (12), 4203–4210.

FOARDE, K.K., VANOSDELL, D.W. and STEIBER, R.S. (1997) Investigation of gas-phase ozone as a potential biocide. *Applied Occupational and Environmental Hygiene*, **12** (8), 535–542.

GRAHAM, D.M. (1997) Use of ozone for food processing. *Food Technology*, **51** (6), 72–75.

GREENE, A.K., VERGANO, P.J., FEW, B.K. and SERAFINI, J.C. (1994) Effect of ozonated water sanitization on gasket materials used in food processing. *Journal of Food Engineering*, **21** (4), 439–446.

GURLEY, B. (1985) Ozone: Pharmaceutical sterilant of the future? *Journal of Parenteral Science and Technology*, **39** (6), 256–261.

HAUFELE, A. and VON SPROCKHOFF, H. (1973) Ozone for disinfection of water contaminated with vegetative and spore forms of bacteria, fungi and viruses. *Microbiology Abstracts*, **157**(1), 53–70.

HOIGNE, J. and BADER, H. (1976) The role of hydroxyl radical reaction in ozonation processes in aqueous solutions. *Water Research*, **10**, 377.

HOLAH, J., BLOOMFIELD, S., WALKER, A. and SPENCELEY (1994) Control of biofilms in the food industry In: *Bacterial Biofilms and their Control in Medicine and Industry*. (eds. Wimpenny, J. *et al.*) Cardiff, Bioline, pp. 163–168.

ISHIZAKI, K., SHINRIKI, N. and MATSUYAMA, H. (1986) Inactivation of *Bacillus* spores by gaseous ozone. *Journal of Applied Bacteriology*, **60** (1), 67–72.

KIM, J.G., YOUSEF, A.E. and CHISM, G.W. (1998) Use of ozone to inactivate microorganisms on lettuce. *Journal of Food Safety*, **19** (1), 17–34.

KIM, J.G., YOUSEF, A.E. and DAVE, S. (1999) Application of ozone for enhancing the microbiological safety and quality of foods: a review. *Journal of Food Protection*, **62** (9), 1071–1087.

KOMANAPALLI, I.R. and LAU, B.H.S. (1998) Inactivation of bacteriophage λ, *Escherichia coli* and *Candida albicans* by ozone. *Applied Microbiology and Biotechnology*, **49** (6),

766–769.

LEWIS, K. (2001) Riddle of biofilm resistance. *Antimicrobial Agents and Chemotherapy*, **45,** 999–1007.

MASAOKA, T., KUBOTA, Y., NAMIUCHI, S., TAKUBO, T., UEDA, T., SHIBATA, H., NAKAMURA, H., YOSHITAKE, J., YAMAYOSHI, T., DOI, H. and KAMIKI, T. (1982) Ozone decontamination in bioclean rooms. *Applied and Environmental Microbiology*, **43** (3), 509–513.

MOORE, G. GRIFFITH, C. and PETERS, A. (2000) Bactericidal properties of ozone and its potential application as a terminal disinfectant. *Journal of Food Protection*, **63** (8), 1100–1106.

STAEHELIN, J. and HOIGNE, J. (1985) Decomposition of ozone in water in the presence of organic solutes acting as promoters and inhibitors of radical chain reactions. *Environmental Science and Technology*, **19** (12), 1206–1213.

VENCZEL, L.V., ARROWOOD, M., HURD, M. and SOBSEY, M.D. (1997) Inactivation of *Cryptosporidium parvum* oocysts and *Clostridium perfringens* spores by a mixed-oxidant disinfectant and by free chlorine. *Applied and Environmental Microbiology*, **63** (11), 1600.

VENOSA, A.D. (1972) Ozone as a water and wastewater disinfectant: a literature review. In *Ozone in Water and Wastewater Treatment*, Ann Arbor, MI, Ann Arbor Science.

XU, L.J. (1999) Use of ozone to improve the safety of fresh fruits and vegetables. *Food Technology*, **53** (10), 58–63.

32

Enzymatic cleaning in food processing

A. Grasshoff, Federal Dairy Research Centre, Germany

32.1 Introduction

When assessing the measures needed to reach a level of sufficient hygienic conditions in food production plants, not only the processes and their results but also the attendant circumstances such as the type and quantity of the required chemicals, water consumption, disposal of chemicals and water, and, last but not least, the costs for the total cleaning process have to be taken into account. Additionally, it has to be considered that the cleaning processes, including the chemicals to be used, must be adapted to the specific requirements. In a dairy, due to different soiling, for example, a staple tank for raw milk or a bulk milk tanker requires a different cleaning process from a pasteurizing plant, a plant for ultraheating consumer milk, a plant for the manufacture of yogurt, or a cheese manufacturer. However, regardless of the difference among the processes, at a certain point the question of disposal and the related environmental pollution arises for all cleaning agents. So long as a substance contained in the waste water is biologically degradable it can be discharged without any problems into the municipal (or, if present, in the company-owned) sewage treatment plant, and be adequately processed. If small quantities of non-degradable components occur in the form of waste-water sludge, most of the remaining components can be discharged in the form of cleaned water into the natural cycle.

The chemicals used in conventional cleaning processes in the food industry are primarily biologically non-degradable substances such as sodium or calcium hydroxide, and nitric and phosphorous acids, which have to be neutralized because of their pH values (12 to 13 or 1 to 2) to a pH value of 8 to 9 by simple merger in neutralizing tanks before being discharged in the waste-water net. Here a considerable quantity of dissolved, biologically non-degradable, inorganic salt concentrations occur, which pass through the sewage treatment plants and subsequently charge the outfall ditch. The idea is to reduce the salt load in the waste water by using agents for scaling-off the soiling, which contain less chemicals than the usual cleaning agents.

An example for this approach comes from the detergent industry for textile cleaning. The detergent industry has been successful at reducing chemicals by using different enzymes (see below), and at lowering the required temperatures in the washing liquor (energy savings). Also for the dish-washing formulae for industrial kitchens, catering service and restaurants, hospitals, etc. enzymes are being successfully used (e.g. in the product neodisher bioClean® of the Chemischen Fabrik Dr. Weigert, D-20539 Hamburg, Germany, www.drweigert.de).

A field where the use of enzyme-based cleaning agents has been largely established is the reprocessing of membranes for micro-, nano- and ultra-filtration as well as for reverse osmosis of various products of the food industry. Membrane processes are, for example, used in the beverage industry for clearing and clarifying juices, wine and beer, in the starch industry for starch cleaning, in slaughter houses for waste-water cleaning (Allie *et al.*, 2003), in egg product manufacture for concentrating albumins and egg, and in dairies for fractionating whey proteins (Arguello *et al.*, 2003), or for concentrating cheese whey (Kessler, 1996). In the course of the filtration process the membrane pores are plugged up with particles from the filtration substrate (preferably with higher molecular protein substances), so that the permeate flux drops down to an acceptable bottom limit at a given maximum trans-membrane pressure. To restore the filtration performance, the pores of the membranes have to be completely cleaned. Here cleaning agents are used that dissolve the substrate molecules accumulated in the membrane pores, or decompose them into smaller components to be subsequently removed. For the hydrolysis of higher molecular proteins, alkalis with a pH higher than 12 are usually used for temperatures beyond 60 °C. These conditions represent no problem for the ceramic or zirconium oxide membranes applied in microfiltration, but may do for the polysulphonic membranes, particularly the cellulose acetate membranes used for ultrafiltration and reverse osmosis. For material-sparing reprocessing of these pH sensitive (and expensive) membranes, enzymatic cleaning agents (e.g. the Ecolab products *P3-ultrasil 53, 54, 60b, 62* and *65*) have been developed and applied.

After the successful use of enzyme-containing recipes in the aforementioned fields of application, comprehensive assays for the use of enzyme-based recipes in further fields of application were primarily performed in the dairy sector. However, despite the positive results obtained, these assays cannot yet be considered as complete.

32.2 Enzyme-based cleaning procedures

32.2.1 Enzyme-based cleaning of the cold milk area

On the way from the producer to the consumer, milk in cartons or bottles, or in milk products such as yogurt or cheese, passes through many stations, where it is exposed to contamination from microorganisms and also to residues of preceding product batches.

Under the presumption of sufficient udder hygiene (cleaning and disinfecting the udder and the separation of the foremilk) the first possible contamination source for milk is the milk production plant, including the cooling container on holding. This area is under the responsibility of the production holding. A daily clean with conventional alkaline chemicals (and with acid chemicals at least once per week), is performed as well as additional, regular disinfections with boiling water. In newer milk production holdings, the use of a two-phase air-fluid flow with extremely high flow turbulence and correspondingly high wall shear stress in the pipelines, allowed considerable reductions in the volume of cleaning solutions used during cleaning by up to 70% (Reinemann *et al.*, 1998, Grasshoff and Reinemann, 1993).

The milk is pumped from the receiving container into the milk tanker. From this point of time the accepting dairy is responsible for its hygiene. The milk tanker, of highly complex geometric construction with inlet tube, pumps, de-aerator, quantity measurement, and automatic sampling devices, bears the risk that milk of irreproachable quality from the production holding may be contaminated by infected residues from a preceding milk batch. To exclude this contamination risk from remaining milk residues it is vital that the milk tanker is thoroughly cleaned.

In the dairy, the milk is filled from the milk tanker into a raw milk receiving tank, whereby milk is once more in touch with inlet tubes, pumps, collecting devices, pipelines, and tank walls, which may be contaminated by residues of preceding batches.

After the subsequent heat treatment of milk (pasteurization, ultraheating), further processes follow, including cooling, storage and filling of consumer milk, cheese making and ice cream production. All these areas distinguish themselves by a similar structure of soiling (in this context product residues are considered as soiling), which contain more or less original proteins, fat, lactose, milk salts and trace elements. For this reason they are an ideal substrate for – undesired – microorganisms, and must be regularly and carefully removed. The conventional way, which is usually successful, is to remove these residues via installed cleaning-in-place (CIP) plants using alkaline and acid chemical formulae.

As enzymes have been regularly used in the detergent industry since the 1960s for textile cleaning, efforts are always being made to enlarge the application fields of enzyme-based cleaning agents. In October 1994, Ecolab North America (www.ecolab.com) presented an enzyme-based cleaning method under the name '*Paradigm*' that has been used worldwide in the dairy, ice cream, meat/egg product industry. At the time of writing, the system '*P3-Paradigm*' is used in more than 10 large food companies in Germany and by a further 15 companies in Poland and in the Netherlands.

The *P3-Paradigm* system includes two separated process steps. In the first step the intrinsic cleaning is performed, followed by a disinfection step with the preparation *P3-paraDES*, necessary to kill microorganisms not removed during the cleaning step, but also to safely deactivate remaining enzyme residues. Before, between and after the individual steps, the plant is rinsed with potable water (Potthoff *et al.*, 1997).

For cleaning, a 0.1% aqueous solution from two concentrate components is manufactured, whereby one component contains buffer salts, complex-forming agents, and surface-active substances, the other the enzyme. The latter is found with a maximum 10% in the concentrate and up to 0.04% in the applied solution. The enzyme is a mixture of Savinase and Alcalase, classified under the proteases, produced by *Genencor International®, Rochester, NY 14618, USA*, and *Novo Nordisk A/S, 2880 Bagsvaerd, Denmark.*

The enzyme mixture operates like a biocatalyst at the decomposition of the whey proteins in water-soluble peptides, and does not consume itself during the process. After the fulfilment of its primary task it is once more available for the protein degradation. Therefore, the *P3-Paradigm* system can be applied either as a single-use cleaner, or as a reuse cleaner over several weeks. In the present case, the phase separation losses amount to approximately 10–20%. Their volume has to be supplemented accordingly. The application conditions typical for *P3-Paradigm* (pH 9.0–9.5, temperature 50–55 °C) are derived from the optimal operating conditions of the enzymes, cf. Section 32.3.2. Parallel to the enzymatic degradation of the proteins, the surfactants (biologically degradable, non-ionic fat alcohol ethoxylates/propoxylates are used) act upon the fat contained in the soiling, cover the fat globules with a bipolar double layer, and transmute them into a form that can be easily transported in the aqueous solution. The complex-forming agents also contained in the solution should finally eliminate, or impede, the deposit of mineral precipitations on the surfaces in touch with the product. Polyacrylates and phosphonates, supplemented by low quantities of NTA (nitrilotriacetic acid), serve as substrates. To avoid inhibition of the enzymes the complex-forming agents are buffered with a borate additive. The *P3-paraDES* solutions for final disinfection of the cleaned plants are generally used only once. After their use they are discarded.

The dosing of the concentrated components of *P3-Paradigm* and *P3-paraDES* to the solutions used and the undertaking of the mixing phase occur exclusively on a volumetric basis via adequate dosing devices. Contrary to the conductivity-monitored dosing of the chemicals in conventional cleaning processes, the electrical conductivity caused by the low substrate concentrations in enzymatic-based cleaning cannot be referred to as a control factor. At pre- and post-rinsing with cold water the temperature can alternatively be used for recognizing the phases, or the mixing phases. The realized water volumes are recorded via inductive flow measurement – as far as the water has a determined minimal water hardness (approx. 4 dH (40.00 mg CaO/l)), thus having a required minimal conductivity for the operation of an inductive flow measurement. If exhaust vapours are used for the mixture of enzymatic cleaning solutions, the volumes have to be measured with mechanically operating systems as the conductivity falls below the minimum conductivity.

P3-Paradigm does not contain phosphates, nitrate or chlorine and the surfactants used are completely degradable. Owing to the low concentrations (approx. 1/10 of the comparable chemical quantity of conventional cleaning processes) the contribution of the contained chemicals to the biological/chemical oxygen demands (BOD/COD) content of the waste water is negligible. The application solutions are not skin-irritating, non-toxic, and have no corrosive effect, which affects the lifespan of the valve seals, particularly in the polymer rotors in the impeller pumps of the milk inlet tubes in the bulk tankers. Furthermore, no negative effects have been observed during the waste-water treatment.

In Germany, the enzyme-based process for cleaning the cold milk zone has been used for between 3 and 7 years. After the required technical re-equipment of the plant, and after the training of the personnel (particularly drivers of the bulk tankers) by the supplier of the enzymatic cleaning agent, there have up to now been no disorders related to *P3-Paradigm*. As a security measure for the long-term use of the enzymatic cleaning process, Ecolab recommends a conventional basic cleaning once a year, both for the product transporting devices, and for the tanks and components of the CIP plants. The cleaning effect should also be regularly monitored by microbiological controls and via ATP measurement. If these requirements are fulfilled, the enzymatic process for cleaning the cold milk zone in the dairy represents real progress allowing cost economics in the purchase and disposal of chemicals, for more secure handling and, above all, enhanced environmental aspects.

32.2.2 Enzyme-based cleaning of the milk heat exchanger

Structure of the deposits

While heating in the temperature range 70–80 °C (for pasteurization and short-term heating of milk) fouling may develop on the surfaces of the heat exchanger plates, which contain between 50 and 70% (mostly denatured) proteins, 30–40% minerals (of which 3–10% is calcium), and 5–10% fat. In the vertical direction, the foulings are non-homogeneous. Immediately after the heating starts, according to Belmar-Beiny and Fryer (1993), in only a few seconds, a thin, 4–10 μm thick, dense and transparent film, mostly from calcium phosphate with organic deposits, is formed on the metallic surface of the heat exchanger. It is only on this film that the visible, measurably thick proteinous foulings develop at a significantly lower velocity. Owing to the calcium content of milk (in non-heated milk, calcium links casein submicelles to casein micelles in the form of calcium phosphate bridges), stable water-insoluble casein whey protein complexes develop during heating by the deposition of denatured whey proteins on the casein micelles. To remove them from the solid sublayer complex, certain measures are required. Up to now the detection of the aforementioned measures occurs fundamentally on an empirical basis. Some detailed issues in the field of surface physics and chemistry remain unresolved, so it is still not possible to do without the empirical level, experimental studies aimed at developing new processes for removing deposits from heat exchanger plates.

Mechanisms of the deposit removal

Deposit removal from the surfaces in touch with the product do not happen spontaneously. Depending on the persistence of the soiling, it may take from a few seconds to several hours to remove the fouling (Grasshoff, 1983, 1988). Contrary to the general assumption, flow mechanics plays a relatively subordinate role in the cleaning of milk heating plants (Grasshoff, 1988, 1992). Its essential task is to convey the cleaning chemicals to the appointed place of their action, to ensure that no depletion of the chemical agent occurs due to chemical or metabolic reactions, and, last but not least, to prevent soiling removed from one spot of the wall attaching at another spot. The requirement for a minimum flow velocity or a minimum wall shear stress of the fluids during the cleaning of plants is not very helpful. The general rule is that fluid flows should display at each spot of the plant a clearly turbulent flow pattern (Hoffmann, 1984). The driving force is the chemical composition of the cleaning solutions, whereby a number of essential functions are transferred to water – apart from its role as carrier of the thermal and mechanical energy. Initially, the firmness of the protection matrix is lowered by integrating water molecules into it to such an extent that a relatively low fluid overflow is sufficient to separate individual soil particles, or also complete layers from their adhesion bonds with the solid walls (Grasshoff and Potthoff-Karl, 1996). As deposits are not water-soluble in the heat exchanger it is the task of chemistry to modify, or destabilize the structure of the soiling to such an extent that finally water can be used as an action medium for soil removal. For example, polarly acting chemical substances can wrap soil particles in an electrical double layer, so that, contrary to the effective direction of the Van der Waals forces, they can be spatially removed by some nanometers from the solid walls (Grasshoff, 1992). Thus, the adhesion bonds of the soiling to the solid wall can be weakened to such an extent that the kinetic energy of the aqueous cleaning solution flowing at a low velocity is sufficient to separate and take away the soiling from the base layer. The proteins of the deposit matrix have to be hydrolysed, and dissolved away by ingredients from the cleaning solution. Mineral components of the soil matrix are generally not attacked by the substances responsible for the degradation of the proteins. The mineral components have to be disintegrated via special ingredients (in conventional cleaning processes, active complex-forming agents are added to the alkaline cleaning solutions; Grasshoff and Potthoff-Karl, 1996). Synergetic effects may occur among the different ingredients of the cleaning solutions (Grasshoff, 1988).

The temperature of the cleaning solutions plays an important role in conventional cleaning processes. When using enzyme-containing chemical formulae, narrow limits are fixed to the temperature parameter by the temperature resistance, or by the temperature optimum of the available enzymes (cf. below). Exceeding the upper resistance threshold, which is different from enzyme to enzyme, results in its spontaneous deactivation, so that the temperature as an independent variable is available only to a very limited extent.

Sakiyama *et al.* (1998) reported the first assays on enzymatic cleaning of protein-soiled surfaces made of stainless steel. They soiled columns, which were packed with stainless steel particles, with β-lacto globulin or gelatin, and used different proteases to remove the soils thus created. Grasshoff (1999) performed practical trials on enzymatic cleaning of stainless steel surfaces on which predominantly proteinaceous foulings had developed as a consequence of the heat application during milk heating.

32.3 Laboratory trials of enzyme-based cleaning

32.3.1 Test soils and experimental procedure

The basis for conclusive trials for detecting the efficiency of cleaning chemicals or cleaning processes is the availability of soiled surfaces, which have to fulfil the following requirements:

- the soil has to correspond to that occurring in practice;
- the soil has to be reproducible; and

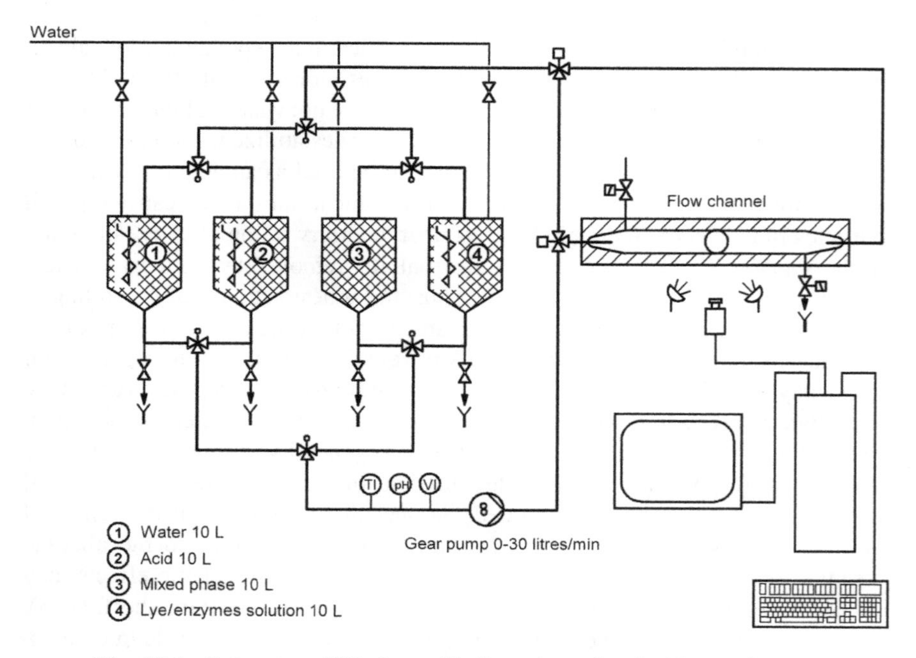

Fig. 32.1 Laboratory CIP plant with flow channel and video equipment.

- the soil removal properties should be more or less similar to those in practice.

Test coupons with soils complying with these requirements were produced in a heating plant on a laboratory scale (Grasshoff, 1999). In a heating trial over four hours, foulings from milk components were formed on the metallic coupons. Their chemical composition was very similar to that occurring in practice in milk pasteurizing plants. At the end of the heating the coupons were removed from the plant, rinsed with distilled water, and air-dried. For a longer period (several weeks), they could be stored without a measurable modification of the scaling-off properties of the fouling in subsequent cleaning processes.

The cleaning trials were performed in a flow channel with rectangular cross-section $3 \times 65\,\text{mm}^2$ and a frontage from acrylic glass, through which the cleaning solutions to be tested under defined conditions (temperature, concentration, flow velocity) were pumped. The test coupons to be cleaned were inserted in the rear ward channel wall. The scaling-off operation of the foulings was observed from the front side with a video camera the output signal of which was connected to a computer. Figure 32.1 displays the experimental set-up of the trial. On individual images, which were saved at regular intervals, the surface deposits were quantitatively detected via electronic image analysis within the 'area of interest' on the test coupons, in which the fouling had been removed up to the metallic surface.

32.3.2 Available enzymes and enzyme-containing ready-made recipes

The enzymes available for laboratory tests as well as their operating data delivered by the enzyme producer are presented in Table 32.1. For details about the enzymes, please refer to Grasshoff (1999).

For practical use, two ready-made recipes with the enzymes Savinase® 16.0 and Properase® L 1600 (Chemische Fabrik Dr. Weigert, Germany) were used. Apart from the enzymes, they contained stabilizers for improving storability and/or for pH regulation. The two recipes contained 5% enzyme and had to be applied in a 1% solution. In the first recipe, the pH value had to be adjusted to the recommended pH optimum for the enzyme by adding Na_2HPO_4 as a buffer substance and NaOH. The second recipe was adjusted to the process water (type and concentration of the water salts) available for the application, and contained, apart from the enzyme stabilizers, special surfactants, so that the solution could be produced directly, without any other measures, except for temperature adjustment, with the available process water.

32.3.3 Description of the enzyme 'Savinase'

A detailed description of the enzyme Savinase and of Savinase-containing recipes will follow as the enzyme had proven its high efficiency at the laboratory scale for the cleaning of milk heat exchangers (cf. results presented below). Technically, it is produced by submerse fermentation of genetically modified *Bacillus* strains. Its efficiency is based on its ability to hydrolyse water-insoluble

Table 32.1 Available enzymes for laboratory tests

Enzyme	Type	Temperature range (°C)	Optimum temperature (°C)	pH range	Optimum pH	Manufacturer
Alcalase®	Protease	<70	60	<11.5	9.0	Novo Nordisk*
Savinase®	Protease	<65	50	<12	9.5	Novo Nordisk
Esperase®	Protease	<75	60	<12	9.5	Novo Nordisk
Purafect® L	Protease	25–65	55	7–12	10.0	Genencor Int.**
Properase® L	Protease	25–65	55	7–12	9.5	Genencor Int.
Flavourzyme™	Endoprotease/ exopeptidase	<50			6.0	Novo Nordisk
CIPzyme™	Protease/ lipase	30–75		7–11		Novo Nordisk
Neutrase®	Protease	45–55			6.5	Novo Nordisk

* Novo Nordisk A/S, 2880 Bagsvared, Denmark (www.novonordisk.com).
** Genencor International, Inc., 925 Page Mill Road, Palo Alto, Ca 94304, USA (www.genencor.com).

proteins to peptides in proteinous soilings such as spots caused by grass, blood, fungi, faeces or food. The peptides thus produced are either completely soluble in the cleaning solution, or are removed with the washing solution as they are decomposed to sufficiently small elements.

Savinase belongs to the serine proteases, which occur in virtually all organisms, and which can be intracellular or extracellular enzymes. These proteins exist as two families, the trypsin-like and the subtilisin-like families. A serine residue, together with one aspartate and a histidine residue, forms a catalytic triad in which the nucleophilic serine O atom attacks the carbonyl C atom of the peptide bond of the substrate to give a tetrahedral covalent intermediate. The aspartate and histidine residues are involved in a hydrogen bond to the serine residue and form a charge-relay system (Lange *et al.*, 1994). Savinase belongs to the subtilisin family of serine endopeptidases, which have molecular masses of 26–29 kDa. Savinase is a highly alcalophilic member of the family secreted by *Bacillus lentus*. It consists of 269 amino acids, and the gene sequence is known (Betzel *et al.*, 1992). In the Chemical Abstract Service it is classified as Subtilisin, CAS No. 9014-01-1, in the catalogue of the International Union of Biochemistry it is listed in the class EC 3.4.21.62. Its substrate specificity is very broad and it displays maximum stability in the pH range 7 to 10 and high activity in the range 8 to 12 (Betzel *et al.*, 1992), thus creating considerable interest as protein-degrading additives to detergents in the washing-powder industry for the removal of proteinaceous soils. The crystallographic structure of Savinase is shown in Fig. 32.2 (Betzel, 2003). There are two calcium ions in the structure, which are responsible for the stabilizing effect of the subtilisin. The relatively high number of salt bridges is likely to contribute to its high thermal stability. The different composition of the S1 binding loop as

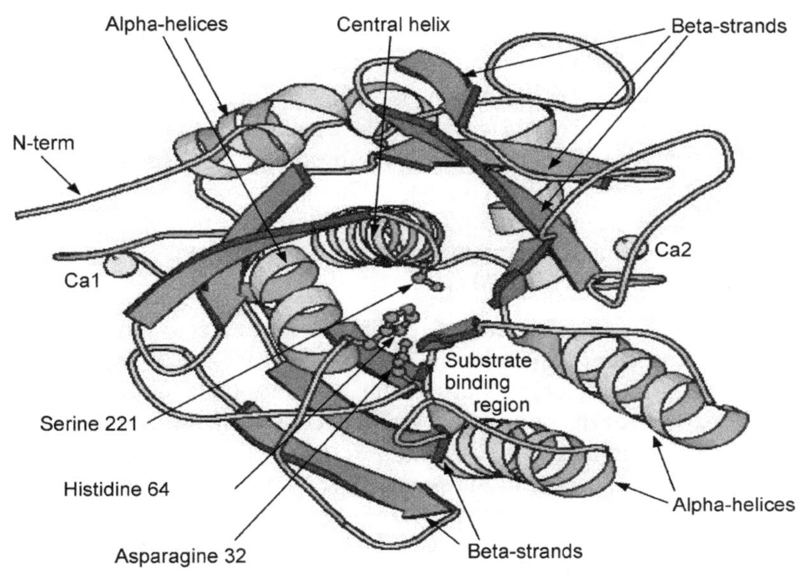

Fig. 32.2 The enzyme 'Savinase' with a view to the central helix. The active centre gets its geometric stability from this helix, together with the Ca2-ion at the surface – close to the active centre. The N-term, stabilized by the Ca1-ion, is important for the enzyme protection. The activity of the enzyme is controlled via the available calcium of the cell or the medium. For cleaning purposes the bonding of calcium, e.g. by complexing agents, must be avoided (Betzel, 2003).

well as the more hydrophilic character of the substrate binding region probably contribute to the alkaline activity of the enzyme (Betzel *et al.*, 1992).

32.3.4 Determination of the enzymatic activity

Enzymes or enzyme preparations are marketed in different forms, as granulates and in liquid form and, depending on the producer, with different active substance concentrations. Owing to the highly complex composition of the enzymes/enzyme preparations, it is nearly impossible to designate the absolute concentration in the individual product, and thus deduce the enzymic quantity required for practical use. During manufacture, standardized processes are applied for detecting the activity of each product batch. Only afterwards is the batch product adjusted to a guaranteed quantity of the active substance. For example, for the enzyme Savinase the producer *Novozymes A/S* indicates the activity in KNPU/g (Kilo-Novo-Protease-Units/gram), which refers to the degradation of dimethyl casein (DMC) at 50 °C, and a pH value of 8.3.

To determine the enzyme activity by the user the enzyme producer *Novozymes* indicates two different methods:

1. The method 2000-14216-01 (*Novozymes A/S, Denmark*, www.novozymes.com) based on the hydrolysis of the substrate *N,N*-dimethyl casein (DMC, delivered

by Novozymes) to low molecular peptides under standardized conditions. The free amino groups formed during hydrolysis react with 2,4,6-trinitrobenzene sulphonic acid (TBNS). The activity is calculated from the optical density photo metrically measured at 425 nm employing a calibration curve.

2. The method Luna no. 2000-10339-01 (*Novozymes A/S*), where the substrate azocasein (trade name, e.g., Sigma A 2765) is hydrolysed to low molecular peptides under standardized conditions. Undigested protein is precipitated with trichloracetic acid, and the quantity of digested protein is determined by spectrophotometry at 390 nm.

For Savinase, the activity under different temperatures and as a function of the pH was investigated at the producer's factory, *Novozymes A/S*, on a laboratory scale, applying haemoglobin as degradable substrate (data sheet 2001-04379-06.pdf at www.novozymes.com). The activity curves displayed in this data sheet expressly refer (according to *Novozymes A/S*) to the trial conditions fixed by the in-house laboratories. Applying substrates other than haemoglobin may result in deviations from this curve. Similar trials were also performed for the other enzymes produced at Novozymes. For the results, please consult the corresponding data sheets.

32.3.5 Results of the laboratory trials

Enzyme type

Trials for removing the test soilings with the enzymes presented in Table 32.1 were performed with the laboratory CIP plant described in section 32.3.1 (Figs 32.3–32.6). Hereby, the enzyme concentration of the applied solutions (10 L) were in a first step uniformally adjusted to 0.05% (related to the concentrate delivered by the producer) – independent of the indications in the data sheets about the individual enzyme activity. 8.0 ml of a surfactant mixture were added to the diluted solutions used. The pH value was adjusted to the optimum value given by the enzyme producer with diluted NaOH, and by adding a phosphate buffer. Starting the trials for scaling-off foulings with the enzymatic solutions with an acid pretreatment (0.5% HNO_3, 60 °C, 15 min) proved to be advantageous and preceded, therefore, all of the enzymatic trials. After an extensive intermediate rinsing (pH control in the rinsing water) followed the circulation of the enzyme-containing solution for a maximum period of 45 minutes. Subsequently, the coupons were removed from the plant, rinsed with distilled water, air-dried and photographed. Marginal fouling residues, which are transparent in a wet state and thus not recognizable, become clearly apparent on the dry metallic surface. Above all the 4–5 μm thin fouling under a layer known as the 'grey veil', neither soluble in NaOH nor in diluted acids – not even after circulation for several hours – is easily recognizable under appropriate lighting in the dry state. Scaling-off trials with 0.5% NaOH without other supplements in the tap water (17° dH) at 65 °C served as a reference.

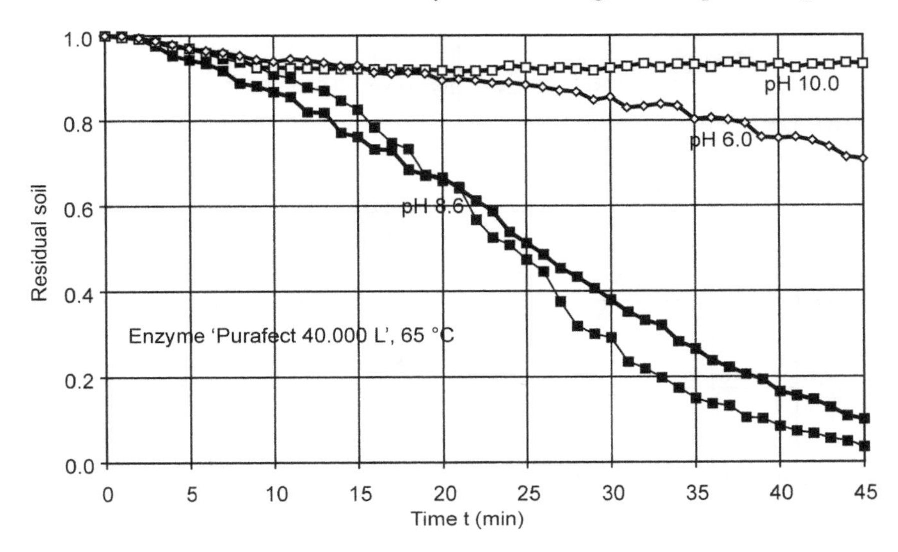

Fig. 32.3 The influence of pH on deposit removal with 0.05% *Purafect 40.000 L*®.

From the eight enzymes included in the study, the substances Savinase, Esperase, Properase L and Purafect L were found to be comparable to the 0.5% NaOH as regards scaling-off of the fouling from the test coupons after a 45-minute cleaning action. Flavourzyme showed no effect at all. With the enzyme Neutrase only 3–6% of the test fouling could be scaled off, while with the

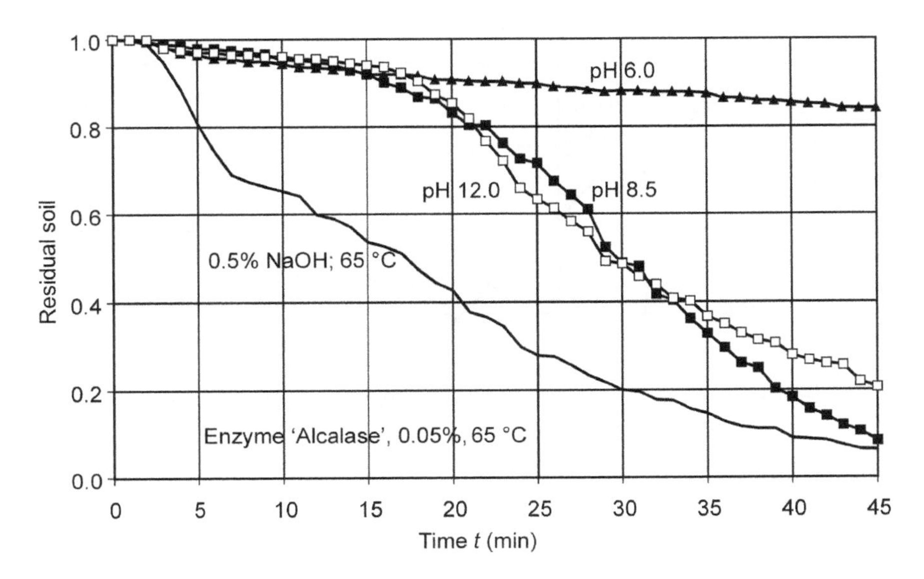

Fig. 32.4 The influence of pH on deposit removal with *Alcalase*®.

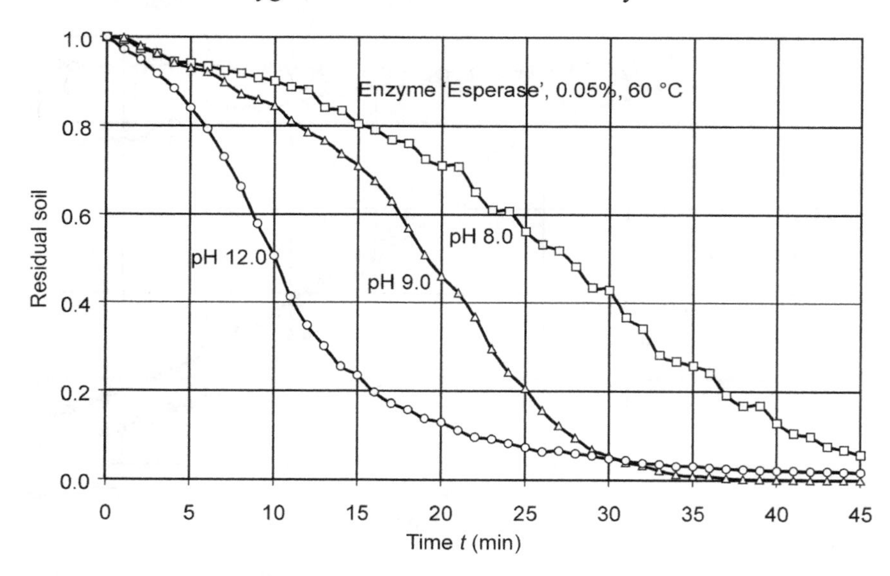

Fig. 32.5 The influence of pH on deposit removal with 0.05% *Esperase®*.

CIPzyme, between 10 and 20% of the test fouling could be scaled off. With the enzymes Savinase, Properase L and Esperase a metallic clean surface of the test coupons was obtained without any visible residues. This was not possible with 0.5% NaOH.

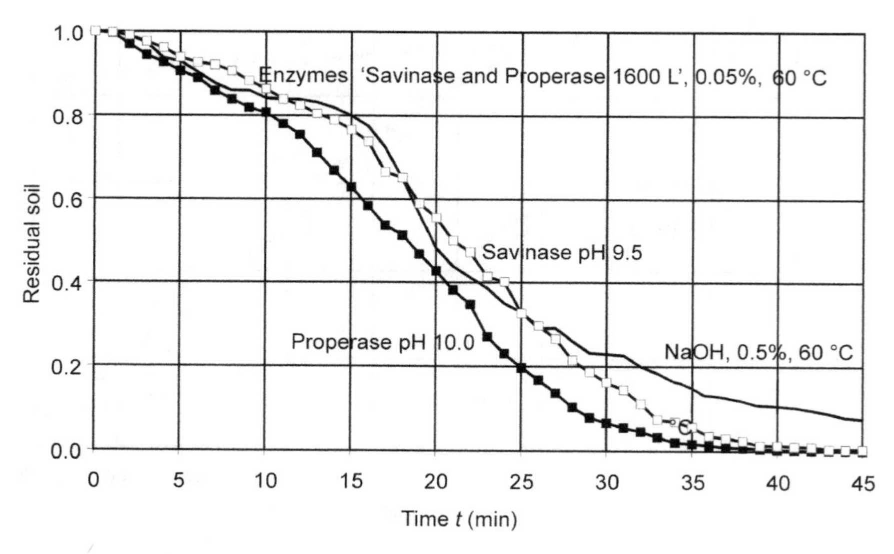

Fig. 32.6 Deposit removal with 0.05% *Savinase®* and *Properase 1 600 L®*, resp.

Enzyme concentration

Owing to the comparability of the preparations Savinase, Properase L and Esperase (concerning their characterization, production and also efficiency) the influence of the enzyme concentration on the scaling-off of the fouling was only analyzed for Savinase. For the trial series displayed in Fig. 32.7, 0.25, 0.50, 1.0, 2.5 and 5.0 mL of the liquid enyzme concentrate (corresponding to 0.0025%, 0.005%, 0.01%, 0.025% and 0.05%) were added to the solution used (10 L, 60 °C, pH 9.5).

A clear dependence of the kinetics upon the deposit removal, or of the remaining soiling, after 45 minutes on the test surface was seen, as long as the test surface did not have a metallic bright appearance before the expiration of a 45-minute trial. Whether an increase exceeding 0.05% brings about a further enhancement in deposit removal has not been investigated as for a possible application of the enzymatic cleaning process in practice, the costs of the chemicals have also to be taken into account, apart from other parameters. For a comparison with a conventional cleaning process, 0.05% as a justifiable upper limit seems to be reasonable.

pH

Figure 32.5 (Esperase) indicates a dependence of the deposit-removing effect of the enzyme upon the pH value of the solution. However, there is no unambiguous proof as to whether the increase of the deposit-removing kinetics can be deduced from the enzyme, or primarily from the NaOH, as the pH value of 12.0 corresponds to a 0.5% NaOH solution, which also has a deposit-removal effect

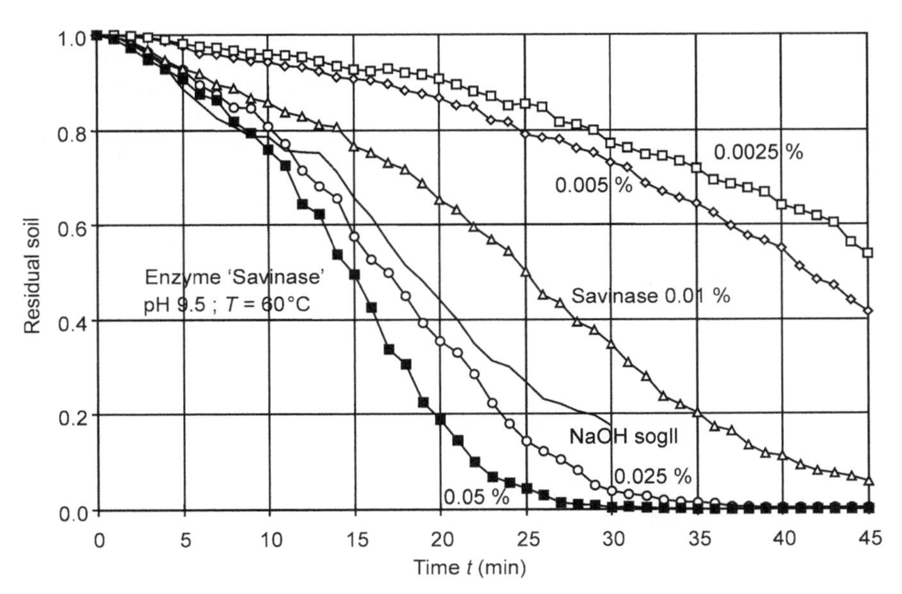

Fig. 32.7 Influence of enzyme concentration on deposit removal with *Savinase*®.

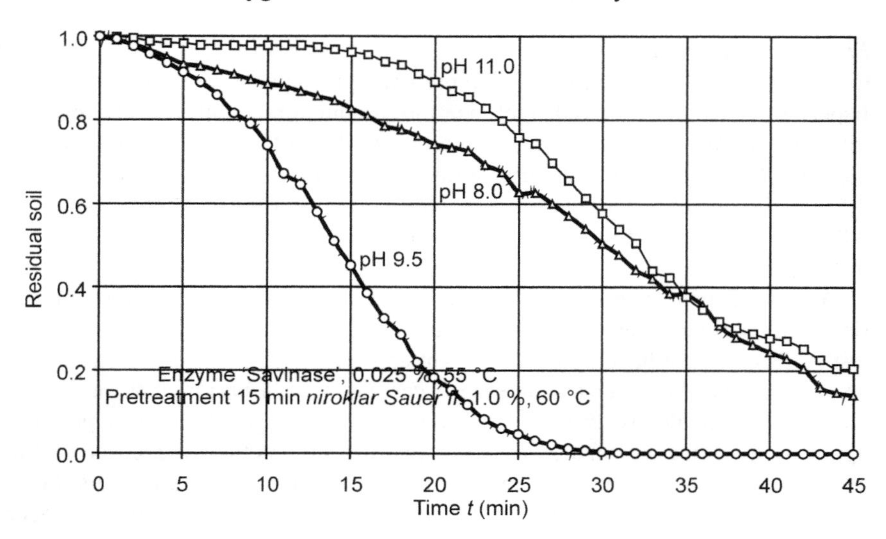

Fig. 32.8 Influence of pH on deposit removal with 0.025% *Savinase*®.

without the enzyme. As the aim of the enzyme input is to reduce the content of biologically non-degradable inorganic substances in the cleaning solution to a level that allows it to be discharged to the sewer without additional treatment, Esperase, with its impact optimum at pH 12.0, is eliminated from the range of enzymes adequate for the heat exchanger cleaning.

Figure 32.8 shows the dependence of the deposit removal for Savinase at an enzyme concentration of 0.025% and a temperature of 55 °C with an unambiguous impact optimum at pH 9.5, thus confirming the information in the data sheet of the enzyme producer. Both undershooting the optimum value by 1.5 pH units and also exceeding it by the same units result in a clear deterioration of the deposit removal. Nearly identical results were obtained for Properase, the impact optimum of which was detected at a slightly higher pH 10.0. Thus the products Savinase and Properase offered by the producers *Novo Nordisk* and *Genencor International* can be considered as being more or less equal.

Reuse of used enzyme solutions and storage of enzyme concentrates

For possible use in practice, it is of interest whether the enzyme-containing solution used can be stored hot after use and be reused for further cleans. To this end, a removal trial with a 0.025% Savinase solution was performed in the laboratory. After the end of the trial the solution was preserved for 1 h at the application temperature of 55 °C. Subsequently, a further test coupon, after an acid pretreatment, was pressurized in the hot-stored enzyme solution: the reused solution was practically inefficient. However, additional dosing of fresh enzyme concentrate after 15 minutes led to a spontaneous development of the deposit removal. Analogous behaviour was registered for the 0.05% Properase solution.

This means that when using enzyme-containing cleaning solutions in practice, their preparation directly before use is required, and also that they can be applied only once – unless a renewed dosing of the enzyme concentrate occurs. Through the addition of fresh enzymes, the solution can be used several times, thus saving water and energy. In this case, pH-regulating substances and surfactants do not need to be added again.

For the ready-made recipes provided by the *Chemischen Fabrik Dr. Weiger,* the influence of extended storage on their efficiency was investigated. Preparations of each of the enzymes Savinase and Properase 1600 L were stored at 6 °C for 21 months. After this period, fresh preparations of the same recipe were produced. Comparison trials were performed with the stored and the fresh recipes: neither the efficiency of the recipe with Savinase nor that with Properase had suffered during storage and the removal curves obtained from the stored and fresh recipes were more or less identical. Additional trials were performed with the aforementioned cool-stored recipes for a further 22 months storage and again, no significant decrease in activity was observed, i.e. the stabilized enzyme recipes can accordingly be held in stock in practice. A restriction is, however, that storage refers only to the concentrates. Diluted solutions have to be used immediately after their preparation.

32.4 Field trials

The laboratory trials have displayed that it is possible – after acid pretreatment – to obtain optically residue-free surfaces in the low-alkaline range pH 9–10 by using enzymatic preparations on the predominantly proteineous foulings of soiled surfaces of heat exchangers. In conventional cleaning processes with highly alkaline solutions (pH 13 and higher) this is only possible after addition of efficient complex-forming agents (Grasshoff and Potthoff-Karl, 1996), or in multistage cleaning processes (alkaline/acid) with a final oxidative step (most efficient: active chlorine in alkaline solution) (Grasshoff, 1988).

Therefore it was logical to undertake a practical field trial of the cleaning of milk heat exchangers with enzyme-containing preparations. Two dairies participated in this trial: dairy 'A' with a milk heat exchanger (VT 60 LOC N, GEA Ahlborn), with a performance of 45 000 L/h, through which 459 000 L cheese factory milk was filtered within a 10-hour operating period, and dairy 'B' with a milk heat exchanger (ZE 23 WP1, Alfa-Laval), with a performance of 25 000 l/h, through which 220 000 L consumer milk was manufactured within a 7½-hour operating period.

32.4.1 Dairy 'A'

Before the cleaning trial, the cleaning programme, which usually ran in the automatic mode, had been converted to the manual mode, and the signals of malfunction and automatic emergency shutdown were turned off. After pre-

Fig. 32.9 High-temperature section of a milk heater (VT 60 LOC N, GEA Ahlborn, Germany), enzymatically cleaned.

rinsing with water the cleaning was started at 60 °C with a 10 minute acid step (1.0% *P3 horolith CIP* a phosphoric acid product, Ecolab Deutschland GmbH, Düsseldorf). (During the acid pretreatment essentially the mineral components were extracted from the soil matrix, whereas the organic components were practically not dissolved. The acid solution, post-acidified if required, can be saved for multiple use. Phosphoric acid as well as nitric acid, or products based on these chemicals, are well suited for this purpose.) After depleting the plant and after a careful intermediate rinsing the plant was filled with water. For adjusting the pH to 9.1, the buffer solution and NaOH had been added to this water, together with surfactant mixture, and ultimately the Savinase containing enzyme concentrate (enzyme concentration in the solution used: 0.05%). The enzyme containing cleaning solution was circulated for 45 minutes at 57–60 °C through the circuit (circular volume approx. 5000 L), and subsequently discharged to drain. To deactivate any remaining residues of the enzymatic cleaner in the circuit, the system was filled with horolith CIP (nitric acid product, Ecolab Deutschland GmbH). After displacement of the acid by cold fresh water the heat exchanger was opened. All the product flow paths of the plate device were irreproachably visibly clean (Fig. 32.9).

32.4.2 Dairy 'B'

As for Dairy 'A', the electronically controlled cleaning programme including the signals for malfunction and the automatic emergency shutdown measures were switched off, and the cleaning was manually controlled (at the PC). After five minutes pre-rinsing with water the acid cleaning was started at 55 °C (15

Fig. 32.10 High-temperature section of a milk heater (ZE 23 WP1, Alfa-Laval, Germany), enzymatically cleaned.

minutes *niroklar Sauer flüssig* 1.0% nitric acid product from the Chemischen Fabrik Dr. Weigert, Hamburg, Germany). After careful intermediate rinsing (pH control in the outflow) the plant was filled with water. A ready-mixed enzyme solution was added to the circulation volume of 1500 L. The pH value, which should be in a range between 9.0 and 9.5 due to the added formula, reached approx. 8.2 and had thus to be slightly corrected by adding a small quantity of NaOH. The enzyme-containing cleaning solution was circulated for 45 minutes through the circuit, and subsequently discharged to drain. After six minutes water rinsing the cleaning was stopped and the heat exchanger opened. All the product flow paths of the plate device were visually clean (Fig. 32.10). Wipe tests with a clean white linen cloth did not hint of possible fouling residues. In the same way, but without daily disassembly, the heat exchanger was operated consecutively for 4 days between 7 and 10 hours, to be subsequently enzymatically cleaned. At the end of day 4, the plate heat exchanger and the separator were opened for control purposes. Once more all the product flow paths and the surfaces in touch with the product of the separator were visually clean. None of the microbiological samples taken from the heated product during the entire trial period revealed abnormal findings.

32.5 Risks

Apart from the function of the enzymes as cleaning agents it is important for the dairy, particularly for the manufacturers of yogurt or cheese, to understand the risks that may arise with the conversion from conventional cleaning to enzymatic cleaning, e.g., whether a change or disturbance of the rennetting ability of the process milk is to be expected if the milk is contaminated by residues of the enzyme solution. To answer this question, the 'worst case in the dairy', namely the intermingling of cheese factory milk with enzyme residues from the heat exchanger cleaning, was simulated in a laboratory trial (Grasshoff, 1999). The trial was expressly done without an acid activation of the enzyme residues, other than in practice. It was presumed that 10 L of a 1% enzyme recipe remained in the heat exchanger system and that these were intermingled with 10 000 L cheese factory milk. This corresponds to a concentration of 1.0 mg of the enzyme recipe in 100 mL milk. In the laboratory trial pasteurized milk was mixed with a differently diluted enzyme recipe, and was left for 30 min at 34 °C for simulation of pre-ripening. Afterwards, rennet enzyme was added and stirred. From the mixture, 1 mL was filled into each of two cuvettes of a lacto dynamograph (device for measuring coagulation; Hellige Thrombelastograph, Germany). With this device operating in a similar way to a rotation viscometer, the periods R_0 up to coagulation, and K_{20} up to an amplitude of 20 mm were detected on the chart recorder as well as amplitude A_{10} in mm 10 min after the beginning of coagulation.

From an enzyme preparation containing 5% *Savinase,* four different dilutions were prepared and added to a given volume of the cheese factory milk. The samples contained 0.1 mg, 0.5 mg, 1.0 mg and 10.0 mg of the enzyme recipe per 100 mL milk. In all the samples the coagulation behaviour was analysed, and compared with the results of a milk sample without enzyme addition that served as a blind sample before and afterwards. Against the zero sample a shortened R_0 period (11.0 min instead of 19.5 min) was measured up to the beginning of the rennet coagulation only in the sample with 10 mg of the Savinase recipe. The K_{20} and A_{10} values displayed no modification against the reference sample. The preparations with 0.1, 0.5 and 1.0 mg enzyme recipe/100 mL milk displayed no significant difference versus the blind sample. Enzyme supplements in this order of magnitude may have no influence on the rennet behaviour of milk.

Coagulation trials were also performed with a preparation containing 5% *Properase 1,600 L*, namely with concentrations 1, 10 and 50 mg/100 mL milk related to the enzyme recipe. Hereby, the first dilution showed no deviation from the zero sample. The addition of 10 mg enzyme recipe to 100 mL milk caused a decrease in the coagulation period R_0 from 16.0 min to 12.5 min, the A_{10} value slightly increased from 14 to 16 mm, and the K_{20} value was minimally shortened from 14.0 min to 12.5 min. The addition of 50 mg of the properase recipe led to a thickening of the sample (before the addition of the proper rennet enzyme), so that the subsequent lacto dynamographic measuring could not be performed. The milk had already coagulated before its filling into the cuvettes of the measuring device. Thus the enzyme, when it comes in high concentration into the milk, can result into serious disorders of the rennet coagulation behaviour.

However, it is unlikely that contamination caused by disorders of the cheese factory milk could occur in practice under the aforementioned order of magnitude. One reason is that before and after the cleaning step with an enzymatic solution a rinsing step with water is performed. The other reason is that the (diluted) enzyme containing solution suffers obvious losses in activity during the 45-minute application. If residues of the solution remain in the system, e.g. over night, it has to be assumed that they have no more enzyme activity the day after. If, as practised in the field trial in dairy 'A' – a nitric acid post-rinsing step is performed the enzyme activity decreases, according to the data sheets for the enzyme *Savinase*, immediately to a low, negligible size.

The conclusion from the laboratory trials with the simulated worst case in the dairy is that the enzymatic cleaning of the milk heat exchanger is unlikely to bear additional, unpredictable risks in the cheese-making process or the yogurt manufacture.

32.6 Future trends

The processes for enzyme-based cleaning of equipment and plants in the food sector, namely cleaning processes in the egg and meat processing industry, in ice cream manufacturing as well as in the cold milk department of the dairy can be considered as sufficiently tested, and established. In these areas the soilings primarily contain non-denatured animal proteins, which are recognized by many enzymes and degraded without any problems. Using cleaning solutions at temperatures of approx. 50 °C and in the pH range of approximately 9.0 means a considerable relief for the environment. Waste waters with this pH can be discharged in a non-treated state into the sewage treatment plant as the waste load of effluents contained in the waste waters as well as possible residues from the enzyme recipes are considered to be completely biodegradable. Additionally, at the time of writing of this chapter, trials with a considerably simplified process are being performed in practice using a ready-mixed one-component recipe instead of the more usual two-component recipe, which is added to the cleaning solution immediately before use via separated dosing devices. If the trials in practice are successful, the enzyme-based cleaning process for the aforementioned application areas will be considerably more user friendly than those currently used. This concerns the expenditure for monitoring measures and the stock-keeping of chemicals and, last but not least, the handling.

The cleaning of the heat exchangers in the dairy is more problematic as the fouling on the heat exchanger plates differs from the soiling in the cold milk area. The fouling of the heating plants in the dairy is tightly structured, it adheres firmly to the solid surfaces, and contains (apart from inorganic substances, which are not enzymatically degradable) primarily denatured proteins. These denatured proteins, are not recognized by most of the specifically acting proteases, limiting the number to only a few known proteases that can be used for the cleaning of milk heat exchangers.

Although the cleaning trials of the milk heat exchanger performed in the laboratory, as well as the field trials in practice achieved excellent results, they have not yet been used in practice. One possible reason may be that in most dairies the cleaning via a central CIP station occurs with 'universal chemicals' – *one* acid and *one* alkaline cleaning agent for all the applications, at best in different concentrations. A prerequisite for the introduction of the enzymatic cleaning is the installation of decentralized cleaning stations with chemicals specifically adapted to the object to be cleaned. For the cleaning of the cold milk area, other processes and chemicals than those applied in the cleaning of heat exchangers have to be used. The cleaning of UHT plants has still to be undertaken conventionally as corresponding laboratory trials were not successful.

Conventional cleaning occurs automatically via computer-assisted programs. This assumes beforehand that the corresponding parameters (periods, temperatures, volumes, flow velocities, concentrations, but also valve functions) are available as measurable values or control factors. Above all, the mixing phases have to be safely recorded. In conventional cleaning the phase separation between product, water and different cleaning solutions occurs via measurement of cleaning solution electrical conductivity. In the conventional method this functions well as the measuring systems are reliable and largely maintenance-free, and as the conductivity of water, product and cleaning solution differ significantly from each other (tap water 0.50 mS/cm; milk 30–60 mS/cm; 0.5% NaOH approx. 25 mS/cm; 3% NaOH approx. 130 mS/cm; 0.5% HNO_3 30 mS/cm, 2% HNO_3 115 mS/cm). In the 1% solution of a ready-made Savinase recipe in tap water, the conductivity is only 0.80 mS/cm. Thus, it is extraordinarily difficult with conductivity measuring systems set up for measuring areas up to approximately 200 mS/cm, to differentiate in the zone near to zero between tap water with 0.50 mS/cm and enzymatic cleaning solution with 0.80 mS/cm, and to guarantee a safe separation of the phases. However, this problem can be solved by the installation of a pH measuring system. pH has its largest spread angle in the neutral or low alkaline area, whereas the electrical conductivity has a roughly linear curve. A clear distinction can be made between pH 7 (tap water) and pH 9 (enzymatic cleaning solution). Contrary to the conductometric measurement, pH measuring systems need regular control, cleaning and calibration. Nevertheless, robust process measuring systems are available on the market. They are equipped with automatically performed maintenance functions, so that they are comparable with more or less maintenance-free systems for electrical conductivity measuring.

The introduction of enzymatic cleaning in the food industry on an broader basis represents a considerable contribution to the recovery of the environment. Most recent trials with new chemicals or enzyme combinations promise an even broader application. Recently, the company *Chemie X 2000* in *Remscheid, Germany*, presented an enzymatic cleaning agent that has an effect comparable to that of a 0.5% NaOH at a pH of 7.25, namely in the absolute neutral range. Although this substrate cannot yet take up the challenge under the cost aspect with the enzyme-based cleaning agents developed for cleaning purposes in the food sector, it shows at least that the development in this field will go on.

32.7 References

ALLIE Z; JACOBS EP; MARTENS A; SWART P (2003), 'Enzymatic cleaning of ultrafiltration membranes fouled by abattoir effluent', *J. of Membrane Sci.* **218** (1–2) 107–116.

ARGUELLO MA; ALVAREZ S; RIERA FA; ALVAREZ R (2003), 'Enzymatic cleaning of inorganic ultrafiltration membranes used for whey protein fractionation', *J. Membrane Sci.* **216** (1–1) 121–134.

BELMAR-BEINY M T; FRYER PJ (1993), 'Preliminary stages of fouling from whey protein solutions', *J. Dairy Res.* **60** (4) 467–483.

BETZEL C (2003) 'Enzyminhibition und Drug-Design wie man die Wirkung von Proteinen modelliert', Arbeitsgruppe für Makromolekulare Strukturanalyse, www.uke.uni-hamburg.de/institute/biochemie/zesi/ag_betzel

BETZEL C; KLUPSCH S; PAPENDORF G; HASTRUP S; BRANNER S; WILSON K S (1992), *J. Mol. Biol.* **223** (2) 427–445.

GRASSHOFF A (1983), 'Modellversuche zur Ablösung festverkrusteter Milchbeläge von Erhitzerplatten im Zirkulationsreinigungsverfahren', *Kieler Milchwirtschaftliche Forschungsberichte* **35** (4) 493–519.

GRASSHOFF A (1988), 'Zum Einfluss der chemischen Komponenten alkalischer Reiniger auf die Kinetik der Ablösung festverkrustete Beläge aus Milchbestandteilen von Erhitzerplatten', *Kieler Milchwirtschaftliche Forschungsberichte* **40** (3) 139–177.

GRASSHOFF A (1992), 'Hygienic design – the basis for computer controlled automation', *Trans. IChemE*, **70** Part C, 69–77.

GRASSHOFF A (1999), 'Laborversuche zur Reinigung von Milcherhitzern mit enzymatischen Reinigungsmitteln', *Kieler Milchwirtschaftliche Forschungsberichte* **51** (4) 295–318.

GRASSHOFF A; POTTHOFF-KARL B (1996), 'Komplexbildner in alkalischen Reinigern', *Tenside, Surfactants, Detergents* **33** (4) 278–288.

GRASSHOFF A; REINEMANN D J (1993), 'Zur Reinigung der Milchsammelleitung mit Hilfe einer 2-Phasen-Strömung' *Kieler Milchwirtschaftliche Forschungsberichte* **45** (3) 205–234.

HOFFMANN W (1884), 'Wandschubspannung als Bezugsgrösse für die Strömungsmechanik beim Zirkulationreinigen von geraden Rohren', *Milchwissenschaft* **39** (11) 645–647.

KESSLER H G (1996), 'Ultrafiltration in der Lebensmitteltechnologie' in *Lebensmittel- und Bioverfahrenstechnik*, 72 ff., München, A Kessler.

LANGE G; BETZEL C; BRANNER S; WILSON K S (1994) 'Crystallographic studies of Savinase, a subtilisin-like proteinase, at pH 10.5', *Eur. J. Biochem.* **224**, 507–518.

POTTHOFF A; SERVE W; MACHARIS P (1997), 'The cleaning revolution', *Dairy Industries International* **62** (6) 25, 27, 29.

REINEMANN D J; WONG A C L; MULJADI A (1998), 'Efficacy assessment of CIP processes in milking machines', in Wilson D I; Fryer P J; Hasting A P M, *Fouling and Cleaning in Food Processing '98*, Office for Official Publications of the European Community, Luxembourg.

SAKIYAMA T; TOYOMASU T; NAGATA A; IMAMURA K; TAKAHASHI T; NAGAI T; NAKANISHI K (1998), 'Performance of protease as a cleaning agent for stainless steel surfaces fouled with protein', *J. Fementation Bioeng* **85** (3) 297–301.

33

Contamination routes and analysis in food processing environments

J. Lundén, J. Björkroth and H. Korkeala, University of Helsinki, Finland

33.1 Introduction to contamination analysis in the food industry

During the past decade, large changes have taken place in food processing, particularly the development of more and more ready-to-eat foods for consumers. The amount of refrigerated processed foods of extended durability (REPFED) also continues to increase. A number of different mechanical production stages are needed during the preparation of these foods, which makes these foods sensitive to contamination. The logistics are very important in order to get the products to consumers as individual packs. Therefore, packaging of REPFED products has been the target for intensive research, and in order to establish long shelf-lives, vacuum and modified atmosphere packaging (MAP) have been employed. Another important approach to increasing the shelf-life of products is cold storage and the maintenance of the cold chain at every production stage.

There is a trend among consumers to desire foods as mildly processed as possible and produced without any preservatives and food additives. This has led to minimal processing that reduces factors inhibitory to bacterial growth during preparation and storage. Heat treatment is associated with preparation of many REPFED products. The heat treatment does not, however, inactivate spores that can later cause safety or spoilage problems. On the other hand, heat treatment inactivates vegetative bacteria present in the raw materials, subsequently making products sensitive to post-cooking contamination and bacterial growth. Many factors associated with modern food processing increase food safety risks.

Changes in food production also influence the contamination and the growth of pathogenic and spoilage bacteria. The use of the cold chain and changes in packaging technology have caused the selection of certain pathogenic and spoilage bacteria. Psychrotrophic, anaerobic or facultative anaerobic bacteria such as *Listeria monocytogenes*, non-proteolytic *Clostridium botulinum* and pathogenic *Yersinia enterocolitica* and *Yersinia pseudoturberculosis* have become the most important pathogenic bacteria, and psychrotrophic lactic acid bacteria have become the most important bacterial spoilage population.

The processing lines of REPFED products are very complex, including different machines and conveyor belts. During preparation, the products are exposed to bacterial contamination. As food contamination is a complicated phenomenon, its cause is not easily traced. When pathogenic or specific spoilage bacteria are found in the final product, the contamination sources should be detected with a thorough contamination analysis.

Contamination research has been markedly developed since 1985. The most important development has been the application of DNA-based methods. In earlier times, the bacterial genus or species was determined, whereas now we can detect certain strains or clones within a species at different stages of processing. We can follow the whole production line and find at which stage the product becomes contaminated with the strains found in the final product. We can also determine the contamination of certain clones of virulent strains or specific spoilage microorganisms. The contamination research has developed and is still developing enormously. In the present paper we focus on the contamination analysis of *L. monocytogenes* and psychrotrophic lactic acid bacteria.

33.2 Different types of contamination analyses

33.2.1 Conventional microbial detection and enumeration analyses and their limitations

Before the genomic era and the use of molecular methods, bacterial contamination in food processing was studied using culture-dependent microbial techniques. Contamination load was estimated by different enumeration techniques and pathogen detection by conventional, usually enrichment-based culturing methods. Today, contamination analyses employ a mixture of culture-based and molecular approaches. This development has been welcome since the limitations of enumeration analyses and the lack of specific information involved in the bare species level detection of a contaminant has clearly been elucidated in association with certain industrial food hygienic problems.

Increasing information on specific spoilage organisms has revealed that total bacterial counts have limitations in showing the relevant sites and sources of bacterial contamination in food processing. Our knowledge has improved our understanding of the microbial ecology of foods. Through this knowledge we can nowadays better identify the relevant spoilage organisms associated with different foods, as not all contaminants have the same ability to cause food spoilage. This can clearly be seen in meat production as it has been estimated that only 10% of the total bacterial load contaminating the carcasses is capable of growing during refrigerated storage (Nychas *et al.*, 1988). Moreover, the meat is usually packaged under vacuum or modified atmosphere (MA), creating a further hurdle for contaminating bacteria. This results in the growth of certain bacterial groups, such as psychrotrophic lactic acid bacteria (LAB). In a recent study (Björkroth *et al.*, 2005) concerning MA packaged, marinated broiler legs, it was shown that the majority of initial LAB contaminants (enterococci) were overgrown during storage at 6 °C by mainly heterofermentative rods spoiling the product. Only 27% of the initial LAB contaminants, forming just a minor fraction of the total bacterial load, would then be relevant to trace during a contamination analysis of this product. These situations show the limitations of the approaches monitoring total bacterial counts or even LAB counts. In a contamination analysis of specific spoilage bacteria, we are quite often facing a 'needle in a haystack' situation, which is challenging our methodology. In the worst case, the main spoilage-causing bacterial group may not even be known.

Monitoring of pathogenic bacterial contamination in food processing sets different challenges from the corresponding analyses of spoilage bacteria. Sometimes detection analyses serve well in tracing contamination sites or source monitoring. However, the nature of pathogen contamination is quite variable depending on the properties of the contaminant. Sometimes traditional or polymerase chain reaction (PCR)-based species detection approaches do not show the whole picture of the problem. In the case of *L. monocytogenes,* persistent strains may cause prolonged product contamination, which should be revealed by strain typing techniques. These aspects are dealt with in more detail in Section 33.3.

Despite the limitations of conventional enumeration and detection, these techniques still form the cornerstone of contamination studies. Since the majority of bacterial typing techniques still rely on culturing of the organisms, however, good knowledge of these methods should be maintained. We are currently quite far from the situation where culture-independent techniques may be applied in routine monitoring of microbial contaminants in industrial food processing.

33.2.2 Molecular-based contamination analysis

The development of molecular-based typing methods has revolutionized the field of industrial contamination analysis. Before the era of these methods, contamination analysis was based on isolation and detection, which left many questions open concerning contamination routes. The characterization of species to strain level has enabled researchers to analyse the relatedness and epidemiology of pathogenic and spoilage microorganisms (Björkroth *et al.*, 1996, 1998; Miettinen *et al.*, 1999; Lundén *et al.*, 2003a). This information has increased our understanding of contamination routes to and within the food processing industry.

Several different molecular typing methods have been used in industrial contamination analysis. Most of them are based on the physical characterization of molecules such as proteins or DNA. There are differences between the molecular typing methods, which the user should be aware of. A good molecular typing method should have a high discriminatory power in order to recognize different subtypes of a wide range of species (typeability) and it should be reproducible. Furthermore it should be rapid and economical. However, these characteristics are not usually all met in a single typing method and the user has to evaluate which characteristics are most important.

DNA macrorestriction analysis by pulsed field gel electrophoresis (PFGE) typing is considered to be a feasible method for subtyping many different microorganisms. This method has a very high discriminatory power, good typeability and reproducibility (Brosch *et al.*, 1996). The method is based on the cutting of chromosomal DNA into large fragments with rare cutting restriction endonucleases. The fragments are run in a pulsed-field gel electrophoresis allowing separation of large fragments (Schwartz and Cantor, 1984). PFGE typing has been successfully used in contamination analysis of e.g. *Listeria monocytogenes*, *Yersinia enterocolitica*, *Escherichia coli* O157 and lactic acid bacteria (Björkroth *et al.*, 1996, 1998; Autio *et al.*, 1999; Fredriksson-Ahomaa *et al.*, 2000; Lahti *et al.*, 2002).

Ribotyping shows a good typeability, but the discriminatory power is not as high as in PFGE typing (Swaminathan *et al.*, 1996; Aarnisalo *et al.*, 2003). An advantage of the method is that it can be automated and large amounts of isolates can be typed in a short time. The method is based on the restriction of DNA, followed by gel electrophoresis of obtained fragments. A subset of fragments is visualized by hybridizing with a labelled ribosomal RNA (Stull *et al.*, 1988). Ribotyping has been used in several contamination analysis studies (Björkroth and Korkeala, 1996; Aarnisalo *et al.*, 2003).

Other molecular typing methods used in contamination analysis are multilocus enzyme electrophoresis (MEE) typing, restriction endonuclease analysis (REA) and random amplification of polymorphic DNA (RAPD). MEE is based on the separation of the soluble metabolic enzymes in a gel electrophoresis, and the method shows good typeability, but poor discriminatory power (Caugant *et al.*, 1996). In REA chromosomal DNA is cut with frequently cutting restriction endonucleases. REA has a high discriminatory power, but may result in numerous fragments, which complicates the interpretation (Gerner-Smidt *et al.*, 1996). RAPD is a PCR-based typing method (Williams *et al.*, 1990), which shows a high discriminatory power (Boerlin *et al.*, 1995). However, obtaining reproducible results is problematic (Wernars *et al.*, 1996), which limits the use of the method in contamination analysis. Recently, amplified fragment length polymorphism (AFLP) typing (Fonnesbech Vogel *et al.*, 2001; Autio *et al.*, 2003; Keto-Timonen *et al.*, 2003) has been used in contamination analysis. AFLP is a PCR-based typing method with a discriminatory power similar to PFGE (Fonnesbech Vogel *et al.*, 2001; Keto-Timonen *et al.*, 2003).

33.3 *Listeria monocytogenes* contamination in food processing environments

Listeria monocytogenes is an exceptionally well-adapted pathogen to food processing environments. The prevalence of the organism may be high in different food processing plants (Table 33.1), and the organism can survive

Table 33.1 Prevalence of *Listeria monocytogenes* in food processing plants

Food processing plant/ sampling site	*Listeria monocytogenes* (%)	Country	Reference
Meat processing plant			
Environment	2.3	France	Chasseignaux *et al.* (2001)
	14	France	Chasseignaux *et al.* (2002)*
	4	France	Chasseignaux *et al.* (2002)*
Equipment	13	Greece	Samelis & Metaxopoulos (1999)
	2.8	Greece	Samelis & Metaxopoulos (1999)
	25	France	Chasseignaux *et al.* (2001)
	53	France	Chasseignaux *et al.* (2002)*
	1.4	France	Chasseignaux *et al.* (2002)*
Poultry processing plant			
Environment	22	UK	Hudson & Mead (1989)
	29	Ireland	Lawrence & Gilmour (1994)
	15	Ireland	Lawrence & Gilmour (1994)
	22	France	Chasseignaux *et al.* (2001)
	55	France	Chasseignaux *et al.* (2002)*
	27	France	Chasseignaux *et al.* (2002)*
Equipment	29	UK	Hudson & Mead (1989)
	20	Finland	Miettinen *et al.* (2001b)*
	16	France	Chasseignaux *et al.* (2001)
	27	France	Chasseignaux *et al.* (2002)*
	0	France	Chasseignaux *et al.* (2002)*
Fish processing plant			
Environment	29	Norway	Rørvik *et al.* (1995)
	11	Finland	Autio *et al.* (1999)
	28	USA	Norton *et al.* (2001)*
Equipment	24	Finland	Autio *et al.* (1999)
Dairy processing plant			
Environment	9.3	France	Jacquet *et al.* (1993)
	4.7	Finland	Miettinen *et al.* (1999)
	0	Brazil	Silva *et al.* (2003)
	14	Brazil	Silva *et al.* (2003)
Equipment	11	France	Jacquet *et al.* (1993)
	5.1	Finland	Miettinen *et al.* (1999)
	0	Brazil	Silva *et al.* (2003)
	0	Brazil	Silva *et al.* (2003)

* The prevalence includes the results of two or more plants

harsh elimination procedures in food processing environments. As these procedures, including different disinfection actions, have not always succeeded in their task there has been a need for novel approaches in the elimination of *L. monocytogenes*. Molecular typing methods have offered new tools in the fight against *L. monocytogenes* as they have enabled researchers to trace the organism in the food processing environments and point out sites of contamination, which in turn has enabled the targeting of preventive measures.

33.3.1 Characteristics of *Listeria monocytogenes* contamination of food processing environments

Listeria monocytogenes contamination can be divided into two major events: (1) contamination of the food processing environment and (2) contamination of the product (Autio *et al.*, 2004). The contamination of the processing environment can further be divided into: (a) contamination originating from the outside of the plant by raw materials and (b) persistent contamination originating from the inside of the plant.

Contamination of the food processing environment can occur through different ways, which are highlighted in Fig. 33.1. Raw materials are frequently contaminated with *L. monocytogenes* (Johansson *et al.*, 1999; Heredia *et al.*, 2001; Miettinen *et al.*, 2001b) and are probably the most important source of contamination (Lawrence and Gilmour, 1995; Berrang *et al.*, 2002). It is therefore not surprising to find *L. monocytogenes* during processing in the raw area of the plant. The presence of *L. monocytogenes* in the raw area requires strict hygiene barriers between the raw area and the high hygiene areas, especially the post-heat treatment area, in order to stop the spread of contamination. Poor separation or no separation has been shown to increase contamination in the post-heat treatment area in meat processing plants (Lundén *et al.*, 2003a).

Although the initial contamination of the raw area in food processing plants appears to be caused by contaminated raw materials, the contamination of the processing environment and especially the post-heat treatment area is more

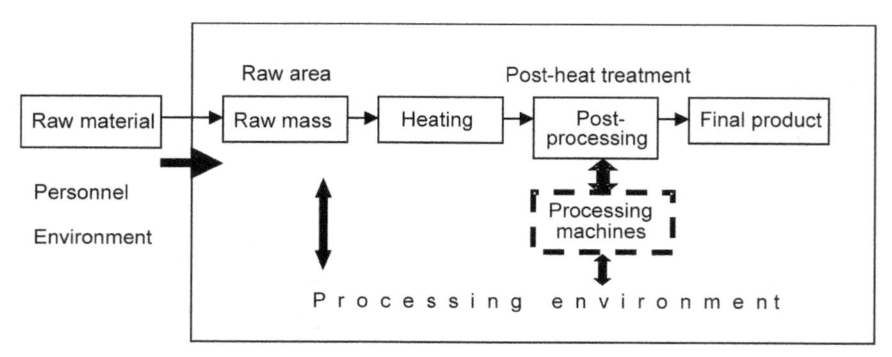

Fig. 33.1 *Listeria monocytogenes* contamination in food processing plants. Block arrows indicate contamination routes of *L. monocytogenes*.

Table 33.2 Properties of *Listeria monocytogenes* associated with persistent contamination

Property	Reference
Enhanced adherence to food contact surfaces	Norwood and Gilmour (1999), Lundén *et al.* (2000)
Disinfectant resistance	Aase *et al.* (2000), Lundén *et al.* (2003b)
Cadmium resistance	Harvey and Gilmour (2001)
Monocin E production	Harvey and Gilmour (2001)
Acid resistance	Lundén (unpublished)
Heat resistance	Lundén (unpublished)

complex. The contamination of the processing environment appears often to be due to *L. monocytogenes* strains already present in the plant environment (Hoffman *et al.*, 2003), and not to continuous ingress of *L. monocytogenes*. There have to be, therefore, other factors causing contamination of the processing environment.

A very particular characteristic of *L. monocytogenes* contamination in food processing environments is the persistence of the contamination (Rørvik *et al.*, 1995; Unnerstad *et al.*, 1996; Miettinen *et al.*, 1999; Lundén *et al.*, 2003a; Autio *et al.*, 2003). *L. monocytogenes* has been observed to occupy food processing plants for months or even years. Even if some of the persistent contaminations can be explained by continual ingress, not all can. The persistent contamination has thus not been eliminated with routine sanitation measures. Molecular typing of *L. monocytogenes* isolates has shown that all or some of the genotypes found in raw materials are not seen in the post-heat treatment area or final products (Rørvik *et al.*, 1995; Nesbakken *et al.*, 1996; Autio *et al.*, 1999; Tkáčiková *et al.*, 2000; Lundén *et al.*, 2002; Norton *et al.*, 2001; Hoffman *et al.*, 2003). This indicates that the persistent strains are true persisters, which survive the elimination procedures in the food processing plant. Therefore, two subpopulations exist in the food processing environment: persistent and non-persistent (or transient) strains. This fact has led researchers to seek further understanding of the development of persistent strains. Some factors associated with persistent *L. monocytogenes* strains are presented in Table 33.2. One very important factor is the ability to adhere to food contact surfaces. This ability is enhanced in persistent strains (Norwood and Gilmour, 1999; Lundén *et al.*, 2000).

33.3.2 *Listeria monocytogenes* contamination of products

The contamination of products with *L. monocytogenes* occurs most typically during post-processing (Fig. 33.1). The processing environment provides several harbourages for *L. monocytogenes*, which in turn serve as the source of contamination of finished products (Autio *et al.*, 1999; Norton *et al.*, 2001; Lundén *et al.*, 2002). Several contamination sites have been identified in molecular contamination analyses.

The contamination sites in different food processing plants for the most part are the same, with processing machines standing out. It has been shown by molecular typing that identical *L. monocytogenes* genotypes can be found in the processing machines and the product that has been manipulated by that machine or in the product waste of such machines. These findings strongly indicate that the processing machines have transferred the contamination to the products (Nesbakken *et al.*, 1996; Autio *et al.*, 1999; Suihko *et al.*, 2002; Lundén *et al.*, 2003a).

Processing machines that are complex and difficult to take apart and therefore sanitize have been found problematic (Table 33.3). In particular, brining machines and particularly recirculating brining machines are difficult to clean (Autio *et al.*, 1999; Greer *et al.*, 2004). Other challenging processing machines are packing machines, freezers and conveyors (Autio *et al.*, 1999; Miettinen *et al.*, 1999; Hoffman *et al.*, 2003) and machines that reduce product size, e.g. slicers and dicers. The difficulty of eliminating *L. monocytogenes* from processing machines is well illustrated in the case where the contamination was transferred with a sanitized dicing machine from one plant to another. The contamination was caused by a persistent strain, which had an increased ability

Table 33.3 *Listeria monocytogenes* contamination sites in food processing plants

Contamination site	Reference
Processing environment	Autio *et al.* (1999), Chasseignaux *et al.* (2001), Berrang *et al.* (2002)
Cutter	Chasseignaux *et al.* (2001)
Tumbling machine	Samelis *et al.* (1998), Samelis and Metaxopoulos, 1998)
Cutting table	Norton *et al.* (2001)
Trimming	Eklund *et al.* (1995), Johansson *et al.* (1999)
Brining machine	Autio *et al.* (1999), Johansson *et al.* (1999), Miettinen *et al.* (2001a)*
Brining solution	Eklund *et al.* (1995), Autio *et al.* (1999), Norton *et al.* (2001)
Skinning machine	Autio *et al.* (1999), Suihko *et al.* (2002), Miettinen *et al.* (2001b)
Slicing machine	Autio *et al.* (1999), Johansson *et al.* (1999), Suihko *et al.* (2002)
Dicing machine	Lundén *et al.* (2002)
Conveyor belt	Salvat *et al.* (1995), Giovannacci *et al.* (1999), Miettinen *et al.* (2001b)
Cold smoker	Norton *et al.* (2001)
Chilling machine	Miettinen *et al.* (2001b)
Cooler	Hoffman *et al.* (2003)
Freezer	Pritchard *et al.* (1995), Lundén *et al.* (2003a)
Mould	Jaquet *et al.* (1993), Salvat *et al.* (1995)
Packaging machine	Johansson *et al.* (1999), Miettinen *et al.* (1999), Lundén *et al.* (2003)
Milk filler	Pritchard *et al.* (1995)
Brine pre-filter	Pritchard *et al.* (1995)

* *Listeria* spp.

to adhere to surfaces (Lundén *et al.*, 2002). The contamination was finally eradicated after thorough disassembly of the machine and harsh alkali–acid–alkali treatments.

33.4 Psychrotrophic lactic acid bacterium contamination in meat processing environments

33.4.1 Lactic acid bacteria and microbial ecology of packaged meat products

Vacuum or modified atmosphere packaging is commonly used for increasing the shelf-life and hygienic handling of meat products. Owing to the anaerobic atmosphere created by these packaging techniques, LAB are the major spoilage-causing bacteria in these products (Borch *et al.*, 1996). Some LAB are also resistant to the effect of smoke and nitrite and tolerate relatively high NaCl concentrations which enables them to grow also in cooked meat products, such as cured ham and sausages (Korkeala and Björkroth, 1996).

Metabolic activity of LAB produces the typical spoilage changes making products eventually unfit for human consumption. Sour flavour and off-odour, formation of CO_2 resulting in bulging of packages, slime formation and discolorations are typical sensory changes involved in LAB meat spoilage. Spoilage changes are usually delayed until the stationary growth phase of LAB (Reuter, 1970; Korkeala *et al.*, 1989), and a product is expected to retain good sensorial quality for several days. In comparison to the aerobic spoilage rate, LAB spoilage is much slower and therefore prolonged product shelf-life is obtained by anaerobic packaging of meat.

There are differences between the ability of various LAB species and strains to spoil meat products (Borch *et al.*, 1996; Korkeala and Björkroth, 1996). On the basis of LAB enumeration, it cannot be determined whether a sample contains species/strains later associated with the spoilage of a product. In the study of Björkroth *et al.* (2005), dealing with LAB spoilage in MAP marinated broiler legs, a clear switch from the initial contaminants to the spoilage-causing LAB was detected during the storage at 6 °C. The results showed that enterococci (35.7% of the initial LAB population) were dominating in the fresh product whereas carnobacteria (59.7%) dominated within the spoilage LAB. Only about 19% of the initial LAB contaminants were carnobacteria and no enterococci were detected in the spoiled products. In general, when the initial LAB population was compared with the spoilage LAB, a shift from homofermentative cocci towards carnobacteria, *Lactobacillus sakei/curvatus* and heterofermentative rods was seen in this marinated product.

Since all LAB detected in the food processing environment cannot be considered as spoilage causing, contamination analyses must be based on the tracing of the right strains. If no prior knowledge exists of the spoilage-associated strains/species, molecular typing methods provide tools for distinguishing spoilage-associated LAB strains from the ‘background noise’.

33.4.2 Detection of LAB contamination in the production of vacuum-packaged, sliced cooked ham

Detection of LAB contamination sites during the manufacture of cooked meat products is difficult. LAB populations are very low in an adequately cleaned production environment (Mäkelä and Korkeala, 1987). Shortly after packaging, LAB populations of the products are usually below the routine detection limit (<10 cfu/g), even if the product later spoils quickly (Björkroth and Korkeala, 1996; Nerbrink and Borch, 1993). The cooking process (internal temperature of 68–73 °C) inactivates LAB and surfaces of the cooked products can be considered sterile (Allen and Foster, 1960; Mäkelä and Korkeala, 1987; Mäkelä *et al.*, 1992). Products mainly become recontaminated with LAB during handling after the cooking process.

The first studies employing DNA typing techniques for bacterial contamination analysis associated with food processing were carried out in order to understand LAB spoilage contamination in the production of a vacuum-packaged, sliced cooked ham product (Björkroth and Korkeala, 1996, 1997). Prior to these studies, molecular typing methods had only been used in the context of molecular epidemiological studies dealing with the spread of pathogenic bacteria within patients and the hospital environment. Molecular typing (ribotyping) was applied in two in-plant LAB contamination analyses in a situation where the product was expected to retain good sensorial quality for 21 days, but sour off-odours and off-flavours were sometimes detected after 14 days. The aim of these studies was to locate the sites and sources of spoilage LAB contamination.

In the first phase of the study, a *Lactobacillus sakei* starter strain was suspected of contaminating and spoiling the product during the slicing and vacuum packaging steps. This species was known to cause spoilage in cooked meat products (Korkeala and Björkroth, 1996) and cross-contamination between a fermented meat product and the cooked ham was suspected. A total of 127 LAB strains from products, packaging room surfaces and the ferment batches were typed using *Eco*RI and *Hin*dIII rRNA gene RFLP analysis (Björkroth and Korkeala, 1996). These patterns were used to recognize the starter strain among the contaminating LAB detected in the product and manufacturing environment. This analysis showed that the fingerprint of the *L. sakei* starter was not associated with the spoilage of ham. Therefore, the second contamination analysis (Björkroth and Korkeala, 1997) was carried out.

A total of 982 LAB isolates from the raw materials, product and the environment at different production stages were screened. LAB typing resulted in 71 different ribotypes, of which 27 were associated with contamination routes. Raw material (macerated pork meat) was distinguished as the source of the major spoilage strains. Contamination of the product surfaces after cooking was shown to be airborne. The removal of the product from the cooking-forms was localized as a major site for airborne LAB contamination. Food handlers and some surfaces in contact with the product during the manufacture were also contaminated with the spoilage strains. Some LAB strains were shown to be able to resist cooking in the core of the product bar. The air in the slicing department and adjacent cold room was shown to contain very few LAB but surface-mediated contamination was detected during slicing/packaging stages.

In this case, molecular typing provided useful information revealing the LAB contamination sources/sites of the product. The airborne contamination site preceding the slicing and packaging phases was not anticipated before these analyses. Instead, the slicing machine, which was difficult to clean, was thought to be the source of LAB contamination. After these analyses it was concluded that the strains, which were detected on the finished product surface were already present before slicing took place. The results also confirmed that following traditional quantification of the LAB associated with the production phases, no conclusions of the relevant contamination sites can be made. Indeed, surfaces were found to contain many strains that were never detected in the product. Apparently these LAB were not able to grow in the product.

These contamination analyses were based only on the ribopattern differences and no species identification was done in the study. Later (Björkroth *et al.*, 1998), an identification study was undertaken in which the most potent spoilage LAB was shown to be a *Leuconostoc carnosum* strain. During this work an LAB identification database employing numerical analysis of *Hin*dIII ribopatterns was established. In addition, macro-restriction patterns separated by pulsed-field gel electrophoresis (PFGE) showed that one major PFGE type was responsible for the most abundant spoilage changes. Depending on the level of information needed, typing techniques possessing high discriminatory power may have to be used for obtaining thorough comprehension of the LAB contaminants and the contamination process.

33.4.3 Tracing spoilage LAB at poultry slaughter house and adjacent product manufacture

Vihavainen *et al.* (2005) enumerated and identified LAB associated with MAP, non-marinated, broiler meat at the end of producer-defined shelf-lives. In order to reveal how spoilage-associated LAB were connected with poultry and processing environment contamination, broiler chicken handled during the early stages of slaughter and air from processing phases related to carcass-chilling, cutting and packaging were studied. The LAB counts in the late shelf-life products varied from 10^4 to 10^8 cfu/g. A total of 447, 86 and 122 isolates originating from broiler products, broiler carcasses and processing plant air, respectively, were subjected to identification using a 16 and 23S rRNA gene RFLP database based on *Hin*dIII patterns.

Carnobacterium divergens and *Carnobacterium maltaromaticum* (*piscicola*) were dominating the developing spoilage LAB populations, forming 63% of the LAB isolated. Other major LAB species detected in the products belonged to the genera of *Lactococcus, Leuconostoc* and *Lactobacillus.* LAB associated with the late shelf-life broiler products were recovered from the production plant air adjacent to the final cutting and packaging stages but not from the broiler carcasses being handled at the beginning of the slaughtering line. According to these results, the incoming broiler chickens were not the major source causing psychrotrophic LAB spoilage. The detection of spoilage LAB in the air, however, suggested that the contaminated processing environment played the major role in product contamination. LAB associated with the developing spoilage population of MAP broiler meat finished products were likely to be introduced to the products from the environment at the late processing operations rather than being indigenous bacteria associated with the microbiome of broiler chicken.

Proper eradication of the psychrotrophic spoilage-causing LAB is not obtained if the preventive hygiene-controlling acts are directed only to the handling of the broiler chicken at the abattoir level since psychrotrophic spoilage-LAB contamination was evident in the air associated with final processing phases.

33.4.4 The origins of psychrotrophic lactic acid bacteria in food processing

When contamination analyses were first undertaken, it was thought that the LAB typically associated with different meat products mainly originated from animal sources. The fact that raw meat material used for the manufacture of products often contained certain LAB species helped support this conclusion. For example, *L. carnosum* was typically associated with ham and *Leuconostoc gasicomitatum* (Björkroth *et al.*, 2000) with poultry. Molecular contamination analyses have, however, shown that this conclusion has not necessarily been correct. It seems to be that psychrotrophic LAB in meat products are actually more likely to originate from the production environment than from the microbiome of the animals.

33.5 Applying knowledge from contamination analysis to improve hygienic food manufacturing

Molecular typing methods have increased the understanding of contamination pathways of pathogens and spoilage organisms in food processing plants and food processes. This approach has made it possible to trace microorganisms in a food processing plant, discover their niches and to target preventive measures. In so doing, it has increased our knowledge of how to manufacture pathogen-free products. It has also provided important information of spoilage organisms in general and on specific spoilage organisms. Molecular contamination analysis has therefore had great implications on the hygiene of food processing plants and the quality and safety of their products. The molecular contamination analyses have not only provided valuable information for individual processing plants but also for the food processing industry in general. Such improved hygiene and food safety has also been recognized in improved public health. The decrease of human listeriosis cases in some European countries at the end of the 1990s is a consequence of improved food manufacturing hygiene.

33.6 Future trends

Molecular typing methods have proven their usefulness in the improvement of food industry hygiene and the use of these methods will also be important in the future. New molecular typing methods will probably be developed and the current culture-independent techniques improved. Microarray technology and sequence-based typing methodology will bring very sensitive and unambiguous typing data into reach. The interpretation of the data and practical use in contamination analyses will be challenging.

Molecular typing methods have increased our knowledge of the epidemiology of individual microorganisms, but a more complete understanding of the microbial ecology in the food processing plants should be acquired. The survival and multiplication of pathogens and spoilage organisms in food processing environments are influenced by several different factors, e.g. other microorganisms. Interactions between microorganisms and their influence on the survival of pathogens are still quite unknown. The study of microbial ecology will provide us with information that should give us again 'a new set of tools' in the elimination and prevention of pathogens and spoilage organisms in food processing environments.

33.7 Sources of further information and advice

Numerous contamination analyses studies, which will provide more information, have been performed since 1995 on different microorganisms in different food industry areas. There are also several studies pointing out weaknesses and strengths of molecular typing methods.

33.8 References

AARNISALO K AUTIO T SJÖBERG A-M LUNDÉN J KORKEALA H and SUIHKO M-L (2003), 'Typing of *Listeria monocytogenes* isolates originating from the food processing industry with automated ribotyping and pulsed-field gel electrophoresis', *J Food Prot,* **66**, 249–255.

AASE B SUNDHEIM G LANGSRUD S and RØRVIK L (2000), 'Occurrence of and possible mechanism for resistance to a quaternary ammonium compound in *Listeria monocytogenes'*, *Int J Food Microbiol*, **62**, 57–63.

ALLEN J R and FOSTER E M (1960), 'Spoilage of vacuum-packed sliced processed meats during refrigerated storage', *Food Res*, **25**, 19–25.

AUTIO T HIELM S MIETTINEN M SJÖBERG A-M AARNISALO K BJÖRKROTH J MATTILA-SANDHOLM T and KORKEALA H (1999), 'Sources of *Listeria monocytogenes* contamination in a cold-smoked rainbow trout processing plant detected by pulsed-field gel electrophoresis typing', *Appl Environ Microbiol*, **65**, 150–155.

AUTIO T KETO-TIMONEN R LUNDÉN J BJÖRKROTH J and KORKEALA H (2003), 'Characterisation of persistent and sporadic *Listeria monocytogenes* strains by pulsed-field gel electrophoresis (PFGE) and amplified fragment length polymorphism (AFLP)', *System Appl Microbiol*, **26**, 539–545.

AUTIO T LINDSTRÖM M and KORKEALA H (2004), 'Research update on major pathogens associated with fish products and processing of fish', in: Smulders J M and Collins J D, *Food safety assurance and veterinary public health, Volume 2, Safety assurance during food processing*, Wageningen Academic Publishers, Wageningen, The Netherlands, 115–134.

BERRANG M E MEINERSMANN R J NORTHCUTT J K and SMITH D P (2002), 'Molecular characterization of *Listeria monocytogenes* isolated from a poultry further processing facility and from fully cooked product', *J Food Prot*, **65**, 1574–1579.

BJÖRKROTH J and KORKEALA H (1996), 'Evaluation of *Lactobacillus sake* contamination in vacuum-packaged sliced cooked meat products by ribotyping', *J Food Prot*, **59**, 398–401.

BJÖRKROTH J RIDELL J and KORKEALA H (1996), 'Characterization of *Lactobacillus sake* strains with production of ropy slime by randomly amplified polymorphic DNA (RAPD) and pulsed-field gel electrophoresis (PFGE) patterns', *Int J Food Microbiol*, **31**, 59–68.

BJÖRKROTH J RISTINIEMI M VANDAMME P and KORKEALA H (2005), '*Enterococcus* species dominating in fresh modified-atmosphere-packaged marinated broiler legs are overgrown by *Carnobacterium* and *Lactobacillus* species during storage at 6°C', *Int J Food Microbiol*, **97**, 267–276.

BJÖRKROTH K J and KORKEALA H J (1997), 'Use of rRNA gene restriction patterns to evaluate lactic acid bacterium contamination of vacuum-packaged sliced cooked whole-meat product in a meat processing plant', *Appl Environ Microbiol*, **63**, 448–453.

BJÖRKROTH K J VANDAMME P and KORKEALA H J (1998), 'Identification and characterization of *Leuconostoc carnosum* associated with production and spoilage of vacuum-packaged sliced cooked ham', *Appl Environ Microbiol*, **64**, 3313–3319.

BJÖRKROTH K J GEISEN R SCHILLINGER U WEISS N DE VOS P HOLZAPFEL W H KORKEALA H J and VANDAMME P (2000), 'Characterization of *Leuconostoc gasicomitatum* sp. nov. associated with spoiled raw tomato-marinated broiler meat strips packaged under modified atmosphere conditions', *Appl Environ Microbiol*, **66**, 3764–3772.

BOERLIN P BANNERMAN F ISCHER J ROCOURT J and BILLE J (1995), 'Typing *Listeria monocytogenes*: a comparison of random amplification of polymorphic DNA with 5 other methods', *Res Microbiol*, **146**, 35–49.

BORCH E KANT-MUERMANS M-L and BLIXT Y (1996), 'Bacterial spoilage of meat and cured meat products. A review', *Int J Food Microbiol*, **33**, 103–120.

BROSCH R BRETT M CATIMEL B LUCHANSKY J B OJENIYI B and ROCOURT J (1996), 'Genomic fingerprinting of 80 strains from the WHO multicenter international typing study of *Listeria monocytogenes* via pulsed-field gel electrophoresis (PFGE)', *Int J Food Microbiol*, **32**, 343–355.

CAUGANT D A ASHTONF E BIBB W F BOERLIN P DONACHIE W LOW C GILMOUR A HARVEY J and NØRRUNG B (1996), 'Multilocus enzyme electrophoresis for characterization of *Listeria monocytogenes* isolates: results of an international comparative study', *Int J Food Microbiol*, **32**, 301–311.

CHASSEIGNAUX E, TOQUIN M-T RAQIMBEAU C SALVAT G COLIN P and ERMEL G (2001), 'Molecular epidemiology of *Listeria monocytogenes* isolates collected from the environment, raw meat and raw products in two poultry- and pork-processing plants', *J Appl Microbiol*, **91**, 888–899.

CHASSEIGNAUX E GÉRAULT P TOQUIN M-T RAQIMBEAU C SALVAT G COLIN P and ERMEL G (2002), 'Ecology of *Listeria monocytogenes* in the environment of raw poultry meat and raw pork meat processing plants', *FEMS Microbiol Lett*, **210**, 271–275.

EKLUND M W POYSKY F T PARANJPYE R N LASHBROOK L C PETERSON M E and PELROY G A (1995), 'Incidence and sources of *Listeria monocytogenes* in cold-smoked fishery products and processing plants', *J Food Prot*, **58**, 502–508.

FONNESBECH VOGEL B HUSS H OJENIYI B AHRENS P and GRAM L (2001), 'Elucidation of *Listeria monocytogenes* contamination routes in cold-smoked salmon processing plants detected by DNA-based typing methods', *Appl Environ Microbiol*, **67**, 2586–2595.

FREDRIKSSON-AHOMAA M KORTE T and KORKEALA H (2000), 'Contamination of carcasses, offals, and the environment with *yadA*-positive *Yersinia enterocolitica* in a pig slaughterhouse', *J Food Prot*, **63**, 31–35.

GERNER-SMIDT P BOERLIN P ISCHER F and SCHMIDT J (1996), 'High-frequency endonuclease (REA) typing: results from the WHO collaborative study group on subtyping of *Listeria monocytogenes*', *Int J Food Microbiol*, **32**, 313–324.

GIOVANNACCI I RAGIMBEAU C QUEGUINER S SALVAT G VENDEUVRE J-L CARLIER V and ERMEL G (1999), '*Listeria monocytogenes* in pork slaughtering and cutting plants: use of RAPD, PFGE and PCR-REA for tracing and molecular epidemiology', *Int J Food Microbiol,* **53**, 127–140.

GREER G NATTRESS G F DILTS B and BAKER L (2004), 'Bacterial contamination of recirculating brine used in the commercial production of moisture-enhanced pork', *J Food Prot,* **67**, 185–188.

HARVEY J and GILMOUR A (2001), 'Characterization of recurrent and sporadic *Listeria monocytogenes* isolates from raw milk and nondairy foods by pulsed-field gel electrophoresis, monocin typing, plasmid profiling, and cadmium and antibiotic resistance determination', *Appl Environ Microbiol*, **67**, 840–847.

HEREDIA N GARCÍA S ROJAS G and SALAZAR L (2001), 'Microbiological condition of ground meat retailed in Monterrey, Mexico', *J Food Prot*, **64**, 1249–1251.

HOFFMAN A D GALL K L NORTON D M and WIEDMANN M (2003), '*Listeria monocytogenes* contamination patterns for the smoked fish processing environment and for raw fish', *Appl Environ Microbiol*, **66**, 52–60.

HUDSON W R and MEAD G C (1989), '*Listeria* contamination at a poultry processing plant', *Lett Appl Microbiol*, **9**, 211–214.

JACQUET C ROCOURT J and REYNAUD A (1993), 'Study of *Listeria monocytogenes* contamination in a dairy plant and characterization of the strains isolated', *Int J Food Microbiol,* **10**, 13–22.

JOHANSSON T RANTALA L PALMU L and HONKANEN-BUZALSKI T (1999), 'Occurrence and typing of *Listeria monocytogenes* strains in retail vacuum-packed fish products and in a production plant', *Int J Food Microbiol*, **47**, 111–119.

KETO-TIMONEN R AUTIO T and KORKEALA H (2003), 'An improved amplified fragment length polymorphism (AFLP) protocol for discrimination of *Listeria* isolates', *System Appl Microbiol*, **26**, 236–244.

KORKEALA H ALANKO T MÄKELÄ P and LINDROTH S (1989), 'Shelf-life of vacuum-packed cooked ring sausages at different chill temperatures', *Int J Food Microbiol*, **9**, 237–247.

KORKEALA H J and BJÖRKROTH K J (1996), 'Microbiological spoilage and contamination of vacuum-packaged cooked sausages. A review', *J Food Prot*, **60**, 724–731.

LAHTI E EKLUND M RUUTU P SIITONEN A RANTALA L NUORTI P and HONKANEN-BUZALSKI T (2002), 'Use of phenotyping and genotyping to verify transmission of *Escherichia coli* O157:H7 from dairy farms', *Europ J Clin Infect Dis*, **21**, 189–195.

LAWRENCE L M and GILMOUR A (1994), 'Incidence of *Listeria* spp. and *Listeria monocytogenes* in a poultry processing environment and in poultry products and their rapid confirmation by multiplex PCR', *Appl Environ Microbiol*, **60**, 4600–4604.

LAWRENCE L M and GILMOUR A (1995), 'Characterization of *Listeria monocytogenes* isolated from poultry products and from the poultry-processing environment by random amplification of polymorphic DNA and multilocus enzyme electrophoresis', *Appl Environ Microbiol*, **61**, 2139–2144.

LUNDÉN J MIETTINEN M AUTIO T and KORKEALA H (2000), 'Enhanced adherence of persistent *Listeria monocytogenes* strains to stainless steel after short contact time', *J Food Prot*, **63**, 1204–1207.

LUNDÉN J AUTIO T and KORKEALA H (2002), 'Transfer of persistent *Listeria monocytogenes* contamination between food processing plants associated with a dicing machine', *J Food Prot*, **65**, 1129–1133.

LUNDÉN J M AUTIO T J SJÖBERG A-M and KORKEALA H J (2003a), 'Persistent and nonpersistent *Listeria monocytogenes* contamination in meat and poultry processing plants', *J Food Prot*, **66**, 2062–2069.

LUNDÉN J AUTIO T MARKKULA A HELLSTRÖM S and KORKEALA H (2003b), 'Adaptive and cross-adaptive responses of persistent and nonpersistent *Listeria monocytogenes* strains to disinfectants', *Int J Food Microbiol,* **82**, 265–272.

MÄKELÄ P and KORKEALA H (1987), '*Lactobacillus* contamination of cooked ring sausages at sausage processing plants', *Int J Food Microbiol*, **5**, 323–330.

MÄKELÄ P M KORKEALA H J and LAINE J J (1992), 'Survival of ropy slime producing lactic acid bacteria in heat processes used in the meat industry', *Meat Sci* **31**, 463–471.

MIETTINEN H AARNISALO K SALO S and SJÖBERG A-M (2001a), 'Evaluation of surface contamination and the presence of *Listeria monocytogenes* in fish processing factories', *J Food Prot*, **64**, 635–639.

MIETTINEN M K BJÖRKROTH K J and KORKEALA H J (1999), 'Characterization of *Listeria monocytogenes* from an ice-cream plant by serotyping and pulsed-field gel electrophoresis', *Int J Food Microbiol*, **46**, 187–192.

MIETTINEN M K PALMU L BJÖRKROTH K J and KORKEALA H (2001b), 'Prevalence of *Listeria monocytogenes* in broilers at the abattoir, processing plant, and the retail level', *J Food Prot*, **64**, 994–999.

NERBRINK E and BORCH E (1993), 'Evaluation of bacterial contamination at separate processing stages in emulsion sausage production', *Int J Food Microbiol*, **20**, 37–44.

NESBAKKEN T KAPPERUD G and CAUGANT D A (1996), 'Pathways of *Listeria monocytogenes* contamination in the meat processing industry', *Int J Food Microbiol*, **31**, 161–171.

NORTON D M MCCAMEY M A GALL K SCARLETT J M BOOR K J and WIEDMANN M (2001), 'Molecular studies on the ecology of *Listeria monocytogenes* in the smoked fish processing industry', *Appl Environ Microbiol*, **67**, 198–205.

NORWOOD D E and GILMOUR A (1999), 'Adherence of *Listeria monocytogenes* strains to stainless steel coupons', *J Appl Microbiol*, **86**, 576–582.

NYCHAS G J DILLON V M and BOARD R G (1988), 'Glucose, the key substrate in meat and certain meat products', *Biotechnol Appl Biochem,* **10**, 203–231.

PRITCHARD T J FLANDERS K J and DONNELLY C W (1995), 'Comparison of the incidence of *Listeria* on equipment versus environmental sites within dairy processing plants', *Int J Food Microbiol,* **26**, 375–384.

REUTER G (1970), 'Untersuchungen zur Mikroflora von verpackten, aufgeschnittenen Brüh- und Kochwürsten', *Arch Lebensmittelhyg*, **21**, 257–264.

RØRVIK L M CAUGANT D A and YNDESTAD M (1995), 'Contamination pattern of *Listeria monocytogenes* and other *Listeria* spp. in a salmon slaughterhouse and smoked salmon processing plant', *Int J Food Microbiol*, **25**, 19–27.

SALVAT G TOQUIN M T MICHEL Y and COLIN P (1995), 'Control of *Listeria monocytogenes* in the delicatessen industries: lessons of a listeriosis outbreak in France', *Int J Food Microbiol*, **25**, 75–81.

SAMELIS J and METAXOPOULOS J (1999), 'Incidence and principal sources of *Listeria* spp. and *Listeria monocytogenes* contamination in processed meats and a meat processing plant', *Food Microbiol*, **16**, 465–477.

SAMELIS J KAKOURI A GEARGIADOU K G and METAXOPOULOS J (1998), 'Evaluation of the extent and type of bacterial contamination at different stages of processing of cooked ham', *J Appl Microbiol*, **84**, 649–660.

SCHWARTZ D C and CANTOR C R (1984), 'Separation of yeast chromosome-sized DNAs by pulsed field gradient gel electrophoresis', *Cell*, **37**, 67–75.

SILVA I M ALMEIDA R C ALVES M A and ALMEIDA P F (2003), 'Occurrence of *Listeria* spp. in critical control points and the environment of Minas Frescal cheese processing', *Int J Food Microbiol*, **81**, 241–248.

STULL T L LIPUMA J J and EDLIND T D (1988), 'A broad-spectrum probe for molecular epidemiology of bacteria: ribosomal RNA', *J Infect Dis*, **157**, 280–286.

SUIHKO M-L SALO S NICLASEN O GUDBJÖRNSDÓTTIR B TORKELSSON G BREDHOLT S SJÖBERG A-M and GUSTAVSSON P (2002), 'Characterization of *Listeria monocytogenes* isolates from the meat, poultry and seafood industries by automated ribotyping', *Int J Food Microbiol*, **72**, 137–146.

SWAMINATHAN B HUNTER S B DESMARCHELIER P M GERNER-SMIDT P GRAVES L M HARLANDER S HUBNER R JACQUET C PEDERSEN B REINECCIUS K RIDLEY A SAUNDERS N A and WEBSTER J A (1996), 'WHO-sponsored international collaborative study to evaluate methods for subtyping *Listeria monocytogenes*: restriction fragment length polymorphism (RFLP) analysis using ribotyping and Southern hybridization with two probes derived from *L. monocytogenes* chromosome', *Int J Food*, **32**, 263–278.

TKÁČIKOVÁ L KANTíKOVÁ M DIMITRIEV A and MIKULA I (2000), 'Use of the molecular typing methods to evaluate the control of *Listeria monocytogenes* contamination in raw milk and dairy products', *Folia Microbiol* **45**, 157–160.

UNNERSTAD H BANNERMAN E BILLE J DANIELSSON-THAM M-L WAAK E and THAM W (1996), 'Prolonged contamination of a dairy with *Listeria monocytogenes*', *Neth Milk Dairy J*, **50**, 493–499.

VIHAVAINEN E LUNDSTRÖM H-S SUSILUOTO T KOORT J PAULIN L AUVINEN P and BJÖRKROTH J (2005), 'Identification of developing spoilage lactic acid bacteria in modified atmosphere packaged broiler meat products and their association with broiler chicken and meat processing environment', submitted.

WERNARS K BOERLIN P AUDURIER A RUSSELL E G CURTIS G D W HERMAN L and VAN DER MEE-MARQUET M (1996), 'The WHO multicenter study on *Listeria monocytogenes* subtyping: random amplification of polymorphic DNA (RAPD)', *Int J Food Microbiol*, **32**, 325–341.

WILLIAMS J G K KUBELIK A R LIVAK K J RAFALSKI J A and TINGEY S V (1990), 'DNA polymorphism amplified by arbitrary primers useful as genetic markers', *Nucleic Acids Res,* **18**, 6531–6535.

34

Testing surface cleanability in food processing

J. Verran, Manchester Metropolitan University, UK

34.1 Introduction

Materials that retain few microorganisms after cleaning would be the hygienic choice, and would present the least risk of cross-contamination. Surfaces that are hard, inert, impervious, easy to clean and difficult to damage are desirable in terms of hygienic status. The ability of these surfaces to actively repel or inhibit the attachment, or reduce viability, of attached microorganisms without engendering any resistance among survivors, would also be desirable. The development of surfaces that present such novel properties may also require the concomitant development of testing methods that retain the rigour of more traditional tests, but that enable the properties of these substrata to be correctly evaluated for their intended use. Furthermore, if hygienic status is focused only on the removal of microorganisms, then the presence of residual organic soil, perhaps from food, inorganic materials, or cleaning applications, is ignored. The cleanability of the surface, as opposed to sanitizing, is also of concern in this context, and the relative removal of the microbial and non-microbial components of the fouling material should be evaluated. Lastly, surfaces that are used repeatedly – clearly an essential property of a hygienic surface – will alter over time, in terms of both chemistry and topography. Rigorous testing methods used to evaluate the effectiveness of a surface, or a cleaning application, should incorporate an assessment of long-term usage, in terms of surface wear. The potential retention of material within surface features over time, and the different behaviours of relevant soil and microorganisms, should be considered.

There are many variables that should be addressed when determining which testing methods to employ in evaluating the hygienic status of a surface and its cleanability. The aim of the test, for example whether to assess in-place cleanability of food processing equipment (Anon, 2003), or to evaluate novel surface treatments or cleaning regimes, will affect its design.

A recognition of the environment in which the surface is to be employed, with a view to its simulation, should be the starting point for development of tests, but there may also be regulations with which a given product must comply. The location of a given process will also affect the type of testing method employed – an aspect that will not be explored in depth in this chapter, but is addressed elsewhere in this volume. The type of equipment (open, closed, robust, delicate, flexible, rigid, mobile, static, clean room) will present different cleanability and access issues if to be sampled *in situ* (or simulated *in vitro*, for example using rigs). Hygienic design should have been implicit in the development of the process, and attention paid to appropriate hygiene monitoring, cleaning-in-place (CIP), personnel training and other issues. This chapter will focus primarily on the development of methods that are rigorous and appropriate for cleanability testing of surfaces primarily used in open systems, but that are also feasible as part of quality assurance (QA) procedures, or in developmental work when considering novel processes, materials or cleaning/sanitation regimes.

The nature of the food, the microorganism and the substratum likely to be used in a given situation may all affect the findings of hygiene and cleanability testing. This chapter addresses some of the issues that might be considered in attempts to improve testing methods and validate any claims as to hygienic status. Three aspects are considered, alone and in combination, namely the microorganism, the substratum and organic soil.

34.2 Microorganisms

34.2.1 Mode of existence on substratum

In general, methods used for monitoring hygiene, evaluating novel surfaces or cleaning regimens tend to soil using microorganisms perhaps combined with some organic material such as milk powder, via single events, followed by cleaning/sanitization and an assessment of the number of cells remaining, but not usually of the amount of organic material present. The simplest means for this assessment is via removal of the remaining cells, and enumeration of viable cells by traditional culture techniques (Wirtanen *et al.*, 2000). Alternatively, the material may be quantified *in situ*, for example using epifluorescence microscopy to provide data on the area of a microscopic field covered, or on total cell numbers. Vital staining of these attached cells, or the application of culture media onto test surfaces to enable viable cells to form colonies, will provide information on the proportion of viable cells present (Barnes *et al.*, 1996). This scenario is appropriate for hygienic surfaces, where the expectation in a food factory would be that regular cleaning processes are carried out, and attached/retained cells are relatively few (compared with the dense populations encountered in biofilms), probably not actively multiplying at this solid–air interface due to lack of moisture and nutrient. The opportunity for biofilms to develop on surfaces would be minimal. The term 'biofilm' is used frequently, and often incorrectly, in this context.

Biofilms are defined, in simplest terms, as 'matrix-enclosed bacterial populations adherent to each other and/or to surfaces or interfaces' (Verran, 2002). The interface most commonly associated with biofilm formation is the solid–liquid interface, with the liquid flowing over the substratum (Costerton *et al.*, 1995). In food processing, the presence of a solid–liquid interface, with an element of liquid flow across the surface, would be limited, for example to production of liquid foods, and water distribution (and disposal) systems, in predominantly closed pipework (Holah and Gibson, 2000; Verran and Jones, 2000). It would be surprising, and undesirable, in the food production/processing context to encounter a stagnant (non-flowing) supply of a liquid which could function as microbial nutrient. Thus the testing of hygienic food contact surfaces for propensity to support biofilm and biofilm development, or for evaluation of biofilm removal by cleaning agents might not be particularly appropriate – depending on the environment of concern.

On many food contact surfaces, the most common interface presented would be solid–air: an open system. The typical solid–liquid interface biofilm could not develop. Cells would be immobilised on a hygienic surface, unable to multiply in the assumed relative absence of food and water. Of course, in a food processing plant, both of these are present in abundance, hence the need to continually clean and sanitize surfaces. (The dynamic accumulation of microorganisms at a solid–air interface is encountered in other environments where the substratum itself is not an inert hygienic material, but provides an element of nutrient. Humidity has a significant impact on the survival and multiplication of these attached microorganisms. Biodegradation of artwork and textiles is a consequence of these 'sub-aerial' biofilms; Dornieden *et al.*, 2000.) The isolation of viable bacteria from such surfaces in the absence of growth indicates their ability to survive. Most species, with the exclusion of spores, die on prolonged exposure on dry surfaces (McEldowney and Fletcher, 1988). Again, organic material may aid survival by protecting organisms from desiccation, physical removal, or cleaning/sanitizing agents (Verran *et al.*, 2002), even if it does not provide nutrient for the microorganisms present.

Solid–liquid–air interfaces could be encountered frequently, as food particles, or cleaning solutions pass across the surfaces on which microorganisms may be attached. The accumulation of cells at these interfaces is a well-recognized phenomenon, and would provide additional confounding factors of variability in maintenance of hygienic status.

34.2.2 Test species

In the literature, fouling and cleanability testing focuses on a relatively limited range of bacteria: for example, *Listeria monocytogenes*, in the meat and fish industries (Frank and Chmielewski, 1997; Fonnesbech Vogel *et al.*, 2001); *Salmonella* spp. in the poultry industry (Gough and Dodds, 1998); *Bacillus* spores and thermophilic streptococci in the dairy industry (Faille and Benezech, 1999; Flint et al., 2000); *Pseudomonas* spp. or *Escherichia coli* and *Staphylococcus aureus* in many general testing procedures as examples of typical Gram-positive and Gram-negative spoilage organisms and hygiene indicators (Wirtanen and Matilla-Sandholm, 1994). The susceptibility of these different organisms to given environments and antimicrobial agents such as detergents and biocides, and their interactions with a given food soil in order to facilitate attachment, survival and even multiplication on a given surface will determine the hygienic status of the appropriate surface.

34.2.3 Attachment or retention: active or passive

The literature on factors affecting the attachment of microorganisms to surfaces is significant. Surface charge effects and hydrophobic properties are used to describe and predict (not always successfully) the way in which approaching microorganisms and surface will interact. A recent review by Strevett and Chen (2003) outlines the general principles in more detail. Modifications may be made to substrata with the aim of reducing this initial stage of colonization, for example by altering the surface hydrophobicity, or topography. Some mention will be made of modifications later in this chapter; more detailed explorations are made elsewhere in this volume. After this 'non-specific' interaction, biological properties of the surfaces assume importance, and a more firm attachment is established. Here, more overt antimicrobial strategies might be implemented, for example by incorporation of antimicrobial agents into surface coatings. These reversible-to-irreversible attachment processes are more likely to occur in a reproducible manner at a solid–liquid interface, where the presence of the liquid enables the surfaces (cell and substratum) to approach one another. Cells will attach at a solid–liquid–air interface, and remain *in situ*, for example as the liquid dries.

Where there is a solid–air interface, the mechanism will be different, with cells brought to the surface either via air, or by transfer from other surfaces. In food processing, there may well be a more forceful and less active encounter between microorganism and substratum, where transfer of cells from foodstuffs to equipment will occur as the food is passed through the plant. This passive transfer may also include food material such as proteins, fats, oils, carbohydrates from the various hot or cold meat or vegetable-derived materials being passed through a particular processing cycle.

Thus, viable microorganisms may be present on a surface, accompanied potentially by a considerable amount of nutrient. Hydrophobic and surface charge effects will have had little impact on this phenomenon, although they may affect the ease of removal of the material from the surface. Indeed, it is the cells remaining on the surface, i.e. their retention after a cleaning and sanitization event, which is of prime importance in the study of hygienic status, rather than the initial attachment. Factors that will enhance retention would include attached cells drying onto the substratum, the presence of organic material, and the presence of surface defects and features in the substratum.

34.3 Hygienic surfaces

34.3.1 Stainless steel

The most frequently quoted example of a hygienic surface is stainless steel, which will therefore be used to address general principles. This material demonstrates all the properties required for hygienic status. High chromium content (over 10.5%) steel resists attack by food acids; engineered surface finishes facilitate cleaning in catering, food processing and surgical applications. All metals react with oxygen and water in the atmosphere to form a surface layer of oxide. On ordinary carbon steel, this is an hydrated iron oxide which, being porous, permits further oxidation, producing rust. By alloying the steel with chromium to produce stainless steel, the oxide formed is chromium oxide. This 'passive' layer a few atoms thick, protects the metal surface from further reaction, and, when scratched, has the ability to reformulate itself within seconds. Aluminium and titanium are other notable metals in this context.

The austenitic stainless steels present good corrosion resistance and toughness properties. Gradings, for example 304 and 316 (ASTM designation), are the stainless steels most commonly used in the food industry, related to the different proportions of chromium, nickel and molybdenum in the formulations (only 316 contains molybdenum, which enhances corrosion performance). There are several other designations, with European national differences in terminologies now being superseded by the EN prefix. The JIS Japanese designation provides the third used to define the global steel market (www.outokumpu.com).

34.3.2 Surface finish

During production, coils of stainless steel are first softened and descaled, in a process known as annealing and pickling. This process uses gas-fired furnaces, cooling, shot blasting and a mixture of hydrofluoric acid and nitric acid to clean the surface of the material to a matt grey appearance, prior to processing to a given finish. After cold rolling, which reduces the thickness of the steel, material may be softened using an inert atmosphere which prevents oxidation and ensures a bright highly reflective finish, called Bright Annealed (BA, or 2R). Typical applications are the production of sinks and washing machine drums. Alternatively, materials can be softened and processed further for enhancement of surface finish. The surface produced by this route is described as 2B, and is the most common 'mill finish' with a smooth, grey appearance. Thus the 2B and 2R finishes are used most commonly in the food industry. Grain boundaries are apparent on steels with a 2B finish. These features serve to entrap microorganisms, and are also less easy to clean, enabling marks due to fingerprints for example, to remain (Freeman *et al.*, 2002a). Thus this material tends not to be used directly in food applications. Special finishes are applied to 316 2B, and to other steels, thus masking the original grain boundary features. These finishes include ground, brushed or polished surfaces and pattern rolled surfaces. Specially polished 'hygienic' finishes are also available. Interestingly, the surface finish is used to describe the *process* rather than the *product*, thus there may well be considerable variability within the properties of steel with a specified finish.

34.3.3 Surface roughness

In the food, beverage and pharmaceutical industries, highly exacting hygienic demands are imposed on the surfaces that come into contact with the product being manufactured (Boulange-Petermann, 1996). These demands are specified in the appropriate food standards, for example, DIN 11850-10/99 specifies a surface roughness R_a max 0.8 μm for tubing, the implication being that surfaces with R_a values below this level are 'hygienic'. The maximum for the weld area is 1.6 μm. The R_a value describes the average departure of the surface profile from a calculated 'centre line'. A probe is scanned across a surface, perpendicular to the lay of the surface features, and a 'centre line' is plotted so that the areas of the profile above and below this line are equivalent. The R_a value is a figure representing the average departure of the profile from this centre line. The R_a value gives an indication of the size (height) of peaks and valleys, but says nothing about the character of the surface in different directions, the occurrence of long waviness or amplitude distributions (Fig. 34.1). Unless otherwise specified the reference length over which the surface is scanned is 0.8 mm. Other parameters have been used to describe surface features, and are defined within the various standards (Anon, 1990), but the R_a value is the most generally accepted. The R_a is deemed of importance because it is assumed, and has been demonstrated, that an increase in surface roughness will cause microorganisms to be more easily entrapped within surface features, hence reducing the cleanability of the surface (Verran and Boyd, 2001). However, this relationship is only demonstrable within certain limits of R_a values. When features are significantly larger than the microbial cells, then the cells are relatively easily removed from the surface.

Various workers have sought to define a minimum R_a value below which there is no relationship between R_a and retention. Furthermore, since the R_a value is derived from a linear trace, information on the three-dimensional character of surface features is lacking. The finish may or may not affect cleanability (Freeman *et al.*, 2002b; Leclercq-Perlat and Lalande, 1994; Steiner *et al.*, 2000). Linear surface features such as those on a brushed finish will be more easily cleaned along the lay of the features than across the features, and presumably also more easily than surfaces where linear features (scratches) occur randomly across the surface. Pitted surfaces present yet another variable, and represent the type of wear more commonly seen in ceramic surfaces (Verran

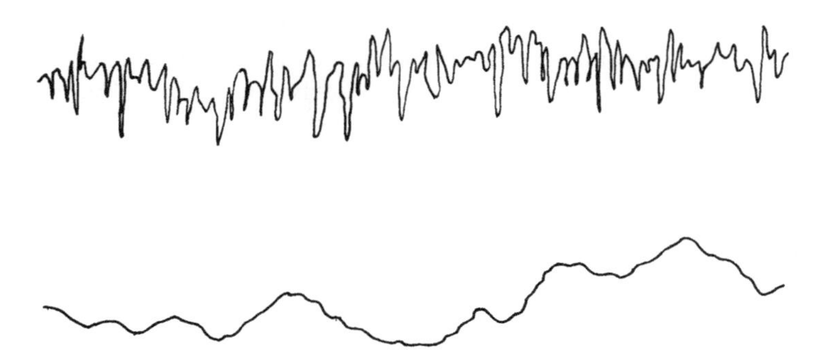

Fig. 34.1 Profiles of two stainless steel surfaces with the same R_a value (0.8 μm) presented along a 0.8 mm length, and at the same vertical scale. The profiles differ considerably; the surface presented by the lower profile would be easier to clean because the features are larger.

et al., 2001a). Surface cleanability has been related to other descriptors such as the depth of inward facing defects (Mettler and Carpentier, 1999), and the area of indentation (Walker *et al.*, 2000), but despite the difficulties encountered in relating R_a to surface cleanability, it is still the descriptor most commonly used to describe a hygienic surface.

While a surface is used in a hygienic setting, wear will inevitably occur. Even hygienic surfaces are susceptible to some degree of wear, perhaps as the peaks of brushed steels are damaged, or as scratches are introduced (Verran *et al.*, 2000; Boyd and Verran, 2001). The effect of this wear on surface cleanability is beginning to be explored.

As more sophisticated profilometers have been developed, information describing surface topography has increased in amount and complexity. Probes of decreasing sizes, from solid through lasers to the nanometre dimension probes of the atomic force microscope (AFM) generate different R_a values, usually describing features within smaller and smaller areas of a surface. The AFM provides R_a data on the nanometre scale, accompanying an image of the area scanned. Interesting differences between measurements and the appearance of the surface emerge. Wear of brushed stainless steel is apparent as small tags occur on the peaks. However, there is little change in the R_a value (Verran *et al.*, 2000). The effect of such wear on surface cleanability in terms of removal of microorganisms is minimal (Fig. 34.2). Studies on a range of stainless steels, all with R_a values below 0.8 μm showed that bacteria were removed (by wiping with detergent) in comparable numbers from all substrata. Using epifluorescence microscopy, it was apparent that the features were retaining cells, thus the pattern of topographic features was affecting the distribution of cells, but not their numbers, or the strength of retention (Verran *et al.*, 2001b,c; Hilbert *et al.*, 2003). Using AFM, cells could be visualized within features, and on top of features whose dimensions were smaller than cells. By increasing the strength of the AFM probe scan across the surface, the relative strength of retention could

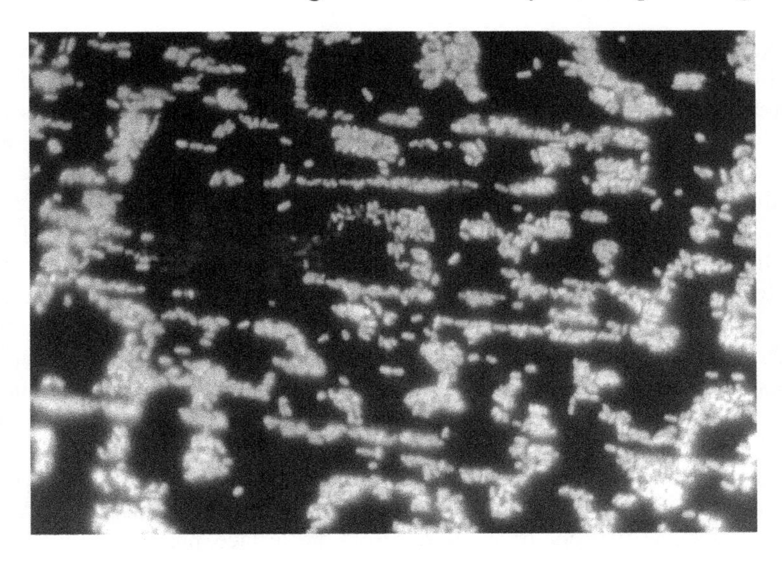

Fig. 34.2 *Pseudomonas aeruginosa* cultures were sprayed onto abraded stainless steel surfaces, and rinsed gently to remove unattached or loosely retained cells. Cells were stained using acridine orange, and viewed using epifluorescence microscopy. Cells are retained in grain boundaries and scratches on abraded stainless steel surfaces. (Image from Deborah Rowe.)

be compared. Cells which were immobilized on features, even if they were smaller than cells, required an increased force (in nanoNewtons) for removal (Boyd *et al.*, 2002). Treatment with detergent reduced the force required (results not presented).

Although perhaps a cleaning force described in nanoNewtons is of little relevance in terms of the more 'macro' business of cleaning and sanitizing surfaces, these findings demonstrate that any topographical change in a surface feature may affect its cleanability. Thus these changes may need some consideration during the development of novel materials and finishes.

34.3.4 Other hygienic surfaces

There are many surfaces deemed to be hygienic other than metals. Ceramic/glass finishes are finding use as chopping boards as well as worktops and wall and floorcoverings. Wear of these surfaces tends to occur via chipping, generating a pitted surface rather than one that is scratched (Verran *et al.*, 2001a). Organic polymers are more malleable and therefore used to fabricate curved surfaces. Softer polymers may also be more easily scratched, which will make cleaning more difficult, thus affecting hygienic status. The use of flexible (more) porous rubbers and similar materials as bungs, seals, may be inevitable: these will require particular attention in any cleaning regime. Surface coatings such as polymers, resins and powders, which may present enhanced cleanability and/or convey protection to the underlying materials, are finding use in a number of applications. Surfaces impregnated with biocides are used in factories and in the domestic market, but there has been some negative reaction to the increased use of biocides in terms of a perceived risk of inducing bacterial resistance, and a possible perception of a reduced need to maintain hygiene standards in the home (Gilbert and McBain, 2003). Photoactivated glass and other ceramics, impregnated with titanium dioxide to produce a 'self-cleaning' and antimicrobial effect are likewise strongly advocated by some users. The unwettable 'Lotus effect' treated surface, from which liquids roll, thus removing any residual material or microorganisms, clearly requires some vertical placement, or a regular directed washing effect to enable 'rolling'. Such treatments must be consistently applied across a surface, and their response to usage over time also requires evaluation.

If intended for use as a hygienic surface, then testing methods appropriate for these novel surfaces and mechanisms of activity may need to be developed (Gibson *et al.*, 1999). If an inhibitory effect needs to be demonstrated on contact between cells and substratum, or if irradiation is required to inactive cells (and remove soil), or diffusion of a material from the surface, presence of liquid over the surface, or activation via a wiping process/novel sanitizing agent, then consideration needs to be paid to the best *in vitro* method appropriate for demonstrating the intended *in situ* effect, thus strengthening the relevance of any in-use claims that might be made by manufacturers, and ensuring adequate in-use performance for users. So far, general criteria and harmonized test methods to evaluate the performance of antimicrobial coatings, as well as criteria for specific applications, are lacking (Hartog, 2004). A number of test protocols have been described which are based on the Japanese Industrial Standard JIS Z 2801:2000. Here, a bacterial cell suspension is held in intimate contact with a coated surface using a sterile cover (e.g. glass microscope cover slip) in humid conditions. After a set contact time, the size of the residual bacterial population is compared with an appropriate control coating using standard microbiological enumeration techniques. ASTM 2180-01 could also be modified to examine coated surfaces (Askew, 2004).

34.4 Organic soil

When a substratum has been used, or exposed to an environment, its surface chemistry will immediately alter. Surface-active materials such as stainless steel will be conditioned with an organic coating, which convey hydrophobic properties to a surface which might have been considered hydrophilic (Boulange-Petermann, 1996; Verran *et al.*, 2001c).

Once a surface has been used, fouled and cleaned, the characteristics of the pristine surface are irrevocably altered (Mettler and Carpentier, 1998). Thus a 'new' surface used in hygienic testing procedures would be encountered only once in use: consideration should therefore be paid to the inclusion of a 'used' hygienic surface in such testing.

An in-use surface will be soiled with organic material as well as micro-organisms: the removal of one component (organic soil) would be a particular focus in terms of cleaning, the other (microorganism) for both cleaning and sanitizing, because of impact on hygiene, spoilage and disease. The importance of a good cleaning regime in the removal of both soil and microorganisms cannot be underestimated. Disinfection/sanitization should be applied to the few cells remaining during or after an effective cleaning process. Thus the combined presence of soil and microorganisms provides a challenging and realistic target for hygiene testing and surface cleanability strategies.

Some fouling and cleanability studies use a soil–microorganism mixture, but then only monitor the microorganisms. There are several recipes for various soils which can be applied to surfaces and their presence monitored (Verran *et al.*, 2002). However, most of these methods tend to lie in the domain of chemists and engineers rather than microbiologists. Fouling and cleaning in food processing may be additionally confounded by the application of heat: for example, both protein and minerals are deposited on the surfaces in milk processing (Changani *et al.*, 1997): thermophilic bacteria and sporeformers are the microorganisms of particular concern (Flint *et al.*, 2000).

34.4.1 Soil and microorganisms

When attempting to develop an appropriate cleaning/sanitizing regime, it should be deemed essential to include some organic soil in addition to microorganisms as the surface inoculum, since the presence of one component may affect the behaviour of another (Barnes *et al.*, 1999; Boyd *et al.*, 2000). It is not easy to compare the relative removal of these two components from a surface, since methods of detection are inevitably not comparable: usually, the number of microorganisms and amount of organic material. Microscopy methods provide potential means for discriminating the components, by employing some differential staining. Image analysis can provide information on the percentage of an area covered by cells, or amorphous material, and compare the relative changes in the two percentage values, but few studies have been carried out. Problems are encountered with intensity of staining, selection of appropriate stains and their differentiation, combined with the presence of an amorphous material (Fig. 34.3). The complexity of the soil is a further confounding issue. If a simple soiling material is used, such as starch or protein, then the selection of an appropriate stain is simplified. Different soils will be encountered in different processes, and will be susceptible to different cleaning regimes and mechanisms, thus the model used need to be selected for a given situation. Complex soils such as the Campden soil, which comprises a mixture of oil, milk and starch, provide a more realistic system, but differential staining in this context is almost impossible, since the different components of the soil will behave differently, in addition to the difficulty of visualizing microorganisms remaining on the substratum.

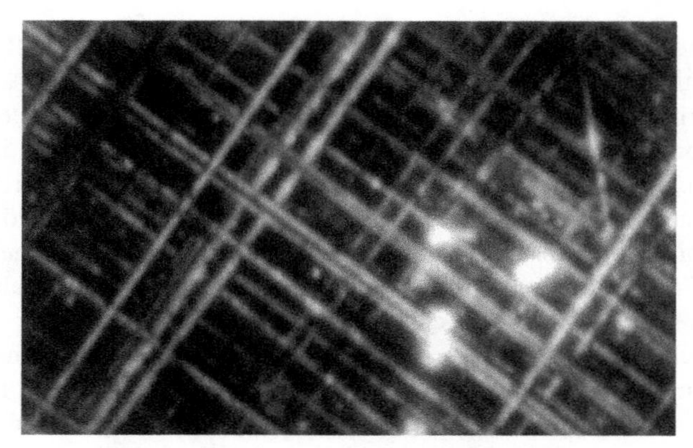

Fig. 34.3 Films of 1% starch plus 0.0002% fluorescein were dried onto abraded stainless steel surfaces and heated at 70°C for 12 h. After light manual brushing using a toothbrush wetted with distilled water, surfaces were viewed using epifluorescence microscopy. Stained organic material is visible within scratches on the surface. (Image from Rob Boyd.)

34.4.2 Repeated soiling and cleaning

In order to monitor the behaviour of organic soil and microorganisms on surfaces subjected to repeated soiling and cleaning, a combination of microbiological culture techniques and epifluorescence microscopy was used to evaluate the effect. Up to 20 soiling (microorganisms plus milk powder) and cleaning (high-pressure spray) procedures were carried out. Half of the area of each test substratum was swabbed to remove cells for culture; the remainder was stained with acridine orange and a percentage coverage measurement of substratum by material (soil and cells) was made.

Viable microbial cells (measure by colony count) were removed by cleaning, and accumulation did not occur, but there was a gradual increase in stained material on the surface, indicating an accumulation of organic soil. This work revealed different behaviours of soil and cells on surfaces (Verran *et al.*, 2001a), but would be very labour intensive if implemented in routine procedures.

34.4.3 Soil, cells and worn surfaces

Stainless steel substrata presenting different types of roughness, all with R_a values of less than 0.8 μm proved comparable in terms of removing retained bacteria from the surfaces (Verran *et al.*, 2001b; Hilbert *et al.*, 2003). However, organic soil was retained in different amounts (measured by percentage area coverage of fluorescein stained material) related to the area of surface presenting defects (not to the R_a value) (Walker *et al.*, 2000). In addition, soil was retained more in deeper, sharper features (Boyd *et al.*, 2001a,b).

In order to characterize the distribution of material across a surface, surface-sensitive analytical methods have been used to provide information about the elements or molecular groups present. Limitations include the lack of depth information, thus only properties of the top nanometre/micrometre layers will be detected. Furthermore, the expense and accessibility of these methods preclude their use in routine testing, although findings may help support the rationale for experimental development. X-ray photoelectron microscopy (XPS) gives information about the distribution of elements across a surface. When a starch soil was mixed with microorganisms, a nitrogen peak could be used to monitor the removal of microorganisms from the surface (Boyd *et al.*, 2001a). The surface topography again had no effect on cell removal in the presence of soil. However, XPS also provided information on the nature of the bonds in which elements are present: thus the presence of markers associated with starch could be detected, showing that starch was retained in larger quantities in the test stainless steel substrata with the larger, deeper surface features. This method was useful because the starch soil lacked nitrogen, enabling discrimination between the two components: other cell–soil combinations would not be so readily differentiated.

34.5 Future trends

Both relatively simple routine methods (epifluorescence microscopy and cell culture) and more complex, expensive and time-consuming surface-sensitive techniques (XPS and AFM) confirmed that the behaviour of soil and cells on surfaces is not necessarily the same when monitoring substrata for hygienic status and cleanability. Several variables should be included in any rigorous testing method, along with the use of surfaces which have been exposed to some element of wear:

1. Intended use, environment, relevant soil and microorganisms.
2. Effect of wear and ageing on substratum chemistry and topography.
3. Mode of action of any antimicrobial/detergent/self cleaning effect.
4. Effect of repeated fouling and cleaning on surface properties and cleanability.
5. Interactions occurring between substratum topographic features, organic soil and microorganisms.
6. Simple to perform, repeatable and valid.

The development of novel food technologies, hygienic materials, finishes, coatings, and cleaning and sanitizing applications should require the concomitant development of realistic, relevant and challenging test methods for evaluation of the surface cleanability and hygienic status.

The methods used to obtain results which support the claims of the manufacturers should be appropriate to the intended mode of action of any novel application. However, once in use, in order for cleanability testing and hygiene monitoring to be carried out with relative ease and reproducibility by the relevant staff, organisation or validating body, then the routine methods ultimately used need to be relatively simple, rapid, cheap and valid. This is a challenge for all concerned.

The overall aim when considering and improving test methods for surface cleanability is to use the information derived from the development of the system, whether novel or well established, to ensure that relevant and reproducible testing methods are employed. Results should confirm the hygienic status and cleanability of the system, and help to ensure the safety of the food produced, and the process by which it is produced, for the consumer.

34.6 Sources of further information and advice.

Some of the organizations involved with development of test methods and standards are noted below. There are many other units which undertake research and development for industry.

- OECD (www.oecd.org). The Organisation for Economic Cooperation and Development comprises an association of countries with interests in a number of areas including biocides and hygiene.
- EHEDG (www.ehedg.org) European Hygienic Equipment Design Group, whose objective is to provide safe and wholesome food and help sustain a positive image of the food industry.
- IBRG (www.ibrg.org). The main objectives of the International Biodeterioration Research Group are in the field of test method development and work to investigate basic principles of biodeterioration. IBRG comprises member organizations which develop testing methods for adoption by standard bodies.
- JIS (www.jsa.or.jp) Japanese standards
- ASTM (www.astm.org) American standards
- BSI (www.bsi-global.com) British standards

34.7 Acknowledgements

The author would like to thank Rob Boyd, Debbie Rowe and Kathryn Whitehead for their involvement with some of the research underpinning this paper, and for the images.

34.8 References

ANON (1990) BS1124: Method for the assessment of surface texture. Part 2. General information and guidance, British Standards Institute, Milton Keynes, UK.

ANON (2003). A method for the assessment of in-place cleanability of food processing equipment. EHEDG, Chipping Campden, Campden and Chorleywood Food Research Association.

ASKEW PD (2004), 'Hygienic control: assessing the role of antimicrobial surfaces' in *Hygienic coatings and surfaces conference papers*, Teddington, Paint Research

Association, paper 6.

BARNES BI, CASSAR CA, HALABLAB MA, PARKINSON NH, MILES RJ (1996), 'An in situ method for determining bacterial survival on food preparation surfaces using a redox dye', *Lett Appl Microbiol* **23**, 325–328.

BARNES L-M, LO MF, ADAMS MR, CHAMBERLAIN AHL (1999), 'Effect of milk proteins on adhesion of bacteria to stainless steel surfaces', *Appl Env Microbiol* **67**, 2319–2325.

BOULANGE-PETERMANN L (1996), 'Processes of bioadhesion on stainless steel surfaces and cleanability: a review with special reference to the food industry', *Biofouling* **10**, 275–300.

BOYD RD, VERRAN J (2001), 'AFM as an in-situ technique for industry', *Microscopy Analysis*, July, 5–6.

BOYD RD, COLE D, ROWE D, VERRAN J, COULTAS SJ, PAUL AJ, WEST RH, GODDARD DT (2000), 'Surface characterisation of glass and poly(methylmethacrylate) soiled with a mixture of fat, oil and starch', *J Adhesion Sci Technol* **14**, 1195–1201.

BOYD RD, ROWE D, COLE D, VERRAN J, HALL KE, UNDERHILL C, HIBBERT S, WEST R (2001a), 'The cleanability of stainless steel as determined by X-ray photoelectron spectroscopy', *Appl Surf Sci* **172**, 135 143.

BOYD RD, COLE D, ROWE D, VERRAN J, PAUL AJ, WEST RH (2001b), 'Cleanability of soiled stainless steel as studied by atomic force microscopy and time of flight secondary ion mass spectrometry', *J Food Protect* **64**, 87–93.

BOYD RD, VERRAN J, JONES MV, BHAKOO M (2002), 'Use of the atomic force microscope to determine the effect of substratum surface topography on bacterial adhesion', *Langmuir* **18**, 2343–2346.

CHANGANI SD, BELMAR-BEINY MT, FRYER PJ (1997), 'Engineering and chemical factors associated with fouling and cleaning in milk processing', *Exptl Therman Food Sci* **14**, 392–406.

COSTERTON JW, LEWANDOWSKI Z, CALDWELL DE, KORBER DR, LAPPIN-SCOTT HM (1995), 'Microbial biofilms', *Ann Rev Microbiol* **49**, 711–745.

DORNIEDEN T, GORBUSHINA AA, KRUMBEIN WE (2000), 'Biodecay of cultural heritage as a space/time-related ecological situation – an evaluation of a series of studies', *Int Biodet Biodeg* **46**, 261–270.

FAILLE C, BENEZECH T (1999), 'Cleanability of stainless steel surfaces soiled by *Bacillus thuringiensis* spores under various flow conditions', in Wilson DI, Fryer PJ, Hastings APM, *Fouling and cleanability in food processing '98*, Luxembourg, Office for Official Publications, EC, 125–133.

FLINT SH, BROOKS JD, BREMER PJ (2000), 'Properties of the stainless steel substrate, influencing the adhesion of thermo-resistant streptococci', *J Food Eng* **43**, 235–242.

FONNESBECH VOGEL B, HUSS HH, OJENIYI B, AHRENS P, GRAM L (2001), 'Elucidation of *Listeria monocytogenes* contamination routes in cold-smoked salmon processing plants detected by DNA-based typing methods', *Appl Env Microbiol*, **67**, 2586–2595.

FRANK JF, CHMIELEWSKI RAN (1997), 'Effectiveness of sanitation with quaternary ammonium compounds of chlorine on stainless steel and other domestic food preparation surfaces', *J Food Protect* **60**, 43–47.

FREEMAN PF, HARGATE N, BARRETT RL (2002a), 'Fingerprint resistant stainless steel', in *4th European Stainless Steel Science and Market Congress*, Paris B5-3, volume I, 283–285.

FREEMAN PF, HARGATE N, MARSH E, CHAO WT (2002b); 'Cleanability of stainless steel', in *4th European Stainless Steel Science and Market Congress*, Paris B1-4, volume I,

173–175.

GIBSON H, TAYLOR JH, HALL KE, HOLAH JT (1999), 'Surface and suspension testing: conflict or complementary', *Int Biodeg Biodet* **20**, 375–384.

GILBERT P, MCBAIN AJ (2003), 'An evaluation of the potential impact of the increased use of biocides within consumer products upon the prevalence of antibiotic resistance', *Clin Microbiol Rev* **16**, 189–208.

GOUGH NL, DODD CER (1998), 'The survival and disinfection of *Salmonella typhimurium* on chopping board surfaces of wood and plastic', *Food Control*, **9**, 363–368.

HARTOG B, KNOL P, STEKELENBURG F, TAP Z (2004), 'Performance evaluation of hygienic antimicrobial coatings', in *Hygienic Coatings and Surfaces Conference Papers*, Teddington, Paint Research Association, paper 9.

HILBERT LR, BAGGE-RAVN D, KOLD J, GRAM L (2003), 'Influence of surface roughness of stainless steel on microbial adhesion and corrosion resistance', *Int Biodet Biodeg* **52**, 175–185.

HOLAH J, GIBSON H (2000), 'Food industry biofilms' in Evans LV, *Biofilms: Recent advances in their study and control*, Amsterdam, Harwood, 211–235.

LECLERCQ-PERLAT M-N, LALANDE M (1994), 'Cleanability in relation to surface chemical composition and surface finishing of some materials commonly used in food industries', *J Food Eng* **23**, 510–517.

MCELDOWNEY S, FLETCHER M (1988), 'The effect of temperature and relative humidity on the survival of bacteria attached to dry surfaces', *Lett Appl Microbiol* **7**, 83–86.

METTLER E, CARPENTIER B (1998), 'Variations over time of microbial load and physicochemical properties of floor materials after cleaning in food industry processes. *J Food Protect* **61**, 57–65.

METTLER E, CARPENTIER B (1999), 'Hygienic quality of floors in relation to surface texture', *Trans I Chem Eng* **77C**, 90–96.

STEINER AE, MARAGOS MM, BRADLEY RL (2000), 'Cleanability of stainless steel surfaces with various finishes', *Dairy Food Env San*, **20**, 250–260.

STREVETT KA, CHEN G (2003), 'Microbial surface thermodynamics and applications', *Res Microbiol* **154**, 329–335.

VERRAN J (2002), 'Biofouling in food processing: biofilm or biotransfer potential?' *Trans IChemE* **80C**, 292–298.

VERRAN J, BOYD RD (2001), 'The relationship between substratum surface roughness and microbiological and organic soiling: a review', *Biofouling* **17**, 59–71.

VERRAN J, JONES MV (2000), 'Problems of biofilms in the food and beverage industry', in Walker JT, Surman S, Jass J, *Industrial Biofouling*, Chichester, John Wiley and Sons Ltd, 145–173.

VERRAN J, ROWE DL, COLE D, BOYD RD (2000), 'The use of the atomic force microscope to visualise and measure wear of food contact surfaces', *Int Biodet Biodeg* **46**, 99–105.

VERRAN J, ROWE DL, BOYD RD (2001a), 'The effect of nanometer dimension surface topography on the cleanability of stainless steel', *J Food Protect* **64**, 1183–1187.

VERRAN J, BOYD RD, HALL K, WEST RH (2001b), 'Microbiological and chemical analysis of stainless steel and ceramic subjected to repeated soiling and cleaning treatments', *J Food Protect* **64**, 1377–1387.

VERRAN J, BOYD RD, HALL KH (2001c), 'The effect of wear on fouling and cleanability of hygienic food contact surfaces', in *32nd Nordic Symposium Proceedings*, Recklam Productcenterna, Enköping, Sweden.

VERRAN J, BOYD RD, HALL KE, WEST R (2002), 'The detection of microorganisms and organic material on stainless steel food contact surfaces', *Biofouling* **18** (3), 167–176.

WALKER JT, VERRAN J, BOYD RD, PERCIVAL S (2000), 'Microscopy methods to investigate the structure of potable water biofilms', *Meth Enz* **333**, 243–255.

WIRTANEN G, MATILLA-SANDHOLM T (1994), 'Measurement of biofilm of *Pediococcus pentosaceus* and *Pseudomonas fragi* on stainless steel surfaces', *Colloids Surf B: Biointerfaces* **2**, 33–39.

WIRTANEN G, STORGARDS E, SAARELA M, SALO S, MATILLA-SANDHOLM T (2000), 'Detection of biofilms in the food and beverage industry', in Walker JT, Surman S, Jass J, Industrial Biofouling, Chichester, John Wiley and Sons Ltd, 176–203.

35

Improving the monitoring of fouling, cleaning and disinfection in closed process plant

A. P. M. Hasting, Tony Hasting Consulting, UK

35.1 Introduction

This chapter aims to summarise and assess currently available monitoring techniques for the food and other relevant industries and indicate some potential future trends in this area. The importance of fouling, cleaning and disinfection as prerequisites to Hazard Analysis Critical Control Point (HACCP) are described together with a description of different types of fouling, their operational consequences and the general approach to cleaning and disinfection processes. Current approaches to monitoring are described together with the key issues, which may limit their application. The systems are also characterised in terms of whether the sensor element is invasive or non-invasive, the measurement made at a point or integrated over the whole system and the data obtained in real time or retrospectively, as these are significant in terms of their potential application in practice. In addition to commercial scale processes, monitoring has a key role in the development of small-scale techniques that allow the fouling, cleaning and disinfection characteristics of new products or processes to be assessed. Such monitoring systems can be used to identify the key parameters that influence fouling and cleaning and offer the opportunity to optimise the process before implementation on a commercial scale. The essential industry requirements, which have to be met before any system is used commercially, are considered and some of the potential benefits from improved monitoring are noted. Future trends in the area of monitoring are suggested based on likely developments within the food industry.

35.2 Background

Fouling, cleaning and disinfection have a critical role in ensuring the product quality and operational performance of food process plant. Cleaning and disinfection are increasingly identified as essential prerequisites, which must be in place before HACCP studies on microbiological hazards can be implemented. Fouling is the unwanted accumulation or deposition of material from the product fluid on to product contact surfaces or stagnant areas of equipment. It can be either heat induced in, for example, a heat exchanger or non-heat induced in, for example, a membrane process, accumulation of material in an unhygienic piece of equipment or growth of a biofilm. The consequences of fouling can therefore be primarily operational; for example the deposit may cause a significant deterioration in the hydraulic properties of the equipment leading to increased pressure drops. Heat transfer may also be adversely affected, leading to heat exchangers being unable to maintain their design performance. Accumulation of product in stagnant areas of equipment can, depending on the process conditions, result in the growth of microorganisms, contamination of the bulk product and loss of microbiological quality.

Cleaning is the process by which the deposits or accumulated soil in the equipment are removed. The process can vary significantly depending on whether the equipment can be cleaned in place or manually. There may be a number of different steps within the overall process but the main steps are:

- removal or recovery of bulk product from the system, generally using water or a product recovery system;
- removal of deposits by contacting them with detergents under predetermined conditions;
- final rinse to remove all traces of detergent residues prior to production.

The process of cleaning may also reduce but not eliminate the number of microorganisms within the system. Disinfection can then be applied after the cleaning process to reduce the number of microorganisms on the product contact surfaces to a level compatible with the level of hygiene required for the product being processed.

Fouling, cleaning and disinfection should be considered as complementary processes since they each can have a significant impact on each other. The type and quantity of fouling will define the challenge for the cleaning process and the optimum strategy for its removal. The efficiency of the cleaning process will strongly influence the disinfection step and, in some cases, failure to clean effectively will compromise disinfection and the safety of subsequent production. Failure to clean effectively may also result in a more rapid build-up of fouling in the subsequent production run.

Effective measurement and monitoring of such processes is critical for food safety and quality. Without measurement it is not possible to monitor and without monitoring, the safety of the product and process cannot be controlled or assured. Monitoring is also essential in the validation of processes to provide the high degree of assurance that the process is being delivered on a consistent basis.

35.3 Current approaches to monitoring

Monitoring is the periodic or continuous determination of the amount of specific material or substances present in a system, the condition of a surface or a process parameter such as flow rate, temperature or pressure. Fouling, cleaning and disinfection data may be produced by a number of direct and indirect methods such as:

- generation and analysis of process plant data;
- controlled studies and analysis at laboratory or pilot plant scale;
- mathematical models.

It is well documented in the literature that fouling is complex and variable, and occurs through a variety of physical or chemical mechanisms. It is rarely uniform or evenly distributed and may vary significantly even on a day-to-day basis. Monitoring the build-up of fouling is important to identify when the plant has reached a condition where cleaning and disinfection is required. All cleaning and disinfection processes require a designated end-point and this may be either defined by protocols based on trials, or, ideally, indicated by sensors, which can monitor the progress of the cycle and determine when cleanliness is reached or disinfection is complete.

Monitoring systems or sensors can be classified as integrated, localised or indirect, depending on the method applied (Hasting 2002):

- Integrated measurements are taken over the complete system or a specific part of it, for example, heat transfer coefficient or pressure drop within a heat exchanger. The sensitivity of these averaged approaches may be limited and there are often sizeable measurement errors. Some techniques, such as heat transfer coefficient, may not be applicable for cleaning when the system is operated isothermally.
- Localised measurements are made at specific points within the system. However, identifying representative locations or relating the information generated to the complete process equipment are key issues.
- Indirect techniques use the measurement of key parameters such as detergent concentration, temperature and valve opening to provide information. The results may be subject to interpretation, particularly where the response may indicate more than one event is occurring.

Sensors may also be classified as invasive if the process equipment has to be modified to allow contact of the sensing element with the process fluid, for example fitting a pressure gauge into a line. A non-invasive sensor is one that does not require the modification of the equipment for installation of the sensor, for example the use of ultrasonic sensors on the outside of a pipe, to measure flow rate within the pipe. The sensor may also provide data in real time during the process or retrospectively in the case where samples are analysed off-line.

Tables 35.1–35.3 summarise the major sensing techniques that have been or could be applied for food processing. Some of the techniques are still at the research or development stage but are based on three main categories:

1. Process parameters, which can be used to provide information on plant process performance.
2. Measurement of surface condition, for example deposit thickness.
3. Determination of quantity of materials within system, for example detergent or biocide concentration.

In current commercial practice the most widely used monitoring approaches are:

- visual inspection;
- predefined process using existing sensors and detergent conductivity;
- heat transfer;
- pressure drop;
- ATP;
- microbiological;
- conductivity.

Table 35.1 Measurement of process parameters

Technique	Measurement principle	Classification	Key issues
Overall heat transfer coefficient, U (Visser *et al.* 1997)	U calculated from basic heat transfer equation using conventional sensors	Real time Integrated Invasive	Assumes uniform deposit properties Accuracy of multiple instruments
Heat flux (Jones *et al.* 1994; Truong *et al.* 1998; Hasting *et al.* 2005)	Heat flux can be calculated directly from power input to sensor	Real time Localised Invasive	Reliability, assumes thermal properties of deposit, representative location Potentially non-invasive
Pressure drop (Corrieu *et al.* 1986)	Increase in pressure drop due to decrease in hydraulic diameter	Real time Integrated Invasive	Accuracy especially when pressure drop low, interpretation of data, assumes uniform deposit thickness and surface roughness known
Existing process sensors, e.g. temperature, time, flow rate	Delivery of minimum acceptable process for cleaning/ disinfection	Real time Indirect Invasive	Assumes that delivery of defined process will assure cleanliness and adequate disinfection

Table 35.2 Measurement of surface condition

Technique	Measurement principle	Classification	Key issues
Ultrasound (Withers 1996)	Attenuation of ultrasound velocity between two sensors	Real time Localised Non-invasive	Sensitive to temperature, particles, gas bubbles. Limited resolution of deposit thickness
ATP	Detection of organic materials on contact surfaces	Retro-spective Localised Invasive	Assumes presence of organic materials indicates microorganisms Calibration required for each soil
Microbiological assay	1. Swabbing to detect microorganisms	Retro-spective Localised Invasive	Limited access, significant time delay, representative samples. Presence of coliforms indicator of poor hygiene/cleaning
	2. End product sampling	Retro-spective Localised Invasive	Significant time delay, unlikely to detect low contamination levels. Difficult to differentiate from other causes
Surface electrical properties (Gale and Griffiths 1995)	Deposit conductivity	Real time Localised Invasive	Calibration, implementation on process areas
Visual	Surface condition	Retro-spective Localised Invasive	Simple, minimal equipment, accuracy and resolution, subjective assessment, non quantitative
Optical (Bartlett *et al.* 1997; Tamachkiarowa and Flemming 1999)	Scattering of light (optical or near infrared) beam or pulses during transmission through film of deposit	Retro-spective Localised Invasive	Access to process surfaces, sensitivity threshold
Pneumatic gauging (Gale 1995, Tuladhar *et al.* 2000)	Approach of air or liquid jet to surface is monitored by change in pressure	Real time Localised Invasive	

Table 35.3 Measurement of material concentration within system

Technique	Measurement principle	Classification	Key issues
Chemical assay	Determines concentration of detergents/biocides	Retrospective Localised Non-invasive	Relatively simple, can be carried out by trained operators
Turbidity (Gallot-Lavallee *et al.* 1981)	Attenuation of incident light as a measure of soil in cleaning solution	Real time Localised Invasive	Assumes that steady state conditions means cleaning has been completed, although system may still contain soil
Liquid electrical properties (capacitance, conductivity, amperometric)	Liquid conductivity, e.g. detergents/ biocides	Real time Localised Invasive	Reliability of measurement, measures material in liquid not on surface
Rinse water analysis (ATP)	Detection of organic matter in circulating rinse water	Real time Indirect Non-invasive	Assumes that steady state conditions means cleaning has been completed, although system may still contain soil

35.3.1 Visual inspection

Visual inspection is still probably the most widely used inspection technique for assessing the impact of fouling and cleaning and in particular confirming that the cleaning results are satisfactory. While a visually clean surface will not guarantee a microbiologically acceptable product, failure to achieve at least visual cleanliness will mean that further cleaning is essential, otherwise the subsequent production process may be compromised. The main practical problem with visual inspection, particularly with closed plant, is access to the product contact surfaces. Ideally, visual inspection should be carried out in areas where the heaviest fouling and hence greatest challenge to the cleaning system occurs. In the case of heat exchangers, the geometry will define the practicality of visual inspection. Some tubular heat exchanger geometries, such as multiple tube in tube and monotube designs, can in principle be inspected by removing the product inlet and outlet connections and exposing the internal surfaces. Other designs such as the concentric triple tube (tube in tube in tube) cannot be inspected in this way. Plate heat exchangers can in principle be dismantled for inspection but, in practice, this could only be considered on an infrequent basis due to the time that would be required. Such a visual inspection should, however, form a key part of the commissioning and acceptance protocol for a heat exchange system.

35.3.2 Predefined process using existing sensors

The majority of fouling, cleaning and disinfection processes are monitored using fairly conventional sensors such as temperature, conductivity, pressure and flow rate. Cleaning and disinfection will usually have the sensors incorporated with the cleaning-in-place (CIP) system rather than the process plant and will therefore be monitoring the end of line conditions. The delivery of the processes is therefore defined by achieving a minimum set of conditions for a predetermined time. While this approach has been shown to be effective for many food applications, the process does not easily lend itself to optimisation, in particular when process lines are processing a far wider range of products than they were originally designed for. It is unrealistic to expect a single cleaning process to be optimal for cleaning a line that might be processing 20–30 very different products.

35.3.3 Heat transfer

For heat exchangers, fouling is one of the major problems that limits operation and requires regular cleaning. Fouling will result in a deposit on the heat transfer surface, which will act as an insulating layer and reduce the heat transfer performance of the exchanger. Typically the performance of a heat exchanger is measured by the overall heat transfer coefficient and the effect of fouling is to reduce this. On-line measurement of heat transfer coefficient therefore provides a simple means of monitoring the condition of the surface during both fouling and cleaning. The heat transfer coefficient can be calculated from the following equations.

Overall heat transferred to the product:

$$Q = M * c_p * (T_{out} - T_{in}) \tag{35.1}$$

where Q is the overall heat transferred, W, M is the mass flow rate of product being heated, kg/s, c_p is the specific heat of the product, J/kg K, T_{out} is the temperature of the product at the exit of the heat exchanger, °C, and T_{in} is the temperature at the inlet of the heat exchanger, °C.

In addition

$$Q = U * A * \Delta t \tag{35.2}$$

where U is the overall heat transfer coefficient, W/m^2 K, A is the heat transfer area of the heat exchanger, m^2, Δt is the log mean temperature difference, °C.

$$\Delta t = \frac{(T_1 - T_{out}) - (T_2 - T_{in})}{\ln[(T_1 - T_{out})/(T_2 - T_{in})} \tag{35.3}$$

where T_1 is the inlet temperature of the heating fluid and T_2 is the outlet temperature of the heating fluid.

If the temperatures, product flow rate, specific heat and heat transfer area of the exchanger are known, the overall heat transfer coefficient can be calculated from equations 35.1, 35.2 and 35.3. The build-up of fouling can then be monitored on-line as a reduction in the heat transfer coefficient and the cleaning process is the reverse with the heat transfer coefficient increasing as soil is removed from the surface.

The accuracy of any technique in providing a quantitative measure of the condition of the surfaces will depend upon both the type and number of instruments used in the calculation. Taking as an example the measurement of fouling based on heat transfer coefficient over sections of heat exchanger in a typical highly energy-efficient dairy pasteuriser with 95% heat recovery. Assume the product is being heated from 5 °C to 72 °C in two stages, firstly by heat recovery and then by heating with hot water. The typical temperatures in such a system would be as follows:

- Heat recovery section – heating medium (hot pasteurised milk)

Product inlet	5 °C
Product outlet	68.7 °C
Heating medium inlet	72 °C
Heating medium outlet	8.3 °C
Δt	3.3 °C

- Final heating section – heating medium (hot water)

Product inlet	68.7 °C
Product outlet	72 °C
Hot water in	75 °C
Hot water out	73.9 °C
Δt	4.0 °C

If the temperature probes were accurate to ±0.5 °C, the overall range of possible temperature differences in the heat recovery section would be 2.3–4.3 °C. Since, when the other process conditions are constant, the overall heat transfer coefficient, U, is inversely proportional to the temperature difference, the accuracy of the probes limits the accuracy of U to approximately ±30%. If the probe accuracy is ±0.1 °C, the accuracy is improved to ±6%. For the final heating section the accuracy of the calculated heat transfer coefficient would be 25% and 5% for probe accuracies of ±0.5 °C and ±0.1 °C respectively. This indicates that such a sensor would be used as an indicator of surface condition rather than an absolute measure. It also tends to be most sensitive at the start of the fouling process and the end of the cleaning process (Hasting 2002). However, since the sensors may well be installed on the heat exchanger as part of the normal instrumentation and control system, it provides a simple and effective way of monitoring the fouling and cleaning processes.

35.3.4 Pressure drop

Pressure drop is another relatively simple method of monitoring the overall condition of a heat exchanger. As a deposit builds up on a surface, the cross-sectional area for flow will be reduced, regardless of the geometry of the heat exchanger, e.g. tube or plate. If the product flow rate is maintained constant, the pressure drop needed to pump the fluid through the heat exchanger will be increased and this change in pressure drop can be used as a real-time monitor. This method has similar limitations to heat transfer measurement in that the pressure drop is the difference between the product inlet and outlet pressure in the heat exchanger. The actual difference and the accuracy of the probes will therefore determine the accuracy of the measured values. The pressure drop can be used to monitor both fouling and cleaning and will tend to have its maximum sensitivity when the surface is heavily fouled and the pressure drop is highest. It is therefore possible to combine heat transfer and pressure drop measurements to provide an enhanced level of information on the conditions within the heat exchanger (Hasting 2002).

35.3.5 Microbiological and ATP methods

Until the 1980s, monitoring the condition of a food contact surface was routinely carried out by conventional microbiology culture methods based on agar plate counts. These methods can detect specific indicator organisms as well as provide information on the numbers of organisms present on a surface. The limitations of such methods are the time taken to obtain a result even with modern rapid analysis techniques and the skilled resource required to ensure the required levels of accuracy.

During the 1980s a bioluminescence assay was developed to detect adenosine triphosphate (ATP), which is always present in viable microbes, as a means of monitoring product contact surfaces. The test uses the enzyme luciferase, which emits light in the presence of ATP. The light is measured in an instrument called a luminometer and results, usually expressed as Relative Light Units (RLU), are available in a few seconds. The degree of light emitted is related to the level of ATP. More recent developments enable the technique to be used for analysing rinse water samples for the assessment of closed plant CIP systems. The limitation for this type of application is that even a consistently low ATP level in the final rinse is no guarantee that there is no residual soil within the plant. ATP techniques are covered more fully in Chapter 36.

35.3.6 Conductivity

The measurement of conductivity, either in-line or off-line, can be used to provide a measure of the concentration of the detergent at the point of measurement, with conductivity being a direct function of concentration. This can then be used to control the concentration of detergent, for example in a CIP recovery system, by dosing additional product into the recovery tank. While this may be acceptable for simple detergent applications, it may not always be appropriate where more complex formulations are used. Different components within the formulation will have specific functions and may not all be depleted at the same rate and a single measure cannot monitor this. An example would be a detergent containing sodium hydroxide for organic soil removal and the sequestrant edta (ethylenediaminetetraacetic acid) for removal of inorganic deposits. In this case, conductivity could be used to monitor and control caustic level but titration would be used to monitor the edta level to ensure that there is always free sequestrant at the end of the cleaning process.

35.4 Laboratory/pilot-scale studies

In addition to commercial scale processes, monitoring has a key role in the development of small-scale techniques that allow the fouling, cleaning and disinfection characteristics of new products or processes to be assessed. Such monitoring systems can be used to identify the key parameters that influence fouling and cleaning and potentially offer the opportunity to optimise the process before implementation at commercial scale.

Other industries have developed portable units that can be transported to the process to be investigated, for example to measure the fouling characteristics of cooling waters (Glen *et al.* 1999, Knudsen *et al.* 1999). They can then be integrated with the process by running them in parallel as a side stream of the main flow and therefore ensuring that representative fluids are being used. Such units are primarily concerned with fouling rather than cleaning. In the food industry the installation of test units in parallel is not widely used.

A number of laboratory or pilot-plant systems for the monitoring of fouling and cleaning have been reported.

35.4.1 Radial flow cell

The importance of fluid shear in fouling and cleaning has been generally accepted and a radial flow cell (RFC) was developed (Fowler and Mackay 1980) to investigate the effects of shear stress on the formation and removal of biofilms. The cell consisted of a pair of parallel discs, one of which had a central hole from which liquid passed radially outwards thorough the narrow gap formed by the plates. As the liquid flowed outwards, its velocity and hence the shear stress on the wall decreased, providing an elegant method of generating a range of shear stresses for a given flow rate. In principle, the critical shear stress for biofilm growth could be calculated from the radius at which a biofilm could be detected. Below this critical radius, the shear stress was sufficiently high to prevent the attachment of the biofilm. Practical problems were encountered with maldistribution of liquid as it moved to the outer edge of the plate, leading to uneven fouling.

A heated version of the radial flow cell was developed (Fryer *et al.* 1985), which permitted the effect of surface shear stress on the initial stages of fouling to be measured under heat transfer conditions. Heating oil was circulated across the top and bottom plates with a plate gap of 0.56 mm. It was found that for mechanical reasons, maintaining such a narrow flow passage over the whole diameter required the use of relatively thick plates. These thick plates incurred significant penalties in terms of heat transfer and were replaced with copper discs, which were plated with stainless steel. Problems were encountered with the deposition of fouling causing localised blockage in the flow cell and distortion of flow within the cell.

35.4.2 Tapered tube system

The tapered tube was developed (Fryer and Slater 1987) to overcome the limitations of the radial flow cell concept. It consisted of a stainless steel tapered tube, 40 cm in length, tapering uniformly from 19.2 mm at the inlet to 12.7 mm at the outlet, thus wide enough to avoid blocking by deposit. The tube was constructed in two halves, with a gasket between to allow examination after each experiment. Shear stresses in the range 0.5 to 8.1 N/m^2 were generated under turbulent flow conditions. The tapered tube could be enclosed in a steam chest to provide the temperature driving force for fouling and a recirculation system used to circulate product through the tube for the required time.

35.4.3 Robbins device

Food process equipment often has surfaces where microbes may attach, grow and develop into biofilms, which can have an adverse effect on the microbial quality of the food. Growth of biofilms can be monitored using stainless steel coupons attached to surfaces within the factory environment. In closed, flowing systems such samples are more difficult to obtain. The Robbins device allowed biofilm development to be monitored at a laboratory scale but under representative conditions. The tubular unit was constructed of 1.42 cm internal diameter, 92.5 cm long, admiralty brass and had replaceable plugs fitted at sampling ports along its length. Individual plugs with a surface area of 0.5 cm^2 could be removed aseptically from the device without needing to drain the system. The build up of biofilms could be monitored by measuring the pressure drop across the unit and the structure could be analysed using scanning electron microscopy. Flow velocities of 1.39 to 2.65 m/s were achieved equivalent to Reynolds numbers of 24 618 to 47 175. Various modifications have been made to the Robbins device (Blanchard *et al.* 1997) to overcome some of the limitations of the original such as the edge effects resulting from locating a flat sample disc at the wall of a cylindrical pipe and its impact on the fluid dynamics. These include using a rectangular flow channel constructed from poly(tetra-fluoroethene) (PTFE), which could be autoclaved prior to use.

35.4.4 Pilot-scale heat exchangers

Small-scale heat exchangers can have the benefit of being able to simulate the commercial size heat exchangers in terms of the geometry of the exchanger and the product temperature profile within the exchanger. Such systems tend to have capacities in the region of 50–150 L/h and hence extended runs to observe realistic fouling build-up can result in substantial quantities of product being used unless the product is recycled. However, recycling can result in unrepresentative fouling characteristics as the fouling-sensitive materials will tend to deposit on the heated surfaces during the early stages of the run and thus reduce the concentration of these species in the bulk. The fouling curve may then tend to indicate a maximum fouling rate at the beginning of the run followed by a reduction in rate eventually reaching an asymptote.

In a pilot-scale plate heat exchanger system for heating products to UHT temperatures (Fryer *et al.* 1996) the capacity of the system was 150 L/h, the total heat transfer area was 1.6 m^2 and the hold-up volume in the heat exchanger was 3.1 L. This meant that for a total product volume of 50 litres each element of fluid is subjected to the full UHT temperature profile three times per hour and hence any heat-sensitive fouling components are likely to be deposited in the relatively early stages of the run. The highly instrumented rig enables the temperatures and pressures over a number of different sections of the machine to be monitored and hence the fouling of product under well-defined temperature conditions.

35.4.5 Heat flux sensor

A number of sensors based on heat flux have been reported with application at both laboratory scale (Jones *et al.* 1994) and on commercial plant (Truong *et al.* 2002). The development and potential application of a patented heat flux sensor (Baginski *et al.* 2002) has been reported (Hasting *et al.* 2005). The sensor can be used to generate as well as monitor both fouling and the subsequent cleaning process. Initial concepts based on the use of a commercial heat flux sensor within the complete system (Jones *et al.* 1994) were predicted to have limited sensitivity to deposit thickness (Hasting *et al.* 2005). A modified design using a heating block directly coupled to the external surface of the flow channel gave significantly greater sensitivity. Initial results using a sensor manufactured according to this design was shown to be capable of monitoring fouling and cleaning with whey protein and skimmed milk fluids under laboratory conditions. The geometry and configuration of the sensor used standard components to incorporate it into the rig and the construction was considered hygienic, robust and compatible with the food industry.

Further work with this sensor design showed the potential value as a small-scale experimental tool. The experimental arrangement used a small volume of fluid, 20 litres, and recirculated this past the sensor and back to a buffer tank. Owing to the small heat transfer area of the sensor, 3.6×10^{-4} m^2, the amount of heat transferred was small and the bulk temperature of the fluid could be maintained at temperatures below which fouling rates would be minimised. The heat transfer area to volume ratio, 0.018 m^2/m^3, was considerably lower than the plate heat exchanger unit described above, 32 m^2/m^3. The system design also meant that the fluid velocity and temperature driving force, which are both considered significant in terms of fouling and cleaning, can be varied independently. The plate heat exchanger system cannot achieve this as a change in flow rate and hence flow velocity will also significantly affect the temperature profile within the exchanger. The benefit of being able to achieve representative fouling and cleaning in small volume equipment would allow the probable fouling and cleaning characteristics to be determined during the product/process development stage, rather than having to determine these during the commissioning phase. It is essential that such equipment can be proven to realistically simulate the performance of full-scale equipment.

35.5 Industry requirements and potential benefits

Any monitoring and sensing system must meet a number of basic requirements if it is to be accepted by the food industry:

- meets a genuine industry need;
- robust and reliable;
- meets claimed specification regarding sensitivity and accuracy;
- simple to install and use;
- low maintenance;
- self-diagnostic;
- cost effective;
- compatible with food industry hygiene requirements.

The installation and use of reliable sensors gives the process operator the opportunity to move from a situation where process data are limited, retrospective and often of doubtful value, to one where the data are real and immediate. This real data may often be very different from that expected and can lead to concerns as to whether the process is functioning correctly. This, however, provides the impetus for identifying whether there is a fundamental process problem or the data do actually represent what would be expected with the layout of the line and its current operation.

A well-designed monitoring system can provide a significant degree of 'peace of mind' that processes critical to the microbiological safety of the final product are being implemented effectively. The use of HACCP in process design and operation studies places an increased demand for monitoring of Critical Control Points (CCPs) in order to define corrective actions to be taken in the case where the process is deemed to be out of control.

Such systems can also provide other potential benefits:

- Validation of the cleaning and disinfection process.
- Monitoring the consistency of the cleaning and disinfection process delivered to the product contact surfaces.
- Provide prediction of maximum time remaining before cleaning required, based on current rate of fouling.
- Optimisation of the cleaning process for different products processed on the same line.

- Evaluation of alternative cleaning approaches or detergents.
- Enable optimised cleaning process to be tailored to particular production needs.
- Proactively highlight potential process problems.
- Identify whether deposits are localised or distributed throughout exchanger.
- Enable fouling reduction strategies to be assessed.

35.6 Future trends

Future commercial trends within the food industry will place continuing emphasis on production flexibility and cost reduction, while requiring greater levels of hygiene assurance and security. This will make process monitoring even more critical in order to enable cleaning to be optimised, while validating that the process delivers the necessary standards required. Process lines will be required to produce an ever-increasing range of products, with more frequent product changeovers. Cleaning will therefore have an increasing impact on plant downtime and reduced operational efficiency. Concerns about potential allergen cross-contamination will place greater emphasis on cleaning performance and validation. There will be a move to a systems-based approach using combinations of sensing elements integrated into a total system. This will require improved techniques for interpretation of the data obtained from such systems. Fouling and cleaning will be treated as complementary rather than separate processes so that the degree of challenge provided by the fouling process can be determined. This can be used in conjunction with the product and hence the expected fouling composition together with the cleaning standards required for the subsequent product to automatically define the optimum cleaning procedure. The development of sensors with self-diagnostic capabilities will increase together with non-invasive techniques that would be particularly relevant for commissioning and troubleshooting applications. For example, a non-invasive flow meter for measuring cleaning flows would be valuable in allowing measurements to be taken at a number of different points and on several lines.

35.7 Conclusions

The monitoring of fouling, cleaning and disinfection is becoming increasingly important in terms of both product safety and operating efficiency. Reliable monitoring is essential to ensure that the appropriate processes are being consistently delivered, without which process validation and optimisation cannot be achieved. The use of commonly available sensors such as pressure and temperature to monitor pressure drop and heat transfer coefficient can provide a valuable indication of the condition of surfaces throughout production and cleaning, although primarily as an indicator rather than an absolute measurement of cleanliness. The most promising approaches to monitoring are likely to consist of a combination of a number of sensors integrated into a total system, which would fulfil the basic industrial requirements of being robust, reliable and cost effective.

35.8 References

BAGINSKI E., BURNS I.W. and HASTING A.P.M. (2002) Monitoring apparatus, Australian patent no. 744109.

BARTLETT G., SANTOS R., BOTT T.R. and GRANT D. (1997) Measurement of biofilm development with flowing water using infra red absorbance. In *Proc. of Engineering Foundation Conf. – Understanding Heat Exchanger Fouling and its Mitigation*, Lucca, Italy.

BLANCHARD A.P., BIRD M.R. and WRIGHT S.J.L. (1997) In Wimpenny J.W.T., Handley P., Gilbert P., Lappin-Scott H.M. and Jones M.V. *Biofilms: Community Interactions and Control*, Cardiff: Biolne Publications, 235–244.

CORRIEU, G., LALANDE, M. and FERRET, R. (1986) On-line measurement of fouling and cleaning of industrial UHT heat exchanger. *Journal of Food Engineering*, **5**, 231–248.

FOWLER H.W. and MCKAY A.J. (1980) In Berkeley R.J. (ed.) *The Measurement of Microbial Adhesion*, Chichester, Ellis Horwood, 143–161.

FRYER P.J. and SLATER N.K.H. (1987) A novel fouling monitor, *Chemical Engineering Communications*, **57**, 139–152.

FRYER P.J., SLATER N.K.H. and DUDDRIDGE J.E. (1985) *Biotech. Bioeng.*, XXVIII, 434.

FRYER P.J., ROBBINS P.T., GREEN C., SCHREIER P.J.R., PRITCHARD A.M., HASTING A.P.M., ROYSTON D.G. and RICHARDSON J.F. (1996) A statistical model for fouling of a plate heat exchanger by whey protein solution at UHT conditions, *Trans IChemE*, **74**, C4, 189–199.

GALE G.E. (1995) A thickness measuring device using pneumatic gauging to detect the sample, *Measurt. Sci. and Tech.* 6, 1566–1571.

GALE G.E. and GRIFFITHS P. (1995) An automatic micrometer for measuring soft electrically conductive materials, *Measurt. Sci and Tech.* 6, 447–451.

GALLOT-LAVALLEE T., LALANDE M. and CORRIEU G. (1981) An optical method to study removing of milk deposits by sodium hydroxide cleaning solution. In Hallstrom B., Lund D.B. and Tragardh Ch (eds) *Fundamentals and Applications of Surface Phenomena Associated with Fouling and Cleaning in Food Processing*, Tylosand, Sweden, 215–224.

GLEN N.F., HOWARTH J.H. and JENKINS A.M. (1999) Fouling monitoring on process plant – Field experience. In Bott T.R., Watkinson A.P. and Panchal C.P. (eds) *Mitigation of Heat Exchanger Fouling and its Economic and Environmental Implications*, July 1999, Banff, Canada, 70–76.

HASTING A.P.M. (2002) Industrial experience of monitoring fouling and cleaning. In Wilson D.I., Fryer P.J. and Hasting A.P.M. (eds) *Fouling, Cleaning and Disinfection in Food Processing*, Jesus College, University of Cambridge, 213–220.

HASTING A.P.M., BAGINSKI E. and BURNS I.W. (2005) A heat flux sensor for monitoring fouling and cleaning in food processing; in press.

JONES A.D., WARD N.J. and FRYER P.J. (1994) The use of a heat flux sensor to monitor milk fluid fouling. In Fryer P.J., Hasting A.P.M. and Jeurnink Th.J.M. (eds) *Fouling and Cleaning in Food Processing*, University of Cambridge, Cambridge, 199–205.

KNUDSEN J.G., EHTESHAMI G.R. and HAYS G.F. (1999) Simulation of fouling in heat exchangers using an annular test section. In Bott T.R., Watkinson A.P. and Panchal C.P. (eds) *Mitigation of Heat Exchanger Fouling and its Economic and Environmental Implications*, July 1999, Banff, Canada, 95–102.

TAMACHKIAROWA A. and FLEMMING H.C. (1999) Optical fibre sensor for biofouling mitigation. In Bott T.R., Watkinson A.P. and Panchal C.P. (eds) *Mitigation of Heat Exchanger Fouling and its Economic and Environmental Implications*, July 1999, Banff, Canada, 343–347.

TRUONG T., ANEMA S., KIRKPATRICK K. and CHEN H. (2002) The use of a heat flux sensor for in-line monitoring of fouling of non-heated surfaces. In Wilson D.I., Fryer P.J. and Hasting A.P.M. (eds) *Fouling, Cleaning and Disinfection in Food Processing*, Jesus College, University of Cambridge, 221–228.

TRUONG T.H., ANEMA S., KIRKPATRICK K. and TRINH K.T. (1998) In-line measurements of fouling and CIP in milk powder plant. In Wilson D.I., Fryer P.J. and Hasting A.P.M. (eds) *Fouling and Cleaning in Food Processing*, 6–8 April, Jesus College, University of Cambridge, Cambridge.

TULADHAR T.R., PATERSON W.R. and WILSON, D.I. (2000) Use of dynamic gauging – a soft film thickness sensor – to probe cleaning-in-place of whey protein deposits, In *Proc. Chemeca*, 200, Perth, Australia, July 2000.

VISSER H., JEURNINK T.J.M., SCHRAML J.E., FRYER P.J. and DELPLACE F (1997) Fouling of heat treatment equipment, *Bulletin of the International Dairy Federation*, **328**, 7–17.

WITHERS P.M. (1996) Ultrasonic, acoustic and optical techniques for the non-invasive detection of fouling in food processing equipment, *Trends in Food Science and Technology*, **7**, 293–298.

36

Improving surface sampling and detection of contamination

C. Griffith, University of Wales Institute Cardiff, UK

36.1 Introduction

Cleaning can be defined as the removal of 'soil' from surfaces and is important in all working and living environments (Dillon and Griffith, 1999). Soil can be described as 'matter out of place' and may be of an organic or inorganic nature, with or without associated microorganisms. In general terms the word soil has become synonymous with dirt and Fig. 36.1 indicates possible sources of contamination for ready-to-eat foods.

Cleaning is important for many reasons, not least of which is human acceptance. While some people are forced, usually through poverty, to live in dirty conditions, some anthropologists believe we have natural tendencies to live in a clean, orderly environment (Curtis, 2001) and there is evidence to suggest consumers avoid unclean food environments (Food Standards Agency, 2004). Clean surroundings are increasingly believed to be important in the prevention of disease transmission, and a dirty environment in the home, hospitals, workplace, etc., can aid the spread of pathogens. Recent experience with SARS and Norovirus outbreaks has refocused attention on the role of the environment in the spread of disease. For the food industry, the adequacy of cleaning may be critical in preventing cross-contamination, especially for pathogens such as *Campylobacter* (Redmond *et al.*, 2004), in the preparation of ready-to-eat foods. However, it is not just microbial pathogens or their toxins in food that can affect consumers' health, but increasingly the presence of small traces of human food allergens, as a result of cross-contamination, can be a cause for concern.

An additional problem for food processors is the presence of food spoilage organisms, which can cause off-odours, flavours or deterioration in food texture,

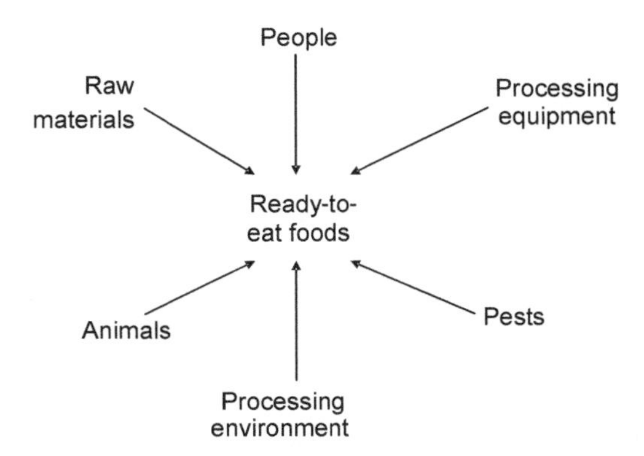

Fig. 36.1 Possible sources of contamination for ready-to-eat foods.

resulting in reduced product shelf-life. Cleaning is essential to minimise microbial build-up and/or presence of biofilms on food contact equipment and surfaces as well as the more general environmental areas of food production/ preparation premises (see Section 35.4.2).

Successful cleaning in the food industry is important for other reasons, including financial ones. Inadequate cleaning can impair equipment performance, reducing efficiency. Cleaning costs money (time, labour, equipment and consumables) and failure to clean properly can literally mean throwing money down the drain! Adequate cleaning can also be a legislative requirement and is mandated as such in the EU for all food products. Within the UK, the Food Safety Act 1990 makes it an offence to contaminate food so that it could be unreasonable to expect it to be used for human consumption. In addition, the Food Safety (General Food Hygiene) Regulations 1995 require the proprietor of a business to identify the steps in the activities of the business that are critical to ensure food safety. There is therefore a clear obligation in law to keep food premises clean, wherever there is a risk to food. The same regulations stipulate premises must be kept clean and maintained in good repair and condition and be designed to allow proper cleaning. It is therefore not necessary to prove that a particular soil or dirt is a risk for an offence to be committed, only that dirt exists (Dillon and Griffith, 1999). Legislation requiring cleaning premises can also be found in other countries (USDA, 2001).

The British Retail Consortium Global Standard – Food (BRC Food Standard) accepted as part of the Global Food Safety Initiative (GFSI) sets out the minimum standards major European retailers expect from their suppliers. Evolving, and likely to become more, rather than less stringent, one section deals with cleaning and cleaning schedules and requires the effectiveness of cleaning to be verified (BRC, 2005). Although not a legal requirement, failure to achieve the standard could be of economic importance and mean considerable loss of business/revenue to a food manufacturer by excluding them from important markets.

In spite of its importance, cleaning could be further improved in manufacturing, retail and food service (Gibson *et al.*, 1999; Griffith *et al.*, 2002; Sagoo *et al.*, 2003).

36.1.1 Cleanliness, microbial growth/survival and cross-contamination

As indicated above, organic matter derived from foods or food-related residues can be associated with microorganisms. When supplied with nutrients and the correct conditions these microorganisms can survive and/or multiply. The ability of both food spoilage and pathogenic organisms to attach to a wide variety of materials used within the food industry, is well documented (Cunliffe *et al.*, 1999). Following attachment, some bacteria can exhibit a variety of physiological and genetic responses to a range of environmental stresses, enabling them to survive in less than ideal conditions (Humphrey *et al.*, 1995). Once attached, some microroganisms can form biofilms (microorganisms plus associated organic matrix), which can be even more difficult to remove with increased resistance to disinfectants and sanitisers (Gilbert *et al.*, 1990). Monitoring cleaning programmes can therefore involve looking for the presence of microorganisms, organic residues or both.

Unlike bacteria, yeasts and moulds, which can grow in, or on, soiled equipment and environmental surfaces, viruses are obligate intracellular parasites, i.e. only grow in other living cells. However, some can survive well outside their hosts and persist in the environment for days or months. Cleaning and good personal hygiene are essential in preventing the spread of viral diseases.

Cross-contamination (defined as the process of contaminating a previously uncontaminated food surface or food) is of particular concern for those microorganisms that have a low minimal infectious dose (e.g. *Escherichia coli* O157, Norovirus). Cross-contamination can occur directly, from contaminated to uncontaminated, e.g. raw ready to eat, or indirectly. Indirect cross-contamination can involve a single event or be much more complex (see Fig. 36.2), involving a complex web of steps involving hands, equipment and surfaces. Within the food service environment given the frequency with which hands touch contaminated

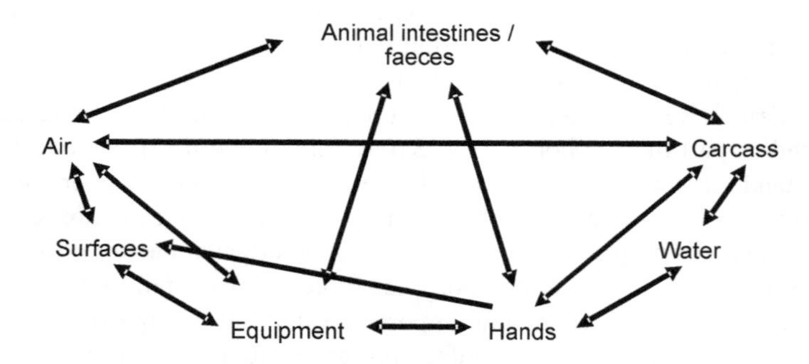

Fig. 36.2 Potential for cross-contamination in an abattoir.

surfaces and raw foods, as well as ready-to-eat foods (Clayton and Griffith, 2004), it is surprising that hand contact surfaces are often omitted from cleaning schedules. Hand contact surfaces are often heavily contaminated (Griffith *et al.*, 2000; Worsfold and Griffith, 2001) and, unless high- and low-risk areas are separated, provide highways by which microorganisms can spread within food environments leading to the contamination of ready-to-eat foods.

36.1.2 Managing cleaning

Ensuring cleanliness starts with the design, construction, operation and maintenance of equipment and premises. Assuming these are appropriately considered, then effective cleaning regimes can be resourced, documented and implemented.

Designing a cleaning regime is best undertaken as the result of a site survey. This considers construction, production flows and type, frequency and sequence of cleaning, facilities available, shift patterns, types of food residues, etc. Once the practicality and potential problems associated with cleaning have been identified, a provisional cleaning plan can be designed, constructed and then validated. After design and validation (proving that the plan works, i.e. is capable of delivering appropriate levels of cleanliness) documentation needs to be considered. Documentation helps to maintain consistency and transparency associated with cleaning methods, is a requirement of certification standards, such as the BRC, and is usually based on standard operating procedures (SOPs). Typical cleaning documentation will include a policy statement, a schedule and procedures, detailed instructions on how to clean each area or piece of equipment as well as record forms. Increasingly, the process is being supported by various software tools. Auditors, e.g. for BRC, may well ask to see both the cleaning programmes, as well as results obtained from monitoring, i.e. the routine assessment of cleaning efficacy. Cleaning regimes need to be current and part of a document control system.

Management responsibility and commitment, in both time and money, are important in ensuring successful cleaning and need to be evident. Unfortunately the process of cleaning is often perceived of low importance, with cleaners poorly paid.

It is said that 'you cannot manage what you do not measure'. Key to managing successful cleaning, both at the time of validation and later during routine implementation, is some means of testing or monitoring cleaning efficacy. Although cleaning practices will vary, Table 36.1 indicates the main stages likely to be involved in most cleaning regimes. The first three stages are designed to reduce surface soil, i.e. cleaning, with stage 4, disinfection, an additional option. This is used to ensure residual surface microbial numbers are reduced to low or acceptable levels.

One stage that is subject to debate is the need for rinsing after disinfection. The European Food Directives are sometimes unclear on rinsing: some state it should be undertaken but others allow it as an option, if it can be assured that there are no residual chemicals that can adversely affect food, people or

Table 36.1 Typical stages in a cleaning programme

Stage	Function	Reason
1 Pre-clean	Remove loose food or dirt, scrape, vacuum, etc. Rinse with water to remove smaller, soluble food particles	Improve efficiency of later stages, allows detergent access to more firmly adhering residues
2 Main clean	Removes more firmly adhering food residue, grease or dirt. Usually detergents used to emulsify food particles and reduce surface tension	Improves efficiency of later stages. Presence of dirt/residue/grease reduce the efficacy of disinfectants
3 Rinse	Removes detergent and emulsified/dissolved dirt and grease	Improves efficiency of disinfection, minimises any reactions between cleaning chemicals. Prevents microorganisms being redeposited on surfaces
4 Disinfect	Further reduction in the number of microorganisms	Minimises risk of cross-contamination, increases product shelf-life and safety
5 Final rinse	Removes traces of disinfectant	Minimises risk of disinfectant contaminating the food
6 Dry	Air dry or use disposable materials to minimise recontamination	Residual moisture provides an opportunity for any remaining microorganisms to grow and survive and increase the risk of cross-contamination (transfer rates)

equipment. The main argument in favour of rinsing is the removal of cleaning chemicals and possibly reducing the chances of developing biocide resistance, but this needs to be weighed against the microbiological quality of available water (at point of use, not entry into the premises), the potential for recontamination of cleaned surfaces and the need to preserve a dry processing environment. In the US, a number of sanitisers have approved limits for non-rinse application. If there are concerns about surface counts after cleaning, the use of these sanitisers first at higher levels, followed by rinsing, followed by their application at a no rinse level has been suggested (Tompkin *et al.*, 1999). An alternative is to use a gaseous bactericide, e.g. ozone, as a terminal disinfectant stage. This can achieve an extra kill before decomposing to oxygen (Moore *et al.*, 2000). The decision is best left to the individual company, bearing in mind the type and concentration of cleaning chemicals used, local water quality, type of product and the level of risk associated with it. It is important to realise, however, that the different stages in cleaning are interlinked and cumulatively help to ensure the overall process is effective. They can also inform how and when monitoring of cleaning needs to be undertaken.

36.1.3 Monitoring cleaning

Cleaning, as stated earlier, is the removal of soil and the process may also reduce the number of microorganisms present. Disinfection is specifically used to further reduce the number of microorganisms present and can be achieved using heat or chemicals. Both cleaning and disinfection can be monitored, although readers are reminded that disinfection is much more difficult and less likely to be achieved if prior cleaning is inappropriately performed.

Figure 36.3 indicates the possible consequences and combinations of surface conditions after cleaning. The reduction in organic residues ensures removal of food debris, allergens, etc., helps reduce the number of microorganisms, as well as preparing the surface for any further disinfection. A low residual microbial surface count reduces the chances of food spoilage and possibly foodborne disease. The presence or absence of residual moisture is also important in helping to prevent cross-contamination both by reducing microbial growth and survival and reducing potential transfer rates. Transfer rates can vary, from less than 1% to nearly 100% depending upon the surfaces, and are greatly increased in the presence of moisture (Harrison *et al.*, 2003). However, drying needs to be performed in a way that will not recontaminate the surface.

Figure 36.4 outlines the various microbiological and non-microbiological methods that could be used to assess the efficacy of cleaning and/or disinfection. Microbiological methods provide only an indication of the numbers of residual surface organisms. Non-microbiological methods primarily assess residual organic surface debris, although some, such as ATP, may, by virtue of its ability to assess microbial ATP, also detect microbial contamination. However, it

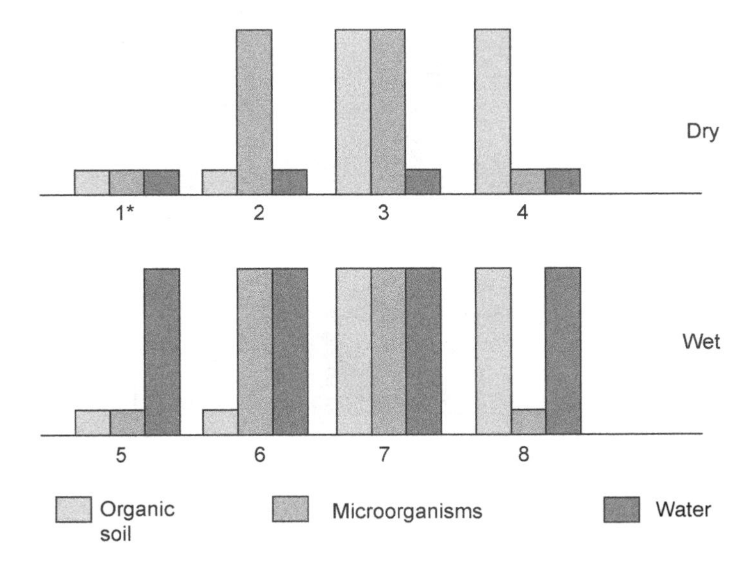

Fig. 36.3 Possible combinations of surface condition after cleaning: *desirable surface condition, dry and free from microorganisms and residual organic soil.

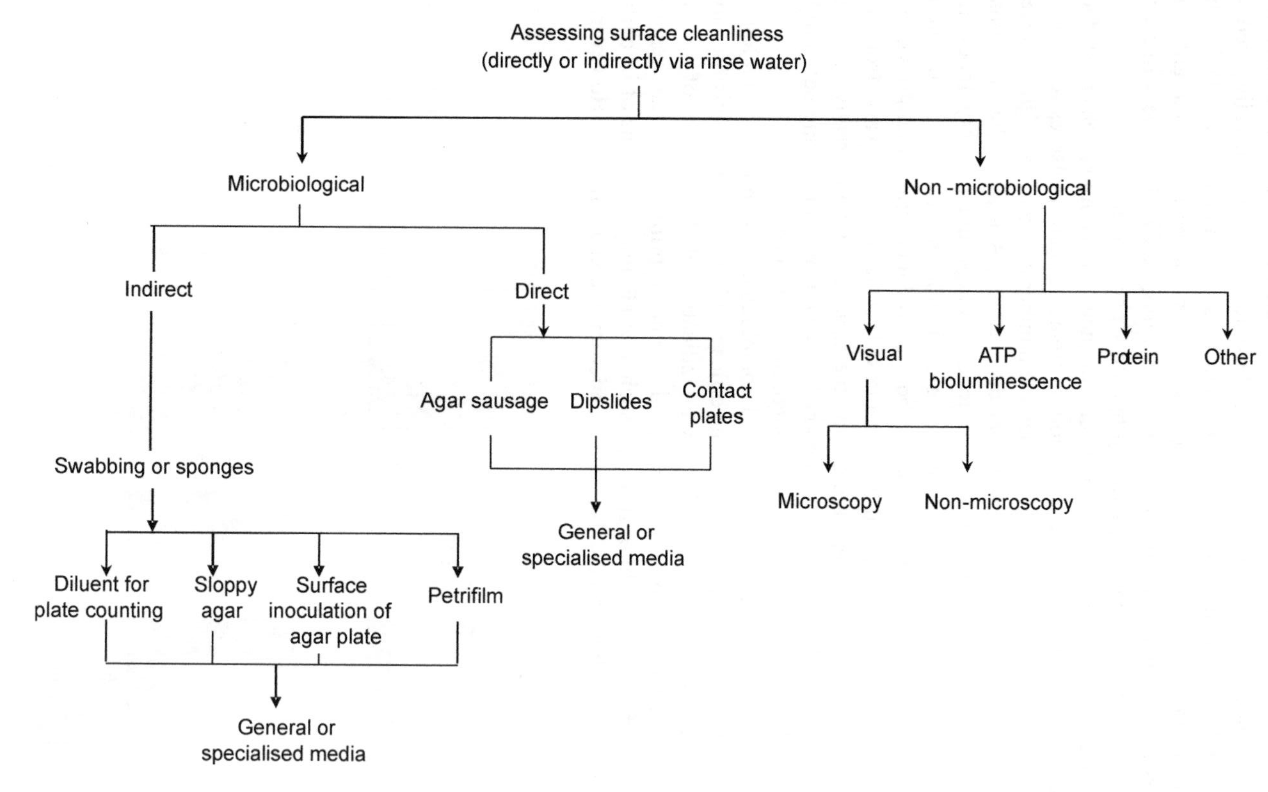

Fig. 36.4 Methods for assessing surface cleanliness.

should be realised that in food environments there is often little value in trying to directly correlate surface counts to ATP readings. For a strong correlation, the ratio between microbial ATP and food debris would need to be constant. This is unlikely to occur in many food premises and sites, with the possible exception of some hand contact surfaces (Griffith *et al.*, 2000). It is possible to have food

Table 36.2 Comparison of microbiological and non-microbiological testing

Attribute	Microbiological (cultivation)	ATP	Protein
Acceptance	Widely accepted by food industry worldwide.	Although not universally accepted, it is widely accepted in the UK and many other countries. Acceptance has increased rapidly since 1995.	Generally less well accepted than ATP but more recently developed technique.
Method/principle testing	Microorganisms derived from surface grow and multiply. Laboratory facilities, or at minimum, portable incubation and safe disposal systems required.	ATP derived from microorganisms and food debris converted into light and detected using a luminometer. No laboratory facilities needed.	Protein from surface food debris transferred onto a swab or equivalent. Protein detected by a colorimetric reaction and compared with colour standards. No instrumentation or laboratory facilities required.
Time for results	18–48 hours.	2 minutes.	5–10 minutes.
Reproducibility for raw milk contaminated surface	CV 84–300%.	CV 9–79%.	Not applicable.
Approximate running costs	60–100 pence (in-house).	95–155 pence.	95–145 pence.
Capital costs	Variable but incubator £130–2000 Autoclave £600–10,000.	Luminometer £900–2500.	Optional colorimeter available for some tests.
Staff requirements	Some level of microbiological training.	Brief training in test protocol. No specialised knowledge required.	Brief training in test protocol. No specialised knowledge required.

surfaces with a high ATP count and low microbial count (e.g. UHT milk), thus a small increase in product residue can increase the ATP count but not microbial numbers. Similarly, depending on the food product, e.g. raw foods, and their level of microbial contamination, it is possible to have a low ATP increase with higher increases in microbial numbers. Factors that may influence the choice of how cleaning is monitored are presented in Table 36.2.

The aims of this chapter are to review microbiological and non-microbiological methods for monitoring cleaning efficacy and suggest ways to manage an integrated programme of monitoring, in an attempt to ensure adequate and cost-effective cleaning.

36.2 Microbiological surface sampling

Microbiological surface sampling cannot be described as new, with reports of its use going back to the 1920s and 1930s (Saelhof and Heinekamp, 1920; Krogg and Dougherty, 1936), although precise methodological details are lacking. However, most of this early work was based on swabbing with direct agar contact methods only developed later.

Advances in microscopy have resulted in the development of methods based on epifluorescent, confocal scanning laser and episcopic differential interference contrast microscopy. These later methods, while providing useful laboratory information, are impractical for routine use in food businesses. The main methods in use within the food industry include the use of swabs and sponges to recover organisms from the surface followed by their cultivation on/in nutrient media (effectively indirect). Methods where the nutrient medium is in direct contact with the surface tested involve impression plates or dipslides. The choice of method will depend on the precise information required and the prevailing circumstances (see Table 36.3). Additional options are whether to test for: an aerobic colony count – a general measure of surface contamination; indicator organisms, some of which can provide a measure of hygiene standards; or specific pathogens. Often the latter may be like looking for a 'needle in a haystack', although is of particular benefit if:

- a specific pathogen has been found in a food sample;
- investigating cases of food poisoning;
- part of a specific pathogen control programme, e.g. controlling *Listeria* in food premises.

36.2.1 Swabbing/sponges

Swabbing in one form or another remains the oldest and probably the most widely used method for surface monitoring (Moore and Griffith, 2002a). It should be noted that although the term monitoring is used, it does not conform to the definition used within HACCP. For the latter, results must be obtained in time for corrective action to be taken and swabbing, like impression plates,

Table 36.3 Comparison of main microbiological methods for hygiene monitoring

Method	Advantages	Disadvantages
Swabbing	Widely used and accepted. Can be qualitative (types of organisms) and quantitative. Any shape, size or surface area can be tested. Relatively inexpensive.	No universally agreed protocol. Methods, media, etc. vary widely between companies and conditions. Incubation and sterilisation facilities needed or external contract laboratory. Staff with some microbiological training needed. Poor recovery especially dry surfaces. Poor reproducibility. Motile organisms can cover surfaces of agar.
Contact plate	Direct contact with surface. Better reproducibility than swabbing. Fixed relatively small area. Can be bought pre-prepared. Available in variety of media. Relatively inexpensive.	Flat smooth surfaces only. Motile organisms can cover surface of agar. Possible agar residue on surface. Lids can become detached in transport, although one make with a lockable lid is available. Incubation and sterilisation disposal facilities needed.
Dipslide	Direct contact with surface. Better reproducibility than swabbing. Fixed area/narrow shape, relatively small surface area. Can be bought pre-prepared in a variety of media. Different media on reverse side of paddle if required. Minimal incubation facilities needed (portable). Can be used to test rinse water. Sealed unit with screw cap. Longer shelf-life. Paddle can be hinged for easier use.	Flat smooth surfaces usually. Motile organisms can cover surface of agar. Incubation and sterilisation disposal facilities needed. Possible agar residue on surface. Semi quantitative depending on numbers.

relies on cultivation, which, depending on the organism can be hours, days or weeks (e.g. for TB).

Most swabbing protocols are based upon the swab-rinse technique originally developed by Manheimer and Yheunez in 1917 (Favero *et al.*, 1968). A sterile swab, consisting of a more or less flexible shaft with a fibrous bud or tip, is pre-moistened in an appropriate wetting agent and inoculated by rubbing over the surface to be tested. The microorganisms transferred to the swab can then be cultivated and counted, either by inoculating the swab directly onto an appropriate solid culture medium or by releasing into a known quantity of sterile recovery diluent, which is then used to prepare pour plates. This description of swabbing also indicates some of the variability in the technique, which can considerably affect the apparent number of organisms recovered (Moore and Griffith, 2002a). If the number of microorganisms on a surface is known (as in laboratory conditions), and compared with the number obtained from swabbing, there is low recovery particularly at low surface population densities below 10^4 cells per cm^2 (Holah *et al.*, 1988). Additionally the swabbing technique lacks reliability, i.e. repeatability and reproducibility are poor (Moore and Griffith, 2002a,b; Moore *et al.*, 2001). Various 'standard' methods are recommended, including by the EU. An ISO Standard (ISO/FDIS 18593) has also been produced but currently there is no universally accepted method of swabbing in use. Some of the possible variables are indicated in Fig. 36.5 and Table 36.4.

Swabbing is widely used in industry to assess surface contamination and as a reference for comparison with other methods. However, basic information is still lacking as to the optimum protocol and the effect that variations may have on recovery rates (Moore and Griffith, 2002a). Overall recovery can be seen as a function of the removal of microorganisms from the test surface, their release from the swab and their subsequent ability to grow. Recovery rates will depend on the technique used but an optimum recovery of 10% for Dacron swabs is not uncommon. Microorganisms can become increasingly difficult to remove from a surface once they have adhered, particularly if associated with a biofilm. Additionally, organism retention within the bud fibres may also lead to poor repeatability and reduced counts. Techniques/variables that improve one element of the swabbing process may adversely affect another. One study (Moore and Griffith, 2002a) showed protocols that improved removal, adversely affected release. Optimum overall recovery may therefore be a trade-off or compromise between different components of the whole process.

The lack of repeatability can make it difficult to interpret the results from environmental swabbing, especially between staff, from different plants and when different protocols are used. An apparent low surface count from a single swab may reflect swabbing technique as much as low contamination levels. This may give a false impression of cleaning efficacy and if guidelines or company specifications have been achieved (see Section 36.5.2). Swabbing, as with other surface assessment techniques, is best used to establish trends in the performance of the cleaning and disinfection programme, where over a period of time, the programme can be seen to be failing or improving. The food manu-

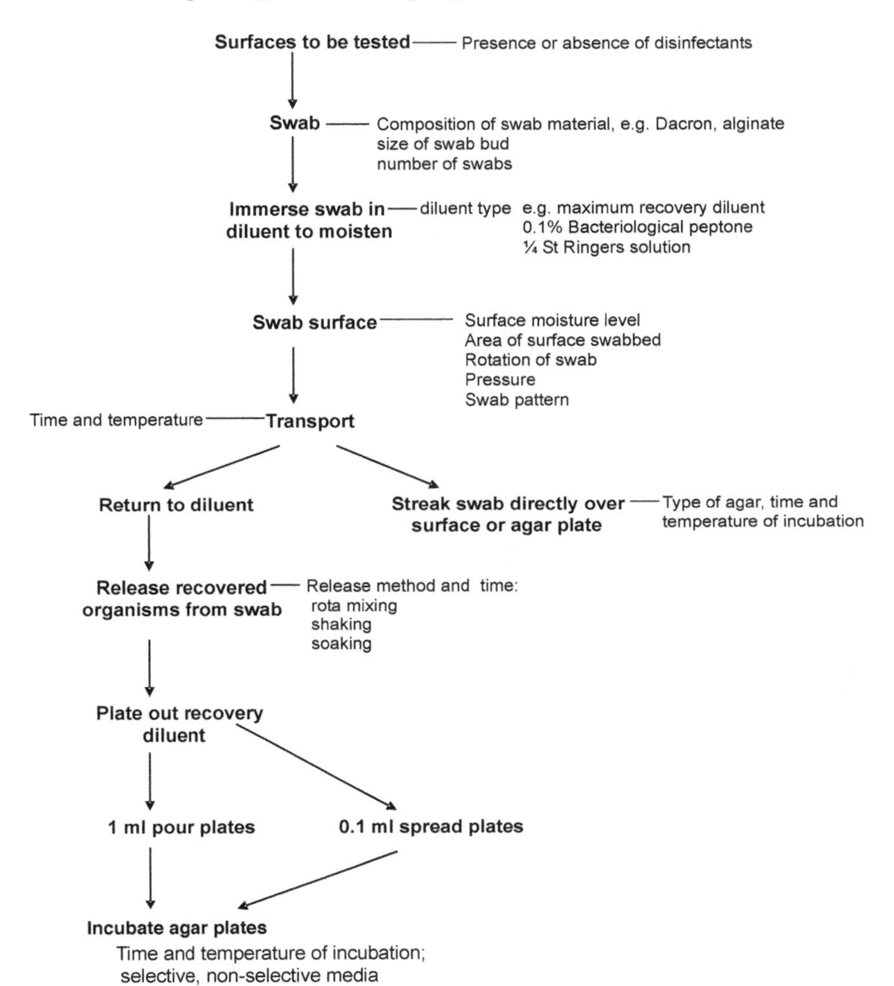

Fig. 36.5 Surface hygiene swabbing variables potentially affecting recovery rates.

facturers' view is that the variability in swabbing per se is not sufficient to prevent the detection of high surface counts on a given day, i.e. the results of a badly implemented cleaning operation.

Understanding the problems associated with overall recovery rates can help to improve and control the process. Sampling wetting solutions, designed to maintain isotonic conditions and reduce physiological stress, can be used to maintain the viability of microorganisms recovered from surfaces (CCFRA, 2003). Care needs to be taken in their selection to ensure they do not artificially increase the count by providing a medium in which recovered microorganisms can grow during transit. Some surfaces may still have residual disinfectant present and neutralising agents, appropriate for the disinfectant being used can

Table 36.4 Selected published surface hygiene swabbing protocols (adapted from Davidson, 2001)

Sampling variable	Collins *et al.* (1989)	Roberts *et al.* (1995)	Harrigan (1998)
Type of swab	Cotton wool or alginate	Cotton wool	Cotton wool or alginate
Area of surface sampled	5 cm × 5 cm	5 cm × 5 cm	Not specified
Swabbing protocol	Swab within 25 cm^2 card or cellophane template area using one swab	Using aluminium template, swab entire area using two swabs, one moist and one dry, rotating swabs. Return both swabs to one diluent tube	Swab predetermined area by rubbing firmly over the surface in parallel strokes, with slow rotation of the swab
Type of diluent	Nutrient broth	MRD or ¼ St Ringers	¼ St Ringers
Release method	Soak and squeeze swab	Shake until cotton wool broken down into fibres	Agitate/shake swab in tube up and down 10 times
Type of culture media	Not specified	Plate count agar	Nutrient agar or trypticase soya agar
Cultivation/plating method	Not specified	Spread/pour plates	1 mL pour plates for cotton swabs, 0.1 mL and 1 mL plates for alginate swabs
Time and temperature of sample storage /incubation	Not specified	Not specified	Not specified
Time and temperature of incubation	Not specified	30 °C, 48–72 hours	Not specified
Expression of results	Count/25 cm^2	Count/cm^2	Count/cm^2

Dacron is now the most widely used 'bud' material. MRD = maximum recovery diluent be added to the wetting solution. These help to prevent organisms, removed from the surface (where they may be more resistant), being killed by residual sanitiser and thereby giving an 'artificially' reduced count.

Ideally swabs should be processed as soon as possible, although this is often impractical, especially when outside laboratories are used. Under such conditions, samples should be transported non-frozen at a low temperature <5 °C, this can result in minimal differences compared with real time analysis (CCFRA, 2003). Times of sampling and processing need to be recorded, as well as delivery temperature, so that any unusual results or significant differences from the norm can be identified and considered when interpreting the results. Variables of time and wetting agent also need to be considered and optimised in sampling for specific pathogens. Appropriate pre-enrichment media should be used, although overgrowth by more rapidly growing non-pathogens needs to be considered.

Some manufacturers may add a surfactant to their wetting solutions to improve 'pick-up' from the test surface. These can, in some cases, artificially increase the number of colonies counted by breaking up clumps of organisms and thereby increasing the number of 'colony-forming units'. Concerns over the inability of swab buds to release recovered organisms have prompted one manufacturer to develop a radically new type of swab. This lacks the normal fibrous bud, which is replaced with short textured flocked nylon in spatula or swab format. This device releases more of the organisms removed from a surface and can yield an approximate 1 log improved overall recovery compared with traditional swabs (Griffith and Moore, 2004). An alternative approach has led to a development of a wet or dry vacuum bacterial collection system. This may be of particular use in pathogen testing as it allows a much larger surface to be tested without the need/use of a swab to lift/remove the organism from the surface being tested.

Other recent variations in swabbing include the use of petrifilms to replace traditional agar plates for cultivation. These are small, thin films coated with nutrients and gelling agents. After wetting the film with approximately 1 ml of de-ionised water to rehydrate the growth medium, it can be used to provide a surface count. Another variation involves self-contained media and hygiene swabs in tubes with the potential to offer more rapid results (Moore and Griffith, 2002b). A traditional swab, after testing a surface, is returned to its accompanying culture tube containing semi-solid agar incorporating an indicator. Microorganisms removed from the surface and retained by the swab grow and, as they multiply in the semi-solid agar, the indicator changes colour. The results are semi-quantitative in that the number of bacteria is not recorded but the time taken for the indicator to change colour is a measure of the original microbial load. Unclean surfaces, depending upon the extent of microbial contamination, can test positive within 12 hours.

Sponges work on a similar principle to swabbing, in that microorganisms are removed, released and cultivated. Recovery is by wiping a compressed sterile sponge (e.g. cellulose acetate) of varying sizes over the test surface. Most have no swab shaft, and in order to avoid contamination, the sponge needs to be held using a sterile glove, usually provided with the sponge. Sponges may be pre-moistened or require the addition of a wetting agent. After inoculation the sponge is returned to a sterile envelope/packet and transported to a laboratory. After the addition of a suitable diluent to the envelope, usually followed by agitation/stomaching, the released organisms can be counted. Similar errors to those encountered in swabbing may occur, and there is some evidence to suggest that the sponge matrix retains even more of the recovered organisms than swabbing, resulting in lower overall recovery (Moore and Griffith, 2002b). Some sponges offer the advantage of greater surface area: being much bigger than conventional swabs they allow larger surface areas to be tested, and may therefore be more useful in testing surfaces for pathogens. Greater pressure can also be applied than with swabs. Other variations include sponges on sticks, and in France, the use of gauze to swab surfaces. Validation data on the effectiveness of some of these alternatives are not widely available.

36.2.2 Replicate organism direct area contact (RODAC) – agar sausages, contact plates, dipslides

All direct agar contact methods, or replicate organism direct area contact (RODAC), involve pressing sterile agar onto a surface to be sampled. A contact time of 10 seconds with a force of $25\,g/cm^2$, without lateral movement, is suggested (ISO/CD 14698-12). Microorganisms are directly transferred onto the agar surface and, after incubation for an appropriate length of time, multiply and form colonies, which are visible and can be counted. In general this approach is best suited to smooth, flat surfaces. The methods vary in how the agar is dispersed. Contact plates resemble small plastic Petri dishes with a lid. The agar is poured into them, leaving a convex contact surface. After removing the lid the agar is pressed onto the test surface. The contact plates are then incubated and examined 24–48 hours later.

Agar immersion, plating and contact (AIPC) slides, more commonly referred to a dipslides or paddles (in the US), were developed from 'dip spoons' used in counting the numbers of organisms in urine samples. They comprise a double-sided hinged paddle with a neutral or selective agar, attached to both sides. The paddle is contained within a transparent cylindrical tube or plastic container. The dipslide is removed, then pressed onto the surface to be tested, replaced back into the tube and resulting colony growth counted, or compared with pictorial estimates/diagrams of surface counts. They can also be used for counting the number of organisms in liquid samples of food, water or rinse water. Recently a flexible hybrid contact plate/dipslide, to test more irregular-shaped surfaces, has become available.

Direct agar contact methods have a number of advantages and disadvantages compared with traditional swabbing (see Table 36.3). Advantages include ease of use, generally lower costs and better recovery and repeatability (Niskanen and Pohja, 1977; Moore and Griffith, 2002b; Moore *et al.*, 2001). Disadvantages include being more suited to flat surfaces and on very contaminated surfaces providing only an approximate or semi-quantitative estimate. This can make statistical analysis of results more problematic. However, if only an indication of cleaning adequacy, i.e. pass or fail is required, rather than the precise number of organisms, this may not be a problem. It is easy to count the individual colonies, obtained from marginally unclean/clean surfaces, based on clean surface counts currently considered attainable (see Section 36.5.2). If a more precise number of colonies, from a heavily contaminated surface, is required then agar contact methods may be inappropriate.

36.3 Non-microbiological surface sampling

Historically, prior to swabbing, visual assessment was the only means to assess the effectiveness of cleaning and is still the most widely used method in food service and the home. Visual assessment still also has an important role to play in food manufacturing as part of an integrated assessment protocol (see Section 36.4.1). In isolation it is not a good method for assessing anything other than gross surface soil. It can be combined with magnification, with or without UV/ black light assistance, as well as touch, dust or powder to detect grease residues. Most auditors will take a torch (flashlight) with them to inspect the visual cleanliness of dark/hidden, out of the way places in food premises.

Various types of microscopy enable individual microorganisms and biofilms to be observed but these are not practical for routine use. More recently a device for visually assessing surface cleanliness, based on detecting fluorescing chemicals, e.g. chlorophyll residues in faeces or meat, has become available. This can be of use in surface assessment in some food processing areas. The advent of swabbing in the early 1900s offered the only major alternative for routine use until the late 1980s. Since then alternative, rapid chemical detection methods, starting with ATP, have been developed (Griffith *et al.*, 1997). These methods detect food/organic debris rather than microorganisms. As cultivation is not required, only a rapid chemical reaction, the test results are available in seconds or minutes, rather than hours or days. These newer tests probably represent a truer assessment of cleanliness (absence of soil), than does a microbial count. Soil can protect microorganisms, and therefore knowledge that the surface is free of soil provides reassurance concerning the potential for microbial growth. Thus the philosophy of their use is different, offering proactive cleanliness management with results available in time for corrective action to be taken (including recleaning) prior to surface use. Microbial enumeration is reactive and proves, by which time the product may have left the factory, that a surface was or was not contaminated after cleaning. Some traditional microbiologists still feel happier with assessing surface microbial contamination, although their approach is further challenged by increased concern over food allergies. If cleaning is inadequately performed, food allergens from one food may remain on a surface and cross-contaminate other foods. Rapid chemical tests are not a

Table 36.5 Characteristics of an ideal assessment method

- Detects microorganisms and food residues with sufficient sensitivity
- Works equally well on wet and dry surfaces
- Good repeatability/reproducibility
- Easy to use
- Rapid
- Cheap
- Foolproof/recordable/tamperproof
- Results can be used in trend analysis

Table 36.6 Considerations in using rapid chemical tests

Universality of test chemical	Residue/moiety detected is found in a wide range of foods
Quantity in food	Amount of the detected chemical contained in different foods
Sensitivity of tests	Lowest level of chemical residue that can be detected by the test
Other	Cost – especially important if many tests undertaken Time – results obtained rapidly to allow corrective action Simplicity – ease of use by all staff with minimum training Documentation – ability to read/record results digitally with time and date
'Horses for courses'!	Choice of test selection varies on individual circumstances and types of food produced

direct replacement for microbiological testing but provide complementary information and should be used as part of an integrated strategy (see Section 36.4.1)

The market for rapid test methods is predicted to increase (Griffith *et al.*, 1997), although it is probably fair to say the ideal test method does not yet exist (see Table 36.5) and their use needs to be considered in relation to the type of business and the food produced (see Table 36.6).

36.3.1 ATP bioluminescence

ATP, or adenosine triphosphate, is the universal energy currency, or donor for metabolic processes, in all living cells. It is present in viable microorganisms (not viruses) and in food stuffs in variable amounts, depending upon their composition. The ATP bioluminescence assay works on the principle (see Fig. 36.6) that ATP in food/food residues and microorganisms, in the presence of an enzyme/substrate complex, leads to light emission. The light is measured quantitatively in a luminometer (light-detecting instrument), with results available in 10–30 seconds. The amount of light emitted is therefore proportionate to

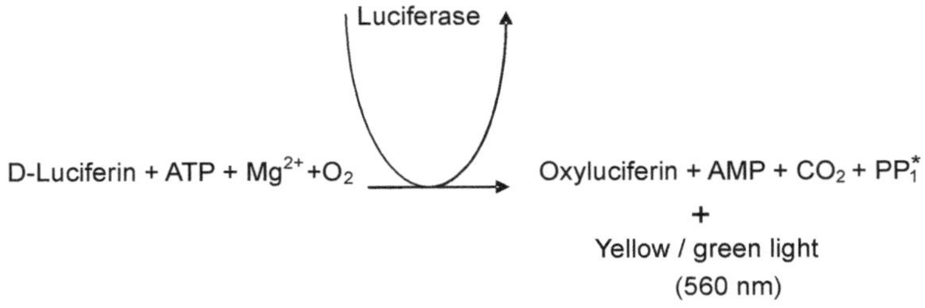

*pyrophosphate

Fig. 36.6 Schematic representation of the ATP bioluminescence reaction.

the amount of ATP on a surface and hence its cleanliness. The level of ATP within cells varies depending upon the type of cell, e.g. animal, yeast, bacteria, and its phase of growth, but the ATP pool in living cells is normally kept consistent by regulatory mechanisms (Davidson *et al.*, 1997). The enzyme-substrate complex luciferin–luciferase converts the chemical energy associated with the ATP into light in a stoichiometric reaction with 1 photon of light produced by the hydrolysis of 1 molecule of ATP. The light emitted is normally measured in relative light units (RLUs), calibrated for each make of instrument and set of reagents. Therefore the readings obtained from assessing food plant cleaning need to be compared with baseline data representing acceptable clean values (see Section 36.5.2).

A range of luminometers and tests are available and major new developments in assays and equipment occur approximately every 4 years. Originally luminometers were large and only suitable for laboratory use. These have evolved over the years into small hand-held models, which can be used anywhere within a plant. Many luminometers use a photomultiplier tube in the light detection system, although as part of the drive to cheaper, smaller instruments, some manufacturers have replaced these with less sensitive photo diode-based systems. This may, depending on the chemistry of the assay system, reduce overall test sensitivity. Luminometers can incorporate a printer, although this need has been obviated in most of the newer instruments with trend analysis software, which can store and then download data to a PC. This software is very useful for comparing data over time and from different sites and plants. It indicates areas frequently improperly cleaned and surfaces that are moving towards loss of control, and allows comparison between cleaning operatives. One manufacturer has added the ability to perform additional checks, e.g. pH and temperature measurement, by adding additional test probes and facilities to the luminometer. This may be useful but can be problematic if one part develops a fault and, at the end of the day, it is how well the luminometer and the test designed for it actually perform that is the most important determinant of choice. Most manufacturers offer calibration and/or positive/negative controls to help ensure accuracy of readings.

Changes in the assays have resulted in the ability for testing to be performed by non-technical staff, with tubes and pipettes replaced by simple, single shot, all in one assays. The exact chemical formulations used in the assays vary with suppliers, but all contain luciferin/luciferase, magnesium ions, buffering, substrates, stabilisers and extractants (to remove the ATP from living cells). They vary in shelf-life, depending on precise composition and the temperature and manner of their storage. Typical refrigerated storage times would be 6 months. A number of manufacturers claim to have developed a test stable at room temperature, although this may vary considerably among countries and climates.

ATP is a universal biochemical found in many, but not necessarily all, foodstuffs. High counts can be found in some fresh foods, e.g. tomatoes, while other foods, especially highly processed foods such as fats, oils or sugar, contain very low amounts. Detergents/sanitisers used in cleaning can have a similar effect to the extractants used in the tests and different studies have demonstrated that commonly used cleaning chemicals can cause either quenching or enhancement of the ATP signal. It is therefore desirable, for consistency of results, to ensure that cleaning agents are removed by rinsing before testing is performed. The repeatability and reliability of the instruments and their tests can vary considerably among manufacturers but is generally superior to microbiological swabbing (Griffith *et al.*, 1997). In the author's experience, problems are more likely to be a reflection of how the cleaning and testing have been performed, rather than the ATP test system. The sensitivity of the instruments and their tests is variable and there has been discussion over exactly how sensitive ATP tests need to be. The key requirement is that they should be able to discriminate 'well cleaned' from 'inadequately cleaned', surfaces important or relevant to a business. There is therefore a demand for a certain minimum sensitivity. Whether a test can be too sensitive is more debatable. Highly sensitive tests, if available, could be adapted to make them more specific, e.g. for microorganisms only or even specific pathogens. Highly sensitive tests could also allow more refined discrimination between clean and marginally unclean.

While test manufacturers will provide guidance on clean benchmark levels, they are usually best determined on site by the food business and then used as the basis for continuous improvement (see also Section 36.4.3). ATP has also been adapted by one test manufacturer for the detection of allergen residues and it is claimed the test detects down to 0.1–5 ppm of allergen food residues.

36.3.2 Protein and other assays

Following the development and application of ATP bioluminescence as a measure of cleanliness, other chemical assays/tests for food residue components have been investigated. The stimulus is to develop a non-instrument dependent test, that is cheap and functional. A range of other chemical residues including protein, reducing sugars, NAD and phosphate, either individually or in combination, are now used as the basis for rapid cleaning tests. Usually the tests lead to the production of a single, or sequence of, coloured end-products within a specified time (1–10 minutes). The colour changes can be qualitatively assessed visually. This can be subjective and the option to use a cheap sample instrument to measure, and/or record the results is available for some tests, if needed. The subjectivity is most variable for marginally unclean surfaces, the clean or very dirty being less subjective. Some tests retain a swab-based format, while others use test strips of plastic or pads of absorbent material impregnated with relevant reagents. Which, if any, of these tests will be of benefit to a food business will depend on a number of factors (see Table 36.6), not least of which is the sensitivity of the assay. Such tests, if cheaper and instrument independent, may find potential use in food service establishments. Often criticised for poor cleanliness, they are the reported location for most outbreaks of food poisoning (Griffith, 2000).

Of the non-ATP assays, protein detection methods offer potential where the food residues, e.g. poultry/meat/dairy products, are high in protein. In some of the assays, other food non-protein, reducing components may also bring about a colour change. Some methods make use of an enhanced Buiret reaction. Under alkaline conditions the peptide bonds of proteins form a complex with the copper II (Cu^{2+}) of the Buiret reagent, reducing it to Copper I (Cu^{+}) ions. These react with bicinchoninic acid, producing an intense purple colour. For this test, the slogan 'if it's green it's clean' is used. Other protein tests make use of different so called 'protein error indicator', dyes (e.g. tetrabromophenol blue), which change colour in the presence of protein at a particular pH. These tests may be swab-based, although some versions use test strips or pads. Depending upon the food examined, protein tests may be more or less sensitive than ATP bioluminescence (Moore and Griffith, 2002c). The intensity of the colour and its speed of production provide an indication of the level of soiling although results are usually just pass/fail.

NAD (nicotinamide adenine dinucleotide) and related forms are chemical residues, also widely distributed in biological materials, including foods and microorganisms. Hence, the level of NAD on a surface provides a measure of organic soiling. NAD is detected in a chemical reaction leading to the production of a pink/purple colour on a test strip, within 5 minutes. As with the other chemical residue detection kits, lack of a positive reaction does not represent lack of microorganisms. This test may be more or less sensitive than other rapid chemical tests and its usefulness needs to be trialled and will depend upon the type of foods produced.

Other swab-based tests can be used to detect either glucose or glucose and lactose, the latter being of practical benefit to the dairy industry. Results are obtained within 60 seconds. Unfortunately, because of the nature of the colour reaction, the results are read 'if it turns green it isn't clean', which can lead to confusion with protein test results! In most cases, the test is less likely to be as sensitive as the equivalent ATP assay but it is claimed that for many food residues, it is nearly as good and is rapid and non-instrument based.

Another option is a test strip containing enzymes that can detect phosphate and carbohydrates. This makes use of capillary action to 'suck' food residues from a wetted surface onto test reagents. A wetting indicator is also incorporated into the test strip, going dark grey when sufficiently wet, the main food residue indicator going deep purple within 120 seconds, if the surface is unclean. The advantages of the strip approach compared with the swab are simplicity and lower costs of production and transport. There is also some indication that these test strips can be useful in assessing the effectiveness of hand-washing. As with all the rapid chemical tests, no conclusion regarding the absence of microorganisms can be inferred from a negative test.

36.4 Monitoring/sampling protocols and strategies

36.4.1 Protocols

It must be recognised that developing a monitoring protocol or strategy is pointless if cleaning itself is poorly implemented and managed. In recognition that no one ideal test exists, the combining of test methods into a coherent protocol, relevant to a business, as part of a consistent approach to plant sanitation is recommended (Figs 36.7 and 36.8). The extent, structure and use of such protocols are likely to be dependent on the plant, the cleaning methods used, the level of risk associated with the product and the sophistication of the

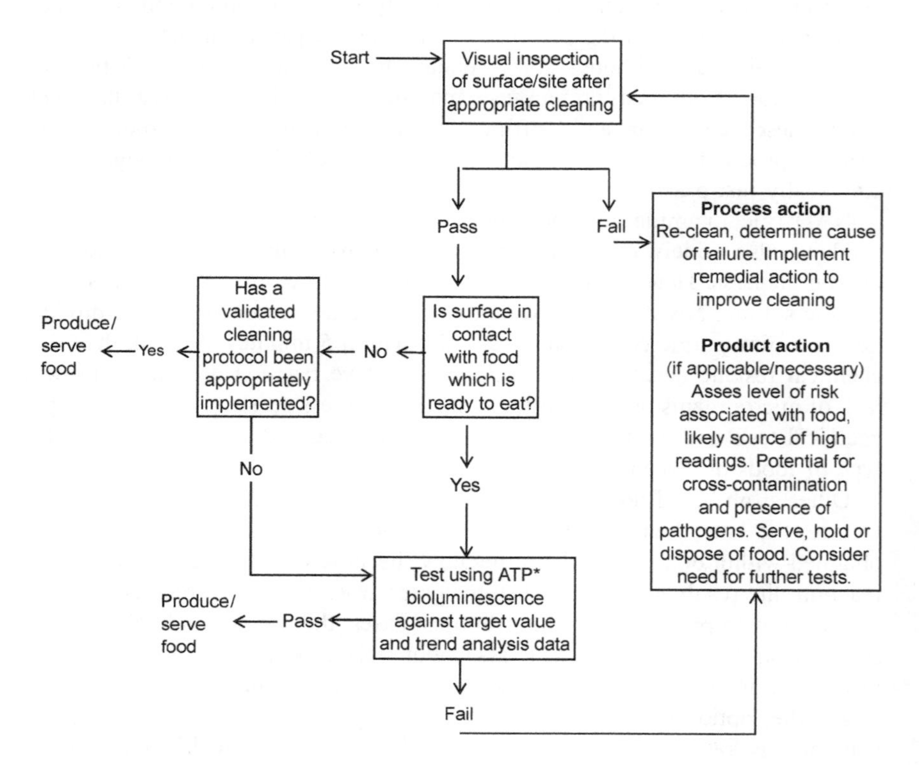

Fig. 36.7 Stages in an integrated cleaning monitoring programme – no microbiological facilities (food service, retail, small processor).

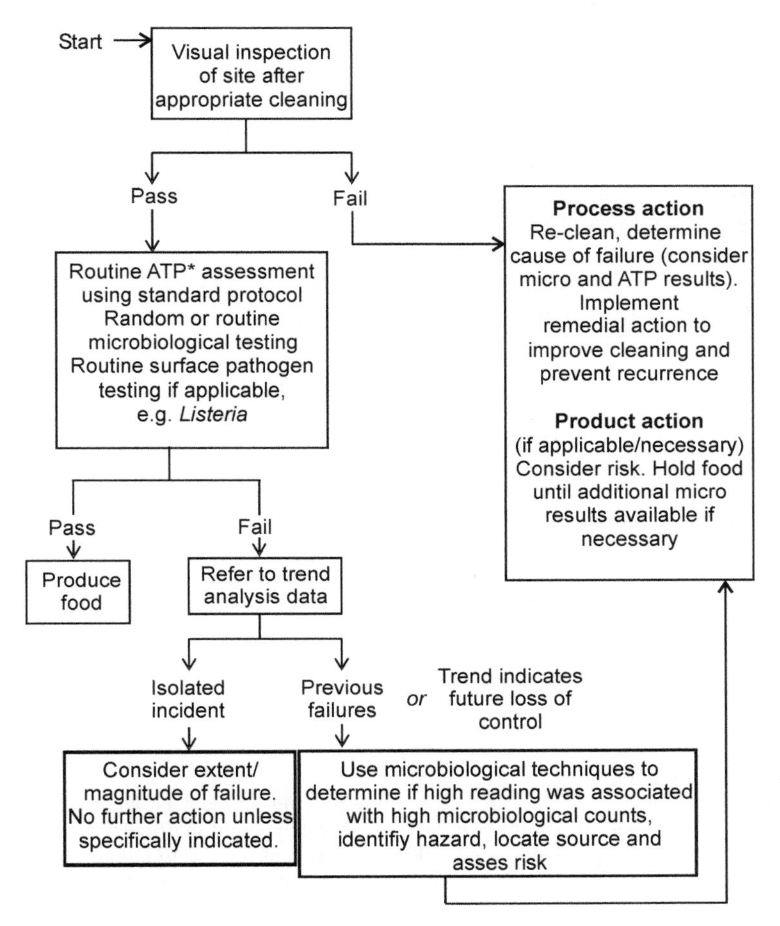

Fig. 36.8 Stages in an integrated cleaning monitoring programme – microbiological facilities available.

quality systems in use. An integrated protocol should recognise the type of information provided by, or the weakness associated with, one test and use another to complement it. This approach has been previously recommended and incorporates corrective actions (Griffith *et al.*, 1997).

The starting point for any protocol should be visual assessment. This is quick and cheap – if a surface is visually dirty then there is likely to be little point in any further testing, additional notes should be made of visual moisture and surface condition/wear. However, in isolation, visual assessment is not a good indicator of surface cleanliness and, therefore, any decision about further testing needs to be considered in relation to product risk as well as the availability of other tests and the types of food soil (see Figs 36.7 and 36.8). One or more types of rapid testing, e.g. ATP, can be combined with microbiological methods to determine the effectiveness of surface cleaning and disinfection. Additionally, both microbiological and non-microbiological tests have value in validating the original cleaning programme and investigating the reasons for any failure to clean effectively (see Table 36.7). Rapid tests can also be used after cleaning, prior to disinfection, to determine whether the surfaces are sufficiently free of soiling to enable successful disinfection (see Table 36.1). They can also help to identify areas difficult to clean or routinely poorly cleaned, thus indicating

Table 36.7 Common reasons for failure to clean and disinfect adequately

Surface not part of cleaning programme	Often occurs with hand contact surfaces. Result may mean surface is not cleaned regularly/efficiently
Non-validated cleaning protocol	Inappropriately designed cleaning regime, i.e. cannot achieve desired results
Work culture	Cleaning not perceived as important, lack of support or motivated workforce
Training	Cleaning staff often poorly trained
Failure to monitor cleaning	Poor cleaning goes unnoticed
Failure to implement cleaning appropriately	
Implementation:	Validated cleaning regime not implemented correctly, resulting in lack of consistency and inefficient cleaning
Equipment:	Dirty water, cloths, equipment, failure to colour code can lead to surface recontamination and/or spread of microorganisms between areas Cleaning equipment, e.g. mops, left wet, act as breeding grounds for microorganisms Failure to change cleaning cloths/equipment frequently enough. As cleaning commences the cleaning equipment can become contaminated
Method:	Failure to adequately remove soil – gross or microscopic Time: cleaning rushed. Disinfectants particularly do not work instantly and require time to destroy microorganisms present Product concentration: chemicals too dilute – insufficient strength to exert full effect – can increase chances of microbial 'resistance'. Too concentrated can be hazardous Product formulation: inappropriate cleaning chemical selected in relation to product composition and soil types, e.g. does product work well if used in hard water area, does product work well if used in high grease areas, does disinfectant destroy viruses if they are a risk, use of high alkaline cleaners on aluminium surfaces

where microbiological testing is most useful.. This type of integrated approach provides a better indication of cleaning efficacy, helps to provide transparency and demonstrates a company's concern for effective cleaning. Additionally, it has the potential to save on cleaning costs by identifying what is, or is not, effective or necessary.

36.4.2 Strategies

Any overall policy to ensure clean surfaces (see Fig. 36.9) should include monitoring surface cleanliness. When and where to sample needs to be considered in relation to risk and the potential for cross-contamination. It is unfortunate that in some companies, sampling concentrates on the centre of large flat surfaces, which are more likely to be well cleaned. Less attention may be given to hand contact surfaces or cracks and crevices where soil and later microorganisms can accumulate. Rinses, especially in CIP, in liquid processing plants

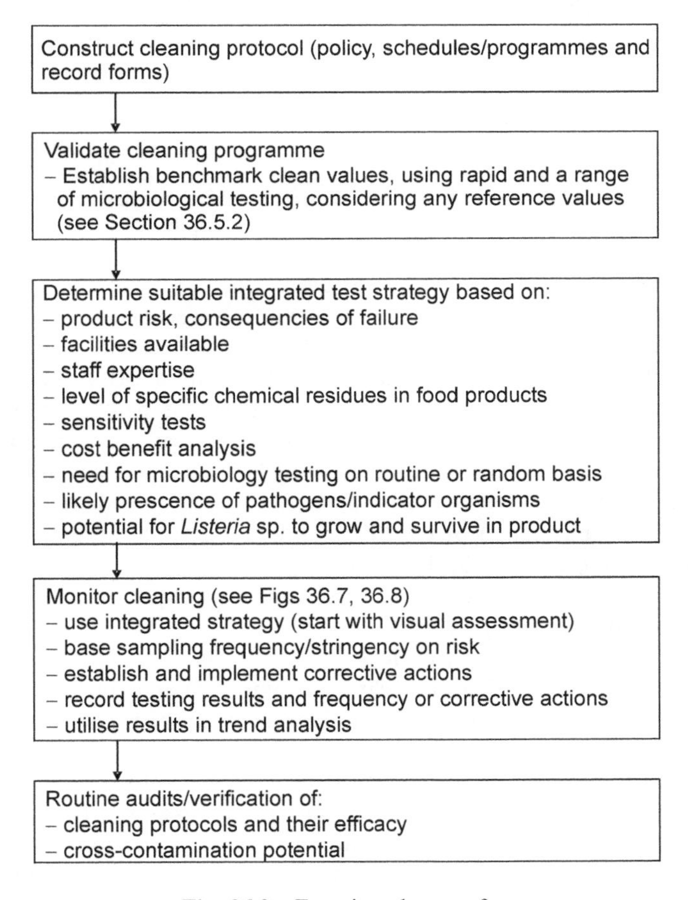

Fig. 36.9 Ensuring clean surfaces.

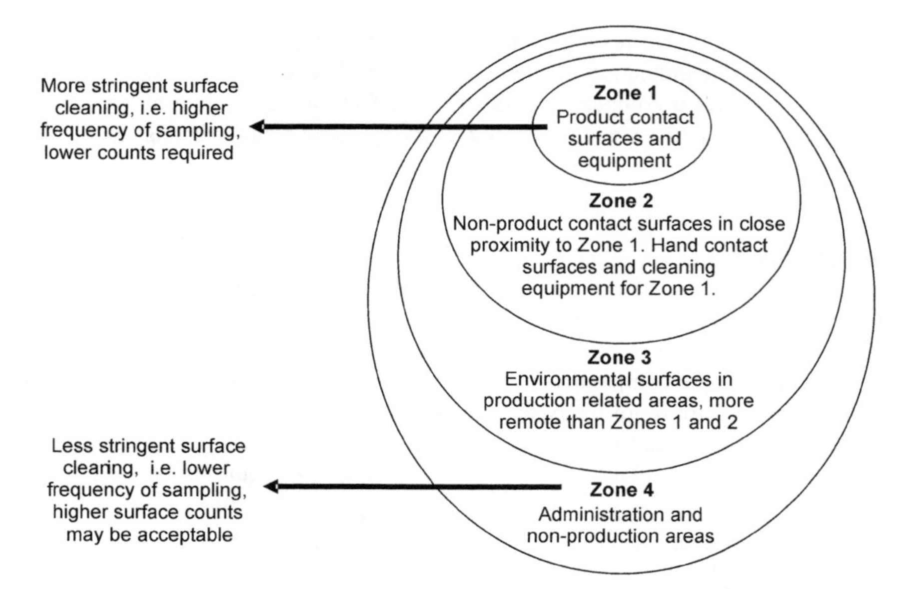

Fig. 36.10 Organisation of areas, based on risk, to determine sampling frequency and stringency. The precise allocation of areas into zones will, to some extent, be product- and plant-specific and the figure is indicative only. Microorganisms can eaily be spread in food premises and molecular sub-typing has shown that pathogens can persist for years in the environment, even after so-called deep cleaning to eradicate them.

can often be tested to provide an estimate of surface cleanliness as can the quality of the first product run after, for example, a weekend shut-down.

One approach is to designate surfaces as food contact, general environmental, hand contact and cleaning (equipment/cloths). The latter need care and attention and can act as vectors causing the zig-zag spread of pathogens within an environment (Harrison *et al.*, 2003). An alternative is to organise areas into high risk and low risk. Greatest emphasis in cleaning being directed towards the high-risk or high-care areas, where ready-to-eat foods are handled. A third option (ICMSF, 2002), a variation of the high risk/low risk, is to arrange areas into zones or shells (see Fig. 36.10). This essentially establishes successively 'cleaner zones' and/or zones of increased sampling frequency and decreasing levels of contamination.

Zone 1 represents the most critical areas of cleaning – mainly surfaces in contact with ready-to-eat products, e.g. conveyor belts and cutters. Filling and depositing heads, spray drying or cream depositers can be particularly difficult to clean effectively.

Zone 2 could include hand contact areas in close proximity to zone 1 and may even include the surfaces used/touched during hand-washing (Griffith *et al.*, 2003). Zone 2 would also include environmental areas in close proximity to Zone 1. The latter may be good locations for the survival of organisms such as *Listeria*. Any *Listeria* control strategy should concentrate on eradication of

Listeria from Zone 2 sites first, before consideration of Zone 1. Failure to do so is only likely to lead to rapid recontamination of Zone 1 surfaces.

Zone 3 includes floors, walls, etc. in areas more distant from Zone 1, and comprises the least critical food handling areas where sampling frequency may be at its lowest and environmental contamination at its highest, e.g. where raw products are received. This is relative, i.e. in relation to Zone 1, and is not an excuse for poor cleaning, or not testing. It is a recognition, based on risk, that less stringent sampling is needed.

However, all three zones need to be considered in terms of product flow and people movement. Depending on where and how it is used, the degree of separation of high from low risk and its potential to spread bacteria, cleaning equipment could be considered as Zone 2 or 3. The use of contaminated cleaning equipment is one of the main reasons for failure to clean effectively (see Table 36.7) and can spread pathogens from low- to high-risk areas. Ideally each area and zone should have hygienically designed colour coded equipment, which should not be used in other areas. Essential also is that this equipment is stored clean and *dry*, or if used on a semi-continuous basis, frequently cleaned and stored in fresh disinfectant solution monitored for concentration levels. Care should also be taken, especially in *Listeria* control programmes, with shoes/boots, tracks, etc. – these need to be cleaned properly as they can spread organisms around premises (Tomkin *et al.*, 1999).

This type of framework fits into the increasing use of cleaning and cross-contamination audits. Cleaning audits (internal or external) should be conducted independently and assess both the quality and adequacy of the cleaning programme and the level of compliance with it. New personal digital assistants (PDA), palm-held auditing tools with appropriate software, are available (http://www.foodsystemsaudit.co.uk), which simplify the whole process and can incorporate data from microbiological or rapid testing. One advantage of the PDAs is that draft reports can be produced, if necessary, before the auditor leaves the areas/premises being audited. Other advantages include greater consistency, overall time savings and greater usability of data for analysing trends and designing corrective strategies. These become even more powerful if combined with cross-contamination audits, which are broader in scope, and used to assess the overall risk or potential for cross-contamination. The latter assess more than just cleaning, and include personal hygiene, facilities available, e.g. hand-washing and drying, traffic and personnel flow. Monitoring surface cleanliness is not without costs but these need to be considered in relation to the costs associated with failing to monitor (see Fig. 36.11).

36.4.3 Using the results

Given the cost of cleaning and the expenditure in time, effort and money that some companies put into surface sampling it is surprising that more use is often not made of the results. It has been said 'if it wasn't recorded it did not happen' and documented results from microbiological and non-microbiological sampling

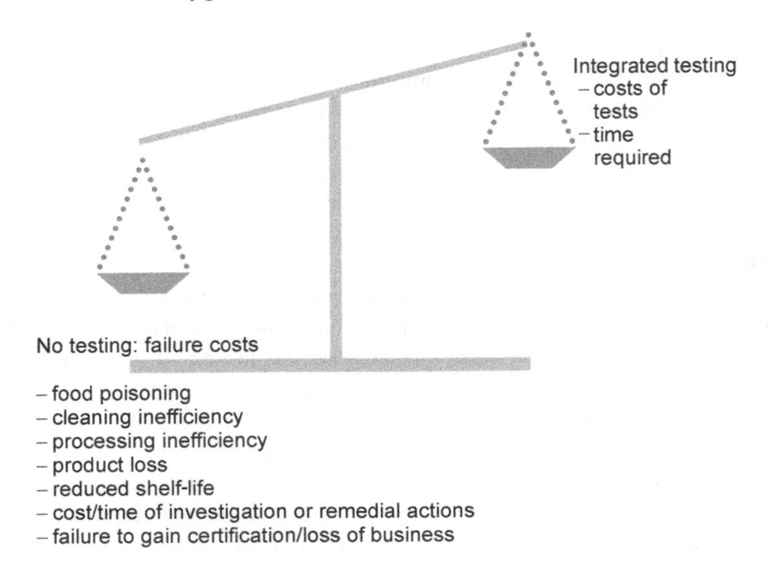

Fig. 36.11 Cost benefits of monitoring cleaning.

can be used to validate cleaning and provide ongoing data for trend analysis. This can be a powerful management tool and identify, for example, areas where cleaning is often/regularly poorly performed, staff who are not cleaning appropriately, effectiveness of changes in cleaning practices. The results can be linked to food end-product counts, staff rotas, shift patterns, etc. Cumulatively this can help to make cleaning and monitoring as cost effective as possible – maximum effectiveness for minimum cost.

36.5 Future trends

With the advent of rapid techniques, the introduction of HACCP, allergen concerns and other factors, the importance and need to assess cleanliness has changed over the years since 1990. This has coincided with the emergence of a number of pathogens with a low minimum infectious dose. It is therefore likely that assessing cleanliness will become more important and evolve to meet changing requirements in the tests that are used, how they can be integrated into an holistic approach, and the results in relation to legislation, reference values or other needs.

An always important factor is cost, especially as the extent of testing needs to be reviewed in relation to the benefits that can be achieved. Such analyses (see Fig. 36.11) should consider the failure costs, i.e. costs associated with poor cleaning as well as the costs of testing. Cleaning should deliver value, i.e. clean surfaces in relation to cost and risk. Future developments are likely to see a further reduction in the costs of rapid test detection instruments and or consumables along with more sophisticated foolproof software. For food service, which currently does little testing, the future is likely to be the introduction of very easy to use, low-cost, non-instrument tests.

36.5.1 Test methods

Changes in test methods are likely to be driven by versatility, speed, specificity, sensitivity and cost. More innovative approaches are likely in the design of flexible agar contact systems, suitable for use on irregular-shaped surfaces. More rapid microbiological tests will be developed. This may be in isolation or in combination with tests for the presence of specific pathogens. The ability to detect lower levels of ATP has developed over recent years with, depending upon the reagents and how they are produced, the ability to detect below 1 femtomole of ATP. ATP or microbiological tests, specific and sensitive enough to detect very low levels of bacteria or ATP, even in dry conditions, could be developed. New formats, other than swabs, may be devised. Currently, no rapid test is well suited for testing surfaces that are high in fats, and tests specific for fats and oils would be useful for some processors.

36.5.2 Guidelines/cleaning standards

When prosecutions for dirty premises/equipment, particularly in the food service sector, take place they are usually based on visual assessment. This may change with the development of proposed reference values for surfaces after cleaning in food premises. However, this depends on improved test reproducibility. Guidelines vary widely (see Table 36.8) and while often their origins are unclear, they

Table 36.8 Some recommended guidelines/standards for clean surfaces

Suggested values	Date and source
80 cfu/cm^2	Herbert *et al.* (1990)
5 cfu/cm^2	USDA (1994)
0–10 cfu/cm^2 for aerobic colony count 0–1 cfu/cm^2 Enterobacteriaecae	EC Decision 2001. Meat (Hazard Analysis Critical Control Point) Regulations 2002
<2.5 cfu/cm^2	Mossel *et al.* (1999)
<2.5 cfu/cm^2	Griffith *et al.* (2000)
<500-RLUs	Applies to use of one specific ATP test/equipment combination (Griffith *et al.*, 2000)
Target 1 cfu/cm^2 Maximum of 3 cfu/cm^2	Swedish Food Agency (1998)

are usually based upon a perception of risk and what is acceptable. An alternative strategy is to decide what is attainable on a surface, after correct implementation of a well-designed cleaning programme. Studies on over 3000 surfaces (Griffith *et al.*, 2000, 2002; Redmond, *et al.*, 2001) indicate that in most cases, levels of <2.5 cfu/cm^2 for a general surface count are attainable and these are relatively close to other suggested standards. Failure to achieve this level of cleanliness or disinfection may mean the cleaning protocol is poorly constructed, is not implemented well, or the surface cannot be satisfactorily cleaned.

Similar standards, using ATP bioluminescence, have also been proposed. However, any standards should always be considered in relation to risk and possible soil types. Crucial to any use of standards is agreement on consistent/approved sampling methods. Cleaning costs money and cleaning needs to deliver, otherwise it is a waste of money and time. Monitoring cleaning efficacy should therefore be part of the work of any food business and if undertaken appropriately can be cost effective (Fig. 36.11), providing a greater confidence in food safety and superior product shelf-life.

36.6 References

BRC (2005) BRC Global Standard – Food, Issue 4 (January). The British Retail Consortium.

CCFRA (2003) *Manual of Hygiene Methods for the Food and Drink Industry*. Guideline No. 45, Chipping Campden, UK

CLAYTON D and GRIFFITH C J (2004) The use of notational analysis to observe the implementation of specific food safety practices in catering. *British Food Journal*, **106**(3): 211–227.

COLLINS C H, LYNE P M and GRANGE J M (1989) *Collins and Lyne's Microbiological Methods*. 6th Edition. Oxford. Butterworth's & Co. (Publishers) Ltd.

CUNLIFFE D, SMART C A, ALEXANDER C and VULFSON E N (1999) Bacterial adhesion at synthetic surfaces. *Applied and Environmental Microbiology*, **65**: 4995–5002.

CURTIS V (2001) Hygiene: How myths, monsters and mothers-in-law can promote behaviour change. *Journal of Infection*, **43**: 75–79.

DAVIDSON C A (2001) *An evaluation of some microbiological and ATP bioluminescence methods for the recovery and detection of bacterial contamination from food contact and environmental surfaces*. PhD thesis, University of Wales.

DAVIDSON C A, GRIFFITH C J and FIELDING L M (1997) ATP bioluminescence and the validation and monitoring of cleaning programmes. *Journal of Bioluminescence and Chemiluminescence*, **12**: 96.

DAVIDSON C A, GRIFFITH C J, PETERS A C and FIELDING L M (1999) An evaluation of two methods for monitoring surface cleanliness – ATP bioluminescence and traditional hygiene swabbing. *Luminescence*, **13**: 1–5.

DILLON M and GRIFFITH C J (1999) *How to Clean: A Management Guide*. Grimsby, M D Associates.

FAVERO M S, MCDADE J, ROBERTSEN J A, HOFFMAN R K and EDWARDS R W (1968) Microbiological sampling of surfaces. *Journal of Applied Bacteriology*, **31**: 336–343.

FOOD STANDARDS AGENCY (2004) *Consumer Attitude Report 2003. www.food.gov.uk*

GIBSON H, TAYLOR J H, HALL K E and HOLAH J (1999) Effectiveness of cleaning techniques used in the food industry in terms of the removal of bacterial biofilms. *Journal of Applied Microbiology*, **87**: 41–48.

GILBERT P, COLLIER P J and BROWN M R W (1990) Influence of growth rates on susceptibility to antimicrobial agents: biofilms, cell cycle and dormancy. *Antimicrobial Agents Chemotherapy*, **34**: 1865-1868.

GRIFFITH C J (2000) Food safety in catering establishments. In: *Safe Handling of Foods*. Edited by Farber and Todd. New York, Marcel Dekker.

GRIFFITH C J and MOORE G (2004) An evaluation of new microbiological surface sampling kits. *International Association for Food Protection Conference in Phoenix, Arizona*. August.

GRIFFITH C J, DAVIDSON C, PETERS A C and FIELDING L M (1997) Towards a strategic cleaning assessment programme: hygiene monitoring and ATP luminometry, an option appraisal. *Food Science and Technology Today*, **11**: 15–24.

GRIFFITH C J, COOPER R A, GILMORE J, DAVIES C and LEWIS M (2000) An evaluation of hospital cleaning regimes and standards. *Journal of Hospital Infection*, **45** (1): 19–28.

GRIFFITH C J, DAVIES C, BREVERTON J, REDMOND E C and PETERS A C (2002) *Assessing and Reducing the Risk of Cross Contamination of Food Stuffs in Food Handling Environments*. London, Food Standards Agency.

GRIFFITH C J, MALIK R E, COOPER R A, LOOKER N and MICHAELS B (2003) Environmental surface cleanliness and the potential for contamination during handwashing. *American Journal of Infection Control*, **31** (2): 93–96.

HARRIGAN W F (1998) *Laboratory Methods in Food Microbiology*, 3rd Edition. London, Academic Press, p. 149.

HARRISON W A, GRIFFITH C J, AYERS T and MICHAELS B (2003) Bacterial transfer rates and cross contamination potential associated with paper towel dispensing. *American Journal of Infection Control*, **31** (7): 387–391.

HERBERT M, DONOVAN T and MANGER P (1990) *A Study of the Microbiological Contamination of Working Surfaces in a Variety of Food Premises Using the Traditional Swabbing Technique and Commercial Contact Slides*. Ashford, Public Health Laboratory Service.

HOLAH J T, BETTS R P and THORPE R H (1988) The use of direct epiflourescent microscopy (DEM) and the direct epifluorescent filter technique (DEFT) to assess microbial populations on food contact surfaces. *Journal of Applied Bacteriology*, **65**: 215–221.

HUMPHREY T J, SLATER E, MCALPINE K. ROWBURY R J and GILBERT R J (1995) *Salmonella enteritidis* phage type 4 isolates more tolerant of heat, acid or hydrogen peroxide also survives longer on surfaces. *Applied and Environmental Microbiology*, **61**: 3161–3164.

ICMSF (INTERNATIONAL COMMISSION ON MICROBIOLOGICAL SPECIFICATIONS FOR FOODS) (2002) Microorganisms in Food 7: *Microbiological Testing in Food Safety Management*. New York, Kluwer Academic/Plenum Publishers.

KROGG A J and DOUGHERTY D S (1936) Effectiveness of the methods of dish and utensil washing in public eating and drinking establishments. *American Journal of Public Health*, **26**: 897–900.

MOORE G and GRIFFITH C J (2002a) Factors influencing the recovery of microorganisms from surfaces using traditional hygiene swabbing. *Dairy Food and Environmental*

Sanitation. **22** (6): 14–24.

MOORE G and GRIFFITH C J (2002b) A comparison of surface sampling methods for detecting coliforms on food contact surfaces. *Food Microbiology*. **19**: 65–73.

MOORE G and GRIFFITH C J (2002c) A comparison of traditional and recently developed methods for monitoring surface hygiene: an industry trial. *International Journal of Environmental Health*. **12**: 317–329.

MOORE G, GRIFFITH C J and PETERS A C (2000) Bactericidal properties of ozone. *Journal of Food Protection*, **63** (8): 1100–1106.

MOORE G, GRIFFITH C J and FIELDING L (2001) A comparison of traditional and recently developed methods for monitoring surface hygiene within the food industry: a laboratory study. *Dairy Food and Environmental Sanitation*, **21** (6): 478–488.

MOSSEL D A A, JANSEN J T and STRUIJK C B (1999) Microbiological safety assurance applied to smaller catering operations world-wide. *Food Control*, **10**: 195–211.

NISKANEN A and POHJA M S (1977) Comparative studies of the sampling and investigation of microbial contamination of surfaces by the contact plate and swab methods. *Journal of Applied Bacteriology*, **42**: 53–63.

REDMOND E, GRIFFITH C J, SLADER J and HUMPHREY T (2004) Microbiological and observational analysis of cross contamination risks during domestic food preparation. *British Food Journal*, **106** (8): 581–597.

REDMOND E C, GRIFFITH C J, SLADER S and HUMPHREY T J (2001) *The evaluation and application of information on consumer hazard and risk to food safety education*. London, Food Standards Agency.

ROBERTS D, HOOPER W and GREENWOOD M (1995) *Practical Food Microbiology*. London, PHLS, pp. 109–110.

SAELHOF J R and HEINEKAMP W J R (1920) Recovery of *Streptococcus hemolyticus* from restaurant tableware. *American Journal of Public Health*, **10**: 704–707.

SAGOO S K, LITTLE C L, GRIFFITH C J and MITCHELL R T (2003) A study of cleaning standards and practices in food premises in the United Kingdom. *Communicable Disease Report*, **6** (1): 6–17.

SWEDISH FOOD AGENCY (1998) *The Swedish Statute Book*. SLV SFS, 10

TOMPKIN R B, SCOTT V N, BERNARD D T, SVEUM W H and GOMBAS K S (1999) Guidelines to prevent post-processing contamination from *Listeria monocytogenes*. *Dairy, Food and Environmental Sanitation*, **19**: 551–562.

USDA (1994) Guidelines for reviewing microbiological control and monitoring programs, part B55, attachment 2. In: *Meat and Poultry Inspection Manual*. Washington DC, United States Department of Agriculture.

USDA (2001) *Food and Drug Administration Food Code*. United States Department of Agriculture, *http://www.usda.gov*

WORSFOLD D and GRIFFITH C J (2001) An assessment of cleaning regimes and standards in butchers' shops. *International Journal of Environmental Health Research*, 11: 257–268.

37

Improving air sampling

H. Miettinen, VTT Biotechnology, Finland

37.1 Introduction

The ability of air to contain and transport liquids, solids and living substances is frequently overlooked or forgotten. After becoming airborne, organisms are able to contaminate products through air as long as they stay viable, which depends on a number of different parameters.

The conditions of sampling and analysis should be considered prior to the choice of the sampler (Ambroise *et al.*, 1999). Microbes are notoriously difficult to assess accurately under variable conditions and no single sampling or assay method is suitable for all applications; rather the method needs to be tailored to the application of interest (Griffiths & Stewart, 1999). Understanding the airborne behaviour of microbes over a range of environmental conditions is vital for recommendation of the handling, sampling and assessment of bioaerosols (Griffiths & Stewart, 1999).

In order to effectively monitor air quality, the nature of bioaerosols as well as the parameters affecting air sampling and assaying techniques have to be understood. Aerosols exhibit complex aerodynamic behaviour resulting from a combination of physical influences that affect success of the air sampling. The bioaerosol sampling parameters and the choice of the sampler must be considered according to the process and aim of sampling. Generally, air samplers must collect a representative sample of the required air fraction with a minimum of stress caused to the microbes to maintain viability. Bioaerosol sampling techniques consist of passive and active methods. The traditional passive air sampling method is the use of sedimentation plates. Active sampling methods are based on different mechanisms such as impaction, centrifugation, filtering and electrostatic precipitation. In addition to bioaerosol sampling there are particle counters for total particle amounts that can be exploited. Bioaerosol samples are analysed with various methods: culturing method is the traditional and widely used method but it has drawbacks. Other assay methods include microscopy, fluorescence, flow cytometry, ATP bioluminescence and molecular methods. For each environment an appropriate sampler and assay method has to be chosen. The bioaerosol results must always be interpreted taking into account sampling place, sampler, assay method and case-specifics. The future development of real-time continuous monitoring of total airborne microbes as well as specific spoilage contaminants is important, but it will be some time before it will be available to the food industry.

37.2 Microbial viability in the air

Traditionally, microbial viability is understood as the ability to divide and multiply. Only viable microbes can cause infection, while both the living and dead ones or their products can be responsible for allergenic and toxic illnesses. An airborne organism may have a very short life, its stability being influenced by relative humidity (RH), temperature, oxygen levels, solar and ultraviolet (UV) radiation, and chemical factors (Parrett & Crilly, 2000). A number of different factors affecting microbes can cumulatively stress them and affect their viability (Griffiths & DeCosemo, 1994). Important factors include organism species, cultivation conditions, method and way of aerosol generation, sampling techniques and the airborne environment. Desiccation, radiation, oxygen, ozone and its reaction products as well as various pollutants can affect the viability of microbes (Griffiths & DeCosemo, 1994).

The growth phase affects the survival of microbes in an aerosol (Griffiths & DeCosemo, 1994). Brown (1953) found that the viability is minimal during the transition from stationary to logarithmic stages for *Escherichia coli*. Better survival rates from the stationary rather than from the logarithmic stages have been observed for both *E. coli* and *Serratia marcescens* (Goodlow & Leonard, 1961; Dark & Callow, 1973). There is little information available on the survival of microbes aerosolized from food environments. Stersky *et al.* (1972) showed that *Salmonella* New Brunswick survived much longer when aerosolized from skim milk than from distilled water.

Bacillus subtilis var. *niger* spores are widely reported to be stable in air. The viabilities of *Salmonella enterica* serovar *enteritidis* and *Salmonella enterica* serovar *typhimurium* were significantly better than those shown by aerosols of *Legionella pneumophila* and *Mycobacterium tuberculosis* studied for two hours in air at 24 °C with 75% RH (McDermid & Lever, 1996). Yeasts are eukaryotes and are likely to be affected differently by aerosolization and sampling than bacteria. Bacterial spores survive better than vegetative cells. Microbes can mutate and adapt to changes in their growth environment, implying that it becomes very difficult to know if a given strain will respond in a consistent way to a stress factor applied over a long period of time (Griffiths & DeCosemo, 1994).

Oxygen slowly kills most airborne microbes through oxidation (Kowalski & Bahnfleth, 1998). The toxic effect of oxygen is related to moisture content; it usually increases with the degree of desiccation, increasing oxygen concentration, time of exposure and also on whether the desiccation is caused by aerosolization or drying. It is important to remember that the combined effects of oxygen and RH should be considered when explaining losses in viability (Griffiths & DeCosemo, 1994). Airborne fungal contamination also correlates with air pollutants, e.g. ozone concentration (Lin & Li, 2000). Airborne bacteria are subject to dehydration caused by evaporation of water from droplet-carrying microbes, as well as evaporation of cellular water. Dehydration of microbes causes osmotic stress and may result in decreased survival. Thompson *et al.* (1994) showed that the total recovery of viable *Pseudomonas fluorescens* was much higher sampled from high than from low RH level air. The decrease in total recovery of microbes with increase in desiccation time was more pronounced at low RH levels of < 50% (Thompson *et al.*, 1994). For yeast cells the survival was four times higher under high RH (> 70%) conditions compared with low RH (2060%). The mould spores of *Penicillum* sp. were not affected by RH of air (Lin & Li, 1999b).

37.3 Why, how and what to sample

Concern is growing in the food industry to determine the importance of the airborne route as a possible source of contamination (Griffiths & DeCosemo, 1994). Many food producers now include bioaerosol monitoring as part of their Hazard Analysis Critical Control Point (HACCP) system (Parrett & Crilly, 2000). Any point at which the product is exposed to air is a possible route for airborne contamination. Bioaerosol monitoring is carried out traditionally for three principal reasons (Griffiths & DeCosemo, 1994): (1) to meet legal requirements in complying with guidelines that often state that air quality may have to be monitored but do not specify the methodology or the acceptable limits to use, (2) to collect epidemiological data, possibly with a view to set occupational exposure limits and (3) for scientific interest to determine how the air affects the products processed. Regular bioaerosol sampling is also important in controlling the effectiveness and hygiene of the ventilation system. Even the filters should have maintenance programmes and advanced planned change programmes, and sometimes the bioaerosol samples are the first sign of the system failure. In addition to mechanical reasons, biological reasons such as the growth on filter materials can also cause problems.

The following criteria should be used to determine a sampling strategy: the sampling method, specificity and level of sensitivity required as well as the speed with which a result is required. Furthermore, information on the importance of total cell counts versus viability of the cells in the sample as well as the particle size range needed put demands on the method chosen. When the reasons for carrying out sampling are fully understood, the correct weighting can be assigned to each of the above-mentioned criteria (Griffiths & DeCosemo, 1994).

It is likely that the overall level of microbes will require monitoring, although the detection of potential pathogens or specific spoilage organisms may also be required. The major concern is probably for the viable microbes contaminating the food products. Viable counts are therefore likely to be more important than the total particle counts. If the sampling method selected uses agar media the choice of the collection media is an important parameter. General agar media for bacteria or fungi support most species. However, if the aim is to isolate and identify specific microbes that can contaminate the product in question it is vital to use a medium that particularly supports those contaminants that are relevant and supported also by the product. The information from a general medium does not indicate whether such strains are able to grow on the relevant product (Kure *et al.*, 2004a). Combination of general and specific media describes the overall situation of the processing. Moreover sensitive methods may be required, since low levels must be detected as in the monitoring of pathogens or potential pathogens. In both the food and pharmaceutical industries the speed with which a result needs to be obtained is a question of economics (Griffiths & DeCosemo, 1994).

The sampling time must be considered according to the process and aim of sampling. In the food industry, monitoring the bioaerosol situation during processing is important but also the situation after cleaning procedures can give additional information. According to Parret and Crilly (2000), important information from the sampling period includes: location and area of the site, date and time of sampling, test temperature and moisture conditions, functioning of the ventilation system during sampling, personnel in the area, volume of air sampled along with sampling period, collection media used and incubation time and temperature used. In addition all unusual operations or events should be recorded.

Choosing the sampling places in the process environment is an important part of accomplishing useful bioaerosol results. It is valuable to look into air flows in the process environment before sampling, since air flows can vary and a sample taken from one place may be totally different from sample taken from a metre away. Smoke generators are available that can be applied to study and visualize air movements. Bioaerosol results are more reliable and give more precise information of the product contamination through an air route if the samples are taken as near the product lines as possible. The number of samples and sample places in one area should also be considered according to air movements, product risk level and the overall knowledge needed.

37.4 Bioaerosols and bioaerosol samplers

Aerosols exhibit complex aerodynamic behaviour resulting from a combination of physical influences that include Brownian motion, electrical gradient, gravitational field, inertial force, electromagnetic radiation, particle density, thermal gradients, hygroscopicity and humidity (Kang & Frank, 1989d). The size of aerosol particles is generally in the range 0.5–50 μm. Particle size is the major factor influencing aerodynamic behaviour (Kang & Frank, 1989a) that affects significantly the sampling effectiveness and yield.

Generally, the performance of an active bioaerosol sampling device is characterized by the device's ability to aspirate particles into the inlet, to transmit them through the sampler's interior and to collect them on the collection surface. In the case of viable bioaerosol sampling, the performance of samplers must maintain the stability of microbial viability as an additional component during sampling (Thompson *et al.*, 1994). The samplers must also collect a representative sample of the required fraction of bioaerosol with a minimum of stress, so that the biological activity of the aerosol is not too impaired (Griffiths & DeCosemo, 1994). The efficiency of a sampler in collecting a particle of a given size is related to the air velocity in the impaction nozzle. Too low a velocity in the inlet will result in failure to collect the particles of interest. Too high a velocity results in a high shear force and may cause serious damage to the microbes, thus decreasing their viable recovery. The larger the aerosol particles, the greater the over- or underestimation of the aerosol concentration is likely to be. If the sampler is not mounted on the axis of the ambient airflow, the measured aerosol concentration may be significantly different from that in the ambient air environment (Thompson *et al.*, 1994).

Samplers differ from each other in flow rate, and optimal sample densities, as well as optimal sampling times, differ considerably from one sampler to another (Nevalainen *et al.*, 1992). Furthermore, the ambient bioaerosol is a mixture of many species of microbes. Spore formers and hardy species may mask the sample and impede the recovery of more sensitive organisms. Thus, sample analysis is severely limited since little information is available on the causative variables that lead to differing colony recoveries (Thompson *et al.*, 1994). Airborne contaminant counts are also dependent on the person operating the measurement device. The more stringent the air purity requirements, the greater the influence of air-sampling techniques (Meier & Zingre, 2000). Contamination from the sampler, especially in clean environments, will decrease if there is a waiting period set for 1–2 min after installation and before starting the air sampler. If measurements are taken in a conventionally ventilated room, airborne contaminant counts may also vary greatly if people are moving around or if work is being performed (Meier & Zingre, 2000).

37.4.1 Efficiency of the bioaerosol samplers

Sampler efficiency is a measure of how well the inlet of a sampler draws in particles without being affected by the particles' sizes, shapes, velocities and direction (Nevalainen *et al.*, 1992). The total efficiency (TE) of a sampler is determined by several factors such as the design of the inlet, collection stage and choice of collection medium, which affect the viability of the collected microbes (Henningson & Ahlberg, 1994). TE can be divided into collection efficiency (CE) and preservation efficiency (PE). Nevalainen *et al.* (1992) determined PE as biological efficiency that refers to the ability of the sampler to maintain microbial viability and prevent cell damage during sampling. CE is usually expressed as the 50% aerodynamic cut-off diameter (D_{ae50}), i.e. the particle size collected to 50% diameter (Henningson & Ahlberg, 1994). For efficient collection it is crucial to choose an impactor with a D_{ae50} below the mean size of the particles being sampled (Jensen *et al.*, 1992). It is not possible to determine the PE of a sampler with the methods on hand. TE can be estimated as (Henningson & Ahlberg, 1994):

$$TE = CE * PE \tag{37.1}$$

where TE = total efficiency, CE = collection efficiency and PE = preservation efficiency.

Nevalainen *et al.* (1992) determined CE as a measure of how well the inlet of a sampler deposits the particles without being affected by their physical properties. Henningson and Alberg (1994) expressed the collection efficiency CE as:

$$CE = IE * (1 - WL) * CSE \tag{37.2}$$

where IE = inlet efficiency, CSE = collection stage efficiency and WL = sampler wall losses.

37.5 Air sampling methods

37.5.1 Sedimentation plates

Sedimentation, also referred to as settle plate technique, has traditionally been used in the food industry. The plates are easy, inexpensive and collect particles in their original state (Kang & Frank, 1989a); however, they have serious drawbacks. The ability of settle plates to collect airborne particles is governed by the gravitational force on the particle which decreases with a velocity dependent on its mass. Hence, settle plates are biased towards collecting larger particles and are sensitive to air movement (Griffiths & DeCosemo, 1994). The method is not quantitative and in high aerosol concentrations the uncountable numbers of colonies can be a problem (Holah *et al.*, 1995).

37.5.2 Impactors and impingers

Impaction is the most common active sampling technique for collection of airborne viable particles (Ljungqvist & Reinmüller, 1998). In impactors the inertial forces are used to collect particles. The inertia of a particle is determined by its mass and velocity. In the collection stage of impactors the air stream is forced to change direction, and particles with too high a level of inertia are impacted on either a solid or liquid surface. Liquid-using impactors are called impingers (Henningson & Ahlberg, 1994). The aerodynamic particle diameter of the target aerosol is one of the most important physical factors determining CE of the sampler of impaction devices (Li & Lin, 1999). Impactors with several collection stages, i.e. cascade impactors, give information in addition on the size distribution of the aerosol (Henningson & Ahlberg, 1994).

There are two types of solid-surface impactors: slit samplers and sieve samplers. The slit sampler usually has a 0.2–1.0 mm-wide tapered slit, which produces a jet stream when the air is sampled by vacuum (Kang & Frank,1989a; Ljungqvist & Reinmüller, 1998). The slit sampler may have a turntable for rotating the agar plate so that aerosol particles are evenly distributed on the agar surface (Kang & Frank, 1989a). The velocity of the air varies according to the slit width of the air sampler used. A linear velocity of 20–50 m/s is typical, since particles with a minimum diameter of 0.5–1.0 μm do not follow the deflecting stream of air but impact against the collection surface. The smaller the size of a particle, the higher the velocity needed in the impaction (Ljungqvist & Reinmüller, 1998).

Sieve samplers are operated by drawing air through a large number of small, evenly spaced holes drilled in a metal plate. The particles are impacted on an agar surface located below the perforated plate (Kang & Frank, 1989a). The impaction velocity is dependent on the size of the perforations, distance to the impaction surface and performance of the vacuum pump (Ljungqvist & Reinmüller, 1998). When the concentration of viable particles in an aerosol is high, one sieve hole may allow more than one viable particle to pass through, resulting in the formation of a single colony from two or more viable particles. This inaccuracy can be corrected by reducing sampling time or by using either the microscopic method or a positive-hole method for enumeration (Kang & Frank, 1989a). Normally the positive-hole correction tables are included in each commercial sieve sampler.

In 1958 a cascade-sieve impactor, the Andersen sampler, was developed and is now probably the best-known sampler for bioaerosols (Andersen, 1958). It is one of the few samplers with a solid collection surface that gives information on the size distribution of the bioaerosol. The sampler collects airborne particles onto a series of agar plates at a flow rate of 28.3 L/min (Griffiths & DeCosemo, 1994). Each sieve has successively smaller holes. This causes increased particle velocity as air flows through the apparatus. Large particles impact at the initial stage and small particles follow the airflow until accelerated sufficiently to impact at a later stage (Kang & Frank, 1989a). Detection of microbes relies on their ability to increase following sampling (Griffiths & DeCosemo, 1994). The CE of the Andersen sampler was found to be greater than 90% for particles with a D_{ae50} of $<2.5\ \mu$m. This efficiency was not affected by air velocity; at air velocities <1 m/s, particles as large as 20 μm D_{ae50} were collected with a near 100% efficiency. The efficiency, however, fell rapidly with increased air velocity. The CE of the Andersen sampler is more affected by air velocity than by particle size (Griffiths *et al.*, 1993). The calculated D_{ae50} value is 0.57 μm (Nevalainen *et al.*, 1992). The Andersen 6-stage sampler is often used as a reference sampler (Henningson & Ahlberg, 1994). The 2-stage Andersen sampler is used to collect and separate respirable from non-respirable particles.

The D_{ae50} value of the first stage is 8.0 μm and the D_{ae50} value for the second stage is 0.95 μm (Andersen, 1958). For this reason viable particles above 0.95 μm in an aerosol can be separated into two size ranges: 0.95 to 8 μm on the first stage and 8 μm and above on the second stage.

There are numerous different 1-stage impactor samplers but only a few are described here. The surface air system (SAS) sampler is a portable, single-stage sieve sampler that collects particles onto a contact plate or standard Petri plate. The sampler operates with a flow rate of 180 L/min (Griffiths & DeCosemo, 1994). The calculated D_{ae50} value is 1.45 μm (Nevalainen *et al.*, 1992). The microbial air sampler MAS-100 is an impactor that aspirates air either horizontally or vertically through a perforated plate with 400 holes 0.7 mm in diameter (Meier & Zingre, 2000). The resulting airstream containing particles is directed onto the agar surface of a standard Petri plate. The impaction speed of the airborne microbes on the agar surface is approximately 11 m/s, which according to the manufacturer corresponds to stage five in the Andersen sampler. This speed guarantees that all particles over 1 μm are collected. At an aspiration volume of 100 L/min and with 400 holes 0.7 mm diameter holes serving as a catch for the lid, the MAS-100 attains a collision speed of 10.8 m/s. The calculated D_{ae50} value is 1.62 μm (Meier & Zingre, 2000) or 1.72 μm (Li & Lin, 1999). Air samplers with a D_{ae50} of less than 2 μm should theoretically be able to precipitate practically any airborne microbiological contaminant-carrying particle (Meier & Zingre, 2000).

Impingement methods are highly efficient for particles greater than 1 μm when high jet velocities are used (Kang & Frank, 1989a). The all-glass impinger-30 sampler is a widely used high-velocity impinger (Kang & Frank, 1989a; Griffiths & DeCosemo, 1994). The sampler operates by drawing aerosols through an inlet tube curved to simulate the nasal-passage respiratory infection potential of airborne microbes (Nevalainen *et al.*, 1992). The jet is held 30 mm above the impinger base and consists of a short piece of capillary tubing designed to reduce cell injury (Kang & Frank, 1989a). The calculated D_{ae50} value is 0.31 μm (Nevalainen *et al.*, 1992). When the sampler is used for recovering total amount of airborne microbes from the environment, the curved inlet tube should be washed with a known amount of collection fluid after sampling, since larger particles (i.e. diameter >15 μm) are collected on the tube wall by inertial force (Kang & Frank, 1989a). The agglomerated microbes are separated into suspension, which increases the number of colony-forming units (CFUs) (Lin & Li, 1999a). The bubbles, rising through the liquid, entrain previously collected particles and create new aerosols by bursting at the liquid–air surface when the impingers operate at a high-level collection fluid and sufficiently high sampling flow rate. The number of reaerosolized particles increased as sampling time increased (Lin *et al.*, 1997). The theoretical overall inlet sampling efficiency including wall loss is close to 100% for 1 μm particles and is significantly reduced for 5 μm and larger particles (Willeke *et al.*, 1992). Hydrophobicity may play an important role in the CE of AGI-30 impingers. Spores from many fungi are hydrophobic and may be lost because they float to the surface of collection suspensions and are re-entered into the exit airflow (Grinshpun *et al.*, 1997; Lin & Li, 1999a). The sampling time and flow rate influence the CE of hydrophobic spores.

There is also a new developed version of AGI-30 impinger, BioSampler, where non-evaporating, higher-viscosity liquids can be used as collection liquid (Lin *et al.*, 1999). The apparatus has equivalent or better microbial relative recovery than the conventional AGI-30, when the collection liquid is water (Lin *et al.*, 2000). The non-evaporating liquid enables longer sampling periods as it does not kill microbes or allow them to grow and does not evaporate even with longer sampling times. The construction difference is that the downward aerosol flow is split into three nozzle flows. Each nozzle has a sonic orifice that is directed at an identical angle toward the curved inner surface. Thus, the aerosol particles are thrown at an angle toward the surface and are removed by the combined forces of impaction and centrifugal motion. The presence of three angular nozzles establishes swirling air motion in the collection vessel. The swirling air flow entrains the liquid and swirls it upward into the region where the aerosol flows from the nozzles reach the inner vessel surface. Thus, the aerosol particles are removed into the liquid film which carries the removed particles down into the liquid reservoir (Lin *et al.*, 1999).

Another application of impinger principle is the multistage liquid impinger with three stages that are intended to correspond to the principal deposition sites in the human respiratory system: the upper respiratory tract, the bronchioles and the alveoli (Griffiths & DeCosemo, 1994).

Impingement is useful for sampling heavily contaminated air, since the liquid samples can be diluted to the appropriate level for subsequent growth culture analysis (Lin *et al.*, 2000). The impinger is inexpensive and simple to operate, but viability loss may occur owing to the amount of shear forces involved in collection. Another limitation is that the sampler should be sterilized before each sampling (Kang & Frank, 1989a). Samples collected by impingement method should be refrigerated and processed as soon as possible to avoid the increase of bacterial culturability (Li and Lin, 2001). Data obtained using the AGI-30 must be used with caution in food-processing environments containing large viable particles, because the agglomerated microbes will be separated into suspension, which increases the cfu number (Kang & Frank, 1989c; Lin & Li, 1999a).

37.5.3 Centrifugal samplers

Centrifugal samplers have a propeller that pulls air into the sampling unit and pushes the air outward to impact on a tangentially placed strip of nutrient agar set on a flexible plastic base. Particles in the incoming air may be thrown out of the air stream by centrifugal force to be caught against the peripheral surface (Ljungqvist & Reinmüller, 1998). Centrifugal samplers do not generate a high-velocity jet flow during sampling, and less stress is imposed on airborne microbes compared with impaction methods. These samplers are simple and easy to operate and can rapidly sample a high volume of air, resulting in more representative sampling (Kang & Frank, 1989a). They, however, demonstrate an inherent selectivity for larger particles and, since larger particles are more likely to include viable particles, there is a tendency towards higher counts than with other types of air samplers (Ljungqvist & Reinmüller, 1998).

The Reuter centrifugal sampler (RCS) is a portable hand-held instrument, much used in the biotechnology and food industry (Griffiths & DeCosemo, 1994). Air from a distance of at least 40 cm is sucked into the sampler by means of an impeller. The air enters the impeller drum concentrically from a conical sampling area. The air leaves the sampling drum in a spiral outside the cone of entering air (Kang & Frank, 1989a). An agar strip is inserted into the drum around the impeller blades. The sampler gives no indication of the size of particles (Griffiths & DeCosemo, 1994). In some investigations the cut-off size of the sampler has been found to be above 3 μm, so single cells pass through the sampler (Clark *et al.*, 1981; Macher & First, 1983).

37.5.4 Filter systems

Several different collecting mechanisms are active in filtration (impaction, interception and diffusion). Usually a single filter is used and all particle sizes are collected with no partitioning into size fractions (Henningson & Ahlberg, 1994). In general, the filters are inexpensive and simple to operate (Kang & Frank, 1989a). The air filtration apparatus consists of cellulose fibre, sodium alginate, fibre glass, gelatine membrane filter or synthetic membrane filters mounted in an appropriate holder and connected to a vacuum source through a flow rate controller (Kang & Frank, 1989a). Microbes can be removed by washing off from the filter to carry out total number rather than viable number enumeration (Griffiths & DeCosemo, 1994). The suspension can also be assessed with appropriate microbiological techniques. Membrane filters can also be directly placed on an agar surface for incubation. The filter methods are good for enumerating mould or bacterial spores (Kang & Frank, 1989a). One of the main disadvantages of using filters in collecting microbes is that they do not protect cells when large volumes of air pass over the filters causing desiccation (Griffiths & DeCosemo, 1994). Shortening of the sampling period for this method may reduce the stress (Kang & Frank, 1989a). The culturability of some fungal spores at low and high RH values of 30 and 85% decreased during the first few minutes, but remained approximately the same for sampling times ranging from 30 min to 8 h. The relative culturability of *Bacillus subtilis* endospores increased with increasing RH as well but decreased with sampling time. Instead *Pseudomonas fluorescens* and *Serratia marcescens* bacterial vegetative cells were culturable only if sampled for 10 minutes or less (Wang *et al.*, 2001).

The large number of pores present in membranes allows a large volume of air to be sampled during a short time (Kang & Frank, 1989a). Prolonged storage of the filters before assay causes death of sensitive vegetative bacteria (Parks *et al.*, 1996). Yeast and mesophilic bacterial concentrations have been observed to decrease significantly in the nucleopore filtration and eluation samples after one day mailing delay before plating, while concentrations of mould and thermophilic bacteria remained steady (Thorne *et al.*, 1994). The material of the membrane does not significantly affect the amount of cells, since equivalent numbers of bacteria and yeasts could be collected on nitrocellulose, polyamide and gelatine filters analysed with 4′,6-diamidino-2-phenylindole (DAPI) counting (Neef *et al.*, 1995).

The gelatine filter membrane is composed of gelatine foam designed to prevent vegetative microbes from being inactivated by desiccation when air is drawn through the filter (Parks *et al.*, 1996). The gelatine membrane is water-soluble so that it can easily be diluted for plating or be solubilized on top of a agar, resulting in colonies on the agar surface (Kang & Frank, 1989a). If gelatine filters are to be placed on agar plates, then the plates should be carefully dried beforehand, because wet plates can make colony counting difficult. The gelatine filters do not melt into the agar if they are incubated at low temperature. Therefore, the filter plates should first be incubated at 26 °C for 1 h to ensure melting of the gelatine filter into the agar (Parks *et al.*, 1996).

37.5.5 Electrostatic precipitation

The airborne microbes can be collected by imparting them with electrical charges and then depositing them in an electrical field onto a growth medium. Electrostatic precipitation can efficiently remove the airborne microbes. Over 70% of viable *Bacillus subtilis* var. *niger* spores and *Penicillium brevi-compactum* fungal spores entering an electrostatic precipitator were enumerated as colony forming units. The sensitive *Pseudomonas fluorescens* vegetative cells were enumerated twice as efficiently in the precipitation as in the impinger BioSampler. For this reason it may be especially useful for collection of sensitive microbial strains (Mainelis *et al.*, 2002).

37.5.6 Particle samplers

Optical particle counters are based on laser-light scattering by a single particle and are widely used for measuring the concentration and size distribution of airborne particles. The crucial response characteristics of a particle counter include the sizing accuracy, counting efficiency and detection limits (Yoo *et al.*, 1996). The detection limit of the smallest detectable particles is a crucial characteristic of the counter. Light-scattering counters are usually calibrated by means of polystyrene latex (PSL) spheres. Owing to the differences between the optical properties of the calibrated PSL spheres and actual particles, errors in the results may occur (Yoo *et al.*, 1996). Optical counters on the market include laser diffractometers, phase-doppler systems, intensity-deconvolution systems, and laser–particle interaction system/image analysers. Other techniques used in particle counters include the electrical mobility techniques used in electrical aerosol analysers and differential mobility analysers as well as light detection and ranging (LIDAR) technology, which uses light waves in the same way that

Table 37.1 Methods including some commercial air sampling devices used in collection of air for microbiological sampling

Sampling principle	Sampler	Collection surface	Comments	D_{ae50} (μm)	References
Cascade sieve impactor	Andersen 6-stage	Agar in Petri dishes	Reliable results, information on size distribution, impractical in industrial use	0.57	Andersen (1958); Nevalainen *et al.* (1992)
	Andersen 2-stage	Agar in Petri dishes	Separates respirable and non-respirable particles	0.95; 8.0	Andersen (1958); Nevalainen *et al.* (1992)
Sieve impactor	Andersen 1-stage	Agar in Petri dish			Nevalainen *et al.* (1992)
	MicroBio MB2	Contact plate			Parret & Crilly (2000)
	Surface Air System, SAS	Contact plate	Practical in industrial use	1.45	Griffiths & DeCosemo (1994); Nevalainen *et al.* (1992)
	MAS-100	Agar in Petri dish	Practical in industrial use	1.62–1.72	Meier & Zingre (2000); Li & Lin (1999)
Slit impactor	Casella MK II	Agar media		0.67	Bourdillon *et al.* (1941); Nevalainen *et al.* (1992)
	Burkard Personal Sampler			2.52	Nevalainen *et al.* (1992)
Centrifugal airstream	Cyclones	Wet or dry surface		> 3	Henningson & Ahlberg (1994)
	Reuter centrifugal sampler; RCS	Special agar strips	Less stress to microbes than in impactors, selectivity for larger particles, practical in industrial use		Kang & Frank (1989d); Clark *et al.* (1981); Macher & First (1983)
Impinger	All glass impinger, AGI-30	Collection fluid	Efficient for collection of bacteria and yeast; impractical in industrial use, glass	0.31	Nevalainen *et al.* (1992); Kang & Frank (1989a,c); Griffiths & DeCosemo (1994)
	Multistage liquid impinger	Collection fluid	Stainless steel		Griffiths & DeCosemo (1994)
	BioSampler	Collection fluid	More gentle and efficient (for spores) than AGI-30, impractical in industrial use		Willeke *et al.* (1992); Lin *et al.* (2000)
Filtration	Sartorius MD8	Gelatine filter	Efficient for spores, decreased desiccation rate of microbes, practical in isolators and for big sampling volumes		Kang & Frank (1989a,c); Parks *et al.* (1996)
Gravitation	Settle plates	Agar in Petri dishes	Simple to use, not quantitative, unreliable		Griffiths & DeCosemo (1994)
Ionization	Electrostatic precipitation	Filter, collection fluids, agar media	Mechanically complex; collection of microorganisms on charged surfaces		Griffiths & DeCosemo (1994); Mainelis *et al.* (1999)
Temperature gradient	Thermal precipitation				Griffiths & DeCosemo (1994)
Specific bindings	Biosensors		Specific, future development; at the moment no industrial use		

radar uses radio waves. Combining particle counting with bioaerosol measurements may allow detection of rapid variations instantaneously and indicate further microbial measurements. No absolute relationship, however, can be established between bioaerosol and particle counts in variable working environments (Parat *et al.*, 1999).

37.5.7 Conclusions regarding samplers

In Table 37.1 microbiological air samplers, collection methods and a few comments on the methods are presented. In addition to the samplers in Table 37.1, there is other commercial apparatus on the market; scientific information, however, is often not available. In the food industry, the settle plates and impactor- or centrifugal-type hand-held samplers are commonly used air-sampling methods in routine monitoring. The Andersen impactor gives the most reliable results, but it is less practical for routine use. The filtration methods may not be optimal for counting vegetative cells due to the stress it places on cells through dehydration during sampling. A glass impinger is inexpensive and simple to operate, but viability loss may occur due to the amount of shear forces involved in collection (Kang & Frank, 1989b,c). The impingers are likely to perform better than the filtration methods for sampling airborne bacterial microbes (Li *et al.*, 1999). Impingement is suitable for yeast bioaerosols (Lin & Li, 1999a) but fungal spores are more reliably collected with filtration than impingement. Lower levels of total bacterial recovery were observed in filter samplers than impinger samplers. This was primarily due to the higher biological stress occurring during the sampling process of filtration. Sampling flow rate did not significantly affect bacterial recovery. Sampling time related to dehydration effect did play a role in bacterial recoveries, especially for sensitive strains. The nucleopore filtration method was recommended for moulds and thermophilic organisms rather than the AGI-30, but for yeasts and mesophilic bacteria the AGI-30 method was more favourable (Thorne *et al.*, 1994).

Each of the proposed monitoring methods has limitations that the user should be aware of. Monitoring effectiveness is dependent to a great extent on the monitoring methods used and the nature of the aerosol present (Kang & Frank, 1989a). Food plant aerosols have not been studied sufficiently to accurately generalize about particle-size distribution. Particle density is influenced by RH and particle composition. It can be stated that an effective separation volume varies more due to the mass of the particle than to the particle size. Bioaerosol results measured with different sampler types are strongly influenced by the sampling environment and the results are sometimes contradictory. The choice of instrument is clearly dependent on the experimental parameters and the secondary means of identification (Neef *et al.*, 1995). Airborne counts recorded from different sampling techniques are not directly comparable but would be expected to show similar trends in level variation (Holah *et al.*, 1995).

37.6 Bioaerosol assay methods

37.6.1 Culturing techniques

Culturing of microbes directly or through a broth on solid agar media with incubation for a certain period and temperature is the traditional method for enumerating the microbial count in an air sample. The culture technique is easy to use and requires no specialized equipment for sampling. The total population of microbes in a sample can, however, be severely underestimated if assessment is carried out with this type of method (Griffiths & DeCosemo, 1994). Microbes in the air may lose their ability to form colonies and still be viable, i.e. be non-culturable. In cases where the airborne microbes are non-culturable, data on cfus does not describe the actual microbial population (Heidelberg *et al.*, 1997). Epifluorescent microscopic methods for counting the total amount of microbes have shown that there is wide variation in the culturability of microbial cells; the culturable amount of total airborne bacteria varied from 0.02% to 10.6% (Lighthart, 1997).

In the development of methods that utilize culture techniques as part of the assessment criteria, growth conditions, diluent and culture plate media must be standardized to be able to estimate microbial counts in the air (Griffiths & DeCosemo, 1994). The culture technique is useful for direct identification of certain pathogens or spoilage microbes in the food-processing air, e.g. spoilage moulds on specific agars (Lund, 1996).

Collection of microbes in the fluid is used in the impingers. The selection of a liquid collection medium is dependent upon the particular organisms to be isolated. In quantitative studies a medium that will minimize both multiplication and death of the organism must be employed (Kang & Frank, 1989a). The usefulness of organic compounds, e.g. di-, tri- and polysaccharides, sugar alcohols, proteins, polypeptides, organic acid salts, antibiotics, chelating agents and culture supernatant fluid in enhancing cell survival has been investigated. The majority of the most effective organic additives are sugars or polyhydric alcohols (Griffiths & DeCosemo, 1994). Marthi and Lighthart (1990) showed that addition of an organic compound, e.g. betaine, to the impinger buffer significantly increased the colony-forming abilities of the airborne microbes. The compound acted in some cases as a protective agent, maintaining the culturability of some bacteria. Heidelberg *et al.* (1997), however, showed that the effect could be the opposite in other cases, owing to strain variability against betaine. Thus, protective agents in routine monitoring may introduce a bias, protecting the culturability of only some parts of the microbial population.

37.6.2 Microscopy and fluorescence techniques

Microscopy is a method with which the total count of microbes as well as morphological data on the microbes can be obtained. The microscopical method is relatively simple and rapid. Automatic counting and size evaluation in the assessment of airborne microbes by means of image processing of fluorescence microscopy data reduce analysis time (Griffiths & DeCosemo, 1994). Fluorescence microscopy can be applied in the evaluation of airborne microbes harvested, e.g. on filters (Kildesø & Nielsen, 1997) and in impinger liquids (Terzieva *et al.*, 1996). However, manual focusing of the microscope is needed due to the impurities, e.g. larger particles, present in the sample (Kildesø & Nielsen, 1997). Phase-contrast microscopy is particularly useful for counting bacterial endospores as they appear phase-bright against the darker vegetative cells (Griffiths & DeCosemo, 1994). The addition of UV fluorescence capability into aerosol counters offers a way to distinguish biological particles from most organic and inorganic particles (Seaver *et al.*, 1999). Viable staining methods have been applied for the detection of viable microbes (Terzieva *et al.*, 1996, Hernandez *et al.*, 1999). The actual viable count may be underestimated when assessed by this technique because microbes injured in the sampling stage may recover later on (Terzieva *et al.*, 1996). Understanding the fluorescence effect on the cell viability, the presence of non-biological particles and the interference from mixtures has not yet been achieved (Brosseau *et al.*, 2000). Technical problems related to fluorescence microscopy include the low-contrast and low-light intensity, rendering difficulties in automatic image processing (Kildesø & Nielsen, 1997).

37.6.3 Flow cytometry

Flow cytometry (FCM) has developed to a rapid screening and enumeration method for airborne particles demanding, however, sampling and sample pre-treatment before analysis. FCM is based on multiple parameters such as forward light scatter (FSC), side light scatter (SSC) and fluorescence emission at wavelengths of interest. It can determine individually a large number of cells in a short time (Prigione *et al.*, 2004). Airborne fungi (Prigione *et al.*, 2004) and bacteria (Lange *et al.*, 1997) have been quantitatively analysed from natural environments. However, the background contaminants cause problems in detection and the sensitivity of the method and it still needs to be improved before being applicable in the food industry.

37.6.4 ATP bioluminescence

Adenosine triphosphate (ATP) is present both in microbial cells and food ingredients and can be measured using the luciferase enzyme complex found in fireflies. The light output of a sample is directly proportional to the amount of ATP present. The detection limit of the method is about 10^4 cells (Wirtanen, 1995). This method is non-specific, i.e. it measures the ATP content of the microbial population in the sample as a whole (Griffiths & DeCosemo, 1994). The ATP content in airborne cells, which are stressed through assessment, can be altered by the effect of aerosolization and this can also affect the detection level of the method. The total concentration of adenylate within the cells may give a better estimate of cell concentration (Griffiths & DeCosemo, 1994). Application of a fully automated ATP monitoring system, AutoTrac, for continuous checking for microbial contamination in air has been developed (Brady, 1999).

37.6.5 Molecular methods

Molecular biology detection methods include polymerase chain reaction (PCR) and gene probes (Griffiths & DeCosemo, 1994) as well as immunological methods. The PCR analysis method permits the detection of DNA regardless of the metabolic state of the cells. The method may therefore be orders of magnitude more sensitive than culture techniques. Applications of this method for the quantitation of airborne organisms are still under development. PCR-based techniques allow the detection and identification of microbes at a group or species level within a background of other microbes. The specificity, sensitivity and reduced processing time of this technique are suitable for the detection of small amounts of target microbial cells in a sample (Alvarez *et al.*, 1995). Specialized equipment and skilled personnel are, however, required for successful applications (Griffiths & DeCosemo, 1994). The air samples may also contain compounds inhibitory to the amplification assay, e.g. high concentration of non-target DNA (Alvarez *et al.*, 1995). In many applications, pre-enrichment of the sample is needed. PCR is at its best a fast and powerful method for the identification of biological air contaminants. It is especially suitable for situations where a particular contaminant or at the most a few contaminant strains in air are searched. The method has to be developed and optimised separately for each microbe and sometimes also for different sampling environments, owing to different inhibiting compounds, in order to have sufficient sensitivity and specificity.

Nucleic acid hybridization has been applied for the detection and identification of microbes in bioaerosols. Each hybridization format is suitable for different aerosol concentrations. It is possible to identify a fast-growing airborne organism within 24 h using the colony-hybridization technique. Colony hybridization can be used to detect amounts as low as even 1 cfu, whereas whole-cell hybridization requires a substantially higher aerosol concentration, e.g. $> 5 \times 10^4$ cfu, from filtered air samples (Neef *et al.*, 1995). A two-step detection strategy, first through selective PCR to amplify a group of the template DNAs in a sample is followed by more specific examination via probe hybridization to a specific target in the PCR products. This can provide better specificity and sensitivity for environmental sample screening (Zeng *et al.*, 2003).

Immunological methods use specific antibodies which detect unique epitopes (antibody-binding sites) expressed by the target organism. The sample is exposed to antibodies that bind to the antigen in the sample and the amount of antibody binding detected. Immunological applications are rare at the moment in bioaerosol monitoring but they have potential in certain cases provided that there is a functional antibody available for the contamination in question (McCartney *et al.*, 1997).

37.6.6 Conclusions regarding assay methods

Traditional and relatively simple bioaerosol assay methods such as culturing and microscopy are most suitable for following general hygienic situations in food industry environments. The regular bioaerosol assays are comparatively cheap, relatively fast to carry out and do not require much work. However, it is important to keep in mind that stresses imposed on the cells in the airborne state may render them non-culturable and the methods that rely on the growth of microbes for their detection can seriously underestimate the number of viable cells within a sample (Griffiths & DeCosemo, 1994). The ratio of the total number of microscopic cells and the viable number of organisms detected may range from 5 to 200 in food industry. The actual microbial count may be biased by several orders of magnitude if concentrations are given only in terms of viable microorganisms (Nielsen & Breum, 1995).

As the information level needed on bioaerosol quality increases or joins together with more effective or targeted sampling techniques, the more advanced assay methods such as molecular and fluorescence methods can help to achieve the precise knowledge needed. The more specific the methods are, the bigger the costs may become. The information gained may, however, be a multiple of that from basic bioaerosol monitoring.

In general, air quality monitoring should be done with one approved sampling and assaying technique to be able to compare the results over the long term. However, as the process, raw materials, employees, equipment and other parameters may change, it is recommended that the bioaerosol quality is checked from time to time with another sampler as well as with another assay method. This helps to prevent unexpected air quality problems despite regular monitoring. No single method can completely demonstrate all the microbes present in bioaerosols.

37.7 Interpretation of bioaerosol results

There are no general limits for microbial concentrations in the food industry air. The bioaerosol results have to be evaluated case-specifically at each processing place. It is normally impossible or at least difficult to compare bioaerosol results from different food industries, from the same industry type or even from different locations in the same factory. In particular one should never directly compare results obtained with different samplers or different assay methods. In addition samples taken with the same sampler and analysed the same way may not be totally comparable if the sampling parameters such as time, collection surface and sometimes even the sample collector have changed between the samples.

The best and fastest way to utilize the bioaerosol results would be a situation where all the product contaminants of air in each space were identified and known. In addition, the concentrations that do not cause product contamination within the product shelf-life should be known. Successful bioaerosol controlling demands knowledge of contaminant concentrations in relation to different product exposure times. This demands much research and is rarely done in the food industry. Without this kind of information, bioaerosol results are only indicative.

Exploitation of the bioaerosol result is usually achieved by analysing the trends and changes in each sampling place results. Careful analysis of long-term results can highlight contamination levels of general or certain contaminants that have caused problems in production. Comparing new results with old results will show whether something abnormal is happening in the process at the moment. In addition the results normally show the seasonal variations that should not be mixed with intrinsic process or process environment created changes caused by potential contamination. The seasonal variations, however, should also be controlled so that they do not present a risk for production or employees. A functional, sensitive and regular bioaerosol monitoring system describes the process's general hygienic situation and production practices as well as the behaviour of ventilation and pressure differences.

37.8 Future trends

The food industry is following the pharmaceutical and medical industries by recognizing that microbial monitoring of air is a must in standard quality control practices. However, in addition to the general contamination power, air can be the major contaminant source of products. There is a published example showing that air has been confirmed as the main contamination source in the production of cheese (Kure *et al.*, 2004b). This kind of example and information will probably become more common as resources are directed on the monitoring and study of air quality and contamination effects of air in the food industry.

In addition to the methods and techniques mentioned in this chapter, the development of real-time continuous monitoring of airborne microbes is important for the food industry. Given the ability to respond quickly to the findings of continuous monitoring, this kind of system would provide a tool to detect abnormalities, problems and trends in processes that are not always found with normal short-term periodic sampling. Continuous monitoring would also offer information on sources and spread of contamination. UVAPS (Ultraviolet Aerodynamic Particle Size Spectrometer) is the first commercial real-time aerosol sampler that monitors viable bioaerosols. It is equipped with a pulsed laser that measures aerodynamic size, scattered light intensity and fluorescence of airborne particles within a size range of 0.5 to 15 μm. It has, however, limitations in its sensitivity to the type of airborne bacteria and especially bacterial spores; in addition, the growth phase and environmental stress level of bacteria reduce the sensitivity level. The counting efficiency of the fluorescent particles depend on particle concentration, with the upper limit of detection being approximately 6×10^7 particles/m^3 and the identification of contaminating species is not possible (Agranovski *et al.*, 2003a, 2003b). Nevertheless this kind of technique will improve and their benefits should be understood and exploited as they develop to functional and practical applications.

37.9 References and further reading

AGRANOVSKI V, RISTOVSKI Z, HARGREAVES M, BLACKALL, P J and MORAWSK L (2003a), 'Real-time measurement of bacterial aerosols with the UVAPS: performance evaluation', *Aerosol Sci*, **34**, 301–317.

AGRANOVSKI V, RISTOVSKI Z, HARGREAVES M, BLACKALL P J and MORAWSK L (2003b), 'Performance evaluation of the UVAPS: influence of physiological age of airborne bacteria and bacterial stress', *Aerosol Sci*, **34**, 1711–1727.

ALVAREZ A J, BUTTNER M P and STETZENBACH L D (1995), 'PCR for bioaerosol monitoring: sensitivity and environmental interference', *Appl Environ Microbiol*, **61**, 3639–3644.

AMBROISE D, GREFF-MIRGUET G, GÖRNER P, FABRIÉS J F and HARTEMANN P (1999), 'Measurement of indoor viable airborne bacteria with different bioaerosol samplers', *J Aerosol Sci*, **30**, Suppl. 1, S699–S700.

ANDERSEN A A (1958), 'New sampler for the collection, sizing and enumeration of viable airborne particles', *J Bacteriol*, **76**, 471–484.

BOER E D and BEUMER R R (1999), 'Methodology for detection and typing of foodborne microorganisms', *Int J Food Microbiol*, **50**, 119–130.

BOURDILLON R B, LIDWELL O M and THOMAS J C (1941), 'A slit sampler for collecting and counting airborne bacteria', *J Hyg*, **14**, 197–224.

BRADY P (1999), 'AutoTrack. The next generation in monitoring', *Brewer*, **85**, 130–133.

BROSSEAU L M, VESLEY D, RICE N, GOODELL K, NELLIS M and HARISTON P (2000), 'Differences in detected fluorescence among several bacteral species measured with a direct-reading particle sizer and fluorescence detector', *Aerosol Sci Technol*, **32**, 545–558.

BROWN A D (1953), 'The survival of airborne microorganisms, III. Effects of temperature', *Aust J Biol Sci,* **7**, 444–451.

CANNON R Y (1970), 'Types and population of microorganisms in the air of fluid milk plants', *J Milk Food Technol*, **33**, 19–21.

CLARK S, LACH V and LIDWELL O M (1981), 'The performance of the Biotest RCS centrifugal air sampler', *J Hosp Inf*, **2**, 181–186.

COX C S and WATHES C M (1995), *Bioaerosols Handbook*, Boca Raton, CRC Lewis Publishers.

CUNDELL A M, BEAN R, MASSIMORE L and MAIER C (1998), 'Statistical analysis of environmental monitoring data: does a worst case time for monitoring clean rooms exist?', *J Pharm Sci Technol*, **52**, 326–330.

DARK F A and CALLOW D S (1973), 'The effect of growth conditions on the survival of airborne *E. coli*', In Hers J F and Winkler K C, *4th International Symposium on Aerobiology*, Utrecht, Oosthoek, 97–99.

GOODLOW R J and LEONARD F A (1961), 'Viability and infectivity of microorganisms in experimental airborne infection', *Bacteriol Rev*, **25**, 182–187.

GRIFFITHS W D and DECOSEMO G A L (1994), 'The assessment of bioaerosols: a critical review', *J Aerosol Sci*, **25**, 1425–1458.

GRIFFITHS W D and STEWART I W (1999), 'Performance of bioaerosol samplers used by the UK biotechnology industry', *J Aerosol Sci*, **30**, 1029–1040.

GRIFFITHS W D, UPTON S L and MARK D (1993), 'An investigation into the collection efficiency and bioefficiency of a number of aerosol samplers', *J Aerosol Sci*, **24**, S541–S542.

GRINSHPUN S S, WILLEKE K, ULEVICIUS V, JUOZAITIS A, TERZIEVA S, DONNELLY J, STELMA G N and BRENNER K P (1997), 'Effect of impaction, bounce and reaerosolization on the collection efficiency of impingers', *Aerosol Sci Technol*, **26**, 326–342.

HEIDELBERG J F, SHAHAMAT M, LEVIN M, RAHMAN I, STELMA G, GRIM C and COLWELL R R (1997), 'Effect of aerosolization on culturability and viability of Gram-negative bacteria', *Appl Environ Microbiol*, **63**, 3585–3588.

HENNINGSON E W and AHLBERG M S (1994), 'Evaluation of microbiological aerosol samplers: a review', *J Aerosol Sci*, **25**, 1459–1492.

HERNANDEZ, M, MILLER, SL, LANDFEAR, D W and MACHER, J M (1999), 'A combined fluorochrome method for quantitation of metabolically active and inactive airborne bacteria', *Aerosol Sci Technol*, **30**, 145–160.

HOLAH J T, HALL K E, HOLDER J, ROGERS S J, TAYLOR J and BROWN K L (1995), *Airborne Microorganism Level in Food Processing Environments*, Chipping Campden, CCFRA, 1–22.

JENSEN P A, TODD W F, DAVIS G N and SCARPINO P V (1992), 'Evaluation of eight bioaerosol samplers challenged with aerosols of free bacteria', *Am Ind Hyg Assoc*, **53**, 660–667.

KANG Y J and FRANK J F (1989a), 'Biological aerosols: a review of airborne contamination and its measurement in dairy processing plants', *J Food Prot*, **52**, 512–524.

KANG Y J and FRANK J F (1989b), 'Evaluation of air samplers for recovery of artificially generated aerosols of pure cultures in a controlled environment', *J Food Prot*, **52**, 560–563.

KANG Y J and FRANK J F (1989c), 'Evaluation of air samplers for recovery of biological aerosols in dairy processing plants', *J Food Prot*, **52**, 655–659.

KANG Y J and FRANK J F (1989d), 'Comparison of airborne microflora collected by the Andersen sieve sampler and RCS sampler in a dairy processing plant', *J Food Prot*, **52**, 877–880.

KILDESØ J and NIELSEN B H (1997), 'Exposure assessment of airborne microorganisms by fluorescence microscopy and image processing', *Ann Occup Hyg*, **41**, 201–216.

KOWALSKI W J and BAHNFLETH W (1998), 'Airborne respiratory diseases and mechanical systems for control of microbes', *HPAC Eng*, **70**, 34–48.

KURE C F, LANGSRUD S and KARLSON I (2004a), 'Mould contamination in cheese production' in Friis A, *Proceedings of 35th R^3 Symposium on Contaminant Control*, Helsingør, R3 Nordic, 258–263.

KURE C F, SKAAR I and BRENDEHAUG J (2004b), 'Mould contamination in production of semi-hard cheese', *Int J Food Microb*, **93**, 41–49.

LANGE J L, THORNE P S and LYNCH N (1997), 'Application of flow cytometry and fluorescent in situ hybridization for assessment of exposure to airborne bacteria', *Appl Environ Microbiol*, **63**, 1557–1563.

LI C-S and LIN Y-C (1999), 'Sampling preformance of impactors for bacterial bioaerosols', *Aerosol Sci Technol*, **30**, 280–287.

LI C-S and LIN Y-C (2001), 'Storage effects on bacterial consentration: determination of impinger and filter samples', *Sci Total Environ*, **278**, 231–237.

LI C-S, HAO M-L, LIN W-H, CHANG, C-W and WANG C-S (1999), 'Evaluation of microbial samplers for bacterial microorganisms', *Aerosol Sci Technol*, **30**, 100–108.

LIGHTHART B (1997), 'The ecology of bacteria in the alfresco atmosphere', *FEMS Microb Ecol*, **23**, 263–274.

LIN W-H and LI C-S (1999a), 'Collection efficiency and culturability of impingement into a liquid for bioaerosols of fungal spores and yeast cells', *Aerosol Sci Technol*, **30**, 109–118.

LIN W-H and LI C-S (1999b), 'Evaluation of impingement and filtration methods for yeast bioaerosol sampling', *Aerosol Sci Technol*, **30**, 119–126.

LIN W-H and LI C-S (2000), 'Associations of fungal aerosols, air pollutants, and meteorological factors', *Aerosol Sci Technol*, **32**, 359–368.

LIN X, WILLEKE K, ULEVICIUS V and GRINSHPUN S A (1997), 'Effect of sampling time on the collection efficiency of all-glass impingers', *Amer Ind Hyg Assoc J*, **58**, 480–488.

LIN X, REPONEN T, WILLEKE K, GRINSHPUN S A, FOARDE K K and ENSOR D S (1999), 'Long-term sampling of airborne bacteria and fungi into a non-evaporating liquid', *Atmos Environ*, **33**, 4291–4298.

LIN X, REPONEN T, WILLEKE K, WANG Z, GRINSHPUN S A and TRUNOV M (2000), 'Survival of airborne microorganisms during swirling aerosol collection', *Aerosol Sci. Technol*, **32**, 184–196.

LJUNGQVIST B and REINMÜLLER B (1998), 'Active sampling of airborne viable particles in controlled environments: a comparative study of common instruments', *Eur J Parent Sci*, **3**, 59–62.

LUND F (1996), 'Direct identification of the common cheese contaminant *Penicillium commune* in factory air samples as an aid to factory hygiene', *Lett Appl Microbiol*, **22**, 339–341.

MACHER J M and FIRST M W (1983), 'Reuter centrifugal air sampler. Measurements of effective airflow rate and collection efficiency', *Appl Environ Microbiol*, **45**, 1960–1962.

MAINELIS G, GRINSHPUN S A, WILLEKE K, REPONEN T, ULEVICIUS V and HINTZ P J (1999), 'Collection of airborne microorganisms by electrostatic precipitation', *Aerosol Sci Technol*, **30**, 127–145.

MAINELIS G, ADHIKARI A, WILLEKE K, LEE S-A, REPONEN T and GRINSHPUN S A (2002), 'Collection of airborne microorganisms by a new electrostatic precipitator', *Aerosol Sci*, **33**, 1417–1432.

MARTHI B and LIGHTHART B (1990), 'Effects of betaine on enumeration of airborne bacteria', *Appl Environ Microbiol*, **56**, 1286–1289.

MCCARTNEY H A, FITT D D L and SCHMECHEL D (1997), 'Sampling bioaerosols in plant pathology', *J Aerosol Sci*, **28**, 349–364.

MCDERMID A S and LEVER M S (1996), 'Survival of *Salmonella enteritidis* PT4 and *Salm. typhimurium* Swindon in aerosols', *Lett Appl Microbiol*, **23**, 107–109.

MEIER R and ZINGRE H (2000), 'Qualification of air sampler systems: the MAS-100', *Swiss Pharma*, **22**, 15–21.

NEEF A, AMANN R and SCHLEIFER K-H (1995), 'Detection of microbial cells in aerosols using nucleic acid probes', *Syst Appl Microbiol*, **18**, 113–122.

NEVALAINEN A, PASTUSZKA J, LIEBHABER F and WILLEKE K (1992), 'Performance of bioaerosol samplers: collection characteristics and sampler design considerations', *Atmos Environ*, **26**, 531–540.

NIELSEN B H and BREUM N O (1995), 'Exposure to air contaminants in chicken catching', *Am Ind Hyg Assoc J*, **56**, 804–808.

PARAT S, PERDRIX A, MANN S and BACONNIER P (1999), 'Contribution of particle counting in assessment of exposure to airborne microorganisms', *Atmos Environ*, **33**, 951–959.

PARKS S R, BENNETT A M, SPEIGHT S E and BENBOUGH J E (1996), 'An assessment of the Sartorius MD8 microbiological air sampler', *J Appl Bacteriol*, **80**, 529–534.

PARRETT F and CRILLY K (2000), 'Microbiological air monitoring', *Int Food Hyg*, **10**, 5–7.

PRIGIONE V, LINGUA G and FILIPELLO MARCHISIO V (2004), 'Development and use of flow cytometry for detection of airborne fungi', *Appl Environ Microb*, **70**, 1360–1365.

SEAVER M, EVERSOLE J D, HARDGROVE J J, CARY W K and ROSELLE D C (1999), 'Size and fluorescence measurements for field detection of biological aerosols', *Aerosol Sci Technol*, **30**, 174–185.

STERSKY A K, HELDMAN D R and HEDRICK T I (1972), 'Viability of airborne *Salmonella newbrunswick* under various conditions', *J Dairy Sci*, **55**, 14–18.

TERZIEVA S, DONNELLY J, ULEVICIUS V, GRINSHPUN S A, WILLEKE K, STELMA G N and BRENNER K P (1996), 'Comparison of methods for detection and enumeration of airborne microorganisms collected by liquid impingement', *Appl Environ Microbiol*, **62**, 2264–2272.

THOMPSON M W, DONNELLY J, GRINSHPUN S A, JUOZAITIS A and WILLEKE K (1994), 'Method and test system for evaluation of bioaerosol samplers', *J Aerosol Sci*, **25**, 1579–1593.

THORNE P S, LANGE J L, BLOEBAUM P and KULLMAN G J (1994), 'Bioaerosol sampling in field studies: Can samples be express mailed?', *Am Ind Hyg Assoc J*, **55**, 1072–1079.

WANG Z, REPONEN T, GRINSHPUN S A, GÓRNY R L and WILLEKE K (2001), 'Effect of sampling time and air humidity on the bioefficiency of filter samplers for bioaerosol collection', *Aerosol Sci*, **32**, 661–674.

WILLEKE K, GRINSHPUN S A, CHANG J-W, JUOZAITIS A, LIEBHABER F, NEVALAINEN A and THOMPSON M (1992), 'Inlet sampling efficiency of bioaerosol samplers', *J Aerosol Sci*, **23**, S651–S654.

WIRTANEN G (1995), *Biofilm formation and its elimination from food processing equipment*, Espoo, VTT Publications.

YOO S H, CHAE S K and LIU B Y H (1996), 'Influence of particle refractive-index on the lower detection limit of light-scattering aerosol counters', *Aerosol Sci Technol*, **25**, 1–10.

ZENG Q-Y, WANG X-R and BLOMQUIST G (2003), 'Development of mitochondrial SSU rDNA-based oligonocleotide probes for specific detection of common airborne fungi', *Mol Cell Probe*, **17**, 281–288.

38

Testing the effectiveness of disinfectants and sanitisers

J.-Y. Maillard, Cardiff University, UK

38.1 Introduction

There has been a dramatic increase in the usage of chemical biocides (i.e. disinfectants, antiseptics and sanitisers) in the food, water and pharmaceutical industries, and in the healthcare and domiciliary environments. The need to reduce and control nosocomial infection (Favero, 2002) and to improve product quality and overall hygiene, for example, in the food industry (Langsrud *et al.*, 2003) has been particularly well reported and advertised among governmental agencies, health authorities and the public. Public knowledge in particular and a better commitment to overall hygiene (Bloomfield, 2002) have contributed to the increased usage of chemical biocides in the home environment, although this has been subjected to controversies (Levy, 2001). With such an increase in the number of products available for disinfection and sanitation, it is essential that the end users can select a product that is appropriate for their needs, but more importantly that the product specification from the manufacturer or supplier, particularly its antimicrobial efficacy, is accurate. Protocols for testing the antimicrobial efficacy of disinfectants and sanitisers are therefore essential to provide reliable information on the efficacy of an antimicrobial product and provide assurance for the end users. However, there are no internationally agreed standard protocols and often countries have their own government laboratory testing with their own standards, although in Europe, CEN/TC216 (the European Committee for Standardisation) aims to produce current and future European disinfectant testing standards (Holah, 2003). Test methodology can range from basic preliminary suspension tests to more complex protocols that simulate conditions in practice. The aims of this chapter are to provide information on the factors that affect biocide efficacy and test reproducibility, to discuss standard protocols for disinfectants and sanitisers, their limitation and possible developments.

38.2 Types of biocidal products

To understand the design and the limitation of antimicrobial test protocols, an understanding of non-antibiotic antimicrobial agents (i.e. chemical biocides) is important. Chemical biocides are defined as agents with disinfectant, antiseptic or preservative activity. Biocides are very diverse in their chemical structure,

Table 38.1 Types and properties of antimicrobial biocides

Biocides	Factors affecting activity		Typical usage
	pH	Soiling	
Aldehydes			
Glutaraldehyde, *ortho*-phthalaldehyde, formaldehyde	>7	–	Preservation (e.g. cosmetics), sterilisation medical equipment, fumigation
Peroxides			
Hydrogen peroxide, peracetic acid	<7	–/+	Surface disinfection, sanitisation, sterilisation of medical equipment, antisepsis
Halogen releasing agents			
Sodium hypochlorite, chlorine, iodine	<7	+	Surface disinfection, water disinfection, sanitisation, antisepsis
Biguanides			
Chlorhexidine, alexidine, PHMB	>7	+	Preservation, antisepsis, surface disinfection
QACs	>7	+	Preservation; antisepsis and surface disinfection, sanitisation
Phenolics			
Bisphenols, halophenols	<7	+/–	Antisepsis and surface disinfection
Alcohols		+	Antisepsis and surface disinfection (often used in combination)
Organic acids	<7		Preservation (sanitisation when used in combination)
Anionic surfactants, sodium lauryl sulphate	<7		Sanitisation when used in combination
Amphoteric agents, betaine, dodecyl-β-alanine, dodecyl-di-glycine	NA	–	

PHMB, polyhexamethylene biguanide; QACs, quaternary ammonium compounds.

Table 38.2 Main desirable characteristics of disinfectants and sanitisers

In relation to antimicrobial activity
Broad spectrum activity including activity against bacteria, fungi, viruses
Rapid antimicrobial activity
Retain stability (product) and antimicrobial efficacy over a wide range of pH
Retain stability (product) and antimicrobial efficacy over a wide range of temperature
Retain activity in the presence of organic load and hard water
Retain activity upon dilution
Residual activity*
In relation with safety
No or low toxicity
Degradable in the environment
In relation with formulation and usage
No or low corrosiveness
Non-staining
No odour
Good wetting and detergency
Easily combined with liquid or powder
Compatible with other chemicals (e.g. surfactants)
Cost-effective

* Residual activity. The presence of remaining low concentration (below the minimum inhibitory concentration, MIC) of a biocide on a surface is the subject of much debate with current evidence on emerging microbial resistance to biocides.

properties and spectrum of activity (Russell, 2001) (Table 38.1). Selecting a biocide for a specific application depends not only on the characteristic of the agent (e.g. toxicity, corrosiveness, spectrum of activity) but also usage conditions (e.g. risk associated with a surface, equipment; types of microorganisms expected, soiling, level of microbial contamination) (Table 38.2).

38.2.1 Biocides, activity and usage

Biocides have been used for centuries, originally for the preservation of foodstuffs and water, and then for the control of infection (mainly through antisepsis). A large number of current antimicrobial agents have been synthesised more recently (Fraise, 2004a). The main applications of biocides are in the food, water and pharmaceutical industry and in the healthcare environment, although some biocides are used in applications in specialised areas, for example the preservation of petroleum products, textile and leather, paint, wood and generally in the construction industry. More recently, products containing low concentrations of biocides have appeared with an increasing frequency in the home environment (Levy, 2001), mainly as a result of a better public understanding of hygiene but also from increased commercial pressure.

Whereas 'disinfectant' is a term used mainly in the healthcare environment (but also increasingly used in the domiciliary environment), 'sanitiser' is used in the food industry, particularly in the USA. A typical definition of a sanitiser is 'a substance that reduces microbial contamination on inanimate surfaces to levels that are considered safe from a public health standpoint' (Grab and Bennett, 2001). Sanitisers are used by the food processing, food handling, preparation and service industries, and milk (i.e. dairy) and beverage-producing (including breweries, soft drink and canning facilities) industries. Sanitisers are usually divided into two groups, depending on whether or not they come into contact with food. This is particularly important, since the testing protocols and the required efficacy differ depending upon their usage. Disinfectants, 'sterilants' and antiseptics are used in the healthcare environment (e.g. hospital, dental, veterinary, nursing, day-care service providers) to prevent and control infection. As for sanitisers, their required microbicidal efficacy depends upon their usage. Three levels of disinfection are usually considered; high-, intermediate and low-level, and they are defined according to the potential risk of microbial infection to patients (Rutala and Weber, 2004).

Biocides used for disinfection and sanitisation

Both disinfectants and sanitisers are used for reducing microbial contamination from inanimate surfaces to an acceptable, safe level. Their antimicrobial efficacy is therefore of prime importance. There are mainly six distinct classes of sanitisers: chlorine, iodophors, quaternary ammonium compounds (QACs), acid–anionic sanitisers, carboxylic acid and peroxy-acid compounds. Additional biocides can be used in disinfectant formulations, mainly aldehydes (e.g. glutaraldehyde, *ortho*-phthalaldehyde), phenolics (e.g. triclosan), biguanides (e.g. chlorhexidine) and alcohols. All these agents are chemically distinct, although some have similar mechanisms of action. For example, QACs, biguanides and phenolics are all membrane-active agents (Russell, 2001). In addition some biocides can be used in combination such as chlorhexidine and alcohols.

Other uses: 'sterilisation', antisepsis and preservation

Biocides are used for a range of applications and play an important role in infection control policies in the healthcare environment. Some highly reactive biocides, such as alkylating (e.g. glutaraldehyde, *ortho*-phthalaldehyde) and oxidising agents (e.g. peracetic acid, hydrogen peroxide) are used for high-level disinfection of critical items (e.g. endoscopes). Usually their use is associated with particular medical equipments such as automated washer-disinfectors (AWD) which combine preliminary cleaning, disinfection and rinsing (Fraise, 2004b).

Biocides are also extensively used for antisepsis. Hand-washing is an essential part of infection control policies in the healthcare environment and is associated with good hygiene and good manufacturing practice in various industries. Appropriate usage of antiseptics and compliance by the end users are essential. In the healthcare environment compliance to hand-washing policies has been associated with a reduction in cross-infection (Boyce and Pittet, 2002).

The preservation of foodstuffs and water by biocides is one of the oldest documented uses of chemical biocides (e.g. the use of copper pots to preserve water) (Fraise, 2004a). In addition, preservation now includes medicines, pharmaceutical and cosmetic products, but also paint, wood, various plastic and textiles, and is used in the petroleum and construction industries. The choice of preservatives has to be balanced between antimicrobial efficacy (now increasingly microbicidal activity) and low toxicity.

38.2.2 Factors influencing biocidal activity

Several factors can affect the antimicrobial efficacy of chemical biocides (Table 38.3). These factors have usually been well characterised for many of these compounds (Russell, 2004). However, their practical significance for the end-product and its usage is rarely discussed. Failure of a disinfection/sanitisation process often reflects the non-respect or lack of understanding of these factors. Hence, it is important to combine the use of a suitable antimicrobial product/formulation for a specific task with the training of the end user. Since compliance to the manufacturer's instructions is particularly important, the efficacy of an

Table 38.3 Factors influencing the activity of biocides

Factors	Comments
Concentration	The main cause for failure of disinfection. Dilution can inactivate biocides, notably those with a high concentration exponent.
Contact time	Non-respect of, or poor compliance with, contact time can result in the survival of microorganisms. Contact time needs to be adapted to the condition of usage.
Organic load	Particularly important in the food industry, and in the clinical context with blood spillage. Can severely reduce the antimicrobial efficacy of biocides.
Temperature	Could be an issue when biocides are used in cold conditions; e.g. cold room, chilled food production, or with the efficacy of preservatives (i.e. products kept at a low temperature).
pH	Can affect both the microorganisms and the agents, especially if it is an acid or a base.
Microorganisms	
Type	Microorganisms vary in susceptibility to biocides, prions and spores being the most resistant to disinfection.
Number	Heavy microbial contamination is more difficult to disinfect/sanitise.
Phenotype	Microorganisms grown as biofilms, or with a low metabolism are more resistant to antimicrobials than planktonic grown microorganisms and microorganisms grown on 'rich' media.
Relative humidity	Particularly important for gaseous disinfectants. Pre-humidification affects the hydration state of microorganisms.
Neutralisation/incompatibility	Can inactivate completely or partially the activity of a biocide. Knowledge of the products is important.

antimicrobial product should be evaluated with standard protocols that investigate a range of conditions. Many antimicrobial tests, notably practical tests, include various parameters in their design, such as concentration, contact time, temperature, soiling, type and number of microorganisms. Generally, these factors can be divided into those inherent to the biocide and those inherent to the microorganisms.

Factors inherent to the product

Concentration is probably the most important factor to consider when antimicrobial efficacy is concerned (Russell and McDonnell, 2000). There have been several published reports of microbial contamination following chemical disinfection, or microbial survival within biocidal products/formulations (Poole 2002; Russell, 2002). For example, many reports concern the failure of QAC disinfectants, although in many cases inappropriate concentrations were used (Prince and Ayliffe, 1972; Ehrenkranz *et al.*, 1980). Holah and colleagues (2002) pointed out that when the concentration of QACs remains high (i.e. $1000\,\text{mg}\,\text{L}^{-1}$), survival of vegetative microorganisms is unlikely. Likewise, failure of high-level disinfectants such as glutaraldehyde to eliminate all microorganisms from endoscope washer disinfectors have been reported (van Klingeren and Pullen, 1993; Griffiths *et al.*, 1997).

The effect of changes in concentration on antimicrobial efficacy can be estimated by the concentration exponent (η) and is given by the equation:

$$\eta = \frac{\log t_2 - \log t_1}{\log C_1 - \log C_2}$$

where C_1 and C_2 represent two concentrations and t_1 and t_2 the respective times to reduce the population to the same level. The concentration exponent varies among biocides (Table 38.4). It gives an indication of the effect of diluting an in-use concentration; i.e. biocides with high concentration exponent will rapidly lose activity upon dilution, whereas those with a low concentration exponent will retain activity upon dilution. This in effect allows the selection of appropriate concentrations to be evaluated with antimicrobial test protocols.

Contact time is an important factor of all antimicrobial testing protocols and the choice of time of exposure usually reflects conditions in practice. There is no simple relationship between activity and contact time, although longer exposure time is usually associated with better activity and might be essential to eliminate the 'resistant' clones of a microbial population. Standard antimicrobial test protocols, for manufacturers' and hygienic guidelines usually specify a set contact time or the minimum contact time required. For example, the European Standard for the testing of surface disinfectants (CEN1276, 1997a) stipulates that 5 $\log_{10}$ reduction in bacterial concentration must be attained within 5 minutes of exposure time. Likewise, the hygienic hand-wash procedure (CEN1499, 1997b) recommends a minimum of 1 minute contact time, which reflects acceptable hand-washing time in practice.

Organic load or soiling (e.g. serum, blood, pus, earth, food residues, faecal materials) contributes to decreasing biocidal activity by either 'mopping up' the

Table 38.4 Examples of concentration exponent (η)

Biocide	Exponent
Phenolics	4–9.9
Alcohol	
Benzyl alcohol	2.6–4.6
Aliphatic alcohols	6.0–12.7
Cationic biocides	
Chlorhexidine	2
Polymeric biguanides	1.5–1.6
QACs	0.8–2.5
Crystal violet	0.9
Aldehydes	
Formaldehyde	1
Glutaraldehyde	1
Peroxygens	
Hydrogen peroxide	0.5
Metallic salts	
Silver nitrate	0.9–1.0
Mercurials	0.03–3.0
Organic acid	
Parabens	2.5
Sorbic acid	2.6–3.2
Potassium laurate	2.3

active concentration or/and offering some protection to the microorganisms. Indeed the antimicrobial efficacy of some biocides can be deeply affected by soiling (Table 38.1). Practical tests now reflect the importance of soiling by stipulating testing under clean and dirty conditions, usually by the addition of serum albumin (e.g. $3\,g\,L^{-1}$ for testing under dirty condition) in the reaction vessel (e.g. CEN1276, 1997a). The effect of soiling also emphasises the necessity of cleaning surfaces and equipment before a biocidal product is used, or combining a disinfectant with a detergent. In the food and dairy industry, a reduction in biocidal activity may occur with the presence of organic matter and effective pre-cleaning prior to disinfection is recommended. Some chemical biocides may exert a detergent action, whereas some detergents exhibit some biocidal activity. In this respect the surface to be treated is important to consider (see below) as it affects the efficacy of a biocide or biocide/detergent combination.

The activity of biocides usually increases with a rise in temperature and this principle is used, when combining biocide and steam sterilisation. Other equipment, such as some automated washer disinfectors, also combine biocides and elevated temperature. On the other hand, low temperature may decrease the antimicrobial efficacy of biocides. Temperature is particularly an issue during storage of a biocidal formulation/product, especially upon preservation, and where chilled food is produced (Taylor *et al.*, 1999). The effect of temperature on activity can be calculated with the temperature coefficient (θ) and more conveniently by the Q_{10} value (change in activity following a rise of 10 °C). The Q_{10} value is given by the equation:

$$Q_{10} = \frac{\text{Time to kill at } T\,°\text{C}}{\text{Time to kill at } (T + 10)\,°\text{C}}$$

Standard testing protocols recommend testing at a temperature of 20 °C ± 1 °C (e.g. CEN1276, 1997a) or around ambient temperature (18–25 °C) (e.g. CEN13697, 2001). However, this does not reflect product usage at low temperature, although the activity of a compound at additional temperature can be tested.

The effect of pH on antimicrobial activity is complex and can affect the microorganism as well as the compound (Russell, 2004). For some biocides, their active state is the non-ionised form (e.g. phenols, acetic acid, benzoic acid) and increase pH decreases their activity. Others (e.g. cationic biocides, glutaraldehyde) show an enhanced activity at an alkaline condition. However, testing for antimicrobial efficacy at different pHs is usually not recommended since the pH is usually set for a given antimicrobial formulation and cannot be altered easily without affecting the stability of the formulation.

Biocides within an antimicrobial formulation can be partially inactivated by the components of the formulation. Surface-active agents, notably non-ionic agents, can affect the activity of antimicrobial compounds. It is therefore not surprising that many non-ionic surfactants are used as neutralising agents and this is considered in more detail below (Section 38.3.3). The potential inactivation of antimicrobial activity by non-ionic agents is particularly a concern for the preservation of pharmaceutical and cosmetic products with biocides such as parabens and QACs. The critical micelle concentration (CMC) of the non-ionic agent is particularly important. Below the CMC it is generally believed that the non-ionic agent increases membrane permeability, but also releases the biocide to produce a highly active solution. Metal ions can also affect the antimicrobial activity of a biocidal formulation, notably when the formulation contains permeabilising agents in the form of ion chelators (e.g. ethylene diamine tetraacetic acid).

Likewise, partitioning of the antimicrobial agent might lead to a decrease in bioavailability. For example, phenol can absorb into rubber, hence reducing its effective aqueous concentration (Allwood, 1978).

The surface to be disinfected is not usually listed as a factor influencing the activity of a biocide as such, but needs to be considered here. The antimicrobial efficacy of disinfectants or sanitisers will depend to some extent on the surface upon which they are used. Surfaces can vary greatly, particularly whether they are porous or non-porous. Porous surfaces will have a tendency to entrap and protect microbial contaminants, whereas non-porous surfaces can reduce bacterial adhesion and facilitate a cleaning or a disinfection process.

Most biocidal formulations are diluted prior to use in hard water. Hard water and notably the presence of divalent cations (e.g. Mg^{2+}, Ca^{2+}) may interfere with the activity of some disinfectants by blocking adsorption sites on the bacterial cell. Water hardness is particularly of concern when the formulation/product needs to be diluted to the appropriate in use concentration with on-site water. To reflect the use of hard water, standard test protocols recommend the use of hard water to dilute the product, and usually state its composition (e.g. CEN1276, 1997a; CEN13697, 2001).

Relative humidity affects the activity of gaseous disinfectants (e.g. ethylene oxide, β-propiolactone, formaldehyde). Pre-humidification and notably the state of hydration of the microorganism is a determining factor for their sensitivity to antimicrobial processes, particularly sterilisation (Russell, 2004).

Factors inherent to the microorganisms

Different types of microorganisms present different levels of sensitivity to a given antimicrobial biocide. Attempts have been made to classify micro-organisms according to their overall susceptibility to biocides (Table 38.5). Usually, such classification relies upon information on the intrinsic property of a microorganism but is not designed to give a definite answer about the susceptibility of a type of microorganism, since variation within species and even strains might occur. Practically, the type of microorganisms expected on a given surface help in the selection of an appropriate disinfectant or sanitiser. Antimicrobial test protocols usually include testing against a range of bacteria and fungi, which are selected depending upon the expected usage of the biocide,

Table 38.5 Susceptibility of microorganisms to biocides

Microorganisms	Level of resistance	Comments
Prions	High	Special sterilisation processes combining heat with biocide. Use of single-use items recommended.
Bacterial spores	High	Sterilisation or high-level disinfection. Often used as biological indicators.
Mycobacteria	High to intermediate	High-level disinfection recommended. Resistance to glutaraldehyde and peracetic acid has been reported.
Small non-enveloped viruses	Intermediate	Enteric viruses might be particularly a problem when low-level disinfection is used.
Fungi	Intermediate to low	Relatively little information is available on the susceptibility of fungi, and particularly fungal spores, to biocides.
Vegetative Gram-negative bacteria	Intermediate to low	*Pseudomonas aeruginosa* is particularly resilient microorganism.
Vegetative Gram-positive bacteria	Low	Generally sensitive to disinfection.
Large enveloped viruses	Low	The lipid envelope is particularly sensitive to biocides.

i.e. food industry, hospital environment, etc. However, the number of test protocols available to evaluate virucidal and mycobactericidal activity (refer to Section 38.4.3) is limited. In addition, there is no standardisation and these protocols tend to vary greatly between countries (Maillard, 2004) notably with the test organisms. In Europe, a virucidal test for food hygiene, domestic and institutional use is available (CEN 13610), although a similar protocol for the healthcare and veterinary environment has not been published yet (Holah, 2003). The choice of the viral indicator is particularly contentious (Maillard, 2004).

The number of microorganisms that should be used in standard tests has long been debated and differs between test protocols. It is generally accepted that the higher the level of microbial contaminant, the more difficult the disinfection. Predicting the level of contamination might be difficult and often the worst case scenario is considered, i.e. a high-inoculum. Most tests work on the basis of reducing the number of microorganisms to an acceptable level (e.g. a 5 $\log_{10}$ reduction on surface), but not to the complete elimination (i.e. sterilisation) of the microorganisms. If this is generally acceptable for most microorganisms, a problem can arise with highly infectious or virulent microorganisms such as the hepatitis B virus, *Escherichia coli* O157 for which a complete elimination would be recommendable.

Bacterial phenotype can affect the activity of antimicrobial biocides. Growth conditions including physical (e.g. temperature, gas) and chemical conditions (e.g. pH), nutrient limitation and diet (i.e. excess of lipids), but also whether the cells are grown as a biofilm or in suspension will produce microorganisms with a different phenotype. The metabolic status of the cell is particularly important since bacteria with a 'low metabolism' or quiescent bacteria are particularly resilient to the antimicrobial effects of biocides (Dodd *et al.*, 1997; Holah *et al.*, 2002; Gilbert *et al.*, 2003). Testing protocols describe meticulously the preparation of the inoculum, including growth medium and physical conditions, but also the number of passage of the strains in broth or solid media. The production of the start-up inoculum is discussed further in Section 38.3.2.

38.2.3 Limitations in the use of biocidal products

The limitations in using biocidal products usually refer to their toxicity, to the alteration of the surface/equipment onto which they are used (e.g. corrosiveness, colour formation), to their incompatibility with other components of a formulation, but also to their overall efficacy against a given predicted microorganism (Table 38.2). For example, high-level disinfectants are needed for the disinfection of critical surfaces in the hospital environment (Rutala and Weber, 1999). Toxicity is also important to consider not only for the end user (e.g. with antisepsis and preservation) but also for the environment. For example, the use of high concentrations might not be acceptable because of the high toxicity for the environment. Within the food industry, consideration must also be given to the potential for any biocide residues to taint or otherwise change the organoleptic properties of the foodstuffs produced.

38.3 Criteria for testing biocidal action

The purpose of antimicrobial efficacy testing is to determine a pass/fail criterion for a given biocide under specific conditions. At first glance, such tests appear straightforward. However, in reality, to obtain appropriate and reproducible results, the design of testing protocols is complex, notably in the number of factors that need to be controlled.

38.3.1 Assessment of antimicrobial activity

Several factors influence the susceptibility of microorganisms to biocides and the strict control of these factors is essential for the standardisation of testing protocols. Past studies have been particularly prolific in showing the variations in growth condition to the sensitivity profile of vegetative microorganisms and spores. The results of many laboratory investigations, notably relating to testing conditions and microbial growth conditions, have helped in the standardisation of modern testing protocols. However, the choice of an appropriate protocol to meet all needs is confusing, mostly because of the multitude of methods available. Indeed several standards have been published by different national organisations: in the USA, the American Association of Official Analytical Chemists (e.g. AOAC, 1990); in Germany, the German Society for Hygiene and Microbiology (DGHM) and the German Veterinary Society (DVG, 1988); in France the French Association of Normalization (AFNOR; e.g. AFNOR, 1989); and in the UK, the British Standards Institution (BSI; e.g. Anon 1988). In Europe, some harmonisation is taking place under the auspices of the European Committee for Standardisation (CEN) by the Technical Committee (TC) 216 (Holah, 2003).

38.3.2 Criteria inherent to the microorganisms

Test strain

Disinfectants and sanitisers are expected to have a broad spectrum of activity at in-use concentration, although some microorganisms are less susceptible, notably bacterial spores and to some extent mycobacteria (Table 38.5). The selection of test strains is crucial but often cause for debate. Ideally antimicrobial testing protocols should use microorganisms that reflect conditions *in situ*. However, most standard test protocols recommend the evaluation of antimicrobial activity against a few selected microorganisms mainly for cost and practical reasons. For example, the CEN1276 (1997a) recommends the use of *Pseudomonas aeruginosa* (ATCC15422), *Escherichia coli* ATCC (10536), *Staphylococcus aureus* (ATCC6538) and *Enterococcus hirae* (ATCC10541), and for specific applications the additional bacterial strains *Salmonella enterica* serovar *typhimurium* (ATCC13311), *Lactobacillus brevis* (DSM6235) and *Enterobacter cloacae* (DSM6234). All microorganisms should be well characterised and readily available from culture collections. Test strains do not include microorganisms that are known to be more resistant to a given product. The inclusion of such strains increases the cost of the testing and might ultimately select for a more toxic biocide, although it would guarantee the efficacy of the biocide in most circumstances. The use of more resistant strains might be recommended for specific usage, for example, mycobactericidal activity. Very few protocols recommend the use of a mixed inoculum and generally these tests are considered to be cumbersome and prone to variability in results. However, there have been attempts to develop biocidal efficacy test using mixed inocula (Best *et al.*, 1994).

Preparation of inocula

Growth conditions (e.g. temperature, media, chemostat or continuous cultures) affect microbial metabolism and physiology (e.g. outer layer, lipid content) and could alter their sensitivity to antimicrobials (Russell, 2004). Water quality can also have a profound effect, for example on the germination, outgrowth and sporulation of *Bacillus subtilis* as reported by Knott *et al.* (1997).

The growth stage of the bacterial population is also essential to control. For example, exponential phase culture of *Listeria monocytogenes* (Luppens *et al.*, 2001) and *Staphylococcus aureus* (Luppens *et al.*, 2002) have been shown to be more sensitive to cationic and oxidising biocides. Nutrient availability and notably limitation can affect growth rate and hence affect also bacterial sensitivity to biocides (Russell and Chopra, 1996). There are considerable differences between bacteria grown in continuous culture *vs.* batch culture, the latter usually predominant in testing protocols. The main criticism of batch culture is that cells of different physiological ages are present. Pre-treatment of a microbial population, for example to increase the dispersion of the individual cells, should be avoided. Pre-treatment with non-ionic (e.g. polysorbate) or cationic surfactants have been shown to alter microbial susceptibility, presumably by altering the cell membrane property. Likewise pre-treatment with permeabilising agents such as ethylenediaminetetraacetic acid (edta), lactoferrin and polycations will alter the microorganism sensitivity following an alteration of membrane permeability. Surprisingly, there is little information on the effect of pH, temperature and anaerobiosis during microbial growth and subsequent sensitivity to antimicrobials (Russell, 2004). It is likely that pH values will change during microbial growth (Messager *et al.*, 2004a). Alterations of phospholipid contents when microorganisms are grown at different pHs or different temperatures has been described (Russell, 2004). Since the alterations of the microbial cell wall might be linked to differences in microbial response to antimicrobials, the strict control of the growth medium and conditions is essential to ensure the production of a reproducible inoculum.

Detection and count of survivors

Antimicrobial testing protocols usually rely on the detection of survivors as visible growth and/or the enumeration of bacterial/fungal colonies or virus plaques. Capacity tests (Section 38.4.2) are based on the detection of growth but do not necessarily quantify the amount of growth following an antimicrobial treatment. In suspension tests where microorganisms are homogeneously dispersed, sampling for survivors is relatively easy. This contrasts with surface carrier tests which make use notably of porous surfaces. For such tests, sampling surviving microorganisms might be challenging and an additional step, such as sonication and the use of enzymes or other compounds for desorbing the microorganisms from the surface, can been used, although this might increase the damage to already stressed organisms.

The size of the inoculum is particularly important for quantitative methods that need to demonstrate a reduction in cell number. In this case, the original inoculum must be high enough to demonstrate the required reduction in number, taking into consideration dilutions caused, for example, by a neutralisation step. However, a large inoculum might not necessarily be representative of microbial contamination in practice, and might cause an 'inoculum effect' (i.e. similar to soiling) on the activity of a biocide. Nonetheless, most standard protocols specify the initial inoculum size, the sampling size after treatment and the volume to be plated on recovery media. The recovery media might play an important role in the survival of damaged microorganisms following an antimicrobial treatment. Most standard protocols recommend the use of complex 'rich' agar (e.g. tryptone soya agar) and optimum temperature for microbial growth. Such growth conditions, although optimum for healthy microorganisms, might be detrimental for stressed and injured ones (Hurst 1977; Dodd *et al.*, 1997). The design of a recovery agar that facilitates the repair of injured microorganisms has been investigated, particularly in relation to food processing (Czechowicz *et al.*, 1996; Farrell *et al.*, 1998; Kang and Fung, 2000; Restaino *et al.*, 2001).

In addition, protocols that rely on visible microbial growth following treatment and therefore long incubation periods to allow the formation of colonies might not provide a rapid response. The use of rapid counting techniques, such as epifluorescence (Pettipher, 1986; Matsunaga et al., 1995; D'Haese and Nelis, 2002; Coma *et al.*, 2003; Yamaguchi *et al.*, 2003), flow cytometry (Shapiro, 1990; McSharry, 1994; Endo *et al.*, 1997, 2001; Auty *et al.*, 2001; Lehtinen *et al.*, 2003), bioluminescence (Stewart 1990; Stewart *et al.*, 1991; 1996; Walker *et al.*, 1994; Hill *et al.*, 1994), impedance (Connolly *et al.*, 1993, 1994) and microcalorimetry (Morgan *et al.*, 2001) have been reported. The measurement of 'optical density' might provide a particularly suitable alternative to plating. Lambert *et al.* (1998) investigated the use of an automated combined shaker incubator-optical reader to measure the kinetics of inactivation following biocidal exposure. Further investigations showed the usefulness of such an approach in terms of rapidity, fast screening and adaptability with the use of different factors such as organic load (Lambert and Johnston, 2001) and inoculum size (Johnston *et al.*, 2000). Furthermore, comparable results to a standard test protocol were obtained (Lambert *et al.*, 1999; Lambert and Johnston, 2000). Nevertheless, standard agar-based recovery protocols are easily standardised and are particularly reproducible and as such they are used as a reference for the development of other alternative counting techniques (Sheppard *et al.*, 1997).

38.3.3 Criteria inherent to the test methods

Quenching antimicrobial activity

The use of a neutralising agent to quench the activity of an antimicrobial or the elimination of the antimicrobial from the recovery medium is essential to determine the lethal activity of the disinfectant/sanitiser. Unfortunately, this crucial step has often been overlooked in many investigations. The 'true' bactericidal activity of a compound then becomes difficult to determine and to distinguish from a residual bacteriostatic effect in the recovery medium. Examples of neutralising agents that quench appropriately the activity of specific biocides are given in Table 38.6. Several 'universal neutralisers' are also available and contain usually a range of chemicals. However, it is unclear as to whether they inactivate the activity of all biocides. Recommended neutralisers in standard protocols may contain several compounds (e.g. Table 38.7).

Table 38.6 Examples of neutralising agents

Antimicrobial agent	Possible neutraliser	Comments
Phenols	None (dilution) Tweens (polysorbate)	Phenolics have high-concentration exponent. The activity readily drops with dilution.
Alcohols	None (dilution)	
Cationic compounds		
Chlorhexidine	Lecithin + tween	
QACs	Lethicin + Lubrol	
Aldehydes Glutaraldehyde *Ortho*-phthalaldehyde	Sodium sulphite/ bisulphite Glycine	Sodium sulphite/bisulphite might be toxic to some microorganisms. Glycine is inefficient in quenching OPA. Higher concentration of sodium bisulphite needed for OPA.
Halogen-releasing agents Chlorine Sodium hypochlorite Iodine	Sodium thiosulphate	Sodium thiosulphate might be toxic to some bacterial species.
Hydrogen peroxide	Catalase or peroxidase Dilution	1 unit catalyses the decomposition of 1 mol of hydrogen peroxide per minute at 25 °C and at pH 7.
Parabens	None (dilution) Tweens	
Metallic salts/compounds		
Mercury compounds Silver compounds Organic arsenical	SH-compounds; e.g. thioglycollate	Thioglycollate might be toxic to some bacterial species. S- and SS-compounds are inefficient to quench silver compounds.

Table 38.7 Examples of recommended neutralising solutions (based on CEN 1276 (1997a))

Number of components	Neutraliser composition
1	• Phosphate buffer (0.25 mol L^{-1}) • Fresh egg yolk (0.5 or 5% v/v) • Phospholipid emulsion (50 mg mL^{-1}) • Glycine (concentration depends upon the product) • Sodium thioglycollate (0.5 or 5 g L^{-1}) • L-cysteine (0.8 or 1.5 g L^{-1}) • Sodium thiosulphate (5 g L^{-1}) • Thiomalic acid (0.075% v/v) • Catalase or peroxidase (for the inactivation of hydrogen peroxide)
2	• Fresh egg yolk (5% v/v), polysorbate 80 (40 g L^{-1}) • Ethylene oxide condensate of fatty alcohol (4% v/v), lecithin (4 g L^{-1}) • Polysorbate 80 (30 g L^{-1}), lecithin (3 g L^{-1})
3	• Polysorbate 80 (30 g L^{-1}), sodium lauryl sulphate (4 g L^{-1}), lecithin (3 g L^{-1}) • Ethylene oxide condensate of fatty alcohol (7% v/v), lecithin (20 g L^{-1}), polysorbate 80 (4% v/v) • Polysorbate 80 (30 g L^{-1}), lecithin (3 g L^{-1}), L-histidine (1 g L^{-1})
4 or more	• Polysorbate 80 (30 g L^{-1}), saponin (30 g L^{-1}), L-histidine (1 g L^{-1}), L-cysteine (1 g L^{-1}) • Lecithin (3 g L^{-1}), polysorbate 80 (30 g L^{-1}), sodium thiosulphate (5 g L^{-1}), L-histidine (1 g L^{-1}), saponin (30 g L^{-1})

Neutralising agents are particularly important for those biocides that interact strongly with the bacterial cell or with those with a low concentration exponent. However, it is important to note that microorganisms that have suffered a certain amount of damage following an antimicrobial treatment might not be able to recover even when a neutraliser is added. Furthermore, some neutralisers might increase the amount of damage already sustained by the microorganisms.

Biocides with a high concentration exponent lose their activity rapidly upon dilution (Section 38.2.2). Although this might be sufficient to overcome any bacteriostatic activity, the effect of sub-inhibitory (residual) concentrations on the bacterial cell remains poorly understood (Maillard, 2002). As mentioned previously, the use of neutralisers might damage further already stressed micro-organisms (e.g. non-ionic surfactants, sodium thiosulphate, thioglycollate). Assessing the toxicity of a neutralising solution (notably when the recommended concentration is changed) against all test microorganisms is of a paramount importance. In addition, the efficacy of a neutralising solution to quench effectively an antimicrobial product against specific test strains and test conditions must also be assessed, notably when several neutralisers can be used

(Table 38.7). Inappropriate quenching of activity or neutraliser toxicity will result in overestimating the antimicrobial activity of a formulation.

If there is no neutraliser available for a particular product, membrane filtration can be used as an alternative. The mixture of biocide and test microorganism is then filtered through a membrane filter which is immediately washed with a rinsing liquid. The membrane is then transferred to a recovery medium to enable survivors to produce colonies after incubation. The advantage of membrane filtration is that neutralisers may not be used, hence reducing the potential toxicity to injured cells. However, the antimicrobial is assumed to be washed off very rapidly. This might not be the case with surfactants or with compounds with a large water–oil partition coefficient. Where neutralisers are used in the rinsing fluid, it is also difficult to verify their neutralisation activity. The use of membrane filtration and notably the rinsing solution might also add further stress to the microorganisms. In addition, the rinsing solution should be compatible with the membrane filter and the filter used should not retain/react with the disinfectant/sanitiser.

Overall, the neutralising or removal of the antimicrobial is of paramount importance to assess the antimicrobial efficiency of a disinfectant/sanitiser. Overlooking the validation of the quenching step might introduce distortions in the results and an overestimation of the killing potency of a biocidal formulation. Standard antimicrobial tests should detail the standardisation of the neutralisation–dilution or membrane filtration protocols.

Physical parameters

Standard testing protocols should clearly describe the experimental conditions, including, for example, the diluent to use, test temperature and pH. Temperature and pH (Section 38.2.2) can affect the antimicrobial efficacy of a biocide. As an example, Taylor and colleagues (1999) studied the effect of temperature on the lethal efficacy of 18 disinfectants against *Ps. aeruginosa* and *E. coli* O157. The number of disinfectants that fail to meet the required lethality against *Ps. aeruginosa* increased at 10 °C (i.e. 11/18 instead of 13/18 at 20 °C). As for the diluent, sterile distilled water, one-quarter strength Ringer's solution, 0.9% w/v saline, peptone water and nutrient broth have been used, and some might increase damage to treated microorganisms. There is no such 'universal diluent' and most standard protocols recommend the diluent to be used.

38.4 Tests for disinfectants and sanitisers

There are a number of protocols available for testing the antimicrobial efficacy of a disinfectant/sanitiser. There is a profusion of information from the literature on exploratory protocols and on the specific advantages and inconveniences of selected tests, but especially on the factors affecting the viability and reproducibility of a standard antimicrobial test. Readers can refer to some excellent reviews and chapters by Ayliffe (1989), Cremieux and Fleurette (1991), Mulberry (1995), Reybrouck (1999) and Lambert (2004).

38.4.1 Classification of disinfectant tests

Disinfectant tests can be classified according to different criteria, such as the test organism (i.e. determination of bactericidal, fungicidal, sporicidal or virucidal activity), the type of action (i.e. 'static' or 'cidal' activity), the test structure (i.e. *in vitro*, practical, in use) and the aims of the test.

Protocols classified according to the test objective can be divided into three main stages: (1) primary testing and screening, (2) laboratory testing simulating possible real-life situations and (3) *in loco* testing (testing in the field) (Table 38.8). The European standard protocols follow the same three-stage classification: in phase 1, only suspension tests are considered with a limited number of microorganisms (e.g. EN 1040; 1275). In phase 2, conditions simulating possible applications are tested. Phase 2, step 1 test refers to more advanced suspension tests that include parameters such as soiling and hard water (e.g. EN1276; 1650). Phase 2, step 2 test concerns practical tests such as carrier tests (i.e. surface disinfection; e.g. EN13697; WI216024). In phase 3, the antimicrobial activity of the compound is tested *in situ*.

38.4.2 Testing the activity of disinfectants and sanitisers

Testing the antimicrobial efficacy of potential disinfectants and sanitisers appropriately is essential to ensure the product will meet the manufacturer's claims and to provide the end user with reliable information. In Europe, the CEN/TC 216 has the role of producing current and future European disinfectant testing standards, whereas in the US the Environmental Protection Agency (EPA) and the Food and Drug Administration (FDA) are involved in their

Table 38.8 Classification of antimicrobial tests according to their objective

Test stage	Phase	Aim	Example of European tests
Preliminary test	Phase 1	Primary testing and screening	EN1040 (basic bactericidal activity) EN1275 (basic fungicidal activity)
In vitro tests	Phase 2	Simulate the conditions encountered for the possible application	
	Phase 2 step 1	Advanced suspension test	EN1656 and EN1276 (bactericidal suspension test) EN1650 (fungicidal suspension test)
	Phase 2 step 2	Advanced surface test	EN13697 (bactericidal and fungicidal surface test) EN13624 (medical instruments disinfectants–bactericidal activity) EN1499 (hygienic hand-wash) EN1500 (hygienic handrub)
In situ test	Phase 3	Testing in real-life conditions	No standard antimicrobial test available

registration, application and use. These institutions have described a range of protocols available for the testing of the antimicrobial activity of disinfectants and sanitisers.

Suspension tests
The most commonly used testing protocols are based on the suspension test. In essence this test involves mixing a microbial suspension with a disinfectant for a set contact time and then checking for survivors after a neutralisation or a filtration step. The suspension test can be quantitative when the number of surviving bacteria is investigated, or qualitative (the most basic option) for which an indication of microbial survival is only required (i.e. a simple pass/fail test).

Most primary tests are in the form of a qualitative suspension test during which the potential activity of a biocide at a given concentration and a set time is investigated. These preliminary tests are useful for a rapid evaluation of antimicrobial activity of a series of concentrations or contact times. There are limitations to the suspension tests, notably when the result is assessed qualitatively. Indeed, the growth and multiplication of a lone survivor are indistinguishable from the growth of part or the entire population. Therefore the antimicrobial effect on the growth and survival of the entire population cannot be evaluated with these tests. Information on qualitative suspension test protocols can be found principally from German institutions (DVG, 1988). The addition of an enumeration step is then highly beneficial and most of the suspension tests used for commercial and research purposes are now quantitative. Microorganisms are counted either by direct agar plating method or by membrane filtration after quenching/removing the activity of the disinfectant/ sanitiser. The lethal effect of the agent (ME) is calculated as follows:

$$ME = \log_{10} N_C - \log_{10} N_D$$

where N_C is the number of colony-forming units in the untreated control(s) and N_D the number of colony-forming units counted after exposure to the biocide.

In Europe, the publication of the European bactericidal suspension test (CEN, 1997a) added much-needed standardisation and supplanted various existing tests, some of which have now been withdrawn. Prior to that, the Dutch standard suspension test and quantitative suspension tests published by AFNOR and DGHM (Reybrouck, 1980) were widely used throughout Europe and required a 5 $\log_{10}$ reduction in microbial number within 5 minutes. The European suspension test does not differ drastically from existing protocols, apart from the addition of soiling, which simulates 'in use' conditions. In addition, more stringency is provided by the precise description of the different stages of the test, notably, the preparation of the inoculum and the test conditions, including biocide concentration, contact time, water hardness, temperature and as already mentioned soiling.

Following the publication of the European bactericidal suspension test (CEN, 1997a), a European-funded research programme 'Andistand' (EU Contract No.

SMT4-CT98-2222) has been undertaken to assess the reproducibility of the method between 16 laboratories in eight European countries. The work showed that the mean precision of the log reduction factor in the test was 3 log orders when a single test was performed, 2 log orders when four tests were performed and 1 log order when nine tests were performed. This work, for the first time, gave an idea of the true precision of suspension tests.

Surface tests

There are many surface tests available from the literature and they cannot all be described here. Surface tests are particularly appropriate for the efficacy testing of disinfectants/sanitisers. These tests attempt to reproduce real-life conditions whereby the antimicrobial compound is added onto an inoculum previously dried on a surface. Survivors are then recovered from the surface and counted, usually after a neutralisation step. Protocols can be easily adapted to represent better surface conditions *in situ*, with the use of a range of porous and non-porous surfaces (e.g. various textiles, ceramic, stainless steel), and specific microbial environments with the use of specific microorganisms associated with, for example, the food or healthcare environments. An alternative to surface tests consists of incorporating the biocidal agent into a surface/product (e.g. textile). The microbial inoculum is then added onto the surface and after a set contact time, survivors are counted, usually after a neutralisation step. In principle, surface tests are easily designed but suffer from a narrow range of applications and their reproducibility as compared with suspension tests is unknown. The inoculum dried on the surface is difficult to standardise, notably since some microorganisms can be severely injured or killed by the drying step (e.g. *Ps. aeruginosa*). This severely limits the practicality of these tests by decreasing their detection threshold limit. Relative humidity then becomes an important factor as well as the medium used to resuspend the cells from the surface. In addition, long drying times may increase the insusceptibility of a microorganism (van Klingeren *et al.*, 1998). The reproducibility of these tests between laboratories is also questionable (Reybrouck, 1986, 1990; Bloomfield *et al.*, 1994).

The development of standard carrier test provides some standardisation for surface protocols. The AOAC Use-Dilution Test is a well-used carrier test, for which a piece of cloth is artificially contaminated, dried and immersed in the disinfectant. It suffered originally from some lack of reproducibility, notably with *Ps. aeruginosa*. This test was modified to result in the AOAC Hard Surface Carrier Test (HSCT) (Beloain, 1993), which was further improved by a standardisation of the inoculum (Rubino *et al.*, 1992; Hamilton *et al.*, 1995; Beloain, 1995).

In Europe, CEN/TC216 published a bactericidal and fungicidal surface test (CEN13697, 2001) and other practical tests are being developed (Holah, 2003). Such practical tests will undoubtedly supersede existing carrier test protocols published by, for example, DGHM and AFNOR. In the EN 13697 a standardised inoculum is dried on the surface of stainless steel disks, onto which the appropriate dilution of the product is added. After a set contact time, the disk is transferred into a neutralising solution and surviving microorganisms enumerated. As for other European tests produced by CEN/TC216, additional microorganisms, contact time and temperature can be used to reflect in-use conditions.

An alternative to inanimate surfaces are the hand-washing tests that have been developed for determining the antimicrobial efficacy of antiseptics. There are several antiseptic test protocols available, (e.g. glove-juice, sterile bag, washing and rinsing), which differ in their basic assessment methodology (Anon, 1997; Sattar and Ansari, 2002). In Europe the main testing protocols were the Vienna and the Birmingham models (Ayliffe *et al.*, 1978; Rotter and Koller, 1991). CEN/TC216 released two antiseptic tests, the Hygienic Handwash test (EN 1499) and the Hygienic Handrub (EN 1500). Basically, in the Hygienic Hand-wash test (CEN, 1997b), fingers and thumb tips are immersed in a non-pathogenic suspension of *Escherichia coli* (NCTC10538) and air dried for 3 min. The antiseptic formulation is then applied and hands rubbed according to a standardised hand-wash procedure. Surviving bacteria are then recovered and enumerated after a neutralisation step.

Other surface test protocols against specific microorganisms such as mycobacteria (Section 38.4.3), spores and *Legionella* have been described. These microorganisms, particularly mycobacteria and spores provide extra challenges for the biocidal product and the test methodology; they are more difficult to grow in the laboratory than standard vegetative cells and they are particularly resistant to biocidal challenges. As a result, more stringent regimens (e.g. high-level disinfectants) are needed to ensure that specific equipment (e.g. endoscopes) have been appropriately chemically sterilised. Standardised practical tests simulating conditions *in situ* (i.e. medical instrument disinfectants – sporicidal/mycobactericidal/virucidal tests) are being developed (Section 38.4.3).

Capacity test

Capacity tests such as the Kelsey-Sykes test (e.g. British Standard BS 6905; BSI, 1987) differ somewhat from the suspension test in that the biocide is challenged with several additions of a bacterial suspension. After each addition, the number of survivors is measured. Such repetitive additions allow the determination of when the biocide activity (i.e. capacity) has been drained. Capacity tests attempt to simulate practical applications in the field (i.e. real-life conditions) and in some instances are used to confirm the efficacy of an 'in use' dilution.

38.4.3 Other testing methodologies

Mycobacteria are generally considered more resistant than other vegetative microorganisms to disinfection (Lambert, 2002) and therefore have to be tested separately (Holah, 2003). Most mycobacteria are slow-growing microorganisms and have a tendency to form clumps (as a result of their high hydrophobicity), which contribute to their resistance profile (Lambert, 2002). The most important species is *M. tuberculosis* and often tests have been labelled tuberculocidal tests. However, *M. tuberculosis* is particularly slow growing and is an important human pathogen, which makes its routine use for disinfection testing difficult and expensive. In addition, there is a marked difference in biocide susceptibility between species, which leads to the question as to which mycobacterial surrogate can be used in antimicrobial efficacy testing. *M. bovis* is recommended in the EPA and AOAC (confirmation) tests, *M. smegmatis* in the AOAC (screening) and AFNOR tests, and *M. tuberculosis* and *M. terrae* in the DGHM. However, *M. smegmatis* was reported to be more sensitive than *M. tuberculosis* and thus is not recommended as a test surrogate. *M. avium-intracellulare* has been found to be generally more resistant to biocides than *M. tuberculosis* and *M. terrae*. Nevertheless, *M. terrae* appears to meet the criteria for an appropriate surrogate for *M tuberculosis* (Griffiths *et al.*, 1998). Mycobactericidal suspension and carrier tests have been published (Ascenzi *et al.*, 1987; AOAC, 1990; Ascenzi, 1991), although concerns have been expressed towards the reproducibility of mycobactericidal suspension tests (Robison *et al.*, 1996).

Disinfectants/sanitisers are usually not required to show antimicrobial activity against bacterial spores, except in specific circumstances, where the risk of contamination/infection by a sporulated species has been established. Specific sporicidal test protocols have been published and the main difference with the test protocols described above is in the preparation of the spore inoculum; vegetative cells have to be removed and it is important that the spore structure and vitality are not damaged during the preparation of the inoculum (Tanimoto *et al.*, 1996). *Bacillus subtilis*, *B. cereus* and *Clostridium sporogenes* have been used as test microorganisms. Several sporicidal tests protocols have been published (Cremieux *et al.*, 2001).

This section would not be complete without mentioning prions, the agents responsible for transmissible degenerative encephalopathies (TDEs). As for spores, disinfectants/sanitisers are not required to show activity against prions, which are probably the most resistant entities. They are highly resistant to biocides and often complete elimination of the agents involve drastic measures such as the use of highly corrosive and toxic biocides or a combination of heat and chemical inactivation (Taylor, 2004). Decontamination studies are particularly difficult to set up since the prion 'inoculum' is often difficult to standardise and is closely associated with host tissues, which might confer some level of protection. In addition, bioassays need to be used to detect reliably TDE infectivity (Taylor *et al.*, 2000). As a result, there is some variability in prion inactivation results in the literature, because of the number of different test protocols used. Nevertheless, it is recommended to use harsh conditions to simulate the worst case scenario (Taylor, 2004).

38.5 Test limitations and scope for improvement

One can argue that there is always scope for improvement, particularly with the reproducibility of these protocols. Reproducibility, robustness and limitations of testing regimens have been reported in the literature (Groschel, 1983; Rutala and

Table 38.9 Test requirements and conditions

Test conditions/parameters	Requirements	Questions/limitations
Test microorganisms	• Readily available from culture collection • Preparation of inoculum clearly specified	• Appropriate surrogate? • Represent microbial diversity *in situ*? • Pathogenicity of the test organisms; e.g. *M tuberculosis*, hepatitis B virus, TDEs? • Inoculum size? • Use of biofilm?
Test conditions		
Water/growth media/diluent	• Composition clearly specified	• Representation of in use conditions (i.e. hard water)? • Toxicity to test microorganisms? • Need to use cells with high metabolic activity? • Appropriate media for cell recovery?
Neutraliser	• Composition clearly specified • Toxicity and efficacy tests need to be conducted	• Detrimental to test organisms? • Use of membrane filtration?
Rinsing liquid	• Compatibility with membrane filter needs to be assessed	• Detrimental to test organisms?
Test parameters		
pH/temperature/contact time	• Need some flexibility to reflect in use conditions	• Choice of appropriate parameters to reflect usage conditions. • Change of experimental procedures? • Associated costs?

Cole, 1984; Reybrouck, 1991; Bloomfield and Looney, 1992; Bloomfield *et al.*, 1994, 1995; Langsrud and Sundheim, 1998; Tilt and Hamilton, 1999; Kampf and Ostermeyer, 2002; Borgmann-Strahsen, 2003; Kneale, 2003). The variability in results on the antimicrobial activity of a given agent as reported in the literature often resides in the differences in protocols used, some tests being less stringent than others (Kampf *et al.*, 2003; Marcheti *et al.*, 2003; Messager *et al.*, 2004b), but also the non-respect of test preparation (notably inoculum) and conditions. Indeed there have been several reports of commercial disinfectants failing a basic suspension test (Jacquet and Reynaud, 1994; Taylor *et al.*, 1999). A drive towards the standardisation of antimicrobial tests, such as with European testing protocols, provides a much-needed improvement in this field. However, the different test requirements and conditions (Table 38.9) need to be clearly thought, defined and respected. Preparation of the inocula is of paramount importance (Bloomfield *et al.*, 1995; Johnston *et al.*, 2000). Standardisation of the inoculum, at least in the European suspension test, has improved reproducibility in results. The reproducibility of antimicrobial surface tests might remain an issue where the inoculum is difficult to standardise. Overall, antimicrobial efficacy tests can be complex (Table 38.9) and it is impossible to standardise all the equipment, glassware and consumables, and thus results will be subject to variability from one laboratory to the next. In addition, the operator will undoubtedly add variability in the test (Bloomfield *et al.*, 1994).

Reproducibility set aside, limitations of antimicrobial test protocols probably are concerned with two levels of testing: practical and *in situ* test. The idea of practical tests is to simulate conditions in the field. The rigidity of some testing protocols does not allow much flexibility, although with antimicrobial tests recommended by CEN/TC 216, additional temperature, contact time and microorganisms can be chosen to reflect conditions in practice more appropriately. Still the determination of mycobactericidal, sporicidal or virucidal activity requires specific testing protocols. Testing for viruses is particularly a problem, notably with the question about the appropriate test virus to be used. Viruses differ tremendously in structure and sensitivity to biocides and a multitude of testing protocols exist (Maillard, 2004), although some standardisation will be introduced (Holah, 2003).

Tests *in loco* are costly and difficult to standardise since parameters cannot be controlled accurately in the field. These tests remain poorly reproducible and their outcomes may be contentious. It is difficult to imagine improvements in the *in situ* tests in the near future, although these tests would provide key information on the antimicrobial efficacy of a disinfectant/sanitiser to the manufacturers and end users.

38.6 Future trends

Suspension and surface tests use microorganisms that have been grown in a planktonic system. Although this might be appropriate for preliminary tests, the preparation of such inocula might be questionable for practical tests, particularly for surface tests. Microorganisms associated with surfaces are often found as a biofilm community rather than single cells. This is particularly important since biofilms are often more resistant than their planktonic counterparts to antimicrobials (Allison *et al.*, 2000). The use of biofilms as the inoculum should add more stringency to the test, and simulate better situations in the field (Fine *et al.*, 2001; Cappelli *et al.*, 2003). As for other practical tests, the standardisation of the 'biofilm inoculum' would be critical for the reproducibility of results. There are several methodologies to produce biofilms (e.g. sedimentation; chemostats and constant-depth film fermentor). Ceri *et al.* (2001) discussed the different methodologies to produce biofilms and their potential application for antimicrobial testing. Other parameters directly relevant to biofilms need to be controlled, such as microorganism type, and biofilm age, depth and structure. The development and standardisation of biofilm testing protocols would be highly relevant for the food and healthcare industries, although it would add more stringency to test protocols and might result in the recommendation for the use of higher disinfectant concentrations.

As mentioned in Section 38.3.2, the development of rapid, sensitive and reliable detection methods to palliate the drawbacks of using agar-based protocols to enumerate surviving microrganisms offers an attractive option. These methods are particularly attractive for those microrganisms that are particularly slow growing, such as mycobacteria, for which survival is difficult to assess (e.g. viruses, prions).

Standardisation of antimicrobial test protocols is beneficial if it allows some flexibility in the parameters and test conditions, especially with practical tests. The establishment of CEN/TC216 has been important in developing a range of tests adapted to specific end users (Holah, 2003). While a number of standard protocols have already been published, a number are still at a development stage. Meanwhile, a variety of test protocols are still being used that are contributing to possible discrepancies in efficacy claims for a given disinfectant/sanitiser. Here, regulatory authorities have to play a role in the development and commercialisation of new biocidal products (Kappes and Rasmussen, 2003; Sobanska *et al.*, 2003). The purpose of antimicrobial tests is to ensure that a product meets its antimicrobial claim and to provide confidence about label claims for the end users (Favero, 2002).

38.7 Sources of further information and advice

There is a wealth of information on testing protocols, factors influencing biocidal activity in the literature and some useful information has already been referred to in this chapter. However, Reybrouck (1999), Lambert (2004), Cremieux and colleagues (2001) and Grab and Bennett (2001) provide detailed and useful information on testing protocols while biofilm test methodologies are described further by Ceri *et al.* (2001). Readers will find useful information on TDEs in the chapter written by Taylor (2004).

The following web sites can be consulted for information on British and European standards (http://www.bsi-global.com), EPA test methods (http://www.epa.gov/epahome/index/), ASTM (http://www.astm.org), ISO (http://www.iso.org) and AOAC (http://www.AOAC.org/) standards.

38.8 References

AFNOR (ASSOCIATION FRANÇAISE DE NORMALISATION) (1989), *Recueil de normes françaises. Antiseptiques et désinfectants*, 2nd edn, Paris, La Défense, Association Française de Normalisation.

ALLISON DG, MCBAIN AJ and GILBERT P (2000), 'Biofilms: problems of control', in Allison DG, Gilbert P, Lappin-Scott HM and Wilson M, *Community Structure and Co-operation in Biofilms*, Cambridge, Cambridge University Press, 309–327.

ALLWOOD M C (1978), 'Antimicrobial agents in single- and multi-dose injections', *J Appl Bacteriol*, **44** (S1), vii–xvii.

ANON (1988), *Method of Test for the Antimicrobial Activity of Disinfectants in Food Hygiene, DD 177*, London, British Standards Institute.

ANON (1997), 'Standard test method for determining the virus-eliminating effectiveness of liquid hygienic handwash agents using fingerpads of adult volunteers', American Society for Testing and Materials, Designation E1838-96, 1–6.

AOAC (ASSOCIATION OF OFFICIAL ANALYTICAL CHEMISTS) (1990), *Official Methods of Analysis*, 15th edn, Arlington, Association of Official Analytical Chemists.

ASCENZI JM (1991), 'Standardization of tuberculocidal testing of disinfectants', *J Hosp Infect*, **18** (SA), 256–263.

ASCENZI JM, EZZELL RJ and WENDT TM (1987), 'A more accurate method for measurement of tuberculocidal activity of disinfectants', *Appl Environ Microbiol*, **53**(9), 2189–2192.

AUTY MAE, GARDINER GE, MCBREARTY SJ, O'SULLIVAN EO, MULVIHILL DM, COLLINS JK, FITZGERALD GF, STANTON C and ROSS RP (2001), 'Direct *in situ* viability assessment of bacteria in probiotic dairy products using viability staining in conjunction with confocal scanning laser microscopy', *Appl Environ Microbiol*, **67**(1), 420–425.

AYLIFFE GAJ (1989), 'Standardization of disinfectant testing', *J Hosp Infect*, **13**(3), 211–216.

AYLIFFE GAJ, BABB JR and QUORAISHI AH (1978), 'A test for "hygienic" hand disinfection', *J Clin Pathol*, **31**, 923–928.

BELOAIN A (1993), 'General Referee Reports. Disinfectants', *J AOAC Int*, **76**, 97–98.

BELOIAN A (1995), 'General Referee Reports. Disinfectants', *J AOAC Int*, **78**, 179.

BEST M, SPRINGTHORPE VS and SATTAR SA (1994), 'Feasibility of a combined carrier test for disinfectants: studies with a mixture of five types of microorganisms', *Am J Infect Control*, **22**(3), 152–162.

BLOOMFIELD SF (2002), 'Significance of biocide usage and antimicrobial resistance in domiciliary environments', *J Appl Microbiol*, **92** (S1), 144–157.

BLOOMFIELD SF and LOONEY E (1992), 'Evaluation of the repeatability and reproducibility of European suspension test methods for antimicrobial activity of disinfectants and antiseptics', *J Appl Bacteriol*, **73**(1), 87–93.

BLOOMFIELD, SF, ARTHUR M, VAN KLINGEREN B, PULLEN W, HOLAH JT and ELTON R (1994), 'An evaluation of the repeatability and reproducibility of a surface test for the activity of disinfectants', *J Appl Bacteriol*, **76**(1), 86–94.

BLOOMFIELD SF, ARTHUR M, GIBSON H, MORLEY, K., GILBERT, P. and BROWN, MRW (1995), 'Development of reproducible test inocula for disinfectant testing', *Int Biodeter Biodegrad*, **36**(3–4), 311–331.

BORGMANN-STRAHSEN R (2003), 'Comparative assessment of different biocides in swimming pool water', *Int Biodeter Biodegrad*, **51**(4), 291–297.

BOYCE JM and PITTET D (2002), 'Guidelines for hand hygiene in health-care settings', *Am J Infect Control*, **30**(8), 1–46.

BSI (BRITISH STANDARDS INSTITUTION) (1987), 'Estimation of Concentration of Disinfectants Used in "Dirty" Conditions in Hospitals by the Modified Kelsey-Sykes Test. BS 6905: 1987', London, BSI.

CAPPELLI G, SERENI L, SCIALOJA MG, MORSELLI M, PERRONE S, CIUFFREDA A, BELLESIA M, INGUAGGIATO P, ALBERTAZZI A and TETTA C (2003), 'Effects of biofilm formation on haemodialysis monitor disinfection', *Nephrol Dial Transpl*, **18**(10), 2105–2111.

CEN (COMITÉ EUROPÉEN DE NORMALISATION, EUROPEAN COMMITTEE FOR STANDARDIZATION) (1997a), 'EN 1276 Chemical disinfectants and antiseptics – Quantitative suspension test for the evaluation of bactericidal activity of chemical disinfectants and antiseptics for use in food, industrial, domestic and institutional areas – Test method and requirements (phase 2, step 1)', London, British Standards Institute.

CEN (COMITÉ EUROPÉEN DE NORMALISATION, EUROPEAN COMMITTEE FOR STANDARDIZATION) (1997b), 'EN 1499 Chemical disinfectants and antiseptics – Hygienic handwash-Test method and requirements (phase 2, step 2)', London, British Standards Institute.

CEN (COMITÉ EUROPÉEN DE NORMALISATION, EUROPEAN COMMITTEE FOR STANDARDIZATION) (2001), 'EN 13697 Chemical disinfectants and antiseptics – Quantitative non-porous surface test for the evaluation of bactericidal and/or fungicidal activity of chemical disinfectants used in food, industrial, domestic and institutional areas – Test method and requirements without mechanical action (phase 2, step 2)', London, British Standards Institute.

CERI H, MORCK DW and OLSON ME (2001), 'Biocide susceptibility testing of biofilms', in Block SS, *Disinfection, Sterilization and Preservation*, 5th ed, London, Lippincott Williams and Wilkins, 1429–1438.

COMA V, DESCHAMPS A and MARTIAL-GROS A (2003), 'Bioactive packaging materials from edible chitosan polymer – Antimicrobial activity assessment on dairy-related contaminants', *J Food Sci*, **68**(9), 2788–2792.

CONNOLLY P, BLOOMFIELD SF and DENYER SP (1993), 'A study of the use of rapid methods for preservative efficacy testing of pharmaceuticals and cosmetics', *J Appl Bacteriol*, **75**(5), 456–462.

CONNOLLY P, BLOOMFIELD SF and DENYER SP (1994), 'The use of impedance for preservative efficacy testing of pharmaceutical and cosmetic products', *J Appl Bacteriol*, **76**(1), 68–74.

CREMIEUX A and FLEURETTE J (1991), 'Methods of testing disinfectants', in Block SS, *Disinfection, Sterilization and Preservation*, 4th ed, Philadelphia, Lea and Febiger, 1009–1027.

CREMIEUX A, FRENEY J and DAVIN-RELI A (2001), 'Methods of testing disinfectants', in Block SS, *Disinfection, Sterilization and Preservation*, 5th ed, Philadelphia, Lea and Febiger, 1305–1328.

CZECHOWICZ SM, SANTOS O and ZOTTOLA EA (1996), 'Recovery of thermally-stressed *Escherichia coli* O157:H7 by media supplemented with pyruvate', *Int J Food Microbiol*, **33**(2–3), 275–284.

D'HAESE E and NELIS HJ (2002), 'Rapid detection of single cell bacteria as a novel approach in food microbiology', *J AOAC Int*, **85**(4), 979–983.

DODD CER, SHARMAN RL, BLOOMFIELD SF, BOOTH IR and STEWART GSAB (1997), 'Inimical processes: bacterial self-destruction and sub-lethal injury', *Trends Food Sci Tech*, **8**(7), 238–241.

DVG (DEUTSCHE VETERINÄRMEDIZINISCHE GESELLSCHAFT) (1988), *Richtlinien für die Prüfung Chemischer Desinfektionsmittel*, 2nd edn, Giessen, Deutsche Veterinärmedizinische Gesellschaft.

EHRENKRANZ NJ, BOLYARD EA, WIENER M and CLEARY TJ (1980), 'Antibiotic-sensitive *Serratia marcescens* infections complicating cardio-pulmonary operations: contaminated disinfectants as a reservoir', *Lancet*, **ii**, 1289–1292.

ENDO H, NAKAYAMA J, HAYASHI T and WATANABE E (1997), 'Application of flow cytometry for rapid determination of cell number of viable bacteria', *Fish Sci*, **63**(6), 1024–1029.

ENDO H, NAGANO Y, REN HF and HAYASHI T (2001), 'Rapid enumeration of bacteria grown on surimi-based products by flow cytometry', *Fish Sci*, **67**(5), 969–974.

FARRELL BL, RONNER AB and WONG ACL (1998), 'Attachment of *Escherichia coli* O157:H7 in ground beef to meat grinders and survival after sanitation with chlorine and peroxyacetic acid', *J Food Protect*, **61**(7), 817–822.

FAVERO MS (2002), 'Products containing biocides: perceptions and realities', *J Appl Microbiol*, **92**(S1), 72–77.

FINE DH, FURGANG D and BARNETT ML (2001), 'Comparative antimicrobial activities of antiseptic mouthrinses against isogenic planktonic and biofilm forms of *Actinobacillus actinomycetemcomitans*', *J Clin Periodontol*, **28**(7), 697–700.

FRAISE AP (2004a), 'Historical introduction', In Fraise AP, Lambert PA and Maillard J-Y, *Principles and Practice of Disinfection, Preservation and Sterilization*, 4th edn, Oxford, Blackwell Science, 3–7.

FRAISE AP (2004b), 'Decontamination of the environment and medical equipment in hopsital', in Fraise AP, Lambert PA and Maillard J-Y, *Principles and Practice of Disinfection, Preservation and Sterilization*, 4th edn, Oxford, Blackwell Science, 563–587.

GILBERT P, MCBAIN AJ and RICKARD AH (2003), 'Formation of microbial biofilm in hygienic situations: a problem of control', *Int Biodeter Biodegrad*, **51**(4), 245–248.

GRAB LA and BENNET MK (2001), 'Methods of testing sanitizers and bacteriostatic substances' in Block SS, *Disinfection, Sterilization and Preservation*, 5th edn, London, Lippincott Williams and Wilkins, 1373–1382.

GRIFFITHS PA, BABB JR, BRADLEY CR and FRAISE AP (1997), 'Glutaraldehyde-resistant *Mycobacterium chelonae* from endoscope washer disinfectors', *J Appl Microbiol*, **82**(4), 519–526.

GRIFFITHS PA, BABB JR and FRAISE AP (1998), '*Mycobacterium terrae*: a potential surrogate for *Mycobacterium tuberculosis* in a standard disinfection test', *J Hosp Infect*, **38**(3), 183–192.

GROSCHEL DHM (1983), 'Caveat emptor: do your disinfectants work?', *Infect Control Hosp Epidemiol*, **4**(3), 144.

HAMILTON MA, DE VRIES TA and RUBINO JR (1995), Hard surface carrier test as a quantitative test of disinfection: a collaborative study', *J AOAC Int*, **78**, 1102–1109.

HILL PJ, HALL L, VINICOMBE DA, SOPER CJ, SETLOW P, WAITES WM, DENYER S and STEWART GSAB (1994), 'Bioluminescence and spores as biological indicators of inimical processes', *J Appl Bacteriol*, **76**(S1), 1295–1345.

HOLAH JT (2003), 'CEN/TC216: its role in producing current and future European disinfectant testing standards', *Int Biodegrad Biodeter*, **51**(4), 239–243.

HOLAH JT, TAYLOR JH, DAWSON DJ and HALL KE (2002), 'Biocide used in the food industry and the disinfectant resistance of persistent strains of *Listeria monocytogenes* and *Escherichia coli*', *J Appl Microbiol*, **92**(S1), 111–120.

HURST A (1977), 'Bacterial injury: a review', *Can J Microbiol*, **23**, 936–944.

JACQUET C and REYNAUD A (1994), 'Difference in the sensitivity to eight disinfectants of *Listeria monocytogenes* strains as related to their origin', *Int J Food Microbiol*, **22**(1), 79–83.

JOHNSTON MD, SIMONS E-A and LAMBERT RJW (2000), 'One explanation for the variability of the bacterial suspension test', *J Appl Microbiol*, **88**(2), 237–242.

KAMPF G and OSTERMEYER C (2002), 'Intra-laboratory reproducibility of the hand hygiene reference procedures of EN 1499 (hygienic handwash) and EN 1500 (hygienic hand disinfection)', *J Hosp Infect*, **52**(3), 219–224.

KAMPF G, MEYER B and GORONCY-BERMES P (2003), 'Comparison of two test methods for the determination of sufficient antimicrobial activity of three commonly used alcohol-based hand rubs for hygienic hand disinfection', *J Hosp Infect*, **55**(3), 220–225.

KANG DH and FUNG DYC (2000), 'Application of thin agar layer method for recovery of injured *Salmonella typhimurium*', *Int J Food Microbiol*, **54**(1–2), 127–132.

KAPPES D and RASMUSSEN K (2003), 'Prioritisation of existing biocidal active substances in the European Union', *Environ Sci Pol*, **6**(6), 521–532.

KNEALE C (2003), 'Problems and pitfalls in the evaluation and design of new biocides for plastics applications', *Polym Polym Compos*, **11**(3), 219–228.

KNOTT A G, DANCER B N , HANN A C and RUSSELL A D (1997), 'Non-variable sources of pure water and the germination and outgrowth of *Bacillus subtilis* spores', *J Appl Microbiol*, **82**(2), 267–272.

LAMBERT PA (2002), 'Cellular impermeability and uptake of biocides and antibiotics in Gram-positive bacteria and mycobacteria', *J Appl Microbiol*, **92**(S1), 46–54.

LAMBERT RJW (2004), 'Evaluation of antimicrobial efficacy', in Fraise AP, Lambert PA and Maillard J-Y, *Principles and Practice of Disinfection, Preservation and Sterilization*, 4th edn, Oxford, Blackwell Science, 345–360.

LAMBERT RJW and JOHNSTON MD (2000), 'Disinfection kinetics: a new hypothesis and model for the tailing of log-survivor/time curves', *J Appl Microbiol*, **88**(5), 907–913.

LAMBERT RJW and JOHNSTON MD (2001), 'The effect of interfering substances on the disinfection process: a mathematical model', *J Appl Microbiol*, **91**(3), 548–555.

LAMBERT RJW, JOHNSTON MD and SIMONS L-A (1998), 'Disinfectant testing: use of the bioscreen microbiological growth analyser for laboratory biocide screening', *Lett Appl Microbiol*, **26**(4), 288–292.

LAMBERT RJW, JOHNSTON MD and SIMONS E-A (1999), 'A kinetic study of the effect of hydrogen peroxide and peracetic acid against *Staphylococcus aureus* and *Pseudomonas aeruginosa* using the Bioscreen disinfection method', *J Appl Microbiol*, **87**(5), 782–786.

LANGSRUD S and SUNDHEIM G (1998), 'Factors influencing a suspension test method for antimicrobial activity of disinfectants', *J Appl Microbiol*, **85**(6), 1006–1012.

LANGSRUD S, SINGH SIDHU M, HEIR E and HOLCK AL (2003), 'Bacterial disinfectant resistance – a challenge for the food industry', *Int Biodeter Biodegrad*, **51**(4), 283–290.

LEHTINEN J, VIRTA M and LILIUS EM (2003), 'Fluoro-luminometric real-time measurement of bacterial viability and killing', *J Microbiol Methodol*, **55**(1), 173–186.

LEVY SB (2001), 'Antibacterial household products: cause for concern', *Emerg Infect Dis*, **7**(3), 512–515.

LUPPENS SBI, ABEE T and OOSTEROM J (2001), 'Effect of benzalkonium chloride on viability and energy metabolism in exponential- and stationary-growth-phase cells of *Listeria monocytogenes*', *J Food Protect*, **64**(4), 476–484.

LUPPENS SBI, ROMBOUTS F M and ABEE T (2002), 'The effect of growth phase on *Staphylococcus aureus* on resistance to disinfectants in a suspension test', *J Food Protect*, **65**(1), 124–129.

MCSHARRY JJ (1994), 'Uses of flow cytometry in microbiology', *Clin Microbiol Rev*, **7**(4), 576–604.

MAILLARD J-Y (2002), 'Bacterial target sites for biocide action', *J Appl Microbiol*, **92**(S1), 16–27.

MAILLARD J.-Y (2004), 'Viricidal activity of biocides', in Fraise AP, Lambert PA and Maillard J-Y, *Principles and Practice of Disinfection, Preservation and Sterilization*, 4th edn, Oxford, Blackwell Science, 272–323.

MARCHETTI MG, KAMPF G, FINZI G and SALVATORELLI G (2003), 'Evaluation of the bactericidal effect of five products for surgical hand disinfection according to prEN 12054 and prEN 12791', *J Hosp Infect*, **54**(1), 63–67.

MATSUNAGA T, OKOCHI M and NAKASONO S (1995), 'Direct count of bacteria using fluorescent dyes – application to assessment of electrochemical disinfection', *Analyt Chem*, **67**(24), 4487–4490.

MESSAGER S, HANN AC, GODDARD PA, DETTMAR PW and MAILLARD J-Y (2004a), 'Use of the "*ex vivo*" test to study long-term bacterial survival on human skin and their sensitivity to antisepsis', *J Appl Microbiol*, **97**(6), 1149–1160.

MESSAGER S, GODDARD PA, DETTMAR PW and MAILLARD J-Y (2004b), 'Antibacterial activity of several antiseptics using two "in-vivo" and two "ex-vivo" tests', *J Hosp Infect*, **58**(2), 115–121.

MORGAN TD, BEEZER AE, MITCHELL JC and BUNCH AW (2001), 'A microcalorimetric comparison of the anti-*Streptococcus mutans* efficacy of plant extracts and antimicrobial agents in oral hygiene formulations', *J Appl Microbiol*, **90**(1), 53–58.

MULBERRY GK (1995), 'Current methods of testing disinfectants', in Rutala WA, *Chemical Germicides in Health Care*, Washington, Association for Professionals in Infection Control, 224–235

PETTIPHER GL (1986), 'Review: the direct epifluorescent filter technique', *J Food Tech*, **21**(5), 535–546.

POOLE K (2002), 'Mechanisms of bacterial biocide and antibiotic resistance', *J Appl Microbiol*, **92**(S1), 55–64.

PRINCE J and AYLIFFE GAJ (1972), 'In-use testing of disinfectants in hospitals', *J Clin Pathol*, **25**, 586–589.

RESTAINO L, FRAMPTON EW and SPITZ H (2001), 'Repair and growth of heat- and freeze-injured *Escherichia coli* O157:H7 in selective enrichment broths', *Food Microbiol*, **18**(6), 617–629.

REYBROUCK G (1980), 'A comparison of the quantitative suspension tests for the assessment of disinfectants', *Zbl Hyg Umweltmed*, **170**, 449–456.

REYBROUCK G (1986), 'Unification of the testing of disinfectants in Europe', *Zbl Hyg Umweltmed*, **182**(5–6), 485–498.

REYBROUCK G (1990), 'The assessment of the bactericidal activity of surface disinfectants.

III. Practical tests for surface disinfection', *Zbl Hyg Umweltmed*, **190**, 500–510.

REYBROUCK G (1991), 'International standardization of disinfectant testing: is it possible?', *J Hosp Infect*, **18**(SA), 280–288.

REYBROUCK G (1999), 'Evaluation of the antibacterial and antifungal activity of disinfectants', in Russell AD, Hugo WB and Ayliffe GAJ, *Principles and Practice of Disinfection, Preservation and Sterilization*, 3rd edn, Oxford, Blackwell Science, 124–144.

ROBISON RA, OSGUTHORPE RJ, CARROLL SJ, LEAVITT RW, SCHAALJE GB and ASCENZI JM (1996), 'Culture variability associated with the US Environmental Protection Agency tuberculocidal activity test method', *Appl Environ Microbiol*, **62**(8), 2681–2686.

ROTTER ML and KOLLER W (1991), 'A European test for the evaluation of the efficacy of procedures for the antiseptic handwash', *Hyg Med*, **16**, 4–12.

RUBINO JR, BAUER JM. CLARKE PH, WOODWARD BB, PORTER FC and HILTON HG (1992), 'Hard surface carrier test for efficiency testing of disinfectants: collaborative study', *J AOAC Int*, **75**(4), 635–645.

RUSSELL AD (2001), 'Types of antimicrobial agents', in Russell AD, Hugo WB and Ayliffe GAJ, *Principles and Practice of Disinfection, Preservation and Sterilization*, 3th edn, Oxford, Blackwell Science, 5–94.

RUSSELL AD (2002), 'Introduction of biocides into clinical practice and the impact on antibiotic-resistant bacteria', *J Appl Microbiol*, **92**(S1), 121–135.

RUSSELL AD (2004), 'Factors influencing the efficacy of antimicrobial agents' in Fraise AP, Lambert PA and Maillard J-Y, *Principles and Practice of Disinfection, Preservation and Sterilization*, 4th edn, Oxford, Blackwell Science, 98–127.

RUSSELL AD and CHOPRA I (1996), Understanding *Antibacterial Action and Resistance*, 2nd edn, Chichester, Ellis Horwood.

RUSSELL A D and McDONNELL G (2000), 'Concentration: a major factor in studying biocidal action', *J Hosp Infect*, **44**(1), 1–3.

RUTALA WA and COLE EC (1984), 'Antiseptics and disinfectants-safe and effective?', *Infect Control Hosp Epidemiol*, **5**(5), 215–218.

RUTALA WA and WEBER DJ (1999), 'Infection control: the role of disinfection and sterilization', *J Hosp Infect*, **43**(S1), 43–55.

RUTALA WA and WEBER DJ (2004), 'The benefits of surface disinfection', *Am J Infect Control*, **32**(4), 226–231.

SATTAR SA and ANSARI SA (2002), 'The fingerpad protocol to assess hygienic hand antiseptics against viruses', *J Virol Methodol*, **103**(2), 171–181.

SHAPIRO HM (1990), 'Flow cytometry in laboratory technology: new directions', *ASM News*, **56**, 584–586.

SHEPPARD FC, MASON DJ, BLOOMFIELD SF and GANT VA (1997), 'Flow cytometric analysis of chlorhexidine action', *FEMS Microbiol Lett*, **154**(2), 283–288.

SOBANSKA MA, ASCHBERGER K, RASMUSSEN K, PEDERSEN F and KAPPES D (2003), 'The review programme in the European Union for existing biocidal active substances – outcome of the notification process', *Environ Sci Pol*, **6**(6), 513–519.

STEWART GSAB (1990), '*In vivo* bioluminescence: new potentials for microbiology', *Lett Appl Microbiol*, **10**(1), 1–8.

STEWART GSAB, JASSIM SAA and DENYER SP (1991), 'Mechanisms of action and rapid biocide testing', in Denyer SP and Hugo WB, *Mechanisms of Action of Chemical Biocides*, Bedford, Society for Applied Bacteriology, Technical Series No. 27, 319–329.

STEWART GSAB, LOESSNER MJ and SCHERER S (1996), 'The bacterial lux gene bioluminescent biosensor', *ASM News*, **62**, 297–301.

TANIMOTO Y, ICHIKAWA Y, YASUDA Y and TOCHIKUBO K (1996), 'Permeability of dormant spores of *Bacillus subtilis* to gramicidin S', *FEMS Microbiol Lett*, **136**(2), 151–156.

TAYLOR DM (2004), 'Transmissible degenerative encephalopathies: inactivation of the unconventional causal agents', in Fraise AP, Lambert PA and Maillard J-Y, *Principles and Practice of Disinfection, Preservation and Sterilization*, 4th edn, Oxford, Blackwell Science, 324–341.

TAYLOR JH, ROGERS SJ and HOLAH JT (1999), 'A comparison of the bactericidal efficacy of 18 disinfectants used in the food industry against *Escherichia coli* O157:H7 and *Pseudomonas aeruginosa* at 10 and 20 °C', *J Appl Microbiol*, **87**(5), 718–725.

TAYLOR DM, McCONNELL I and FERGUSON CE (2000), 'Closely similar values obtained when the ME7 strain of scrapie agent was titrated in parallel by two individuals in separate laboratories using two sublines of C57BL mice', *J Virol Methodol*, **86**(1), 35–40.

TILT N and HAMILTON MA (1999), 'Repeatability and reproducibility of germicide tests: A literature review', *J AOAC Int*, **82**(2), 384–389.

VAN KLINGEREN B and PULLEN W (1993), 'Glutaraldehyde resistant mycobacteria from endoscope washers', *J Hosp Infect*, **25**(2), 147–149.

VAN KLINGEREN B, KOLLER W, BLOOMFIELD SF, BÖHM R, CREMIEUX A, HOLAH J, REYBROUCK G and RÖDGER H-J (1998), 'Assessment of the efficacy of disinfectants on surfaces', *Int Biodeter Biodegrad*, **41**(3–4), 289–293.

WALKER AJ, STEWART GSAB, SHEPPERD F, BLOOMFIELD SF, HOLAH JT and DENYER SP (1994), 'Bioluminescence imaging as a tool for studying biocide challenge upon planktonic and surface attached bacteria', *Binary-Computing Microbiol*, **6**(1), 16.

YAMAGUCHI N, ISHIDOSHIRO A, YOSHIDA Y, SAIKA T, SENDA S and NASU M (2003), 'Development of an adhesive sheet for direct counting of bacteria on solid surfaces', *J Microbiol Methodol*, **53**(3), 405–410.

39

Traceability of cleaning agents and disinfectants

D. Rosner, Ecolab GmbH & Co., Germany

39.1 Introduction

Tracing of base materials is an important safety factor in modern food production. With a fast growing network of suppliers, on a global base, it becomes increasingly important to know who the producer of a raw material is, how it is produced and that the supplier complies with the users' quality and safety standards. In the case of any problem or question, the manufacturer of foodstuffs should be able to trace any raw material used in a preparation back to its sources. This requirement is easy to understand and logical – but what does traceability mean regarding cleaners and disinfectants? Of course, they are absolutely not allowed to become an ingredient in a food product but they play a very important role in food production! Without a strict hygiene regime there is no quality production of food. A poor quality or wrong composition, or expired shelf-life, can have an adverse effect on hygiene quality – in the worst case an unsuitable product could leave tainting substances in production equipment with a negative impact on the food quality. But this is only one aspect with such products. Proper labelling with the product name, main ingredients, manufacturer information and *batch coding* will deal with that regard. Nowadays cleaners and disinfectants are not sold without safety data sheets that give even more detailed information not only about safe handling, but also about the composition of the products.

39.1.1 Manufacturers' responsibilities in tracing cleaners and disinfectants back to their origins

Manufacturers of cleaners and disinfectants are responsible for the constant quality of their products. They must know the requirements of their customers and select the raw materials for cleaners and disinfectants to avoid or at least to minimise any potential risk. Tainting or toxic substances should not be used; raw materials should not contain dangerous by-products, e.g. heavy metals. Production equipment should be dedicated exclusively to products for the food industry and separated from equipment used for janitorial products that might contain dyes and perfumes. Responsible care involves an efficient quality management system dealing with the aforementioned subjects. Batch records and batch samples have to be retained for a sufficient period to clarify any question and complaint the customer might have regarding the product bought. If this is done properly, the product can be traced back by using the batch number as far as to the raw materials and equipment that were used in the manufacturing process.

39.1.2 Additional aspects

Another point of interest – and this is definitely the more complex part – is the tracing and identification of these products during handling and logistics in a production site. Filling other containers, spillages, old stocks with damaged labelling, receiving of bulk deliveries, etc. require tracing and identification methods.

Tracing of cleaning solutions and hygiene products is above all a matter of food hygiene and safety. Even with fully automated systems, it is important to identify hygiene products before they are filled from road tankers into bulk storage tanks. When handling hygiene products manually, operator safety and plant safety are even more of an issue. Unintentional blending of products or even product solutions may cause severe accidents. In the case of accidents and spillage, or simply with old stocks with damaged labels, it can be important to identify the nature of the products involved and take adequate countermeasures. Being able to identify the nature of all products used in a plant is a part of good housekeeping practices.

The intention of this chapter is to show solutions for this relatively complex subject of tracing hygiene products.

39.2 General issues in tracing of cleaning solutions and hygiene products

Today analytical tracing and detection of most chemicals, even in very small concentrations, no longer seems to be a physical problem. When looking at product mixtures or traces of substances in a matrix, chemical analysis becomes much more problematic. Considering the costs of sophisticated analytical equipment, cost–benefit comparisons, personnel costs and time factors, many scientifically possible methods turn into theoretical options of little practical value.

However, there should be ways to control hygiene products under affordable conditions focusing on practical demands, necessities and cost–benefit relations. A closer look at the composition of cleaners and disinfectants will provide a better understanding of analytical options and challenges.

39.2.1 The nature of industrial cleaners and disinfectants

In most cases, cleaners and disinfectants are multiple-component preparations. Tables 39.1–39.4 give an overview regarding the possible ingredients and their purpose in common formulations of hygiene products. The list of components is intended only as an informative overview with examples, and not as a complete list of all possible ingredients in hygiene products and their nature.

39.2.2 The challenge of analysing cleaners and disinfectants

Tables 39.1–39.4 demonstrate the complexity of cleaner and disinfectant formulations. There is always more than one chemical compound to analyse. An additional factor to be considered is the fact that nearly all industrial hygiene products are of technical grade. This means there must be certain allowances for variations *within acceptable limits* regarding the percentage of an ingredient in a formula (quality control). Furthermore, the majority of raw materials used in industrial hygiene products are of technical grade, too. By-products and variations have to be accepted *within agreed specifications*. This is no problem

Table 39.1 Acid descalers and cleaners

Ingredient	Example	Purpose
Inorganic acids	Nitric acid	Owing to its oxidising properties at higher concentrations nitric acid cannot be used in more complex formulations, e.g. with surfactants. Its use is limited to descalers for removing inorganic residues as water scale, milk stone, beer stone, etc. from surfaces.
	Phosphoric acid Sulphonic acids Sulphuric acid	These acids permit a combination with surfactants, foam-controlling agents and other materials supporting a cleaning process and may be used in formulations for removing inorganic residues as well as organic residues at the same time from food contact surfaces.
Organic acids	Formic acid Citric acid Lactic acid Gluconic acid Sulphamic acid	
Inhibitors	Phosphonic acids	Protect materials against chemical attack.
Surfactants	Non-ionic or anionic surfactants	Provide cleaning efficacy regarding organic soil; enhance scale-removing properties and are the choice for removing fat residues.
Defoamers	Hydrophobic non-ionics	Suppress foam occurring from formula components and/or removed soil.
Stabilisers	Hydrotrophic substances	Stabilise liquid formulations at high and/or low temperatures

Table 39.2 Neutral cleaners (including mildly acidic or mildly alkaline products)

Ingredient	Example	Purpose
Builders	Phosphates Phosphonates Citrates	Enhance soil-removing and suspending properties as well as the effects of surfactants.
Surfactants	Non-ionic or anionic surfactants	Provide soil penetration, soil emulsification, surface wetting and low surface tension.
Defoamers	Hydrophobic non-ionics	Suppress both foam occurring from formula components and/or removed soil.
Enzymes	Protease types	Support (enable) protein removal at nearly neutral pH from sensitive surfaces.
	Lipase types	Support (enable) fat removal without using surfactants.
	Specific enzymes	Support (enable) removal of difficult-to-remove substances from surfaces without using aggressive chemicals.
Stabilisers	Hydrotrophic substances	Stabilise liquid formulations at high and/or low temperatures.

regarding the functionality of the hygiene product, but it is a problem regarding precise analytical data and physical parameters for product identification.

Exceptions to this rule are *specific good manufacturing practice (GMP) manufactured products* for specific industries, e.g. the pharmaceutical industry. Such products have much smaller tolerances regarding standard deviations and by-products. They are also supplied with an elaborate analytical background.

A possible and practical solution is the determination of *summary parameters of the main ingredients* and the definition of *acceptable deviations*. In most cases a qualitative rather than a quantitative result is sufficient for quality control of the products or product identification.

For these purposes, there are various quick tests, physical instruments and analytical methods available that can be adapted to the required level of control.

39.3 Particular issues in tracing of hygiene products

Tracing of hygiene products may be divided into five segments:

1. Control of incoming goods, positive identification.
2. Automatic control of preparing use-solutions.
3. Control during manual handling and refilling.

Table 39.3 Alkaline cleaners

Ingredient	Example	Purpose
Bases	Sodium hydroxide Potassium hydroxide	Solve, peptise, soften or decompose organic soil.
Chelates	Ethylenediamine-tetraacetic acid (edta) Nitrilotriacelate (NTA) Imidodisuccinate (IDS) Gluconate	Chelates not only suppress the negative impact of water hardness on cleaning efficacy and provide scale prevention, they also can remove fresh and thin layers of inorganic deposits, attack inorganic soil and assists in removing organic soil when combined in a matrix with inorganic scale.
Builders	Phosphates Phosphonates Citrates Silicates	By chemical nature most of the builders are chelates. They enhance soil-removing and suspending properties as well as the effects of surfactants.
Surfactants	Non-ionic or anionic	Provide soil penetration, soil emulsification, surface wetting and low surface tension.
Defoamers	Hydrophobic non-ionics	Suppress foam occurring both from formula components and/or removed soil.
Sequestering agents	Polyphosphates Phosphonates	Prevent scaling, especially in rinse phases.
Corrosion inhibitors	Silicates	Protection of soft metals against chemical attack.
Oxidising cleaning boosters	Hypochlorites	Assist in removal of tenacious and insoluble soil. By means of oxidation they can break down larger molecules into smaller fractions or render soil soluble by introducing functional groups.
Stabilisers	Hydrotrophic substances	Stabilise liquid formulations at high and/or low temperatures.

4. Validations of clean surfaces free of residues from hygiene products (within legal requirements).
5. Identification in case of accidents and emergencies, e.g. unidentified spills, unlabelled containers.

39.3.1 Control of incoming goods, positive identification

Hygiene products are usually delivered to the consumer in two ways, either filled in a range of small containers often termed 'jerry cans' with sizes from usually 5, 10, 20 and 30 L, drums with 200–220 L, transit tanks with 800 to 1000 L or as bulk chemicals in road tankers.

Table 39.4 Disinfectants

Ingredient	Example	Purpose
Disinfectants	Hypochlorites Peroxides Quaternary ammonium compound (QAC) Ampholytes	Killing of microorganisms by complex reactions on either the outside or inside of the microbial cell.
pH-regulators, buffers	Bases Acids Salts	Provide optimum pH for the active biocide, stabilise the pH during application, e.g. to control a corrosion risk with oxidising disinfectants or to provide product stability in solution or in concentrated form.
Surfactants	Non-ionic or anionic surfactants	Improve wetting, enhance biocidal efficacy or enable foam applications.
Defoamers	Hydrophobic non-ionics	Control foaming during application.
Stabilisers	Hydrotrophic substances	Stabilise liquid formulations at high and/or low temperatures.

The containers must be properly labelled including relevant product information, safety symbols and phrases and batch identification, and the bulk delivery has to be accompanied by proper documentation giving the required information for identification, safe storage and handling as well as potential hazards.

An additional analytical identification of the products may serve only quality assurance reasons in case of the products in containers, drums and transi-tanks. Regarding bulk deliveries that are pumped into customer-owned holding tanks still containing certain amounts left from the last delivery, it is also a safety measure to identify the product before permission for pumping is given. Especially in the second case it will not be possible to run time-consuming analysis while keeping the transport driver waiting.

Reliable, rapid methods for the identification of deliveries (incoming goods)

Measuring pH

Solution pH can be measured correctly only in highly diluted solutions, because only in this case is is possible to measure the hydrogen ion concentration properly, because the physical laws regulating pH measurement can only be applied to rather diluted solutions. At higher concentration, especially in very concentrated products, the measured pH becomes increasingly unreliable because overlapping physical effects occur. Therefore, it is important to measure the pH in solutions diluted using water of a consistent water quality. Preferably, demineralised water should be used.

With less sophistication, such a test can also be done with pH-paper. Testing with pH-paper may also be used in case of spills as an informative test to determine whether a product is alkaline, neutral or acidic.

Measuring the conductivity
What is true for pH measuring applies also for conductivity measuring: only in diluted solutions can correct values be obtained. At higher concentration the degree of dissociation and ion mobility decreases, thus altering the proportionality between concentration and conductivity. Conductivity is only a measure of the physical property of conducting electrical current but does not give any other information about the character of a product when used as a single method. Carefully adapted, this method might be used for the identification of concentrates in combination with a second method.

Combining pH and conductivity
Two different products can easily have the same conductivity but still have a very different composition. Finding two concentrates in one place with the same pH **and** conductivity is more unlikely. Therefore, a combination of both methods will dramatically improve reliability as a rapid and simple product identification method.

Refractive indices
This is a very fast and simple method with a highly discriminating character; therefore it is important to define a range, not a standard value. Deviations that occur due to the technical grade of the products question will be detected and might cause uncertainties. When used as a third parameter along with pH and conductivity the identification is nearly 100% safe.

The aforementioned methods are available as automated on-line systems as well as 'hand-held' manual procedures or laboratory procedures. The required time, even for all three methods together, is less than 10 minutes, including evaluation. Alone or in any combination, they can be used to safeguard bulk tank receptions or as simple but highly reliable quality check, i.e. a check of specification – quality is here more used in a sense of properties to ensure the material really is what it should be, and not in a sense of good or bad quality.

Density
Measuring the density of a product is also a relatively simple and fast method to collect information. In combination with conductivity and pH it is a very secure method to identify a product very precisely. If not using the most accurate methods and evaluating the third figure after the decimal point, measuring a product's density can be more practical than the refraction index.

Measuring the velocity of sound in a product
This is done by employing piezo-electronic elements. A disk-shaped emitter is forced into fast oscillations thus emitting short clubbed pulses (like a radar

system). The pulses travel a fixed distance to a receiver that converts sonar energy to electric energy. The time the pulse travels between emitter and receiver is measured and the sonar velocity calculated from the fixed distance between emitter and receiver and the lapsed time.

Also this physical property of a material can be automatically measured even with in-line probes and may serve in combination with one or more of the other methods described as a rapid and easy means of identification for product concentrates.

Tracers
Some products may contain tracers that can be detected with a dedicated method. This is of course an indirect method that can only be used for the identification of a product type, but not as a quality control measure because only the presence of the tracer can be proven but no physical property of the product.

Checking other product qualities
It is even possible to identify compounds of cleaners with simple tests, e.g. using test strips similar to pH-paper. Available chlorine and peroxides, as well as QACs can be identified with such tests, and when pre-diluted under controlled conditions, the tests may be quantitative. However, quality control of incoming goods might be more useful if proper concentrations of the main ingredients in the hygiene products can be determined.

The following chemical ingredients are common and relatively easy to determine: total alkalinity, total acidity, available chlorine, total peroxides, hydrogen peroxide and peroxyacetic acid, anionic surfactants and cationic surfactants. In addition, with relatively inexpensive photometric equipment, there are quick tests for phosphates and non-ionic surfactants. These simple methods permit a sufficient level of identification and quality control for incoming hygiene products.

According to a previous statement about the technical grade of the discussed products, it is always essential to set and agree upon appropriate ranges with the product supplier rather than discrete values. The tested sample has to be within the range of agreed specifications.

Conclusion
The methods described may well serve as product identifiers to avoid a road tanker filling the wrong holding tank, but also to measure product properties for quality testing or even to identify products from old stocks with missing or damaged labelling.

39.3.2 Automatic control of preparing use solutions

In automatic cleaning systems, it is very unlikely that products are mistaken, especially when the concentrates are taken from bulk storage tanks with satisfactory control of incoming goods. Here the human factor is still the highest risk when connecting small containers such as jerry cans, drums or transi-tanks to cleaning systems. Transponder systems, bar code readers, etc. are under discussion but cannot guarantee total safety.

The proper dilution rate and maintenance of a certain concentration in the cleaning system can often be monitored by using online conductivity meters. In addition, there are a few direct online methods to measure certain ingredients in use solutions.

Manual or semi-automatic control of use solutions

This is a task that is usually not easy to achieve with automated equipment. The concentration of products with a suitable conductivity can be controlled by using conductivity meters. For a very few chemicals there is a dedicated automatic measuring method, e.g. available chlorine or peroxyactic acid. However, in most cases the use of summary titration of the main ingredients or direct titration or analysis of the main components by titration, photometric determination or even using tracers is required.

29.3.3 Control during manual handling and refilling

Industrial cleaners and disinfectants are not dangerous *when handled properly*, but one should not forget that any manual handling of cleaning chemicals bears several potential risks. There are direct risks regarding the person handling the chemicals. Industrial cleaners have the purpose of attacking and dissolving fat and proteins – and these are main components of human body tissue! In addition there are risks of filling cleaners into containers with wrong or missing labelling. The problems that could arise from these mistakes can range from insufficient cleaning via equipment damage and food contamination to full-size chemical accidents. Food contamination may occur when filling chemicals into food or food additive containers or when filling tainting materials such as fuel into detergent containers.

A good example is the case of a dairy worker who filled an empty 1000 L transi-tank with nitric acid from a bulk tank. He brought the transi-tank safely with a forklift truck to the place where it was needed and left, because his shift was finished. About 15 minutes later the top lid of the transi-tank was blown off and a dark brown smoke column erupted 20 m into the air as if a miniature volcano. In very little time, the environment was contaminated with highly toxic nitrous gases travelling with the wind towards a nearby shopping centre. Only the immediate action of joint security forces including several fire brigade teams, paramedics, helicopters and police forces prevented a disaster. What had happened? The worker had selected an empty container that had the proper labelling for an acid cleaner. What he did not know is that this cleaner contained organic material that must not be blended with concentrated nitric acid. The residues of about 1000 mL in the empty tank had been sufficient to start a time-retarded but accelerating reaction leading to a complete decomposition of the nitric acid into nitrous gases and steam. Because of this experience, the dairy management stopped any refilling of chemicals into other containers.

Of course, the world is not ideal and there are things we have to do and accept a compromise. To make refilling of chemicals from larger containers or bulk tanks into smaller containers as safe as possible there are professional refilling stations available. Such stations can be designed to fill from drums into jerry cans, to fill from transi-tanks into jerry cans or to fill from bulk tanks into smaller containers.

There are also so-called 'user pack solutions' to avoid misuse of chemical container for transporting and storing other goods. The user-packs have a closed tab that cannot be removed without destroying the container. This tab fits only to a valve on the filling station. Including personnel protection gear and hazard information, ergonomic designs for easy handling and a safety retaining area to retain spills, these refilling stations provide much more security but still cannot completely eliminate the factor of human error and failure.

39.3.4 Validations of clean surfaces to be free of residues from hygiene products (within legal requirements)

A complete control of a confined surface (inside machinery, tanks and pipework) is not possible, but there are well-accepted approximation methods that have proven their reliability, even under the stringent requirements of pharmaceutical production. First, equipment should be designed to permit proper drainage and there should always be sufficient time for draining. Research into the behaviour of common cleaning solutions has shown that after proper draining, 20–40 mL/m of the solution are left on vertical equipment walls. This equals approximately 0.2–0.4 mL of a product when used as a 1% solution. Secondly, dependent on the food product category (in Europe all chemicals must be rinsed of dairy processing surfaces but not necessarily from meat products) or relevant national legislation, these remaining solutions may need to be rinsed off with potable water. Where it is acceptable to leave disinfectants on surfaces, they must be non-toxic and non-tainting and may need to be specially approved.

Rinsing is relatively easy to control. In most cases the conductivity or the pH of the cleaning solution is different from the pH and conductivity of the rinse water. In this case the rinse process can be controlled by comparing the conductivity and/or the pH of the incoming rinse water to the used rinse water leaving from the rinsed equipment. When there is no longer a difference, the surface has been sufficiently rinsed. This method can be used as a permanent installation or in a more or less frequent auditing process to validate the cleaning process.

In the case of particular ingredients one can collect rinse water samples during an extended rinse cycle and analyse the samples for critical ingredients. This more complex method is used during validation of cleaning methods in the pharmaceutical industry but may also serve as a tool in other industries to set CIP parameters. However, even this method can only show and prove a sufficient rinse time. There is still no information about what might still be on the surface in the equipment.

It is a physical law that any substance having had contact with a surface leaves at least a very thin molecular layer adhering to the surface. It is a questions of physics involving the affinity surface of a surface material to a substance or vice versa as to how much of a residue has to be expected. Here only swab tests with suitable solvents or a placebo production with later analysis of traces will give further information.

In normal food production, this would be impractical and not economical. A simple risk assessment would answer many questions. Such a risk assessment can be based upon a worst case study calculation as to how much of a cleaner is left without rinsing, using the fact that 20–40 mL/m will remain on a surface after draining. Considering the use-concentration and the filling volume of the equipment the maximum possible concentration of a cleaner or disinfectant in the final product can be calculated. Evaluating toxicity data and threshold levels shows the impact of the worst case scenario on a food product.

Ingredients of cleaners and disinfectant for the food industry do not have a high toxicity level. Their risk potential derives mainly from their acidity or alkalinity as a concentrate and in use solution. Even this risk decreases exponentially with further dilution. If there are any other substances that might be undesirable, such as nitrates, surfactants and defoamers, their application should be limited to the absolute required level.

Cleaning is normally not a single step process; alkaline cleaning is often followed by an acidic cleaning step and/or disinfection. Here the influence of the following step has to be considered and often this step can provide a safeguard. A good example for this is the elimination of proteolytic enzymes by using an oxidising disinfectant. Responsible manufacturers of enzymatic cleaners use only proteolytic enzymes that can be easily destroyed by oxidation. This provides additional security in the case of rinse failures. Lipolytic enzymes on the market are much more stable and there is no simple safeguarding method. If non-ionic surfactants are regarded as a problem, a surfactant-free acidic disinfectant, following the use of a surfactant-containing cleaner, will further reduce the minimal residue potential.

Plant design (hygienic design) with regard to smooth surfaces providing sufficient rinse water contacts is essential. Further, there should be no dead legs and sagging pipe sections. In combination with a choice of adequate cleaners and disinfectants used in a proper hygiene, sequence will virtually eliminate the last risk of a food contamination by properly applied cleaners.

39.3.5 Identification in case of accidents and emergencies, e.g. unidentified spills, unlabelled containers

This last point can be mainly handled in the same fashion as the first point, when dealing with containers having lost their labelling. All methods described may be used to identify the product for proper elimination.

In most cases, as with spills, it is sufficient to find some basic information such as:

- Is it alkaline or acidic?
- Does it contain available chlorine?
- Does it contain peroxides?
- Does it react with organic material (such as concentrated sulphuric acid or peroxides)?

To answer these questions it is sufficient for most cases to use pH-paper, iodine–starch paper, peroxide test stripes, etc.

Potential hazards regarding contact with organic material can be checked by the immersion of a cotton swab into the spill. A brown discoloration of the cotton or any visible reaction is an indication not to treat a spill with organic adsorbents. As a general rule no organic material such as sawdust or cotton should be used in areas where chemical spills might occur. Also sand is not a genuine absorbent. Products such as diatomateous earth or special absorbents sold for absorbing chemicals should be available in any chemical storage to be at hand when needed.

39.4 Conclusion

Tracing chemical agents and good housekeeping of chemical products is important in the light of food safety and personnel safety, and can be attained by:

- installing and maintaining good manufacturing practices;
- training of operators and safety officers;
- safety by design, not by coincidence.

Methods have been described to allow the positive identification of chemicals not only before receiving, handling and use of chemicals but also when discharging, and dealing with old stocks or spills.

39.5 Future trends

Nowadays the rapid development on the electronics market leaves prediction difficult. Electronic equipment for analysing chemicals becomes less expensive and even today chemical companies have already started to use on-line near-NIR spectroscopy to analyse raw materials on delivery from road or rail tankers. Other spectroscopic methods might follow and become available at lower price levels. Multi-array technologies combining automated analytical sensors might enable 'fingerprint' identification of products. The near future might bring auto-sensors that control and manage cleaners and disinfectants during reception and when connecting to a dosing system, thus preventing dangerous blending of concentrates or the application of wrong products. Simultaneously such systems can generate reports for quality management systems about the products used for plant hygiene in food processing.

40

Improving hygiene auditing

P. Overbosch, Kraft Foods, Germany

40.1 Introduction

Auditing for hygiene improvement may be seen as an oxymoron. ISO 9001 defines an audit as a 'Systematic, independent and documented process for obtaining audit evidence and evaluating it objectively to determine the extent to which audit criteria are fulfilled.' Audits are being carried out for purposes other than ISO certification, but there is a general perception that audits are for determining compliance, with or without certification of some kind. Audit reports point out non-conformances, supporting the auditees in their effort to meet all requirements of the predefined standard they were audited against.

Where improvement is the primary purpose, a far more ambitious, though less precisely defined goal, than compliance, the term 'assessment' is often used. For our present purposes we will continue to use the word 'audit' in the context of various possible purposes and design characteristics of this type of exercise.

The subject matter of our audit for improvement is hygiene, or rather the hygiene management system (HMS), i.e. the management system that establishes and maintains hygienic conditions in the facility in question (our primary focus will be on food manufacturing sites, although most of the considerations will equally apply to warehouses, kitchens, etc.).

According to the Codex definition, 'Food hygiene' comprises 'all conditions and measures to ensure the safety and suitability of food at all stages of the food chain',[1] but that is too wide for our present purposes because it would include HACCP into the HMS.

The relation between hygiene and Hazard Analysis Critical Control Point (HACCP) is often explained in terms of hygiene being a prerequisite programme. Early in the HACCP decision tree, a distinction needs to be made between significant and non-significant hazards. Those that are realistically unavoidable and require active control at all times are subsequently managed within HACCP. The function of a hygiene programme is to help limit the list of manufacturing equipment and food production environment-related hazards through relatively (in comparison with HACCP) simple methods that are still sufficiently robust to be consistently effective. Sperber *et al.* note that: 'occasional deviation from a prerequisite program requirement would not by itself be expected to create a food safety hazard or concern'.[2] The position of good manufacturing/good hygiene practice (GMP/GHP) within an overall quality management approach is illustrated in the ILSI report on Food Safety Management Tools[3] (Fig. 40.1). For the purposes of this chapter, we will not deal with HACCP itself, which is an entirely separate subject.

A second part of the HMS job description is that it should provide adequate protection against negative influences related to the same sources (manufacturing equipment and food production environment) that are not 'hazards' in the strict sense of the word, but do impair quality (spoilage organisms, non-toxic taints, non-hazardous foreign matter including hairs, etc).

The success of an HMS can therefore be evaluated on the basis of its:

- relevance (have all relevant hygiene concerns and potential routes of contamination been identified and addressed and is there a mechanism to keep the system up to date?);
- effectiveness (can the success of the system be demonstrated?);
- robustness against irregularities and disturbances;
- efficiency in terms of time, resources and costs involved.

This chapter will provide some fundamental and practical considerations regarding audits for hygiene improvement, but does not intend to introduce any new hygiene standards or requirements per se.

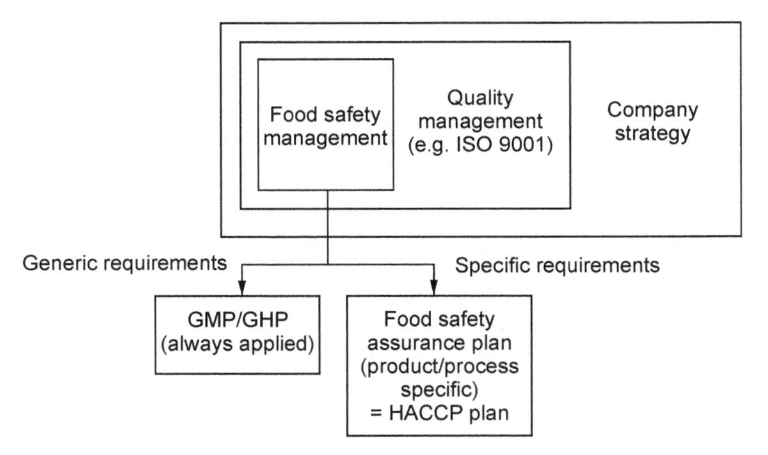

Fig. 40.1 The elements of a food safety management system as part of a quality management system (QMS).

40.2 Why have a hygiene improvement audit in the first place?

Before we get into some of the technicalities of auditing for improvement, the question why and under what circumstances a facility should seek a hygiene improvement audit should be addressed. First of all, compliance to local legal/regulatory (hygiene) standards (European Union, US Food and Drug Administration, US Department of Agriculture, Food Standards Australia New Zealand) is simply a *conditio sine qua non* for any food manufacturer and authorities will normally come to inspect facilities. Therefore, the first priority must be to meet basic legal/regulatory (usually Codex Alimentarius-derived) hygiene requirements.

Beyond that, customers may have additional proprietary standards and will audit against them. Assuming that a facility demonstrably complies with all locally applicable standards, why would anyone be interested in an exercise that will take time and resources to dig much deeper into hygiene issues and come up with improvement proposals that will probably again involve time, resources and costs? The reasons lie in the criteria mentioned in the introduction.

Although the facility may have passed various compliance focused inspections, may operate an internal (usually also compliance focused) audit system and may never have suffered a hygiene-related incident up to now, management would still not know:

- what parts of their system are only barely adequate;
- which represent costly overkill;
- whether there may be accidents waiting to happen;
- whether all hygiene-related standards and procedures are understood and practised in the same way by all involved;
- whether the system needs to be updated now or in the near future because of new emerging issues, existing development plans (site, equipment, product portfolio) or problems with ageing infrastructure.

After complying with basic applicable standards, and having that confirmed and recognised through auditing, there are therefore excellent reasons to take a closer and deeper look and seek a hygiene improvement audit.

40.3 Auditing and the hierarchy of a controlled system

Quality systems for food manufacturing operations are meant to ensure a consistent flow of safe, legal and in-specification materials all the way from raw materials to finished product at the point of sale, but the entire system can schematically be reduced to four levels applying to every unit operation (Figs 40.2–40.4):

1. *Basic design.* This includes all aspects of equipment and infrastructure design, as well as standard procedures including scheduled maintenance,

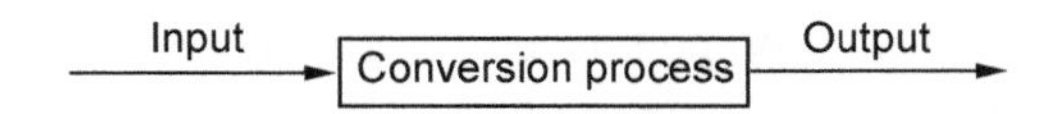

Fig. 40.2 Uncontrolled process.

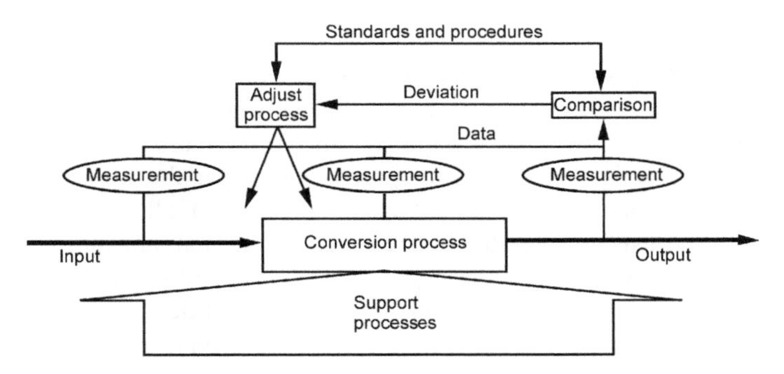

Fig. 40.3 Primary process and first control loop.

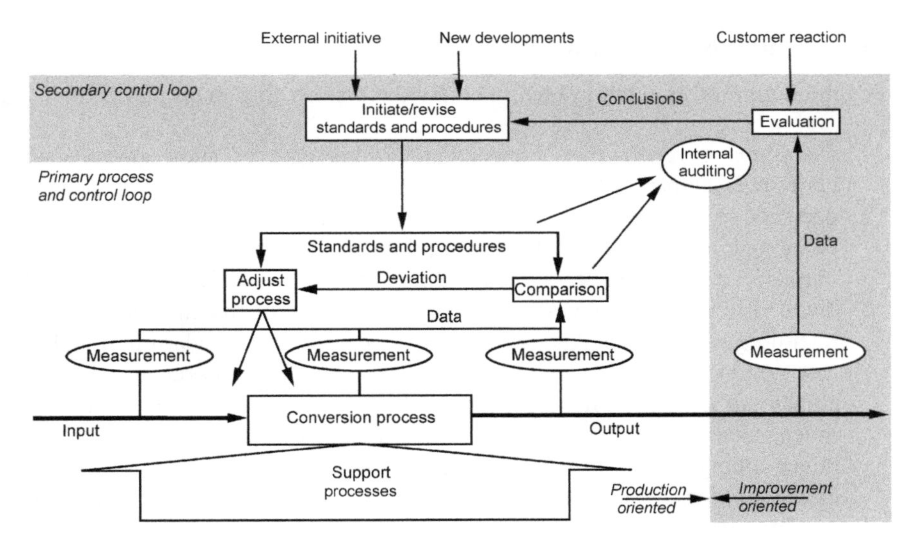

Fig. 40.4 A controlled process (two loops).

sanitation etc. Using basic design elements only, including predefined regular maintenance, a process will run, but it is not equipped to notice or respond to out-of-specification situations of any kind.

2. *Control*. This includes all physical/chemical measurements, including temperature, pH and microbiological swabs of environment/equipment, etc. that are being taken, with results being compared with predefined specifications and corrective action being taken to bring the process back into specification when appropriate.

3. *Incident management.* While this term is usually reserved for incidents that may lead to public consequences (such as product withdrawals and recalls), within this context we will use it to refer to mechanisms to address situations where normal procedures have failed and the process in question is 'out of control'.
4. *Audit.* For our purposes, the definition of an audit system may now be expanded a bit. While audits may concentrate on one or more of the above levels (e.g. look at established control mechanisms but never open a pump to assess basic hygienic design or the effectiveness of a standard maintenance/sanitation programme), it is clear that the definition should cover all relevant aspects: a critical assessment of all the above levels against all considerations that are relevant for the fundamental purpose of the system (as detailed at the beginning of this section). For most audits carried out today, this is much too wide but we will need both the depth in terms of levels and the breadth in terms of 'all relevant considerations' for our current 'auditing for hygiene improvement' purposes.

40.4 Purposes of an auditing system

For the design of an audit system, we may distinguish four main purposes:

1. *Compliance, verification.* Fundamentally this type of audit assesses an entire management system according to the above-mentioned 'basic design' perspective: have all prescribed elements been put in place and are they being operated according to the requirements of the standard? For this purpose we will need pre-agreed standards that are sufficiently precise and detailed to allow clear yes/no (or graded in terms of major/minor) scores regarding compliance. Auditors will have to know the standard thoroughly (but no more than that) and, because of the potential consequences for the auditee of non-compliance, need to be seen as strictly impartial and independent. Assuming that the standard is relevant and sufficient (the never-ending stream of new standards seems to indicate that this remains a challenge) and that the facility in question does not yet meet all requirements, a compliance audit may lead to improvement. That, however, is not the main purpose of this type of audit and improvement potential is anyway limited to what is foreseen by the standard itself.
2. *Effectiveness, capability to perform to specifications, validation.* This type of audit normally builds on the compliance perspective, extending the question whether everything is in place according to the standard to: 'does it work as intended?'. The focus is shifted from 'basic design' to 'control'. An intimate knowledge of 'the standard' is no longer sufficient, the auditor will have to be able to make an informed decision regarding the effectiveness of the control system, based on quantitative historical information. Where process capability issues are found, some indication of the root cause is

normally expected. While an expectation of impartiality remains, the focus will shift somewhat to technical expertise, experience and the ability to make appropriate judgments in situations that are not black and white.
3. *Efficiency*. Audits for efficiency are mentioned here because efficiency is a necessary part of the next category (improvement). In isolation these audits are usually based on an assumption of compliance and effectiveness and targeted at reducing identifiable 'losses' in terms of time, materials, utilisation and other cost categories. Because of these assumptions and focus, auditors are usually from a financial background.
4. *Improvement*. An audit for improvement essentially takes all of the above levels (basic design, control) and aspects (compliance, relevance) and seeks to identify, explain and prioritise items for potential improvement against the background of the above-mentioned 'all relevant considerations'. This will include internally and externally established standards but will go beyond those where appropriate. Compliance to formally applicable and adopted standards remains highly relevant as such, and for the audit team to be able to judge the organisation's ability to improve – if the facility cannot implement what they have formally adopted, the credibility of further improvement initiatives is dubious.

An improvement audit is much more demanding on the auditor, who must be able to make an informed judgement regarding the appropriateness of basic design, control system and incident management specifications and practices (one of the reasons why auditors usually come in teams). As the outcome of this type of audit will be closer to a blueprint of an improvement programme, without any formal approval or certification consequences, the above-mentioned technical expertise combined with communicative and teaching abilities will be more important than assurances of formal independence and impartiality.

Apart from the main purposes of an audit system, there are secondary purposes to be considered:

- *Education of auditors-in-training or of the auditees*. Within each of the main purposes, the common format of an expert assessment of systems and practices is an ideal opportunity for training purposes. In some third party certification audit formats this is precluded by strict requirements of independence and avoidance of any potential legal liability, but an improvement audit by its very nature should be more like a master class.
- *Sharing of best practice*. Where manufacturing sites are in a position to share best practice with each other (e.g. because they belong to the same company), audits are potentially excellent opportunities to exchange experiences.

40.5 Designing a system for improvement audits

Focusing on improvement audits, we can address a number of audit-system design elements:

- *Frequency*. All audits are supposed to be cyclical, either to ascertain that an operating unit still complies to the same standard, or in our current case to ensure that the improvement momentum is refreshed and maintained. General experience would suggest that a 3-year cycle is a minimum.
- *Reporting and scoring*. Whereas the audit team are necessarily the sole authors of compliance-oriented audit reports, the improvement audit format with its emphasis on understanding, ownership and commitment implies that the auditees capture their learnings and improvement plans themselves, supported and checked by the audit team. Scoring may be very helpful for prioritisation purposes, for tracking a facility's progress over time, for condensed communication to senior management, and for internal and external benchmarking. The basis of any scoring system should be a categorisation of the opportunity (in our case this may take the form of the categorisation mentioned in the introduction: relevance, effectiveness, robustness and efficiency), possibly combined with a magnitude scale.
- *The use of a questionnaire*. Normally, questionnaires are derived from standards, and the 'all relevant considerations' frame of reference for improvement audits would seem to preclude the use of questionnaires for this purpose. Nevertheless, a questionnaire that evolves over time can be a great way to maintain consistency, capture learnings, support the education process and assure that potentially relevant aspects are not accidentally forgotten. As long as it is clearly understood that not all questions necessarily apply to the particular site being audited and that the quest for relevant improvement opportunities may at any moment lead to explorations beyond the current questionnaire, the use of questionnaires may have more advantages than disadvantages.

40.6 Performing the audit

40.6.1 Scope of a hygiene improvement audit

Apart from restrictions placed on the scope of the audit by the expertise of the audit team, the scope of a hygiene audit may extend in principle to include all aspects and elements of the facility's basic design, control systems and incident management preparedness relevant to the microbiological, chemical, physical and toxicological integrity of the product. This is meant to cover the facility itself (may or may not include external production and logistics of raw and packaging materials as well as all downstream logistics), include all hazards and risk factors normally associated with food production, but excludes the HACCP system and food security issues such as malicious tampering in any form.

40.6.2 Putting a team together

For an improvement audit to be successful, the auditees must come away with a much deeper understanding of the strengths, weaknesses, threats and opportunities associated with their HMS, and have fully participated in the identification and prioritisation of improvement opportunities. Full participation implies that they act as members of the audit team together with the external auditors.

Qualifications of the external members have been briefly discussed above. For our current purposes we will need external members that have no direct conflicts of interest, represent a set of fresh eyes, have expertise in all relevant areas – hygienic design, microbiology, sanitation systems, statistical sampling and data analysis, etc., observational skills, curiosity and willingness to go out of their way to uncover issues/root causes, work at all hours, have excellent people skills at all levels, including the ability to explain complex issues and be great communicators and teachers. This is a tall order and in practice there needs to be an understanding between auditees and the audit team what the actual scope of the HMS improvement audit will be, based on the team's actual skills and expertise.

Internal members will have to represent the various disciplines and departments involved in the HMS, be willing and able to contribute a detailed knowledge of local conditions and practices to the exercise, be eager to learn and improve and have the ability to think outside their own local box.

The size of such a mixed team will depend on the size of the operation, but sizes between four and eight members (50/50 internal/external) usually work well. Other internal sources may be required on a temporary basis for specific purposes. The audit is led by the external team leader, who is also responsible for the appropriate use of audit related methods (data and root cause analysis, brainstorming sessions, etc.).

40.6.3 Preparation for the audit

For an audit to be both effective and efficient, both sides will have to do some pre-work. The focus should be on understanding the various elements of the current hygiene management system, and gaining some insight into indicators of past performance of the system. This may include the following:

- Lay-out of the site, including all zoning.
- Description of all processes, products and raw materials used.
- Trend analysis of micro-data (e.g. swabbing data and trends before and after sanitation, environmental swabs, drains, etc.) and pest control reports.
- Recalls/withdrawals. (Anything potentially related to hygiene concerns?)
- Complaints history.
- Reports/citations/fines by visiting authorities.
- Previous external audit reports.
- Internal standards (is there an expectation that equipment be designed according to EHEDG/3A standards?, sanitation procedures including chemicals used).
- Management structure and supervision (cleaning during the night shift? By third parties?)
- Internal training/refresher programmes.

40.6.4 Some key attention points during the audit

Keeping in mind that we intend to assess the hygiene programme against the background of its primary job description, i.e. to keep potentially negative environmental influences well away from the food and from HACCP consideration, and to support later prioritisation of improvement opportunities we may want to make a distinction between items in direct contact with the food, and primary, secondary and tertiary environments.

Items in direct contact with the food include the following:

- All aspects of food contact equipment (materials used, welds, dead ends, drainability, etc). Following the breakdown introduced in section 40.3, the audit team will need to assess whether appropriate design standards have been applied, what type of maintenance (regularly scheduled, preventive, breakdown) and sanitation are used, whether records confirm the effectiveness and appropriateness of the maintenance programme (if an unsanitary situation has arisen and persisted for any length of time before repairs were made, this should be raised as an issue and possibilities for preventive action should be considered).
- Chemicals used for maintenance and sanitation, and the residues that may remain on food contact surfaces should also be included.
- Workers' hands in direct contact with the food; bare hands or glove policy, glove replacement, hand washing procedures etc.

Primary environment includes the following:

- Immediate surrounding of the food production. The audit should assess the 'zoning' policies (access rules and GMP requirements applying to people, materials, utensils, vehicles, pallets) applied to the immediate surrounding of (partially) open food production areas. Sensitivity of the food materials and processes determines zoning rules.
- Potential contact materials; safety status and management of those materials (lubricating oils/grease, heating/cooling materials) that could inadvertently come into contact with the food.

Secondary (the rest of the plant) and tertiary (direct vicinity of the plant) environment includes the appropriateness and management, including zoning, of designated areas outside the primary environment. Consideration of 'dirty' areas inside or in the direct vicinity of the manufacturing site. The general layout of the facility, in terms of being supportive of appropriate zoning, and potential routes and mechanisms of contamination should be assessed from food contact surfaces to the various environments, including the use of historical data pointing at actual possibilities.

Some topics for special attention are mentioned below. The list is not meant to be a quick and balanced guide through all applicable hygiene management or equipment and infrastructure design standards, or a reliable approximation of 'all relevant considerations' for a HMS. Rather it is meant to provide a summary of this author's experience with potential HMS weaknesses in food processing and storage facilities around the globe. The emphasis needs to be on asking the right questions – and for the audit team to keep digging until the issue is clear – and to directly observe where possible.

- Management example and attention: as day-to-day hygiene management is to a large extent an issue of keeping workers' attention and commitment, is management seen to actively support and uphold the hygiene management system in all its aspects? Important elements include personal behaviour (respect for zoning standards) and management attention to the HMS as expressed through the internal audits and management review, which are normally part of ISO type QMSs. While ISO has no specific focus on hygiene (though it is a topic in ISO 15161[4] and the upcoming standard for food safety management systems: ISO 22000), hygiene needs to be part of these internal audits and management reviews.
- Change management: do internal management procedures and records ensure that relevant hygiene conditions will be preserved or improved as a result of changes to infrastructure, equipment, environment or service providers (maintenance, sanitation or building)?
- People management: questions here include personal cleanliness, training, discipline, the role of employees in the continuous improvement process and the role of people during hygiene-related issues in the past (if any). A very important aspect here is the management of illnesses and injuries. Are the plant's policies and practices sufficiently specific and clear to all involved, are they optimally conducive to maintaining sanitary conditions on site and are they being followed?
- Contractors/visitors: hygiene management systems are inherently vulnerable to contractors' activities, especially in more sensitive zones. How are these situations being managed? Is there any indication that contractors have caused unsanitary conditions in the past?
- Maintenance and repair work may threaten hygienic conditions and may even be completely incompatible with ongoing production in some areas. How is this managed?
- Hand-over procedures between shifts, and their members' interpretation of internal standards and procedures. There should be consistency between shifts, but that is not always the case, and for the audit team there is no other option than to be there and observe.
- Catering: while many companies have outsourced canteen facilities and catering, maintaining the appropriate food safety including hygiene conditions in the canteen remains of the utmost importance. How is this managed, including the caterer's employees?
- Uniforms (relation to zoning, location of pockets, repair, replacement and laundry policy, is there any way that uniforms can carry contaminants from the outside into the facility?) and personal items/jewellery.
- Hand-washing facilities – drying/foot baths/air curtains, hand-swabs/toilets, are all these facilities appropriately designed as such, are they appropriately

situated on site, are they being maintained, what is their current sanitary situation? Are eating/chewing/drinking/smoking policies appropriate to the zone, in place and implemented?

- Cleaning/sanitation procedures and the use (including storage) of water/chemicals/utensils. Designing and maintaining a sanitation system to cope with different conditions (dry/wet environment, targeting to be microbiologically clean, ready for an allergen related change-over, removing particulate matter or colours, etc.) is a challenge that increases with the complexity of materials, processes and tight production schedules. The sanitation system should be one of the primary topics for the audit team. This includes everything from direct food contact surfaces to general areas. For effectiveness of cleaning of non-visible food contact areas the team will have to open up some equipment immediately after a cleaning cycle and inspect it directly. Particular care should be taken with difficult to clean areas, e.g. pumps, poorly designed equipment.
- Temperature control is not always necessary (one of the items for the audit team's consideration), but in cold conditions people's protective clothing in itself may present a hygiene challenge, and in very hot conditions there may be an issue with keeping doors/windows consistently closed.
- Air quality and filtering for direct product contact (air may be used as an ingredient). Air flow, including aerosols and dust, will be a potential concern where zoning is concerned. Ventilation systems provide a conduit in/out of the facility and may be difficult to clean internally.
- Water (including steam and ice) quality can vary significantly depending on its source (well, river, municipal system) and from place to place and time to time. While useful distinctions can be made to ingredient water and non-ingredient water (cleaning, heating, cooling), the ongoing potable status of water in the plant for all purposes should be a concern of the audit team.
- Pallets are well known for being difficult to track, clean and control and for turning up in places in the plant where they should not. Pallet policies and practices, and their role in zoning requirements, may be key elements in managing contamination routes.
- Silos and tanks that are continually topped up. Are they being cleaned on a regular basis? Is there evidence to demonstrate that the methods and frequency are appropriate?
- Forklift trucks, where they are allowed to go, what they carry from one zone to another, what they spread around as exhaust fumes and oil leakages, and where they refuel or reload their batteries.
- Incoming trucks, trains or ships and the issues they bring with them, in terms of carry-overs of previous cargoes, cleaning chemical residues, pests, potentially unsanitary pallets and worker-related hygiene issues.
- Drains, their design, general state of cleanliness and the integrity of the drainage system. The status of the underground part of the drainage system is often unknown, especially in older plants. The audit team should investigate what is known about this aspect and to what extent it could potentially impact

the hygiene status of various areas (or have environmental consequences).

- Overhead piping and (false) ceilings, as these are often difficult to clean but may harbour dust, and pests and suffer from leakages and peeling paint.
- All entrances (doors, windows, other holes in the walls) to the facility, their physical status and their status in relation to the zoning principles (e.g. where hand-washing is required, are all entrances covered by hand-washing stations?)
- Walls and floors, their basic design and current condition.
- Pest control: minimising pest-related risk often requires a careful balance between keeping pests out while at the same time keeping the use and availability of potentially toxic pest control materials to a minimum. What type of pest controls is the facility using (traps, fumigation), can the audit team assess effectiveness and is there an appropriate balance between specific prevention (various types of pest traps) and general prevention (keeping doors closed, protecting the immediate environment of the facility, etc.)? The team should also check records of findings and corrective actions. For our current purposes, pest control also applies to the management of domestic animals on site.
- Packaging materials usually come on pallets (see above) and may in themselves present hygiene risks (e.g. mould on cartons).
- Waste, rejected materials, rework and materials on hold will all require different storage conditions, which should be consistent with appropriate requirements for their respective purposes.
- Lubricants may be potential food contact materials, and as such have a high priority as (usually non-HACCP) contamination concerns, but not all lubricants used on site will fall into that category.
- All equipment specifically used for protecting the hygienic status of the food materials or the general environment (sieves, magnets, de-stoners, sorters); are they relevant, appropriately positioned, effective, robust and efficient?

40.6.5 Follow-up after the audit

As argued above, the report of a hygiene improvement audit should be written by the auditees, demonstrating their learning by capturing all items for potential improvement, an assessment of their priority and the pertinent follow-up plans. These plans should clearly indicate what needs to be done, timing, responsibilities and what the measure of success will be. The external auditors will approve the audit report, confirming the observations and conclusions, prioritisation and appropriateness of the proposed improvement projects.

Decisions will then have to be taken by management in terms of accepting (or not) the audit proposals and providing resources.

40.7 References

1. Codex Alimentarius, Supplement to Volume 1B, General requirements (Food Hygiene), second edition, 1997, FAO, Rome.
2. W. H. Sperber *et al.*; The role of prerequisite programs in managing a HACCP system, in *Dairy, Food and Environmental Sanitation*, July 1998, 418–423.
3. J.L. Jouve *et al.*: *Food Safety Management Tools*, ILSI Europe Report Series, 1998, International Life Science Institute, Brussels.
4. ISO 15161:2001; 'Guidelines on the application of ISO 9001:2000 for the food and drink industry', ISO, Geneva.

Index

CPSIA information can be obtained
at www.ICGtesting.com
Printed in the USA
BVHW04*0729090818
523967BV00004B/125/P